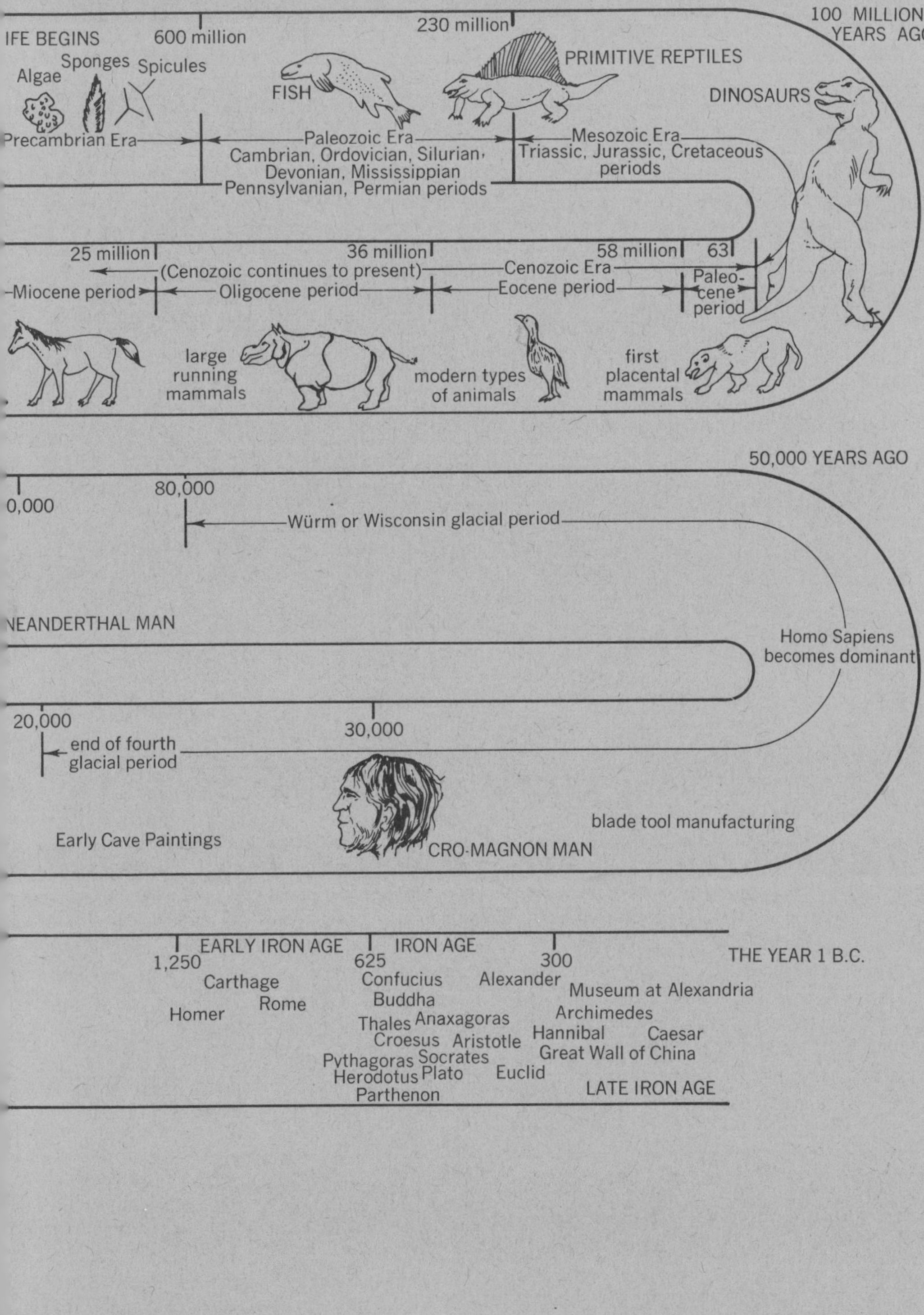

100 MILLION YEARS AGO

LIFE BEGINS — 600 million — 230 million — PRIMITIVE REPTILES — DINOSAURS

Algae, Sponges, Spicules — FISH

Precambrian Era — Paleozoic Era (Cambrian, Ordovician, Silurian, Devonian, Mississippian, Pennsylvanian, Permian periods) — Mesozoic Era (Triassic, Jurassic, Cretaceous periods)

25 million — 36 million — 58 million — 63

Miocene period — Oligocene period — (Cenozoic continues to present) — Cenozoic Era — Eocene period — Paleocene period

large running mammals — modern types of animals — first placental mammals

50,000 YEARS AGO

0,000 — 80,000 — Würm or Wisconsin glacial period

NEANDERTHAL MAN — Homo Sapiens becomes dominant

20,000 — end of fourth glacial period — 30,000

Early Cave Paintings — CRO-MAGNON MAN — blade tool manufacturing

EARLY IRON AGE — IRON AGE
1,250 — 625 — 300 — THE YEAR 1 B.C.

Carthage, Confucius, Alexander, Museum at Alexandria
Homer, Rome, Buddha, Archimedes
Thales, Anaxagoras, Hannibal, Caesar
Croesus, Aristotle, Great Wall of China
Pythagoras, Socrates
Herodotus, Plato, Euclid
Parthenon

LATE IRON AGE

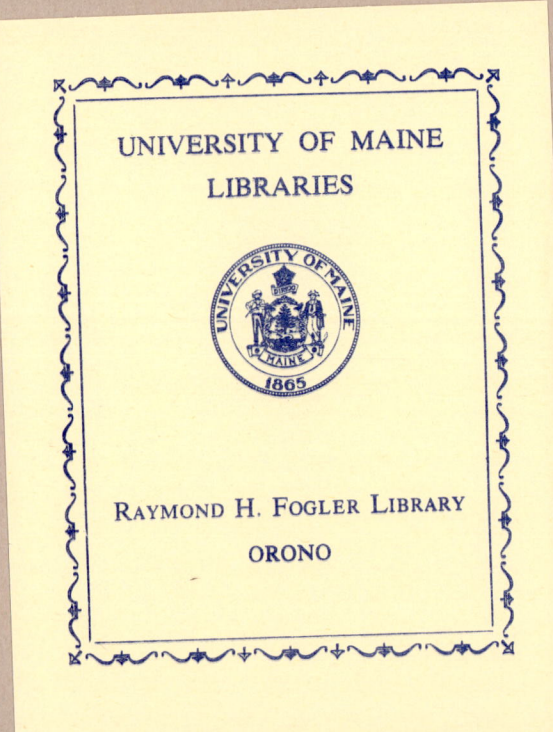

Introduction to Natural Science

V. LAWRENCE PARSEGIAN
Rensselaer Polytechnic Institute

ALAN S. MELTZER
Rensselaer Polytechnic Institute

ABRAHAM S. LUCHINS
State University of New York at Albany

K. SCOTT KINERSON
Russell Sage College

Introduction to Natural Science
Part One:
The Physical Sciences

Academic Press
New York and London

Copyright © 1968, by Academic Press Inc.

All rights reserved.
No part of this book may be reproduced in any form,
by photostat, microfilm, or any other means, without
written permission from the publishers.

ACADEMIC PRESS INC.
111 Fifth Avenue, New York, New York 10003

United Kingdom Edition published by
ACADEMIC PRESS INC. (LONDON) LTD.
Berkeley Square House, London W.1

Library of Congress Catalog Card Number: 68-14657

Second Printing, 1971

PRINTED IN THE UNITED STATES OF AMERICA

To the end that Science and Technology may contribute to the total progress of man.

Preface

THIS VOLUME constitutes the text material for the first-year portion of *Introduction to Natural Science*. When used together with the *Laboratory and Mathematics Supplement* (and with the *Teacher's Guide*), the volume meets the needs for either a 6-credit hour or an 8-credit hour course. When followed by Part II for the second year (entitled *The Life Sciences*), the entire course of 12- to 16-credit hour equivalent provides a comprehensive introduction to science for the future teacher, lawyer, economist, government official, artist, theologian, social scientist, anthropologist. The future of society, as well as of science itself, demands that these potential leaders of society understand the historical, philosophical, and social significance of science; its esthetic and cultural values, its bounty of goods and services, the hazards to society that accompany misuse of science, as well as a knowledge of the subject itself. This volume is addressed first to these students.

This volume has a second purpose, derived from the untoward effects that have come from specialization of disciplines. While specialization is unavoidable, it has been accompanied by separation and isolation of disciplines to the point of being detrimental both to the interests of society and of the disciplines. Cooperative effort among disciplines suffers when specialists scarcely understand each other. For this reason it is believed that the present course provides an integrated view which is necessary also for students who plan careers in the physical and life sciences.

The intent of the course is to give some historical perspective, with special effort directed to the identification of the common elements and interrelationships that give pattern to the institutions called *science* and *society*. A major theme is that of systems or cybernetics, including concepts of feedback influences and of stability. The probabilistic and statistical character of natural phenomena, difficulties attending measurement and understanding of natural phenomena, energy transformations and the nature of matter, philosophic and social implications of modern science, and questions of determinacy-indeterminacy constitute other themes.

The project called Science Courses for Baccalaureate Education which produced the new course was initiated in September, 1964 with the financial support of the Charles F. Kettering Foundation. Although centered at Rensselaer Polytechnic Institute under a resident chairman, the project had the benefit of a distinguished Project Advisory Board and of faculty members from many schools. Some helped to write text material, others served as reviewers and critics, while others taught pilot classes. Their contributions were necessary for the success of the project. It is fair to say, however, that the most remarkable influences that determined the character of the course came from the students who were in pilot classes at several institutions. We sought their influence by organizing "feedback committees" of students, by use of questionnaires, and by inviting criticisms and suggestions on every aspect of the course.

Considerable flexibility has been incorporated into the organization of the text in order for it to be used in a wide variety of colleges and junior colleges, usually at the freshman and sophomore levels. For this purpose there are *optional* sections of the text (printed in color) which treat given topics in greater depth, but which can be omitted from lectures and reading assignments with no loss of continuity. The optional sections are suitable for students who have had some calculus and good physics courses in their high-school years; without these sections the course has been found to be within the reach of students who previously have not had physics or more than three years of mathematics. The *Teacher's Guide* offers many suggestions on how to adapt the course to each class.

Laboratory work is included, with the recommendation that there be at least ten experiments each year (out of nearly twice that number included in the *Laboratory and Mathematics Supplement*). The mathematics portion of the *Supplement* is designed for both review and instruction, although the text requires only a minimum use of mathematical reasoning. The science and mathematics covered in this course meet the requirements for teacher certification in most states.

The course is designed to be taught economically. When given as a 3-credit-hour series for four semesters, it is recommended that there be three 1-hour sessions each week with large classes (of which two are for lecture and the third for quizzes, showing of films, discussion of experiments, and so forth) plus a 2-hour session for laboratory or recitation with small sections. The laboratory and recitation sessions can then be safely entrusted to graduate students when necessary to do so.

Because the scope of topics of the full course is so broad, the first-year portion (with Part I) is designed to be taught by a physical scientist, while the second-year portion is designed to be taught by a biological scientist. It has been possible, nevertheless, to integrate material from many disciplines within each portion. There can be occasional visiting lecturers to treat special topics when they are available.

The many faculty members who have participated in the development of this course hope that the student and general reader will derive some of the excitement and satisfaction that have accompanied the project effort.

The authors

Troy, New York

ACKNOWLEDGMENTS

While it is not feasible to list all the individuals (they number well over one hundred) whose encouragement and help have aided this work, I want to identify at least some whose contributions gave direction and continuing help to the preparation of the text material.

The Advisory Board that functioned during the first two years of the Project included the following:

DAVID G. BARRY, *San Jose State College, California*
WALTER H. BAUER, *Rensselaer Polytechnic Institute*
LOREN EISELEY, *University of Pennsylvania*
HARRY W. JONES, *Columbia University School of Law*
ADOLPH LOWE, *New School for Social Research*
HENRY MARGENAU, *Yale University*
RONALD A. H. MUELLER, *Rensselaer Polytechnic Institute*
ERNEST NAGEL, *Columbia University*

Henry Margenau has continued to give valuable advisory help.

Other contributors to the development of the *first-year text* include:

M. BRYAN BAYLY, *Rensselaer Polytechnic Institute*
KURT BING, *Rensselaer Polytechnic Institute*
RICHARD L. CARTER, *Rensselaer Polytechnic Institute*
A. S. COTERA, JR., *Allegheny College*
ROBERT R. DONALDSON, *State University of New York at Plattsburgh*
I. G. FOSTER, *Florida Presbyterian College*
GEORGE GOE, *Rensselaer Polytechnic Institute*
CAROL M. GOULD, *Rensselaer Polytechnic Institute*
ROBERT D. HARRIS, *Rensselaer Polytechnic Institute*
EDWIN J. HOLSTEIN, *Rensselaer Polytechnic Institute*
F. H. KNELMAN, *York University*
K. J. KOOYOOMJIAN, *Rensselaer Polytechnic Institute*
T. H. LEITH, *York University*
EMANUEL LEVINE, *Rider College*
THOMAS S. MENDENHALL, *Colgate University*
FLOYD V. MONAGHAN, *Michigan State University*
LEONARD MULDAWER, *Temple University*
GUY C. OMER, JR., *University of Florida*
JOHN C. PALMQUIST, *Monmouth College*
JULIAN A. RIPLEY, JR., *Stanford University*
RICHARD SCHLEGEL, *Michigan State University*
ROLAND F. SMITH, *Russell Sage College*
HARRY SOODAK, *The City College of New York*
K. M. THOMAS, *Jackson State College*
LEONARD W. WEIS, *The University of Wisconsin*
MICHAEL J. ZENZEN, JR., *Rensselaer Polytechnic Institute*

I would like to express special thanks to the Charles F. Kettering Foundation for the financial support of this project at a time when other financial help for such projects was not easily available. The early contacts with that Foundation were with Edward H. Vause and Charles F. Kettering III, whose enthusiastic support has continued to this day.

I am grateful also to the Trustees of Rensselaer Polytechnic Institute, to President Richard G. Folsom and Provost Clayton O. Dohrenwend for their foresight in establishing a Chair of Rensselaer Professor for myself with complete freedom to pursue multidisciplinary studies that explore the social as well as the technical progress of man.

Troy, New York

V. L. PARSEGIAN
Chairman,
Science Courses for Baccalaureate Education
Rensselaer Polytechnic Institute

Contents

Preface, vii
Acknowledgments, x

1. Objectives and Approach

1.1 Natural philosophy and natural science, 1
1.2 The systems aspects of nature, 3
1.3 New organization brings new functions, 6
1.4 Role of chance phenomena, 6
1.5 The limitations of our approach, 8
1.6 Developing concepts, 9
1.7 Handicaps along the way, 10
1.8 General approach for the studies, 11
1.9 The miracle worker, 12
1.10 The problem of language, 13

2. The Beginnings of Science

2.1 Introduction, 16
2.2 "In the beginning," 17
2.3 Radioactivity dating techniques, 21
2.4 A brief history of the earth, 26
2.5 The beginnings of Homo sapiens, 28
2.6 The beginnings of civilization, 34
2.7 Urbanization and the bronze age, 35
2.8 Writing, 36
2.9 Numbers and mathematics, 38
2.10 Astronomy, 40
2.11 Medicine, 41
2.12 The Greeks and the beginnings of science, 41
2.13 Thales and the Ionian school of cosmologists, 42
2.14 The Pythagorean concept of numbers, 44
2.15 Plato and Aristotle, 45
2.16 Summary, 47

3. The Development of Astronomy

3.1 Introduction, 54
3.2 The celestial sphere as seen by the naked eye, 56
3.3 The movements of the sun as seen by the eye, 59
3.4 The apparent movements of the moon, 64
3.5 The apparent motions of the planets, 66
3.6 What is the shape of the earth?, 69

3.7 The illumination of the moon and planets and the nature of an eclipse, 72
3.8 Models of the universe that account for planetary motion, 73
3.9 Science from the Hellenistic period to the Renaissance, 77
3.10 Revival of learning and the Renaissance, 82
3.11 Copernicus and the heliocentric universe, 84
3.12 Tycho Brahe and Johannes Kepler, 86
3.13 Galileo and astronomy, 89
3.14 Other scientific contributions of Galileo, 92
3.15 Summary, 93

4. Motion and Change in Nature

4.1 Introduction, 99
4.2 Motion of a particle, 101
4.3 Motion in a straight line, 102
4.4 Instantaneous velocity, 107
4.5 The application of mathematics, 108
4.6 Acceleration, 109
4.7 More on acceleration, 111
4.8 Extension to other variables, 112
4.9 A little more on instantaneous rates of change, 116
4.10 Summary, 118

5. The Coming of Age of Science: The Newtonian Period

5.1 Science in the seventeenth century, 124
5.2 Force, mass, acceleration: the first two laws of Newton, 128
5.3 The third law of Newton, 133
5.4 Conservation of momentum, 135
5.5 Kinetic energy and work, 136
5.6 Applying the method of the calculus, 140
5.7 Rotational motion, 141
5.8 Gravitational forces, 148
5.9 Simple harmonic motion, 151
5.10 Forces involved in simple harmonic motion, 153
5.11 Equations for simple harmonic motion, 155
5.12 Energy in simple harmonic motion, 157
5.13 Some properties of waves, 158
5.14 Doppler effect, 166
5.15 Light waves and interference phenomena, 168
5.16 Elastic vibrations in solids, liquids, and gases, 171
5.17 Sound waves, 172
5.18 Supersonic and shock waves, 175
5.19 Summary, 176

6. Systems, Feedback, Cybernetics

6.1 Extension of "systems," 185
6.2 Cyclic character of natural phenomena, 186
6.3 How oscillations increase despite restoring forces, 190
6.4 Modifying cyclic changes: controls, 192
6.5 Introduction to on-off control, 192
6.6 Negative versus positive feedback, 194
6.7 Driving an automobile, 196
6.8 Some control principles—nomenclature, 199
6.9 More on on-off control, 200
6.10 Characteristics of proportional control, 202
6.11 Feedback, 204
6.12 The elements of control systems, 205
6.13 Control concepts: cybernetics, 207
6.14 Some examples of systems, 209
6.15 Functional relationship: notations, 212
6.16 "Black box" approach, 213
6.17 The closed-loop amplifier system, 215
6.18 The nature of "living" systems, 218

7. Introduction to the Space Sciences

7.1 History of space exploration, 224
7.2 Why explore space?, 225
7.3 The mechanics of space travel, 227
7.4 Rockets, 232
7.5 Guidance, control, and communication systems, 234
7.6 Life support in space, 239
7.7 Reentry problems, 241

7.8 Some results of space research 1957–1967, 243
7.9 Distance scales in the universe, 246
7.10 The galaxies, 252
7.11 Radio astronomy, 260
7.12 Summary, 261

8. Probability and Statistical Concepts in Human Affairs

8.1 The importance of statistical reasoning, 265
8.2 Reasoning from sample data to population expectancy, 268
8.3 Errors of sampling, 273
8.4 Probability: its early beginnings, 274
8.5 The basic notions of probability, 277
8.6 Two axioms concerning combinations of events, 279
8.7 Some applications to combinations of events, 281
8.8 The stability of statistical ratios—the law of averages, 285
8.9 The normal distribution, 287
8.10 Range, average, standard deviation, 288
8.11 The central limit theorem, 293
8.12 Inference—mostly statistical, 294
8.13 Summary, 297

9. Observation, Measurement, Evaluation

9.1 The role of measurement in knowledge, 301
9.2 Effect of errors in the planning of experiments, 303
9.3 Errors of observation and of judgment, 304
9.4 Measuring lengths, 306
9.5 Measuring velocity and time, 309
9.6 Temperature measurement, 310
9.7 Calibration of instruments, 313
9.8 Altering the length being measured, 315
9.9 Significant characteristics of measurement schemes, 315
9.10 Psychological measurements, 316
9.11 Vision, 318
9.12 Visual acuity, accommodation, intensity discrimination, depth perception, 322
9.13 Extension of the senses through instruments, 324
9.14 Microscope and telescope, 325
9.15 Concepts of time, 327
9.16 Summary, 329

10. Heat and Thermodynamics

10.1 Early concepts of heat and temperature, 332
10.2 Early theories: the caloric fluid theory of heat, 334
10.3 Mechanical equivalent of heat, 335
10.4 Brownian motion, 336
10.5 Microscopic versus macroscopic analysis, 337
10.6 Atomic numbers and dimensions, 338
10.7 Kinetic theory of heat, 339
10.8 Distinction between heat and temperature, 341
10.9 The pressure of an ideal monatomic gas, 341
10.10 Actual gases, 344
10.11 The phases of water, 345
10.12 The evaporation of a molecular crystal, 349
10.13 Average properties of a large number of atoms, 350
10.14 Zeroth law of thermodynamics, 352
10.15 The first law of thermodynamics, 352
10.16 Order and disorder, 353
10.17 Second law of thermodynamics, 354
10.18 Systems and conservation of energy, 357
10.19 Entropy and time, 358
10.20 Summary, 358

11. Electricity and Magnetism

11.1 The discovery of electric charge, 363
11.2 Magnets and magnetic fields, 367
11.3 Electric currents, 369
11.4 The electron, 373
11.5 Conservation of electric charge, 374
11.6 The electronic properties of matter, 375
11.7 The concept of field, 377
11.8 Moving charges and magnetic field, 381
11.9 Motion of charged particles in electric and magnetic fields, 382
11.10 Accelerating charges, 385

11.11 Summary, 391

12. The Theory of Relativity

12.1 Introduction, 394
12.2 The concepts of absolute space and absolute time, 395
12.3 Michelson-Morley experiment, 399
12.4 Einstein's special theory of relativity, 402
12.5 The relativity of simultaneity, 406
12.6 Time dilation, 408
12.7 Lorentz contraction, 410
12.8 Relationship of time and space, 411
12.9 Mass change with velocity, 412
12.10 Mass-energy equivalence, 413
12.11 General relativity, 414
12.12 Summary, 421

13. Transition from Determinism to Indeterminacy

13.1 The miracle of history, 426
13.2 The approaches of Plato and Bacon, 427
13.3 Technology, rationalism, and the Industrial Revolution, 429
13.4 Energy and modern civilization, 430
13.5 The limitations of Newtonian mechanics, 433
13.6 Philosophy of cause and chance phenomena, 436

14. The Earth and Its Atmosphere

14.1 Man and his environment, 443
14.2 The earth's interior, 445
14.3 The earth's crust and atmosphere, 449
14.4 Energy conversions in the earth and atmosphere, 450
14.5 Self-adjusting systems in the earth and its atmosphere, 453
14.6 Interactions between the inanimate world and human affairs, 456
14.7 The inanimate world in modern technology, 457
14.8 Water, earth, and man, 460
14.9 Time and geological processes, 464
14.10 The processes that build mountains, 467
14.11 Cyclic processes of geology, 473
14.12 Summary, 477

15. The Biosphere as Environment for Populations

15.1 What is the biosphere? 481
15.2 The influence of geography on man, 482
15.3 Influence of geography on political and economic development, 484
15.4 The carbon and oxygen cycles, 485
15.5 The nitrogen cycle, 488
15.6 The energy requirements for life, 490
15.7 Some observations on growth of populations, 491
15.8 Population controls in nonhuman societies, 493
15.9 Population increase in human societies, 498
15.10 Summary, 498

16. Transition from the Classical to the Atomic Period

16.1 The microscopic picture thus far, 502
16.2 Atomic science in the early nineteenth century, 503
16.3 The Mendeléev periodic table of elements, 505
16.4 The "fourth state of matter" and the electron, 508
16.5 Force and energy: concept of causality, 511
16.6 Energy as matter or motion, 513
16.7 Duality of matter and waves, 513
16.8 Determination in macroscopic and microscopic phenomena, 515

17. The Birth of Modern Physics

17.1 Radiation from charged particles, 517
17.2 Radiation from hot bodies, 519
17.3 The birth of the quantum hypothesis, 524

17.4 The photon and the photoelectric effect, 526
17.5 Some further thoughts on the "quantum," 529
17.6 Summary, 531

18. The Bohr Atom

18.1 Experimental evidence from the spectrometer, 535
18.2 The Rutherford scattering experiments, 539
18.3 The Bohr atom, 542
18.4 Bohr's problem, 546
18.5 Success and shortcoming of the Bohr model, 547
18.6 Extensions of Bohr theory, 548
18.7 Explanation of the periodic table of elements, 552
18.8 The covalent bonding of atoms, 554
18.9 The ionic bond: electrolysis, 556
18.10 Summary, 559

19. Wave Mechanics

19.1 Difficulties with the Bohr model, 563
19.2 The early concepts of matter versus energy, 564
19.3 Wave properties of particles, 564
19.4 "Eigenvalues" for atomic processes, 568
19.5 Confirmation of the wave nature of electrons, 569
19.6 Some further considerations of matter waves, 570
19.7 The uncertainty principle, 572
19.8 Probabilities and atomic structure, 574
19.9 Summary, 577

20. Radioactivity and the Atomic Nucleus

20.1 Introduction, 580
20.2 Phosphorescence; discovery of X-rays, 581
20.3 Discovery of radioactivity, 583
20.4 Alpha, beta, and gamma radiation, 584
20.5 The Curies, 585
20.6 The nature of radioactive decay, 586
20.7 The laws of radioactivity transformations, 588
20.8 Measurement and applications of radioactivity, 589
20.9 What is the structure of the atom? 590
20.10 Isotopes and transmutation of an element, 591
20.11 Discovery of the neutron, 593
20.12 The mass spectrograph and nuclear masses, 594
20.13 Structure of nuclei; binding energy, 596
20.14 Mass-energy equivalence, 597
20.15 Some nuclear reactions, 598
20.16 Beta decay; the positron and the neutrino, 599
20.17 High-energy physics and elementary particles, 601
20.18 Summary, 605

21. The New World of Nuclear Power

21.1 The world of 1939, 608
21.2 The fission process, 610
21.3 The power of nuclear devices, 612
21.4 Discovery of plutonium, 613
21.5 Getting a chain reaction, 614
21.6 "Manhattan district" and the bomb, 616
21.7 Following Hiroshima, 619
21.8 The decades since Hiroshima, 621
21.9 The future for atomic power, 623
21.10 Nature of nuclear power reactors, 628
21.11 The world of new atoms and of ionizing radiation, 633
21.12 Effects and products of ionizing radiation, 638
21.13 Radiography and medical therapy, 640
21.14 The hazards from nuclear bombs, 641
21.15 The sun and fusion reactions, 644
21.16 Solar energy utilization, 648
21.17 Summary, 649

22. What Is Scientific Method?

22.1 Limitations of models, 652
22.2 Failure of the Bohr model, 654
22.3 Deterministic versus probability approach, 656
22.4 Science is a human enterprise, 656

22.5 Experience, knowledge, and the domain of science, 657
22.6 Perceptual and conceptual planes, 658
22.7 Experience and concept of temperature, 662
22.8 Experience and concept involving light, 662
22.9 What makes a concept acceptable? 663
22.10 A "systems" approach to concepts, 665
22.11 The concept of time, 668
22.12 Induction, deduction, intuition in birth of concepts, 670
22.13 Deduction, induction, intuition in the progress of science, 671
22.14 A creed for the scientist and the layman, 673

23. Science and the Progress of Man

23.1 The panorama before us, 676
23.2 Science and technology in national affairs, 678
23.3 The contributions of science and of technology, 682
23.4 Problems created by advancing technology, 683
23.5 Science and human values, 685
23.6 The impact of science on religion and philosophy, 688
23.7 Impact of science on education, 692

Appendix, 697
Index, 713

CHAPTER ONE

Objectives and Approach

It is given to man to dream dreams, to search the ties that make him one with the farthest reaches of time and space. The story of that search—varied, personal, purposeful and always fascinating—we sometimes call science.

V. L. P.

THE DICTIONARY definition of philosophy (as commonly used) might read, in part: "The science which investigates the most general facts and principles of reality and of human nature and conduct; a search for the underlying causes and principles of reality." The breadth of this definition makes human knowledge and philosophy almost synonymous, as indeed they were throughout most of history.

1.1 Natural philosophy and natural science

The body of knowledge concerning the world in which we live, the flora and fauna which cohabit this world and the universe of which they are a part, is termed *natural philosophy*.† The term *science* is often used as being synonymous with natural philosophy. About a century ago, owing to the increased volume and specialization of human knowledge, natural science broke away from philosophy. The modern subdivisions of natural science into biology, chemistry, physics, followed in turn. Within the past hundred years the body of knowledge and the degree of specialization have increased to the point that not only have scientists tended to forget their philosophic origins, but also the various sciences have grown apart from each other to such a degree that cooperative effort among them has been made difficult. While the clock cannot (and should not) be turned back, there is great need to correct the harmful effects caused by the separation of disciplines. This will be one of the objectives of this volume.

† The reader should note that the commonness of science and philosophy is still evident in the fact that most doctorate degrees awarded to scientists are Doctor of Philosophy (Ph.D.) degrees.

An introduction to the science of nature tends to become also an introduction to the science of man, since the universe has meaning only to the extent that we comprehend it through human experience, perception, and contemplation. We wish that we might dare to pursue this study with the objective of unifying the whole of man with the whole of the universe, not excluding spiritual values and the designs of nature. But this is quite beyond us. We must leave such vision and daring to men like Pierre Teilhard de Chardin (1881–1955), the French paleontologist and explorer. For ourselves, we must set humbler goals while identifying the spirit of our approach with those given by Pierre Teilhard†, to give the reader:

A sense of spatial immensity, in greatness and smallness, disarticulating and spacing out, within a sphere of indefinite radius, the orbits of the objects which press around us;

A sense of depth, pushing back laboriously through endless series and measureless distances of time, which a sort of sluggishness of mind tends continually to condense for us in a thin layer of the past;

A sense of number, discovering and grasping unflinchingly the bewildering multitude of material or living elements involved in the slightest change in the universe;

A sense of proportion, realizing as best we can the difference of physical scale which separates, both in rhythm and dimension, the atom from the nebula, the infinitesimal from the immense;

A sense of quality, or of novelty, enabling us to distinguish in nature certain absolute stages of perfection and growth, without upsetting the physical unity of the world;

A sense of movement, capable of perceiving the irresistible developments hidden in extreme slowness—extreme agitation concealed beneath a veil of immobility—the entirely new insinuating itself into the heart of the monotonous repetition of the same things;

A sense, lastly of the organic, discovering physical links and structural unity under the superficial juxtaposition of successions and collectivities.

Awareness of these features and distinctions constitutes major elements of our study of nature and of man, a study that builds on a long history of human effort and human experiences. But as we shall see, neither natural phenomena nor human activities ever occur as single events unrelated to other events. A stone starts to move because you throw it. Events have causes. Even more than the details of throwing, there is often need to explore the *causes* and *relationships* of events, and perhaps the mental processes that lead to throwing. We seem to find that not only was there an almost infinite sequence of causes that found expression in the throw, but also that the throw initiated a host of new reactions. For example, the throw may have been the result of mental reasoning and an aggressive posture on your part. But if the stone strikes someone, he may react with enough violence to knock you down and thereby drastically revise your aggressive posture. We can call this reaction a "feedback" effect on the earlier posture. We shall have much to say about such influences.

Our study of man and of nature must take into account the close interrelationship that exists between the two. Man's whole existence requires careful accommodation to his environment, in the cloth-

---

† *The Phenomenon of Man.* New York: Harper & Row, 1965. Reproduced by permission of Harper & Row.

ing he wears for protection, the shelter he builds, the food he cultivates, and the relationship he establishes with his fellowmen. The relationship is not "one way," however, for he may also drastically change his environment. He may burn the forests, improve or ruin the productivity of the soil, or contaminate the atmosphere for many years.

In its simplest form, an interrelationship is present even in relatively static situations, as when we push hard against a solid wall (action) and find that the wall resists just as hard (reaction). In this case the effort seems to be wasted, since the wall does not come down. When we sit astride a bicycle and push backward with our feet against the ground, the results are more satisfactory; the earth remains firm, but we do start forward on the bicycle.

The engineer, being a practical person, is likely to carry the idea further. He designs an electrical system that opens a door when he pushes a button. His mechanical pressure triggers a change in an electromechanical circuit in which electric energy (possibly from a chemical battery) produces a magnetic field which is converted to mechanical energy and releases the latch of the door. Conversion of energy from one form to another is very common in our daily life, whether we ride a bus or train or automobile, or merely sit quietly and eat our breakfast. Actually, a rather complex conversion takes place in our first examples of pushing the wall or riding the bicycle, while from the simple act of ingesting food, the food becomes muscle and nerve energy by a complex metabolic process. Even these elementary examples illustrate how intimately mechanical, electrical, biological, and mental processes become interwoven in almost every situation involving man.

1.2 The systems aspects of nature

Let us pursue this interrelation theme a little further. Most processes have more to them than a simple reaction to every action. There can be a variety of factors or causes that lead to the decision to ride a bicycle. As the rider starts to move on the bicycle, a complex control system—including inner and outer forces—comes into play. His body senses tell him whether he is balanced and steady on the cycle. To maintain balance he manipulates his position around a balance point. If we could examine the motion carefully we would find that he constantly adjusts the cycle back and forth across that balance point. That is, perfect balance is not maintained very long; he "hunts" back and forth across the balance, automatically and often unconsciously. In fact, we find balance, equilibrium, and stability to be the result of a continuous push-pull and hunting about some intermediate point.

How does the rider know what movements to initiate to maintain balance? Visual perception tells him what is going on in his path; for example, that he is headed toward danger. That is, as the cycle moves ahead, vision becomes "feedback" that returns information related to the forward motion. (Of course he had to learn from earlier experience what things were dangerous.) The very *feedback* signals that constitute vision pass from his eyes to his brain centers, where he may interpret what he sees as dangerous and as requiring action. It is only as a result of this feedback of information that his body nerve and muscle systems come into play and steer the cycle away from that hazard. The light energy that enters the eyes produces nerve discharges which initiate cerebral and muscle activity on the part of the cyclist. The source

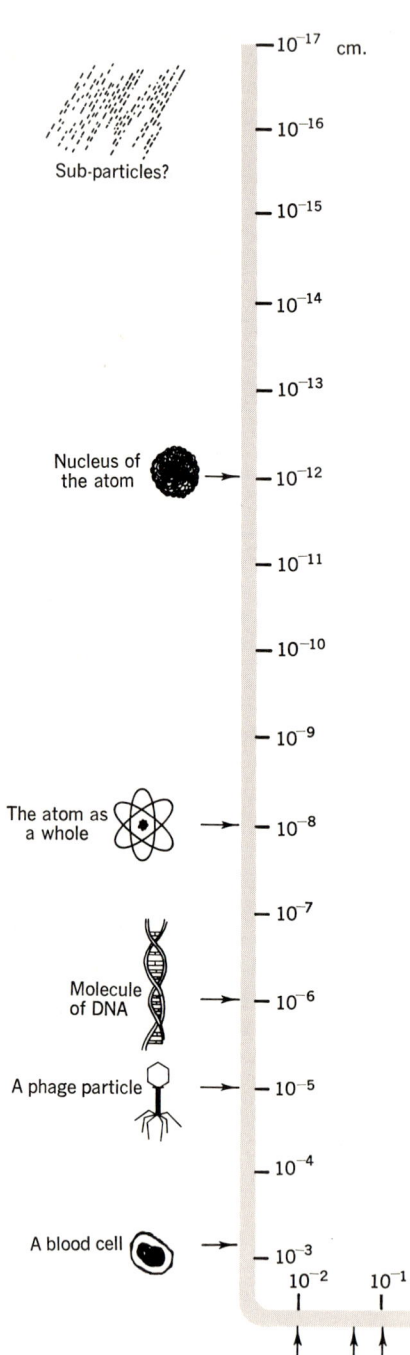

Fig. 1.1. Dimensions of objects, in centimeters, ranging from the smallest subnuclear particles to the distances to the farthest galaxies yet revealed.

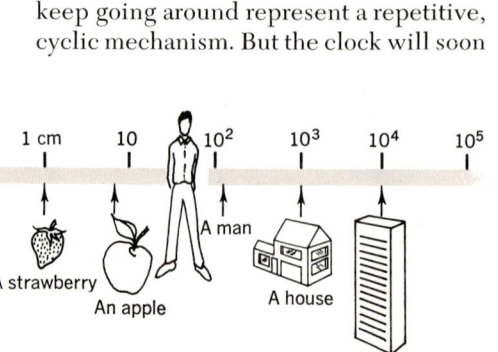

of energy for the balancing movements is the food which is converted to energy by the metabolic processes of the body (except for the negligible light energy which comes directly from sunlight).

This example illustrates an aspect of natural (including human) phenomena which is as important as it is common. We may call it the *systems* aspect of nature. Very often, as in the example of the cyclist, there is a control feature to maintain balance, which is part of the system. Control is usually a result of the presence of feedback information in the system. The feedback information may have many forms. For our purposes we can think of it as being the result or product of some initial change within the system, which becomes directed so as to feed back information that initiates action which can oppose or even reverse the initial action. Thus, in the case of the cyclist, every side motion that affects equilibrium requires quick restoration. Similarly, every move forward gives rise to a new situation (with respect to position, speed, or danger) which causes the eye to send back new information (feedback) as a signal either to continue the same motion or to initiate some change from that motion.

A system may appear to have only cyclic character without the control function's being too evident. The mechanical clock that ticks away and the hands that keep going around represent a repetitive, cyclic mechanism. But the clock will soon run down and fail to perform unless the owner winds the spring again. Here we see another characteristic of systems: The mechanical clock is only a small *sub-system* of a larger one, which includes the one who winds the clock, and a still larger system, which includes the people who built and sold the clock, not to mention the industrial operations that produced the metals and the parts.

Examples of systems and feedback go beyond machines and body nerve and muscle energy. They also involve human experience and understanding, human thinking, and behavior. They often involve community behavior, which in turn evolves around complex codes of ethics that are the products of historical development and of time. Perhaps the most significant example of feedback to modern man has to do with *information that derives from books and news media.* This form of feedback may topple governments during elections, spell the life and death of nations, and give rise to radically new religious movements. The principles will be discussed in detail in due time, and we shall see that the same principles seem to apply whether one analyzes a mechanical system, a living organism, or the larger cycles of nature. This approach to systems and to control is often called *cybernetics*, from the Greek *kybernes* meaning *steersman*.

Systems vary enormously in their physical and time dimensions. Cycles that transform ocean bottoms to mountain tops may take many millions of years. The conversion of plant and animal life to coal and oil deposits may take nearly as long. Large bodies of ice glaciers, requiring tens of thousands of years to accumulate,

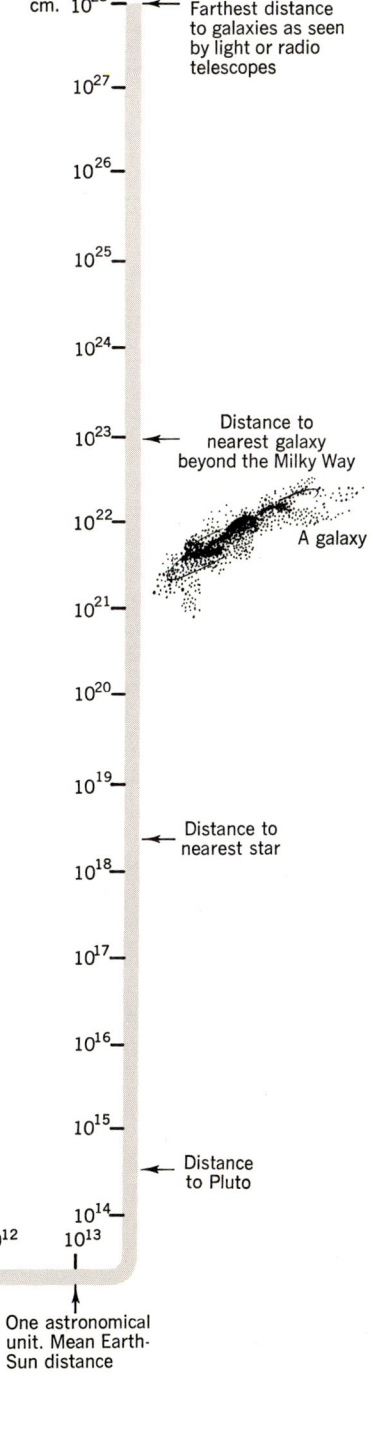

have periodically moved over vast land areas to affect the climate and geology of the earth. Life on earth, the evolutionary processes of earth and sky, and the "birth" and "death" of stars involve even longer periods of time, measured in billions of years. There are shorter cycles of nature as well: The yearly changes of the seasons, the daily repetition of night and day, the microscopic biological processes that produce order and growth out of random motion of atoms, all are also nature's cycles.

1.3 New organization brings new functions

Another characteristic of natural phenomena is illustrated nicely by the systems theme. A system comprises parts which, when working together, may have functional properties that are dramatically new and different from anything that any one part by itself may have when studied in isolation. Certainly, neither the metal wire and rubber that make up the wheel of a bicycle, nor the metal frame and leather seat, nor the rider standing alone could conceive of the possibilities that lie in the combined, articulating system of the whole.

Consider, for another example, the functional properties of the much smaller entity—the atom. As we shall see, sodium is one of the metals whose atoms react quite violently with water, and chlorine is a gas that is quite toxic to the human body; yet when they combine they form ordinary table salt (NaCl) which is very necessary for rather than harmful to the proper functioning of the cells of the body. More than that, the salt crystals need not remain combined in the body but may separate and move in and out of membranes to carry both electric charges and chemical properties to other molecular assemblies. In fact the sodium atoms are found to be very much involved in every nerve discharge experience that sends a signal from a sensory organ to the brain and from the brain to muscle fibers.

1.4 Role of chance phenomena

In some of the examples given thus far (throwing a stone, pushing a wall) there would be no difficulty in describing the phenomena in terms of the force applied, or in terms of cause and effect, action and reaction. In Chapter 5 we shall dwell on such phenomena in some detail in connection with the work of Newton. It was a tremendous achievement when, after long centuries of uncertain groping, it became possible in the seventeenth century to express in Newton's three laws of motion such diverse phenomena as the movements of a stone and the forces that hurled it, as well as to explain the movements of the planets in their courses. With what appeared to be new and undeviating laws, the processes of nature seemed to be quite clear and *determined* to many philosophers and scientists (Descartes, Spinoza, Leibnitz, Laplace). What appeared to be due to chance lacked only a more complete knowledge of causative factors in any situation.

The confidence that was created by these developments did not last. For one thing, it was not easy to apply the new laws to situations involving many interacting bodies. The difficulty becomes magnified when they are applied to very large numbers of very small particles. For example, a cubic centimeter of air (at the freezing point of water, 32°F, or 0°C, and at ordinary barometric pressure) contains about 30,000,000,000,000,000,000 molecules,† all bouncing and colliding at a vigorous rate. The large numbers make it impossible to attempt any meaningful analysis of the behavior of even one molecule while it is subject to collisions from all the others. What is more serious, however, is the appearance of utter randomness about the whole thing, with each molecule subject to *chance encounters* with other molecules.

Such a seemingly hopeless situation (as far as possiblities for rigorous analytic processes are concerned) appeared to suggest a return to allowing *indeterminacy* in nature—a belief prevalent in the period that preceded Newton—were it not for one important additional feature, namely: *The molecules in their total, statistical behavior followed patterns that could be just as precisely defined as were the laws of Newton for single particles.* Despite the haphazard experiences that seemed to be the lot of every molecule, and because the numbers were so large, it was possible to predict quite accurately the spread of kinetic energies of the molecules at any interval of time no matter how small. Apparently, while there was the appearance of complete randomness in the experience of each molecule, there could nevertheless be defined a certain *probability* for each type of experience. The *statistical summation* of all these probabilities, in turn, gave a net result that could be defined just as accurately as could be the behavior of a single stone subjected to a force. Indeterminacy existed for the single member of the class, but not for the class. We shall see examples of this in very many situations, such as the behavior of atoms and molecules in gases, in chemical and biological processes, in radioactivity, as well as in the behavior of individuals in social organizations.

While we are assured that determinacy and complete order does exist in a statistical sense, we continue to feel uneasy as to how such order can emerge out of situations that appear so chaotic. There can be many philosophical arguments on this question, not the least of which concerns the question of how life came into existence. Does chance really rule nature as the Greek atomists claimed? Or is the appearance of chance merely an indication of our inability to conceive of alternative explanations? Did life begin when random collisions fortuitously brought together the right set of atoms? An affirmative answer would deny the need for purpose or special creation. What was the probability that this might have happened in the long stretch of time since the formation of the earth? The probability appears to be very slight, according to some calculations, unless there is in nature a "natural" trend toward greater complexity. We shall see, however, that nature seems to tend toward ever more complete randomness of motions and agglomerations in the molecular world. How, then, is it possible that in the living world of plants and animals there is continuing change in the direction of more complex and specialized organization and function in the living organism? Why and how do the changes in the living world go in the opposite direction from that of the universe as a whole, as defined in the concept we shall call *entropy*?

It is in the light of these questions and phenomena that we view our task. Surrounded as we are with a very intimate relationship between our bodies and the natural and social environment from which we draw sustenance, and with which we interchange action and reac-

---

† As we shall see, this number is more easily written as 30×10^{18}. See also Fig. 1.1.

tion, love and hate, information and misinformation, we must somehow identify the *parts* and *functions* that make up the systems relationship. From the confines of daily experience we reach for comprehension of the vastness of outer space as well as an appreciation for the minuteness of molecular dimensions. From what amounts to a moment in time in the span of life, we try to reach back to the eternity of the past, and contemplate the eternity of the future, to locate that moment of time. From the phenomena that are before us we must sense the flow and the direction of time, as well as the trend of nature's processes, into and out of that brief span. In this effort we continually seek to give order and meaning to the world through concepts that are our own creation. How realistic or true to nature can such concepts be? The question will remain with us throughout this course of study.

1.5 The limitations of our approach

What are the dimensions of time and of space that we shall meet? Suppose we begin with the first hour of the 24 that make up the day. When we go to geologic time and geologic cycles, we must extend our thinking to as many as billions of years, which amounts to writing the one hour with at least 14 zeros after it (10^{14} hours). The time for some astronomical cycles seems to be many times that in duration. Going in the other direction toward events of shorter and shorter duration, we ultimately reach atomic frequencies which may require placing 18 zeros in front of the one hour, or 10^{-19} hour.

For spatial dimensions, our ordinary experiences are measured in feet or in miles. The mile is a common measure for anyone who drives an automobile. But nature's dimensions reach out to 10^{23} miles or more at the astronomical level, and down to as small as 10^{-20} mile at the atomic level. We see that with respect to both time events and spatial dimensions, our daily experiences cover a very small segment of nature's magnitudes, and this segment lies roughly in the middle of the ranges of time and of space.

As we examine the cyclic processes of nature we shall learn something of the geologic and chemical processes that produce oxygen, hydrogen and carbon dioxide, which also at one time established conditions by which the sun's energy could support biological systems through photosynthesis and metabolic processes. As implied earlier, we shall find that the process of living seemingly represents a remarkable uphill struggle that in a few significant respects goes contrary to conventional chemical processes. We shall explore some details of this process in discussing biological processes. Although the biological processes that involve thinking, reasoning, and judgment are at present the center of much discussion (see Hebb, 1966) they remain beyond the scope of this text.

In most respects, our approach must be anthropocentric in that we know the world mainly through our senses of sight, hearing, touch, taste, and smell. We know the significance of hot and cold, of pain, sickness and well-being, through personal experience. There are very elaborate instrumental methods for analyzing the features of the universe, but they are all extensions and refinements of our basic personal sensations and reasoning processes. *In view of this it will be useful to discuss the capabilities and limitations of human sensory mechanisms.* We would like to do this quantitatively wherever possible, since exact knowledge is a goal of science. Unfortunately, many of the aspects of sensory experience that are most important to us as individuals do not permit exact measurement; for example, there are no generally accepted, precise, and valid measures for attractive or unattractive personalities, for joy or sorrow, for anticipation or disappointment, although these constitute the most meaningful reactions that we experience as individual human beings. Nevertheless, in the development of their disciplines, social and behavioral sciences such as anthropology, psychology, sociology, and economics have devised methods and concepts by which to deal with such sensations and human reactions. Just how to fit their findings into a meaningful structure that includes the physical sciences is one of the contemporary problems of science and philosophy. Unfortunately, their progress is handicapped by an inappropriate emphasis on the physical sciences as being "exact sciences" in contrast with the others.

We shall find to our dismay that even in the physical sciences *few phenomena are so well understood that we can apply the term "exact sciences" to them.* To begin with, every measurement we attempt is subject to error, sometimes gross error. To be sure, the mathematical expressions that describe such phenomena as gravitational attraction and electromagnetic phenomena seem to be highly accurate. But we have difficulty in explaining the basic causes of these phenomena despite our ability to measure their effects to a high degree of precision and to utilize them in a thousand applications. Often we are driven to accept basic concepts that run counter to human experience, both in the realm of small, atomic dimensions and in the expanse of the universe. Our discussion of the concepts of quantum mechanics, relativity, astrophysical phenomena, and many others will leave us deeply puzzled. We are likely to find it difficult to comprehend the existence of an outer limit to the universe, and yet it will be equally difficult to assume that there is no outer limit to the universe.

A critical analysis of the things we know and the things we do not know may remove some of the aura of infallibility from the physical sciences. Nevertheless, great credit is due to the physical sciences for the progress that has been made. Despite inadequate understanding of basic facets of nature, we shall find that careful measurements and observations have led to hypotheses which have been very useful for guiding us toward new information and for developing improved concepts.

1.6 Developing concepts

What methods are useful for going from observation to concepts? We can identify three:

(1) Everyday experience with specific situations sometimes reveals a pattern of behavior that is so general as to suggest a general trend, which we formulate into a law of nature. For example, the frequent observation that objects that are hot (say, a cup of coffee) always lose heat to the colder surroundings, and never (of themselves) gather heat to become hotter than their surroundings, eventually suggested to physicists the laws of thermodynamics. In the same way, what we learn by playing billiards can be very instructive in revealing more general laws of mechanics. These are examples of the first, or *inductive* approach.

(2) Having achieved a general concept or theory by some process, one may de-

duce from it the results that apply to specific situations, such as to the behavior of a particular cup of coffee, or to more complex situations involving flow of heat. The generalities drawn from observing the interactions between billiard balls might be useful for playing the next time, or for analysis of other colliding bodies. This second method is called the *deductive* process. The solving of mathematical and engineering problems most often pursues this approach, since they begin with general principles and theories of good practice.

(3) Sometimes a situation does not yield to the formal experience and logic of either the inductive or deductive approach, yet finds solution and understanding through a process called *intuition*. This third approach, to which the term *hunch* is also applied, has been very important.

Actually, any case history of a scientific or mathematical discovery is likely to reveal that important achievements of science reflect the use of all three approaches. We shall go into these matters to some extent in Chapter 22, where we describe how experiences and observations become *concepts*.

1.7 *Handicaps along the way*

The examples presented previously emphasize the fact that science and scientific effort are not cold, logical pursuits, but in fact are very human experiences, subject to human virtues and human weaknesses. To the limitations due to sensory mechanisms and to the procedures for problem solving, which utilize processes of intuition, induction, and deduction, we must add the tendency of human beings to become set in attitude. This is observed in the influence of the science of one generation on the thinking of succeeding generations. New observations often tend to provide data that support existing theories and concepts. Sometimes the concepts become so woven into society's thinking that they are accepted as the "common sense" of philosophies and of religious beliefs, to an extent that society develops a rigidity of attitude that resists concept modification, thus delaying further progress. The distinctions between responsibility and stability as compared to inertia and rigidity are not easily drawn in human affairs. Those who in seeking stability and order give the names unstable or irresponsible to those who seek change are in turn called conservative, rigid, and like terms. There have been many occasions when the achievements of one generation have shackled the freedom of pursuit of later generations. For example, the tremendous achievements of Aristotle and of Ptolemy posed obstruction to innovation many centuries later (in Galileo's time) because the intervening generations had given theological significance and the label of "common sense," "ultimate truth," to the findings of the earlier period. The weight of Newton's authority similarly discouraged development of the wave concept of light. Certainly, the weakness lay not with the findings of Aristotle or Ptolemy or Newton, but with the use made of them by succeeding generations. We cannot be sure that present-day theories are not serving to hold back new and better concepts that explain the notion of man and his place in nature. An example in our day is the difficulty of pursuing unhindered such topics as the origin of life, evolution of species, the significance of man in nature, or any of a dozen other themes without encountering complications that have theological or moral implications to some church or community.† Nor is it unusual for scientific and professional societies to resist changes that threaten an existing structure.‡

1.8 General approach for the studies

We noted earlier that the study of natural phenomena has become so complex as to require specialization on the part of those who go into professional studies such as physics, chemistry, biology, geology, medicine, engineering, to name a few. While specialization cannot be avoided, overspecialization tends to isolate one scientist from another, one profession from another. Someone has said that with specialization, a person learns more and more about less and less. There have been important moves in the reverse direction recently; for example, biologists, mathematicians, and physical scientists have been cooperating in new areas called biochemistry and biophysics, with the result that there has been much faster progress on problems which could not be resolved by any one of them individually.

In this book we shall try to disregard as much as possible rigid division between disciplines. There must be, of course, a systematic approach to the various topics. In general this makes the studies of the first two semesters predominantly concerned with physical science topics (physics of particles, earth science, atomic physics, and so forth). The third and fourth semesters are similarly oriented toward chemistry, biology, and ecology. But we shall find that from the beginning to the end, the discussions and examples will attemp to view the unity of nature. The first semester emphasizes several important concepts, the first being the systems theme mentioned in connection with the bicycle rider. The second concept relates to the randomness and probability aspects of natural phenomena.

An important characteristic of nature is its state of continual change and movement. The heavens, the waves of the sea, the depths of the earth, the grass of the field, the cells of the human body, the thought of the moment, all are in a state of continual movement and change. *All these involve energy transformations of nature,* and so the uses and control of energy will constitute a main theme. The text will continue to reflect this state of flux and of energy exchange as being the characteristic of nature. In this connection we shall have occasion to identify the transition from the way of life which we enjoyed until the 1940's, with conventional chemical-energy sources, to the

† Until 1967 the teaching of the principles of evolution was illegal in Tennessee. The famous "Monkey Trial" of Tennessee brought world attention to the situation in the 1920's. The experience of the Jesuit Father Pierre Teilhard de Chardin, who was also a distinguished paleontologist, offers another example of imposition of restraints for religious reasons. In 1938, he was appointed Director of the Laboratory of Advanced Studies in Geology and Paleontology in Paris, but the outbreak of World War II kept him in China, where he had been on a paleontological mission. Prior to this appointment, he had been denied the right to teach or to publish his major works because of his different approach to the theological concept of original sin and its relation to evolution. When he finally returned to Paris in 1946, he was enjoined by his superiors not to write any more on philosophical subjects. When he moved to the United States in 1951, he was persuaded to leave his manuscripts with a friend so that they could be published after his death, when the permission of authorities would no longer be required. The reader is urged to read as many as possible of his remarkable works.

‡ See Luchins and Luchins in references.

dramatic political, social, and technical revolution that came with the "New World of Atomic Energy."

The path leading from the advent of life and of man to the current pages of science journals has been long and tortuous. The slow, uncertain progress of early centuries gave way to a faster pulse, faster rate of change, faster rate of progress. Strangely enough, *the present stage of scientific development is the most knowledgeable, the most active, the most productive,* and yet in some respects, of all the periods of science since early history it is the *least certain* of its foundations. We shall review some of this history at appropriate points, with the hope that the reader will achieve some feeling for the excitement of this period of history. We, as did our ancestors, are still searching for the fundamental forces that hold the tiny fragments of the atom together, that hold the heavenly bodies to their courses, that suddenly give something called life to molecules, and that make up a human being who thinks and wants to learn the secrets within himself. We are living in a period of technological development in which man has achieved the power of atomic energy, and the skill to transform the earth into paradise or into a hell that matches the conjectured abode of the damned. The progress of science has always been an intimate part of the commerce, religion, military conquest, and social structure of its day. We shall want to note the character of forces and influences that bear on us in the present decade.

1.9 The miracle worker

While a formal definition of science and of scientific methods is not necessary at this point, we offer a simple illustration from the experience of Helen Keller.

In the play, *The Miracle Worker,* the author William Gibson portrays the heart-rending experiences of Helen Keller, who became deaf and blind as an infant. As a handicapped child, Helen was permitted to live a life of anarchy, entirely free of rules of behavior. Her behavior was animal-like. She utilized the sense of touch, of smell, of taste. She knew affection, and knew the reactions of her own body to heat and cold and discomfort. Her world was exceedingly small and chaotic. The new Irish governess, Anne Sullivan, on being given the task of helping the child, sought to develop an awareness of self-discipline, of law and order. How Helen resisted and fought back! How intensely and passionately did she cling to her state of anarchy! Those who were privileged to see Anne Bancroft's performance as the nurse in the play or film are not likely to forget the struggle between the two.

There was delight in the home when Helen Keller conceded a little by folding her napkin at the dinner table. But the supreme moment came when she suddenly realized that each of the things she touched and lived with had a name. By giving attention to learning the names of *doll, water,* and the *tree,* she could give meaning to the unseen chaos that surrounded her. By exercising self-discipline, her world suddenly became alive with meaning and activity. She could add a new speech facility to her expression and use new capabilities for analyzing and understanding the world around her. Animal-like noises and behavior gave way to disciplined, intelligent search, which developed a most notable example of human progress against severe handicaps.

While we would discount the implications of the word "miracle" in this connection, the history of science has had some close parallels to the experience of Helen Keller. The early civilizations observed the heavenly bodies above and touched the earth and water below, but feelings of fear, awe, and reverence pervaded all their knowledge. Many centuries passed until there came awareness of guiding laws and principles. Once born, science continued to be subjected to disciplined analysis, using new knowledge to analyze the old. As a result, what we call "science" has become not knowledge in any fixed or enduring sense but is rather a search for new ideas and new explanations. The process has brought many new developments, new avenues of thought, new means for communication, transportation, and opportunities for living. For some, science has destroyed the mystique of nature and of man; for others, it has deepened the mystique by opening up uncharted realms of ideas.

1.10 The problem of language

Although the means for communication have never been so favorable as they are now in the form of radio, television, journals, books, newsprint, and highly sophisticated private and government news organizations, it seems at times that the world has rarely experienced so many barriers to the progress of understanding as those that exist at the present time. There is, first, the problem of language. While with Helen Keller we can feel great satisfaction in the philosophic insight and flexibility that accompany the use of symbols and words, the problems of communication and of understanding remain.

There are the differences of national language and dialects that separate peoples into a thousand and more noncommunicating segments. Of course, dictionaries are available to translate words such as "legal," "serene," "democracy," "peace-loving," "aggressor," "truth," from one language to another. But how can a dictionary assure that these words will evoke identical understanding and emotions in New York, Moscow, Peking, or Bombay? Symbolism and language fall far short of communicating complete understanding, for their true significance derives not simply from dictionaries but from individual perceptions, which are themselves the product of particular events and experiences and of social relationships that converge uniquely on each individual. It is this difficulty that makes it possible for nations to establish "curtains" of separation and misunderstanding, and to develop almost any interpretation they wish to ascribe to words and concepts.

Nor is the problem of symbols and language absent in the sciences. In fact each scientific discipline tends to develop its own terminology, to such extent that specialists from different disciplines have great difficulty in understanding each other. It is not altogether out of place to recall the answers which the several blind men gave to the question "What is an elephant?" as they explored various parts, one the trunk, another the legs, another the tail.

Fortunately, the sciences have achieved a strong ally in the use of mathematical symbols. Mathematical symbols and equations at once permit a degree of preciseness of meaning, generality, and manipulative flexibility, which far exceed the capability of words. A symbol X (or Y, or any other letter of the alphabet or graphical notation) can be chosen to represent an object or phenomenon; thereafter one can proceed to multiply or divide that symbol without loss of meaning or clarity. Indeed the progress of science would have been reduced to a slow crawl even during our own century had there not been developed the conciseness of expression and of manipulation made possible through mathematical formulations.

Most of us, unfortunately, fail to take advantage of the virtues and capabilities of mathematics. For this reason the present text also will depend more on explanation through physical and graphical means than through mathematics. We hope, nevertheless, that such use as we do make of mathematics will give the reader some feeling for its inherent beauty, simplicity, and flexibility, qualities that enable him to translate simple experiences into concepts and relate these concepts to other experiences.

The reader will therefore understand our difficulties as we try to develop some understanding of the very substantial body of information and concepts called science which is based on demonstrable experience. Even more important than the communicating of information, however, we must probe the *processes* by which science obtains and continuously improves on this information before our very eyes, since the latter is more representative of the progressive nature of science. The reader will not become a scientist, and is not likely to learn how to fix an electric bell or the carburetor of his automobile from this exposure. He can nevertheless gain new insights and fresh interest for the world of which he is a part, and combine a spirit of search with tolerance toward the imponderables and foibles of nature and of his fellow man. We hope also that he will learn to view the progress of science with a sense of personal involvement and to view the concepts, philosophy, and methods of science as an intimate part of his cultural heritage.

Questions/Discussions

Although the contents of this chapter present very little on which to base problem assignments, nevertheless they do pose questions that the reader can pursue to some advantage. (You will realize very soon that neither the following questions, nor most questions of life, permit single, unqualified answers.)

1. Give examples of the most isolated systems (with respect to interactions with outside bodies) that you can imagine. If there continue to be interactions with bodies or systems outside your chosen system, describe their nature.

2. Continuing with Question 1, can you identify a system in which the interaction with other bodies does not involve energy exchange?

3. Explain in your own words and with examples the meanings encompassed in each of the paragraphs quoted in Sec. 1.1 from Teilhard de Chardin. (It would be a very good experience to write out your answers and then to make them the basis for exchange among students and for class discussion.)

4. Think over your personal experiences of a typical day and identify them as one of the following categories:

(a) Some act or decision on your part which was based on the *probability* that something would (or would not) happen.
(b) Some act or decision on your part that was based on an *inductive* process; a *deductive* process.

(c) Some act or decision based on a "hunch" or *intuition*.

5. What would you say is the difference between acts based on logic involving probability considerations and acts involving intuition?

6. Can you list specific new questions and ideas that occurred to you with the reading of this chapter?

7. What factors lead an individual to accept or to reject a new idea? Can you give examples that have to do with each of the following factors: (a) economic, (b) political, (c) philosophical, (d) religious, (e) scientific, (f) institutional, (g) personal?

References

There are very many scientific and semiscientific books that relate to the topics briefly noted in this introduction. The following books, many of them available in paperback, can make good additions to the student library.

Bronowski, J., *The Common Sense of Science*. New York: Random House (Vintage book). *This little book is both informative and thought provoking.*

Du Nouy, Lecomte, *Human Destiny*. New York: New American Library (Mentor book), 1947. *This book also will appeal to many readers for its blending of scientific and philosophical thought.*

Hanson, N. R., *Patterns of Discovery* (Paperback). New York: Cambridge Univ. Press, 1965. *An excellent analysis of scientific discoveries.*

Hebb, Donald Olding, *Textbook of Psychology*, 2d ed. Philadelphia: Saunders, 1966.

Luchins, A. S. and Luchins, E. H., *Rigidity of Behavior*. Eugene, Oregon: Univ. of Oregon Press, 1959. *A comprehensive treatment of personality, and of social and educational conditions that foster rigidity of attitudes.*

Newman, James R. (ed.), *What Is Science?* New York: Washington Square Press, 1961. *A series of twelve essays covering natural science.*

Snow, C. P., *The Two Cultures and the Scientific Revolution*. New York: Cambridge Univ. Press, 1959. *Professor Snow created very deep controversy with this little volume of lectures.*

Teilhard de Chardin, Pierre, *The Phenomenon of Man*. New York: Harper & Row, 1965. *The interested reader will find this and other writings of this sensitive thinker to be very worthwhile.*

Weisskopf, Victor F., *Knowledge and Wonder*. New York: Doubleday (Anchor book), 1963. *This little volume will be useful throughout the year.*

Whitehead, Alfred North, *Modes of Thought*. New York: Putnam (Capricorn book), 1938, 1958. *Clearly for the deep thinker.*

CHAPTER TWO

The Beginnings of Science

They that know the entire course of the development of science will, as a matter of course, judge more freely and more correctly of the significance of any present scientific movement than they who, limited in their views to the age in which their own lives have been spent, contemplate merely the momentary trend that the course of intellectual events takes at the present moment.

MACH [1900]

THE ENGLISH word "science" is derived from the Latin *scientia*, meaning knowledge. Current use of the word often implies the study of natural phenomena and sometimes also the study of technological developments that accompany new ideas or useful knowledge. Our preference is for the broader definition, which is also closer to the Latin meaning. Science is more than a body of knowledge or information, however. It encompasses the processes or *methods* by which knowledge comes into being. It also includes attitudes toward existing knowledge and toward the pursuit of new knowledge. Indeed, the very *attitude* with which questions are posed and pursued may determine the effectiveness of science.

2.1 Introduction

An introduction to science should provide some understanding of the circumstances and the often tortuous mental processes by which man has achieved present-day science. Science is a human activity. In fact, we shall see that science is an activity in which social situations and personality factors come strongly into play. Over the centuries its foundations and progress have been intimately associated with the progress of technology, art, literature, social and political innovations, as well as with the introduction of new philosophic ideas. Paraphrasing Carlyle, a history of science often becomes the history of individuals whose activities encompass many fields of activity. Progress in one area often introduces a stimulus for developments in other areas. Each new application of metallurgy, each new excursion into space or across the oceans, each significant communication from one community to another community, each new literary contribution represents and invites potential values for the progress of science. The invention of a lens made possible more careful study of the movements of the heavenly bodies as well as of microbes. Even the invasion of one nation by another sometimes influences scientific progress by extending and combining the influences of differing civilizations.

In the course of our brief historical review we shall have occasion to mention ideas held by people in the past and which appear rather naive in view of "sophisticated" contemporary concepts. It is tempting to adopt an attitude of condescension and even disbelief of the naive notions of the ancients. Some reflection on the great changes in concepts that have occurred over the past seventy years suggests that we, too, may appear to our grandchildren as having been quite naive.

In this chapter we shall attempt a bird's-eye view of the beginnings of earth and of man. Aside from the information that will be gathered along the way, our purpose is to gain a sense of time in terms of the magnitudes with which geologists and paleontologists count time. We shall see how science provides a means for extending our thoughts beyond the immediate confines of daily life into the far reaches of the past and of the future.

2.2 "In the beginning . . ."

Much of our study will be concerned with the relationship of man to the earth environment of which he is a part and to which, until recently, he has been confined. There is good justification, therefore, for beginning our studies with an analysis of the earth environment.† But

† The theories of the origin of the earth and of the solar system constitute a branch of

just as a study of the events of a particular day in the life of an individual may appear meaningless without some knowledge of the events that led to that day, so a study of the present structure of the earth requires knowledge of the history and the processes that gave rise to this structure. How, we might ask, is it possible for us, living as we are in the present, to probe into the dim past? We can go back into the past for some 5000 years, through the records of previous ages that remain in the form of books, scrolls, cuneiform tablets, and hieroglyphic inscriptions. The records become increasingly sparse and incomplete as we go farther back, however. Archaeological discoveries are constantly adding tidbits to this information, in some cases reaching back 15,000 years. Even this extended period appears to be of trivial duration in comparison with the much longer periods of geologic processes. In our investigations we are faced with the difficulties of tracing events back to origins that are completely beyond the scope of what now constitute human memory and human experiences.

But just as the habits and activities of people in a day or in a year provide an index to the activities of previous days and years, so we can expect that the geological processes that are taking place at the present time may give some clue to

astronomy known as *cosmogony*. Until relatively recent times, the creation of the earth was considered one and the same with the creation of the universe. More recently, however, with man's growing knowledge of the nature of the stellar universe, cosmogony has assumed the more restricted meaning given above. The creation and evolution of the universe is the subject of *cosmology*. Although we might logically begin our story with cosmological theories rather than cosmogonical, we shall make a "long story" short by starting with the earth.

Fig. 2.1. Examples of a slow process changing the earth's surface. (a) (Above) Aerial view of a glacier (U.S. Forest Service). (b) (Below) A glaciated valley. Notice U-shaped cross section. (Courtesy U.S. Geological Survey.)

the events of earlier ages. We observe that nature is in a state of continual change. There are earthquakes, tornados, hurricanes, volcanic eruptions that on occasion dramatically modify the face of the earth. Even these dramatic changes seem to follow a characteristic pattern. The seasonal changes that occur more gradually also leave characteristic effects. We can observe how seasonal rains steadily erode land areas, and how hills of dust that are blown about by the wind and the silt that becomes deposited by rainwaters and overflowing lakes and rivers are capable of covering the earth with thick layers of sediment. We can surmise that a combination of these processes will erode mountains into valleys and sea beds, layers of sediment eventually into beds of rock, and that upheavals will elevate the bottom of a sea into high ground.†

It was thinking somewhat along these lines that led James Hutton, a medical doctor in Edinburgh, to present a paper to the Royal Society of Edinburgh in the year 1785 titled, "The Theory of the Earth." In this paper, he pictured the earth as being in a state of continuous change, undergoing repair as well as wasting away much as the cells of the body simultaneously suffer both destruction and renewal. He was convinced that:

(1) The history of the earth can be explained in terms of natural forces that are observable and acting today.

(2) The principal changes in the earth have been more the result of slow, gradual transitions than of catastrophic upheavals.

† For an excellent photographic essay on erosion and geologic change see *The Earth*, by Arthur Beiser, Life Nature Library, New York, Time, Inc., 1963. Figures 2.1(a) through 2.1(e) in these pages illustrate some of the phenomena that change the face of the earth.

These propositions, called *uniformitarianism*, eventually gave life and direction to the science of geology.† They also prepared the way for explaining the evolution of biological life as propounded by Darwin in *Origin of Species* in 1859. The concept of evolution directed by natural forces became one of the guiding principles by which to understand many of the changes affecting both earth and man.

Present-day experience still offers two alternatives with respect to the major causes of geologic changes, namely, (1) through catastrophic events at infrequent, irregular intervals, or (2) by slow processes extended over long periods of time. The supporters of the catastrophic concept point to earthquakes and volcanic action, while the supporters of uniformitarianism depend on slow erosion and sedimentation. How can we reconcile the two views or determine which of the two types of processes has been the more influential through the history of the earth? Actually a conflict exists only if one attempts to see all the phenomena from one theoretical position. In this case there need be no conflict between the two views, since the action of the one process does not exclude the action of the other. We shall make the assumption that the surface characteristics of the earth have been determined by both the slow evolution emphasized by the uniformitarianists and sudden, catastrophic changes superimposed at random intervals. In pursuing geohistory, we shall note that the slow evolutionary process proposed by uniformitarianism appears to dominate the overall picture, although catastrophic events may affect small regions.

† The idea was not wholly new with Hutton. Others before him had begun to think along these directions, but Hutton succeeded in giving direction to the ideas.

It is worth noting that the concepts (of catastrophic versus slow, "uniform" processes) considered above in connection with the evolution of the earth can be considered also in connection with other features of nature such as the evolution of life forms or of human social or political institutions. In the latter, wars, disease, and famine are catastrophic events that may greatly affect the populations concerned and sometimes alter the pattern of social or political trends. On the other hand, the slow, gradual changes that occur as a result of population growth, increasing knowledge and technology, or the changing needs of a population may develop even greater pressures and long-term effects on social patterns. Genetic changes presumably can be of slow, evolutionary nature without precluding more sudden influences. In this connection we must recognize that the study of human development cannot isolate man from his environment. A change in the earth environment, such as a slow increase in the concentration of oxygen in the atmosphere, may promote slow but eventually large-scale changes in biological development without producing observable effects in the biological systems in any one generation. When we discuss the biosphere in a later chapter, we shall learn that the presence of life has greatly modified the surface and atmosphere of the earth, which in turn has resulted in even larger modifications in the nature of life. (We shall refer to the interaction as having the nature of *feedback* interaction in a *systems* context.)

To continue our story of the earth, the geologist probes into the layers that make up the crust of our planet. He notes that the erosion of river beds, such as produced the Grand Canyon, and the shifting and cracking of the earth's crust

Fig. 2.1. (c) *Aerial view of a portion of the Grand Canyon. Note the layered nature of the rock, which denotes the sedimentary processes that formed these rocks.* (From Time, Space and Matter, *copyright 1964, Princeton University.*)

in many areas often reveal many layers of rock piled one on top of the other. He traces and analyzes seams of coal and veins of ore and minerals. He compares the chemical composition of one vein with another and finds that some appear to have had similar origin. He measures the rate with which rocks and earth erode *at the present time* under the relentless pounding of the ocean waves or the flow of a river. Sometimes he even duplicates in the laboratory the conditions that he observes in nature. For example, he observes in the laboratory that some of the compounds present in rocks require high temperature or high pressure, or both, for their formation. This suggests that such compounds were probably subjected to equivalent temperatures and pressures at some time in the past.

From such observations he eventually develops a picture of what appear to be the continuing processes of nature. The eroded materials of high ground that flow away with rainwater and rivers to settle in the seas or other lowlands as sedimentary layers have chemical compositions of the materials from which they were derived. Each layer may, in time, enclose the trees and foliage or bones and shells of its period and thus give evidence of things that lived and died during the time the sediment was being deposited. Eventually, such layers of sediment turn into consolidated rocks without altogether destroying the shape of the dead flora and fauna that became enveloped. Volcanoes may pour their molten rock (lava) over the sedimentary layers, or wind-laid volcanic ash may form younger layers above the water laid materials. The earth may split open, or new mountains may be pushed into the air. What was once the bottom of a lake may eventually find itself the side of a mountain, and undergo erosion anew from the rains. Fertile populous areas can change to arid desert, as appears to have taken place in the area of the Sahara.

The geologist, however, can often decipher the *order* of the layers and the sequence of events, and estimate the probable time of duration for the formation of each layer and event. The paleontologist, whose interests are in the organic remains that are preserved in the sedimentary rock and dust, joins in the search. He is interested in the deposits of limestone (calcium carbonate, $CaCO_3$) because many of these were formed on ocean bottoms from the shells of tiny marine animals. As he uncovers evidences of plants and animals, each type may become associated with a specific geologic or sedimentary layer, and the total may offer considerable detailed information about the earth and its early history and inhabitants. The influences and paths of moving glaciers of the ice ages can also be traced by observing the grinding and mixing of rock and soil and by tracing the debris they left behind. Such glaciers once flowed from the Scandinavian highlands to cover much of Europe, and others once moved southward from Canada to cover roughly half of North America, reaching as far south as Kansas and Kentucky.

2.3 Radioactivity dating techniques

While geologic and paleontological studies reveal the sequence of layers of sediments, erosion rates, chemical composition, and the relative duration of time involved in geological processes, until recently there was little to go on for establishing a precise time scale in terms of some common unit, say, the year. But

Fig. 2.1. (d) (Below) *The Alaska earthquake of 1964 caused substantial damage to the city of Anchorage.* (From Physical Geology, 3d ed., by Leet and Judson, copyright 1965. Used by permission of Prentice-Hall, Inc., Englewood Cliffs, N.J.)

Fig. 2.1. (e) (Below) *Nearly all the islands in the Atlantic Ocean are the tops of volcanoes that have risen from the ocean bottom. Helping confirm this hypothesis was the emergence of this volcanic island in 1957 in the Azores.* (From Fundamentals of Physical Science by Krauskopf & Beiser, copyright 1966. Used by permission of McGraw-Hill Book Company.)

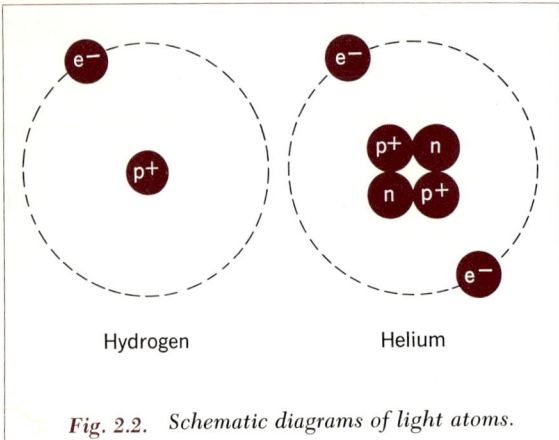

Fig. 2.2. *Schematic diagrams of light atoms.*

within the twentieth century there has evolved a most remarkable technique which utilizes the new knowledge of atomic processes.

As we shall learn in detail, all materials consist of *atoms*. Each atom has a tiny but dense central core, or *nucleus*, and outer electrons.† The nucleus itself is a complex structure consisting of electrically positive charged particles called *protons*, and electrically neutral particles called *neutrons*. In normal atoms the number of protons in the nucleus of the atom equals the number of electrically negative charged electrons. The chemical properties of an atom are determined by this number, called the *atomic number*. Thus, if an atom were to lose a proton from its nucleus (and consequently also an outer electron), there would be a decrease of one in the atomic number and a change in the chemical properties of the new atom. A chemical *element* would be *transformed* into another *element* by a change in the nucleus of the atom.

Toward the end of the nineteenth century, it was discovered that certain atoms spontaneously undergo such transformation—a process called *radioactivity* because the emission of particles and radiation that accompanied the transformation was discovered first, before the changes in the atoms had been analyzed. It is this property of some elements (to be naturally radioactive) that provides us with

---

† These statements on the atomicity of matter and the structure of the atoms are introduced briefly without any "proof." All this material will be reintroduced in a more detailed manner in later chapters. At this point, the intention is to give the student an awareness of the nature of the radioactivity dating technique and its values for determining geologic time. Figures 2.2, 2.3, and 2.4 illustrate some of the atomic characteristics that will be discussed in greater detail in later chapters.

a valuable method of dating over extremely long periods of time.

To understand how radioactive materials are used as a "clock," let us assume that we have a certain amount of radioactive material, of an element A. Assume that we have an extremely pure sample of A (that is, no other elements are present) and that we take the necessary precaution to prevent contamination. Let us assume that the sample of A contains a number, N_0, of atoms.†

Since this material is radioactive, the nuclei of the A atoms can spontaneously change by giving off a particle of matter, or gamma radiation, and be transformed into another element, say B. Although during any short interval of time the transformations will appear to be quite random, we would find that over a longer period *each radioactive element would follow a very well defined time rate of change.* Suppose that a long time, t,

elapses, and that by some chemical means we separate all new B atoms from the original A atoms in our sample. We now find that there are less than the original N_0 atoms of A, while the number of atoms of B has increased from its original value of zero so that the number of atoms of A plus the number of atoms of B equals our original N_0. Let us repeat the same experiment many times, keeping a record of the time when we make the number determinations. If we now make a graph of number versus time for each of the elements A and B, we get the curves shown in Fig. 2.3. We note that with increasing time, the number of A atoms decreases continuously, while the number of B atoms increases correspondingly. Notice that after some time the number of A atoms has dropped to one-half of its original value (in Fig. 2.3 it took two time units). We call this the *half-life* of the radioactive element. After one additional half-life (two additional time units) has passed, we find that the number of A atoms has again dropped by one-half; that is, it is now one-quarter of its original value. *Indeed, the number of A atoms decreases by a factor of two during each half-life.*

---

† We can determine the number of atoms in our sample by weighing it and dividing this weight by that of a single atom of the element. The weight of a single atom can be ascertained by techniques that will be discussed later in this course. In this notion we can think of the subscript 0 (zero) to designate time $t = 0$.

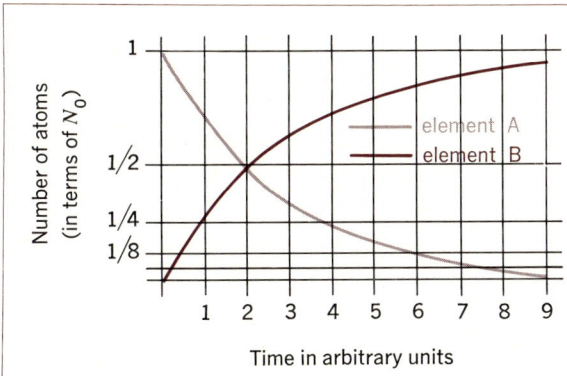

Fig. 2.3. Plot of the number of atoms of original element A, and of decay product, B, as a function of time.

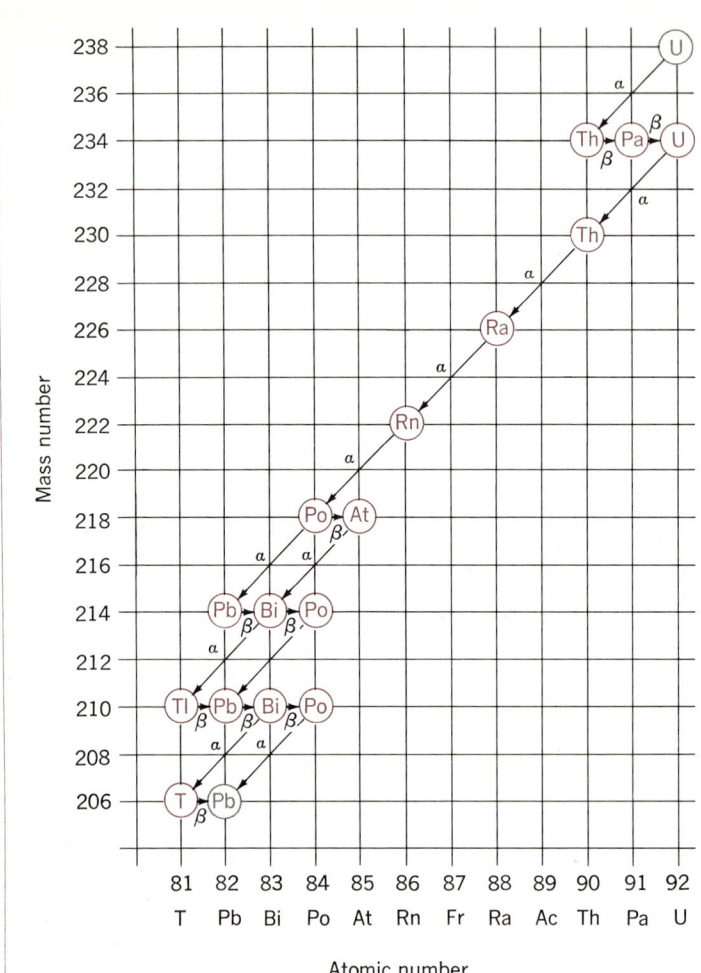

Fig. 2.4. *Uranium-238 radioactive series.* $_{92}U^{238}$ is the "parent"; $_{82}Pb^{206}$, nonradioactive, is the "end product." The radionuclei emit alpha particles or beta particles; occasionally, a fraction of the nuclei of an element (for example, Po^{218}) emits alpha particles while the remainder emits beta particles.

Similarly, the number of B atoms increases in such a way as to keep the total number of $A + B = N_0$.

Now suppose we find a rock sample containing elements A and B in nearly equal amounts. If we make the assumption that originally (when the rock was formed) no element B was present, then from Fig. 2.3 we would conclude that about one half-life had passed since the rock was formed.

In our story so far we have tacitly assumed that element B is itself stable, but in nature this is often not true. B may transform into C, C into D, and so on, each with a different half-life. An example of such a series of decays (or transformations) is involved in the conversion of uranium to lead. This complicates the picture somewhat, but it does not alter the underlying theory that has just been presented. All elements heavier than lead are naturally radioactive. The half-lives of these elements range from fractions of a microsecond to tens of billions of years. Most of these elements are quite rare in nature. However, a few, such as uranium, occur frequently enough to be useful as a clock. Common uranium, U^{238}, has a half-life of 4.5 billion years.† It decays, via many steps, to lead, Pb^{206} (see Fig. 2.4). None of the substeps has a half-life anywhere near as long as 4.5 billion years. (For example, U^{234} and thorium, Th^{230}, are the longest lived and they have half-lives of about 100,000 years.) Thus, a rock that originally contained pure U^{238} in time would have a measurable amount of lead but only negligible traces of the other members of the transformation series.

Suppose that a rock is discovered in which there is a mixture of both uranium and lead, and that after chemical separation the two are found to be about equal in weight. Since the uranium becomes lead with a half-life of 4.5 billion years (neglecting the intermediate steps), we would say that that particular uranium concentration was formed about 4.5 billion years ago. There is a basic assumption here that originally the uranium

gradually came out of solution and became concentrated at that spot without any lead being mixed with the uranium. This appears to be a good assumption because uranium and lead have quite different chemical properties, and the two are not likely to come together in significant amounts under these original conditions.†

Using the U^{238}/Pb^{206} transformations, as well as other transformations of heavy nuclei, rock samples have been dated as far back as 3 billion years.

For shorter times of the order of a few thousand years, which are more common in archaeology, a radioactive isotope of carbon, C^{14}, whose half-life is 5600 years is useful. C^{14} is produced when a high-energy particle interacts with a common nitrogen atom in our atmosphere. The high-energy particles result from the bombardment of the atmosphere by cosmic rays from outer space, and produce C^{14} at a certain rate. The ratio of radioactive C^{14} to normal carbon, C^{12}, in the atmosphere is known. While alive, all plants ingest carbon in the form of carbon dioxide, CO_2, from the atmosphere, and hence some C^{14} is also introduced along with C^{12} into the plant. Each living plant, therefore, has a known fraction of C^{14} mixed with the C^{12} that makes up the body of the plant (or a known C^{14}/C^{12} ratio). But when a plant ceases to live and no more C^{14} is taken from the air, the ratio C^{14}/C^{12} decreases with time as C^{14} experiences radioactive decay. The concentration of C^{14} is determined by measuring the rate of its emission of radiation. This dating scheme, for which Willard Libby received a Nobel Prize, is a valuable tool for archaeological research.

---

† The symbol U^{238} means that there are 238 nucleons in the nucleus. Since uranium must have exactly 92 protons, there must be 146 neutrons. Changing this latter number does not change the element—for there still remains 92 protons—but gives nuclei of different weights, called *isotopes*. For example, if there were only 143 neutrons in our uranium nucleus, we would have U^{235}. This will be discussed more fully in Chapter 20.

† Question: What would have been our conclusion as to the age when the uranium became deposited if the lead we found constituted one-tenth the amount of the uranium?

With the help of these analytical aids it has been possible to piece together some details of the history of the world. There are still very large gaps in our knowledge, but how much more information we have, compared with the information available to people of four thousand years ago! The people of earlier periods struggled for survival against the violence of natural phenomena. Although catastrophes were common and there was little understanding of their causes, in terms of modern physics, it was nevertheless difficult for thinking man to accept chaos as the necessary or dominant characteristic of nature. He sought to find answers to the inexplicable, the social and natural catastrophes, in his search for order in life and in nature. He found it possible to develop tools and weapons with which to improve his living conditions and feeling of security. It was within this context that the gods of early civilizations served, although there were times when the gods were no less capricious than were the forces of nature which they were supposed to control. Nevertheless, it was this continual need to search for order that found majestic exaltation in the words:

"In the beginning God created the heaven and the earth. And the earth was without form, and void; and darkness was upon the face of the deep. And the Spirit of God moved upon the face of the waters. And God said, Let there be light: and there was light. And God saw the light, that it was good: and God divided the light from the darkness. (Genesis 1:1-4)"

2.4 A brief history of the earth

We saw in Sec. 2.3 that radioactivity dating techniques yield ages of the order of 3 billion years for the oldest rock on the earth's surface. Thus the *crust* of the Earth from which we obtain specimens containing uranium has been solid for at least 3 billion years. This still does not tell us how much longer than this the Earth has been in existence, or where it came from and what it was like in its early years. To these questions we now turn.

How the Earth began is a matter of conjecture. Many theories have been proposed and discarded over the past 200 years. The most widely accepted theory considers the entire solar system to have a common origin in a cloud of interstellar dust and gas. About 5 billion years ago this cloud may have started to contract as a result of its own gravitational attraction. With contraction the center of the cloud could have become heated by the great energy generated in the collapsing cloud. (See Fig. 2.5.) At some point in time, perhaps a few hundred million years after the contraction began, the central temperature of the cloud may have reached about 10 million degrees. At that high temperature, nuclear reactions produce new energy† and the cloud becomes a star, which in our case was the proto-Sun.‡

During the collapse of the cloud to form the proto-Sun, some material present in the cloud could have formed rings of dust and gas about the central cloud. These rings, in turn, could have condensed to form the original protoplanets.

The original chemical composition of the proto-Earth was probably the same as that of the present-day Sun: by mass about 70% hydrogen, 27% helium, and 3% of all other elements in the same relative amounts that we find on Earth today. At

† Chapter 21 discusses these reactions more fully.
‡ The prefix *proto* as applied to the Sun and later the Earth and other planets denotes that certain large-scale changes would yet occur to bring these bodies to their present state.

the time the proto-Sun began shining, the proto-Earth was a cold, dense, solid body surrounded by a thick gaseous atmosphere. The light that radiated from the proto-Sun probably heated the gases sufficiently to drive them away, leaving the surface of the planet dry and airless. During the subsequent millions of years, internal changes within the earth (perhaps further gravitational contraction) could have caused the interior to be heated. Volcanism and earthquakes may have freed water and gases trapped below the surface, and hence replaced the lost atmosphere and filled the seas with water. The new atmosphere was probably rich in the lighter gases such as hydrogen, methane, and ammonia, unlike our present atmosphere of nitrogen and oxygen. Since the light gases do not seal out the Sun's ultraviolet light, as our present atmosphere does, high-energy radiation from outer space could have reached deep into the Earth's atmosphere to promote chemical reactions between its constituents. This could have resulted in the formation of complex organic molecules, giving rise to the first living organisms.

Meanwhile, because the Earth's gravitational attraction was insufficient to hold onto the lighter components, the composition of the atmosphere gradually changed (with the help of living things and photosynthesis on Earth) to the present composition of 78% nitrogen and 21% oxygen with a trace of carbon dioxide and other gases. The new mixture has been responsible for the nature of the living organisms that now populate the earth. All of this occurred more than 3 billion years ago. Within the intervening period the continents have formed and movements of the Earth's crust have changed the topography of the Earth's surface to the present form.

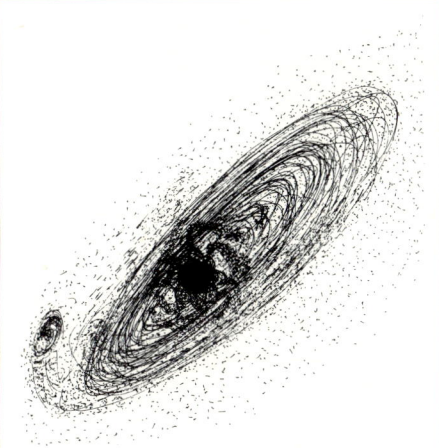

Fig. 2.5. A cloud of interstellar dust and gas may have condensed to form the solar system. An artist's view of the origin of sun and earth.

Figure 2.6 summarizes the various eras of geologic history. The names applied to the various eras and epochs—Mesozoic or Pleistocene, for example—are traditionally employed in geology and are listed here merely for convenient reference. (See also the illustration on the inside front cover.)

2.5 The beginnings of *Homo sapiens*

Since the story of science is part of the story of man, it is therefore proper to ask at this time, "When and where did man first appear on earth?" There are several conjectures about where man originated as well as whether man originated in one place. The popular belief is that he originated in the forest and was a tree dweller. His advance to manhood, it is said, was initiated by his leaving the forest for the plain. We really do not know—he may have originated on the sea coast. Moreover, man may have evolved in several places and even at different times. We shall learn about these problems from the biology part of the book where we shall deal in greater detail with the evolution of species, including man.

Earlier in this chapter we mentioned the importance that evolutionary processes appear to have had both in the formation of the earth as we now see it, and of living things. We wish there were as clear evidences and understanding of the processes of organic evolution as there are of the erosion of mountains. There simply is not available the same degree of understanding for the principal mechanisms or events by which man has developed. We shall continue to use the concept of organic evolution, despite the ambiguity and vagueness of the concept. By so doing we run the risk of appearing to know more than we actually do. The gaps in knowledge are today very wide in this particular area, and overdependence on the general concept of evolution sometimes leads to a tendency toward simply adding more years of chronological time to account for serious gaps in knowledge.

We believe that primitive life forms existed on earth at least a billion years ago. How and when life originated on this planet is a fascinating subject to speculate on and will be discussed in the second volume of this text.† Our knowledge of the development of prehistoric life forms is based on discovery and analysis of fossil remains. From a fossil skull or bone (or, more frequently, only a fragment of such a skeletal member) the paleontologist attempts to reconstruct the basic anatomical features of a prehistoric animal. The total picture that emerges from these studies is far from complete, but it does offer a glimpse of what may have been the history of the development of man.

The history of man revealed by these fragments is neither unambiguous nor universally accepted. In some cases, species appear only to become extinct. When the fossil remains of a hitherto unidentified species are discovered, it is difficult (sometimes impossible) to ascertain whether it belongs to the main line of some evolutionary chain. Because of this uncertainty we cannot state definitely how modern man (*Homo sapiens*) evolved. We do know that about 70 million years ago primates began to appear. These early primates probably included the ancestors of our modern lemurs,

†References at the end of this chapter contain titles of several books on the origin and evolution of life. These books were selected because they are of a level of sophistication appropriate to our course. The interested student is urged to make use of outside readings to supplement the material in this text.

2.5 The Beginnings of Homo sapiens 29

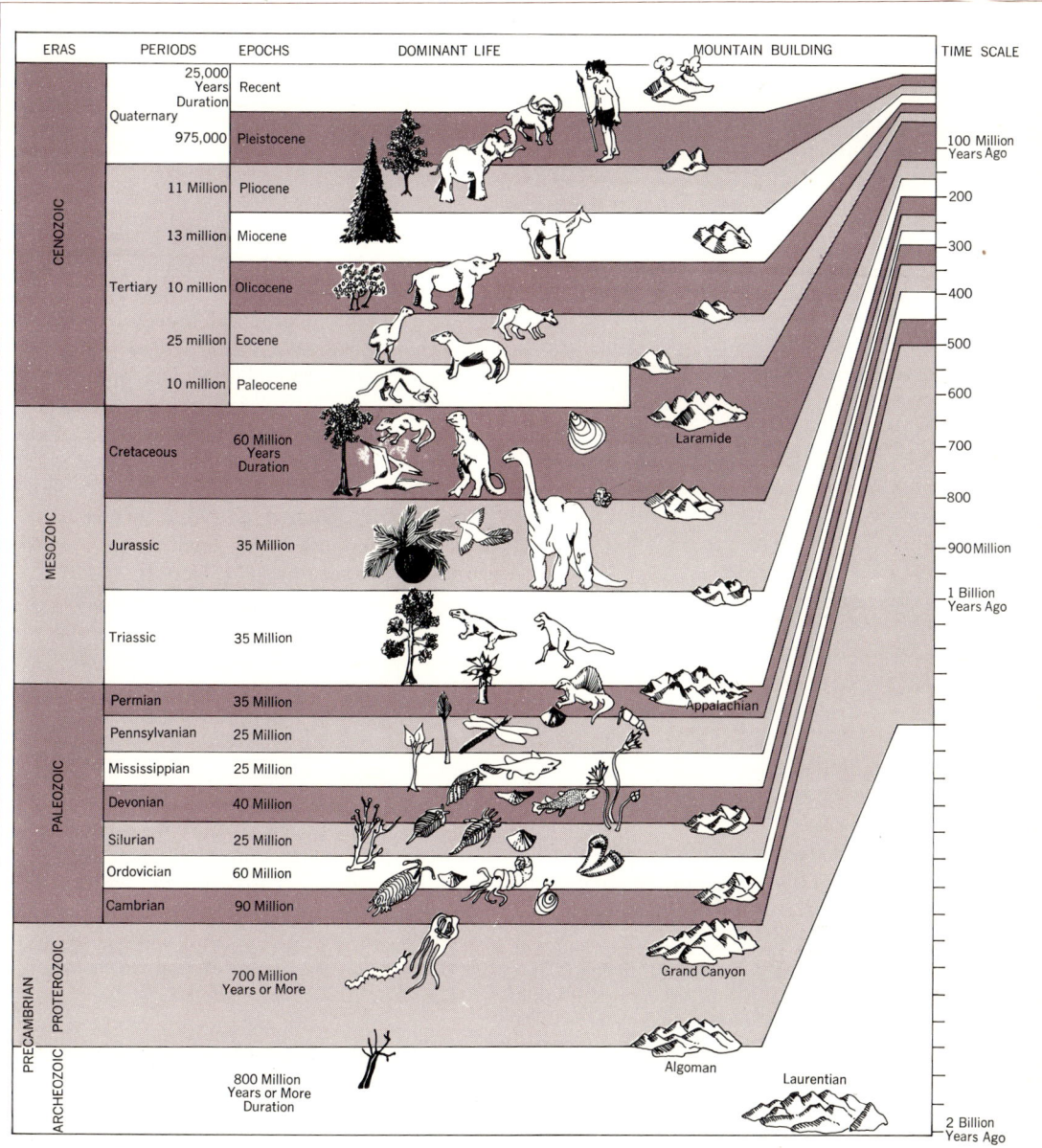

Fig. 2.6. A chart showing the geological history of the earth divided into eras, periods, and epochs. (Courtesy University of Colorado Museum.)

monkeys, apes, and man. By 30 million years ago a subdivision of the primates gave rise to the anthropoids. Soon thereafter the ancestors of the great apes (and possibly of man) became distinct from those animals that evolved into present-day monkeys.

Leaving aside the question of dates, the time and manner by which man became distinguished from the great apes depend in large part on definitions. Even today, man bears many anatomical resemblances to the great apes—the gorilla, orangutan, chimpanzee, and gibbon. If we follow the definition suggested by most paleontologists and anthropologists, we distinguish man by his *ability to manufacture and to use tools.*† Thus, by definition, man is set aside from the lesser animals by these particular abilities of his mind and hands.

The earliest evidence of man, according to the paleontological definition, appears between 1,000,000 B.C. and 500,000 B.C.‡ The remains of Australopithecus (of South Africa), which date from this period, have been found in connection with pebble tools (see Fig. 2.7). Similar development may have also taken place in southern Asia.

By the beginning of the second ice age (about 500,000 B.C.) a new species of man inhabited parts of southern Asia and Africa (and perhaps southern Europe). Grouped together under the single genus Pithecanthropus, they included the so-called Peking man and Java man. These men also produced tools of chipped stone (see Fig. 2.7). In addition to tool making, they took shelter in caves and learned the use and control of fire.

The slow evolution of the human race continued with the appearance of *Homo neanderthalensis* about 100,000 years ago, between the third and fourth ice ages. *Homo neanderthalensis* was a short, heavy figure with a brutish face. He possessed a large brain and was a skilled toolmaker. In general, he was a cave dweller, and his remains have been found throughout Europe and western Asia. Despite his apelike appearance, the Neanderthal man seems to have been capable of having human emotions. He buried his dead ceremoniously in a way that indicates belief in some form of afterlife, perhaps a religion that made ancestors of the family the object of veneration.

The earliest examples of our own species, *Homo sapiens,* seem to date back to the second interglacial period, some 250,000 years ago. It is likely that at various times the different species of man—*Pithecanthropus, Neanderthal,* and *Homo sapiens*—coexisted on earth and that the cultures of these different populations at any given time were comparable. It is still a matter of dispute whether *Homo neanderthalensis* was a direct ancestor of *Homo sapiens* or whether he represented instead the termination of a species. Evidence indicates that these two types coexisted during the last ice age, but by 40,000 B.C., *Homo sapiens* had become dominant and Neanderthal man was extinct. Why did he become extinct? Was he exterminated by *Homo sapiens?* Did the two species inter-

---

† There can be considerable disagreement with the choice of this particular definition for man. A strong argument can be made for other distinctions, such as in man's use of symbolic language for communication or man's ability to plan for future needs. Sherwood L. Washburn has presented the thought (*Scientific American,* September 1960) that tools antedate man and that the use of tools by prehuman primates gave rise to *Homo sapiens.*

‡ Recent discoveries by Prof. Leakey in southern Africa indicate that humanoid remains may date as far back as 2,000,000 B.C.

breed?† What implications are there as to the nature of *Homo sapiens* if the former or the latter question is answered in the affirmative?

Numerous different cultures of *Homo sapiens* have been identified as existing from 40,000 B.C. until about 10,000 B.C., which was the end of the Paleolithic or Old Stone Age. Among these were the various cultures of Cromagnon man, who are purported to be the direct ancestors of contemporary Europeans. In addition to an increased use of stone and bone tools, these people produced some forms of art and religious traditions. They constructed dwellings of wood, stone, and clay. By now, man had not only made material objects but also had developed languages and systems of beliefs which he passed on to his offspring. Moreover, he had developed systems of social relations and group norms and standards, which were passed on to succeeding generations. In short, he had developed a culture and a society that was maintained from generation to generation by various formal and informal methods of teaching, which not only preserved what had been learned and discovered by former generations but which also enabled each generation to apply such knowledge in a productive and even a creative manner to its own uses. Some believe that the tempo of social and cultural evolution quickened at the end of the ice ages. Man's skill as a toolmaker increased. He learned to fashion dwellings and artifacts of skins, wood, stone, and clay. In so doing, man

† It is of interest to note here that although a species is often defined as a reproductive entity, there is some evidence that this may not be entirely correct. That group X does not interbreed may not be due merely to genetic reason but to other factors that keep or cause members of a species to prefer their own kind.

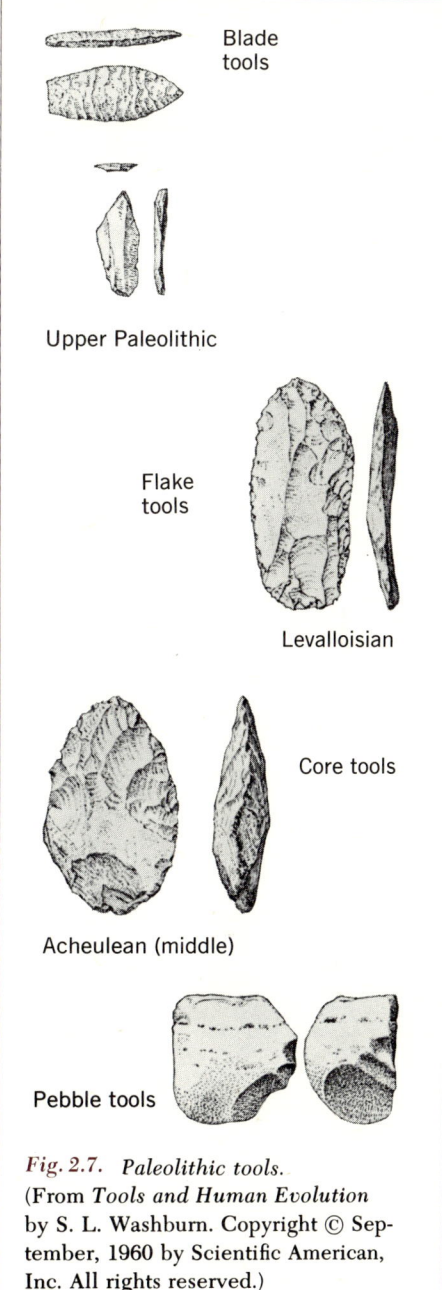

Fig. 2.7. Paleolithic tools. (From *Tools and Human Evolution* by S. L. Washburn. Copyright © September, 1960 by Scientific American, Inc. All rights reserved.)

Fig. 2A. Bronze Iranian arrowheads of the early First Millenium B.C. (Courtesy of the Metropolitan Museum of Art. Gift of Jerome M. Eisenberg, 1962.)

Fig. 2B. Old Akkadian (Near Eastern) clay tablet, ca. 2200 B.C.: contract for the purchase of a house. (Courtesy of the Metropolitan Museum of Art.)

entered into a new stage of cultural development.

Man learned the use and control of fire, which not only affected his diet but influenced his way of hunting and toolmaking. It also gave rise to metallurgy. The fact that man was no longer dependent on the Sun and bright Moon "to see" may have also influenced his social and personal relations. He could now work by the light of the fire, and could sit around it and report past experiences as well as plan new experiences. He domesticated the dog, probably first as a scavenger to help keep his settlements clean but later as a hunting companion. Some evidence exists that the goat was the first animal domesticated by man; by living among them and protecting them from predators, man developed a symbiotic relation with the herd. By the end of the last glaciation (about 20,000 years ago), as the Paleolithic age drew to a close, a relatively sophisticated society of mammoth and reindeer hunters existed in Europe, western Asia, northern Africa, and North America. The later stone-age cultures possessed certain of the nonmaterialistic aspects we attribute to "civilization." They evolved religious beliefs, as evidenced by the ritualistic burials given their dead, their painting, and their sculptured and clay figures. Clay figurines fashioned by the mammoth hunting people of the Balkans and south-western Russia imply the existence of "medicine men" even in this period, as do the cave paintings of France and Spain.

Early man did not, as it is claimed we do, draw a sharp distinction between himself and nature as it acted on him and he in turn acted on it. Natural events were dynamic and living experiences, not the abstractions of modern science and modern religions, which secularized man's habits as these philosophies became the common sense of the masses of people. In short, the effects of the changing weather conditions, the variety and sometimes catastrophic nature of natural phenomena, and the vagaries of the hunt, among other events of his experience which were beyond his control, may have suggested to man that supernatural agencies were at work.

A summary offered by Jacquetta Hawkes is interesting in this connection:

"This ends the account of the emergence, speed and differentiation of man as a zoological breed. It is a history that begins with rival genera and species, then narrows to the races of Homo sapiens, the single species which is mankind. Through the tens of millions of years of the Cenozoic era the human frame, so familiar a possession of each one of us, has been seen slowly shaping among the primates, until by Pleistocene times a large-brained, upright biped by beginning the creation of culture has won human status. Throughout this vast stretch of time the increase in the size and complexity of the neo-pallium or New Brain makes the central theme; in the fossil skulls which are our principal record for the human epic we see the forehead and vault rising, their capacity swelling. Whether or not he is inclined to indulge in the modern name-calling of highbrow or egghead, no one can question that man is distinguished as the highbrow or egghead of the primates. Here, housed within the curved bone plates of the skull, is the most subtle and complex instrument in the world, which at the command of the whole man, has created the rich and varied cultures, the superb individual works of art, the inspiring if never final systems of thought, that make the history of mankind."†

Fig. 2C. Gold ewer, Anatolian metalwork of the Pre-Hittite period, ca. 2100 B.C., height 7 in.; reputedly found near Amasia. (Courtesy of the Metropolitan Museum of Art.)

† See Fig. 2.6 for relationship of the periods referred to.

2.6 The beginnings of civilization

The Paleolithic age saw an increase in the sophistication and diversity of tools produced by some groups of men. Throughout this period man was a food gatherer, a hunter, and a forager rather than a food producer. This demanded that he follow the animal sources of food as they moved about, resulting in a pattern of living that often discouraged the erection of permanent settlements or the acquisition of personal property on any large scale. The fact that children were nonproductive and required protection was a handicap to this nomadic existence. These conditions may have tended to limit the population as well.

After 10,000 B.C. the situation began to change. Man learned how to grow large crops of food, and acquired herds of domesticated farm animals to provide meat, hides, and bone for tools. There was less need for migration; therefore permanent settlements could be established. This community living seemed to increase man's interest in acquiring material possessions. Children could now stay with parents and become heirs to the material possessions of their parents. As a result, larger family groups began to appear.

This new stage of development, which we call the *Neolithic Age* or *New Stone Age*, did not occur as a single period of time, but rather at different times among different peoples. Indeed we have several examples of Neolithic cultures on earth even today, among them the Australian aborigines and some of the peoples of New Guinea. Also, although the name Neolithic (*new stone*) is given to this stage because the use of polished stone implements had replaced the cruder chipped-stone tools of the Paleolithic period, some cultures had introduced the use of metals even before they settled down to a food-raising economy.

The Neolithic period did not spring fully developed from the Paleolithic, but rather followed a transition period called the *Mesolithic* or *Middle Stone Age*. It was during this transition period that the techniques of toolmaking were improved and new techniques were introduced, which led to the polished stone implements of the Neolithic stage.

The oldest Neolithic settlement of which we have knowledge at the present time was at Jericho,† and dates to at least 7800 B.C. according to carbon-14 dating techniques. The search by archaeologists goes on, however, for as they uncover a town such as Jericho, they find many different levels of habitation, each corresponding to a different period of time. The first or lowest level corresponds to the period of earliest occupancy, since new settlements were built over the old whenever natural or man-made disasters destroyed or covered an area.‡

In the earliest settlement at Jericho, the inhabitants built houses of sun-baked brick and used mill stones for grinding grain into flour. The tools found at this level resemble those of the Mesolithic cultures of the area, however. By 6000 B.C. the use of clay pottery was introduced. Multistoried buildings were used, probably for the storage of grain. We can see in this town of six to eight thousand years ago many of the features of community life that are familiar aspects of modern civilization. The second level revealed that the town had been populated by a totally new people and had grown to a population of over 3000. These

---

† Recent archeological findings indicate that Catal Hüyük in what is now Turkey (Anatolia) may actually be the oldest known Neolithic settlement.

‡ James Michener's novel, *The Source* (Random House, 1965), offers an interesting reconstruction of the history of such a site in a fictional approach to "what might have been."

people constructed elaborate dwellings and made bowls and dishes cut from stone. Tools of obsidian began to be used. From clay figures of animals—goats, cattle, sheep, and pigs—found in the temple, we may assume that these people practiced some form of "animal worship." The development of such town-oriented agricultural societies occurred much later in Europe and eastern Asia. It is thought by some that farming and animal husbandry were first discovered in western Asia (Asia Minor) and from there slowly spread throughout Europe and Asia and North Africa. During this period the cities of Mesopotamia, Anatolia, and Egypt had their beginnings, cities that played such a large role in the history of man.

Primitive man must have been aware of many of the natural phenomena that surrounded him—the rising and setting of the sun, drought and flood, disease and death. He did not view them as mere "happenings," but as events of significance and meaning to him and his life. Man wondered the "How"? and "Why"? of them. We have seen that Neanderthal man, some 50,000 years ago had ceremonial burial rituals, denoting a belief in some form of afterlife and hence in some form of religion. These early practices, whether demonistic or totemistic in nature, were possibly an attempt on his part to explain the otherwise inexplicable phenomena of nature. Perhaps they also mark man's earliest attempts to develop a natural philosophy, an attempt that continues in present-day science.

At some stage in his early history, man developed language. Language, it is believed, shapes and determines thinking and perception. It helps clarify ideas and fixes them in memory. It is not only an aid to memory but also a means of communicating ideas, personal thoughts, feelings, and the social traditions and beliefs. Perhaps the use of language and verbal expression for communication could be considered as important as the manufacture and use of tools for distinguishing man from other living things. (The ability to engage in abstract thoughts and use symbols has been equated to "human nature.") When man developed the means to express himself in written language, he was further able to clarify and record his thoughts, beliefs, and feelings. With the written records man enters the period of history.

2.7 Urbanization and the Bronze Age

The villages of the early Neolithic age were primarily the abodes of a farming society. The entire population was occupied with the raising (or obtaining) of food, although different people may have specialized in different ways of obtaining food. But as the tools and techniques of agriculture improved, some people could specialize in activities other than food production; for example, some specialized in the exchange of produce. Since the storage and distribution of goods was centered around the city's temple and involved religious rites, the temple priests grew in importance. As demands for their goods and services increased in populous areas, various other professions appeared (for example, "healers" or physicians, smelters, perfumers, carpenters). As the villages grew, so did specialization, and as the villages became *cities,* some men no longer secured their sustenance directly. Rather they applied their own specialized skills to service for others and depended on the specialized skill of others to secure other essentials for their own use. By 3000 B.C. many cities existed in Asia Minor. The city-states of the "Urban Revolution" had come to Mesopotamia, resulting in such centers of civilization as Sumer, Erech, Ur, and Akkad.

Fig. 2D. Fragment of Urartean bronze belt, about Seventh Century B.C.; found near Lake Urmia in N.W. Persia. (Courtesy of Metropolitan Museum of Art.)

For some time man had been using metal, but it was metal that had been found in nature; for instance, natural copper. At the time when the city-states were flourishing, man developed ways of smelting poor-grade copper ores. Since man had been firing his pottery, perhaps it suggested to him the possibility of smelting copper and the ores that yielded copper. Now man had a material suitable for fashioning better tools of war and industry. This development led man into an industrial and social era, called the Age of Metals. We do not know when man also discovered that if molten copper is alloyed with tin, the even more useful metal bronze is produced, but by 3000 B.C. the art of alloying appears to have been well known in Sumeria. It is the history of these bronze-age cultures in such areas as Egypt, Sumeria, Babylonia, Assyria, and elsewhere that give the essential features of the early civilizations.†

Since our purpose in this course is to study the development of ideas rather than to reconstruct the early history of mankind, we shall at this point abandon the historical approach in favor of examining a few of the key aspects of human knowledge on which man's future progress was most dependent.

2.8 Writing

The art of writing probably grew out of the need for records which resulted from the growth of industry and trade in the Urban Revolution. Something more per-

---

† While attention has been limited to the civilizations that developed along the "fertile crescent" of the Mediterranean, from which our Western culture and science evolved, we must note that there were probably comparable civilizations such as the Mayan, Hindu, African, and especially the Chinese civilizations in these early periods.

manent than man's memory was needed. The earliest writings were probably symbols that were used as aids to memory (for example, the knotted ropes, quipus, which the Incas used for records), but in time these symbols became conventionalized. The earliest form of "written" material was the personal seal, which denoted ownership of the object in question (much as branding of cattle at the present time). The first system of writing that we now know of was that of the Sumerians. Since the Sumerian cities were the domain of gods, the temple was the economic center of the city, supervising the planning, production, distribution, and exchange of goods and services. With the increase of economic activity, the attendants at the temple who were involved in economic activity of the city had to have written records to keep track of things. Thus it is speculated that the origin of writing is related to the need for business record keeping and legal (commercial) transaction. Many such business records have been found in baked clay tablets.

Pictographs were used to represent sounds of the spoken language. This appears to have developed independently in both Sumeria and Egypt. The Sumerian writing was in many ways similar to the Chinese, not in form of the pictographs but rather in their relation to the words they represent. It is conjectured that the cuneiform writings of Sumeria, Babylonia, and Assyria, and the hieroglyphics of Egypt, required great effort to master; therefore the ability to read and write was possessed by only a small fraction of the populace—scribes who may have been priests. However, it is important to realize that written language which uses an alphabet does not make for universal literacy. Japan has had a higher literacy rate than most alphabet-

Fig. 2E. Sumerian gypsum stela, of the early Dynasty I period (ca. 2850 B.C.), perhaps from Umma; height of $8\frac{3}{4}$ in. (Courtesy of Metropolitan Museum of Art.)

using people. To learn to read and write may therefore need more than an "easy" form of written language.

Writing was not confined to keeping the numerous fragments of ancient writings, but included also religious texts, historical records, letters and scientific records, usually of astronomical observations (see Fig. 2.8). Writing offered a means for communicating with others who might be far distant in both space and time. As Galileo phrased it three centuries ago:

"But surpassing all stupendous inventions, what sublimity of mind was his who dreamed of finding means to communicate his deepest thoughts to any other person, though distant by mighty intervals of place and time. Of talking with those who are in India; of speaking to those who are not yet born and will not be born for a thousand or ten thousand years; and with what facility, by the different arrangements of twenty characters upon a page."†

2.9 Numbers and mathematics

It is impossible to say when the concept of numbers first appeared. The Paleolithic hunter could probably distinguish between one mammoth or two, but may have been content to regard a herd as containing *many* animals rather than explicitly counting 47 members.

We do know, however, that by the time of the Pyramid builders of Egypt (around 2700 B.C.) the concept of numbers and their manipulation was firmly established. The Sumerians used a sexagesi-

---

XII 11	31 29	21 49	♋	ush	43 45 30
XIII 22	29 41	21 30	♌	,,	41 57 30
II 4	28 38	20 8	♍	,,	40 32
III 16	30 26	20 34	♎	,,	42 20
. 1	32 14	22 48	♏	,,	44 8
V 17	34 2	26 50	♐	,,	45 56
VII 5	35 50	2 40	♒	,,	47 44
VII 25	37 38	10 18	♓	,,	49 32
IX 13	36 38	16 56	♈	,,	48 54 30

Fig. 2.8. (a) Cuneiform text of Jupiter table. (b) Transcription of text. (From A. Pannekoek, A History of Astronomy, John Wiley (Interscience), New York, 1961.)

---

† *Dialogues Concerning the Two Chief Systems of the World (End of First Day)*. Translation by Stillman Drake, Univ. of California Press, 1962, p. 105.

mal notation (that is, one based on 60†) in addition to the use of the base 10. This system persists today in our measurement of time (60 seconds = 1 minute, 60 minutes = 1 hour) and of angle (360 degrees). The Egyptians, on the other hand, favored the base 10 (possibly because they originally counted on their fingers). Arithmetical operations—addition, subtraction, multiplication, and division—were carried out in these number systems, albeit somewhat cumbersomely. We also find that by the beginning of the Second Millennium B.C. the mathematicians of Babylon were familiar with square roots and cube roots of numbers. The architecture of the period indicates some knowledge of geometry. We also know that geometry had some development in Egypt, where the annual flooding of the Nile made it necessary to have some means of surveying in order to reestablish property boundaries. The scribe Ah-mose (around 2000 B.C.) worked out the areas of many figures. In one of the few mathematical texts of this period, he gave good rules for finding the volumes of cylinders, spheres, and other solids. In this work (today known as the Rhind papyrus) the author uses pi (π) equal to 256/81, or 3.16, which is close to our present value of 3.1415. . . . ‡ It is relevant to note that for all their correctness, the formulas or rules given in this work are all *stated without proof* of any kind. Is this because they were rules of thumb derived from trial and error by people who had certain pragmatic need for them, or is it that the proofs were not recorded? It must be remembered that in ancient times much knowledge was handed down orally. In their oral discussions, these early mathematicians might have wondered and proved what they recorded as a rule of thumb for those who needed useful knowledge. One would, for example, not know the wealth of theories and speculations of the ancient Hebrews if they had not finally begun to record the then "oral" learning in order to preserve it.

The early culture of the Indian subcontinent gave us two of our present-day aspects of mathematics: the form of our present-day numbers (called Arabic because they were introduced to the Western world by the Islamic culture at a much later time) and the concept of zero. There is, however, some evidence that aspects of the concept of zero were known to the Babylonians.

Before leaving this discussion of numbers it should be noted that numbers and counting imply that a system of units exists. For example, if I lend you two bags of sugar and expect you to return two bags to me some time later, we must agree on what constitutes a *bag* of sugar. I cannot give two of the small, individual portion bags containing 1 teaspoonful of sugar and expect you to return two 5-pound bags to me. Thus, some standard system of measurements has to be introduced before we can make full use of our numerical abilities. As early as 2500 B.C. a royal edict in Babylon set forth the standard measures that had to be employed in commerce. The importance of maintaining true weights and measurements is seen in references in the Bible and other religious beliefs. It was sinful not to give exact measure. This need to guarantee the standard is seen in our modern governmental efforts and industrial scientific endeavor to improve and

---

† The base 60 has the convenient property that it is divisible by 1, 2, 3, 4, 5, and 6.

‡ The Greek symbol π (pi) is used to denote the ratio of the circumference of the circle to its diameter.

2.10 Astronomy

Among all the natural phenomena that surrounded primitive man, celestial phenomena were in a unique category. It is not easy to believe that Paleolithic man was unaware of the rising and setting of the sun, the phases of the moon, and even the existence of the brightest planets. The phenomena relating to the Sun and to the orderly progression of the seasons were of prime importance to an agrarian society. Moreover, since man engaged in trade and travel over uncharted seas, deserts, and jungles, he must have used the heavenly bodies as guides in his journeys. The time for planting and harvesting had to be known to ensure a good crop, and these times had to be determined from the sun. It is not surprising, therefore, that one of the first uses of astronomical observation was for setting up a calendar.

No sophisticated equipment is needed to determine that the year lasts approximately 365 days† or that the time required for the moon to go once through all its phases (a lunation) is a convenient subdivision of the year, namely the month. The calendar used by the Sumerians was based on the moon. The Jewish calendar, while based on the solar year to conform to the seasons, continues to utilize the lunation as a unit of measure.

The importance of the calendar went beyond its usefulness to the farmer. One of its prime functions then, as now, was the determination of dates for religious festivals. It is to be remembered that the religious rituals were intimately related with the economic and social lives of the people and that it was important that the rituals be performed not only in an exact manner but also at the proper (auspicious) time. Because of such concern, those responsible for the performance of the ritual (the priests) played a major role in the formulation of the calendar.

Interest in the calendar led the Sumerians and Babylonians as well as the Egyptians into more detailed study of astronomy. They gave names to the five brightest planets: Mercury, Venus, Mars, Jupiter, and Saturn. They became aware of solar eclipses. Perhaps it was due to the influence of the priests that the development of astronomy in the Mesopotamian civilizations leaned toward mystical explanations and toward use of astronomical phenomena for predictive purposes. Astronomy became astrology, whose main purpose was the casting of horoscopes for the prediction of the future. This was especially true of the Hellenistic period in Egypt. The practice has continued with varying degrees of respectability to our own period, even though the Old Testament forbade astrology, divination, and soothsaying.

However mystical their explanations and use of celestial phenomena, we know that by 500 B.C., the Babylonians were making highly complex calculations of the times of various lunar and planetary configurations. The recent explanation of the Stonehenge structure in England as a well-designed architectural calendar inspires a great deal of respect for the competence of the early astronomers. We shall return to this story of astronomy in the next chapter.

---

† Chapter 3 will discuss various celestial phenomena in more detail.

† One could argue that astrology became astronomy.

2.11 Medicine

Although the practice of medicine in all the bronze-age cultures often included mystical elements (somewhat like the folk or native doctor in India or China), records from that period indicate that there was in existence a well-developed pharmacopoeia. Numerous plants and minerals were considered to have healing powers. The physicians of Nineveh knew of opium, belladonna, and other narcotics as well as many herbal remedies still in use today. Mingled with very sound medical advice one often finds in ancient records statements that appear to us, and may really be, nonsense. A Mesopotamian record describes a drug (probably marigold) that was reputed to be useful for curing scorpion stings, toothaches, ear and eye troubles, jaundice, snake bites, stomach troubles, and venereal disease, and could be used also to exorcise ghosts that had taken possession of a human body. (Such claims seem like the advertising claims of modern medicine peddlers and manufacturers.)

Although the records do not inform us about their knowledge of physiology, surgery was a highly developed art in Sumeria, Babylonia, and Egypt. Moreover, in Egypt the study of anatomy and surgery reached its early heights. This was due in part to the complex process of embalmment (mummification) used in Egypt, which allowed for careful study of the human anatomy. However, from existing surgical texts of the period it must be noted that although experience had taught the Egyptian surgeon a great deal about how to perform surgery on the brain and to amputate a limb, he seems (from the records) to have been rather uninformed on the functions of the various parts of the human body.

At the beginning of the Third Millennium B.C. in the cities of India such as Mohenjo-daro, medicine and surgery appear to have reached a state comparable with that of the Western nations by the end of the First Millennium B.C. Records show that surgical operations for hernia and cataracts were being performed.

The briefness of time allotted to cover the bronze age does not permit discussion of the art and architecture of the period, the fine work with clay and metals, and the various technological innovations of the time, which were of very sophisticated nature and showed great skill. In all probability the growth of technology and art affected science. It seems that the knowledge (or science) and the technology of any age go hand in hand, even though there may be periods when there is little or no communication between them. A new idea may result in the building of new "machines," while the use of these same machines may make possible the acquisition of more knowledge.

2.12 The Greeks and the beginnings of science

Along with the development of large city centers, science, language, writing, and technology, the period we have discussed witnessed an enormous increase of commerce and trade. There were also numerous invasions by armies that traveled large distances to plunder or to impose their culture, as well as their rule, on other nations. Both trade and war became influential vehicles for exchanging and promoting science and technology and for the more frequent assimilation of peoples and cultures. It is therefore not surprising that despite the many differences in the peoples of the known world of the ancients, certain similarities of

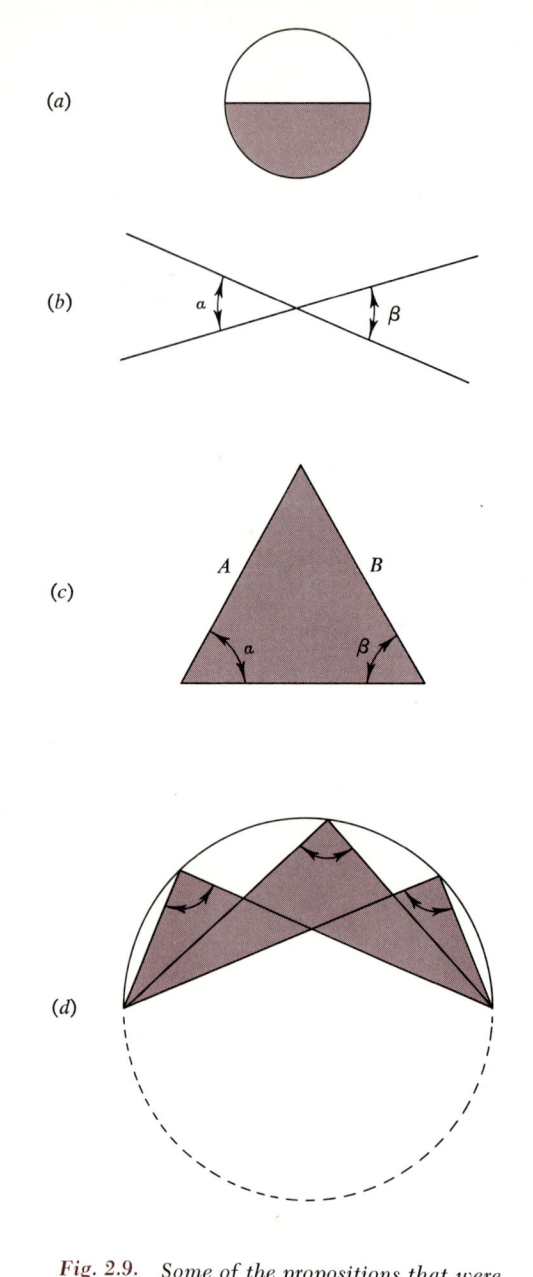

Fig. 2.9. *Some of the propositions that were proposed and proved by Thales.*

culture and technology developed among the Egyptians, the peoples of Mesopotamia, and the Hittites of Anatolia. The influence of these cultures extended from North Africa through western Asia and perhaps to Crete and even Greece. But, the early Semitic culture of Crete and Greece was, in the Second Millennium B.C. overrun by a new group of uncivilized people from the North—a people who spoke a new language and who had developed the use of iron tools and weapons. These people are commonly called the Greeks.

By the middle of the First Millennium B.C. most of the eastern Mediterranean world except Greece was under the political domination of the Persian Empire. We are familiar with the warfare that ensued between the Greeks and the Persians over the next several centuries, culminating with Alexander's overthrow of the Persian Empire. It was a time of flux of ideas as Alexander the Great encouraged exchange and interactions of the ideas of "East and West." Much of ancient science and technology was interwoven with mysticism in the sense that explanation of reality was not sought for or made in terms of ordinary material phenomena but in terms of myths and magic. It is the story of this period that we pursue in the remaining sections of this chapter.

2.13 Thales and the Ionian school of cosmologists

Our story of Greek natural philosophy begins with Thales, a Phoenician born in Miletus, an Ionian city in what is present-day Turkey, about 620 B.C. We are told that he traveled extensively and that he studied in Babylon and Egypt. Such traveling in pursuit of knowledge was not unusual among Asia Minor people in ancient times. Although none of his work in natural philosophy has come to us directly, he is widely quoted in later Greek writings and it is through them that we have learned of his views on science.

Thales believed that natural phenomena occur according to fixed *natural laws* and not according to the capricious will of the gods. He further conjectured that all material objects were formed of a single "element," which he took to be water. This search for the laws of nature was to dominate Greek philosophical thought for several centuries and, indeed, is still the underlying principle of science today. Thales' proposal of water as the basic building block of all matter was a *model* which, with the application of the laws of nature, could be used to explain the world around us. This use of a model is an essential tool of science to this very day.

Thales is also credited with another significant step in the progress of scientific thought—the introduction of the formal (deductive) *mathematical proof*. We noted earlier that the Egyptian geometers had discovered mathematical formulas for the areas and volumes of various geometrical forms. These were presented without derivation or formal "proof." Thales is credited with giving geometrical proofs of a number of simple geometrical theorems, some of which are familiar to us from Euclid's later assembly of many theorems (Figs. 2.9 and 2.10).

The Ionian school of philosophy that started with Thales was among the first to assume that the entire universe is potentially explicable by ordinary knowledge and rational inquiry. Natural events were seen as following a cyclical course. The idea of cyclic change, which will play an important role in the continuation of our discussion, was stressed by the Ionian philosophers.

Among the other members of the Ionian school, the best known is Anaximander (610–545 B.C.) who is said to have been the first to note that the heavens revolve around the pole star and that the visible sky is but half of the complete sphere. Anaximenes (d. *ca.* 526 B.C.) noted that moonlight is due to reflected sunlight. We shall return to these ideas in Chapter 3.

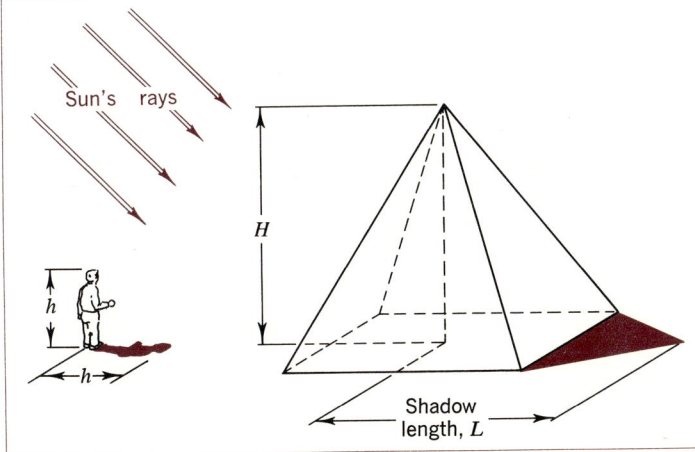

Fig. 2.10. When the sun is in such a position that a man's shadow is exactly equal to his height h, the length of the measured shadow of the pyramid L must be equal to the height of the pyramid H.

2.14 The Pythagorean concept of numbers

Against the seemingly concrete and materialistic view of the Ionian school, the Pythagoreans emphasized what is abstract and seemingly mystical. Pythagoras (ca. 582–500) was a native of Samos but moved to Crotona in southern Italy in 530 B.C. Here was founded a school of philosophy that was to have great influence (through Plato) on philosophy and on the course of science. The Pythagoreans formed a brotherhood devoted to a life of mathematical speculation and, what one would call it today, mystical contemplation. Their community admitted women on equal footing with men and held all property and ideas in common among the members. They, however, preserved their philosophic ideas as secret mysteries to outsiders, and adopted a highly ritualistic mode of living. After the death of Pythagoras, the brotherhood broke up into a religious wing and a scientific wing, the latter led by Philolaus.

To the Pythagoreans, *numbers* played a central role in the description of all natural phenomena. They thought of numbers as physical and geometrical entities made up of unit points. They introduced the notion of squares and cubes of numbers, giving to them geometrical shapes as well as quantitative size. But in each case they sought general theorems of form or structure and underlying relationships, rather than limiting their concepts to describing specific examples.

For example, it was known that a right triangle which has sides of length 3 and 4 units also has a hypotenuse of length 5 units. But the famous theorem attributed by Euclid to Pythagoras states that *every* right triangle has a hypotenuse of a length which is given by the square root of the sum of the squares of the other two sides (or $c^2 = a^2 + b^2$). The same theorem did cause some consternation among the Pythagoreans, who considered the natural numbers (integers) and their ratios (rational numbers) to play a central role in their philosophy. For consider a right triangle with two equal sides of length 1; then the hypotenuse has length $\sqrt{2}$, a number that cannot be expressed as the ratio of any pair of integers (that is, it is irrational). Yet $\sqrt{2}$ can be shown to exist geometrically, which is contrary to the belief that numbers (integers) themselves have fixed geometrical properties.

It is interesting to show that the number $\sqrt{2}$ cannot indeed be represented as the ratio of two integers. We may accomplish the proof of this by assuming the opposite; that is, the number $\sqrt{2} = p/q$, where p and q are two integers having no common divisor. Then, assuming

$$\sqrt{2} = \frac{p}{q}$$

squaring both sides gives

$$2 = \left(\frac{p}{q}\right)^2$$

$$= \frac{p^2}{q^2}$$

or

$$p^2 = 2q^2$$

and hence p^2 must be an *even* number because it is divisible by 2. But this means that p itself must be even, for the square of an odd number is always odd. We may write

$$p = 2r,$$

where r is another integer. Our last equation relating p and q becomes

$$p^2 = (2r)^2 = 4r^2 = 2q^2$$

or, in lowest terms,

$$2r^2 = q^2,$$

and hence q^2 is even, making q itself even. But if both p and q are even, we violate our initial assumption that they contain no common divisor. Thus, there cannot exist any pair of integers p and q such that $\sqrt{2} = p/q$, and therefore $\sqrt{2}$ must be irrational.

Note that our proof assumes the converse of what we sought to prove and arrives at a contradiction or absurdity. Such a proof is termed a *reductio ad absurdum* and is a valuable technique in mathematical logic.†

The Pythagoreans did not accept the Ionian idea that the universe consisted of a single element, but held instead that there were four elements: earth, water, air, and fire. They introduced the abstract concept of number as divorced from the objects being counted. They recognized the sphericity of the earth and explained the apparent rotation of the heavens by considering a moving earth! But to the Pythagoreans, the earth moved not about the sun, but rather about an unseen central fire, balanced by an also unseen "counterearth."

2.15 Plato and Aristotle

The philosophy of the Greeks reached its highest point in systematization with the work of Plato and Aristotle. But Plato looked to the possibility that the mind (soul, or the Greek *Nous*) of man shall ultimately free itself from overwhelming preoccupation with immediate sensations and shall find through a pattern of liberal education the clues that can yield truth and reality from these limited shadows and impressions.

Plato (428–348 B.C.)† and his teacher Socrates differed from the Ionians in that they stressed the importance of knowing man and not necessarily the cosmos. In this respect they may be said to represent an anthropological school. "Man is the measure of all things," preached the Sophists. The slogan was reiterated by Socrates and Plato, but they sought to find out what is true reality and how one gains knowledge of it. They rejected the idea that knowledge comes from sense experience. For the senses of man are easily deceived and do not offer a direct approach to reality. In an allegory given in Book VII of his *Republic*, Plato (in the name of Socrates) describes man as being prisoner in a cave with only his sensory impressions to guide him as he watches the shadows on the wall and tries to discern reality from these shadows:

"And now, I said, let me show in a figure how far our nature is enlightened or unenlightened:—Behold! human beings living in an underground den, which has a mouth open towards the light and reaching all along the den; here they have been from their childhood, and have their legs and necks chained so that they cannot move and can only see before them, being prevented by the chains from turning round their heads. Above

---

† We should note that some mathematicians (the Intuitionists) do not consider proofs that depend on *reductio ad absurdum* to be acceptable. Accepting only direct proofs, they reject the principle of the *excluded middle of logic*, which is the basis of such indirect proofs.

† In jumping from Pythagoras to Plato we omit a century filled with the work and ideas of many eminent men. We are compelled to do this because of lack of time only. Plato and his successors made abundant use of the work of the Pythagoreans who preceded them.

and behind them a fire is blazing at a distance, and between the fire and the prisoners there is a raised way; and you will see if you look, a low wall built along the way like the screen which marionette players have in front of them, over which they show the puppets."†

The dependence of the senses on the mind for understanding of the fundamental structure of reality led Plato to stress the world of *ideas*, the world of *ideals* and of *forms*. These offered more valid knowledge than that derived from sense experience. (Our later chapters will be concerned with the reservations of modern science concerning the reliability of the senses.) To Plato, the Pythagorean concepts of numbers and knowledge of astronomy offered a link between the world of ideas and forms and the phenomenological world. In a dialogue titled *Timaeus* (after a Pythagorean) the

". . . world of the senses is described . . . as a world of happenings or becomings. Since whatever becomes is the product of an agent, there is an artisan or creator who makes this world. The model or archetype that the creator uses, is exact, eternal and unchanging, but its copy, the world of the senses, is variable and changeable. Plato stresses that discourse about the model can be exact and final, as is the object it deals with, whereas discourse about its copy must be approximate, tentative, and subject to correction."‡

Plato, like most Greek philosophers, belittled experiment as a base mechanical art.§ Mathematics, a deductive sci-

---

† (From the translation by C. B. Jowett. New York, Bozok, 1943.)
‡ (Luchins and Luchins, 1965, see references.)
§ This was in line with the Greeks' ideas that work coarsened ones character and corrupted the soul. Technology and the artisan's way of

ence, he held in high esteem. Plato himself formulated the concept of negative numbers. Observation perhaps could suggest concepts which, when examined by the mind and subjected to logical analysis, would lead to valid knowledge.

Aristotle was born in 384 B.C., the son of the physician to the court of Philip of Macedon. He, too, served Philip as tutor to his son Alexander. Although he was a disciple of Plato, he showed some independence of thinking in rejecting the mysticism and certain of the social views of his teacher. Aristotle founded a school of philosophy known as the "Peripatetic" from the custom of master and pupils walking in the gardens of the Lyceum at Athens. His aim seemed to be to systematize existing knowledge, just as Euclid had systematized geometry.

Aristotle became the most important collector and classifier of knowledge of the ancient world. Indeed the systematic approach that he developed and recorded became the method and text from which the West drew its knowledge of science and of nature right up to the Renaissance period.

In his *Physical Discourse*, Aristotle envisions an "Unmoved Mover" to be the acting cause that gives existence, matter, form, motion, time, and space to nature and to the universe. (In contrast, Plato appeared to assume that a cause is needed only to deflect the course of nature from a straight path.) In his book, *On the Heavens*, he pictured the heavenly bodies to be made of aether, moving in circles; but in the space between these perfect and incorruptible bodies and the earth there develop opposing principles of hot and cold, wet and dry, which react in pairs to produce the four elements of

---

life corrupted both body and soul, the contemplative life being the more desirable.

fire, air, earth, and water (Fig. 2.11). He rejected the atomic theory of Leucippus and Democritus, who believed all matter to be divisible into a multitude of indivisible bits of various shapes and sizes called *atoms,* with *empty space* (the void) between the atoms. In general, his ideas of physics, the properties of matter, and of astronomy, while not acceptable in terms of current ideas, nevertheless comprised the dominant theory for a longer time than any other philosophical or scientific system.

His contributions to biology were much more in accordance with some modern ideas, but were not so influential in the past because biological topics did not mature and become a part of the world view of the times. Life was "the power of self-nourishment and of independent growth and decay." He mentions about five hundred different animals, of which fifty are illustrated in such manner as to have required dissection on his part. In embryology he repudiated the widespread idea that the father is the only real parent, and assigned to the mother an equally significant role for generation of the embryo. He urged biologists to make careful observation of anatomical structure before they framed their views; he also urged that they recognize that there is a "final cause" for whatever nature did. But his denial of the sexuality of plants caused a long delay in the acceptance of that concept.

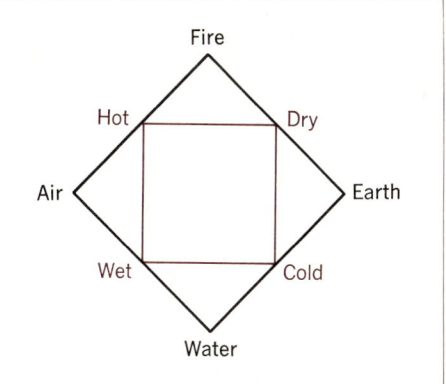

Fig. 2.11. Aristotle's four elements.

2.16 Summary

We have quickly scanned some 5 billion years of geological and human history, and have arrived at the periods of early civilizations that were within a few centuries of the Christian era. It has seemed to us important to reach backward in time in order to achieve some perspective and

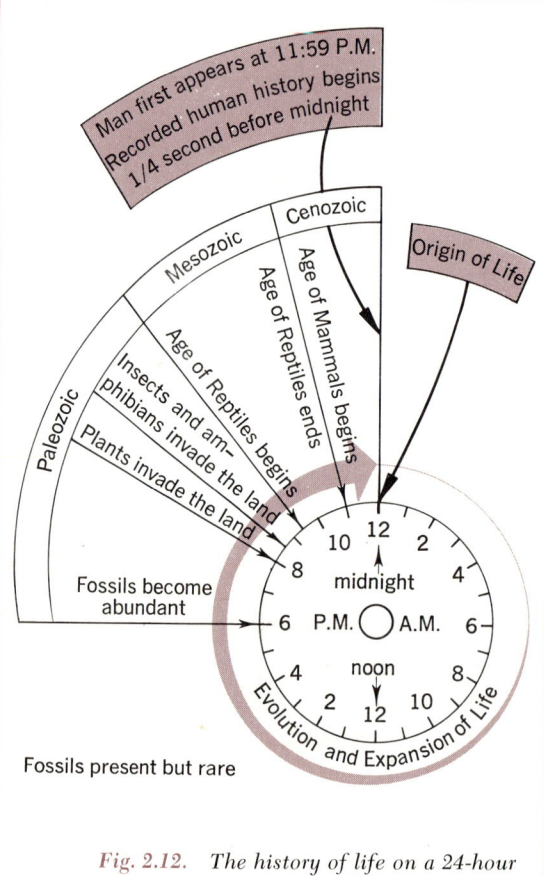

Fig. 2.12. *The history of life on a 24-hour scale.*

understanding of the processes and chronological periods that have brought us to the present stage of our earth and of man.

For this backward look into history the techniques of the geologist, chemist, paleontologist, and archaeologist have been very useful. Even the apparently unrelated nuclear science gave us the radioactivity dating technique that has proved to be especially effective in this connection.

The principle of *uniformitarianism* has been useful in relating to the evolutionary processes of past ages the processes or erosion, chemical reaction, physical change, and laboratory experiences that are currently observable. Indeed these slow, evolutionary processes seem to have had greater influence in bringing the earth to its present form than did the more catastrophic phenomena.

An interesting analogy illuminates the results of our findings. Suppose that life appeared on earth just after midnight and existed for exactly one day or 24 hours to the next midnight. During the course of this one day, which represented only 2 billion years of the 5 billion years of earth's existence, the first mammals do not appear until 11:00 P.M., and man enters the story at 11:59 P.M. All of written history would occupy only the last $\frac{1}{4}$ second of this fictional day.

This bird's-eye view of the geologic periods and of the beginnings of civilization seemed to find man to be at the mercy of the physical elements that surrounded him in a chaotic world of hot and cold, day and night, life and death. But the features that distinguished man from other living things included a spirit that sought an alternative to the concept of chaos as the unalterable character of nature. The sense of total chaos sought relief in the early concepts of a world ruled by God, as exemplified in the

Rabbinical traditions of the Hebrews. At the time of the Greeks there was even stronger acceptance of the concept that nature is orderly and directed, and the purpose of man is to seek out the laws that govern nature. (Perhaps we should note that even to this day science has been searching for clarification of the laws that determine this orderliness, with only limited success.)

For the moment at least, the earth itself seems to have settled down to a state of considerable stability with respect to the environment it offers to man, although this stability may be largely the result of the shortness of our view. Catastrophic volcanic eruptions, glacial movements, fall of meteors, and other destructive natural phenomena occurring in only restricted local areas have very little effect on the overall stability of the habitable portions of the world. The patterns of temperature and weather change seem to be slow enough to permit living organisms to become adapted to the changes, especially since man has achieved considerable control over his environment.

This chapter has traced the emergence of agricultural and then large urban civilizations, and the influence of the new tools and techniques, including ceramic arts, metallurgy, the healing arts, and commerce. Observation, generalization, trial and error, and an empirical approach characterized man's progress. Careful recording of observations of cyclic phenomena on earth and in the heavens also gave beginnings to the science of astronomy, although the observations became more used for purposes of astrology than for developing a model for the universe. Meanwhile, conquests of armies and commercial travel tended more and more to exchange knowledge and experience among the regions of the early civilizations.

It was from the eastern shores of the Mediterranean, and later from among the Greeks, that man went beyond astronomical data and technological rules of thumb to identify principles and abstract concepts from which to deduce the world of everyday life. This resulted in a diversity of models or principles that contended for the basis on which to found philosophical systems. Observations, to the Greeks, had value only in so far as they could be placed in larger systems or beliefs about phenomena. Structure, order, and unity were the characteristics that the Greeks sought for in man and in nature.

Throughout the history of Greek thought there was a drive toward unification of events and generalization of ideas. They achieved extraordinary connections and associations, and explored concepts of proof. Whereas their predecessors, the Babylonians and Egyptians, seemed (as far as their records show) to be content with recording and acting on the regularities they observed, the Greeks, from Thales through Pythagoras to Plato and Aristotle (and later Euclid), generalized their observations into systematic deductive proofs. They sought to distinguish what is only sensed from what is postulated as ideal in a systematic fashion. They glorified a certain kind of mathematics, especially geometry and geometric forms for numbers, to reach for the ideal. Under the influence of Aristotle they sought a synthesis of what is observed with mathematical or other explanations; algebra and arithmetic, which had been developed by the Sumerians and Babylonians, were largely ignored in favor of geometry. One may find within the scope of their explorations a wide range of themes, some of which demanded the use of rigorous logic while others gave base for the mysticism of later centuries.

Questions/Discussions

The chapters of a textbook on science are ordinarily followed by a list of questions and problems that are designed to develop skill in problem solving, or at least to require a careful review of the main points of each chapter. The contents of this chapter are designed neither for developing problem-solving skill nor for eliciting firm answers. Instead, we want to introduce a sense of time and of history. But both time and history ordinarily have significance only as they reveal *change,* whether it be measured by the ticking of a clock or the succession of events in human experience. (We shall probe the more philosophic question of the *direction of time* in a later chapter.) The early thinkers began to ask difficult questions about the nature of existence and of reality. Each age has continued to ask the same questions, each within the limitations of its own experiences and observations, even to our present decade. In that same vein, instead of asking the usual questions about the specific contents of this chapter, we have chosen to invite discussion of a wider range of topics. The following questions and answers are almost extreme, polar alternatives to each other; yet they find proponents in our very day.

The reader is invited to study each of the six categories of questions and then to formulate a statement of about 300 words which best represents his own preferred answers to each of the six categories of questions. It will be evident to the reader that many of the topics have not been discussed as yet, nor will all be discussed in the course of time. But the posing of the questions should nevertheless prove useful for revealing how widely our attitudes can differ on some common questions. Some "bull sessions" on these topics can be very illuminating.

How did the universe originate and what is its nature?

(1a) It was specially created for man pretty much as it is now.
(1b) It evolved from some primordial matter through a long series of changes.
(2a) It is closed and finite.
(2b) It is open and infinite in extent.
(3a) It is static and fixed in size.
(3b) It is expanding, increasing in size.
(4a) Its development is completely determined in structure and process by the properties of its constituent matter. Objects and events that appear to be novel and due to chance are not.
(4b) Its development is not completely determined. Novelty and chance are real and not just illusions.

What do you believe about the nature of the solar system?

(1a) It was specially created for man and unique (the only one of its kind anywhere).
(1b) It has evolved as a normal event in the history of a certain type of star and is one of many of its kind.
(2a) It will exist until the day of judgment and then will end.
(2b) It will eventually become uninhabitable, or will be destroyed as a result of nuclear changes in the sun.
(3a) It is a vast machine like a clock operating according to natural laws.
(3b) It is like some vast organism, moving and changing in response to its own inner life in order to achieve its purposes.

What do you believe is the past history of the earth?

- **(1a)** It was specially created pretty much as it is now except for some minor differences.
- **(1b)** It was specially created but subjected to many more or less large changes on the way to its present appearance.
- **(2a)** The changes in the past were much more violent than now: violent earthquakes, vast floods, great and rapid upheavals and subsidences of parts of the surface.
- **(2b)** The changes in the past were much the same as those now going on. Slowly, almost imperceptibly over a very long time, the changes produced the rocks and the surface we now know.
- **(3a)** The earth was made for man and all the changes in it have been directed for man's benefit.
- **(3b)** The earth has evolved, like the solar system, without any special connection with man.

What is the nature of man?

- **(1a)** He was specially created and is unique in the universe, the capstone of creation and only a little lower than the angels. The plants and animals were made specially for his use.
- **(1b)** He has evolved, as have other living things, from lower, less organized forms and is thus one with all organic life.
- **(2a)** He is basically a mechanical system operating according to chemical and physical laws which are, for the most part, well known and well understood.
- **(2b)** He is a living soul, distinctly non-mechanical, having freedom of will and choice.
- **(3a)** Since man is a creature of evolution, he is king in his own right (cf. Swinburne, *Hymn to Man*).
- **(3b)** Man's knowledge of evolution gives him control not only over his fate but over all evolutionary process.

What is the nature of matter; what is it really like?

- **(1a)** It is continuously divisible, without any least unit of structure.
- **(1b)** It is discontinuous, made up of small separate individual units of structure.
- **(2a)** It is made up of solid substantial material parts held together in some way.
- **(2b)** It is mostly empty space with a few electric charges whizzing around in it.
- **(3a)** It is permanent and enduring forever. It cannot be created nor destroyed.
- **(3b)** It is interconvertible with energy. It can be created and it can be destroyed.

What is the nature of natural law?

- **(1a)** Nature obeys certain laws; its behavior conforms to natural laws.
- **(1b)** Natural laws do not control nature; they describe only what we observe nature to do.
- **(2a)** Natural laws are the laws of God, imposed on natural events when He made the world.
- **(2b)** Natural laws do not exist apart from man as a discoverer. They are man-made statements.
- **(3a)** Natural laws are exact statements of relationships. They are true everywhere and always.

(3b) Natural laws are only approximate statements of relationships. They hold good only within certain well-defined limits.

(4a) Natural laws are exact statements about the exact conditions under which certain events or occurrences will take place.

(4b) Natural laws are statements about the probability of the occurrence of certain events if the prescribed methods of observation are used.

(5a) To say that the laws are of divine origin is to say that they can be changed at any time.

(5b) To say that the laws are of divine origin is to say that they are eternal and unchanging.

References

The treatment of many of the topics has been, of necessity, very brief. This list is intended to suggest further reading material. It is not complete—for many good references to this material exist—but should serve to acquaint the reader with some of the work in the fields covered.

For Secs. 2.2, 2.3, and 2.4:

Faul, Henry (ed.), *Nuclear Geology*. New York: Wiley, 1954. *This volume contains the proceedings of a symposium of "Nuclear Phenomena in the Earth Sciences." Most of the papers are of a specialized nature, however. Chapter 9 contains an excellent survey of the various techniques used and a list of significant results.*

Gamow, George, *A Planet Called Earth*. New York: Bantam Books, 1965. *This delightful little book can serve as a reference for Secs. 2.2, 2.3, and 2.4. It is written in the lucid, informal style that marks all of Gamow's popular works, and is highly recommended. Pages 13–19 treat the techniques of geochronology.*

Gillispie, C. C., *Genesis and Geology* (New York: Harper & Row) (Torch book), 1951. *This volume contains a very thought-provoking discussion of the impact of discoveries in geology on religious beliefs in England before Darwin.*

Kummel, Bernhard, *History of the Earth—an Introduction to Historical Geology*. San Francisco: Freeman, 1961. *An interesting textbook on historical geology. Chapters 1.4 and 1.5 introduce the notions of geologic time scales and how they are determined. Chapter 1.2 gives a brief review of the contributions of James Hutton and others to the modern science of geology.*

Libby, Willard F., *Radiocarbon Dating*, 2d ed. Chicago: University of Chicago Press, 1965. *A somewhat technical account of this valuable archaeological tool by the man whose work in this field earned him the Nobel Prize. Nearly one-half of this book is devoted to a list of "dates" obtained by this technique.*

Mather, Kirtley, and Mason, Shirley (eds.), *A Source Book in Geology*. New York: Hafner, 1964. *This book contains excerpts from the writings of many eminent scientists that help follow the history of geology as a science. Of special interest to our treatment are the sections as follows:*

Hutton, James, "Theory of the Earth" (pp. 92–100). *This is Hutton's original presentation of his concept of uniformitarianism.*

Playfair, John, "Proofs of the Huttonian

Theory" (pp. 131–137). *Excerpts from Playfair's book which popularized Hutton's theories.*

Lyell, Sir Charles, "Uniformitarianism" (pp. 263–267).

Thomson, Sir William (Lord Kelvin), "On the Secular Cooling of the Earth" (pp. 472 ff.). *The first section (numbered 8 in the original paper) discussed uniformitarianism and catastrophism.*

Rapport, Samuel, and Wright, Helen, *The Crust of the Earth*. New York: New American Library (Signet Book 2083), 1955. *The chapter "Age of the Earth" (pp. 159–167 are to be noted especially).*

FOR SEC. 2.5:

Hawkes, Jacquetta, "Prehistory" in Vol. 1, Part 1, *History of Mankind, Cultural and Scientific Development*. New York: Harper & Row, 1963. (Also available in paperback as Mentor Book MQ632, New American Library, New York, 1965). *This volume was used as source for the material used in this section and is highly recommended in its entirety. The treatment is comprehensive, and includes good bibliographic listings, so that further references will not be given by us. The book was sponsored by UNESCO toward a complete history of mankind.*

FOR SECS. 2.6 THROUGH 2.11:

Childe, V. Gordon, *Man Makes Himself*. New York: New American Library (Mentor Book MP384), 1951. *An excellent, easily read history of mankind from paleolithic hunter through the Neolithic and bronze ages. Highly recommended.*

Neugebauer, Otto, *Exact Sciences in Antiquity*, 2d ed. Providence, R.I.: Brown University Press, 1957. *Also came out in paperback in 1962 (Harper).*

Woolley, Sir Leonard, "The Beginnings of Civilization," in Vol. 1, Part II, *History of Mankind, Cultural and Scientific Development*. New York: Harper & Row, 1963. (Also Mentor Book MY633, New American Library, New York.) *This excellent treatment of the bronze-age cultures of the world up to 1100 B.C. contains a considerable amount of material used in the text. An excellent bibliography is included.*

FOR THE REMAINING SECTIONS OF THIS CHAPTER:

Dampier, Sir William C., *A History of Science*. Cambridge: Cambridge Univ. Press, 1929. (Paperback edition published 1966.) *See pp. 9–35 for an easily readable account of this period in the history of science.*

Farrington, B., *Greek Science*, 2 vols., London: Pelican Books, 1944 and 1949. *A comprehensive review of Greek study.*

Mason, Stephen F., *A History of the Sciences*. New York: Collier Books, 1962, pp. 25–47. *This fine history briefly covers the principal contributions to science made by the various Greek philosophers.*

Windleband, Wilhelm, *History of Philosophy*, 2 vols. New York: Harper & Row (Torch book), 1961. *This is a rather thorough survey of Greek philosophical thought.*

Luchins, Abraham S., and Luchins, Edith H., *Logical Foundations of Mathematics for Behavioral Scientists*. New York: Holt, Rinehart, 1965.

CHAPTER THREE

The Development of Astronomy

*Read not to contradict
and confute, nor to
believe and take for
granted, nor to find
talk and discourse,
but to weigh and
consider.*

FRANCIS BACON [1561–1626]

LIVING AS WE DO in an age that has seen space probes launched into the orbit of Venus and to Mars and has observed soft landings of instrument packages on the Moon, we are likely to accept as obvious some facts about our solar system. We know that the sun is a typical star, similar to the myriad of stars that illuminate our night sky. Nine major planets (Mercury, Venus, Earth, Mars, Jupiter, Saturn, Uranus, Neptune, and Pluto, in order of increasing distance from the sun) orbit the sun. In addition, numerous minor bodies (comets, asteroids, and so forth) also orbit our sun. Their orbits are usually ellipses, with the sun at one focus.† From our own experiences we know that an attractive force, which we call gravity, exists between our bodies and the earth. Through Newton's contributions we believe a similar gravitational attraction holds the heavenly bodies to their courses.‡

3.1 Introduction

The cycle of day and night we know to be due to the rotation of the earth about an internal axis, while the progression of the seasons is due to the changed angle at which the sun's rays strike the surface of the earth as our planet completes the orbit about the sun in a year of $365\frac{1}{4}$ days.

We shall learn§ that the multitudinous stars which fill our night sky are all at vast distances from us, but all are constrained by gravitational attraction to form one huge system, the Galaxy, or Milky Way. Finally, we shall learn that

† These will be discussed presently. See also the Laboratory and Mathematics Supplement for the characteristics of the ellipse.
‡ A fuller account of gravitation will be given in Chapter 5.
§ See Chapter 7.

our galaxy is but one of millions of such systems, all of which appear to be receding from one another as if hurled by a tremendous explosion (the "big bang" theory is one of the theories offered to explain this phenomenon).

From our vantage point in this century it is sometimes easy to overlook how difficult has been the process of developing these concepts. How many centuries passed until it was generally acknowledged that the earth revolved about the sun, and not the sun about the earth!

In this chapter we shall discuss a few of the historical events and the tortuous path that marked the development of our present understanding of the celestial universe from the early days of civilization. From our review of the development of astronomy we shall discover much that is true of all of natural science, for until the nineteenth century the history of science was primarily the history of astronomy.

We have recounted how the early civilization of the Sumerians, Egyptians, and Babylonians accumulated a great deal of observational data about the movements of the heavenly bodies. These data provided a base for the development of a calendar which was useful for anticipating the seasons, and was also used by the astrologers. A model, or theory of the system of heavenly bodies, did not emerge until the time of the Greeks, however.

A number of models did emerge from Greek thought, some of them quite complex. These ultimately converged to a form presented by Ptolemy†; a model that lasted until the middle of the sixteenth century. These models were finally discarded, but lest we underestimate their

† See Sec. 3.8.

significance, we shall ourselves explore the heavens as they did with unaided eye and find that the Greeks' approach was logical if not correct. Our purpose is threefold: To show (1) that the historical features of this development are closely allied with the progress of society, (2) how much one can learn from careful observation alone, and (3) how important it is to corroborate what one observes by comparing with other sources of information and correlating these with existing models. This avoids misunderstanding the observed data and leads to development of an acceptable theory.

We shall also learn how the Ptolemaic concepts served through many centuries for navigation purposes (not to mention astrology), how they became entrenched in religious and political thought, and how astronomical models became the focus for religious conflict as well as the gates through which new science emerged at the time of Copernicus, Galileo, and Newton. The drama of science was exceeded only by the intensity of human passions during this critical period.

The reader may ask why these models played such an important role that they could become a cause of human conflict. To answer this question, one must realize that spiritual explanations of natural phenomena were deeply entrenched in religious teachings at that time. The conflict arose because scientists, in trying to establish interrelationships among the various phenomena, necessarily had to relate random tidbits of individual observations, some of which had been already assigned a religious connotation. It was the latter that incited dispute, for with their inclusion the non-scientists condemned logical scientific theory to an inferior role in the process

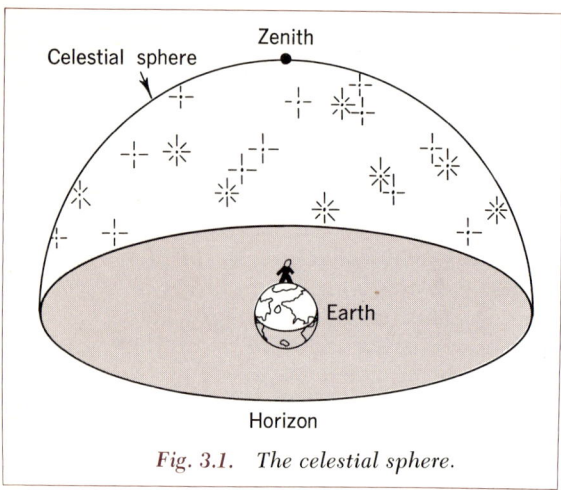

Fig. 3.1. The celestial sphere.

of developing a pattern to substantiate their theories. To see a pattern more clearly, we usually require an image, or a model, which may remain in our mind or find expression on a piece of paper or in a material form. Models are rarely complete or accurate, but they do nevertheless help our minds to focus more clearly on complex phenomena and serve to develop better models and better understanding.

With these purposes in mind, and realizing that the eye is today, as in early centuries, the first instrument for observation, what are the phenomena that one can observe in the sky with the unaided eye? Only by answers to this question can we evaluate the difficulties and the achievements of the early civilizations.

3.2 The celestial sphere as seen by the naked eye

On a clear night, a look to the heavens reveals a great number of points of light. Some are bright, some faint; some are red, others are yellow or blue. From early childhood we have learned to call them *stars*, and before we are subjected to more rigorous instruction it is rather natural to envision these stars to be embedded within a huge hemisphere, with ourselves at the center. We call this hemisphere the *celestial sphere*† (see Fig. 3.1). The point directly overhead is the *zenith* and the imaginary circle on the celestial sphere which is everywhere 90 degrees from the zenith is called the *horizon*. Thus the horizon divides the celestial sphere into two hemispheres,

---

† Use of the name "celestial sphere" at this point is getting a little ahead of our story in that it requires the assumption that the hemisphere of the sky continues into a hemisphere beneath us (which we cannot see) to complete a sphere.

one of which is above us and is visible, the other below us and not visible to us.

As we continue watching the night sky for several hours, another phenomenon is observed. When we are in the United States and looking north, we may see a configuration of stars as in A (open circles) in Fig. 3.2. This view includes the Little Dipper, the Big Dipper (in the "9 o'clock" position)† and the W shape of Cassiopeia (at approximately "4 o'clock" position). Some 2 hours later we would note that although the relative position of the three configurations remained unchanged, their position relative to the horizon had changed as in B in Fig. 3.2 (shaded circles); that is the Big Dipper would be at about 8 o'clock and Cassiopeia would be at 3 o'clock). At a still later time, they would appear as C in Fig. 3.2 (dark circles) with the Big Dipper near 7 o'clock and Cassiopeia at about 2 o'clock. Thus, during the few hours we observe the sky, the celestial sphere seems to rotate by nearly 60 degrees about an axis that runs from us (as observers) to the Little Dipper. Does the entire celestial sphere partake of this rotation, or are only the three groups of stars we have noted being affected? If, when we began our watch, we had also noted the stars directly overhead (that is, near the zenith) and those about 30 degrees above the eastern horizon, by the time the star groups in the northern sky had changed to position C in Fig. 3.2, the stars originally overhead would have been found to be about 30 degrees above the western horizon, while those originally near the eastern horizon would be almost overhead. Thus the entire celestial sphere appears involved in the observed rotation.

† Note that it is very useful to imagine one's range of vision to be divided according to the positions of the numbers one observes on the face of a clock.

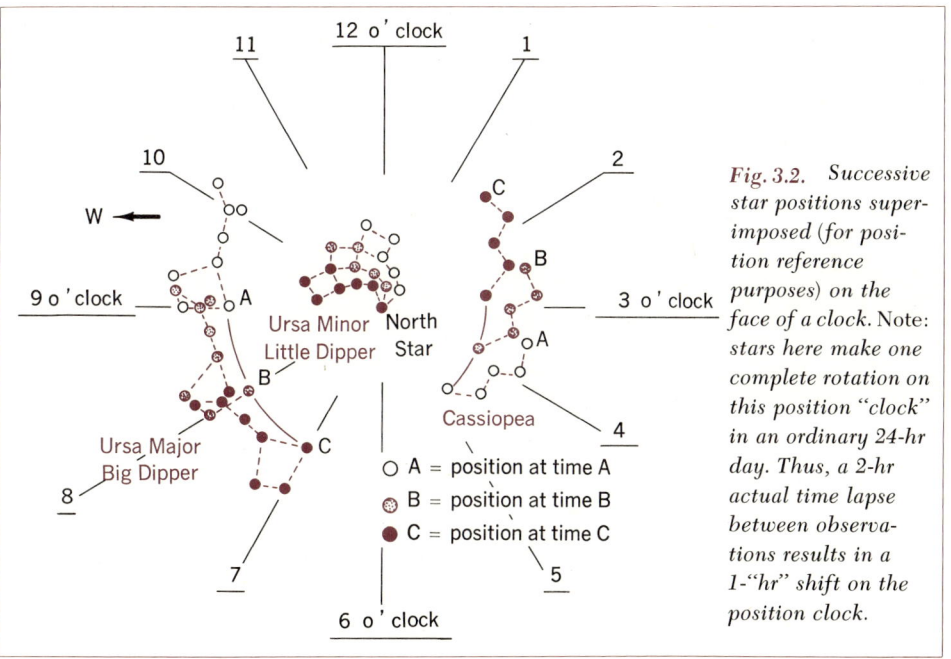

Fig. 3.2. Successive star positions superimposed (for position reference purposes) on the face of a clock. Note: stars here make one complete rotation on this position "clock" in an ordinary 24-hr day. Thus, a 2-hr actual time lapse between observations results in a 1-"hr" shift on the position clock.

Fig. 3.3. Trails left by noncircumpolar stars as a result of the apparent rotation of the celestial sphere. (Lick Observatory photo.)

Fig. 3.4. Star trails near the north celestial pole. (Lick Observatory.)

If we had taken note of a prominent star near the western horizon early in the evening, it would have "disappeared" by the time we concluded the observations. On the other hand, stars would now appear at the eastern horizon that were not there when we began. If we would picture ourselves at the center of a hollow sphere (see Fig. 3.1), plotting the position of a single star as we observe it at different times during the night, we would find that it traces the arc of a circle whose plane is perpendicular to the axis that runs from us to the Little Dipper.

The stars that drop below the western horizon are said to *set;* those appearing on the eastern horizon are said to *rise.* Hence stars rise in the east and set in the west, being carried by the apparent rotation of the celestial sphere. If we plotted the position of each star through the course of several hours, we would obtain a picture such as that shown in Fig. 3.3. A similar plot for the stars around the Little Dipper would give the picture in Fig. 3.4, in which the stars *do not* rise or set, but make complete circles about some central point. In the Northern Hemisphere we call that point the *north celestial pole.* The people of South America similarly identify a *south celestial pole* for their hemisphere, but this is unseen in the north.

The north celestial pole is at the present time located close to the relatively bright star Polaris, or the North Star, which marks the end of the handle of the Little Dipper. The daily rotation of the celestial sphere about an axis joining the celestial poles is called the *diurnal motion.*

A great circle can be drawn on the celestial sphere passing through the north celestial pole and the zenith. This imaginary circle, called the *celestial meridian,* cuts the horizon at the two points, north and south. A star rising on the eastern horizon reaches its highest point at the celestial meridian, after which it drops lower and lower until it sets in the west. The angular distance between the star and the horizon is called its *altitude*.

Since (by definition) the celestial meridian passes through the north celestial pole, the North Star is always on the celestial meridian and does not experience the general diurnal motion of the celestial sphere.†

At Troy, New York (where this text is being written), the North Star has an altitude of about 42 degrees. Since the apparent rotation of the celestial sphere causes every star to follow a circular path around the celestial pole, those stars less than 42 degrees from the North Star trace out circles that do not intersect the horizon (that is, they never rise or set when viewed at Troy). We call these stars *circumpolar*. Observations from more northerly locations, say from Hudson Bay in Canada where the altitude of the North Star is about 55 degrees, would include correspondingly more circumpolar stars. From a southerly position, say from Mexico City, the North Star appears to be only 19 degrees above the horizon.

3.3 The movements of the sun as seen by the eye‡

The sun, the most prominent object in the sky, the source of light and heat and of life itself, will be considered next.

---

† This assumes the North Star to be exactly at the north celestial pole. It is actually a small angular distance away from the pole, and hence this statement is not strictly true. It is close enough, however, so that our analysis introduces no major errors.

‡ In this section we assume throughout that our observer is in the middle latitudes of the Northern Hemisphere.

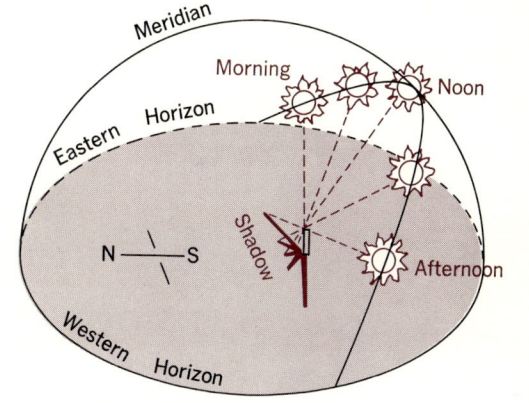

Fig. 3.5. Apparent path of sun for a given day, as seen in the Northern Hemisphere. The varying position and length of the shadow cast by a vertical stick are shown.

The sun rises in the east and continues to climb in altitude until it reaches our meridian (high noon), after which it sinks towards the horizon to disappear in the west. Between sunrise and sunset we have *day;* until it rises again we have *night*. As seen from the United States, when the sun reaches the meridian (high noon) it is always somewhat south of the zenith. The time interval between successive noons we call the *day*.† With the sun shining, let us place a vertical stake in the ground and watch the shadow it casts as the day progresses. On a winter morning when the sun is a little above the horizon in the southeast, the shadow of the stake points northwest. As the sun moves toward the meridian, the shadow moves northward until at noon it points due north. While the sun is sinking toward the southwestern horizon, the shadow moves eastward until just before sunset, when it points northeast (see Fig. 3.5). By subdividing the path of the shadow, we can use the position of the shadow at any moment to indicate the *time* of the day. The stake becomes a clock—the sundial, which was useful to early civilizations.

Let us repeat these observations every day for, say, 365 days. Let us begin, for a reason that will soon become apparent, in the winter season on December 21 of our present calendar when the sun rises in the southeast and sets in the southwest. As days pass, the sun's position on the eastern horizon at sunrise (and similarly its position on the western horizon at sunset) slowly moves northward, reaching its most northerly position near the end of June. During this period when the sun's position on the horizon moves north, its altitude at noon increases and the length of the shadow of our stake, at noon, decreases (see Fig. 3.6). After June 21 the shadow lengthens again, reaching its maximum length about December 21, when the sun at noon is at its lowest altitude. During the period required for these observations, we would become

† Unfortunately the English language uses the same word, day, with the two definitions given above. However, this seldom causes any misunderstanding.

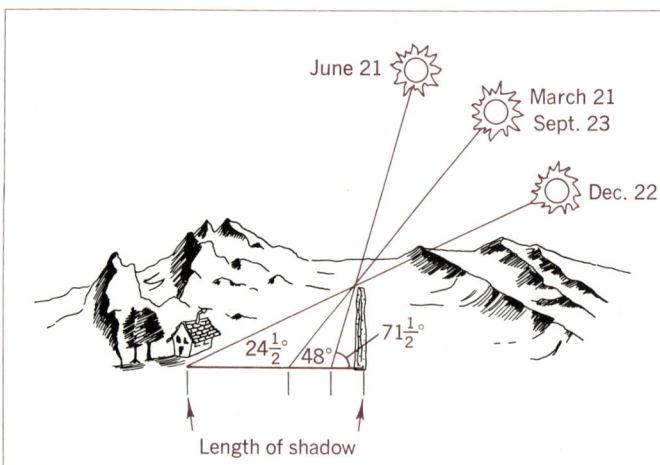

Fig. 3.6. Variations with time of year in the length of the shadow cast by a vertical stick.

very much aware of another phenomenon—the seasons. When we started in December, it was winter, but this gave way to spring, then to summer and fall, and back to winter again. This orderly cycle of the seasons can also be used as a measure of time, namely, the *year*. How many days make a year? If we choose as the length of the year the number of days in the interval from the time the shadow of our stake is longest at noon until it returns to its greatest length, and repeat the observations over many years, we would find that $365\frac{1}{4}$ days make a year.

We learned in our childhood that night is longer than day during the winter months and shorter in the summer months. As the sun moves northward (going from spring to summer) it reaches its highest altitude in the sky about June 21, when the night is the shortest (and the shadow of our stake, the shortest). This is called the *summer solstice*. Going in the reverse direction (from autumn to winter), the sun drops to lower and lower altitudes and the nights become longer until about December 21, when the sun reaches its lowest altitude (or *winter solstice*) and the night is the longest.

Between the times when nights are longer than days and days are longer than nights, there must be two occasions when the periods of the night and of the day are the same. Going from December 21 (when the longest night occurs) toward the shorter nights of spring, we note that around March 21 the night and day become equal in duration; this occasion is called the *spring*, or *vernal equinox*. When going in the reverse direction (away from the longest day of the summer solstice), the days become equal to night about September 21, which is called the *autumnal equinox* (see Fig. 3.7).

In the earlier discussion we noted that the celestial poles marked the end points of the axis of (apparent) rotation of the celestial sphere. We can draw a great circle on the celestial sphere halfway between the two poles (or everywhere 90 degrees from the north celestial pole). We call this circle the *celestial equator*. It is interesting to note that the two equinoxes occur when the sun is on the celestial equator. From vernal equinox to autumnal equinox, the sun lies above (or north of) the celestial equator. From Fig. 3.6 we see that (at Troy, New York) the sun has an altitude (at noon) of 48 degrees at both equinoxes and that this is just the angular distance between the celestial equator and the horizon. At the summer solstice (June 21) the altitude of the sun is $71\frac{1}{2}$ degrees, or $23\frac{1}{2}$ degrees above the equator, while at the winter solstice its altitude is $24\frac{1}{2}$ degrees, or $23\frac{1}{2}$ degrees below the equator.

Is the sun's motion through the sky exactly the same as the rotation of the celestial sphere? We can try to answer this question by focusing attention on the relationship of some one star to the movements of the sun. Consider a star that is on the eastern horizon just before sunrise today. If the entire motion of the sun is due to the rotation of the celestial sphere, the relative position of this star and the sun should remain fixed. Therefore this star should rise above the eastern horizon just before the sun's rise every morning. But if we should repeat our observations two weeks hence, we would find the star to be almost 15 degrees above the eastern horizon at sunrise! That is, the sun would appear to be falling behind the star in its rising. Thus the sun is moving eastward with respect to the celestial sphere by almost 15 degrees in two weeks (14 days) or about 1 degree per day. (Actually this would be 360 degrees in 365¼ days.) Thus, if the stars were visible during the daytime, we should observe the sun to move eastward (by 1 degree per day) among the stars.†

The apparent path of the sun among the stars is called the *ecliptic* (see Fig. 3.8). Since the sun may rise 23¼ degrees above the celestial equator, the ecliptic plane must be inclined by this same angle, 23½ degrees, to the plane of the celestial equator. This angle is called the *obliquity of the ecliptic.*

The ecliptic is a great circle through the stars on the celestial sphere. It can be arbitrarily broken up into 12 arcs of 30 degrees each. The star groupings in these twelve divisions of the ecliptic (covering an area of about 20 degrees above and below the ecliptic) have been identified by a group of 12 constellations‡ known as the zodiac (Aries, Pisces, and so forth) because their arrangements of stars mostly suggest forms of animals. The constellations of the zodiac played an important role in the astrology of the ancients and are familiar to us for the same reason even today.

Before completing our observations of the sun we should note also its appearance—an exceedingly bright, yellowish

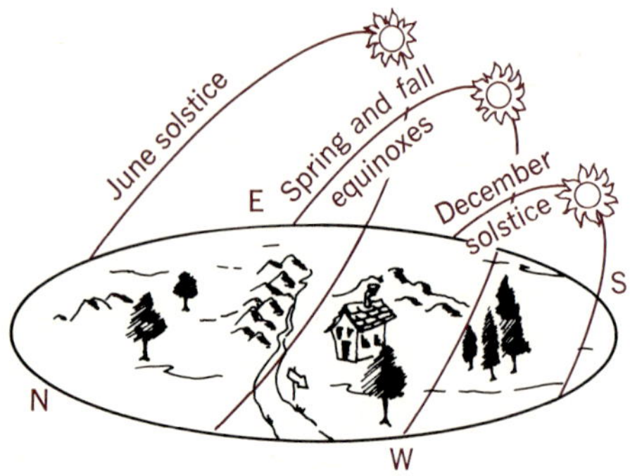

Fig. 3.7. Apparent path in the sky of the sun as seen by an observer in the Northern Hemisphere.

---

† Although we cannot see the stars when the sun is shining, we can observe the antisolar point (that is, the point in the sky 180 degrees away from the sun or halfway around the sky from the sun). Whereas the sun is on the meridian at noon, the antisolar point reaches the meridian at midnight. By observing which stars are on the meridian at midnight (at the proper altitude) we can trace the path that the sun will follow among the stars 6 months later.

‡ A constellation is a somewhat arbitrary grouping of stars on the celestial sphere and does not imply any actual physical connection between these stars. Most of the constellation names used by astronomers today derive from Greek mythology.

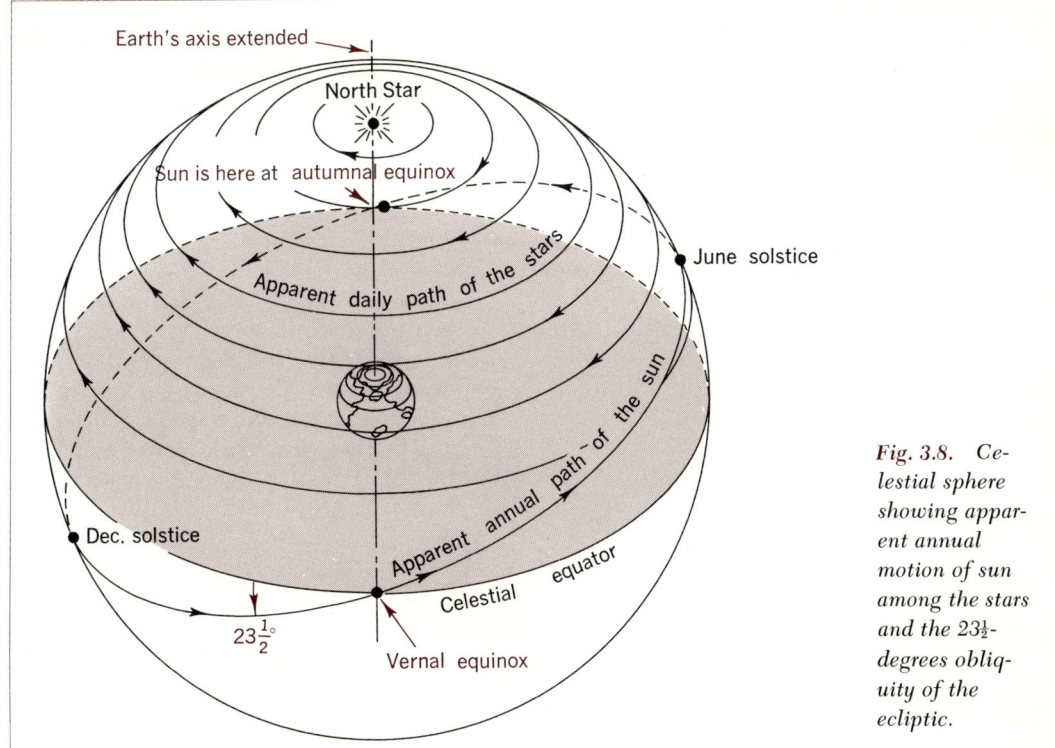

Fig. 3.8. Celestial sphere showing apparent annual motion of sun among the stars and the $23\frac{1}{2}$-degrees obliquity of the ecliptic.

circle in the sky. The diameter of this circle can be easily measured (without the use of sophisticated equipment) to be about 0.5 degree. (*Question:* How would you go about making the measurement?)

Finally, we note that we can use the stars instead of the sun to give us a unit of time. The period of one rotation of the celestial sphere is denoted as the time needed for a given star presently on the meridian to return to the same position on the meridian. This time unit we call the stellar, or *sidereal*,† day. Consider a star that reaches the meridian today at noon (that is, at the same time as the sun).

† From the Latin *sidus*, meaning constellation, star.

One sidereal day later, this star will again be on the meridian, but now the sun, which is moving eastward with respect to the stars by 1 degree per day, will be 1 degree east of the meridian. It will require nearly 4 minutes additional time for the sun to reach the meridian (noon). Thus the day as measured by the sun (*solar day*) is approximately 4 minutes longer than that measured by the stars (*sidereal day*).

Both solar and sidereal days are subdivided for convenience into 24 hours of 60 minutes each, each minute containing 60 seconds. Thus the sidereal second is shorter than the solar second by about 4 minutes in 24 hours (1440 minutes) or 1 part in 360 (1 solar second equals

1 + 1/360, or 1.00277 . . . sidereal seconds).

The preceding discussion of time assumes that the solar day has a constant, fixed length. This is not true. The length of the solar day varies with the season. The time we use (told by clocks) is based on a fictitious "mean" solar period that has all its days of equal length. Mean solar time may differ from that of a sundial (which uses the true solar time) by as much as 15 minutes at certain seasons.

3.4 The apparent movements of the moon

Next to the sun, the moon is the most prominent celestial object. Like the sun, the moon appears to be a yellowish disc whose diameter is approximately 0.5 degree. Unlike the sun, however, the moon shows *phases*. Suppose tonight the moon is rising at the eastern horizon just as the sun is setting in the west. It will appear as a full circle or disc (full moon). Tomorrow night the moon will rise nearly 1 hour *after* the sun sets, and its disc will appear slightly flattened on the westward (leading) edge. Each night thereafter the moon will rise later and appear more flattened (gibbous) until about a week after our original observations it will rise at midnight and appear as a semicircle, the western half of the disc being invisible (third quarter moon). During the next week the moon continues to rise 1 hour later each day. Its

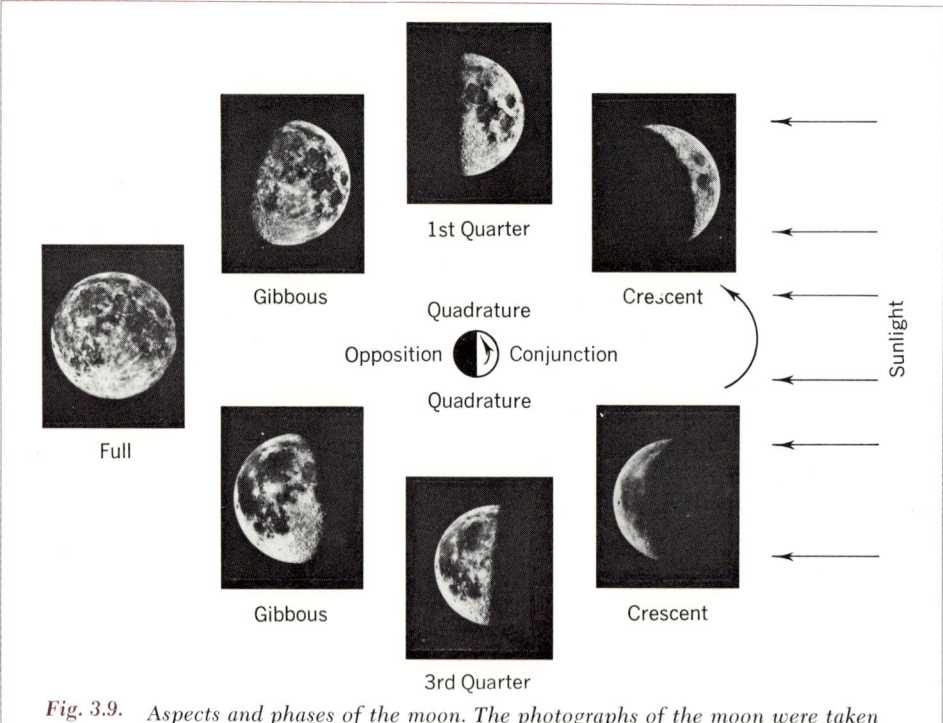

Fig. 3.9. Aspects and phases of the moon. The photographs of the moon were taken through the 36-in. refracting telescope at the Lick Observatory.

appearance changes from a semicircle to a "crescent"† which decreases in size, day by day, until at the end of the week, no moon is visible (new moon). During the following week the process is reversed, the moon rising after the sun in the form of a crescent. At the end of this third week the moon would rise at noon and set at midnight. Its appearance would again be a semicircle, now with the eastern edge missing (first quarter moon). At the end of about 4 weeks (actually, $29\frac{1}{2}$ days) the moon will be full again, rising at sunset (see Fig. 3.9).

Several other observations about the moon can be noted. First, on a dark night with the moon in crescent phase, it is just possible to discern the remainder of the circle of the lunar disc (sometimes romantically referred to as "the new moon in the old moon's arms"). Secondly, the moon can be seen during daytime and the phases can be observed. And finally, even without a telescope, the physical appearance of the moon is always approximately the same (that is, we always see the same side of the moon).

A phenomenon that appears to involve both the moon and the sun is the solar eclipse. For those of us who have been fortunate enough to view it, a *total eclipse* of the sun can be an awe-inspiring experience. As viewed through some suitable filter (to cut down the sun's light and prevent eye damage), the sun appears as a circular disc, shining brightly. But at some moment the circular appearance of the sun is marred by a small "bite" taken out of its westward edge. Slowly this "bite" becomes larger and larger until

Fig. 3.10. Eclipse of the sun on June 30, 1954. A series of photographs by Roy Swan in Minneapolis, Minn. Eclipse starts before sunrise (lower left) and ends with the sun high in the sky. (From Struve, Lynds and Pillans, Elementary Astronomy, *Oxford University Press.)*

† The word "crescent" has as its root a word meaning "to grow" and should be applied only to the moon between new moon and first quarter. Popular usage calls both the phases preceding and following new moon a crescent.

only a thin crescent of the sun remains visible. Even when 99% of its disc is gone, the sun is still an extremely bright object. The disc then disappears entirely and day becomes night. The stars appear in the sky and around the spot where the sun was, there appears a pearly white luminescence called the *solar corona*. The total eclipse lasts only a few minutes and then the process reverses itself. First, a thin crescent sun reappears at its westward edge, followed gradually by growth of this crescent until the entire sun reappears (see Fig. 3.10). A total eclipse of the sun is visible only along a narrow band of sites on the earth; outside this band observers see only a partial eclipse. Not all solar eclipses reach totality to observers on earth. Some start as described above and reach the crescent phase only when the process reverses. These are called *partial eclipses*. The amount of the sun eclipsed in the partial eclipse decreases as one gets farther from the band of totality until, for distant observers, no eclipse is seen at all.

In a similar fashion, the moon can be eclipsed (see Fig. 3.11). All lunar eclipses are visible over the entire region of the earth where the moon would normally be visible at the time of the eclipse.

Observations show that solar eclipses take place only at or near the new-moon phase, whereas lunar eclipses occur only at or near the full-moon phase. But as we well know, not every new moon sees a solar eclipse nor does every full moon become eclipsed.

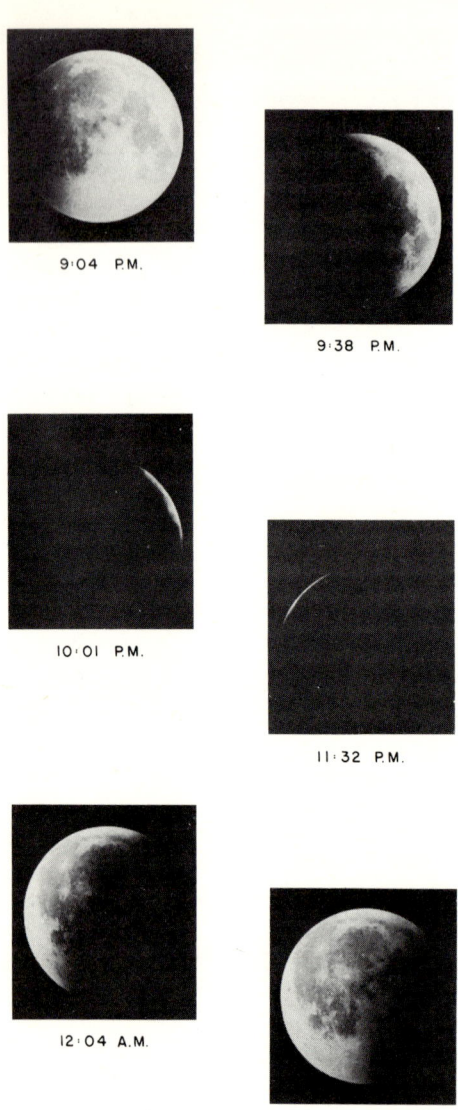

Fig. 3.11. Sequence of photographs of the total Lunar eclipse of Nov. 17 and 18, 1956. (Photographed by Paul Rogues, Griffith Observatory.)

3.5 The apparent motions of the planets

Since the sun and moon move with respect to the celestial sphere, we can assume that they are not rigidly a part of

3.5 The Apparent Motions of the Planets   67

the system of stars. Several other objects in the sky, which we call *planets,* from the Greek word meaning "wanderers," appear to deviate from the motion of the celestial sphere. Three of them (Mars, Jupiter, and Saturn) are always found within the band of the twelve constellations called the zodiac; that is, always close to the ecliptic. When viewed over a period of many nights they are seen to move relative to the stars on the celestial sphere, moving generally eastward among the stars.†

When a planet rises in the east just as the sun sets in the west, it is said to be in *opposition* (that is, at the opposite side of the sky from the sun), and when it rises and sets at the same time as the sun, it is said to be in *conjunction* with the sun. When the angle between the planet and the sun, as seen from earth, is 90 degrees, the planet is said to be at *quadrature* (either rising or setting at noon).

Although the planets, as do the sun and moon, appear to move eastward with respect to the stars, near the time of opposition the motion seems to reverse itself and become westward instead. We call this *retrograde motion* (see Fig. 3.12).

Two different periods may be noted for the planets Mars, Jupiter, and Saturn.

† We shall return to this point below.

One is the measure of the time it takes the planet to go from opposition to opposition (or conjunction to conjunction) and is called the *synodic period* (from the *synodos,* meaning meeting).† The second period is a measure of the time needed for the planet to make one complete circuit of the celestial sphere and return to the same place among the stars. This is called its *sidereal period.* Table 3.1 lists the synodic and sidereal periods of Mars, Jupiter, and Saturn. When the observed motion of these three planets is plotted over their entire sidereal period, we get Fig. 3.13.

Note that the synodic period for Mars exceeds its sidereal period, while for Jupiter and Saturn, the synodic period (which is slightly more than a year) is far shorter than the sidereal period. We also note by simple observation that these planets are much brighter at opposition than at conjunction.

An observer will see what appear to be four additional heavenly objects that do not completely follow the general motion of the celestial sphere. Two of these are the evening star and the morning star, called Hesperos and Eosphoros, respectively, by the ancients. On a given even-

† The period of the moon determined by its phases (29½ days) is its synodic period.

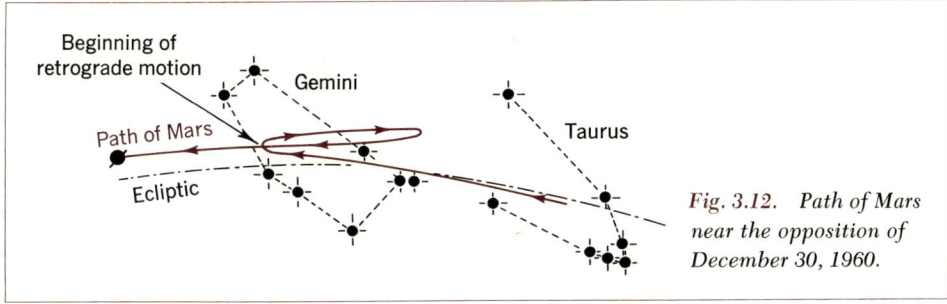

Fig. 3.12. *Path of Mars near the opposition of December 30, 1960.*

TABLE 3.1

SYNODIC AND SIDEREAL PERIODS

Planet	Synodic Period, years	Sidereal Period, years
Mars	2.14	1.88
Jupiter	1.92	11.86
Saturn	1.35	29.65

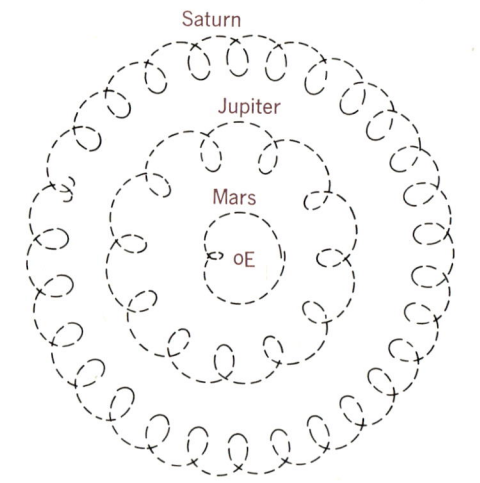

Fig. 3.13. Apparent orbits of Saturn, Jupiter, and Mars as seen from the earth.

TABLE 3.2

ASPECTS OF MERCURY AND VENUS

Planet	Synodic Period, days	Sidereal Period, days	Greatest Elongation, deg
Mercury	116	88	28
Venus	584	224	48

ing Hesperos may appear a little more than 45 degrees above the western horizon at sunset, setting some 3 hours after the sun. On succeeding nights it will appear closer and closer to the western horizon at sunset until some 220 days after our initial observation it is so close to the sun as to be no longer visible. A few days later, Eosphoros will appear in the eastern sky, rising slightly ahead of the sun. During the next 220 days its *elongation* (that is, apparent distance from the sun) will increase until it rises about 3 hours ahead of the sun. Then the process will reverse and Eosphoros will start to decrease its elongation, appearing closer and closer to the rising sun until it, too, becomes so close to the sun as to be invisible. The time needed for Eosphoros to go from greatest elongation to conjunction during this phase is about 72 days. Soon thereafter Hesperos will appear again, setting soon after the sun. Another 72 days will have to pass before it again reaches maximum elongation (of about 45 degrees) and sets 3 hours after sunset.

Whenever Eosphoros is visible in the morning sky, Hesperos is not visible in the evening, and vice versa. For a moment we shall depart from the approach of this section (of presenting only what one may observe of celestial phenomena without recourse to models or explanations) and note that some 2500 years ago it was recognized that Eosphoros and Hesperos are one and the same object—the planet we call Venus. Similarly the second pair of objects mentioned with Eosphoros and Hesperos at the beginning of this discussion are just the evening and morning aspects of the planet Mercury. Table 3.2 lists some pertinent data about these planets.

Mercury and Venus, as do Mars, Jupiter, and Saturn, exhibit retrograde motion at times. However, Mercury and Venus are never seen at opposition. In their cases, retrograde motion occurs at the conjunction when the planet is going from being the "evening star" to becoming the "morning star."

These observations of the stars, sun, moon, and planets by no means exhaust the knowledge we can obtain by patient study of the heavens, using only our eyes and occasionally a stake in the ground, and by being willing to travel to different places to make observations. However, this was the kind of information available 3000 years ago, from which the models of the universe were constructed. We shall now turn to these models.

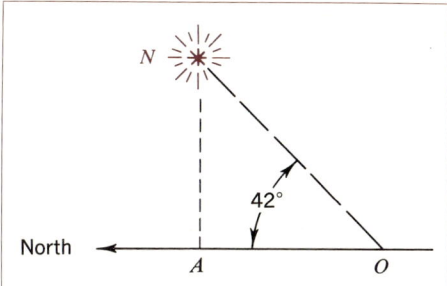

Fig. 3.14. Altitude of the North Star in the "flat earth" model.

3.6 What is the shape of the earth?

It is time now to analyze our observations and to develop some models that explain the phenomena we have thus far observed with our eyes.

To the casual observer, the earth upon which we live appears flat, marked here and there by mountains and hills.

A sufficiently long journey from any point on Earth will eventually lead to the sea. Hence, the picture of the earth as a flat disc surrounded by water is quite consistent with this simple model. We noted that the altitude of the North Star

(1) was about 42 degrees at Troy, New York;
(2) increased as we traveled north;
(3) decreased as we traveled south, being about 19 degrees at Mexico City.

On the model of the flat earth, (1) and (2) can be explained by a sketch such as shown in Fig. 3.14, where N is the North Star and O the observer. As the observer moves to the north (left) the angle NOA,

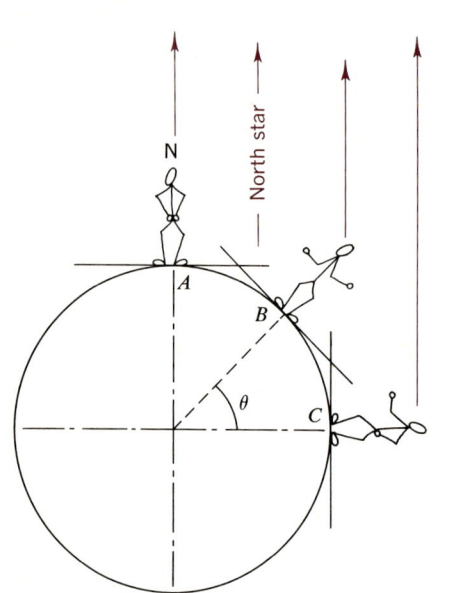

Fig. 3.15. *Elevation of North Star for an observer on a spherical earth.* Note: Distance to North Star is assumed to be so great that sight lines to it from widely separated points A, B, and C are still parallel.

representing the altitude of the North Star, increases in agreement with (2). From (1), the distance *NA* must be almost equal to *AO*, since the tangent of 42 degrees is nearly 1. In traveling north to Hudson Bay we find the altitude of the North Star to be 55 degrees. A simple calculation shows the distance *OA* to be about 4000 kilometers (since Hudson Bay is about 1500 kilometers north of Troy) and the height *NA* of the North Star above the surface of the earth to be about 3500 kilometers. As we head south (to the right in Fig. 3.14) the angle *NOA* decreases. At Mexico City (about 3000 kilometers south of Troy) the tangent of this (*NOA*) is 3500/5800, or tan *NOA* = 0.603 and the angle is 31 degrees. But, according to (3), this angle should be only 19 degrees! Our model is therefore inconsistent with observation.

An alternative model that we might propose is that the earth is a sphere. In Fig. 3.15, suppose that the distance to the North Star is large compared with the size of the earth. To an observer *A*, the North Star appears overhead, whereas to the observer *C* it appears on the horizon. To any other observer, say *B*, at some intermediate geographic latitude† θ, the altitude of the North Star will be θ also. (Can you prove this statement?)

The new model of a spherical earth is consistent with our observations of the altitude of the North Star. This does not mean that it is the only possible model, but we shall consider it in a little more detail before deciding whether we need look for other models.

Anyone who has watched a ship sail out to sea has observed that it disappears bottom first—that is, first the hull disap-

---

† Latitude is the angular distance on the earth's surface, measured in degrees, north or south from the equator.

pears, then the superstructure, and finally the smoke. Fig. 3.16 illustrates this phenomenon.

Another, more subtle but most convincing, argument for the sphericity of the earth comes from a study of eclipses of the moon. We shall jump ahead in our story and note without further comment at this time that a lunar eclipse is due to the passage of the shadow of the earth across the moon. In Fig. 3.11 we can see that the edge of the eclipsed area is roughly circular; thus the earth's shadow is roughly circular in cross section. This phenomenon is true in all lunar eclipses, no matter at what time of night they occur or from what point on earth they are viewed. Thus the earth must be of spherical form to cast a circular shadow in any direction.†

Today we take the sphericity of the earth for granted.‡ The last word on the subject might come from a quick look at a photograph taken from an earth satellite such as shown in Fig. 3.17 or from a statement of one of the Gemini 10 astronauts who, after orbiting the earth in July 1966, declared: "Columbus was right; it is round."§

Fig. 3.16. Ship disappearing below horizon on a spherical earth.

---

† Actually the earth is an oblate spheroid, being slightly smaller in the north-south direction than in the east-west direction. This is, however, only a minor deviation from a true sphere, and need not be considered at this time.

‡ It may come as a surprise that even today in the United States there are those who believe the earth to be flat, but they are few in number. See Martin Gardner, *Fads and Fallacies in the Name of Science*, New York, Dover Publications, 1957, for an interesting account of such beliefs today.

§ Columbus must have been stating a view held by many educated persons of his time when he said that the earth was round. The Greek philosophers and other learned people were well aware of the sphericity of the earth over 2000 years ago.

Fig. 3.17. This photograph was taken from 22,300 miles above the earth by NASA's Application Technology Satellite-1. (NASA, 1966.)

3.7 Illumination of the moon and planets and the nature of an eclipse

Let us explore models of the solar system that can explain *all* of our observations without exception. What model will explain our observations of the phases of the moon, noted in Sec. 3.4? We shall start by making a few assumptions, namely:

(1) The earth is a sphere.

(2) The moon revolves around the earth.

(3) The sun is more distant than the moon.

(4) The sun illuminates the moon.

Of these assumptions, (1) was discussed in the preceding section and requires no further comment. Assumption (2) seems to be what we observe, although we have to admit that not only the moon, but the sun and stars also appear to revolve about the earth. As-

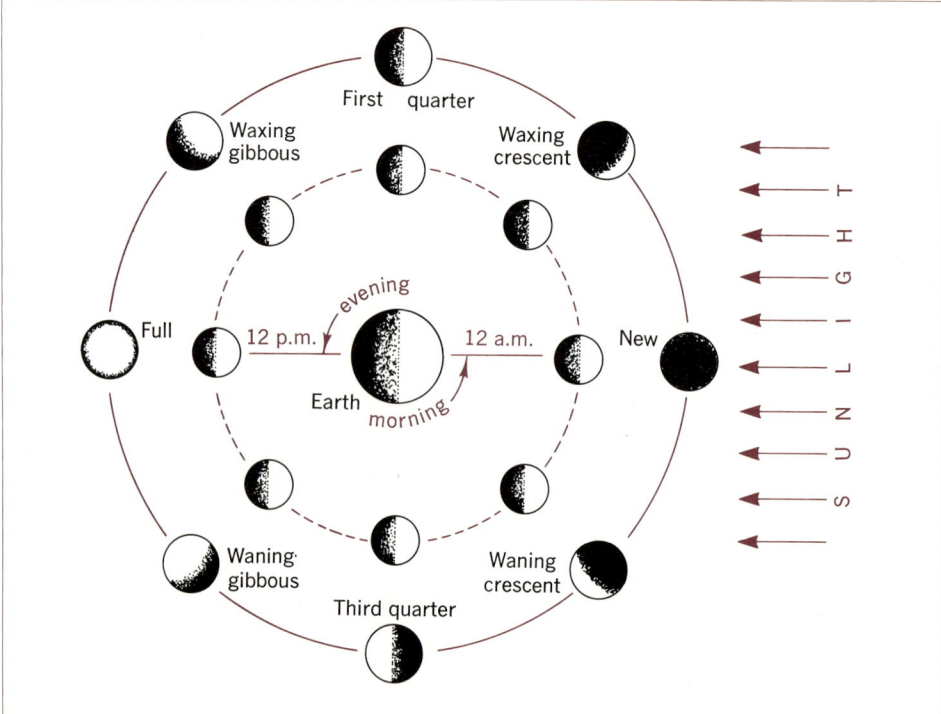

Fig. 3.18. *An explanation of the phases and aspects of the moon. The outer circle shows phases of the moon as they appear to an observer on earth when the moon is in each of the positions shown in the inner circle relative to the sun and the earth.*

sumption (3) seems acceptable, since if the moon were more distant than the sun, then the new moon would be behind the sun, and it would be more difficult to understand the moon's role in a solar eclipse. As for (4), it would be difficult to explain the phases of the moon on the basis of its being a self-luminous body; hence the sun would be a logical choice as a light source. We have not proved these assumptions to be true, but have attempted to make them seem plausible, pending the results of further analysis.

Figure 3.9 illustrates the illumination of the moon at different times during one synodic period with the sun in a fixed position. (By this we cannot yet imply that the sun is not also rotating about the earth.) Figure 3.18 illustrates how the phases of the moon are accounted for by this model.

We know that the earth is illuminated by the sun and have assumed, by (4), that the moon is similarly illuminated. We know also from common experience that illuminated objects cast shadows that have the shapes of the objects themselves; therefore the earth and moon should each cast a shadow that extends into space away from the sun. From the fact that a solar eclipse is not seen everywhere on the daylight portion of the earth, we can conclude that the moon's shadow, at the moon-earth distance is smaller than the earth in cross section. On the other hand, a lunar eclipse is usually total, the earth's shadow covering the entire moon. Since the size (diameter) of the shadow is dependent on the distance and size of the object casting the shadow, and since the distance from moon to earth is the same as that from earth to moon, the earth must be larger than the moon!

Eclipses do not occur during every synodic period of the moon, and therefore the moon's shadow must generally pass above or below the earth; similarly, the earth's shadow must pass above or below the moon. This implies that the moon's orbit is inclined to the plane of the ecliptic (the sun's apparent path around the earth). See Fig. 3.19. Eclipses can take place only when all three bodies—sun, moon, and earth—lie in the same plane. The plane formed by the moon's orbit around the earth intersects the ecliptic plane in a line known as the *line of nodes* of the lunar orbit. The moon must lie on this line of nodes for an eclipse to occur. Knowing this and knowing when previous ones have occurred, it is a matter of simple arithmetic to predict a future eclipse.

We have shown that our knowledge of the phases of the moon and of eclipses of the moon and the sun is consistent with the sun as the source of illumination of both earth and moon. It is a small step to the assumption that the planets are also illuminated by the sun.

3.8 Models of the universe that account for planetary motion

We must try to expand our tentative models to include the planets and the stars. Based on our eye observations alone it appears that all the stars are fixed to the inside of a "large hollow sphere" (the celestial sphere) with the earth at its center. But since the motions of the sun and moon and the other planets† (Mercury, Venus, Mars, Jupiter, and Saturn) differ from the motion of this

---

† To the ancients, the sun and moon, although brighter and more important than the other five planets, were also considered to be planets. These were the seven planets of the ancients.

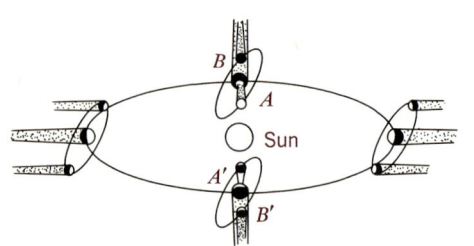

Fig. 3.19. Eclipses of the sun at A and A' and of the moon at B and B' occur only when the moon crosses the plane of the ecliptic at such a time that the sun, moon, and earth lie on the same straight line. The inclination of the plane of the moon's orbit relative to the ecliptic is actually only about 5 degrees, but is exaggerated here for clarity.

starry "hollow sphere," does this mean that we must add seven additional independent spheres, one for each body, to account for their quite independent movements?

The "sphere" of the stars rotates about an axis passing through the celestial poles. The "sphere" containing the sun rotates about an axis inclined by $23\frac{1}{2}$ degrees to that of the stars, while those of the moon and other planets have rotation axes inclined by a small amount to that of the sun. With a model of eight "spheres," all rotating about different axes, we seem to get a rough approximation to what is observed in the sky.

The model is not quite enough, however, for we have not yet accounted for several observed phenomena—the difference in brightness of a planet at various times, the varying length of the day in different seasons, the retrograde motion of the planets or the fact that Mercury and Venus are never found far from the sun. To improve our model we must allow each planet more freedom of movement than that of a simple rotating sphere. Eudoxus (ca. 370 B.C.), a disciple of Plato, constructed a model that utilized 27 spheres, all concentric with the earth.

There are other models possible which do not depend on concentric spheres. One such model was put forth by Claudius Ptolemaeus (Ptolemy) of Alexandria in 140 A.D. This model, known to this day as the Ptolemaic system, was put forth by him in a treatise we call the *Almagest* literally, the "greatest" composition. Like Euclid's *Elements* and the other textbooks of the Hellenistic era, Ptolemy's book is not entirely his own work. Ptolemy very frankly gives credit to his predecessors, particularly Hipparchus. The Ptolemaic model is basically one of common sense in which the movements of the heavenly bodies are represented exactly as they appear to move to the observer. Ptolemy proceeded with astronomy (as Euclid began his book on geometry) with five postulates or axioms that could be accepted without requiring proof beyond that of observation, namely:

(1) The heavens are spherical and rotate.
(2) The earth is spherical in form.
(3) The earth is at the center of the universe.
(4) The earth is of infinitesimal size compared with that of the universe.
(5) The earth is stationary, without any local motions whatsoever.

There was one more assumption, not listed as such but fully as effective in directing Ptolemy's thoughts. This arose from the belief of some Greek philosophers that the heavenly phenomena represented nature in its state of perfection. For this reason the movements of the heavenly bodies were required to be in circles and at uniform velocities, since the circle represented the most perfect of plane figures and a sphere of solids.

Building on his five postulates, Ptolemy proceeded to the details of the model. While the daily movement of the celestial sphere accounts roughly for the appearances of one day, continued observation for many days shows that the sun, moon, and planets move slowly about on the celestial sphere. There is, superimposed, a slow motion eastward which is contrary to the apparent daily westward rotation of the celestial sphere.

Also the motion of the sun is not uniform along the ecliptic but is at slightly different rates each day during the year, so that the apparent times required for the sun to move through successive 90 degree intervals of its annual path vary noticeably. Ptolemy showed that two different geometrical constructions of

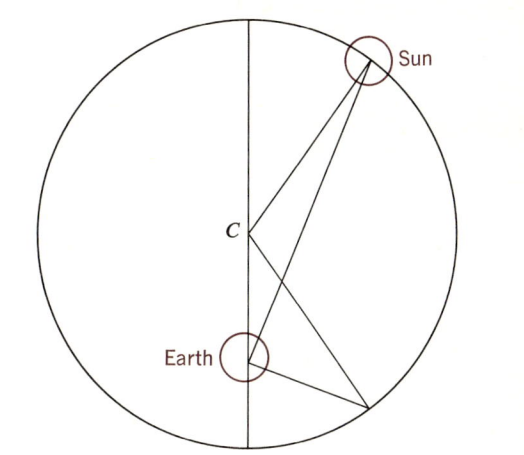

Fig. 3.20. *Eccentric hypothesis. The earth is displaced from the center C of the sun's orbit, assumed circular.*

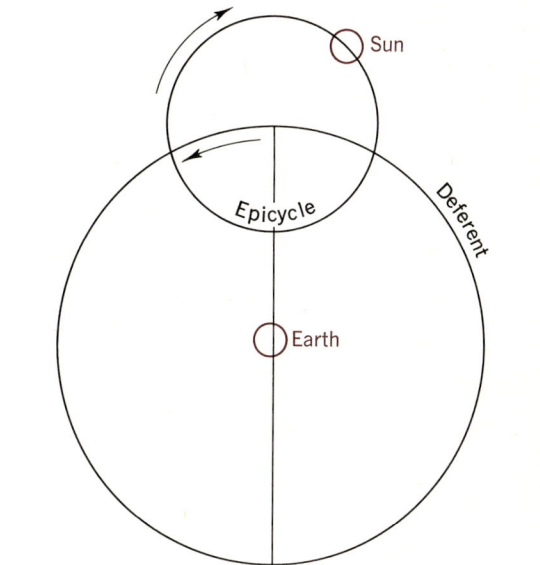

Fig. 3.21. *Epicycle hypothesis. Sun moves on circle (epicycle) centered on a second circle (deferent), which is centered on the earth.*

circles could each account for the observed irregular motion of the sun. In the first, which is the *eccentric hypothesis* (Fig. 3.20), the sun moves uniformly on a circle that is not centered on the earth. In the second, which is the *epicyclic hypothesis* (Fig. 3.21), the motion of the sun is the result of the superposition of motion in two different circles. One circle, the deferent, is centered on the earth and bears another rotating circle, the epicycle, on its circumference. The apparent motions of the sun in the sky can be reproduced by placing the sun on the epicycle and properly adjusting the periods of revolution and the diameters of the two circles. To account for the planetary appearances, Ptolemy used circles with displaced centers and epicycles along with a new hypothesis of the "equant." This proposed that a circle does not turn uniformly about its center but about quite a different point, the equant. Ptolemy continued to assume that all heavenly motions could be explained in terms of some arrangement of circles, since circular motions were "natural" for heavenly bodies.

The motion proposed for Mercury by Ptolemy was even more complicated, since the equant itself moved upon a small circle. However, with a suitable combination of enough circles, almost any observed apparent motion could be accounted for. The reader is referred to the highly simplified drawings of the Ptolemaic system, Figs. 3.20 through 3.23.

Since the earth is the most important "fact" in the life of a man and all observations of the universe are made from earth it is not surprising that two of the models were geocentric. In fact it was not until the twentieth century that we learned that there can be no absolute motion; that all motion is relative to an arbitrary reference frame.

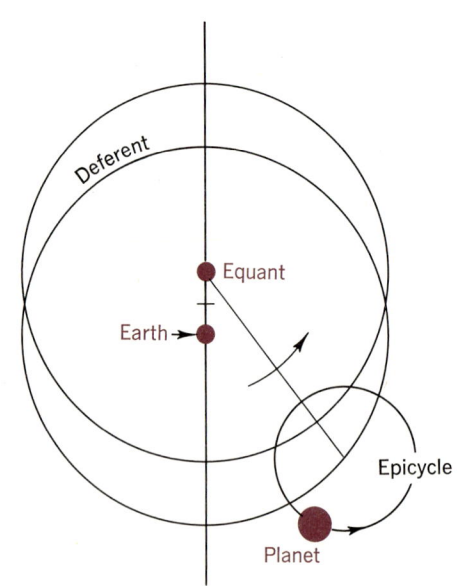

Fig. 3.22. Ptolemy's model of planetary motion, including both epicycle and equant hypotheses. The planet moves uniformly around circle denoted as epicycle, the center of which moves along one lower large circle (deferent). The motion along the deferent is not uniform with respect to earth, but rather with respect to a noncentral point called the equant.

These geocentric models (and there were others) seemed to account for everything known about the universe, within the accuracy of the data. The fact that complexities had to be introduced did not make the models incorrect, for nature need not be simple. We recognize, however, that the models reflected no interest or awareness of forces and energy sources to make the motions possible, beyond ascribing all motions and energy sources to a *prime mover*.† Also we must repeat that the models were subject to the philosophical requirement that the heavens demonstrate unchanging perfection.

The astronomy we have introduced briefly will, in part, become a major thread with which to continue our story of the development of scientific thought through the ages. We shall find that a most important revolution in scientific thought and scientific approach came about when it became desirable in the sixteenth and seventeenth centuries to develop quite different models to explain observations of the movements of heavenly bodies. But let us now turn to a review of some other aspects of the science of the early civilizations.

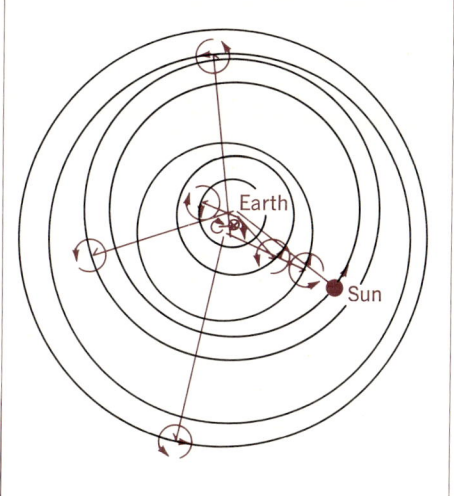

Fig. 3.23. A schematic view of the Ptolemaic model of planetary motion.

3.9 Science from the Hellenistic period to the Renaissance

During the period‡ when Aristotle was producing his works on natural philosophy that were to influence the world of science for many centuries, a former pupil of his was exerting an influence of a different kind on the entire known world.

---

† The concept of the *prime mover* as deity developed in the centuries following the Greeks.

‡ See Fig. 3.31 for a summary of major events and individuals discussed in this section and their temporal relationship to other selected historical epochs.

Alexander (called the Great) of Macedon accomplished in a few years what the Greek city-states had not accomplished in centuries—dominion over Greece and the conquest of the Persian Empire. And where Alexander's army marched, Greek language, culture, and philosophy went apace. From the Indus River on the east to Egypt on the west, the ideas of the Greeks were mingled with those of the local cultures and introduced a new era for art and science. This age, from Alexander's conquests until the establishment of the Roman Empire some 350 years later, is called the Hellenistic period.

Greek science, leavened with ideas that had emanated earlier from Egypt and Babylon, flourished during this period. Increased knowledge of the earth led to greater curiosity and a more probing attitude toward nature. Philosophic systems and encyclopedic surveys gave way to more circumscribed studies, which in effect encouraged specialization in the search for knowledge. This same period also saw the spread of Babylonian astrology into Greek astronomy.

One of the centers of Hellenistic learning was Alexandria, the city Alexander founded on the Nile delta to commemorate his conquest of Egypt. It was here that Ptolemy (the general appointed by Alexander to rule Egypt, not the astronomer who lived 500 years later) founded the famous Museum (literally, House of Muses) which contained research facilities for a hundred professors plus a library of a half-million manuscripts. This city became the mecca for scholars of the ancient world.

Among the great scholars who worked in Alexandria was Euclid (*ca.* 300 B.C.) whose collection of theorems on geometry is used to this day. Aristarchus (*ca.* 310–230 B.C.) put forth a model of the universe in which the sun was at the center. This heliocentric view was too revolutionary to be acceptable to the common-sense approach of the Greeks, however. The greatest astronomer of antiquity, Hipparchus (190–120 B.C.) worked in Alexandria from 160 B.C. to 127 B.C. He is credited with invention of both plane and spherical trigonometry and with making the first accurate catalog of star positions containing a scale of relative brightnesses. He was also the innovator of epicycles to explain planetary motion. The astronomical researches started by Hipparchus culminated in the work of Ptolemy (*ca.* 140 A.D.), which we have already discussed. Eratosthenes (273–197 B.C.), who was librarian of the Museum, used the angles of the sun at two different latitudes to compute the circumference of the earth (see Fig. 3.24).

During this same period, Archimedes of Syracuse (287–212 B.C.) was laying the foundations for our modern science of mechanics. He discovered the principle of bouyancy, which still bears his name, and introduced the concept of "center of gravity" in his discussion of the lever. His chief interest lay in the field of geometry, however, where he made significant contributions.

The Hellenistic Age also saw great technological achievements. Hero of Alexandria built the first steam engine sometime around the beginning of the present era. Archimedes constructed devices of pulleys and levers.

Whereas earlier Greek natural philosophy was of a speculative nature that sought to generalize ordinary experiences into all-embracing world models, the scientists of the Hellenistic period were content to solve more specific problems. There was now greater emphasis on the classification of knowledge, with the work of the men and women of the Museum being not unlike the interests of modern university professors. Thus, Archimedes studied the lever and wrote a mathematical treatise on its operation without seeking to extend his findings into a theory applicable to all nature.

The last centuries of the Hellenistic Age also saw a spurt in the growth of superstition and mysticism. By the end of the period, occult science overshadowed natural science. An example of this is that Ptolemy was more remembered during the First Millenium A.D. for his tract on astrology and how to cast horoscopes than for his world model. This reemphasis of the mystical in science was accompanied by similar changes in theology, for example, the Neoplatonism of Plotinus. The Olympian deities were abandoned in favor of mystical cults concerned with ideas about fate. In philosophy we find the growth of Stoicism.

The Hellenistic period also saw the rise of the political power of Rome. By the end of this era, much of the known world was under its domination. Indeed, the Alexandria of 140 A.D. where Ptolemy lived and worked had been ruled by Rome for nearly 200 years.

The Roman Empire was the cultural heir of Greece. The Romans were well known for their administrative and engineering skills, but they did little to extend the natural sciences. Some great collec-

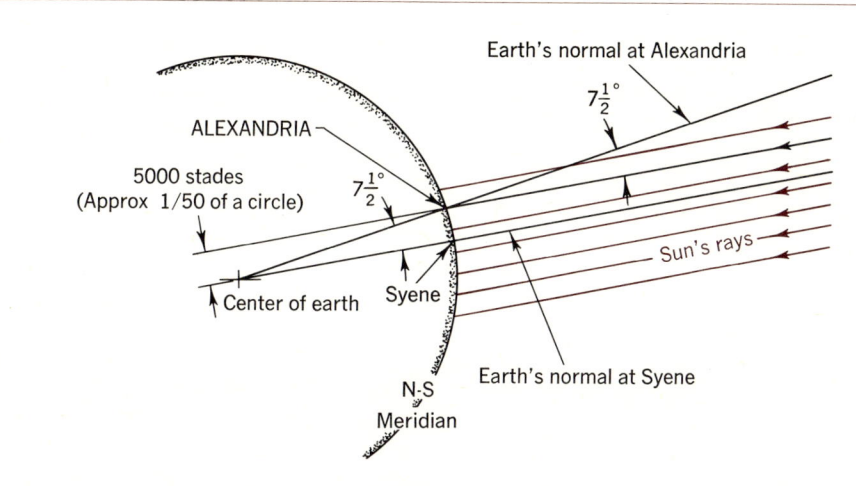

Fig. 3.24. Measurement of the circumference of the earth according to Eratosthenes. At the time when the sun was directly overhead at Syene, it was $7\frac{1}{2}$ degrees (or about $\frac{1}{50}$ of a circle) away from that position at Alexandria. The distance between the two points was 5000 stades (approximately 500 miles). Therefore, the circumference of the earth was estimated at about 250,000 stades, or 25,000 miles.

tions of Greek philosophic thought were compiled during this time (for example, Pliny the Elder's *Natural History*), but little original work was accomplished. One exception was the work of Galen in physiology and medicine.

The Roman people were temperamentally well suited to accept the mysticism of the late Hellenistic period. Astrology was very popular among them, and alchemy had its birth during this period.

It was during the era of the Roman Empire that Christianity arose and spread. With religious fervor, some of the more zealous proselytes to Christianity sought to rid themselves of all vestiges of their pagan past, sometimes destroying temples and works of art along with philosophical and scientific works of the Classical period. The destruction of one section of the library at Alexandria by the Bishop Theophilus in about 390 A.D. was an example of such actions.†

Another force, which was to play a major role in European history, was the successive waves of westward migration of peoples from eastern Europe and western Asia into the heartland of the Continent. Some of these invaders were brought into the Roman Empire to serve as buffers against future onslaughts, others were hired as mercenaries, while others entered the Empire with fire and sword. In 476 A.D., Rome itself fell to the invaders. By the time of the fall of Rome, the Empire had been divided into two parts—a western part ruled from Rome and an eastern part governed from ancient Byzantium, renamed Constantinople.

† Also, in the year 415, a woman mathematician of renown, Hepatia, the last professor of the Museum, was the victim of a mob that was encouraged by the Christian authorities to destroy her. She was pulled by the mob from a carriage into a church, and torn to pieces.

Christianity was the accepted religion in both parts of the Empire.

Following the fall of Rome, the western Empire disappeared as a political entity. The people who were heirs to the cultural legacy of Rome had been raised in a society with different traditions and different desires. The leisure class of the society that had given birth to Greek art and philosophy was absent. The economic decline of western Europe, which followed the fall of Rome, was not conducive to the development of a society dedicated to science and the arts.

The spread of Christianity throughout western Europe had a profound influence on the history of the centuries that followed the fall of Rome. Official religion defined the character of this period, and natural philosophy as well as theology were circumscribed by contemporary prescribed doctrines.

The Eastern Roman Empire, centered at Constantinople, fared somewhat better during this period. Containing, as it did, Greece, Egypt, and the Middle East, it was closer to the sources of Classical and Hellenistic thought. For a while the spirit of scholarship as well as the ancient philosophies were maintained there. This spirit was greatest, perhaps, in the Byzantine monasteries in Armenia and Syria, where the works of the Greek philosophers were translated into Syriac, a semitic language akin to Hebrew and Arabic. Eventually, however, the philosophic discussions of the Byzantine Empire developed into theology, and the Greek traditions fell into disrepute. The scholars of Armenia and Syria moved eastward into Persia. At Junishapur on the border between present-day Iran and Iraq, the Persian monarch established a great center for medical research which became an important center for learning.

Mohammed, born in 570 A.D. near the

Red Sea, brought into being a rival religion that inspired the Arab world to conquer much of the Middle East, parts of India, and to reduce the Mediterranean Sea to a Moslem lake in the course of the seventh and eighth centuries. Some of the Arabs showed keen interest in Greek learning, while making their own contributions to the number system and to algebra. For the next several centuries the caliphs of Baghdad were among the patrons of science in the Western world, in contrast with the Christian attitudes of the period toward science. Arabic translations of Aristotle and Ptolemy as well as of other Greek philosophers were made and studied.

After 1000 A.D., the center of Arabic learning shifted to Spain, then ruled by the Moors who accepted Christian and Jewish scholars. By 1200 A.D. the political power of Islam began to decline, and Arab culture also began to stagnate under the pressures of orthodoxy. The Moslems had destroyed libraries, such as the famous Museum of Alexandria in 640 A.D. In 1140 the caliph of Baghdad ordered the destruction of all the philosophical books of Avicenna, one of their scholars who was highly respected and studied by the Christian world. The scholar Ibn Habib was put to death by his emir for studying philosophy. The coming of the Turks brought persecution of the Christians and prevented travel to the city of Jerusalem, which was a holy city to three faiths. The failure of the crusades to free Jerusalem only prefaced the capture of Constantinople by the Turks in 1453. Meanwhile, Latin translations (from the Arabic) of Greek works began to appear in Europe. The philosophy of Greece had traveled from Greece and Alexandria via Rome and Constantinople to Baghdad and Andalusia and thence back to Rome.

In the founding of Christianity the Early Church Fathers had drawn upon the beliefs and philosophic traditions of their times. Thus it is not surprising that the works of pagan philosophers, especially Plato, should have played a large role in the development of early Christian beliefs. Saint Augustine (354–430 A.D.), the most influential of the early Church Fathers, used Platonism in combination with the Pauline Epistles to achieve the first great synthesis of Christian philosophy. But, a millenium hence, his successors were shocked by the realization of the influence that pagan thought had had on the foundations of Christian doctrine. W. C. Dampier (1929) sums up the Classical influence on the Church with:

"The Roman Empire died, but its soul lived on in the Catholic Church. Philosophically the Catholic Church was the last creative achievement of Hellenistic civilization; politically and organically it was the offspring and heir of the autocratic Roman Empire."

During the eleventh century the Persian al-Ghazzali accomplished a synthesis of Aristotelian natural philosophy with Moslem theology. A century later, Rabbi Moses ben Maimon (Maimonides), working in Spain, accomplished the same synthesis with certain Jewish traditions. The same synthesis of Aristotelian science with Christian theology was begun by Albertus Magnus (1206–1280 A.D.) and completed by his student, Saint Thomas Aquinas. Thus, by the end of the thirteenth century, the great theological systems of the Western world had thoroughly embodied the natural philosophy of Classical Greece and had determined the character of the scholasticism of the period.

To summarize, the period from 476 A.D. until about 1200 A.D.—the so-called Middle Ages—was characterized by religious orientation of all intellectual endeavors. The rise of Christianity in Europe and of Islam in the Middle East played a major role in controlling the minds of men. This period, which lasted for nearly a millenium, did not give birth to any new science, although there were the beginnings of chemistry in the form of alchemy, some developments in mathematics and medicine, and in the eighth century the beginnings of technology in Europe. Few, if any, significant changes occurred in knowledge of astronomy and indeed by the end of this period much of Alexandrian astronomy had been forgotten. Religious interpretation seemed to satisfy whatever desire men had for the understanding of the universe. The spirit of scientific inquiry was decidedly dormant. (On a latter occasion a pope was to tell Galileo that men should be more concerned with how to go to heaven than with how the heavens go.)

Before leaving this period of history we should note that several significant developments had taken place in the Far East. Among the technical achievements of the Chinese, the manufacture of porcelain, improvement of ship design, and the invention of block printing were important. Gunpowder was invented during the ninth century. Chinese mathematics became highly developed during this period. In India a major astronomical center was located at Ujjain, but the astronomical theories proposed there were based largely on those of Greece. Hindu mathematics was largely algebraic, although they did introduce the sines of angles into trigonometry. They also introduced the concept of *zero* into mathematics, and with it the idea that division by zero gives infinity. It was not until the end of the thirteenth century, however, that contact between East and West was reestablished.

3.10 Revival of learning and the Renaissance

The twelfth century saw a revival of learning in Europe and the beginnings of the secular universities which were to play a large role in future education. This was the *Age of Scholasticism* when Greek and Arabic writings were studied and interpreted at both church and secular schools throughout Europe. The works of Aristotle were studied, first from Latin translations from the Arabic, but later directly from the Greek by scholars such as Robert Grosseteste, Chancellor of Oxford and Bishop of Lincoln. Grosseteste's pupil, Roger Bacon, wrote a book on Greek grammar so that more scholars could go back to the original sources of Greek science and philosophy. This was the period when St. Thomas Aquinas was completing the amalgamation of Greek philosophy and Christian theology.

The Scholastics sought knowledge within the works of past ages. In a way, they, like ancient Greek philosophers, believed that true knowledge was to be obtained only through contemplation. Roger Bacon's was one of the first voices raised against this method of learning. He felt that the truth or falsity of theories must depend on how well they account for what we observe. Although Bacon is generally credited with reviving the "experimental method" in science, with the exception of a few experiments in optics, he personally did little to promote the method.

William of Occam (Ockham) (*ca.* 1295–1349) held that a body in motion does not necessarily require continuous physical contact with the mover. This attack on a basic postulate of Aristotelian physics presaged the development of the concept of inertia 300 years later. William is also remembered today for enunciating *Oc-*

cam's Razor which, in modern terms, says "When choosing between rival theories explaining the same phenomena, pick the simplest one (that is, the one with fewest assumptions)."†

Despite the change in outlook inspired by these men we are reminded that Dante's *The Divine Comedy* served as a mirror of the philosophy of the age in which it was written. (See Fig. 3.25.)

The fourteenth century marked the beginning of a new period in the history of Europe, namely the Renaissance.‡ There was rebirth of interest in art and science in Italy, perhaps because only there, among all the nations of Europe, did the upper classes live in the cities and were close to the people. By the fifteenth century, interest in classical literature on the part of western Europe attracted Greeks from the east, a process hastened by the fall of Constantinople to the Turks. Once more the spirit of free inquiry spread throughout Europe. Such men as Johann Müller (known as Regiomontanus) sought to combine science and humanism. Even the church shared in this new spirit until the capture of Rome in 1527 by troops of the Holy Roman Empire, after which the church officially tended to oppose search for new knowledge and to become an obstacle in the path of learning.

The Renaissance was also the age of exploration. The fifteenth century saw the voyages of De Gamma (1497), Columbus (1492), and in the early sixteenth century, Magellan (1516). It also produced such artistic masters as the Italians Leonardo Da Vinci (1452–1519), Michelangelo

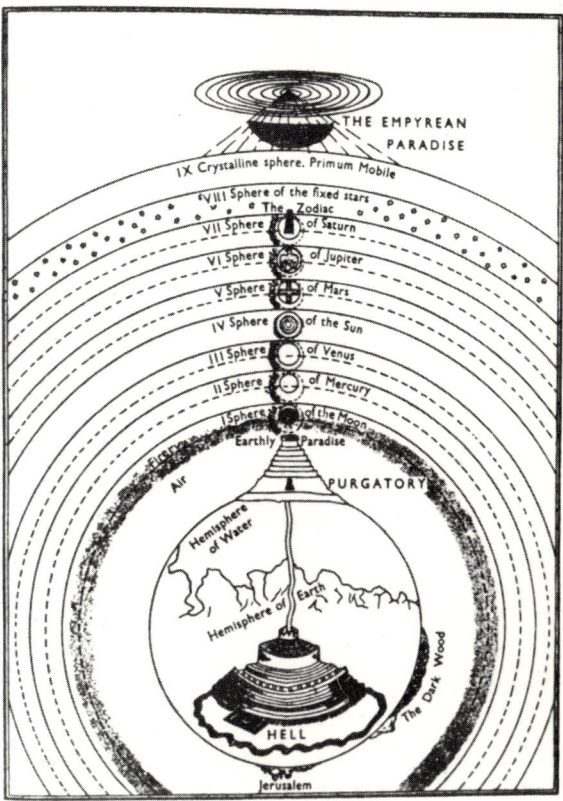

Fig. 3.25. Dante's scheme of the universe. (There was the impression that the Copernican theory, by replacing the earth with the sun as the center of the universe, tended to demean the position of man and of the earth. Note, however, that like Dante scholastic philosophers of the period regarded the farther distances from the earth as the more favored locations for paradise.) (From The Fabric of the Heavens, *copyright 1961. By permission of Harper & Row, New York.)*

† The original said, "It is vain to do with more what can be done with fewer."

‡ We should note that an earlier renaissance in the twelfth century may have paved the way for this one.

(1475–1564), Botticelli (1444?–1510), and the German painter, Dürer (1471–1528). These men introduced the ideas of perspective into Renaissance art. They studied anatomy, and Dürer exhorted his colleagues to study mathematics in order to understand geometrical relationships. Da Vinci's interests embraced all facets of human knowledge and many of his inventions are known even today. He worked at mathematics and astronomy as well as his better-known endeavors in the world of art.

The spirit of the Renaissance was halted for a time by the political and religious upheaval of the Reformation and the wars that followed it. It was this period of unrest that gave impetus to the development of modern science.

3.11 Copernicus and the heliocentric universe

Niklas Kippernigk, better known as Nicholas Copernicus (1473–1543) was born in the city of Thorn on the Vistula and educated at the universities at Cracow and at Bologna where he studied secular and canon law, medicine, mathematics, and astronomy. His education was accomplished in a period marked by Renaissance liberalism of thought and discussion. He returned to his native land with the degree of Doctor of Canon Law and was appointed canon at the Cathedral of Frauenburg by his uncle, then Bishop of Ermland. Copernicus retained this appointment for the remainder of his life, but he is remembered today for his contributions to the science of astronomy.

According to Rheticus, his disciple, Copernicus disliked the Ptolemaic idea of equants, to a large extent because of the aesthetic incongruities he saw in that approach, and sought a better geometry to explain astronomical observations. He preferred his circles to turn on their centers rather than eccentrically, and was able to eliminate the equants by placing the sun at the center of the system. Nevertheless, in his early work he still required 34 circles for the solar system (as compared to 79 circles required by Fracastoro to fit the Ptolemaic system).

With keen understanding for the age in which he lived, Copernicus worked quietly on his new system without making his thoughts known to the authorities. But there were those who were aware of his work, and in response to a direct inquiry, Copernicus wrote a letter, now known as the *Commentariolus*, outlining his system. The letter was copied by hand and found its way to a small audience around Europe. In it he objected to the Ptolemaic model and set forth a "more reasonable" system in a list of seven assumptions,† namely:

1. There is no one center of all the celestial circles or spheres.
2. The center of the earth is not the center of the universe, but only of gravity and of the lunar sphere.
3. All the spheres revolve about the sun as their midpoint, and therefore the sun is the center of the universe.
4. The ratio of the earth's distance from the sun to the height of the firmament is so much smaller than the ratio of the earth's radius to its distance from the sun that the distance from the earth to the sun is imperceptible in comparison with the height of the firmament.
5. Whatever motion appears in the firmament arises not from any motion of the firmament, but from the

† Nicholas Copernicus, "The Commentariolus" from *Three Copernican Treatises.* Translated by Edward Rosen, Columbia University Press, New York, 1939.

earth's motion. The earth, together with its circumjacent elements, performs a complete rotation on its fixed poles in a daily motion, while the firmament and highest heaven abide unchanged.

6. What appear to us as motions of the sun arises not from its motion but from the motion of the earth and our sphere, with which we revolve about the sun like any other planet. The earth has, then, more than one motion.

7. The apparent retrograde and direct motion of the planets arises not from their motion but from the earth's. The motion of the earth alone, therefore, suffices to explain so many apparent inequalities in the heavens.

Planet	Copernicus (au)	Modern (au)
Mercury	0.3763	0.3871
Venus	0.7193	0.7233
Earth	1.0000	1.0000
Mars	1.5198	1.5237
Jupiter	5.2192	5.2028
Saturn	9.1743	9.5388

Copernicus also assumed that the planets move in some combination of circles, namely a deferent and several epicycles. Figure 3.26 is a simplified drawing of the Copernican system (compare with Fig. 3.23). The exterior planets —Mars, Jupiter, and Saturn—are assumed to move in deferents having various centers, with none centering on the sun. The planets move on epicycles which are themselves carried on other epicycles. The theoretical paths produced by these various combinations of circles are oval orbits not centered upon the sun.

While it cannot be said that the Copernican scheme was dramatically simpler than the Ptolemaic system, it did make two significant departures from the Ptolemaic, namely:

(1) The sun rather than the earth became the center of the planetary system, making the earth simply another planet along with Mercury, Venus, Mars, Jupiter, and Saturn.

(2) The fact that equants were no longer needed was the most impor-

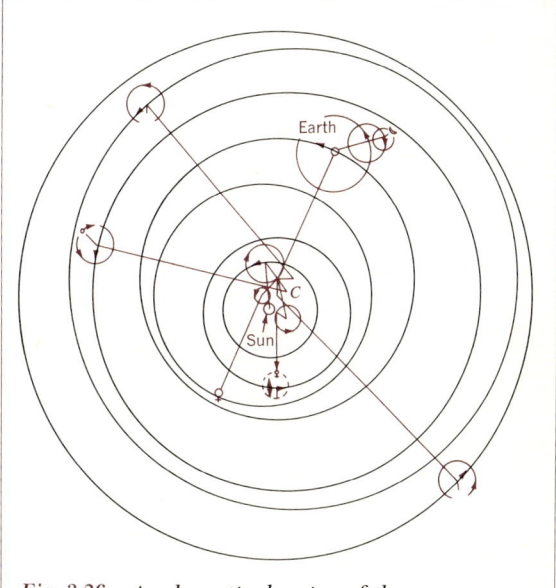

Fig. 3.26. *A schematic drawing of the Copernican model of the solar system.*

tant advantage in the mind of its creator.

Copernicus was able to calculate the mean distances of the other planets from the sun (that is, the radii of their deferents), whereas Ptolemy had not been able to reach any decision even as to their relative order. The table on page 85 compares Copernicus' calculations (taken from the second half of his *Commentariolus*) with the modern figures. The unit of distance used in the table is the astronomical unit (au) which is the average distance between the earth and the sun.

The second astronomical "publication" of Copernicus was also a personal "Letter against Werner," sent to a former classmate. This was the extent of his astronomical publications throughout his lifetime, in spite of strong pleas from scientists all over Europe who had heard about his work. Finally, near the end of his life, Copernicus relented and sent his manuscript to be printed. The first copy of this book reached him on the day that he died, May 24, 1543.

The book carried the title of *De Revolutionibus Orbium Coelestium Libri VI* ("Six Books Concerning the Revolutions of the Heavenly Spheres"). Copernicus had changed many details of his planetary system in this volume, going to more than 34 circles to account for the motions of the planets.

The *De Revolutionibus* was to be the center of a controversy for a century to come. In the middle of the 16th century the Roman Catholic Church was fighting to preserve its position. The success of St. Thomas in integrating and harmonizing Aristotelian physics, Ptolemaic astronomy, and Christian theology made an attack on any one of these an attack on them all. In 1616 the Catholic Church put the *De Revolutionibus* on the index of banned books. Copernicus' work did not fare any better in the new Protestant church. Luther referred to Copernicus as the fool who wished to reverse the entire science of astronomy despite the statement of Holy Scriptures to the effect that Joshua commanded the sun to stand still, not the earth (Joshua 10:13).

The hundred years following Copernicus' death were marked by an increasingly dogmatic approach to natural philosophy on the part of both Catholic and Protestant churches. New ideas, whether scientific or religious, could earn their advocates the stamp of "heretic." Indeed, Giordano Bruno was convicted of heresy and burned at the stake as a result of preaching the infinite extent of the starry sky and the plurality of worlds.

Despite the continuous wars waged for the control of the bodies and minds of men, the fifteenth and sixteenth centuries were times of great creativity in all fields of human endeavor. The magnificent creations of Michelangelo, Rafael, and Cellini date from this period. Nor was science stagnant during this period. William Gilbert published his book *Concerning the Magnet* in 1600. Vesalius published his great work on human anatomy in 1543, the same year that Copernicus' *De Revolutionibus* was published. Vesalius' book differed from others of his time in that it was not a rewriting of Galen, but rather was based primarily on his own observations. The sixteenth century also witnessed the work of three men who were to have a profound effect on all of science, namely, Tycho Brahe, Johannes Kepler, and Galileo Galilei.

3.12 Tycho Brahe and Johannes Kepler

Tycho Brahe, born in Denmark in 1546, became one of the outstanding observers of the skies of all time. Despite the inaccuracy (compared to modern observational instruments) of Tycho's instruments his observations far surpassed the others of his day in precision and accuracy. He measured the positions of over a thousand stars with an accuracy of about 1 minute of arc, and compiled extensive tables of planetary positions.

Although Tycho's greatest accomplishments were in the field of observational astronomy, he, too, speculated on the nature of the universe. He pointed out that if Copernicus was correct and the earth does move about the sun in a circular orbit, then we should observe stellar *parallax*, that is, an apparent cyclic change in the star's position in the sky due to its finite distance from earth and the change in our observation point caused by the earth's motion.† Tycho was unable to observe such parallactic motion, and for that reason rejected Copernicus' idea of a moving earth as being inconsistent with observation (much as Ptolemy rejected the heliocentric view of Aristarchus for the same reason). Actually, it was not until 1837 that the first evidences of stellar parallax were measured, and they turned out to be less than 1 second of arc! This explained why Tycho could not observe them with his limit of 1 minute of arc. Tycho's error lay in his inability to conceive the enormity of the distance to even the nearest stars.

Although he rejected Copernicus' model of the solar system, Tycho realized that the Ptolemaic system needed modification. He proposed yet another model, the Tychonic system in which the planets orbit around the sun, but the sun and moon still orbit around the earth. Geometrically, the Tychonic system was equivalent to that of Copernicus.

During the last years of his life, Tycho

† Parallax is discussed more fully in Sec. 7–9.

served as court mathematician to Rudolph II, the Holy Roman Emperor, at Prague. In February of the year 1600 he hired a new assistant named Johannes Kepler, then 29 years old. Kepler showed little interest or aptitude for the observational aspects of astronomy. He was a gifted mathematician and it was to the theoretical principles of astronomy that he was to make his greatest contributions.

Kepler believed in the Copernican system. Yet, like many scholars of his time, he shared the common ideas of the time. His earliest published work, predating his employment at Prague, dealt with an unsuccessful attempt to fit the orbits of the six Copernican planets (Mercury, Venus, Earth, Mars, Jupiter, and Saturn) to the surfaces of the five Platonic solids. On arrival at Prague, he had at his disposal the accurate planetary observations of Tycho. He was armed also with the conviction that these observations had to agree with numerical predictions if a model or hypothesis was to have value. He felt that the model had to explain the positions along the entire orbit, not just at selected, critical points. Moreover, he sought a physical force that would account for the motion of planets in their orbits, rather than depending on a divine agent for this force.

With these data and convictions, Kepler set out to prove the circular (hence perfect and divine) nature of the planetary orbits. His major work concerned the orbit of Mars, which he found not be circular in shape. Kepler, in keeping with the spirit of his times, was loathe to abandon the perfection of the heavens. When the circle would not fit the orbit, he tried an "egg shape" or ovoid, but finally was forced to conclude that Mars followed an elliptical orbit. He found the earth's orbit to be elliptical also. From the time needed for these planets to traverse the various parts of their orbits, Kepler found that a planet travels fastest when it is closest to the sun and slowest when it is farthest away. If a line is drawn from sun to planet, it will cut out equal areas of the ellipse in equal times (Fig. 3.27).

Although Kepler wrote on astrology and earned money by casting horoscopes, he had a mystical spirit, akin to that of the Pythagoreans, that made him seek "harmonies" in the celestial universe. He sought numerical relationships between orbit sizes and periods for the known planets in order to discover the "music of the spheres." He sought to identify the sun with God the Father, the sphere of fixed stars with God the Son, and the intervening ether with God the Holy Ghost. What he found was that the square of the sidereal period, P, of a planet was proportional to the cube of the semimajor axis of its orbital ellipse.

Kepler's findings are summarized as follows:

1. Planets move in plane elliptical orbits with the sun at one focus.
2. A line drawn from sun to planet sweeps out equal areas of the ellipse in equal time.
3. The square of the ratio of sidereal period P for any two planets equals the cube of the ratio of the semi-major axes of their orbital ellipses a, or

$$\left(\frac{P_1}{P_2}\right)^2 = \left(\frac{a_1}{a_2}\right)^3$$

According to Kepler's three laws of planetary motion, the variation in the sun's apparent motion with the season of the year can be explained by the fact that the earth moves in an elliptic orbit about the sun, being closer to the sun and hence traveling faster in January than in June. Epicycles were no longer needed to account for this phenomena. Likewise, the fact that planets show retrograde motion at opposition could be explained on the Copernican model by the fact that both earth and planets are in motion. Referring to Fig. 3-28, let the earth have its normal period of one year to go around the sun, and the planet Jupiter its period of 12

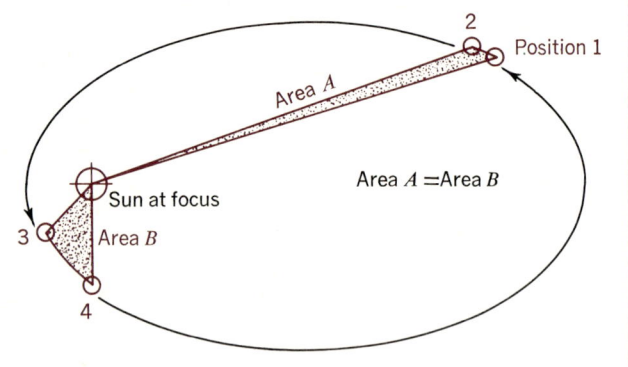

Fig. 3.27. The radius vector of Kepler's second law is the line from the sun at the focus to the planet. This line sweeps across equal areas, such as areas A and B, in equal times. Thus, since the distance from position 3 to position 4 is much larger than that from 1 to 2, then the planet must travel fastest in its orbit when nearest the sun.

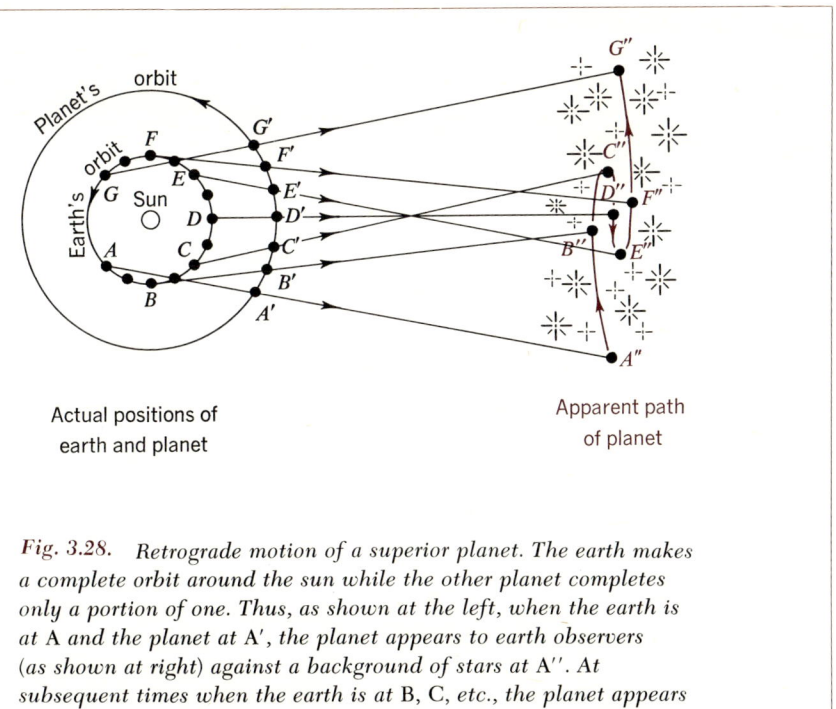

Actual positions of earth and planet

Apparent path of planet

Fig. 3.28. Retrograde motion of a superior planet. The earth makes a complete orbit around the sun while the other planet completes only a portion of one. Thus, as shown at the left, when the earth is at A and the planet at A', the planet appears to earth observers (as shown at right) against a background of stars at A''. At subsequent times when the earth is at B, C, etc., the planet appears to move in the same direction to B'' and C''. However, as the earth at C and D overtakes and passes the planet at C' and D', the apparent motion of the planet reverses, as shown at the right.

years for one revolution around the sun. The earth therefore has 12 times the angular velocity of Jupiter. When the earth is at point A of its orbit, we would see Jupiter at some point, A', in relation to the star background. As the earth continues in the lower part of its orbit to point B, its vertical position will have changed downward, whereas Jupiter will have moved upward; therefore Jupiter will appear to have moved much higher against the star background. But as the earth moves through positions C, D, and E, its rate of *vertical* distance change will now be in the same direction and faster than Jupiter's, and Jupiter will actually appear to have backtracked to position E' against the star background. Our present-day model of the solar system is really Keplerian.

3.13 Galileo and astronomy

The most dramatic and poignant experience of sixteenth century science revolved around the person of Galileo Galilei.

Galileo was born in Pisa in 1564, the year of Michelangelo's death and the year of the birth of Shakespeare. While studying medicine at the University of Pisa, it is said, he chanced on a lecture on Euclid and thereafter pursued mathematics. He developed an early interest in the mechanics of moving bodies, including bodies that move under the force of gravity. He soon was at odds with those who held to the Aristotelian ideas on falling bodies. It became desirable for him to move away from Pisa, where, although he had many friends, he had also made enemies through his lack of tact, to assume a post at the University of Padua. He became interested in the new theories of Copernicus and espoused them after careful analysis.

Galileo became interested in the telescope, just invented by a Dutch spectacle maker. He built his own, using a weak convex lens for the first lens and a concave lens for the eyepiece, which gave a magnification of only three times. While studying the moon with the telescope, he discovered the mountains and craters of the moon and measured the height of some of the mountains. Revealing such earthly characteristics as mountains on the moon brought wide disagreement with the Aristotelian concept for the characteristics of heavenly bodies (that is, perfect spheres). With improved telescopes, going up to magnifications of 30 times, he was able to recognize the Milky Way as made up of myriads of stars. He discovered four "stars" near Jupiter (Fig. 3.29) and identified them as moons of Jupiter. This gave strong support to the Copernican scheme because Jupiter and her moons gave clear demonstration of how a planetary system can work.† He discovered dark spots on the sun (called sunspots) and used them to show the rotation of the sun. He also was the first to note that Venus has phases like those of the moon (see Fig. 3.30). Galileo, in 1610, published many of his telescopic observations in a little book entitled *The Sidereal Messenger*.

But multiplication of proofs for the Copernicus theories were of no avail. The church had taken a contrary position and would have none of it. It was during this period, when all who defended the Copernican scheme were under suspicion, that Galileo proceeded to stronger arguments to define the relative roles of science and religion. He declared, as had some ancient theologians, that where science is involved, the language

† The publication *Sky and Telescope* prints, each month, diagrams indicating the changing positions of Jupiter's four largest satellites. These are the "stars" observed by Galileo.

Fig. 3.29. *The motions of the four brightest satellites of Jupiter. (Yerkes Observatory photo.)*

of the Bible must be taken metaphorically and not literally. Scripture and nature are both works of the same divine author and therefore cannot be in conflict.† His defense was strong and logical, but to no avail. A group of theological experts, who had been appointed by the church to examine the teachings of Copernicus, declared that the idea that the sun does not move was false and absurd in philosophy and formally heretical. The idea that the earth both moves and spins was declared to be false and absurd, and at least erroneous in faith. Galileo was reproved and in some disgrace.

Copernicus' book was suspended in 1616, pending its "correction," after which Catholics might read it. On the election of a friendlier pope, Galileo received honors and permission to write a noncommittal book to explain the arguments for the competing systems of Ptolemy and Copernicus. In 1632 he published *The Dialogue on the Two Chief Systems of the World*, written in Italian rather than Latin so that all might read it. The dialogue includes Salviati, a philosopher who presents the ideas of Copernicus as Galileo might present them. Sagredo serves as attorney to Salviati. Simplicio, who represents the views of Aristotle and Ptolemy, is given a difficult time and made to appear a fool. The pope ordered the Inquisition to forbid the book and to reexamine Galileo, who was summoned to Rome for the purpose. He was forced to recant his views as false and to sign a document of apology. He was sentenced to perpetual confinement, and died in 1642.

Galileo's book was placed on the Index Expurgatorius, where it remained until 1835 along with Copernicus' *De Revolutionibus* and one of Kepler's works.

† See Stillman Drake.

Fig. 3.30. (a) *Five phases of Venus. These five photographs of Venus illustrate the phases and relative size of the disk of Venus during its synodic period. (Lowell Observatory photograph.)*

3.14 Other scientific contributions of Galileo

Although Galileo's contributions to the science of astronomy were extensive and profound, by no means did they represent his entire scientific output. He is credited with the invention of the first thermometer and of the pendulum clock. But the greatest achievements of Galileo were in the field of mechanics. In 1638 he published his *Discourses on Two New Sciences* in which he set forth in a clear and lucid manner his thoughts on statics and kinematics.

Galileo took exception to the Aristotelian belief that a force must be continually applied to a body in order to keep it in motion. He held, along with his predecessors William of Occam and Stevinius of Bruges, that a body in motion would remain in motion unless acted on by a force.† He also disagreed with Aristotle on the matter of falling bodies. Aristotle had stated that when two bodies of different weight are dropped from the same height, the heavier one will reach the

† It should be noted that Galileo usually had in mind motion around a center of gravity rather than linear motion, since linear, inertial motion would disperse the stars and planets and thus disintegrate the universe.

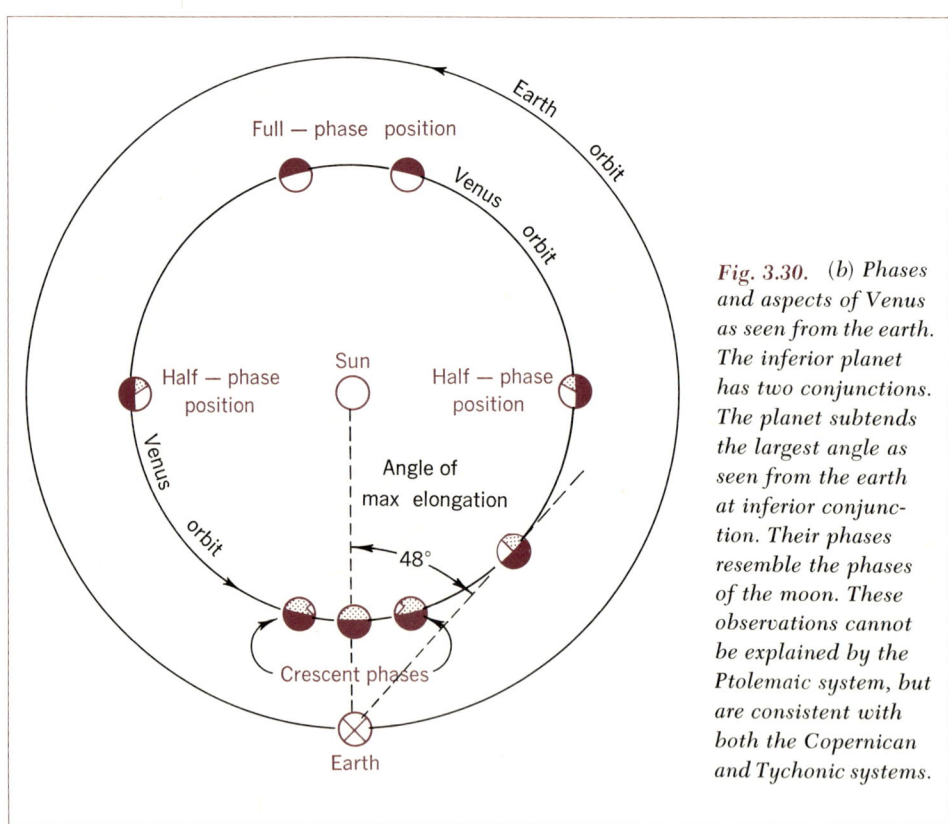

Fig. 3.30. (b) *Phases and aspects of Venus as seen from the earth. The inferior planet has two conjunctions. The planet subtends the largest angle as seen from the earth at inferior conjunction. Their phases resemble the phases of the moon. These observations cannot be explained by the Ptolemaic system, but are consistent with both the Copernican and Tychonic systems.*

ground first. If we drop a lead ball and a feather together we can see the basis of the Aristotelian idea. Galileo stated that, in a vacuum, both objects would fall at the same rate, the observed difference in travel time when not in a vacuum being due to air resistance affecting the feather more than it affected the lead ball. In the *Two New Sciences* he described in detail an experiment involving the rolling of balls down various inclined planes all having the same height, and extrapolated his results to free fall. He made quantitative measurements of time and distance during the experiment and stated clearly how these measurements were to be made, so that any other experimenter could reproduce his results.

Galileo's concern over experimentation is a forerunner of our present-day experimental method in science. In this he differed from the Platonic attitude about obtaining knowledge. Galileo did not invent experimentalism, but he popularized it in such a way as to make a profound impression on all future science. Despite his belief in experimentation, Galileo did not forget the importance of analytic thought. He combined experiment with mathematics so that each enhanced the other. It is this attitude towards scientific inquiry that is Galileo's legacy to the modern world.

3.15 Summary

The science of astronomy offers a remarkable path through which to trace the development of many scientific and technological ideas back to Babylonian times. There are many reasons why it is natural to find the origin of science in astronomy. It was in the stars, planets, sun, and moon that the most marvelous regularities were found. In contrast with the mutability of earthly events, the capriciousness of life, and the inevitability of decay and death of all organisms, the events in the sky exhibited eternal regularity and stability. Divinity was often ascribed to the course of the heavenly bodies, and the development of astrology went hand in hand with the task of producing calendars. The great ancient astronomer Ptolemy wrote not only the *Alamagest,* but also the astrological treatise *Tetrabiblos* which for many centuries was widely read and was more influential than his astronomical work.

The Greeks sought a geometric model that permitted all the motions of the heavenly bodies to be understood in terms of the perfection of the circle and the sphere, with the earth at the center of all circles.

It is all too easy for an observer to conceive of the stars as embedded in a huge spherical dome moving uniformly about the earth, although the observed variations in the speed with which the sun and moon move against the background of the stars argue against assuming the same spherical uniform motion for all. More puzzling is the motion of the five observable planets. Not only is there a variation in their brilliance, which implies that they are first closer to and then farther away from the earth, but still more puzzling is the observation that whereas they normally move in an easterly direction relative to the stars, there are periods when they reverse this motion (the retrograde motion). To account for all these variations within a consistent geometric model taxed the imagination and the mathematical ingenuity of the Greeks to the utmost. The culmination of Greek cosmology is found in the work of Ptolemy. Whether or not the universe was to be regarded as earth-centered, or the earth as rotating on its axis, was not so important as preserving the uniform and circular motion of all the celestial objects. It was natural to start with the premise that the earth was stationary. It was natural also that Aristotle and his followers tried to keep their first principles in accord with what was observed, and daily observation easily convinced them of a stationary earth.

In this chapter the major observations that can be made without a telescope were outlined. These give the reader the picture of a celestial sphere carrying the fixed stars in a diurnal rotation. Each star has a fixed position on the celestial sphere, and we can picture the sphere itself as having a celestial equator which can appear to be quite fixed in relation to our earth position. Against this fixed background of stars, the motions of the sun, moon, and planets are then de-

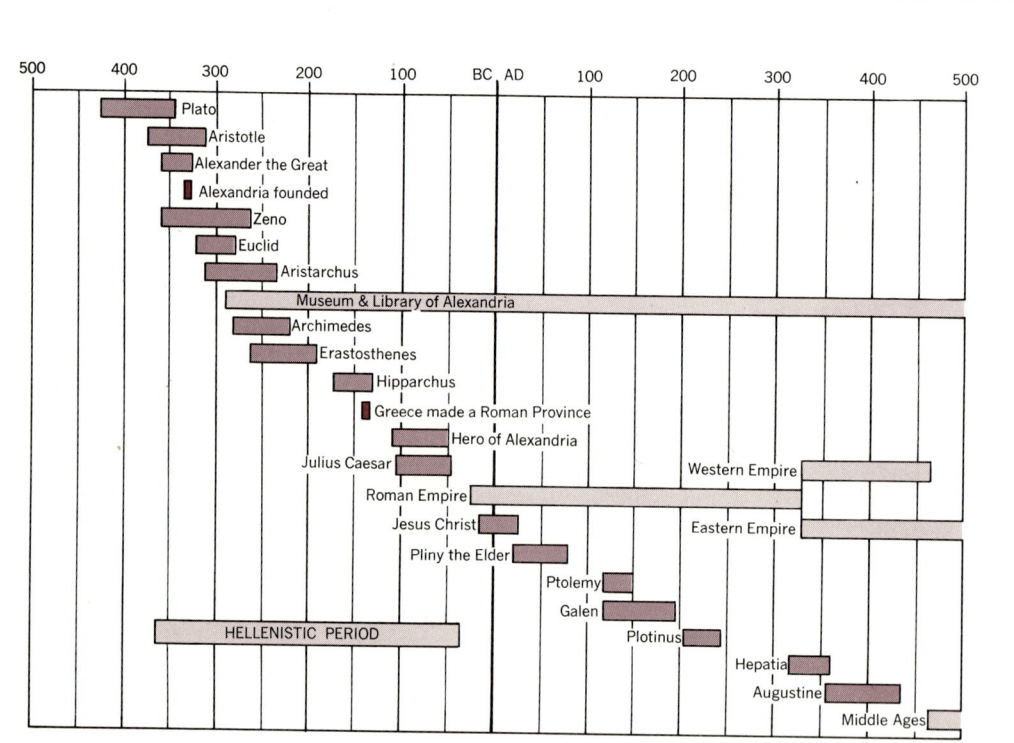

Fig. 3.31. *Cultural achievements from 500 B.C. to 1500 A.D. Chronology of important people and events.*

scribable. In general, all move in a path very close to what appears to be the path of the sun as traced against the background of the stars. This path is called the *ecliptic*.

If the ecliptic and the celestial equator coincided, there would be no seasons. But they are inclined at an angle of $23\frac{1}{2}$ degrees. In a year's time the sun "traverses" the celestial sphere, crossing the celestial equator twice, once at the vernal equinox and once at the autumnal equinox. At these crossings it lies, of course, directly on the celestial equator. The stars that lie along the ecliptic are grouped in constellations known as the twelve signs of the zodiac.

(We must emphasize that this "observed" picture, which is also a very convenient picture for describing motions, is

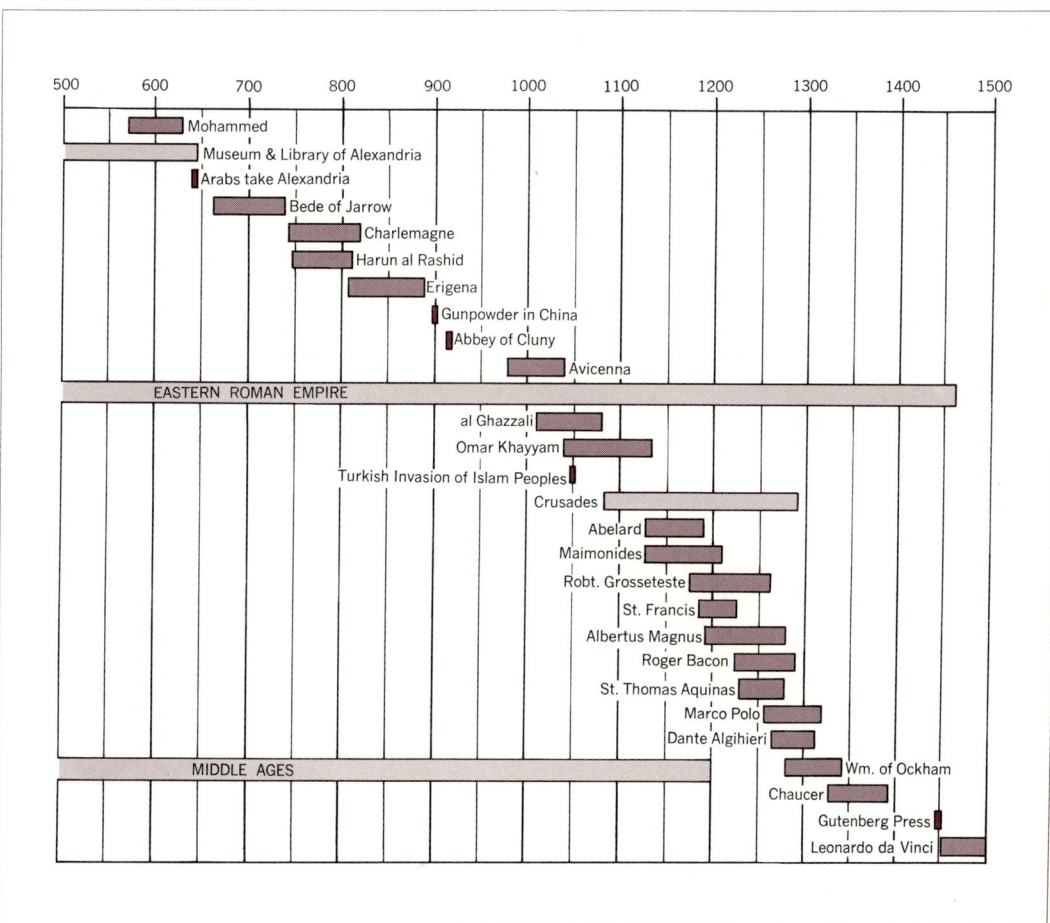

useful only because the stars are so far away appear to be so fixed in position. The actual reason for the change in seasons is the change of relationship between the earth and the sun with the yearly rotation of the earth around the sun.)

The elements of the Ptolemaic system which preserved circular and uniform motion have been presented. The most important of these is the system of deferents and epicycles, which explains the retrograde motion of the planets and the appearance of Venus and Mercury in the west as evening stars and a few months later in the east as morning stars. To account for variations in motion and variations in the distance from the earth, Ptolemy used a system of eccentrices and equants, epicycles, and deferents.

It was the Ptolemaic theory which St. Thomas Aquinas incorporated together with Aristotelian cosmology into his theology of the Roman Catholic Church. It is true that the theory had to be "tinkered with" to keep in accord with observations (or to use the ancient Greek phrase to "save the appearances"), but the theory of Ptolemy was in general eminently successful in accounting for observation, in making predictions, and in being amenable to inclusion in the theology of the times. No alternative theories were seriously proposed until the publication of Copernicus' *De Revolutionibus* in 1543. By this time the Renaissance was well begun.

The shift from the Ptolemaic system to that of Copernicus is often called "the Copernican revolution" because it represented such a profound change in the way man regarded the universe and such a break from traditions, religious as well as scientific. It is worth studying carefully as a paradigm. Successful scientific theories again and again tend to freeze the thoughts of man into a pattern that is difficult to break. To accomplish anything in science requires theory building, but the theories always carry with them a necessity to be revised as new observations are made. Such is the dynamism of science which was almost forgotten at the end of the nineteenth century before the new scientific revolution in which we live engulfed us.

Much can be learned about how alternative scientific theories are judged. It is not uniformly the case that one theory is obviously "proved" and the other rejected. Copernicus was guided by his conviction that the workings of the "divine" laws of nature should be simpler than the scheme presented by the Ptolemaic system, but he was forced to retain many of the same complexities because he, like his predecessors, could not abandon the notion that all heavenly motions are circular.

In fact one aspect of the Ptolemaic system was much closer in agreement with observation than was that of Copernicus. The Ptolemaic system implied the absence of parallax; the Copernican system implied its presence. None was discovered for 300 years. Copernicus could account for the absence of parallax only by removing the stars to infinitely greater distances than had previously been considered. This in turn opened a Pandora's box of speculation as to the place of man and his distance from God.

Nevertheless, and in spite of good scientific as well as theological reasons to the contrary, Galileo's and Kepler's instincts and feeling for simplicity caused them to accept the Copernican theory. The telescope had been discovered, the moons of Jupiter had been seen, the "imperfections" on the surface of the moon and spots on the sun had been revealed, the phases of Venus had been observed.

While none of these "proved" the correctness of the Copernican theory, they did cast elements of doubt upon the Ptolemaic system. Finally, with Kepler's enunciation of the elliptical orbits which explained so many of the puzzling variations that required the vast complex machinery of epicycles and eccentrics of both the Ptolemaic and Copernican theories, a new, simpler, and more precise theory of the motion of heavenly bodies was at last developed. Copernicus' thinking was still half-grounded in the Middle Ages. Kepler and Galileo were children of the Renaissance, but their discoveries marked the beginning of modern science, laying the ground work for Newton and his contemporaries.

Questions/Discussions

1. What similarities and differences do you see between the experiences of Galileo and of modern science teachers with respect to restraints by religious beliefs? (Recall the experiences of Teilhard de Chardin and the Tennessee biology teacher, noted in Chapter 1.)

2. Explain the concept of: Celestial sphere; equinox; ecliptic; sidereal day; phases of planets; synodic period; retrograde motion; Hellenistic period; orthodoxy in science; orthodoxy in religion.

3. Develop three questions which in your opinion are suitable for a 1-hour examination (open-book type) of your class on the significant contents of this chapter. Outline your answer to each question.

4. Look up and write down the dictionary (unabridged) definitions for *induction* (logic), *deductive method*, and *intuition* (philosophical). Now analyze the following and describe what part inductive and deductive processes, or intuition, played in the development of the following:

  (a) Use of the North Star for navigation
  (b) The Ptolemaic model of planetary motion
  (c) The findings of Kepler on planetary motion.
  (d) Astrology

(*Note:* Throughout the course there will be considerable emphasis on the role of sensory information and the human learning processes involving induction, deduction, intuition.)

5. We shall make considerable reference to "feedback" influences in nature and in human society. An invention of a new machine, for example, may completely alter (for good or for bad) the habits and way of life of the inventor. How would you analyze and give examples of the chain of influences that include

  (a) The influence of scientific and technological progress on social progress.
  (b) The influence of social progress on religious thought and religious institutions.
  (c) The influence of religious thought on technological and scientific progress.

6. The accumulation of information and ideas on the behavior of the heavenly bodies led both to the development of the science of astronomy and to astrology. Discuss the significance of the latter with respect to its "feedback" influence on the further development of astronomy.

References

The astronomical data presented in Secs. 3.2 and 3.5 may be found in many books, only a very few of which are noted here.

Abell, George, *Exploration of the Universe.* New York: Holt, Rinehart and Winston, 1964. *This recent, general astronomy text surveys in a clear and lucid manner the entire science of astronomy. Chapters 2 and 3 are most pertinent to this chapter.*

Armitage, A., *The World of Copernicus* (orig.: *Sun, Stand Thou Still*). New York: New American Library, 1952. *The life and times of Nicolas Copernicus and his scientific works; an enjoyable little book.*

Dampier, W. C., *A History of Science.* New York: Cambridge Univ. Press, 1929. (Reprinted in a paperback in CAM 366 in 1966.) *A highly readable account of the history of scientific thought and its relation to philosophy and religion. Chapter 1 covers the period up to the fall of Rome; Chapter 2, the period of the "Middle Ages"; and Chapter 3, the Renaissance. These 165 pages carry science up to 1660.*

Di Santillana, Giorgio, *The Crime of Galileo.* Chicago: Univ. of Chicago Press, 1955.

Drake, Stillman, *Discoveries and Opinions of Galileo.* Garden City, N.Y.: Doubleday (Anchor books), 1957. *Contains the complete translation of Galileo's Sidereal Messenger.*

Dreyer, J. L. E., *A History of Astronomy from Thales to Kepler.* New York: Dover Books, 1953. *A comprehensive treatment of the subject in some detail. Rather difficult reading but rewarding to the interested reader.*

Galileo, *Dialogues Concerning the Two Chief World Systems.* Translated by J. Drake. Berkeley, Calif.: Univ. of California, 1962. *Galileo's historic defense of the Copernican system is enjoyable reading even today.*

Kuhn, T. S., *The Copernican Revolution.* New York: (Vintage books V-164) Random House, 1959. *Highly recommended.*

Mason, S. F., *A History of the Sciences.* New York: Collier Books, 1962. *An easy to read, comprehensive history of all sciences. The first 191 pages deal with the subject matter of this chapter.*

Pannekoek, A., *A History of Astronomy.* New York: Wiley, 1961. *A detailed history written by an eminent astronomer.*

Rogers, Eric M., *Physics for the Enquiring Mind.* Princeton, N.J.: Princeton Univ. Press, 1960. *Offers details on most of the topics covered in this book.*

Thiel, R., *And There Was Light,* New York: New American Library (MT 290), 1960. *A delightful history of astronomy, written for the general reader. The first 157 pages cover the history up through Galileo.*

Toulmin, S., and Goodfield, J., *The Fabric of the Heavens.* New York: Harper & Row (Torchbook TB 579 L), 1961. *Another highly readable history of astronomy.*

CHAPTER FOUR

Motion and Change in Nature

Nothing in nature is more ancient than motion, and the volumes that the philosophers have compiled about it are neither few nor small; yet have I discovered that there are many things of interest about it, that have hitherto been unperceived.

GALILEO

THE BRIEF REVIEW we have attempted of the history of science, and especially of astronomy, holds many lessons for us. Certainly one of the more important lessons is that while intuition, contemplation, and speculation constitute necessary first steps toward formulating hypotheses and theories, effort must be made to test all hypotheses that are accepted lest they cease to represent a current and viable view of nature. The Greek philosophers were aware of the difficulties of obtaining "truth" through the senses and did not realize the possibilities that the fallible senses afford for experimentation and testing of observations. We have already mentioned the Greeks' antiempiristic bias. They used a logical, geometrical pictorial approach rather than the quantitative numerical approach so characteristic of modern science and of aspects of Babylonian science. The Greeks were "visual image worshipers, the Babylonians were magic number proteges." Their attitude toward science is characterized by their dealing with the square root of 2. "The Greeks proved it irrational; the Babylonians computed it to high accuracy." (See *Science Since Babylon* by de Solla Price, Yale University Press, for a detailed discussion of these attitudes.)

4.1 Introduction

In what follows we shall see the fruits of quantifying observational data and studying their "numerical properties" in order to understand phenomena.† The calculation of the actual movement of objects in space, in place of logical debate about

---

† The reader will do well to read the mathematics portion of the Laboratory and Mathematics Supplement in parallel with Chapters 4 and 5.

the nature of space and of motion, led to understanding of the dynamics of nature.

The hands of man had hurled stones with great skill for countless years. His life often depended on the choice of design of the material to be hurled and on the accuracy of throw. Yet until the time of Galileo there was little exact knowledge of the basic laws that govern the motions of objects in space. People intuitively knew that an object thrown into the air would in due time always return to the earth, and that with experience one could be successful in throwing an object to a certain place. The principles about motion that had been promulgated by the Greeks, especially Aristotle, involved philosophers in debates about the nature of motion but could not be used to specify the course of a moving object. Moreover, even these principles were in dispute. Through his ideas, and emphasis on testing ideas, Galileo set an example to would-be scientists for pursuing ideas. His ideas and experiments concerned motion, but this led naturally into other fields. In contrast to his predecessors, who generally favored a logical or rational approach to the attainment of truth by sheer reason, he brought whatever *physical and mental process* he had at his command to test his ideas and conjectures.

Perhaps the most significant aspects of our observations and experiences have to do with the dynamic, continuously changing character of nature. We have seen in an earlier chapter that the earth itself, and all biological systems, have experienced one evolutionary change after another. Within our own lifetime, and indeed within every hour of our daily lives, there is a state of movement and change. Some changes have only momentary significance, like the movement of an arm or a stroll across the room. Other changes have greater consequence, such as the processes within our bodies, the growth of plants, the life of our community, and the industrial output of the nation. But whether important or not, the world of nature and of man is a world of motion and changes. It is desirable, therefore, to think of nature in terms of *magnitude* of change, *direction* of change, *rate* of change, as well as *type* of change. This subject is the key to understanding all science, for it bears alike on the movements of our bodies and of the planets, on the flight of birds, on the blood corpuscles that flow in our veins, and on the evolutionary changes that continue for billions of years.

First we shall introduce the concepts of change and rate of change by considering the subject called *kinematics,* which deals with the motion of bodies without taking into account the cause of motion. (Historically similar considerations led to the new concepts of the Newtonian world which then dealt with the causes. Thus kinematics is limited to describing motion.) This procedure can be compared to watching the course of an automobile along a road without taking into consideration the behavior of the motor or the driver; or following a bird in flight with only the path of flight in mind. We can, continuing to abstract our examples, consider the moving object (automobile or bird) to be represented by a single particle or point. It is often possible to simplify situations in this manner. Thus, in discussing the motion of the earth about the sun we may consider the earth to be a single point in space. However, we must remember that when we examine the rotation of the earth we can no longer use such a simplifying assumption. A little later we shall expand the same ideas to apply to changes of variables other than motions of particles, reaching in fact that the motion of a particle is only a single example of the dynamic and changing character of all aspects of nature—physical, biological, and social.

4.2 Motion of a particle

Suppose we watch an automobile as it travels down a road as shown in Fig. 4.1, and for some reason we must describe its movements to someone else. What kind of information must we gather? The vehicle has a *direction* of travel, a *speed* of travel. There can also be identified a *time* when it arrives at each *location* or position along the road. Figure 4.1 gives only the path (direction) of the vehicle, without any graphical indication of speed or time. It is fairly clear that, to give the exact direction of travel and location of the auto at any instant of time, one would have to have all the information that was used to prepare the road map, plus the exact location of the car in relation to the road as a function of time.

If we define the position of the vehicle at the beginning, (call this location A) and its position at a later time (call this location B), we can measure the distance between the two locations in terms of feet or of miles.† Let us assume that this dis-

---

† The meter is the standard of length for scientific work. The original standard meter was a bar of platinum kept in France, with working standards in use elsewhere. Recently the meter has been defined more precisely in terms of the wavelength of the light emitted by atoms and molecules. 1 meter = 100 centimeters = 1000 millimeters = 1,650,763.73 wavelengths in vacuum of the 6056 A light emitted by the element krypton, which is one of the noble gases. In this chapter we use the English system of units (foot, mile, and so forth), since they are conventionally used in the United States in the situations we are discussing.

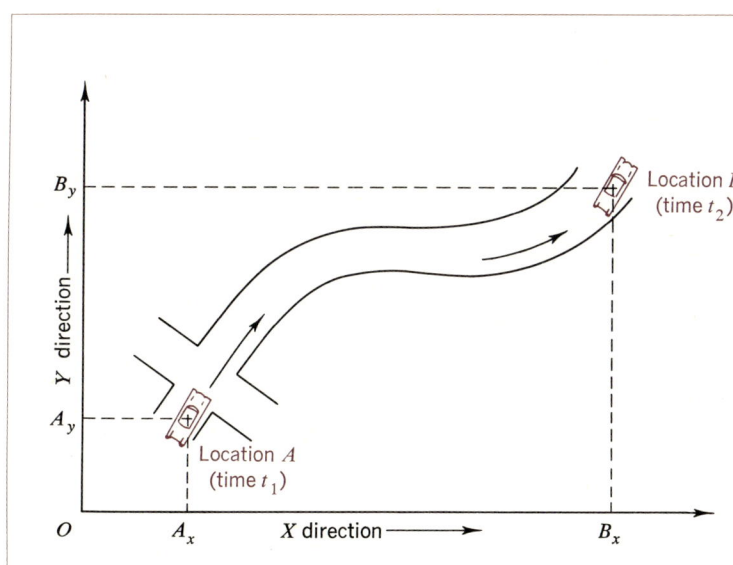

Fig. 4.1. To analyze movements of an automobile, we must have reference axes to determine location, and a measure of the time involved in going from one place to another.

tance of $(B - A)$ is just 1 mile. We can also measure the time that was required to cover that mile.

Clocks and our everyday experience tell us that time is a constantly increasing variable which has a fixed rate of change.† Everything else may vary its rate of change, but time increases in the same way for all changes, by assumption. We can refer to a particular time of day and day of year, such as November 11, 1965, 11:45 A.M., Eastern Standard Time. That instant of time would be well defined for everyone. We could also say that in going from location A to location B, a full minute of time had elapsed. Over that particular distance, this would make the *speed* of the vehicle an *average* of just 1 mile per minute, or 1 mile per 60 seconds, or 60 miles per hour (mph). If the vehicle starts from a stopped position at A, its initial speed will be zero miles per hour. After it begins to move, we could be certain that along some portions of that mile the vehicle will travel at a rate in excess of 60 mph if the average speed ends up at that value. The driver may have started and stopped many times between points A and B, however.

We see, therefore, that the unit of time and the progress of time are well defined for practical purposes while everything else may vary. It becomes useful, therefore, to express very many variables in terms of their *change with time*, which is another way of saying that we want to express the variables as *functions of time*.‡ If we have a record of the time when the vehicle arrived at each point along the road, we can draw additional graphs to show more detail about the motion of the vehicle. Of course we can identify functional relationships between many other types of variables. The variable might be the salary of an individual as it changes with time. An OY axis (ordinate) could then be plotted directly in dollars per year, and the abscissa would be a time axis. Or the graph could illustrate the melting of a piece of ice with time, and the OY axis could be calibrated to represent either the amount of melted ice or the amount that is still not melted. Time, then, is the *independent variable,* not depending on any other change, and the other variables are referred to as *dependent*.

These simple examples illustrate how *time* (the relativity theory will tell us that it is not independent) can be utilized as a reference axis to express such features as the motion of a particle, or the change in condition or state of a substance as *functions of time*. If a displacement in position (or other change) is referred to the interval of time that elapses for the change to take place, we can determine the *rate of change* over that interval of time. We must distinguish between the *average rate* and the *instantaneous rate,* however. We shall illustrate this by considering the motion of a particle.

4.3 Motion in a straight line

We are now prepared to discuss the velocity of a vehicle (or particle) in motion. The arguments will apply equally

---

† The philosophic aspects of the time concept will be treated in later chapters. The concept of rate of change will be developed in this chapter.

‡ A brief and mathematically nonrigorous definition of a functional relationship between two variables is: If two variable quantities are related in such a way that when a value of one (the independent variable) is given, there exists a systematic means (equation, graph, table of numbers) for finding the corresponding value of the other (the dependent variable), then the latter is said to be a function of the former.

4.3 Motion in a Straight Line

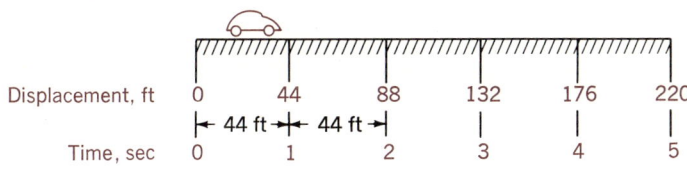

Fig. 4.2. Description of a moving object. The time of its arrival is indicated under each point of observation.

well to rate of change of other variables. In this section we shall discuss motion in a straight line, with a vehicle moving along a level, straight road (see Fig. 4.2).

Suppose that the car is at point 0 when we begin to observe it. For convenience we can designate that time as zero on a clock. In this particular case we chose the reference axis representing *displacement,* calibrated in feet, and another axis representing the *time* when the vehicle arrives at each point. For each value t of the time, there will be one value of the displacement, s. That is, s in this case is a function of time t.

A function can be represented in several ways: (1) by a table of numerical values, (2) by a graph that utilizes the data in the table, or (3) by a mathematical equation. For our particular function, Table 4.1 gives a tabulation of numbers.

Figure 4.2 is a pictorial way of describing a moving object. Figure 4.3 shows the same function in graphical representation.† Notice that equal displacements (+44 feet) occur during each second. A motion in a straight line that is of this kind (that is, in which equal displacements occur during equal time intervals) is called *uniform motion.*

In Table 4.1, Δs represents the incre-

† Usually the observations on which a graph is based are limited in number. One must draw the most suitable curve based on the available points. One may then utilize the curve to obtain data lying *between* the points. This is called *interpolation.* If one makes predictions on the basis of a curve extended beyond the given limits of the curve, the process is called *extrapolation.* The user of a graph or an equation should know the conditions under which the information was obtained and use only with caution.

TABLE 4.1 OBSERVED DATA FOR A MOVING OBJECT

Displacement from Position 0, ft	Increment of Change in Displacement, Δs	Time of Travel from Position 0, sec	Increment of Time intervals, Δt
0	−feet	0	−sec
44	44	1	1
88	44	2	1
132	44	3	1
176	44	4	1
220	44	5	1

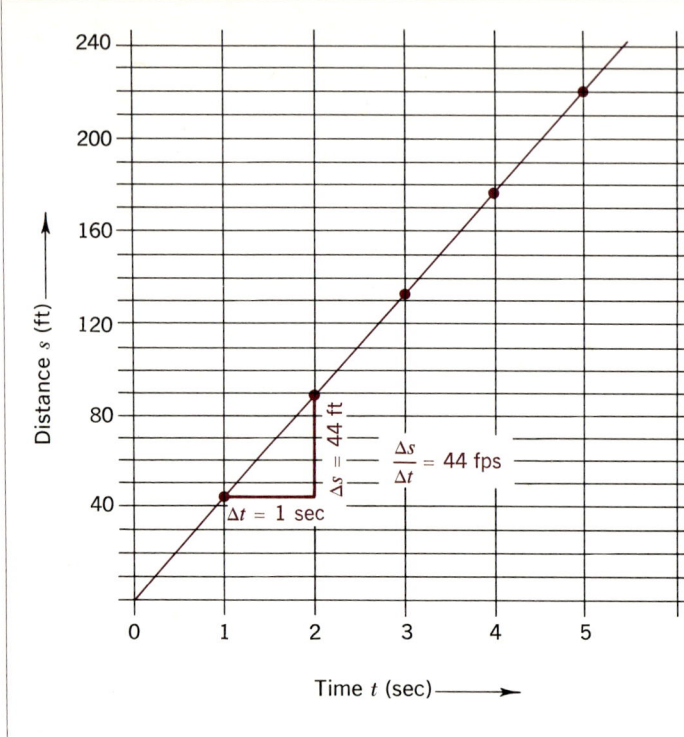

Fig. 4.3. Distance-time graph for motion with constant speed. (If we were plotting the melting of ice, and an equal amount of ice melted in each minute of time, we would refer to it as having uniform rate of change.)

ment of distance change during each increment of time, Δt.† To give a simple mathematical expression to this particular function, we note that a displacement of +44 feet takes place during each second of time. Thus

Distance traveled
= (distance traveled per second)
× (number of seconds)

or, much more simply,

$$s = 44t \qquad (4\text{-}1)$$

† The Greek letter Δ (delta) will be used to indicate "a small change in . . . ." Thus Δs does not mean "delta times s," but reads "delta s," that is, an increment in distance, a small displacement covered during the corresponding short time interval, Δt. Therefore we can write $\bar{v} = \Delta s / \Delta t$.

When t doubles, then s doubles, and so on. We say that s varies with t, and indicate this by the relation†

$$s \propto t$$

or

$$s = ct \qquad (4\text{-}2)$$

where c is called the *constant of proportionality*.

In our example, the constant of proportionality is given by

$$c = \frac{s}{t} = 44 \text{ fps}$$

(read 44 feet per second)

As mentioned earlier, the ratio of the

† The symbol $\propto$ is used to denote relatedness in the sense of proportionality.

displacement *s* of an object in uniform motion to the (elapsed) time *t* is, by definition, the *velocity v*, of the object.† It is the displacement per unit of time. We have, in words, (velocity) = (displacement) ÷ (time required for that displacement), or more simply,

$$v = \frac{s}{t} \qquad (4\text{-}3)$$

This also gives

$$s = vt \qquad (4\text{-}4)$$

or expressed in words,

Displacement = (velocity
 or distance traveled per second)
  × (time of motion, seconds)

In Table 4.1 we have listed, in addition to displacement and time, the increments of displacement, Δs, and the corresponding increments of time, Δt. Within each interval we can state the velocity to be

$$v = \frac{\Delta s}{\Delta t} = \frac{44}{1} \text{ fps}$$

Note in Fig. 4.3 that we have arbitrarily called the origin of the two axes zero for both displacement and time, and have assumed the two directions to be positive when moving in the direction of the arrows. Thus this vehicle is said to increase its distance if the displacement is to the right and therefore positive, and reduce its distance if it moves in the reverse direction. The term *displacement* therefore has a directional aspect, being either positive or negative according to its direction. In going from the 88-ft position to the 132-ft position, we say that the vehicle has a displacement of +44 ft, whereas if it were to go from the 132-ft position to the 88-ft position, we would say that the vehicle displacement

† We shall presently distinguish between velocity and speed.

is −44 ft. In both cases the numerical value of the distance is 44 ft.

Similarly, the term *velocity* has a directional aspect, and may be positive or negative. Referring to the vehicle displacement given above, travel from the 88-ft position to the 132-ft position in 1 sec represents a *velocity* of +44 fps. Travel in the reverse direction, from the 132-ft position to the 88-ft position in 1 sec represents a velocity of −44 fps. But in both cases the *speed*, represented by the numerical value only, irrespective of the ± sign, is 44 fps.

Whenever the *direction of change* also has significance in addition to *magnitude of change,* such as is the case with these variables, we say that a variable has the properties of a vector†. Therefore, in discussing them, we must give attention to both their magnitudes and direction. As a crude analogy, all this says is that when a person boards a fast train intending to head for Chicago, he had better be careful not to board a fast train going in the opposite direction.

In most cases variables do not change at a constant rate. In the case of our automobile the motion may be as indicated in Figs. 4.4 or 4.5. In Fig. 4.4, the velocity was constant for 10 sec, then abruptly increased and remained constant for 10 sec more, and finally abruptly decreased to another constant value for the last 20 sec. In Fig. 4.5, the rules for uniform motion do not apply, and we need a more general concept of velocity. The dashed lines of the figures represent the equi-

† It is recalled that vectors represent quantities that have both magnitude and direction (such as a force applied in a specific direction), whereas a scalar has magnitude only (such as a mass). The concepts of force and mass are used here only with their intuitive meaning; they will be defined more specifically in Chapter 5. Vectors are distinguished from scalars either by bold type or by an arrow above the symbol.

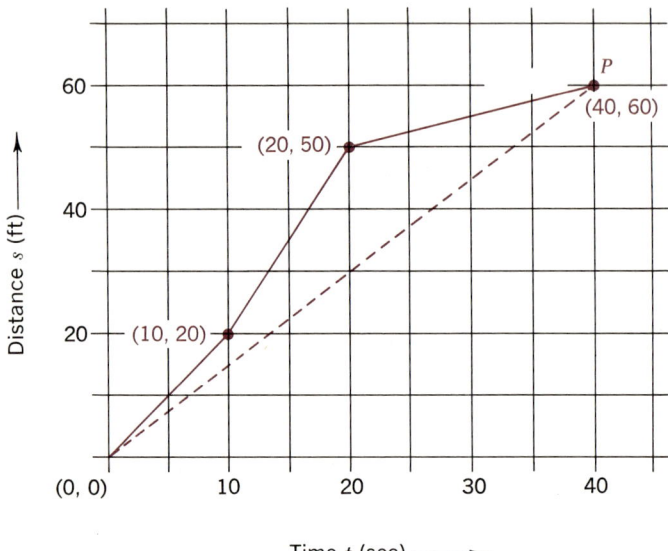

Fig. 4.4. Motion involving two abrupt changes of speed.

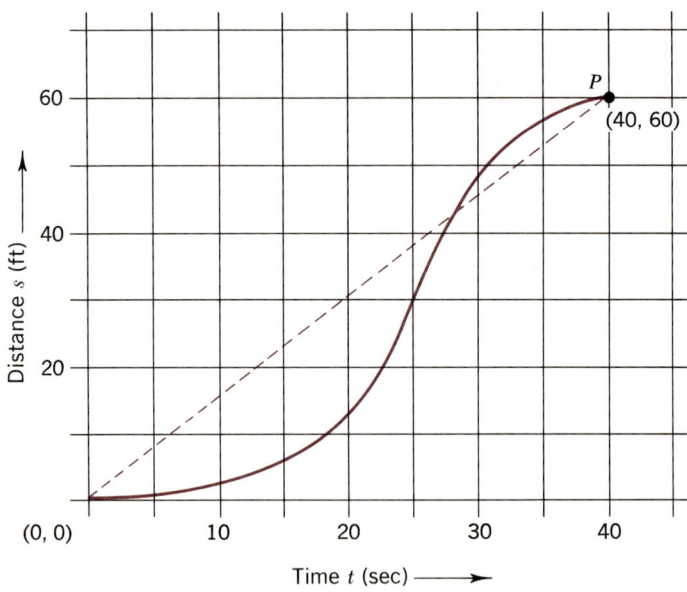

Fig. 4.5. Motion involving gradual changes of speed.

valent motion, with constant velocity, which would have displaced an object by the same number of feet in the same time.

We must now distinguish between three different concepts: *constant velocity* (Figs. 4.2 and 4.3), *average velocity*, and *instantaneous velocity*. The average† velocity $\bar{v}$ of a body moving along a straight path during a period of time t is defined as the displacement s that has taken place in that period of time, divided by t:

$$\bar{v} = \frac{s}{t} \qquad (4\text{-}5)$$

We note that this is the same as the definition of v for a body moving with constant velocity, since in that case v and $\bar{v}$ are the same. In Figs. 4.4 and 4.5, the average velocity is $+60$ ft/40 sec, or $+1.5$ fps.

As another example of average velocity, consider a train that travels $+125$ miles in 2.0 h, stops 0.5 h to take on freight, and then traverses $+120$ miles in the next 3 h. What is the average velocity for the first 2.5 h and for the entire trip? To answer the first part of this question, we say

$$s = 125 \text{ miles} \qquad t = 2.5 \text{ h}$$

Then

$$\bar{v} = \frac{s}{t} = \frac{125 \text{ miles}}{2.5 \text{ h}} = 50 \text{ mph}$$

The second part of the question is left for the student to answer.

4.4 Instantaneous velocity

Thus far, all our measurements of velocity have involved somewhat extended time intervals (which resulted in an averag-

† A bar above the symbol represents average value, whatever the variable may be.

ing process in each case) and have disregarded the changes in velocity along the way. However, we are often required to know the velocity of a moving object *at a particular instant* without any averaging process. For example, looking at Fig. 4.5, what was the velocity at the instant when the vehicle had traveled 10 sec? Of course the driver could read the speedometer at that point, since it indicates the rate the wheels are turning *at that instant*. We obtain an approximation to the instantaneous velocity at any chosen time by measuring the displacement that takes place over the smallest period of time that we can measure. We again go through an averaging process in practice, but the averaging is computed at the smallest possible time interval at the point in question. In the case of the speedometer reading, the averaging could be taken for a single rotation of the wheel instead of many rotations. Thus, to find the velocity at a particular instant of time, $t = 10$ sec, we divide a small but finite displacement Δs by an interval of time Δt; but we can choose an interval of time that is so small that the average velocity *approaches* the actual velocity at $t = 10$ sec.

More precisely, the instantaneous velocity (symbolized by v) at any time t is defined by

$$v = \lim_{\Delta t \to 0} \frac{\Delta s}{\Delta t} \qquad (4\text{-}6)$$

This simply says that the instantaneous velocity at any point is found by dividing an increment of displacement by an increment of time while the increment of time and the corresponding increment of displacement are made smaller and smaller so as to approach zero at the point in question. We took a preliminary step in this direction when we listed the increments Δs and Δt in Table 4.1. At each interval along the way, the *average velocity within that interval* would be given by $\bar{v} = \Delta s/\Delta t$. But as the intervals Δs and Δt become smaller and smaller, the velocity approaches the true instantaneous velocity, **v**. But while the increments Δs and Δt approach zero in magnitude, their *ratio* remains finite. That is, even if the increments of time of Table 4.1 were to be reduced from $\Delta t = 1$ sec to 0.000001 sec, and the corresponding increment of displacement were also divided by a million to become 0.000044, the ratio $\Delta s/\Delta t$ would still be

$$\frac{0.000044}{0.000001} = 44 \text{ fps}$$

Intuitively, the velocity at time t may be described as the instantaneous rate of change of the displacement at t. This was in fact the step taken by Leibnitz and by Newton when they invented the method of the calculus to cope with the variable aspects of natural phenomena.

4.5 The application of mathematics

From the specific question of how the velocity v is to be found in practice, one is led to the more basic question of how mathematical concepts such as the system of real numbers, the notion of a function, and similar deductions are related to the physical world. This is a difficult philosophical question that we shall not try to answer. We recognize, however, that when we use mathematical concepts as we have done, we make the assumption that they are a suitable means for describing the physical world. Also, expressing the result of a measurement (say, of a distance or a time interval) by a real number is an idealization, since no measurement is exact. Similarly, describing the motion of a car in a straight line by the function $s = 44t$, when the values of t are the real numbers between 0 and 5, is an idealization. We justify the use of such idealizations (models) simply because experience tells us that they are useful for describing the physical world.

When we discussed instantaneous velocity, we performed the algebraic equivalent of finding the number representing the slope of the straight line that is tangent to the curve at the point P in Fig. 4.6. Briefly, the situation is as follows: Say that

$$\frac{\Delta s}{\Delta t} = \frac{s_1 - s_0}{t_1 - t_0}$$

and that this ratio is the slope of the chord connecting the points on the curve P [corresponding to the pairs (t_0, s_0)] and Q_1 [corresponding to (t_1, s_1)] for which Δs and Δt are determined. The tangent at P is defined as the line through P whose slope is the limit of the slopes

$$\frac{\Delta s}{\Delta t} = \frac{s - s_0}{t - t_0}$$

of the chords PQ [where Q corresponds to (t, s)] as $\Delta t = t - t_0$ approaches 0. Thus by definition, the slope (that is, the tangent of the angle of inclination) is the $\lim_{\Delta t \to 0} (\Delta s/\Delta t)$ or the instantaneous velocity **v** (at the time t_0 to which the point P corresponds), assuming that the units of the ordinate and abscissa are designed to give this information.

Thus, in order to find the instantaneous velocity at a given time t_0, instead of calculating $\Delta s/\Delta t$ we may find the tangent to the curve representing the function at the corresponding point P and calculate its slope. It is this concept that we shall find most useful.

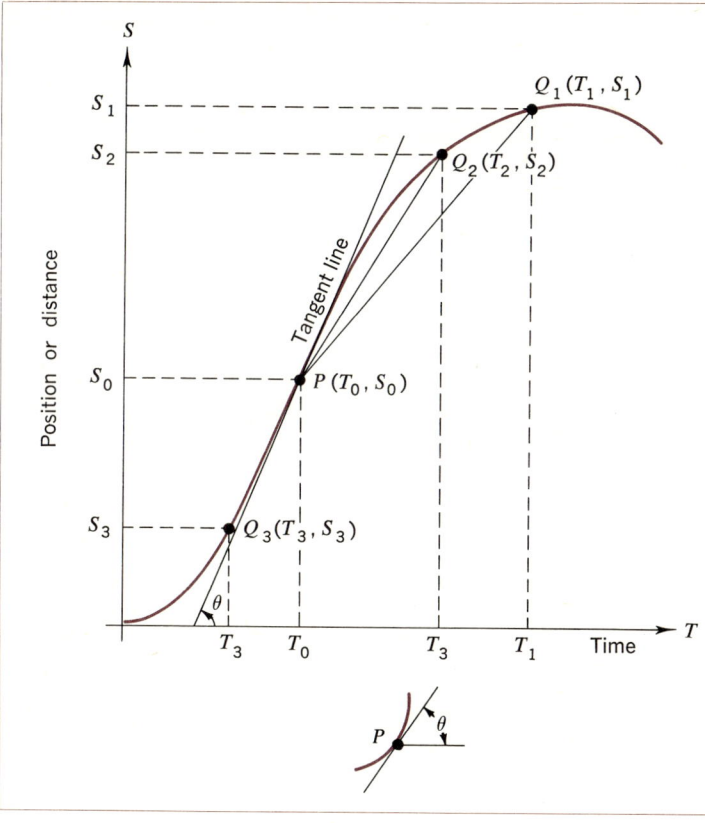

Fig. 4.6. *Instantaneous velocity. When the speed with which an object moves is variable, the velocity at any instant of time is given by the tangent drawn at that point of the location-time graph. The lower sketch illustrates the method of determining the slope or tangent in the neighborhood of the point P, by taking the ratio of a small interval Δs to the corresponding interval Δt. The slope equals $\Delta s/\Delta t$. As Δt is made smaller and smaller, approaching zero, the ratio $\Delta s/\Delta t$ approaches the tangent at the point P.*

4.6 Acceleration

Since velocities need not remain constant, there can be frequent *acceleration* to higher velocity of *deceleration* to lower velocity.

An important motion is that which has *uniformly changing* velocity; that is, motion with constant acceleration (going faster and faster at a constant rate of increase) or constant deceleration (going slower and slower at a constant rate of decrease). Galileo found this to be the behavior of freely falling bodies and of objects rolling down a smooth inclined plane. In more modern experience we are aware of the acceleration of an automobile each time we start from a stopped position or when we "schuss" down a steep slope.

For all such cases, we define the concept of accleration as the ratio of the change of velocity $(\mathbf{v} - \mathbf{v_0})$ to the time interval $(t - t_0)$ during which this change occurs, or

$$\mathbf{a} = \frac{\mathbf{v} - \mathbf{v_0}}{t - t_0} = \frac{\mathbf{v}}{t} \qquad (4\text{-}7)$$

Acceleration, like velocity, is a vector

quantity. To illustrate this new concept, let us consider one example.

Example A car has been traveling on a straight, level road at a speed of 26 mph. The driver presses down on the accelerator and in 2 min the car reaches a speed of 50 mph, with uniform acceleration. What was the acceleration during the 2-min interval?

We may write $v_0 = 26$ mph, $v = 50$ mph, $t - t_0 = 2$ min $= 120$ sec; $a = ?$ By the definitions of a, we have

$$a = \frac{50 \text{ mph} - 26 \text{ mph}}{120 \text{ sec}} = +0.20 \frac{\text{(mph)}}{\text{sec}}$$

The positive value of the acceleration means that v increases with time. Notice also that the units for acceleration are (length/time)/time, which can also be written as length/time² when the time units are the same (that is, both are expressed in seconds or in hours).

For later use, it is advantageous to note that Eq. (4.7), for the case where $t_0 = 0$, can be written as

$$v = v_0 + at \tag{4-8}$$

which says that the velocity v of an object at the end of a time interval t is equal to the sum of the original velocity v_0 and the added velocity at due to acceleration. In particular, if the object starts from rest, $v_0 = 0$, at time $t_0 = 0$, then

$$v = at \tag{4-9}$$

Equation (4-8) defines v as a function of t, just as Eq. (4-4) defines s as a function of t. Since these two equations are of similar form, the graph of the function defined by Eq. (4-8), is like that of the function defined by Eq. (4-4), that is, a straight line. (See Fig. 4.7.)

We can easily find the displacement s of a body moving with *constant acceleration* as a function of the time t in which the body acquires equal increments of velocity in equal time intervals.

Galileo's experiments led him to believe that all objects, no matter what their chemical constitution or surface condition, if sufficiently heavy to overcome air resistance, would fall with the same acceleration. This is actually the case.† This

† It would be more proper to say that different types of bodies do show some variation in their rate of fall, but that the variations can be explained in terms of differences in drag caused by air friction.

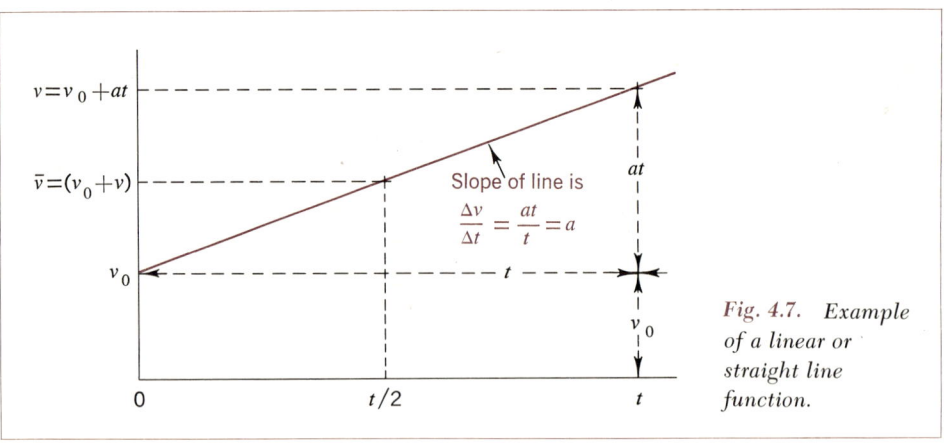

Fig. 4.7. Example of a linear or straight line function.

acceleration is a constant of nature that is designated by the symbol g and is approximately equal to 32 ft/sec² or 9.8 m/sec² at sea level at 45 deg latitude. When these relations deal with falling objects, g is usually written in place of the symbol a for acceleration.

4.7 More on acceleration†

When a particle experiences constant acceleration and the motion begins from a state of rest at time $t_0 = 0$, the following relations hold among the quantities s, v, a, and t:

$$v = at \qquad (4\text{-}9)$$

$$s = \tfrac{1}{2}vt = \tfrac{1}{2}(at)t = \tfrac{1}{2}at^2 \qquad (4\text{-}10)$$

$$v^2 = (at)^2 = a^2 t^2 = a^2 \frac{2s}{a} = 2as \qquad (4\text{-}11)$$

These equations can be easily adapted to the case in which the velocity v_0 and displacement s_0 are not 0 at $t = 0$. (That is, when the particle is already in motion before we give it faster or slower motion.) We now have the six quantities s, s_0, v, v_0, a, and t. We find four relations, in each of which four of these magnitudes appear and s, v, a, or t is missing. Thus, proceeding as above, but using Eq. (4-8) instead of (4-9), we obtain

$$v = v_0 + at \qquad (4\text{-}8)$$

$$\begin{aligned} s &= s_0 + \bar{v}t = s_0 + \tfrac{1}{2}(v_0 + v)t \\ &= s_0 + \tfrac{1}{2}(v_0 + (v_0 + at))t \\ &= s_0 + \tfrac{1}{2}(2v_0 + at)t \\ &= s_0 + v_0 t + \tfrac{1}{2}at^2 \end{aligned} \qquad (4\text{-}12)$$

$$\begin{aligned} v^2 &= (v_0 + at)^2 \\ &= v_0^2 + 2v_0 at + a^2 t^2 \\ &= v_0^2 + 2a(v_0 t + \tfrac{1}{2}at^2) \\ &= v_0^2 + 2a(s - s_0) \end{aligned} \qquad (4\text{-}13)$$

† The use of colored pages indicates that this is an optional study section.

Note that Eq. (4-12) may be interpreted as follows: The total displacement is the sum of the initial displacement plus the displacement due to the initial speed v_0 and the displacement due to the constant acceleration a.

Since s, v, and a could be positive, zero, or negative, Eq. (4-12) can be used as it stands, even if the moving object slows down, stops, and reverses the direction of its motion. Care must be taken to attach the proper sign to each quantity and then to interpret the results in accord with the original assignment of signs to indicate direction.

Before leaving the study of freely falling objects, let us look at some examples of the usefulness of these equations in practical cases.

Example Suppose a stone is thrown vertically upward at a speed of 64 fps by a person standing at the edge of a cliff. Where (how far from the thrower) will the stone be after 3 sec have elasped? Replacing words by symbols, we write that $v = 64$ fps, $t = 3$ sec, and (from data not given in the problem) $a = -32$ ft/sec²,† and $s = ?$ These values may be substituted into Eq. (4-12). Thus

$$\begin{aligned} s &= (64 \text{ fps} \times 3 \text{ sec}) \\ &\quad + \tfrac{1}{2}(-32 \text{ ft/sec}^2)(3 \text{ sec})^2 \\ &= [192 \text{(ft/sec)sec}] \\ &\quad + [-144 \text{(ft/sec}^2) \text{sec}^2] \\ &= 48 \text{ ft} \end{aligned}$$

Since the displacement is a positive quantity, the stone is 48 ft above the thrower. (Note how the units of time have been handled.)

† If v is positive upward and v decreases with time, as we know it does for an object thrown upward, then the acceleration a is positive in the downward direction or negative in the direction of the initial v.

Example An object is thrown vertically downward with an initial speed of 40.0 fps. After it has fallen 50 ft how fast is it moving? Translating the words into algebraic symbols, we can write $v_0 = 40.0$ fps, $s = 50.0$ ft, $a = 32$ ft/sec^2, and $v = ?$† Equation (4-13) contains v_0, v, s, and a, and can therefore be used to find the unknown final velocity. Thus

$$v^2 = (40.0 \text{ fps})^2 + 2(32 \text{ ft/sec}^2 \times 50.0 \text{ ft})$$
$$= 1600 \text{ ft}^2/\text{sec}^2 + 3200 \text{ ft}^2/\text{sec}^2$$
$$= 4800 \text{ (fps)}^2$$
$$v = 69.3 \text{ fps}$$

Notice that the square root sign applies to the units (ft^2/sec^2) as well as to the numerical quantity. Note too that although the solution of the equation $v^2 = 4800$ ft^2/sec^2 includes two roots, one a positive quantity and one a negative quantity, only the positive root 69.3 fps is adopted as a solution. The negative root is physically meaningless because the downward direction was chosen as the positive one and we have no reason to think that the object can reverse its direction of motion.

So far we have dealt with accelerated motion in a straight line to which we can ascribe constant acceleration. Before closing this section we should note that a more general discussion would have to include a concept of instantaneous rate of change of acceleration.

4.8 Extension to other variables

The concepts and equations that have just been presented for motion of a particle apply in the same way to many other variables in nature. In every metabolic process by which food is converted to energy in the body, there is a *rate* for the metabolic process which corresponds to the velocity of the particle. There is a rate for every chemical process, as well as a rate for the increase of the gross national income. The national income can go down as well as go up, and a chemical reaction may act in the reverse direction as well.†

Whenever the rate of change or reaction itself changes, there must be acceleration or deceleration similar to that observed in the motion of particles. The data for the metabolic process, chemical reaction, or gross national income can be plotted as if they were displacement values. The average rate over a long interval of time can be determined by taking the ratio of the amount of change to the time interval for that change. Similarly, the instantaneous rate can be determined for a specific point by taking very small intervals of change and the corresponding small interval of time at the point in question, or better still by taking the tangent of the curve at the point in question.

In this discussion the graphs have all been drawn with time as one of the reference axes or variables. We could just as well have drawn a graph to illustrate how the acceleration varied as a function of the velocity of the vehicle. In that case the abscissa (horizontal reference axis) could have been calibrated in terms of feet per second or meters per second instead of time, and the ordinate would have been calibrated in terms of feet per second squared or meters per second

---

† Note that we have chosen downward as the positive direction in this example. We may arbitrarily select any direction as positive as long as we are consistent.

† The belief existed even in early days with such men as Laplace and Poisson that the findings from celestial mechanics could be applied to "social mechanics."

squared. A graph may be drawn in any manner that illustrates the relationship to be shown.

The mathematical techniques discussed in this chapter are also used by some social scientists to construct theories that explain social and economic behavior. This is especially true of economists, who have used the $\Delta y/\Delta x$ technique for almost a hundred years. Let us consider some concepts of economic theory involving rates of change of one variable with respect to another.

A central notion of one economic theory (that of the "London" school of economics) is the *law of diminishing marginal returns,* or the law of nonproportionality.† Suppose, reasons the economist, we add units of labor (such as man-weeks of a given grade and skill) to a bicycle factory of a given fixed size. What kind of line (function) will trace the

† Although this theory is not accepted by all economists, it is used here because it provides a clear example of the point we wish to illustrate.

output of bicycles as a function of labor? Referring to Fig. 4.8, will it be a straight line such as *OA* or *OB*? Will (Δ bicycles/Δ labor) be a constant value? The economist thinks not. He postulates that there is some level of labor (L_1, Fig. 4.8) that will produce the maximum number of bicycles, B_1, from a given factory. More labor (beyond L_1) would actually cut down on "bike" production; a case of too many cooks spoiling the broth.

Furthermore, the economic theorist would maintain that since the factory is designed with a specific amount of labor in mind, the curve that traces the number of bicycles produced as a function of labor is *not* a straight line, but a curved line such as *ODC*. In other words, the ratio (Δ bicycles/Δ labor) is not a constant, but has a continually changing value. Hence we have the name "law of nonproportionality." (Note the similarity between Figs. 4.8 and 4.6. The curve is of the same type; only the axes are different.)

To this ratio (Δ bicycles/Δ labor) the economist gives the special name "mar-

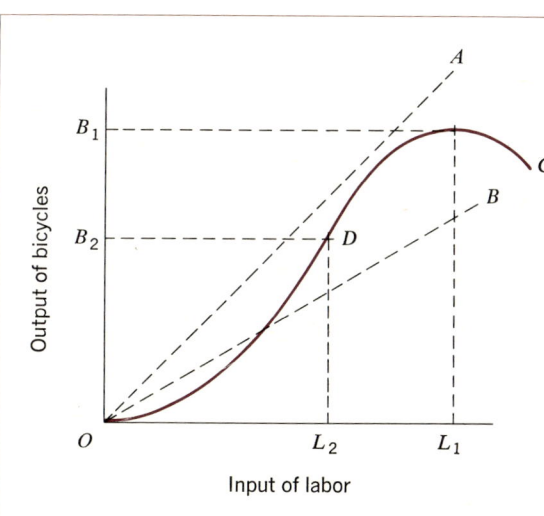

Fig. 4.8. *An example of the law of diminishing marginal returns. Note that at labor level L_2, the slope of the curve has its greatest value. At that labor level a small increase in the labor force will therefore result in the largest increase in bicycle output. However, at labor level L_1, an increase in labor will actually cause a decrease in production.*

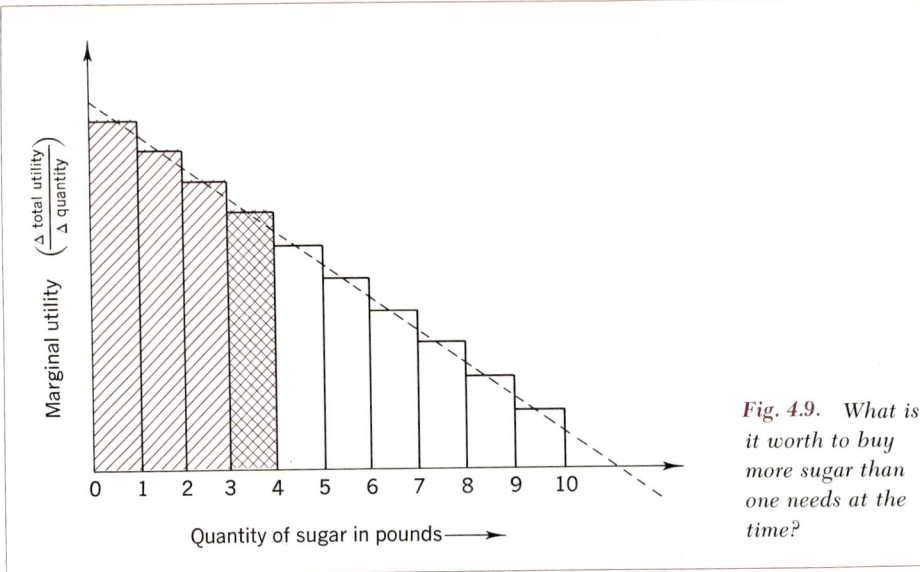

Fig. 4.9. What is it worth to buy more sugar than one needs at the time?

ginal product." Note that for various inputs of labor the value of (Δ bicycles/ Δ labor) for small values of Δ labor (corresponding to instantaneous velocity) is greatest at point D on curve ODC, which is equivalent to labor amounting to L_2 and bicycles amounting to B_2. In other words, a unit of one man's labor added to bring the level of labor to L_2 yields more *additional*, or *incremental*, or *marginal* bicycle production per man than any other. Beyond L_2, added increments of labor yield diminishing *marginal* returns.

The implications of this concept for the determination of the price of bicycles and the level of wages in the bicycle industry, which are developed at length in this type of economic theory, go far beyond the scope of this illustration. Suffice it to say here that this $\Delta y/\Delta x$ or (Δ bicycle/Δ labor) approach is central to the attitude of the manufacturer and his theories of wage determination.

Another instance of the use of the $\Delta y/\Delta x$ approach in economics is the *law of diminishing marginal utility*.† In fact, this "law" is one of the first fruits of very early attempts to apply incremental or marginal analysis in economics.

Suppose that you go to a store to buy a pound of sugar. However, when you ask for a pound of sugar, the clerk urges you to buy 10 lb. What is your reaction? Would you not ask how much cheaper per pound the sugar is when you buy 10 lb rather than 1 lb? Such a probable reaction indicates that when contemplating the purchase of 10 lb rather than 1 lb of sugar, each additional pound is less desirable to you, or, as the economist would say, has less utility for you. The money you give up for the extra pounds is no longer available to you for other purchases. Such a situation in exaggerated form is represented in Fig. 4.9.

Notice in Fig. 4.9 that the *total* utility, or satisfaction, afforded by 4 lb of sugar

† This, again, refers to an economic theory that is not universally accepted.

is the sum of rectangles 1 + 2 + 3 + 4, or the whole shaded area. (Why?) But the marginal or incremental satisfaction afforded by *a* fourth pound is rectangle 4 only.

If you are bothered by the steplike appearance of the curve representing marginal utility as a function of quantity of sugar, just imagine that on the quantity scale (*x* axis) the amount of sugar is increasing not by pounds but by half-pounds, or even by grains or half-grains. In fact, imagine that the *increases* along the "sugar" scale get smaller and smaller. As they do, the rectangles become straight lines, the tops of which would give a smooth curve like the dotted line in Fig. 4.9. What this curve says in effect is: "As a person contemplates acquiring larger and larger quantities of some good, the satisfaction from the possession of the good increases at a decreasing rate." It is this concept that is central to one explanation of the *law of demand,* namely, why people individually and collectively will buy more of an article or product as the price goes down. Furthermore, according to this economic theory, it is the balancing of marginal or incremental utilities per dollar that determines how a person will, at some particular time, allocate a limited income to the purchase of several different goods—a situation with which many of us are confronted every day.

Another more recent use of the $\Delta y / \Delta x$ technique in economics relates changes in spending for consumption purposes (such as food, shelter, and clothing) to changes in the level of income. Both income Y and consumption C are measured in billions of dollars per year. That is, they are "flows," like gallons of water per minute. The central concept in this analysis is $\Delta C / \Delta Y$, or the *marginal propensity to consume* (see Fig. 4.10, which depicts a hypothetical "consumption function," line CC).

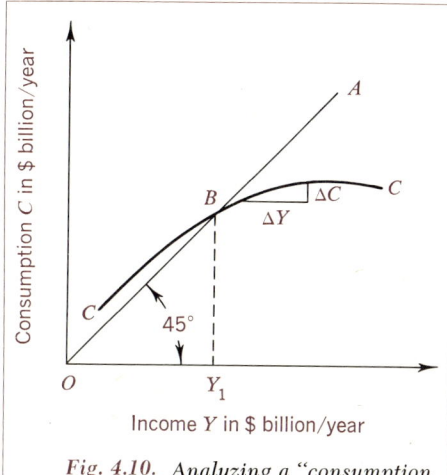

Fig. 4.10. *Analyzing a "consumption function" line C-C.*

The consumption function, line CC in Fig. 4.10, is hypothetical because economists are still not in complete agreement about its probable shape, but as it is drawn here it assumes two things. The first assumption is that at income level Y_1, consumers will spend all their income for consumption purposes. This is so because the 45-deg line OA is the locus of all points equidistant from the C and the Y axes. Therefore line BY_1, which is the level of consumption at income Y_1, must equal line OY_1, which is the level of income at Y_1.

The consumption function CC in Fig. 4.10 also assumes that if income rises above level Y_1, not all the incremental income will be spent for consumption. This is true because the value of $\Delta C/\Delta Y$ is less than 1. That is, as income increases, ΔC is always less than ΔY. Furthermore, if the consumption line CC "bends over," or rises more slowly at higher incomes, as it does in Fig. 4.10, the value of the ratio $\Delta C/\Delta Y$ gets smaller as income increases. When we draw the consumption function as shown in Fig. 4.10 we are hypothesizing that "as the income in an economic system rises above the level where it must be all spent for the necessities of life, consumers will not spend all the additional or incremental income and in fact will spend a decreasing percentage of additional income as income rises." Note that all ideas in this rather long and involved sentence can be conveyed by simply drawing four appropriately labeled lines as in Fig. 4.10.

Again, as in previous illustrations, the full implications of this concept of the marginal propensity to consume, as it is used together with other similar concepts to explain the level and growth of economic activity, are beyond the purpose of this brief section. A probing analysis requires more formal treatment in courses in economics. We must be content at this point with the general assertion that the actual value of the marginal propensity to consume is an important determinant of the income-generating power of new business or government spending.

The three illustrations of the $\Delta y/\Delta x$ technique presented here are only samples of the many mathematical techniques used by economists. The purpose of introducing these concepts at this point is simply to demonstrate that the usefulness of the techniques you have been studying in this chapter is by no means confined to the study of the natural sciences.

4.9 A little more on instantaneous rates of change†

By now the reader has undoubtedly realized that most variables in nature do not change at a uniform rate. When we start a ball rolling over an ordinary road, not only does its position change, but also the distance it covers each second of time will vary from start to the end of its motion. Determining just how fast it is moving *at a particular point* of its path becomes difficult. In our treatment thus far we used the delta, Δ, or increment notation, in which we tried to measure the *increment of distance,* Δs, centered at a given spot, traveled by the ball during an increment of time Δt. For a large increment Δt that we might choose, the measured Δs yields a value of velocity $\Delta s/\Delta t$, which is an average over the large Δt. We therefore reduced the time intervals (and hence, in general, the incre-

† Again note that the colored pages represent optional study material.

ments of distance) to smaller and smaller values until Δt became the shortest time interval in which we could make meaningful measurements. We then had a sequence of numbers $\Delta s/\Delta t$ for continually decreasing values of Δt. This sequence tended toward a limiting value as Δt approached closer and closer to zero. We called this limiting value $\Delta s/\Delta t$ the instantaneous velocity at a given point.

Galileo understood and used this concept intuitively, although the full answers in the form of the methods of the calculus did not come until much later. He saw that "each time interval however small may be divided into an infinite number of instants which correspond to the infinite (number of) values of velocity."

It will help us at this point to give a few illustrations of the functional relations we have discussed already. However, we shall state these in the terminology that is associated with the calculus and which constitutes the simplest way to determine instantaneous rates of change.

The general approach is to think of the increment Δ (whether Δs, Δt, or Δx) as becoming smaller and smaller until it approaches the value zero (or some infinitesimally small value). We designate this very small increment by the letter d instead of the symbol Δ. The increment Δs becomes differential ds, increment Δt becomes differential dt, and so forth. Now, although the differentials ds and dt are infinitesimally small, their ratio continues to be a finite number, as was the ratio of the increments Δs to Δt.†

This transition from Δs to ds is expressed mathematically as

$$\frac{ds}{dt} = \lim_{\Delta t \to 0} \frac{\Delta s}{\Delta t}$$

† This may not be so if the observations occur at a point where the function happens to be discontinuous.

or in words: The ratio of the differential ds to differential dt is given by the ratio of the increment Δs to the increment Δt as the increment Δt is made smaller and smaller, approaching zero.

Illustration 1 Suppose it is found that in a certain situation the displacement s increases with time at a constant rate, which we call k. Thus $\Delta s = k\, \Delta t$, or $\Delta s/\Delta t = k$ feet per second. In the notation of the calculus, this would be $ds/dt = k$ feet per second. From the differential form, the exact relationship between displacement s and time t can be determined immediately (by the rules of the process called *integration*) to give $s = kt$, where k is the proportionality constant. This is the result based on the change we "observed" in the interval of time dt. Of course the particle may have started from some position s_0 before we started to "observe" the increment or differential change; therefore we should add s_0 to this equation to give the actual $s = kt + s_0$.

Illustration 2 But now suppose that while watching for a change in s, we note that the increments of change keep getting larger with time. That is, $ds/dt = kt$. This would be the case when a particle is dropping under the influence of gravity and is being accelerated at a constant rate. Here, again, by the rules of *integration* of the calculus, we immediately obtain

$$s = \tfrac{1}{2}kt^2 + v_0 t + s_0$$

In this case we have added the last two terms to account for the fact that the (falling) body had an initial position s_0 and an initial velocity v_0 *before* we started watching to see what happened in the next interval dt.

Illustration 3 Sometimes the increase in s may vary, not with time but with the value of s itself. This might be the case when the number of children born in a year is proportional to the population of that year. The equation would be

$$\frac{ds}{dt} = ks$$

and the integration of this equation by the rules of the calculus† gives $s = ce^{kt}$. Here e is the base of the system of natural logarithms, a number equal to 2.178 . . . , while c is the value of s at the time $t = 0$.

Illustration 4 If the rate of change of the rate of change (acceleration) of s is negatively proportional to s, it can be represented mathematically by

$$\lim_{\Delta t \to 0} \frac{\Delta[\lim(\Delta s/\Delta t)]}{\Delta t} = -ks$$

(where k is *positive*). Using the calculus, this is more elegantly written as

$$\frac{d^2s}{dt^2} = -ks$$

Two implications follow when this equation is solved according to the rules.

(1) The relationship between s and t can be expressed in terms of a sine or cosine function. If $s = 0$ when $t = 0$ and if the acceleration is zero when $t = 0$, this may be mathematically represented by

$$s = A \sin (\sqrt{k})t$$

where A is a constant known as the amplitude, or the greatest value that s takes on periodically.

---

† These "rules" are available in tables of integration and are very simply applied in most cases. See the Laboratory and Mathematics Supplement to this volume, for example.

(2) The change of s is periodic; that is, the value of s recurs to the same value at regular intervals. The period depends upon k. (The period is given by the equation $T = 2\pi/\sqrt{k}$.)

Now the four mathematical equations, given as the preceding four illustrations of rates of change, can be variously combined. A tremendous variety of phenomena of changes in nature can be analyzed precisely or approximately in terms of these four expressions.

4.10 Summary

In this chapter we take cognizance of the fact that to understand natural and social phenomena which are in a continuous state of change, we must think in terms of *increments* of change, *rates of change with respect to time* or with respect to *other variables*, as well as rates of change *of the rates of change*. We began with analysis of motion of objects, but very quickly we were forced to distinguish between *average velocity* (or average rate of change) and *instantaneous velocity* (or the actual rate of change at some instant of time.) To define average velocity as "the displacement of an object in a given direction taking place over a given length of time" results in a definition easily understood, since this simply means that we measure the total displacement and the time it takes to make that displacement. But as soon as we ask "What is the velocity and how is it changing at a specific *point* of the motion?" we are forced to the mathematical notion of limits.

To grasp the analytical idea of *instantaneous velocity* and *instantaneous changes in velocity*, geometric models are helpful and have been given. If distance traveled is proportional to time (that is, $s = kt$), we can plot a straight line running through the origin to represent this relationship; instantaneous velocity is then the same as *average* velocity and is represented by the constant k. If, however, the velocity is *changing*, the result is a curved line. At any given point the instantaneous velocity (expressed crudely in Galileo's terms as an "infinitely small" change in displacement divided by an "infinitely small" duration of time) is represented by a *slope of the tangent line at that point* of the graph of the displacement-time relationship. The same procedure can be followed by representing a change of velocity by means of a graph on which the instantaneous velocities are plotted against time durations. *Acceleration* now can be defined as the rate of change of velocity, that is, *as a rate of change of a rate of change*.

Finally, in this chapter it has been shown how some of these same basic methods of analyzing change can be generalized. Rates of change—whether in physics, chemistry, biology, economics, or even sociology—can be specified by the same type of definition first developed in physical science. In some situations, two factors considered dominant in a particular system (mathematically called *parameters* or *variables*) are related by a mathematical function such that a change in one implies a change in the other. For emphasis we cite one example from geometry.

We know that the area of a circle is proportional to the square of the radius, the proportionality factor in this case being π. How, then, does a *change* in radius affect the *change* in area? This change in area relative to change in radius is symbolized as the limit of $\Delta A/\Delta r$ as Δr is made smaller and smaller. Using mathematical methods now named "the calculus" and which were anticipated in "the method of exhaustion" used by the Greeks and others in ancient times, the change in radius is proportional to the radius itself, the proportionality factor now being 2π.[†] This idea of *relative change of one variable with respect to another variable* when they are functionally related is capable of a marvelous generality so as to include analyses of a tremendous variety of phenomena. A few illustrative examples from economics were given in this chapter. The important points to keep in mind are: (1) the generalized notion of rates of change (and rates of change of rates of change) and (2) the importance of precise and clear definitions.

[†] That is, $A = \pi r^2$ and its derivative with respect to r is $dA/dr = 2\pi r$ (by the rules of the calculus for finding the derivative).

Questions/Discussions

1. Table 4.2 gives the population census for a particular city. We would like to analyze these numbers to develop some idea of the growth or decline of the population. For many analyses, a graphical representation of the data permits some useful information to be drawn easily. Using linear graph paper ($8\frac{1}{2} \times 11$ in.), carefully plot the population of this city as function of time. (Beginning the or-

TABLE 4.2

Year of Census	Population to the Nearest 500
1860	41,000
1870	46,000
1880	53,000
1890	61,000
1900	66,000
1910	70,000
1920	74,000
1930	77,500
1940	80,000
1950	81,500
1960	82,000

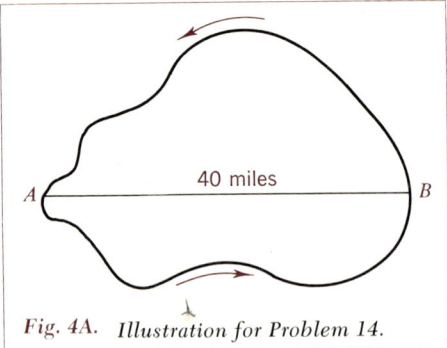

Fig. 4A. *Illustration for Problem 14.*

dinate with 41,000 and the time with 1860 offers a convenient choice of coordinates.) We can think of Table 4.2 and your graph as having the relationship found between Table 4.1 and Fig. 4.1. If you have difficulty in drawing a good curve through the points of your graph, consult your instructor.

2. From Table 4.2 the change in population from 1860 to 1870 is seen as +5000. (Notice the use of the plus sign to indicate an increase.) There is a similarity here to our earlier use of the concept of *displacement*, which we now identify simply as *change*. From the table or from your graph, what were the changes in the periods 1880 and 1890? 1885 to 1890? 1890 to 1895? 1895 to 1900? (We can *interpolate* the data for periods shorter than a decade by means of the graph. In determining these figures, would it be more accurate to take the average number between 1880 and 1890, or the average number between 1890 to 1900, or to interpolate from your graph?)

3. A change as a function of time gives the *rate of change* (corresponding to some features of our velocity concept). Using only Table 4.2, give the rate of change of population for the periods 1860 to 1870, 1870 to 1880, 1900 to 1910, 1950 to 1960. Now using the graph, can you suggest a different value for the rate of increase of population in the period near 1890 (by the method of drawing a tangent at that point). See discussion of Fig. 4.6.

4. The rates of change given in Problem 3 differ and therefore the rates themselves must experience effects that may be compared to the acceleration and deceleration of a particle in motion. From your graph identify the year when the rate of population increase experienced

a change from acceleration to deceleration (that is, when the curve changed from an increasing slope to a decreasing slope at a point of inflection).

5. Over what period was there no change in the rate of population increase (that is, no acceleration or deceleration of population)?

6. From Table 4.2, the increase of 5000 people from 1860 to 1870 represents an increase of about $5000/41,000 = 12+\%$. What was the percentage increase between 1950 and 1960?

7. The census figures for the United States were as follows:

Year-of Census	Population of United States
1790	3,929,000
1900	76,212,000
1950	151,326,000
1960	179,326,000

How does the percentage increase for our city (Problem 1) compare with the percentage increases in the United States for the period 1900 to 1950? For the period 1950 to 1960? What thoughts are suggested by the differences and similarities?

8. At the instant of observation of an airplane, it is traveling with a velocity of 300 mph on a straight course subjected to an acceleration of 1 mile/min². What will be the velocity of the airplane at the end of 15 min?

9. Assuming that efficiency in bicycle production is defined as the maximum number of bicycles per unit labor, is the labor level at point L_2 (Fig. 4.8) the most efficient? Why or why not? [*Hint:* For determining the efficiency of production per unit labor, would we take the ratio of (total bicycle production/total labor) or (Δ bicycle production/Δ labor)?]

10. Referring to Fig. 4.8: When the total bicycle production is at a maximum, what is the value of the ratio (Δ bicycles/Δ labor)? What general principle does your answer suggest?

11. Draw a graph based on Fig. 4.9, which shows the behavior of *total* utility as a function of increasing quantities of sugar. Can (Δ total utility/Δ quantity) ever be negative? If so, give an example.

12. Referring to Fig. 4.11: Starting with income Y_1 what would the consumption function look like if $\Delta C/\Delta Y = 1$ at all higher levels of income?

13. Assume that any part of income Y which is *not* spent for consumption purposes is saved; that is, $Y - C = S$. Referring to Fig. 4.10: From the information furnished about the consumption function (CC), show how you would construct a savings function, that is, a line showing the level of savings as a function of income. On your graph show the marginal propensity to save $\Delta S/\Delta Y$. Do you sense any constant relationship between $\Delta S/\Delta Y$ and $\Delta C/\Delta Y$?

14. A car is driven from site A to site B, located 40 miles away. In going from A to B the lower path (on the accompanying figure) is traversed (a distance of 50 miles) in 1 hour. The return trip (along the upper path in the figure) covers 60 miles in 1 hour. Assuming no time is lost by stopping at B, (a) what is the average speed at which the car travels during the total trip? (b) what is the average velocity of the car for the round trip? (Fig. 4A.)

15. A rock is thrown vertically upward with an initial velocity of 64 fps. How far above its starting point is the rock at the highest point in its trajectory? How long does it take to reach this height? What is the total time elapsed between the time

it is thrown and the time it returns to its original position? What is the velocity of the rock when it returns to its initial position? Use $g = 32$ ft/sec² or the acceleration of gravity.

16. Rework Problem 15, assuming the rock was thrown upward at an angle of 30 deg to the horizontal with an initial velocity of 64 fps. How far from the thrower will the rock return to its original level?

17. In Problem 16, if the initial velocity is 64 fps, at what angle should the rock be thrown in order to obtain the maximum horizontal distance of travel?

References

General

Atkins, K. R., *Physics*. New York: Wiley, 1965, pp. 26–28. *Provides an excellent introduction to vectors and the motion of bodies.*

Galileo, Galilei, *Dialogues Concerning Two New Sciences*. Translated by Crew and de Salvio. New York: Dover, 1637. *This translation of Galileo's historic work still provides an excellent introduction to kinematics.*

Greenberg, D. A., *Mathematics for Introductory Science Courses: Calculus and Vectors*. New York: Benjamin, 1965. *An excellent introduction to the mathematical tools needed in this course.*

Hill, W., *Mathematics for the Layman*. Patterson, N.J.: Littlefield, Adams, 1964. *Chapter 23 introduces some ideas from the calculus.*

Lagemann, R., *Physical Science*. Boston: Little Brown, 1963. *Chapters 8 and 9 treat Galileo's work and projectile motion.*

Ore, O., *Graphs and Their Uses*. New York: Random House (New Mathematical Library 10), 1963.

Resnick, R., and Halliday, D., *Physics for Students of Science and Engineering*, 2d ed. New York: Wiley, 1966. *A more analytic treatment of kinematics, especially Chapters 3 and 4. Requires elementary calculus.*

Mathematics. *The mathematical concepts needed, including an introduction to the calculus, may be found in the following books:*

Kleppner, D., and Ramsey, H., *Quick Calculus*. New York: Wiley, 1965. *A short self-instruction manual.*

Ritow, I., *Capsule Calculus*. New York: Doubleday (Dolphin book C336), 1962.

Rogers, E., *Physics for the Inquiring Mind*. Princeton, N.J.: Princeton Univ. Press, 1960.

Sawyer, W. W., *What Is Calculus About?* New York: Random House (New Mathematical Library 2), 1961.

The reader should refer also to the Laboratory and Mathematics Supplement to this volume.

CHAPTER FIVE

The Coming of Age of Science: The Newtonian Period

Does not nature move or stop in response to causal force; and therefore is there not determined cause for every act of nature and of man?

V. L. P.

THE ADVANCE of science that marked the sixteenth century in Europe continued apace—or even quickened—during the seventeenth century. The achievements of Galileo were in part due to the difference in his attitude from that of his predecessors. Galileo did not seek an all-embracing natural philosophy embodying causal relationships for phenomena. Unlike many philosophers who asked *"Why does nature work as it does?"* Galileo asked *"How does nature work?"* In this attitude he was more akin to Archimedes and the early Hellenistic philosophers than to Plato and Aristotle who provided the basis for the natural philosophy of the sixteenth century. Table 5.1 lists a few of the notables of that century. Thus far we have mentioned only Kepler and Galileo. Francis Bacon (1561–1626), a contemporary of Galileo, was a strong protagonist for the experimental method and for an inductive approach toward making discoveries. He was conscious of the need for careful procedures in observation and proposed methods of gathering and analyzing of data, which were systematic although not very practical in our current view. He did not undertake experimental work himself, nor did he propose or seek to learn the grand designs of nature, his attitude being that designs emerge from experience with phenomena. But his emphasis on the need for a systematic approach in science was to be influential in guiding the Royal Society of London. His publication of *Novum Organum* (The New Instrument) in 1620 played an important role in promoting careful experimentation as the main interest for science and was one of the first attempts to devise a logic of discovery. But as we shall note later, Bacon sought *useful* knowledge as compared with Plato's search for *true* knowledge that represented the essence of the good in nature.

5.1 Science in the seventeenth century

René Descartes (1596–1650), a younger contemporary of Galileo, who is considered by some to be the prophet and popularizer of science, stressed a mathematical approach toward obtaining knowledge because of the "certainty of its reasoning." He maintained that since man's opinions change and the senses so often deceive, it is necessary to doubt them. We must doubt everything that can be doubted. "The power of forming a good judgment and of distinguishing the true from the false, which is properly speaking what is called Good Sense or Reason, is by nature equal in all men . . . the principal matter is to apply them well" (Descartes, *Discourse on Method*, 1637). In his *Rules for the Direction of the Mind*, he describes methods that will lead men not to assume what is false to be true. His advice as to how to initiate an inquiry is rationalistic. One starts by doubting everything that can be doubted and from what remains, the undoubtable ideas, one deduces one's knowledge. His focus on a deductive formal procedure is in contrast with Bacon's focus on inducing principles from instances of the phenomenon under consideration. He was so convinced of the superiority of his approach that he did not think much of Galileo because Galileo experimented. Although Descartes' methods are not considered to have had a great influence in experimental science, he did devise analytic geometry, which has had important effects on mathematics and science. Descartes is said to have extended the work of Greek, Hindu, and Arab mathematicians by bringing algebraic analysis into geometry (analytic geometry). In this connection he introduced the familiar rectilinear coordinate system which still bears his name (Cartesian coordinates). In mechanics he is credited with introducing the concepts of work and energy, and some aspects of modern determinism were formulated by him. Incidentally, the logical culmination of his ideas is seen in the work of Spinoza. His books were widely read, especially by the upper classes of France. This did a great deal toward encouraging a more receptive public opinion and freedom of thought with respect to science and new ideas.

Christian Huygens (1629–1695) investigated the subjects of gravitation, mechanics, and optics. He is most remembered for his development of the wave theory of light. He showed on a mathematical basis that Galileo's discoveries about the pendulum were a natural consequence of the uniformity of the gravitational force. He also derived the law of centripetal force, which will be discussed presently.

The most important figure of seventeenth century science, and indeed the science of many centuries, was Isaac Newton (1642–1727). At the age of nineteen he entered Cambridge University where he studied, among other subjects, mathematics. The Great Plague reached Cambridge in 1666 and the student body was dispersed. Newton, then 24, returned to his birthplace at Woolsthorpe in Lincolnshire, where he spent the next two years. It was here that he set himself to investigating the problem of planetary motion. In the light of Galileo's discoveries involving the motion of falling bodies, it seemed evident to Newton that a body in motion without the action of external forces will remain in (straight-line) motion. For a body to be in circular (or elliptical), motion required that a central force operate in such a manner as to cause the body to deviate from its straight-line motion by the proper amount each second. Balliani in Italy and Hooke in England reached a similar conclusion at about the same time, but it was Newton who placed the theory on a firm mathematical basis.

Newton considered the motion of the moon about the earth. What was the nature of the force that caused the moon to cling to this orbital path instead of going off in straight-line motion as it would do if no force were acting on it? It was at this point, as reported by Voltaire, that Newton, while sitting under a tree, was struck on the head by a falling apple. Newton's thoughts then turned to falling bodies. He knew that a body near the surface of the earth fell toward the earth's center (that is, downward) with an acceleration of 9.8 m/sec², owing to the force called "gravity." Newton asked the question, "Does the force due to gravity extend out to the moon, and if so, how large is it at the moon's distance from earth?" According to Eq. (4-10), the gravitational acceleration of 9.8 m/sec² causes the apple to fall 4.9 m in the first second. How far does the moon fall in 1 sec? Figure 5.1 depicts a portion of the moon's orbit about the earth. If at time $t = 0$ the moon is at point A in its orbit, it will be at point B 1 sec later. Had there not been some central force in operation, it would have traveled instead along the straight line AB'; hence this force caused the moon to fall inward, toward the earth, by an amount $B'B$ in 1 sec. If we consider for the moment that the center of the moon moves in a circular orbit of radius 3.8×10^8 m about the center of the earth, the value of $B'B$, the distance fallen by the moon each second, is 0.0013 m.

Example Let us call the angle subtended at earth by A and B the angle θ. Note that

Fig. 5A. Isaac Newton, 1642–1727. (From Physical Science, Men and Concepts, *by Omer, Knowles, Mundy, and Yoho, copyright 1962. Used by permission of D. C. Heath, Inc.*

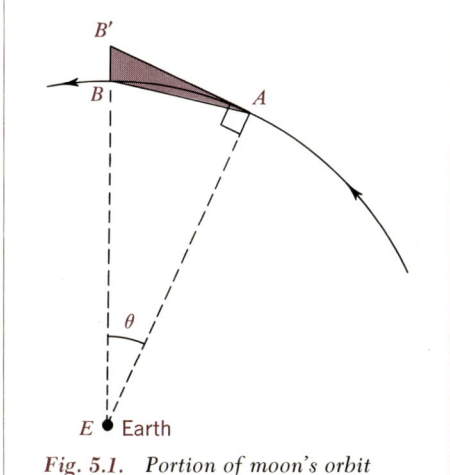

Fig. 5.1. Portion of moon's orbit about the earth.

the angle $B'AE$ is 90 deg and angle EBA = angle EAB. (Why?) By the law of sines

$$\frac{\overline{B'B}}{\sin(\text{angle } BAB')} = \frac{\overline{BA}}{\sin(\text{angle } BB'A)}$$

Where notation $\overline{B'B}$ should be read "the length of line segment $B'B$."

But, since angle EBA = angle EAB and since the sum of the angles in a triangle is 180 deg, angle $EAB = 90° - \frac{1}{2}\theta$ and angle $BAB' = \frac{1}{2}\theta$. Likewise, angle $BB'A = 90° - \theta$. Thus our equation becomes

$$\frac{\overline{B'B}}{\sin(\theta/2)} = \frac{\overline{BA}}{\cos\theta}$$

To find θ, recall that the moon travels 2π radians (rad)† in 27.3 days‡ (2.4×10^6 sec), or 2.6×10^{-6} rad/sec. Thus θ, the angle moved in 1 sec, is 2.6×10^{-6} rad. This is a very small angle and we may use the approximations§

$$\sin(\theta/2) \sim \theta/2 \text{ in radians}$$
$$\cos\theta \sim 1$$

Thus

$$\overline{B'B} = (\overline{BA})\left(\frac{\theta}{2}\right)$$

Similarly, arc length $BA \sim \overline{BA}$, which is equal to θ times the distance R from earth to moon, where

$$R = 3.8 \times 10^8 \text{ m}$$

Thus

$$\overline{BA} = \theta R$$

† Radian measure will be discussed later in this chapter.
‡ This is the sidereal period, a measure of the actual time required for the moon to make a complete orbit of the earth, as contrasted with the synodic period of 29.5 days mentioned in Sec. 3-4.
§ The single ($\sim$), or sometimes a double ($\approx$), wavy line is used to denote "is approximately equal to."

and

$$\overline{B'B} = (\theta R)\left(\frac{\theta}{2}\right) = \frac{1}{2}\theta^2 R$$

$$= \frac{1}{2}(2.6 \times 10^{-6})^2(3.8 \times 10^8) = 0.0013 \text{ m}$$

The ratio of the distance the moon "falls" in 1 sec to the distance the apple falls in the same time is

$$\frac{0.0013}{4.9} = 0.00027 \sim \frac{1}{3600}$$

The ratio of the distance of the apple from the center of the earth (6.4×10^6 m) to the distance of the moon's center from the center of the earth is

$$\frac{6.4 \times 10^6}{3.8 \times 10^8} \sim \frac{1}{60}$$

The number 3600 is 60×60, or 60^2. From this we see that under the influence of the earth's gravity, the ratio of the distance a body falls when near the surface of the earth compared with its fall when it is at the distance of the moon is equal to the inverse square of the ratio of its respective distances from the center of the earth.

Newton reasoned that the same force that caused the apple to fall also holds the moon in its orbit—the force called *gravity*. He hypothesized that this force acts between any pair of bodies that possesses mass† and that the magnitude of this force is proportional to the product of these masses and inversely proportional to the square of the distance separating their centers; that is,

$$\text{Force} = \text{constant} \times \frac{\text{mass}_1 \times \text{mass}_2}{\text{distance}^2}$$

$$= G \times \frac{m_1 \cdot m_2}{r^2}$$

† The concept of mass as well as force will be treated in the next section.

Although Newton's work on the *universal law of gravitation* dates back to 1666, he did not publish this work until nearly 20 years had passed. His book *The Mathematical Principles of Natural Philosophy* (more commonly called *The Principia*, from its Latin title) appeared in 1686 and provided the base for the science of mechanics up to the present time.

Newton also made substantial contributions to mathematics, being concerned with such matters as the binomial theorem and infinite series. He originated the "method of fluxions," the forerunner of modern calculus,† and wrote on optics and heat transfer. Much of the creative effort of his early years went into alchemy, while his later years were devoted to writing on theological topics.

Newton was elected a member of parliament in 1686. Later he became Master of the Mint, and for his services in this capacity he was knighted in 1707.

A parallel development of the seventeenth century was the founding of scientific academies, predecessors of modern scientific societies. The first of these was founded in Naples in 1560. The Academia dei Lincei, founded in Rome in 1603, counted Galileo among its members. The Philosophical College, which began meeting in London in 1645, was officially incorporated in 1662 as the Royal Society by Charles II, under which name it meets to this day. The Academie des Sciences was founded in 1666 by Louis XIV. The scientific periodical *Journal des Savants* was first issued in Paris in 1665, followed three months later by *The Philosophical Transactions of the Royal Society*. These learned societies and journals provided

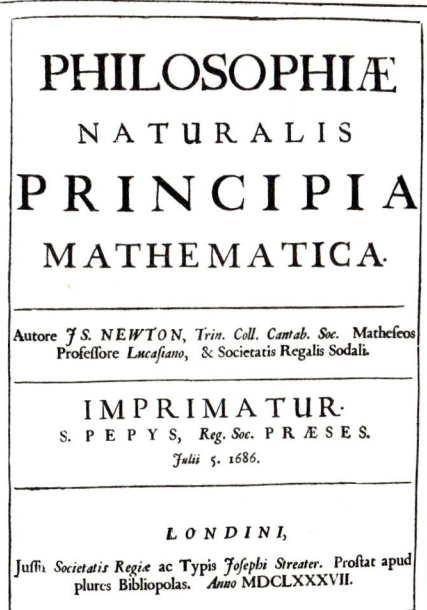

Fig. 5B. (*From* Foundations of Modern Physical Science *by Holton and Roller, copyright 1958. Used by permission of Addison-Wesley Publishing Co.*)

---

† This same method was independently introduced by Leibnitz at about the same time.

> As in Mathematicks, fo in Natural Philofophy, the Inveſtigation of difficult Things by the Method of Analyſis, ought ever to precede the Method of Compoſition. This Analyſis conſiſts in making Experiments and Obſervations, and in drawing general Concluſions from them by Induction, and admitting of no Objections againſt the Concluſions, but ſuch as are taken from Experiments, or other certain Truths. For Hypotheſes are not to be regarded in experimental Philoſophy. And although the arguing from Experiments and Obſervations by Induction be no Demonſtration of general Concluſions; yet it is the beſt way of arguing which the Nature of Things admits of, and may be looked upon as ſo much the ſtronger, by how much the Induction is more general. And if no Exception occur from Phænomena, the Concluſion may be pronounced generally. But if at any time afterwards any Exception ſhall occur from Experiments, it may then begin to be pronounced with ſuch Exceptions as occur. By this way of Analyſis we may proceed from Compounds to Ingredients, and from Motions to the Forces producing them; and in general, from Effects to their Cauſes, and from particular Cauſes to more general ones, till the Argument end in the moſt general. This is the Method of Analyſis: And the Syntheſis conſiſts in aſſuming the Cauſes diſcover'd, and eſtabliſh'd as Principles, and by them explaining the Phænomena proceeding from them, and proving the Explanations.

Fig. 5C. (From Foundations of Modern Physical Science *by Holton and Roller, copyright 1958. Used by permission of Addison-Wesley Publishing Co.)*

the scientists with a forum for discussion and debate, and did much to promote the rapid advance of science during the seventeenth and subsequent centuries.

In 1703, Newton became president of the Royal Society, a post that he held until his death at age eighty-five.

5.2 Force, mass, acceleration: the first two laws of Newton

The most important scientific development up to the seventeenth century, and indeed of the intervening centuries up to our own, had to do with the branch of science called *mechanics*. Mechanics has to do with the behavior of bodies that are subjected to forces such as the gravitational forces just discussed. Every movement of our bodies, every motion of every vehicle and indeed of the multitudinous heavenly bodies, and every movement of an atom is subject to the laws given by this science. It is necessary, therefore, that we delve into the subject in a little more detail.

Common experience seems to tell us that when a body such as a billiard ball is at rest, it continues in a state of rest unless some force is applied to move it. Similarly, if a ball starts rolling at one end of the table, it will continue to roll in a straight line until it strikes another ball or the end of the table. The velocity it has when it reaches the end of the table will be reduced somewhat from the velocity it starts with; reduced because of the slight but constant friction with the surface of the table. A smoother table introduces less friction and less reduction in speed. Less significant is atmospheric "friction" created by virtue of the ball's motion. From these observations we are led to believe that without the interference from friction a body in motion would continue in motion without loss of velocity.†

These two simple observations constituted the essence of a discovery by Galileo, which later became expressed as the first law of Newton, to the effect that:

Law I Every body persists in its state of rest, or of uniform motion in a straight line, unless it is compelled to change from that state by forces impressed upon it.

If we apply to a ball at rest [see Fig. 5.2(a)] a very slight, steady push with our finger or with the cue, the ball will move along slowly. The small push would be just enough to overcome the frictional forces of the table top which tend to keep the ball from moving. When we apply a larger push [Fig. 5.2(b)] the ball moves with some velocity. Of course, to reach that velocity, the ball must have experienced an acceleration. We saw in Chapter 4 that acceleration is the time rate of change of velocity and that it can be measured by the use of meter sticks and clocks.

If we were to apply a series of larger forces to the ball and determine the acceleration that results from each force, we would find that in each case the acceleration a is directly proportional to the force F. This is given as the second law of Newton which in current speech might read as follows:

Law II An object of mass m, which is subjected to a net force† F, will experi-

† This statement assumes that the ball does not suffer collision with other balls or with the end of the table.

† Since a body may be subjected to several forces, some of which oppose and therefore cancel each other, only the resultant or net force will be effective in moving the body.

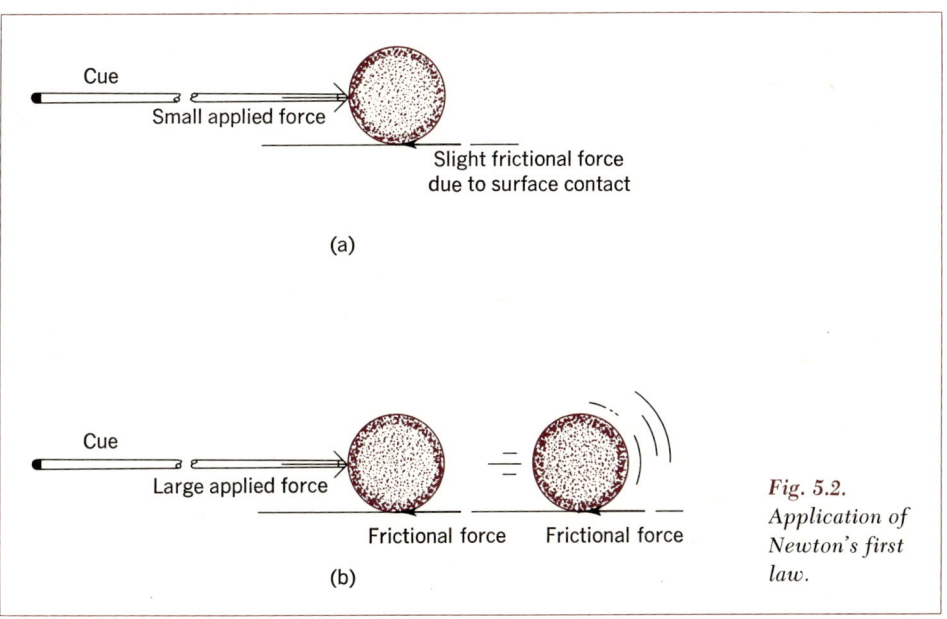

Fig. 5.2. Application of Newton's first law.

ence an acceleration (in the direction of the force) which is proportional to the force and inversely proportional to the mass.

This is expressed by the relationship[†]

$$\mathbf{F}_{net} = m\mathbf{a} \qquad (5\text{-}1)$$

or, in words, Force = mass times the acceleration of the mass.

What do we mean by the terms *mass* and *force*? Mass is a fundamental property of matter. Matter can change form, as when a block of ice becomes water and then steam on heating. But that same quantity of steam behaves exactly as did the ice as far as Eq. (5-1) is concerned. That is, whether the force is applied to the block of ice or to its liquid or gas equivalent, the mass will experience the same acceleration. In contrast with our ability to make direct measurement of length with a meter stick, there is no direct measure for mass. Mass can be measured only through its properties, such as the property of *inertia* which is revealed in experiments that involve Eq. (5-1). From such experiments a standard of mass has been chosen in the form of a cylinder of platinum kept at the Bureau Internationale des Poids et Mesures at Sévres, near Paris, with working standards elsewhere. This standard is named the kilogram-mass, or kilogram (kg).

The convenient way to measure mass, once a standard is defined, is to measure one of its properties, namely, the pull of gravity or the weight of the kilogram-mass.[†] The standard kilogram-mass can be placed in the arm of a balance in order to find a second 1-kg mass; this second mass then becomes a working standard against which to calibrate various other masses, weighing springs and scales. This, of course, is based on the assumption that two 1-kg masses have twice the property of weight (or pull of gravity) that characterized the 1-kg mass.[‡]

Fortunately our appreciation for the concept of *force* has the benefit of personal experience that comes from pushing and pulling objects. To standardize the concept and methods for measuring force, we utilize experiments based on Eq. (5-1), which include a measuring stick and a clock or stopwatch. We place a 1-kg mass on a surface that is sufficiently smooth to reduce frictional forces to a negligibly small value, and arrange to pull this mass

---

[†] When it is desired to call particular attention to the fact that the quantities symbolized are vectors, a boldface roman letter indicates the vector quantity. Sometimes we use a small arrow over the letter when writing it. To avoid complicating the symbols we shall not often designate symbols with vector notation except as required to make a point. When solving problems, the *directional* features as well as *magnitude* must be kept in mind, however.

[†] Note that mass and weight, although closely related, represent quite different aspects of matter. The mass of an object is a *constant*, which is quite independent of the pull of gravity on it, whereas its weight is a *variable* that does depend on gravity. For example, a 1-kg mass weighs 9.8 newtons on earth. If this were taken to the moon, it would still be a 1-kg mass, but it would then weigh only about 1.6 newtons because of the smaller gravitational pull of the moon.

[‡] This procedure can be justified for mass determination, but we are reminded that doubling things does not necessarily double properties (take, for example, the property of surface area; doubling the mass does not necessarily double the surface area). The use of balances also assumes that the two weighing pans experience identical gravitational pull, although they are not in identical positions. Newton made many tests involving gravity until he was satisfied that mass was proportional to weight. It was discovered later that the pull of gravity does vary from place to place on the earth. Of course the gravitational pull would be quite different at the surface of each of the planets.

through a calibrated scale, as illustrated in Fig. 5.3. If the mass is pulled with such force as to give it an acceleration of 1 m/sec², the indication on the scale at that pull represents one unit of force, which is called the *newton*. The pull can be increased until it imparts an acceleration of 2 m/sec², at which point the scale can be marked as 2 newtons, and so on to higher forces. The scale can then be given divisions and subdivisions and thereby made suitable for measuring unknown masses.

At the beginning of the present century the concept of inertial mass became considerably modified with the introduction of the theory of relativity. As we shall see, the characteristic of a mass we call *inertia* increases with its velocity, although the increase becomes measurable only when the velocity increases to a very high value. Theoretically at least, this inertia characteristic becomes infinitely large as the velocity of the mass approaches the velocity of a light as a limit. But for all applications that involve ordinary velocities of daily experience, we can assume that the inertial mass is constant.

Through experiments of the type indicated by Fig. 5.3, we can calibrate the spring scale to read in newtons. If now we raise that same kilogram-mass and spring scale to a vertical position, as illustrated in Fig. 5.4, what will be the reading on the scale? Obviously, this reading represents the pull of gravity on the 1-kg mass, and would be approximately 9.8 newtons. If the mass is released from the scale, this same gravitational pull ($F = 9.8$ newtons) will force the mass downward with an acceleration

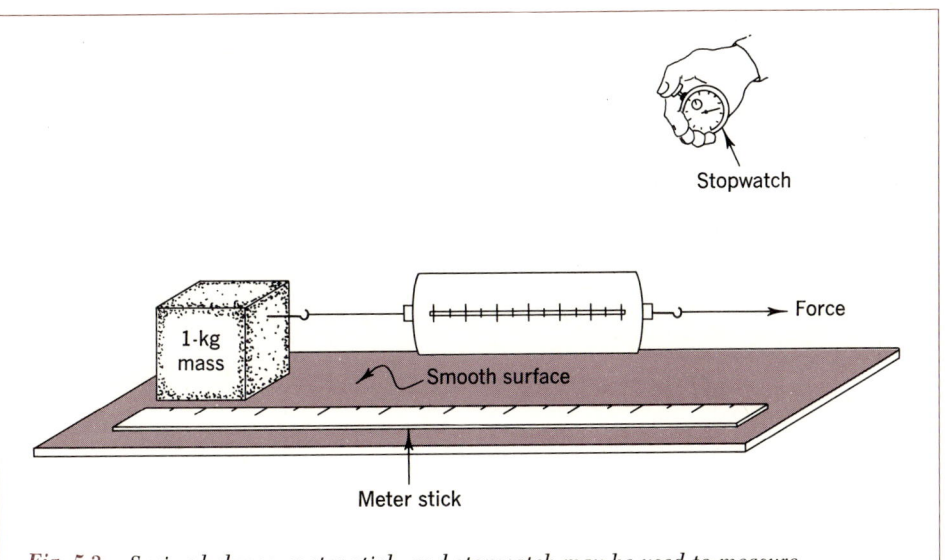

Fig. 5.3. Spring balance; meter stick, and stopwatch may be used to measure accelerating force in newtons.

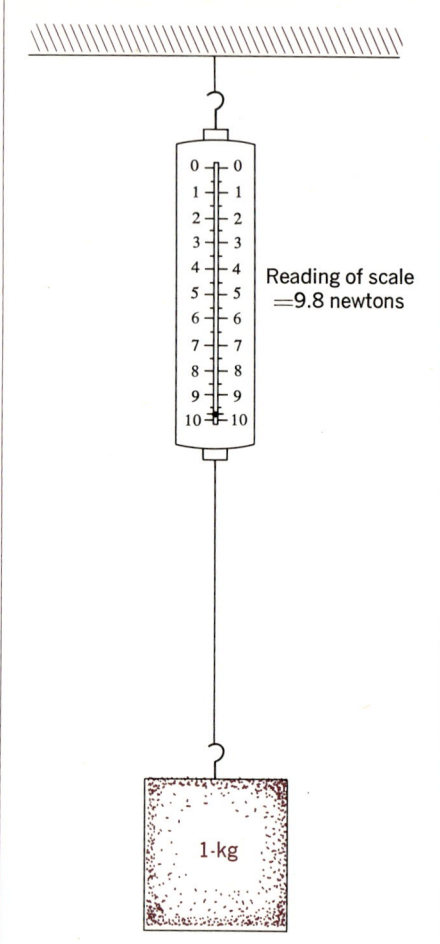

Fig. 5.4. Pull of gravity on 1-kg mass extends the spring of the balance to indicate 9.8 N.

of 9.8 m/sec², according to Eq. (5-1). This value for acceleration due to gravity is called g. Doubling the mass to 2 kg would also double the weight or gravitational pull (19.6 newtons), so the acceleration would remain the same.†

The second law was stated in somewhat different terms by Newton, and we shall repeat it in another form in order to reveal one additional concept. He defined "quantity of motion" as the product of of mass times velocity (which we now call *momentum*) and stated the second law as follows:

Law II (Alternative Statement) *The time rate of change of "quantity of motion" (or momentum) is proportional to the motive force impressed, and is in the direction in which that force is impressed.*‡

† Mass has been variously defined and used. In the metric system it was originally defined as the gram, or the mass of water contained in 1 cubic centimeter (cc) at a specified temperature and pressure. This made the density of water 1 gram per cc. The gravitational attraction on this mass, equal to 980 dynes, gave an acceleration of 980 cm/sec². The dyne is the unit of force in the CGS system (centimeter-gram-second system) and is defined as the force needed to accelerate a 1-g mass by 1 cm/sec². There is also an English system in use in which the unit-pound is sometimes used as the gravitational pull and sometimes as a unit of mass. For our purposes we shall use the relationship that a kilogram-mass weighs 2.2 pounds (lb), and that a force of 1 pound gives a mass of 1 slug an acceleration of $g = 32$ ft/sec².

‡ This may be expressed in the notation of the calculus as

$$F = \frac{d}{dt}(mv)$$

When m is a constant, this reduces to

$$F = m\frac{dv}{dt} = ma$$

However, when m varies with time, as is the case when v is close to the velocity of light

This version makes clear that the impressed force, acting for a given time, produces a change in the momentum (mv). When the mass m remains constant, the change becomes a change in v only, or simply acceleration. But we shall see that according to the theory of relativity, the aspect we call mass, m, need not remain constant at very high speeds. Therefore it is more proper to speak of the impulse† as causing a change in momentum.

Let us analyze the experience of pulling a wagon from a standing position (Fig. 5.5). We know that some effort is required to get the wagon to start moving because of the frictional drag due to friction in the bearings and from the earth, represented by F_f. As the man's pull (F_m) increases, it will reach a value that is exactly equal to F_f. The wagon still remains unmoved.‡

However, any further increase of F_m will cause the wagon to move with an acceleration a, which is proportional to that increase of the force or pull. That is,

$$F_m - F_f = Ma \qquad (5-2)$$

where both F_m and F_f are measured in newtons when M is in kilograms and a is in meters per square second.

5.3 The third law of Newton

Newton's third law is stated as follows:

Law III To every action there is an equal reaction; or the mutual actions of two bodies upon each other are always equal, and in opposite directions.

Whatever pulls or pushes an object is pulled or pushed by that object by an equivalent amount. (See Figs. 5.6 and 5.7.) If you press a stone with your finger, the finger is also pressed by the stone. If a horse pulls a stone tied to a rope, the horse is pulled back toward the stone with equal force; if the earth pulls a sky diver downward, the sky diver pulls upward on the earth an equal amount.

The significance of the law is easily demonstrated by the examples cited by Newton and by common experience. Ordinarily we substitute the term *force* for *action* and *reaction*, and say that a force F_1 applied to an object gives rise to

---

(special relativity), the alternate form remains the correct form for the second law.

† Impulse is defined as the product of the impressed force times the time interval during which it acts.

‡ We should note that there is no frictional resistance F_f until someone tries to pull the wagon, at which time the frictional drag makes itself felt and reaches a starting point value of F_f just before the wagon moves.

We should note also that in most cases of frictional drag, the value of F_f tends to drop to a lower value as soon as the wagon starts to move, since the lubricants and rubbing surfaces experience less frictional drag when moving.

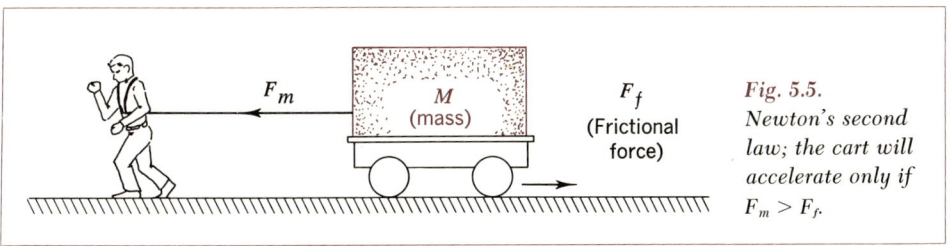

Fig. 5.5. Newton's second law; the cart will accelerate only if $F_m > F_f$.

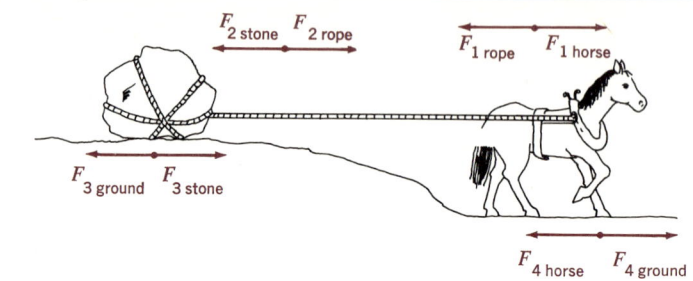

Fig. 5.6. Horse pulling on stone—action-reaction forces (Newton's third law). Each of the action-reaction pairs of forces are equal to each other, for example, $F_{3\,ground}$ (the force exerted by the ground on the stone) must equal $F_{3\,stone}$ (the reaction force exerted by the stone on the ground).

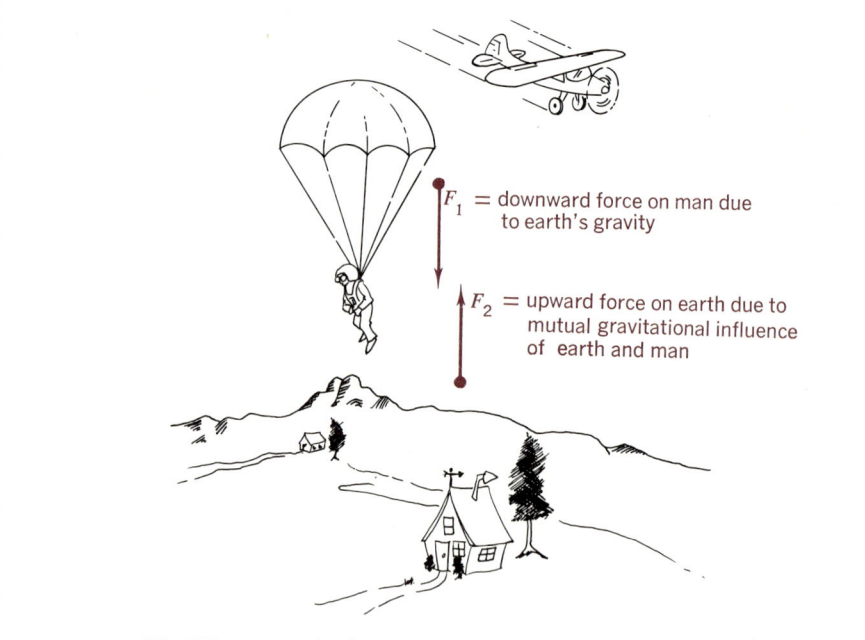

Fig. 5.7. Newton's third law, action-reaction forces. Man and earth experience equal forces. Which one will be accelerated by the greater amount?

a force F_2 that is exactly equal to F_1 but has opposite direction and acts on the other body. When a book or other object is held on the palm of the hand, it presses down because of the pull of gravity, and the hand must exert an exactly equal upward force to keep in a fixed position. Indeed, if a mosquito suffers a head-on collision with a fast-moving locomotive, the force of impact of the mosquito on the locomotive has exactly the same magnitude (but opposite direction) as that exerted by the locomotive on the mosquito! Thus we can write

$$F_1 = -F_2 \quad (5\text{-}3)\dagger$$

or, in words, we can say

Force of object A on object B =
 $-$(force of object B on object A).

In a dynamic system where the parts move, we can substitute for F from Eq. (5-1), for a body of mass m_1 and a body of mass m_2. If the encounter of the two bodies produces motion, we would have

$$m_1 a_1 = -m_2 a_2 \quad (5\text{-}4)$$

Thus, according to the third law, if a person on roller skates tries to push against an object such as a wall, he finds himself pushed back instead. This is demonstrated in a more subtle way in the rocket engine, which we shall discuss in Chapter 7.

5.4 Conservation of momentum

Let us go back to Eq. (5-1) and recall from earlier discussions that acceleration

† This equation may be rewritten

$$\mathbf{F}_1 + \mathbf{F}_2 = 0,$$

which implies, since $\mathbf{F}_1$ and $\mathbf{F}_2$ are not generally both equal to zero, that they are in opposite directions. We can say that a state of *equilibrium* exists when the opposing forces balance each other exactly.

is change of velocity over time, or $a = \Delta v/\Delta t$. Equation (5-1) can therefore be written as

$$F \, \Delta t = \Delta(mv) \quad (5\text{-}5)$$

where we have included the mass m in the parentheses as Newton did. We call the product of force and the time interval during which it is applied the *impulse*. As stated earlier, mv is called the *momentum*, often designated by the letter p. Equation (5-5) can be read as

(Force) × (the time interval during which the force acts)
 = (the change of momentum)

or

Impulse = change of momentum

If the force is expressed in newtons, time in seconds, mass in kilograms, and velocity in meters per second, the units for momentum become kilogram-meters per second.

This leads to an important principle of mechanics, which states that the *total momentum of a closed system is conserved*.† In a system in which two bodies collide, the momentum lost by one of the two colliding bodies is always gained by the other colliding body. The total momentum of the system after the collision is exactly equal to the total momentum of the system prior to the collision (for example, see Fig. 5.8). This conservation law will be useful whether we analyze the collisions between billiard balls, between automobiles, or between atomic particles.

Suppose we bounce a ball of mass m against a larger ball of mass M, causing the larger ball (originally at rest) to move.

† This statement, known as the *law of conservation of (linear) momentum*, is an empirical rule that plays a very important role in physics.

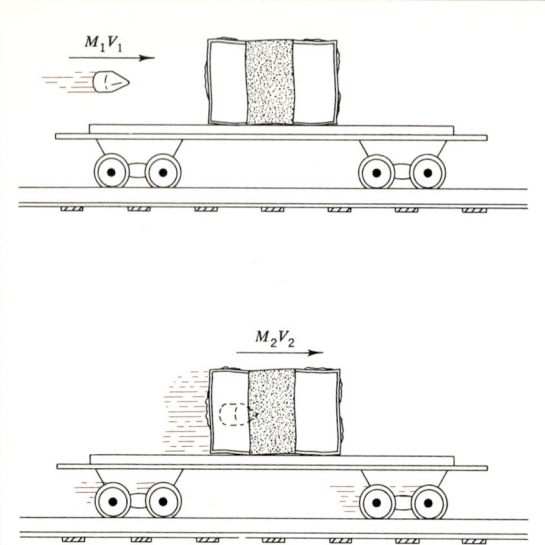

Fig. 5.8. *Conservation of momentum, $M_1V_1 = M_2V_2$. Since the mass of the projectile M_1 is much less than the mass M_2 of the freight and car, it follows that V_2 is much smaller than V_1.*

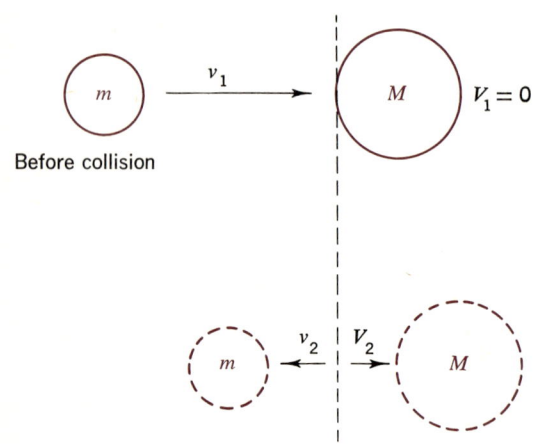

Fig. 5.9. *After collision, the small ball goes backward with a smaller velocity $v_2 < v_1$, while the large ball acquires a velocity V_2.*

How fast does it move? We shall assume that the collision is along the central line, so that the motion of both balls is along a straight line. In Fig. 5.9 we look down on the motion of the small mass m, having initial velocity v_1 (therefore initial momentum mv_1). This collides with a larger ball of mass M which has an initial velocity and momentum of zero. After collision, suppose the larger ball moves with velocity V_2 as a result of the impact, having absorbed some momentum, and the small ball moves with velocity v_2. The principle of conservation of momentum requires that

$$mv_1 = mv_2 + MV_2 \quad (5\text{-}6)$$

By rearranging, we obtain

$$mv_2 = mv_1 - MV_2$$

which says that the momentum of the smaller ball after collision is reduced by the amount of the gain in momentum of the second larger ball. It must be remembered throughout this discussion that momentum is a vector quantity and that a reduction of momentum does not necessarily imply a reduction in the *magnitude* of the momentum but rather may mean that the initial momentum was reversed in *direction*, as is the case in Fig. 5.9.

5.5 Kinetic energy and work

To obtain actual values for the sharing of momentum, we need to introduce a second relationship between the two colliding bodies, since Eq. (5-6) contains too many unknowns. For that purpose we now introduce the concept of *kinetic energy* of moving objects. A ball or other object of mass m traveling with velocity v is said to have kinetic energy of

$$K = \text{kinetic energy} = \tfrac{1}{2}mv^2 \quad (5\text{-}7)$$

The definition is useful because, as thus defined, kinetic energy represents both the *amount of work* that was done to get the object up to that velocity from rest and the amount of work that the rolling ball can perform before it is stopped.† The rolling ball can be stopped by causing it to exert some force on an obstacle in its path until its kinetic energy is dissipated, much as an automobile is stopped by applying the brakes for a certain distance. If the brakes alone do the stopping, the kinetic energy of the traveling auto becomes dissipated by the braking force F (assumed to be kept at a constant value), times the distance S that is required to stop the auto. Similarly, the force F applied over the same distance S would be required to get the auto rolling again at the original speed, if we assume there are no other frictional losses.

This gives us a definition of *work*, namely: For a constant force F, the work done is the product of F times the distance S over which F is applied. If the force is doubled, the distance can be halved to be equivalent to a particular value of work done. If the brakes of the auto are applied lightly, the vehicle will continue over a correspondingly longer distance before coming to a stop.

Suppose that we are pulling on a 30-kilogram wagon with a constant force of 5 newtons, and continue to pull with this force over a distance of 60 meters. How much work is done? The work done is $5 \times 60 = 300$ newton-meters. Note that the fact that the wagon has a mass of 30 kg does not enter into the *work done on the wagon*. The work done *on the body* is the applied force times the displacement or distance.

† Question: Is all the energy of the "rolling" ball in its forward velocity? What have we glossed over in this statement?

That is,

$$\text{Work} = F \times S \quad (5\text{-}8)$$

This is the total work done to move the wagon. Some of this work becomes lost in friction (against the frictional force F_f), but the difference, $F - F_f$, goes into accelerating the wagon, which begins to move according to the relation

$$F - F_f = ma \quad (5\text{-}9)$$

The actual work that goes into kinetic (moving) energy of the wagon becomes†

$$W = (F - F_f)S = (ma)S$$
$$= ma(\tfrac{1}{2}at^2) \quad (5\text{-}10)$$

or

$$\text{Kinetic energy} = m\frac{v^2}{2} \quad (5\text{-}11)$$

The last two steps result from substituting $S = \tfrac{1}{2}at^2$ and $a = (v - v_0)/(t - t_0)$ (where v_0 is taken to be zero at time $t_0 = 0$; see Chapter 4) into Eq. (5-9). Notice that the kinetic energy, Eq. (5-11), now contains the mass of the vehicle.

Just as momentum is conserved, so the total energy of an isolated system is conserved.‡ When two particles collide, energy may be exchanged or become translated into other forms of energy (such as heat), but the total remains the same.

We now can go back to Fig. 5.9 and the collision of a ball with a larger ball. Equation (5-6) is based on the principle of the conservation of momentum. Applying the principle of conservation of energy, we

† A more "elegant" approach, using the methods of the calculus, is applied to these discussions in Sec. 5-6.

‡ It should be noted that the system may possess other forms of energy besides kinetic energy (for example, potential energy) and that it is possible to transform the various forms from one to another. It is the total energy that is conserved. In our special case, we have only kinetic energy.

obtain

$$\tfrac{1}{2}mv_1^2 = \tfrac{1}{2}mv_2^2 + \tfrac{1}{2}MV_2^2 \quad (5\text{-}12)$$

Combining Eqs. (5-6) and (5-12) gives

$$v_2 = \frac{m-M}{m+M} v_1 \quad (5\text{-}13)$$

or, since M is larger than m,

$$v_2 = -\frac{(M-m)}{(M+m)} v_1 \quad (5\text{-}14)$$

and

$$V_2 = \frac{2m}{m+M} v_1 \quad (5\text{-}15)$$

Equation (5-14) says that the smaller ball will bounce back with a reduced velocity, since $-(M-m)$ is negative and also smaller than $(M+m)$. When M is very large, say when a ball bounces against a wall, Eq. (5-14) indicates that $v_2 = v_1$. That is, the ball bounces back with the same speed (but opposite direction) it had at impact. When the mass M of the stationary ball is less than m, both will move forward in the direction of the oncoming ball because $-(M-m)$ becomes positive. What happens when the two balls are of identical mass? From Eq. (5-13) when $M = m$, $M - m = 0$, and v_2 becomes zero, while from Eq. (5-15), $V_2 = v_1$. This means simply that when the two balls are identical, the ball that is struck takes on the full velocity, momentum, and kinetic energy of the initially moving ball, while the initially moving ball comes to a complete stop. This is exactly the phenomenon that is observed with billiard balls (see Fig. 5-10). (We should perhaps note that a billiard ball has rotational energy as well as kinetic energy due to its linear velocity, but this does not change the argument in substance.)

We shall see in later sections that collision phenomena, and exchanges of

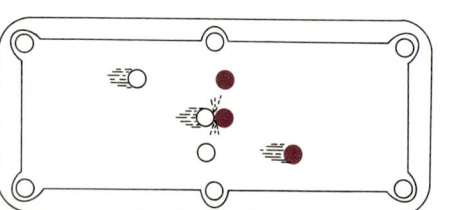

Fig. 5.10. *Conservation of momentum. A sketch of two balls of equal mass in three different positions. After collision the white ball comes to rest and the black one moves off with a velocity equal to the original velocity of the white one.*

energy among particles through collisions, are very common in nature. The characteristics of all gases and liquids, of chemical processes, and of nearly all biological processes are determined to a large degree by the collisions of their molecules with one another.

Let us note the consequences of the conservation principles in connection with an extreme situation, namely, hurling a tennis ball against a stone wall.† It is obvious that the wall remains unmoved. How are momentum and energy conserved? Figure 5-11 illustrates the situation. The ball reaches the wall with a momentum of mv_1 and returns with identical mass and velocity except that the direction is reversed. That is,

$$mv_1 = -mv_2 \quad \text{and} \quad v_2 = -v_1 \quad (5\text{-}16)\ddagger$$

The ball has the same amount of momentum following the collision as that existing before the collision, but with reversed direction. The net change in momentum of the ball is

$$\Delta(mv) = mv_1 - mv_2 = -2mv_1 \quad (5\text{-}17)$$

This is the situation with the ball bounc-

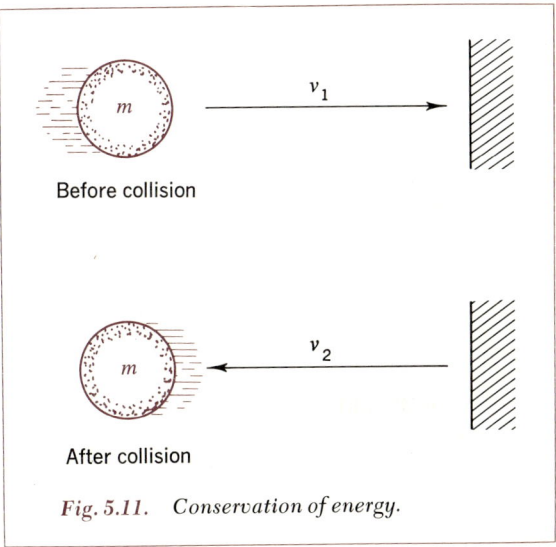

Before collision

After collision

Fig. 5.11. Conservation of energy.

---

† A collision in which the kinetic energy is conserved is said to be *elastic*. If all or some of the kinetic energy is lost, the collision is said to be *inelastic* (as, for example, throwing a lump of putty at a wall and having it stick). Occasionally the system can gain kinetic energy, as in the case where an electron collides with an excited atom, going off with more kinetic energy than it had before (see Chapter 18). Such collisions are termed *superelastic*. We shall deal only with elastic collisions at this time.

‡ Here, again, as for Eq. (5-1), we call attention to the fact that the velocities v_1 and v_2 are vector quantities and hence they are symbolized as $\mathbf{v}_1$ and $\mathbf{v}_2$. Since the product of a scalar such as m and a vector is also a vector, the momentum mv is also a vector and so it can be symbolized $\mathbf{mv}$ by the boldface type to emphasize its vector characteristic.

ing at right angles to the wall. In the process, there is actually a change in momentum of the ball, amounting to $-2mv_1$ because (mv_1) becomes $(-mv_1)$. This change must be supplied by the wall, since the total momentum of the system (ball and wall) must be conserved.† The total kinetic energy also remains the same. One can visualize the kinetic energy of the ball as becoming converted to potential energy in the compression of the ball and wall materials, ready to spring back and transform again to kinetic energy.

5.6 Applying the method of the calculus‡

In the preceding discussion we restricted ourselves to considering only forces that remain constant. This may not always be the case. Consider, for example, that we pull the wagon in Fig. 5.5 with a force of 5 newtons for the first 30 meters, and then abruptly double the force to 10 newtons for 30 additional meters. The work done on the wagon will now be $5 \times 30 = 150$ newton-meters in the first 30 meters plus $10 \times 30 = 300$ newton-meters in the final 30 meters, for a total of 450 newton-meters during the entire trip. The work done over the entire trip is equal to the sum of the work done during each subinterval. We can extend this idea to a force that is continuously varying. Let us divide the total distance traveled into a large number (N) of intervals, each of length Δs (in our example we might

---

† Since the wall has such a large mass compared with the mass of the ball, its momentum after the collision, although twice that of the ball, is accounted for by the product of its large mass and an imperceptibly small velocity.

‡ Note again that the colored pages represent optional study material.

choose $N = 30$, then $\Delta s = 1$ m; or when $N = 3000$, $\Delta s = 0.01$ m $= 1$ cm, and so forth). For each of these small intervals we can determine the average value of the force $\bar{F}(s)$ acting during that interval Δs. By summing over all N intervals the product of $\bar{F}(s) \times \Delta s$, we find that the total work done is

$$W = \sum_{\text{all } N} \bar{F}(s) \, \Delta s$$

where the symbol Σ (Greek sigma) stands for "sum of." If we now take larger and larger values of N (and therefore smaller and smaller values of Δs), the average value of the force acting during the interval becomes closer and closer to the actual force at, say, the beginning of the interval. By letting N become infinite, we get

$$\lim_{N \to \infty} \sum \bar{F}(s) \, \Delta s$$

or

$$\Delta s \to 0$$

In the notation of the calculus, this may be written

$$W = \int_{s_1}^{s_2} F(s) \, ds \qquad (5\text{-}18)$$

where the symbol $\int_{s_1}^{s_2}$ is called the definite integral over the path starting at distance s_1 and ending at distance s_2. Equation (5-18) is a valid equation for the work, even if our force is continuously varying.

It should be further noted that if F is constant, then all the $F(s)$ are the same and may be factored out of the sum or integral:

$$W = \int_{s_1}^{s_2} F(s) \, ds = F \int_{s_1}^{s_2} ds$$

But the integral $\int_{s_1}^{s_2} ds$ simply represents

the sum of all the distance intervals Δs between our starting point s_1 and end point s_2; thus the total distance traveled is

$$\int_{s_1}^{s_2} ds = S$$

and then

$$W = F \times S$$

which is Eq. (5-8)!

The reader can refer to mathematics texts for more details concerning the integral and integration (summing up) concepts and processes. By applying these concepts to derive the relations between work W and kinetic energy K for the general case of a continuously varying force, Eq. (5-5) becomes

$$F = m \frac{\Delta v}{\Delta t}$$

and our sum becomes

$$W = \Sigma F \times \Delta s$$
$$= \Sigma m \frac{\Delta v}{\Delta t} \times \Delta s$$
$$= m\Sigma \, \Delta v \times \frac{\Delta s}{\Delta t}$$

where, by Eq. 4-7, $\Delta s/\Delta t = v$.

In the notation of the calculus, these become

$$W = \int_{s_1}^{s_2} F(s) \, ds$$
$$= \int_{s_1}^{s_2} m \frac{dv}{dt} \, ds$$
$$= m \int_{v_1}^{v_2} dv \cdot \frac{ds}{st}$$
$$= m \int_{v_1}^{v_2} v \, dv$$

where v_1 and v_2 are the velocities at s_1 and s_2, respectively. Evaluating this integral gives

$$W = \tfrac{1}{2}mv_2^2 - \tfrac{1}{2}mv_1^2$$
$$= K_2 - K_1$$

or the *work done is equal to the change in kinetic energy.*

5.7 Rotational motion

Thus far we have only discussed motion in a straight line. Motions in a plane or in three dimensions are mathematically more complex to treat, but do not present new phenomena. There is, however, another class of motion, namely, rotational or circular motion, which is both common and important. Many machines are designed to include parts in rotary motion. The planets and satellites move in a rotational pattern under the influence of gravitational forces.

Consider a mass m attached to one end of a string and whirled around above our head, the other end of the string being held fast by our hand. What observations can we make of this motion? We can keep the rotation uniform, that is, with such periodicity that the same number of seconds is required to complete each rotation. Call this period T seconds. This is the time required to cover one turn of 360 degrees of a circle, as shown in the plane view of Fig. 5.12. We could express the period also in terms of the frequency f or the number of times that the mass rotates in each second or in each minute. That is, the frequency is the reciprocal of the period T. If the frequency is 2000 rpm, the period (time for one revolution) is simply $1/2000 = 0.0005$ min.

The outward direction along the string is called the *radius vector*, which sweeps out a certain angle of turn (or area) each second as it rotates. We know that a circle has 360 deg of angle. The circumference

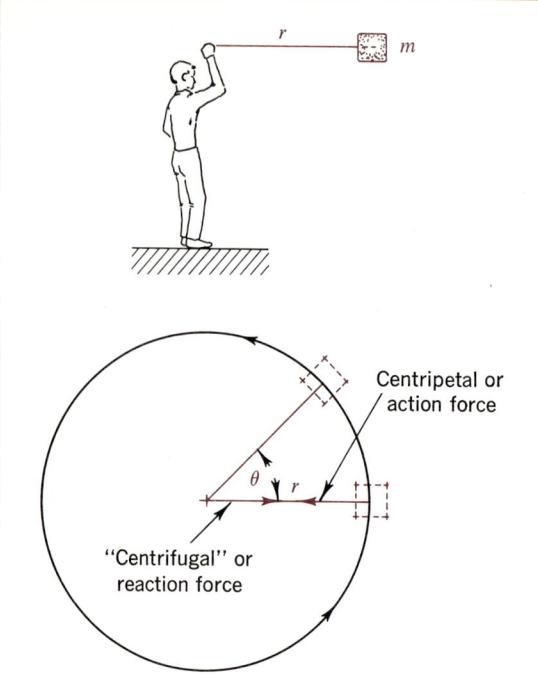

Fig. 5.12. *Rotational motion requires a central accelerating (centripetal) force.*

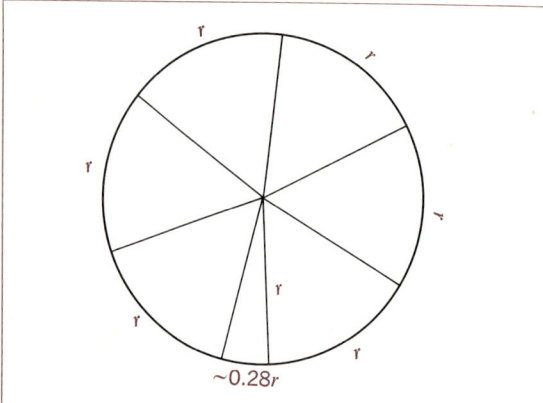

Fig. 5.13. *A circle divided into radians. There are 2π (or about 6.28) radians in a circle of 360 degrees.*

of a circle has a length of $2\pi r$, where r is the radius. It is very useful to divide the 360 deg by 2π, giving 57.3 deg, which is called a *radian* (rad). The radian is therefore the angle that subtends a portion of the circumference of the circle equal to the length of the radius. Expressing rotation in terms of the radian has some advantages of simplicity. A right angle, or 90 deg, equals $\pi/2$ rad. Figure 5.13 illustrates the division of the 360 deg of a circle into 2π radians.

We need a term to express angular velocity and which will correspond to linear velocity v. Angular velocity, usually expressed by the Greek letter ω,† is frequently expressed in radians per unit time. Since there are 2π rad in each circle, a frequency of rotation, f, would represent an angular velocity ω of

$$\omega = 2\pi f = 2\pi \frac{1}{T} \qquad (5\text{-}19)$$

expressed as radians per unit time.

The preceding terms represent the rate of rotation during a single turn. How do we add up the repeated rotations to determine how far the mass m travels in a certain period of time? We can start at some arbitrary time and add up the number of radians traversed during t. In that time, t, there will be t/T complete revolutions, and $2\pi t/T$ rad of rotation. What does this mean in travel distance to the mass at the end of the string? We know that for each rotation, the mass covers the length of the circumference (which is $2\pi r$); therefore, in the time t, the mass travels a

---

† ω (omega) like v, is a vector quantity. The direction of ω is chosen to lie along the axis of rotation, that is, perpendicular to the plane in which the mass travels and therefore perpendicular to both the radius and the velocity. It is defined by convention as having the direction in which a right-handed screw travels when rotated in the same manner.

distance $2\pi rt/T$.† The velocity (assumed constant) of the mass is then given by the distance traveled divided by the time elapsed, or

$$v = \frac{2\pi r}{T} \qquad (5\text{-}20)$$

or

$$v = r\omega \qquad (5\text{-}21)$$

[from Eq. (5-19), which holds only when ω is measured in radians].

We see that for any angular velocity ω, the linear velocity, v, of a mass at the end of a string increases as the length of the string increases.

The foregoing relations define circular motion. As we whirl the string and its mass above our heads, we feel a stronger

---

† Note that the longer the length of radius r, the greater the distance traveled by the mass. Also the shorter the period of a single rotation T, the greater the distance traveled by the mass.

and stronger pull of the string as we increase the frequency of rotation. The mass m "wants" to pull away, and the force with which it pulls increases with angular velocity (or frequency of rotation per unit time). If the string should break, the mass would fly off at a tangent to the circle as illustrated in Fig. 5.14. The mass goes in that direction simply because that is the direction of its motion when the string breaks and it "wants" to continue in that direction, as stated by the first law of Newton. This means that in each succeeding instant of time, the force transmitted by the string pulls on the mass and *changes* the direction of its velocity. Since velocity is a vector (that is, it consists of a direction as well as a magnitude), the continuous change in direction represents a continuous change in velocity. But this is, by definition, an acceleration a. The mass m therefore experiences a constant acceleration under the influence of an accelerating force as it whirls around.

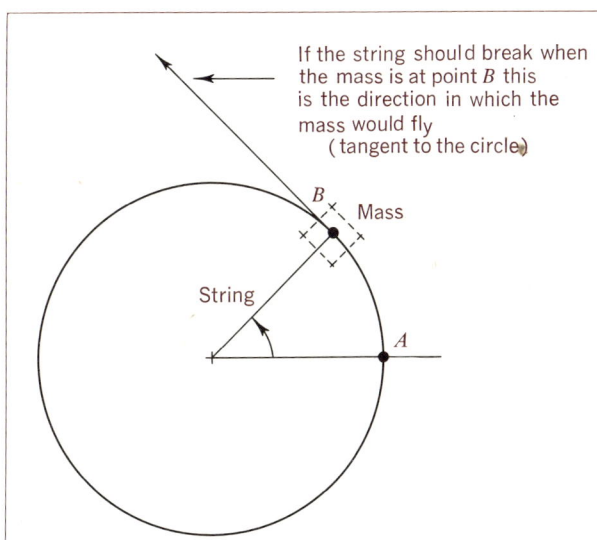

Fig. 5.14. If the string breaks, the moving mass will obey Newton's first law of motion.

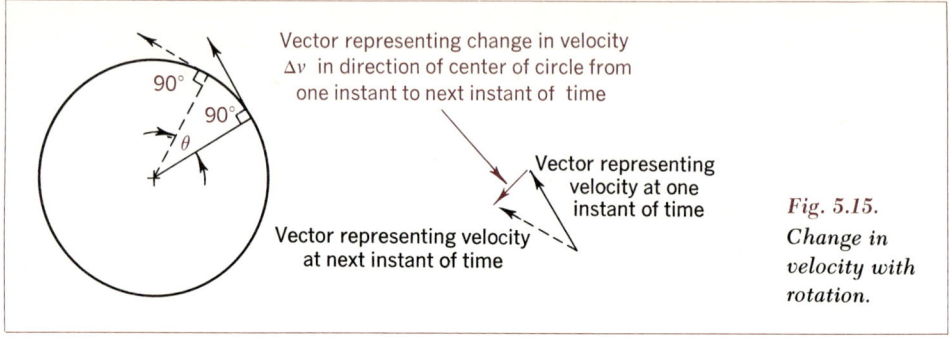

Fig. 5.15. Change in velocity with rotation.

In what direction is the acceleration? The pull of the string on the mass is obviously along the direction of the string, toward the center of the circle. From Newton's second law, the acceleration must be in the direction of the pulling force, that is, toward the center of the circle. Figure 5.15 illustrates the vector diagram for the change in velocity.

In other words, in circular motion there is a continual acceleration toward the center of the rotation. The force along the radius is $F_r = ma = m(\Delta v/\Delta t)$. This inward force is called *centripetal* force, and its resulting acceleration is called *centripetal acceleration*. By further analysis† of the circular diagram and the displacement vectors one can show that this force is equal to

$$F_r = m \frac{v^2}{r} \qquad (5\text{-}22)$$

which, on substituting for r from Eq. (5-21), becomes

$$F_r = mr\omega^2 \qquad (5\text{-}23)$$

† See one of the texts listed in References at the end of this chapter, or see the experiment on centripetal force in the Laboratory and Mathematics Supplement to this volume.

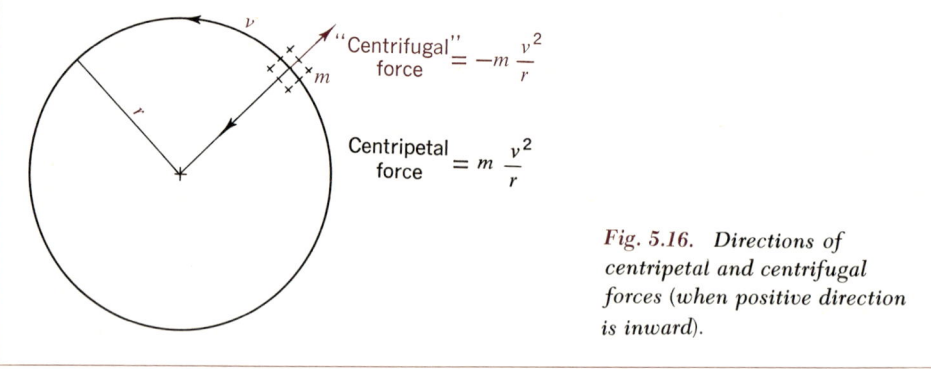

Fig. 5.16. Directions of centripetal and centrifugal forces (when positive direction is inward).

This inward centripetal force is, of course, equal and opposite to an outward force that the string transmits to the hand. This *outward* force is sometimes referred to as *centrifugal* force.† The *centripetal* force is the force that the string exerts on the point where it is attached to the mass m, while the *centrifugal* force at least appears to be a force exerted by the mass on the string. They are equal in value, but opposite in direction. Figure 5.16 illustrates the relative directions.

We note one important consequence of this acceleration toward the center. Along with acceleration, we can assume that there is a displacement toward the center, a kind of "falling" toward the center. We have made use of this already in our discussion of the orbit of the moon under the earth's gravity.

We have found that conservation of momentum and conservation of energy are exceedingly important concepts in connection with linear motion. How are these affected in rotational motion? Indeed they remain just as important in rotational motion, but now we must take into account the fact that the moving mass is at the end of a string.

In linear motion the momentum was given as $p = mv$, and kinetic energy = $\frac{1}{2}mv^2$. The *angular momentum* is defined as equal to the linear momentum times the radius, or

Ang. mom. $= mv \times r = mr^2\omega$ (5-24)‡

The *direction* of this angular momentum (which is also a vector) is at right angles to both radius and velocity, and in the direction of the advance of a right-handed screw, as it is turned in the direction of rotation.† The angular momentum and the energy of a rotating system are conserved in a closed system, as is linear momentum.

Angular momentum is described above for the special case where all of the mass m is concentrated at a distance r from the axis of rotation. Although this applies reasonably well to the case of satellite or planetary motion, it is seldom encountered in other rotating objects. More often we find solid bodies rotating in such a manner that some of the mass is very close to the axis of rotation while other portions are at a considerable distance from it. The amount of energy required to impart a given angular velocity depends on just how the mass of that body is arranged relative to the axis of rotation. The term *moment of inertia*, I, is used to describe this aspect of rotational motion, and it is conveniently defined by reference to Fig. 5.17.

If any irregularly shaped body is to be rotated about an axis at any arbitrarily located position, its moment of inertia around any particular axis of rotation may be found by considering the body to consist of a large number, N, of tiny bits of mass, each at its own distance, r, from the axis of rotation. Then the moment of inertia of each bit is equal to mr^2, and the sum of all such products is given as I, where

$$I = m_1 r_1^2 + m_2 r_2^2 + \ldots m_n r_n^2 \quad (5\text{-}24\text{a})$$

By considering larger and larger numbers, n, we may write

$$I = \lim_{n \to \infty} \sum_{i=1}^{n} m_i r_i^2 \quad (5\text{-}24\text{b})$$

as the moment of inertia of that body when rotated about that axis.

Using the integral calculus, Eq. (5-24b)

---

† Actually, the centrifugal force is a fictitious force because, strictly speaking, the mass at the end of the string "wants" to go at a tangent to the circle, not radially outward.

‡ See Mathematics Supplement for meaning of **v** × **r**.

† It should be noted that the angular momentum vector is parallel to **ω**.

reduces to

$$I = \int r^2 dm \qquad (5\text{-}24c)$$

where the integral is taken over the entire mass of the body. Using this definition, one may determine formulas for computing the moments of inertia of masses M with geometric configurations and rotated about axes as in Fig. 5.18.

For a hoop rotating about an axis through the center and perpendicular to the plane of the hoop Fig. 5.18(a),

$$I = MR^2$$

For a solid cylinder rotating about the cylinder axis, Fig. 5.18(b),

$$I = \tfrac{1}{2}MR^2$$

For a solid sphere rotating about any diameter, Fig. 5.18(c),

$$I = \tfrac{2}{5}MR^2$$

For a thin rod rotating about an axis through its center and perpendicular to its length, Fig. 5.18(d),

$$I = \tfrac{1}{12}ML^2$$

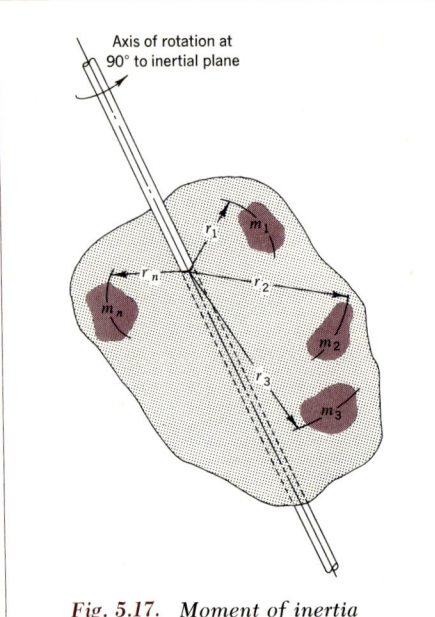

Fig. 5.17. Moment of inertia (Eq. 5-24(a)).

The moment of inertia (or rotational inertia) I bears the same relationship to rotational kinetic energy as does mass to linear kinetic energy. Thus, in the linear case [recall Eq. (5-7)],

$$K = \tfrac{1}{2}mv^2,$$

whereas in the rotational case,

$$K = \tfrac{1}{2}I\omega^2 \qquad (5\text{-}25)$$

Similarly, moment of inertia is related to angular momentum for rotating bodies in the same way that mass is related to linear momentum for bodies with linear motion.

Thus, in the linear case [Eq. (5.5)],

$$p = mv,$$

whereas in the rotational case,

Angular momentum $= mr^2\omega = I\omega$ (5-26)

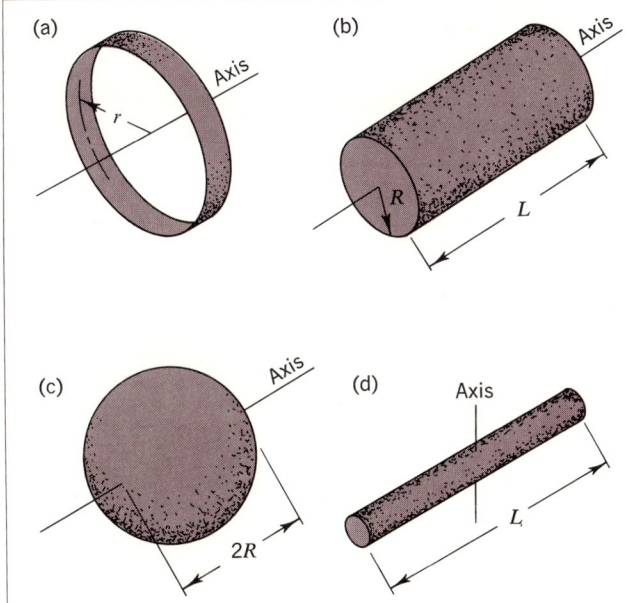

Fig. 5.18. *Determining moments of inertia for various geometric configurations. (a) For a hoop rotating about an axis through the center and perpendicular to the plane of the hoop* $I = MR^2$; *(b) for a solid cylinder rotating about the cylinder axis* $I = (\frac{1}{2})MR^2$; *(c) for a solid sphere rotating about any diameter* $I = (\frac{2}{5})MR^2$; *(d) for a thin rod rotating about an axis through its center and perpendicular to its length* $I = (\frac{1}{12})ML^2$.

An interesting illustration of the principle of conservation of angular momentum is provided by the figure skater of Fig. 5.19. In the first case, the skater is holding his arms well away from his axis of rotation. This gives him a relatively large moment of inertia combined with a modest rotational speed, ω_1. In the second case, as his arms and legs are held close to his body, the value of I is reduced, but since the product $I\omega_1$ must remain constant, one observes an immediate increase in his angular speed, which becomes ω_2 where $\omega_2 > \omega_1$.

The importance of the conservation principles described in this section cannot be overstressed. They have a very significant bearing on a great portion of modern science and particularly in atomic science as we shall see later in Chapters 16 through 20.

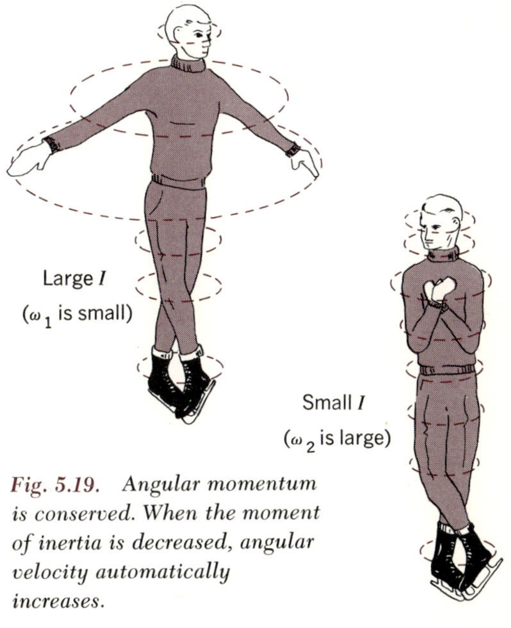

Fig. 5.19. *Angular momentum is conserved. When the moment of inertia is decreased, angular velocity automatically increases.*

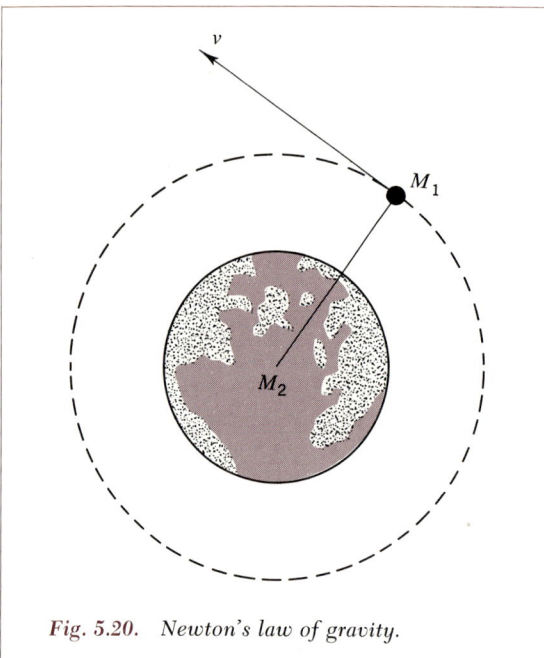

Fig. 5.20. *Newton's law of gravity.*

5.8 Gravitational forces

We return now to an analysis of the forces and motions that are associated with gravity. Newton came to the conclusion that the force that causes an apple to drop from a tree to the ground is related to the forces that hold the moon and the planets to their orbital motion. An apple falling from a tree at the surface of the earth, and which is one earth radius from the center of the earth, experiences an acceleration of 32 ft/sec², or 9.8 m/sec². Since the moon is (roughly) in circular motion around the earth at a distance of 60 times the earth's radius, with a period of 27.3 days, he could calculate the centripetal acceleration of the moon. Newton found that the acceleration of the moon is about 3600 times smaller than the acceleration of the falling apple at the surface of the earth. The square of the distance of the moon from the center of the earth is 3600 times the square of the distance of the apple from the center of the earth. These facts enabled him to conclude that the acceleration due to the earth's gravity at a point in space should vary inversely with the square of the distance of that point from the center of the earth, and therefore the force due to gravity on a body should also vary in inverse proportion to the square of this same distance.

Combining reasoning with a good deal of intuition, Newton proposed the law of universal gravitation. According to this law, F, the force of natural gravitational attraction between two particles of masses, M_1 and M_2, with their centers separated by a distance r, is given by

$$F = G \frac{M_1 M_2}{r^2} \qquad (5\text{-}27)$$

where G is a constant that requires experimental determination. When the masses are expressed in kilograms, the distance in meters, and the force in newtons, G has the value 6.67×10^{-11} newton-m²/kg². Newton estimated the mass of the earth by guessing the average density of the earth and obtained a rough figure for G.

A later experiment in 1797 by Henry Cavendish employed four masses to determine G. His experiment is illustrated by Fig. 5.21. How constant is the value of G? Experiments with various types of material, even including meteoric material, give the same value of G. When the relationship is applied to the orbital motion of the planets, according to Kepler's laws, the same value of G seems to be valid. Indeed, the same value of G seems to hold even in treating the dynamics of galaxies. At least for our purposes, G may therefore be accepted as a *universal* constant for gravitation, although there have been contrary opinions.

According to the *law of universal gravitation*, the force of gravity on a mass m on the surface of the earth is given by

$$F_m = G \frac{M_e}{r_e^2} m \qquad (5\text{-}28)$$

where M_e is the mass of the earth and r_e is the radius of the earth. From Newton's second law, $F_m = ma$, where a is the acceleration resulting from the gravitational force F_m. The acceleration due to gravity is usually designated by the letter g. Therefore it follows that

$$mg = \frac{GM_e}{r_e^2} m \qquad (5\text{-}29)$$

Dividing both sides of the Eq. (5-29) by m, we obtain

$$g = \frac{GM_e}{r_e^2}$$

Since G, M_e, and r_e are all constants, the

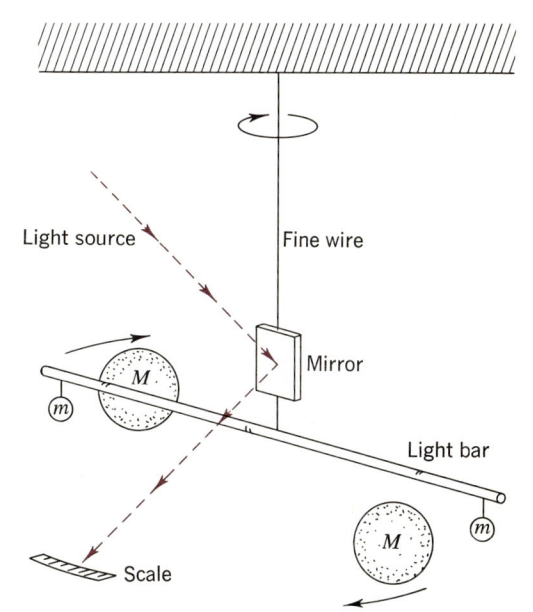

Fig. 5.21. Cavendish experiment for determination of gravity. Two small lead balls, (m), were attached to a light metal rod which in turn was supported by thin metal wire suspension from a rigid support. The apparatus was enclosed in a glass case (not shown in diagram) to eliminate disturbance due to air currents. Two large balls, M, were then brought near the small ones in such a manner that the gravitational attractions of the heavy balls for the small ones would tend to twist the bar in the same direction. The resulting twist on the suspension wire is measured accurately by means of a lamp and scale arrangement.

acceleration g due to gravity on the surface of the earth is also a constant.†

The significance and preciseness of the laws of gravitation become revealed even more vividly as we consider the gravitational effects of the planets on each other and on other large masses in the solar system. Figure 5.22 illustrates the effects of the moon which produce the tides on the earth.

By applying the law of universal gravitation to the motion of the planets, Newton showed that Kepler's three laws were predicted by this theory, on the assumption that only the gravitational attraction of the sun need be considered. The gravitational attraction of Jupiter, Saturn, and even the earth and moon, do cause the orbit of Mars to deviate slightly from the Keplerian ellipse. In fact, each of the planetary and satellite orbits includes some *perturbation* due to the other members of the solar system. For over 200 years the calculations of the perturbed orbits of the various planets and of the moon have posed severe problems in the branch of astronomy known as celestial mechanics, and the interest increases as we draw closer to the day of interplanetary travel.

An impressive result of Newton's laws came 140 years after their formulation. In 1781, William Herschel of Bath, England, a musician and "amateur" astronomer, discovered what he first thought was a comet, but which he later identified as a new planet, Uranus. The new planet was twice as far from the sun as Saturn. Its

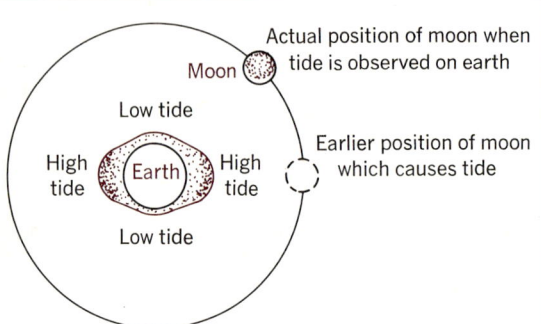

Fig. 5.22. Tides are caused by differences of the moon's attraction. The side nearest the moon experiences a greater pull (as compared with that at the earth's center) so that water pulls up on that side. The side farthest away experiences a smaller pull, so that in effect the earth is pulled away from the water, thus producing another high tide. Because of inertia, tidal friction, and rotational effects the tides are delayed by several hours so that they are not highest opposite the moon. The pull of the sun on the earth is about 175 times greater than that of the moon on the earth, but because the sun is so much farther away, the difference of the pull on the two sides of the earth is much less significant for producing tides.

† This result is valid only if we make the assumption that earth is a perfect sphere of radius r_e. In reality, the earth is not a perfect sphere and therefore g varies slightly from point to point, depending upon the point's distance from the center of the earth. Also, since the earth is spinning, the "centrifugal" acceleration tends to reduce the effective gravitational acceleration, the effect being greatest at the equator and zero at the poles.

period of 8.4 years and elliptic orbit agreed with the laws of Kepler, but careful analysis of the orbit seemed to suggest that there might be some perturbation caused by still another planet. This possibility led a young student at Cambridge University, John C. Adams, to undertake careful mathematical analysis of the orbit. After a few years of work he discovered the direction from which the additional attraction had to come, and wrote to the Royal Observatory at Greenwich to suggest that they search for a new planet in the predicted direction. He failed to arouse their interest. A young Frenchman, Leverrier, had similar interests and got a quicker response from an observatory in Berlin just a few months later. On receiving Leverrier's suggestion, the director of the observatory discovered the planet Neptune in the predicted direction on the very first night of observation in the year 1846!

5.9 Simple harmonic motion

Our studies of cyclic phenomena must now come down from planetary motion to delve into phenomena that are closer to personal experiences. We shall discuss simple harmonic motion (abbreviated SHM) because it is easy to reproduce and to analyze and also because it approximates cyclic changes that are common in natural and man-made systems with which we shall be concerned.

We begin by observing the simple rotating motion of the motor shown in Fig. 5.23, which has an arm and a shiny ball mounted on the motor shaft. When we look at the motor from the right, along the direction of the motor shaft, we see that the ball executes circular motion. But when we view the motor in a direction at right angles to the motor shaft, the ball executes a simple up-and-down motion. We shall see that this up-and-

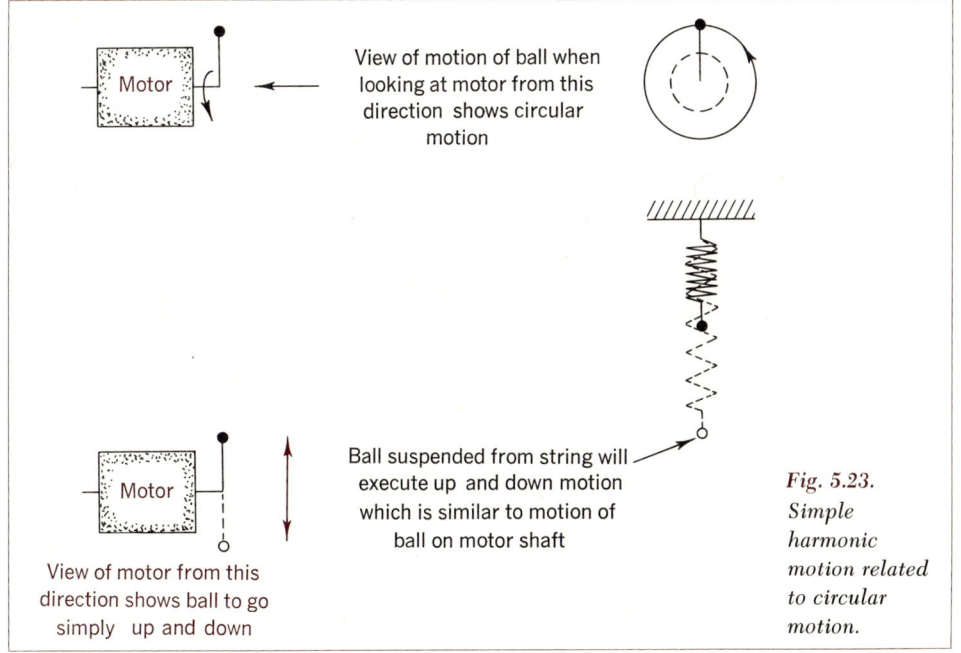

Fig. 5.23. Simple harmonic motion related to circular motion.

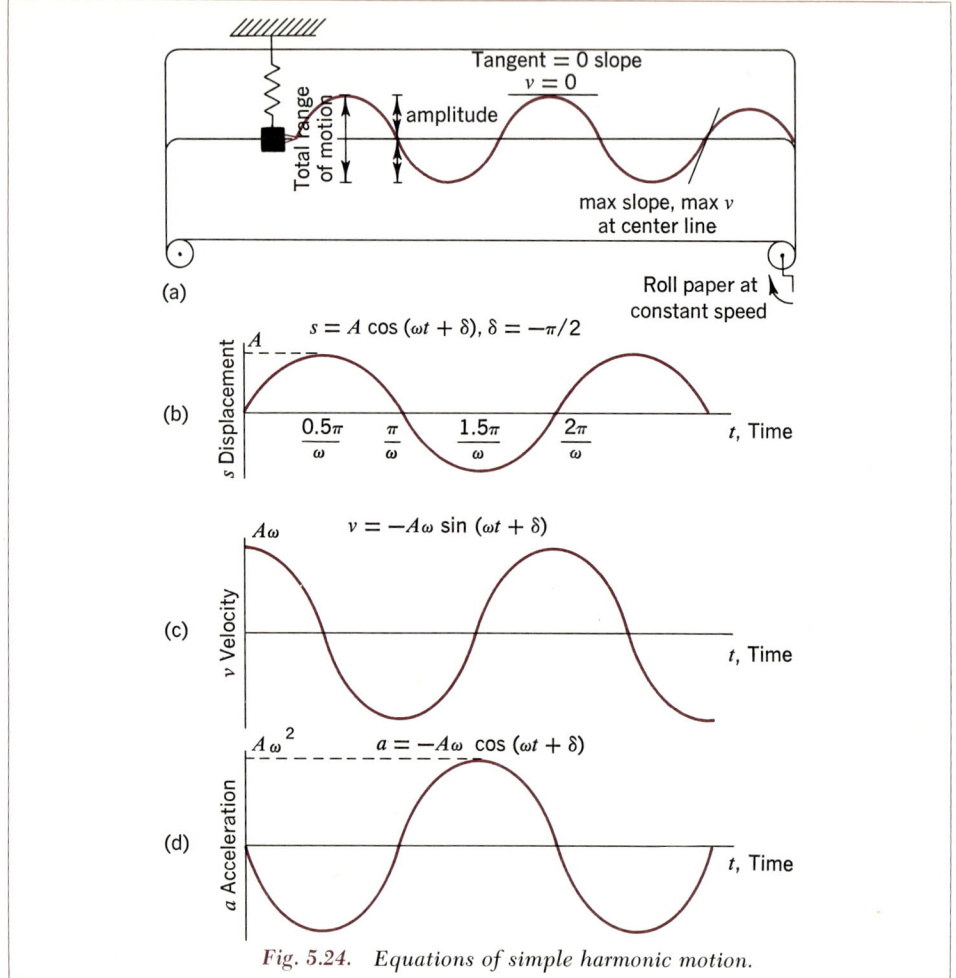

Fig. 5.24. *Equations of simple harmonic motion.*

down motion is simple harmonic motion, and is exactly of the form that an ordinary weight hanging from a spring might execute. Of course, when we view the motor from other angles, the ball will appear to execute elliptic movements.

Several simple observations of the characteristics of SHM can be made by reference to a weight suspended on a spring, as illustrated in Fig. 5.24(a). When the bob is at rest, a pointer on it will be in zero position. When we raise the bob to some deflection A (amplitude) above the zero line and then release it, the bob will drop back and beyond its zero position to a lower position as shown. The total change in position is then 2A. Of course the bob will gradually settle back to the zero position as the energy we gave to the system (by raising the bob) dissipates as air friction and heat.

If we let the pointer act as a pen to

write on a roll of paper, and pull the paper along while the bob is going up and down, the pointer will draw a curve as shown in Fig. 5.24(a). This is the well-known sine curve (or cosine curve) that represents simple harmonic motion.

What observations can we make by watching the bob? Since the bob reverses direction at the upper and lower extremes of the motion, it must be at rest for a very brief instant at each end. At each end of the stroke the bob reverses direction, picks up speed in the new direction to develop maximum velocity as it passes the zero rest point (or center point for the stroke), but immediately on passing this center point the bob slows down again so that it comes to rest for an instant at the opposite end of the stroke.

The graph of the motion, Fig. 5.24(b), represents the displacement s as a function of time. (It is a function of time because we pulled the paper along at a constant rate to show the changes of bob position with time.) We can also obtain the velocity of the bob at each position by drawing tangents to the displacement curve and calculating their slopes, as in Fig. 5.24(c). Since tangents to the curve at the crest of the wave and at the trough of the wave are horizontal, their slopes are zero, which means that the velocities are zero as well. But the tangent is steepest at the center-line position, which means that its slope, and hence the velocity, is a maximum. But the maximum velocity, like the rest period at the ends of the stroke, lasts only for a very brief instant. The velocity is constantly changing, increasing as the bob moves toward the center and decreasing as soon as it passes the center. But if the velocity is changing constantly, there must also be continuous acceleration and deceleration, and we know from Newton's first and second laws that a change in velocity is always the result of a force acting on the body. We shall now analyze these forces.

5.10 Forces involved in simple harmonic motion

What are the forces that bring about simple harmonic motion? Since the motion reverses direction, we must look for forces that also reverse direction and increase and decrease like the motion itself. Let us examine the spring and the bob of mass m that hangs from it (Fig. 5.25). *When the bob is still* (at the "zero" position), the pull of gravity will be equal to mg. The spring must pull up with an equal force for the bob to be at rest, and of course the spring becomes extended in the process. That is, without the weight of the bob, the spring would have a length L_0, but when the bob is added to the spring, it becomes extended to a length $L_0 + \Delta L$. The amount of this extension, ΔL, depends on the stiffness of the spring; a very stiff spring would have to stretch only a very little to develop the added pull equal to the weight of the bob, while a soft spring might have to stretch a great deal farther down.

The general relation between pull (or force) and stretch for a spring is given by

$$P = ks \qquad (5\text{-}30)$$

where P is the pull, k is the stiffness coefficient, and s is the amount of stretch. This relationship is known as Hooke's law. If P is measured in newtons and s in meters, the stiffness coefficient k will have the unit of newtons per meter. The stiffness coefficient is also referred to as the *force constant* of the spring, or spring constant. For our case, this relation is $P = k\,\Delta L$.

If now we disturb this balanced situation by pulling the bob farther down by

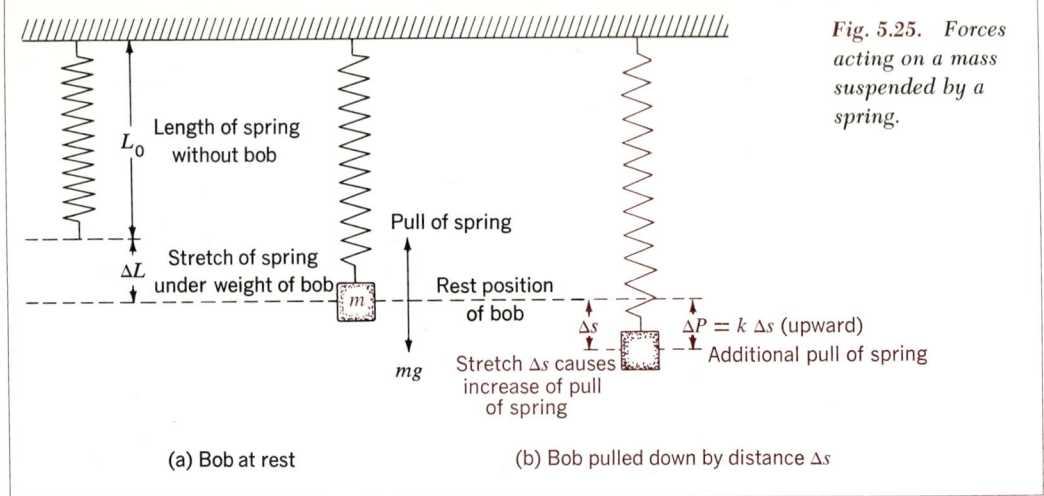

Fig. 5.25. Forces acting on a mass suspended by a spring.

hand, we increase the stretch of the spring and the spring pulls upward with increased force to equal the pull of our hand. Suppose we pull down a distance Δs from its equilibrium length, or $L_0 + \Delta L$. The increase in the spring pull will be given by

$$\Delta P = k\, \Delta s \qquad (5\text{-}31)$$

On the right side of Fig. 5.25 we see that the added stretch Δs will cause the spring to pull upward, to *restore* the original position, with an *additional* pull $\Delta P = k\, \Delta s$. (We must remember that the pull of gravity is still present, balanced by the original force and stretch of the spring.)

The characteristics of **SHM** evolve from this one fact, namely, that *displacement of the body develops a force that is proportional to the displacement and which is in a direction to restore the body to zero-displacement position.* Since the displacement is greatest at the two ends of the motion, the restoring forces are greatest at the two ends also. As the body approaches its maximum displacement position, this force first exerts maximum effort for decelerating the body to a halt, and then exerts maximum effort to accelerate the body in the opposite direction.

The reader may well ask at this point: "I can understand that pulling the bob *down* will increase the length of spring and therefore the restoring force. But what develops a similar situation of maximum force in the reverse direction, when the spring is at its shortest length and the bob is at its highest level?"

To answer this question we must note that two forces act in opposition on the bob, namely, the pull of the spring, ks, and the constant pull of gravity, mg. The two forces are equal at the initial zero point. Below this zero point the spring pull is greater, while at all positions of the bob above zero the gravity pull is greater. The *change* in the forces as the bob goes from the lowest position to the highest level is dictated by the characteristics of the spring only. If the spring is linear in its characteristics, a displacement of the bob of 1 cm below or 1 cm above the zero line would give exactly the same *change* in its pull, and always in a direction to restore the bob to the zero or intermediate position.

We shall take an actual example to illustrate the principles just discussed. When a force of 6 newtons is applied to a long, coiled spring,† the spring is seen to extend 15 cm (=0.15 m). The spring constant k [Eq. (5-30)] is then given as

$$k = \frac{\Delta P}{\Delta L} = \frac{6}{0.15} = 40 \text{ newtons/m}$$

If we were to hang a 2-kg mass from this spring, the pull of gravity downward on this mass would be $F = mg = 2 \times 9.8 = 19.6$ newtons, and this force would extend the spring by

$$\Delta L = \frac{\Delta P}{k} = \frac{19.6}{40} = 0.49 \text{ m}$$

The position at which the restoring force of the spring is equal to the weight of the mass may be referred to as the rest point, zero point, or balance point. If now we pull down this mass by hand to increase the length of the spring by another 0.10 m, the upward pull of the spring increases by $\Delta P = k \, \Delta s$, or $\Delta P = 40 \times 0.10 = 4$ newtons. This added force of 4 newtons gives the 2-kg mass an upward acceleration of 2 m/sec² (from $F = ma$, $a = F/m$; in our case, $a = 4$ newtons/2 kg $= 2$ m/sec²).

As the mass begins to move back toward the zero position with this acceleration of 2 m/sec², several things happen: As the displacement decreases, the restoring force and the acceleration also decrease. Meanwhile the velocity increases to a maximum value at the zero point. But now deceleration begins as soon as the mass moves above the zero point, and the gravity pull downward becomes

---

† The mass of the spring itself is not taken into consideration in these discussions.

greater than the spring pull upward. The deceleration continues, but the mass does not come to a stop until it has reached a point about 0.10 m above the zero point. At that instant the spring pull is *decreased* by 4 newtons (just as it *increased* by 4 newtons when we pulled it down 0.10 m), and this same difference of force, $\Delta P = 4$ newtons, starts the motion downward.

5.11 Equations for simple harmonic motion

The graphs of displacement, velocity, and acceleration of a particle executing SHM as functions of time are shown in Fig. 5.24(b), (c), and (d). All have the general shape of sine and cosine functions.

A usual way of representing displacement s of a particle in SHM as a function of time t is

$$s = A \cos(\omega t + \delta) \quad (5\text{-}32)$$

where A is the amplitude, ω is the angular frequency, and δ is the phase constant. The term $(\omega t + \delta)$ is usually referred to as the *phase*. The magnitude of the maximum possible displacement from the equilibrium position of the particle is the amplitude of the motion; ω, being the angular frequency, $2\pi/\omega$ gives the period of the motion T. For a particle of mass m executing simple harmonic motion with a force constant k, ω can be shown to be equal to $\sqrt{k/m}$. Therefore

$$T = 2\pi \sqrt{\frac{m}{k}} \quad (5\text{-}33)$$

From Eq. 5-33 we see that the period of simple harmonic motion is dependent only on the mass of the particle and the force constant of the spring, and is independent of the amplitude of the motion. Since frequency of the motion, f, is the reciprocal of its period, it is given by

$$f = \frac{1}{2\pi}\sqrt{\frac{k}{m}} \qquad (5\text{-}34)$$

For a fixed mass m, the greater the value of k, the greater will be its frequency and the smaller will be its period.†

Two SHMs may have the same amplitude and angular frequency, but may differ in phase. For example, if $\delta = 0$, then $s = A \cos(\omega t)$ and the displacement at $t = 0$ is a maximum. If $\delta = \pi/2$, then $s = -A \sin \omega t$, and the displacement at $t = 0$ is zero. Other phase constants lead to other initial displacements. If we were to represent the displacement in SHM shown in Fig. 5.24 by an equation of the form of (5.32), we would see that $\delta = -\pi/2$ in that case. The equations for the velocity v and the acceleration a of a particle in SHM, whose displacement is expressed by the Eq. (5-32), are obtained by use of the calculus in the following manner:

$$v = \frac{ds}{dt}$$

and

$$a = \frac{dv}{dt} = \frac{d}{dt}\left(\frac{ds}{dt}\right) = \frac{d^2s}{dt^2}$$

From the rules of the calculus, the derivatives of the functions $\sin(\omega t + \delta)$ and $\cos(\omega t + \delta)$ are

$$\frac{d}{dt}\sin(\omega t + \delta) = \omega \cos(\omega t + \delta)$$

$$\frac{d}{dt}\cos(\omega t + \delta) = -\omega \sin(\omega t + \delta)$$

Thus

$$v = \frac{ds}{dt} = \frac{d}{dt}[A \cos(\omega t + \delta)]$$

$$= -A\omega \sin(\omega t + \delta) \qquad (5\text{-}35)$$

† $\omega = 2\pi f = \sqrt{k/m}$.

$$a = \frac{dv}{dt} = \frac{d^2s}{dt^2}$$

$$= \frac{dt}{dt}[-A\omega \sin(\omega t + \delta)]$$

$$= -A\omega^2 \cos(\omega t + \delta) \qquad (5\text{-}36)$$

which are the formulas given in Fig. 5.24(c) and (d).

Let us analyze the motion of the spring given in our example in the notation of the calculus. The *total force* acting on the bob at any moment is (neglecting gravity, which acts in the same direction with a constant value throughout)

$$F = -ks$$

Here, $s = 0$ at the unstretched spring position, and is assumed to be negative when the spring is compressed and positive when it is extended. By Newton's second law,

$$F = ma = m\frac{d^2s}{dt^2}$$

Thus

$$-ks = m\frac{d^2s}{dt^2}$$

or

$$\frac{d^2s}{dt^2} + \frac{k}{m}s = 0$$

By Eq. (5-34),

$$\omega = 2\pi f = \sqrt{\frac{k}{m}}$$

so

$$\frac{d^2s}{dt^2} + \omega^2 s = 0 \qquad (5\text{-}37)$$

Equation (5-37) is referred to as the *differential equation* of simple harmonic motion. To show that $s = A \cos(\omega t + \delta)$ is indeed a solution to this equation, substitute for s in the differential equation:

$$\frac{d^2s}{dt^2} = -A\omega^2 \cos(\omega t + \delta)$$

$$\frac{d^2s}{dt^2} + \omega^2 s = -A\omega^2 \cos(\omega t + \delta)$$
$$+ A\omega^2 \cos(\omega t + \delta)$$
$$= 0$$

5.12 Energy in simple harmonic motion

The oscillating mass has a kinetic energy given by $K = \frac{1}{2}mv^2$ [Eq. (5-11)]. The kinetic energy has its maximum value (of $K = \frac{1}{2}kA^2$) when the mass is passing through the intermediate point of the motion,† and a value of zero at the extremes of the motion when the velocity is zero.

We learned earlier that energy is conserved in a system. Where, then, does the energy go when it is no longer kinetic energy? The energy is transformed from kinetic energy to *potential energy* $(u = \frac{1}{2}ks^2)$.‡ At the lower end of the stroke the spring is extended. Work was needed to extend a spring, and the spring is at this instant ready to accelerate the mass to transform its potential energy into kinetic energy of the oscillating bob. At the upper end of the stroke the bob is raised a distance A above its point of balance of forces, and has stored in its height a potential energy $(u = \frac{1}{2}kA^2)$ equivalent to the maximum kinetic energy. The total energy E of the system is given by the sum

---

† At the zero point, from Eqs. (5-32) and (5-35), the sin $(\omega t + \delta)$ equals sin $(\pi/2) = 1$, and $v = -A\omega$. From Eq. (5-32), $\omega = \sqrt{k/m}$. Therefore $K = \frac{1}{2}mv^2 = (\frac{1}{2}mA^2)(k/m) = \frac{1}{2}kA^2$.

‡ When the force is proportional to the displacement of $-ks$, as in SHM, increasing the displacement against a constantly increasing force builds up the work done to the potential energy $u = \frac{1}{2}ks^2$ (from integral calculus).

---

of the potential and kinetic energies at any given instant. For example, at the instant when $K = 0$ and $u = \frac{1}{2}kA^2$, then

$$E = K + u = \frac{1}{2}kA^2 \tag{5-38}$$

We see that the energy of a body in simple harmonic motion is determined by the spring constant k and the square of the amplitude of oscillation, A^2.

Example We shall discuss one more example of simple harmonic motion, namely, the simple pendulum made up of a bob of mass m and suspended by a string. The bob will normally hang vertically and at rest. When pulled to one side and released, the pendulum will execute oscillatory motion. For small amplitude oscillations the motion of the simple pendulum is simple harmonic (see Fig. 5.26).

T must equal $mg \cos \theta$ in magnitude, to keep the string taut. The restoring force $mg \sin \theta$ acts on m, tending to return it to the equilibrium position. Therefore the restoring force F is equal to $-mg \sin \theta$. In case the angle θ is small, $\sin \theta$ is approximately equal to θ in radians, and we can write the equations of restoring force as

$$F = -mg\theta$$

Since $\theta = s/L$,

$$F = -mg\frac{s}{L}$$

The force is therefore directly proportional to the displacement and acts in a direction to restore the bob to its equilibrium position. Thus, for small amplitudes, the motion is simple harmonic. Comparing $F = -mg(s/L)$ with the restoring force for the spring, $F = -ks$, we find that mg/L for the pendulum has the same physical significance as k does for the spring. Thus the period of the pendulum

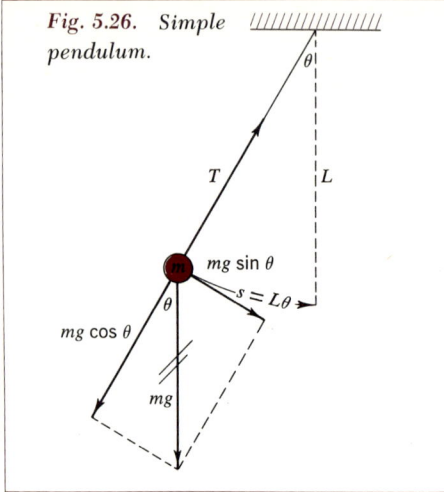

Fig. 5.26. Simple pendulum.

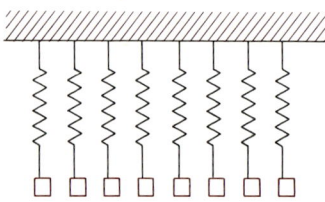

(a) Identical springs and bobs, independent of each other, can independently execute SHM

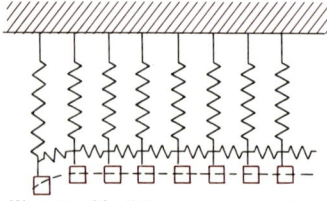

(b) We now add a light spring connecting adjacent bobs, and start the end bob in a vertical motion by pulling it downward

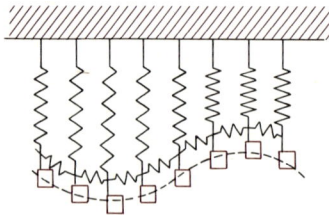

(c) Presently all bobs will be in motion, and will define a traveling wave

[from Eq. (5-33)] is

$$T = 2\pi \sqrt{\frac{m}{k}}$$

$$= 2\pi \sqrt{\frac{m}{mg/L}}$$

$$= 2\pi \sqrt{\frac{L}{g}} \qquad (5\text{-}39)$$

which shows the period to be determined by the length of the string, L, and not by the mass of the bob.†

5.13 Some properties of waves

Figure 5.24 illustrates how a particle that is in simple harmonic motion may be used to draw a sine or cosine wave.‡ Suppose we take a large number of such springs and bobs, as shown in Fig. 5.27(a). Each will execute simple harmonic motion without influencing the others.

But now let us connect each bob with its neighbor by means of light coiled springs [Fig. 5.27(b)]. Let us also pull down the end bob, and release it. The bob will execute simple harmonic motion, and in the process it will pull the adjacent bob along. The adjacent bob will follow behind somewhat, but will cause its next neighbor to go into motion. Before long, all the bobs will be going up and down, each a little behind or ahead of its neighbor. The net result is that the

---

† This relation was first noted by Galileo, supposedly after he observed the oscillation of a chandelier made by Benvenuto Cellini that hung in the cathedral at Pisa. Christian Huygens derived this relation mathematically in his consideration of the effects of gravitation. This relation applies to the pendulum clock.

‡ We note that when their amplitude and angular velocity are the same, a sine wave and a cosine wave are identical in shape except that one is a quarter of a wavelength (one-fourth of the period) displaced from the other.

Fig. 5.27. (Left) Wave propagation in a series of connected springs.

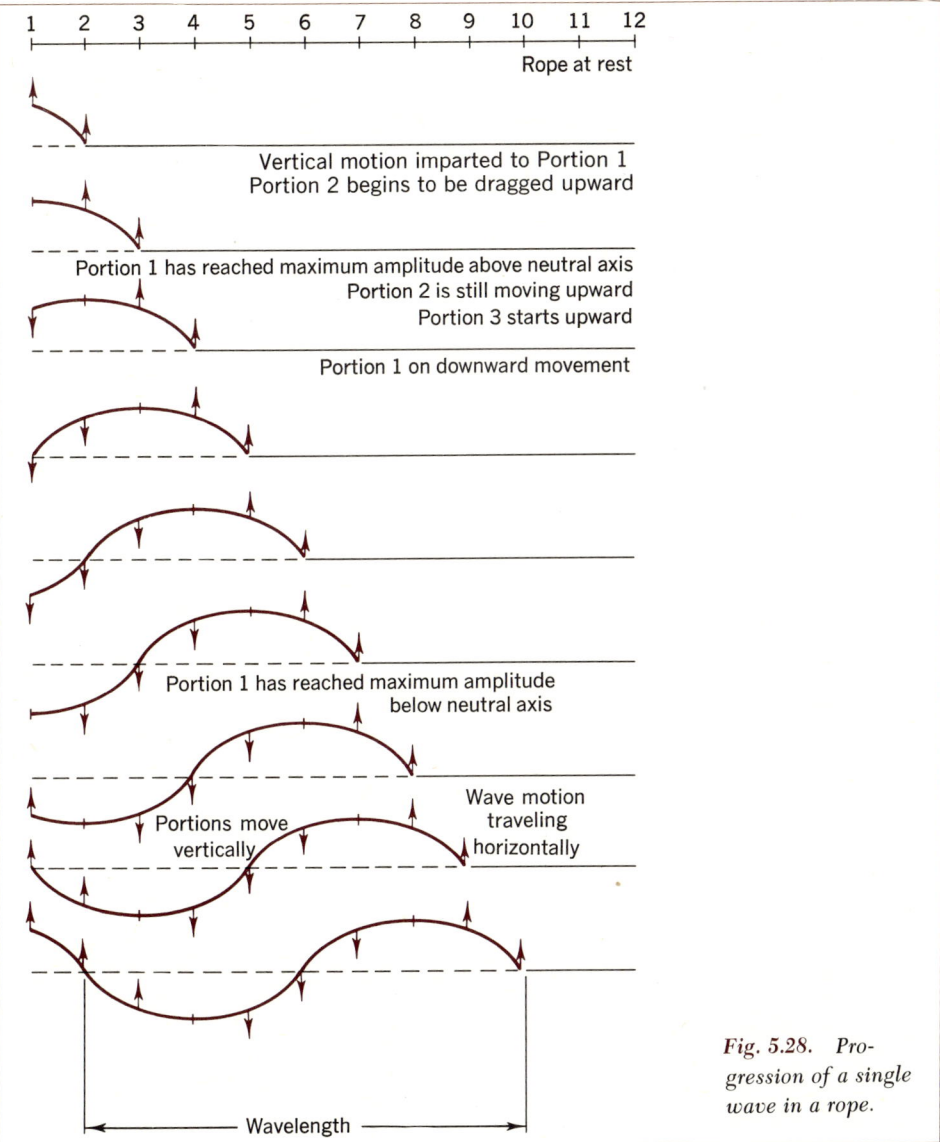

Fig. 5.28. Progression of a single wave in a rope.

bobs will, by their relative position, define a wave as illustrated in Fig. 5.27(c). The wave will appear to be traveling along (moves transverely), although each of the bobs executes only up-and-down motion (moves vertically).

The behavior is illustrated in a more realistic manner by Fig. 5.28, where the bobs become now only the weight of the cord. Motion upward imparted to the end of the cord drags along the adjacent portion and develops a traveling wave as shown.

These examples then illustrate the motion that is characteristic of *waves:* A disturbance becomes transmitted along

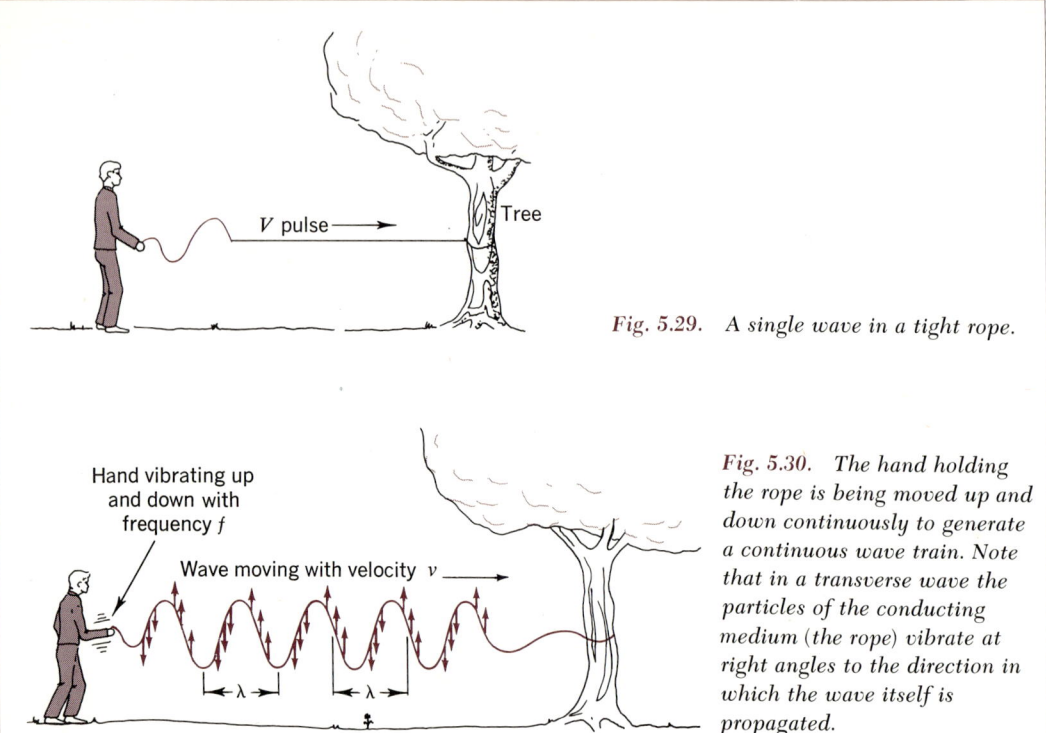

Fig. 5.29. *A single wave in a tight rope.*

Fig. 5.30. *The hand holding the rope is being moved up and down continuously to generate a continuous wave train. Note that in a transverse wave the particles of the conducting medium (the rope) vibrate at right angles to the direction in which the wave itself is propagated.*

the cord while the portions of string, or the bobs and springs, execute only vertical or transverse motion. A single pulse can be transmitted along a tight rope by giving the end of the rope a single, sharp, up-and-down motion (Fig. 5.29), while a continuous up-and-down motion will establish waves along the whole length (Fig. 5.30).

If the person in Fig. 5.30 moves the end of the rope with a frequency of, say, $f = 5$ times each second, the waves that pass any particular point along the length of the rope will number 5 per second. The time for each wave to pass this point will then be $T = 1/f = 1/5$ sec. If the length of each wave is $\lambda = 3$ m, then at the end of one second, five 3-m waves will have passed our viewing point, and it should be evident that the first wave will have traveled a distance of $f\lambda = (5)(3) = 15$ m in 1 sec. Therefore the velocity v of the wave may be found by multiplying the frequency by the wavelength (λ), or by the following

$$v = f\lambda \qquad (5\text{-}40)$$

The vertical displacements of different portions of the string are independent of the velocity with which the wave moves along the string. Also, the velocity of the wave is entirely determined by the properties of the conducting medium and is not affected by changes in frequency. However, the velocity is related to both wavelength and frequency by Eq. (5-40). Thus, since v is constant, if the frequency of the oscillation is doubled, there will be twice the frequency, or $2f$ in the same second, with each wave now being half of its original length.

We shall now consider several properties of waves which are characteristic of *all* waves, namely, *reflection, refraction, interference,* and *diffraction.* Perhaps the most interesting aspect of these phenomena has to do with the wide variety of situations in which their effects can be demonstrated. These include very colorful demonstrations involving visible light, prisms and lenses, found in the works of Newton. They are effectively demonstrated by water waves and by the strings of musical instruments. As we shall see in Chapter 6, interference phenomena constitute a main interest when we discuss the effect of feedback in more complex mechanical, socioeconomic, or behavioral situations.

Let us first examine water waves produced in a ripple tank (Fig. 5.31). The apparatus for these demonstrations resembles a large glass pan containing water to a depth ranging between $\frac{1}{2}$ and 1 in. A small ball on the end of a rod (point source), or a horizontal bar (line source), is made to oscillate up and down, just touching the water surface with each downward motion. The frequency of oscillation can be varied so that the ripples generated by contact with the water travel out in waves of varying wavelength corresponding to each selected frequency. When a stroboscope† is used to illuminate the waves, and the frequencies of light pulses and rod are appro-

† The stroboscope is an instrument used to synchronize light pulses of known frequencies with the movement of a rotating mechanism. A light is flashed intermittently on the part whose frequency of rotation is to be determined. The pulse frequency is adjusted until it appears to be constant rather than intermittent, at which point the frequency of the rotating member and that of the light pulse are synchronized; that is, are equal.

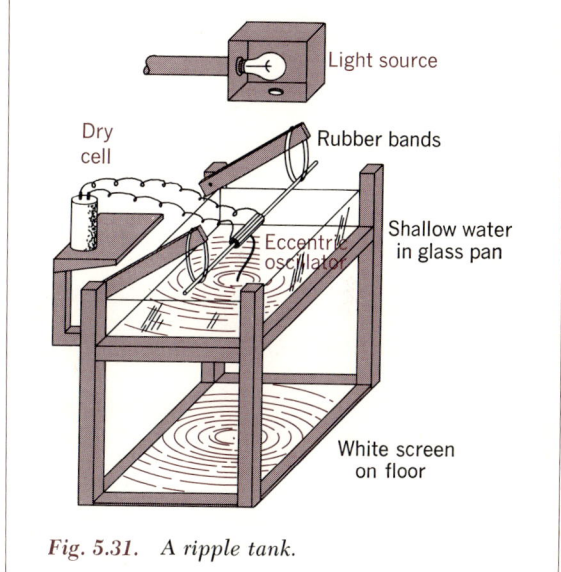

Fig. 5.31. A ripple tank.

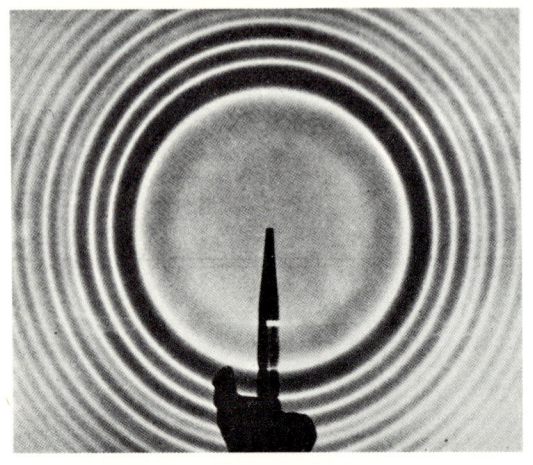

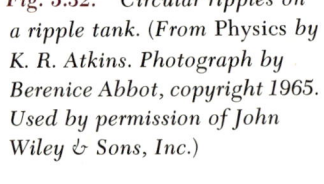

Fig. 5.32. Circular ripples on a ripple tank. (From Physics by K. R. Atkins. Photograph by Berenice Abbot, copyright 1965. Used by permission of John Wiley & Sons, Inc.)

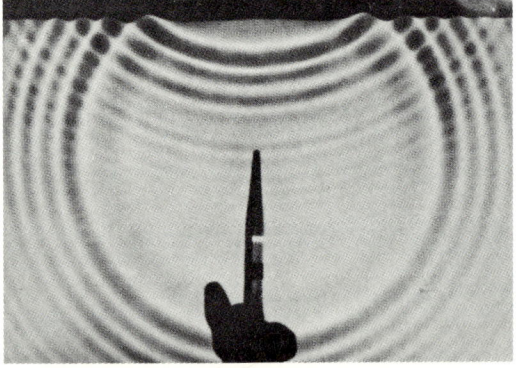

Fig. 5.33. Circular wave reflected from a straight barrier. (From PSSC Physics, 2d ed., copyright 1965 by Educational Services Inc. Used by permission of D. C. Heath, Inc.)

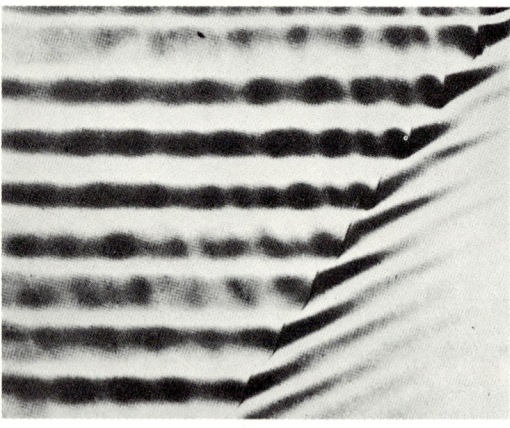

Fig. 5.34. Plane waves entering shallower region of tank obliquely (from right) are changed in direction of motion (bent) due to decreased velocity. (From PSSC Physics, 2d ed., copyright 1965 by Educational Services Inc. Used by permission of D. C. Heath, Inc.)

priately matched, the wave pattern will seem to be standing still (see Fig. 5.32).

When a straight barrier is placed in the ripple tank, the waves striking this barrier are "turned back" or *reflected* (see Fig. 5.33).

If a water wave obliquely encounters a change in depth in the tank, its path is bent or *refracted* (Fig. 5.34). At the same time, its wavelength, λ, changes (Fig. 5.35).

Reflection and refraction at a barrier, although characteristic of waves, are not properties of waves alone. Particles also would show reflection (for example, bouncing a ball off a wall) and bending of their paths when entering a new medium (refraction). But two important phenomena that we shall consider (interference and diffraction) are found to occur only with waves.

In Fig. 5.36 we see what happens when two wavefronts, originating at different points, interact. The pattern that results is marked by the absence of any wave along certain directions. Clearly there is *interference* between the two waves. A similar pattern is obtained when a wave is allowed to pass through a pair of narrow slits.

The *diffraction* of a wave normally incident on a single slit is illustrated in Fig. 5.37. In Fig. 5.37(a), the slit width is comparable in size to the wavelength of the incident wave. Notice that the wave spreads out after passing through the slit. In Figs. 5.37(b) and 5.37(c), shorter λ's are used for the incident wave, so that the slit width is larger than λ. The diffraction effect is still observed, but to a lesser extent than in Fig. 5.37(a).

The phenomena of interference and diffraction can be understood by considering an essential property of waves—that two waves can be added to produce a single resultant wave. This process is

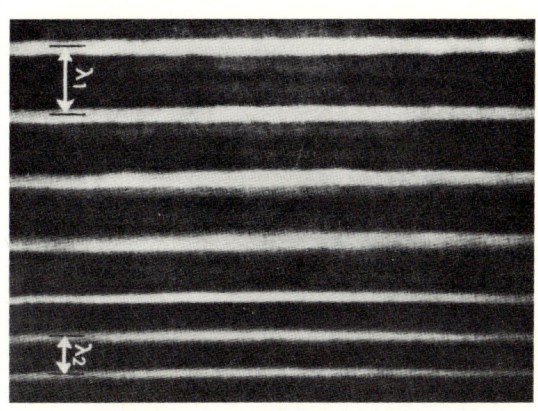

Fig. 5.35. Plane waves normally (perpendicularly) incident on shallower region. (Wave moves upward in picture.) Notice λ_2 in shallow region is shorter than λ_1 (incident). (From PSSC Physics, 2d ed., copyright 1965 by Educational Services Inc. Used by permission of D. C. Heath, Inc.)

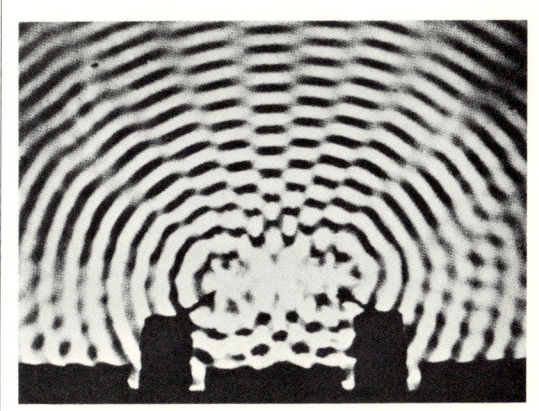

Fig. 5.36. Interference of waves produced by two separate sources. (From PSSC Physics, 2d ed., copyright 1965 by Educational Services Inc. Used by permission of D. C. Heath, Inc.)

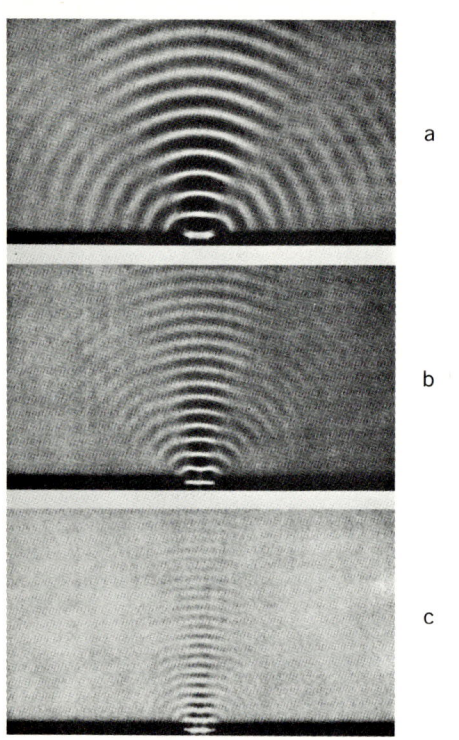

Fig. 5.37. *The diffraction of water waves in a ripple tank. Slit is at bottom and wave proceeds upward. (From PSSC Physics, 2d ed., copyright 1965 by Educational Services Inc. Used by permission of D. C. Heath, Inc.)*

termed *superposition*. We can illustrate superposition by considering two waves of equal wavelength (and hence of frequency f) but out of phase (see Fig. 5.38). It is left for the student to show graphically that two waves of equal *wavelength* and *amplitude*, but 180 deg out of phase, will add up to exactly zero.

Example Let us treat this phenomenon analytically. We can represent the heights of wave 1 and wave 2 (y_1 and y_2, respectively) of Fig. 5.38 by an equation of the type (5-32), where the amplitudes are chosen to be equal, for simplicity.

$$y_1 = A \cos(\omega t + \delta)$$
$$y_2 = A \cos(\omega t + \delta)$$

If the phase constant δ in the equation for y_1 is arbitrarily taken as zero, then the δ in the equation for y_2 is just the phase difference between the two waves:

$$y_1 + y_2 = A\{\cos \omega t + \cos(\omega t + \delta)\}$$
$$= A\{\cos \omega t + \cos \omega t \cos \delta$$
$$- \sin \omega t \sin \delta\}$$
$$= A\{\cos \omega t (1 + \cos \delta)$$
$$- \sin \omega t \sin \delta\}$$

When $\delta = 0$ (both waves in same phase),

$$y_2 + y_2 = 2a \cos \omega t = 2y_1 = 2y$$

and we have *constructive* interference.

When $\delta = 180$ deg (out of phase by 180 deg),

$$y_1 + y_2 = A[\cos \omega t (1 + \cos 180°)$$
$$- \sin \omega t \sin 180°]$$
$$= 0$$

and the waves cancel one another (*destructive* interference).

Consider two sources of waves having the same amplitude and wavelength. Assume them to be in phase at their respective points of origin but separated

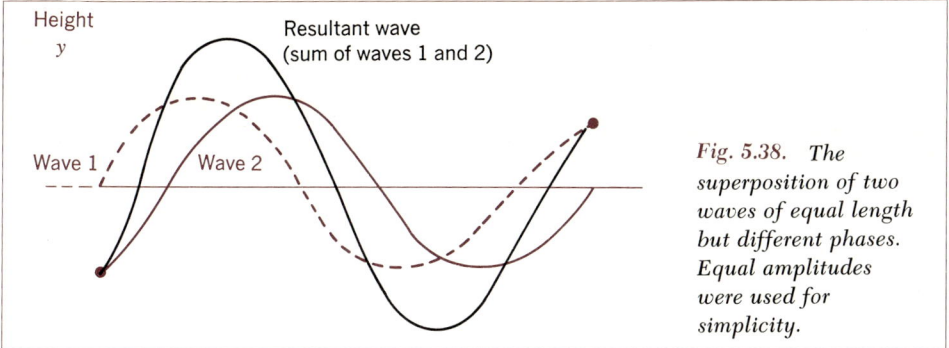

Fig. 5.38. *The superposition of two waves of equal length but different phases. Equal amplitudes were used for simplicity.*

by a linear distance a (see Fig. 5.39). Consider first the point P_0 on the line perpendicular to the line connecting the sources and passing halfway between them. The path length x_1 for waves from source 1 to the point P_0 is equal to x_2 (the path from source 2 to P_0). Hence, x_1 and x_2 require the same number of wavelengths between the source and P_0; thus, as the waves are in phase at P_0, they add *constructively*; that is, the resultant wave at P_0 has twice the amplitude of either initial wave.

Now consider the point P_1, which lies a distance d above P_0 along the line through P_0 parallel to the line joining the sources, and let x be the distance from the line joining the sources to the line $P_0 P_1$. The distance from source 1 to P_1 is X_1; that from source 2 to P_1 is X_2. Since the waves traveling along X_1 and X_2 are both of wavelength λ, the number of wavelengths covered along X_1 is X_1/λ, and along X_2 is X_2/λ. If initially both waves were in phase, the phase difference δ between them at P_1 is

$$\delta = \left(\frac{X_2}{\lambda} - \frac{X_1}{\lambda}\right) 2\pi$$

$$= \frac{2\pi}{\lambda} (X_2 - X_1) \qquad (5\text{-}41)$$

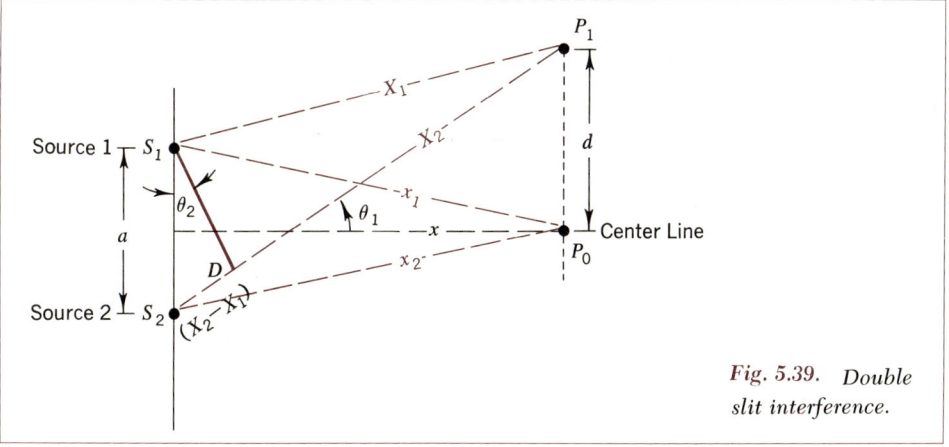

Fig. 5.39. *Double slit interference.*

If the distance x is very much greater than the source separation a, then $\angle \theta_1 \cong \angle \theta_2$ and the triangle formed by S_1, S_2, and D is essentially a right triangle. Thus

$$(X_2 - X_1) = a \sin \theta_2$$

Therefore, from Eq. (5-41) we have

$$\delta = \frac{2\pi}{\lambda}(X_2 - X_1)$$
$$= \frac{2\pi}{\lambda} a \sin \theta_2$$

Also

$$\sin \theta_2 \cong \sin \theta_1$$
$$\cong \frac{d}{\sqrt{d^2 + x^2}}$$
$$\cong \frac{d}{x}$$

Thus

$$\delta = \left(\frac{2\pi a}{\lambda}\right)\left(\frac{d}{x}\right)$$
$$= 2\pi \frac{ad}{\lambda x} \quad (5\text{-}42)$$

Constructive interference occurs whenever $\delta = 0$ or any even multiple of π radians ($\delta = 2\pi$, 4π, and so forth).† That is, $\delta = 2\pi n$, where $n = 0, 1, 2, \ldots$. Thus we have constructive interference whenever

$$n = \frac{ad}{\lambda x}$$

or

$$d = \frac{n\lambda}{a} x \quad (5\text{-}43)$$

Whenever $\delta = (2n + 1)\pi$ radians, $n = 0, 1, 2 \ldots$, we have destructive interference and hence no wave occurs at

$$d = \frac{(2n+1)\pi\lambda x}{2\pi a}$$

or

$$d = \left(n + \frac{1}{2}\right)\frac{\lambda}{a} x$$

for

$$n = 0, 1, 2 \ldots$$

Diffraction is an interference phenomenon that occurs when a wave passes through a narrow slit, as in Fig. 5.40. If x is again much larger than a or d, the difference in phases, $\delta = (\Delta X/\lambda) \cdot (2\pi)$, between waves passing through the bottom of the slit and those passing through the top is given by

$$\delta = \left(\frac{2\pi}{\lambda}\right)\left(\frac{ad}{x}\right)$$

The energy of a wave is proportional to its amplitude squared. It can be shown that at any d, the energy E_d in the resultant wave is

$$E_d = E_m \frac{\sin^2 \theta}{\theta^2} \quad (5\text{-}44)$$

where E_m is the maximum energy (that is, for constructive interference with $\delta = 0$) and $\theta = \delta/2$. Figure 5.41 shows the effect on E_d of varying a or λ. Note this effect in Fig. 5.37.

5.14 Doppler effect

A phenomenon known as the Doppler effect has been exceedingly useful, not only for applications to conventional, earthbound systems but also for analyzing the motion of stars and galaxies that are at vast distances from the earth. The principle is rather simple. If a source of light, or of sound of a given frequency, moves toward us, the waves coming in our direction will be squeezed and will

† 2π rad $= 360°$, $4\pi = 720°$, and so on.

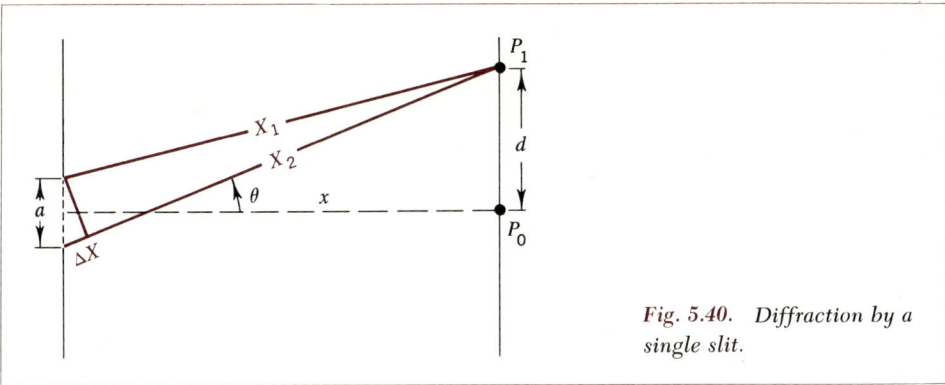

Fig. 5.40. *Diffraction by a single slit.*

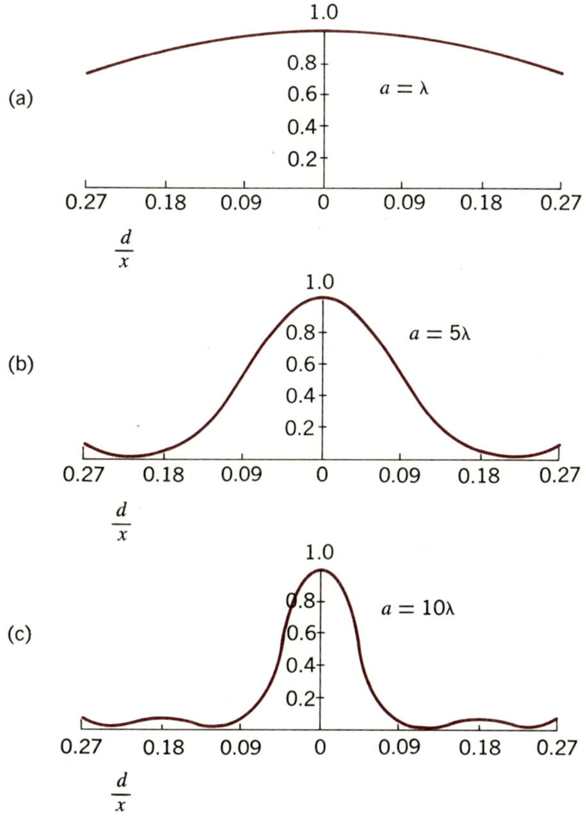

(a)

$a = \lambda$

(b)

$a = 5\lambda$

(c)

$a = 10\lambda$

Fig. 5.41. *The relative intensity in single-slit diffraction for three values of the ratio a/λ, with respect to d/x.*

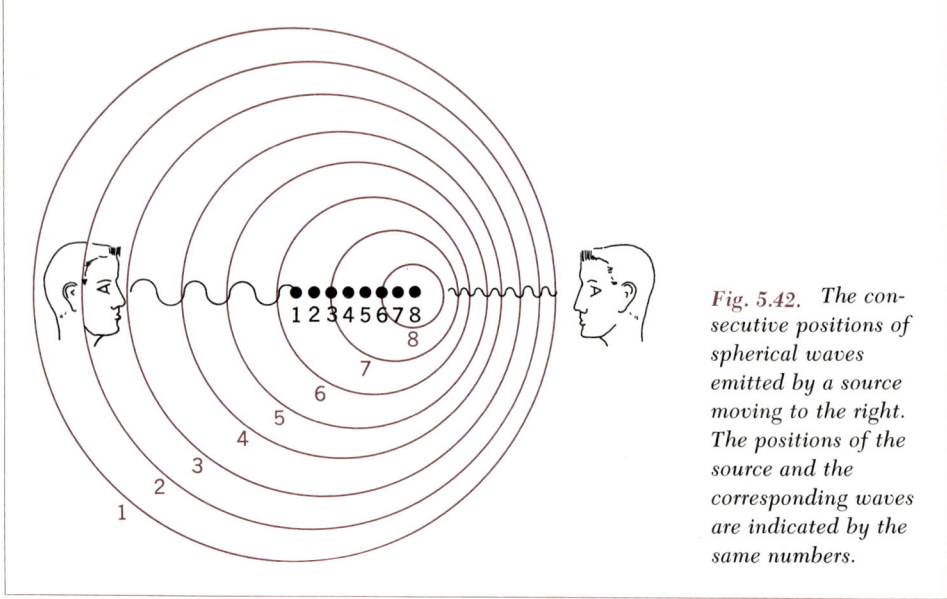

Fig. 5.42. The consecutive positions of spherical waves emitted by a source moving to the right. The positions of the source and the corresponding waves are indicated by the same numbers.

appear to us to be shorter than those from a stationary source. In the case of a receding source, the situation is the opposite; the arriving waves will appear to be longer. Figure 5.42 demonstrates how the waves seem to shorten as the source approaches us.

In the case of sound, the approaching source will appear to have a higher pitch and the receding source a lower pitch. A common example of this is the sudden change in the pitch of the railroad whistle as the locomotive rushes toward us and then away from us at a high speed. In the case of a moving light source, we would see a shift of spectral lines toward the higher frequency (toward the violet) when the source approaches us and toward lower frequencies (toward the red) when the source recedes from us. We can show that the relative change of wavelength, $\Delta\lambda/\lambda$, due to the Doppler effect equals the ratio of the velocity of the source to the propagation velocity of the waves (sound or light velocity), c. Thus

$$\frac{\Delta\lambda}{\lambda} = \frac{v}{c} \qquad (5\text{-}45)$$

where v = source velocity.

Through this simple relationship, by measuring the *change of wavelength*, we can easily calculate the *velocity of the source*. This method is widely used in astronomy for estimating the line-of-sight velocities of stars and galaxies from their observed spectra, and is the basis for several theories on the nature of the universe.

The Doppler effect also occurs in the case of a stationary source and a moving observer.

5.15 Light waves and interference phenomena

In later chapters we shall see that the wave concept is very useful in explaining certain observed properties of light

(Chapter 11) and even of electrons and other atomic particles (Chapter 19). Here we shall limit ourselves to a brief description of interference phenomena with light waves and other more familiar examples of wave motion.

The fact that light can produce beautiful interference phenomena, as shown in Fig. 5.43, was well known, but the full realization that these were due to the wave properties of light escaped even the genius of Isaac Newton. In his great work on light, *Opticks: or a Treatise on the Reflexions, Refractions, Inflexions and Colours of Light,* which was published in 1704, he gave certain wave properties to light. He did not accept light as being altogether a wave phenomenon, however, for he was convinced that on passing through a small opening it would not spread and exhibit diffraction effects as described below.

Figures 5.44(a) and (b) show the patterns that would be revealed with monochromatic light (light having only one wavelength). White light, on the other hand, consists of many different wavelengths, or colors. If white light were incident on such a double slit, the line of nodes for the longer red waves would be diffracted (bent) more strongly and the shorter blue waves less sharply than

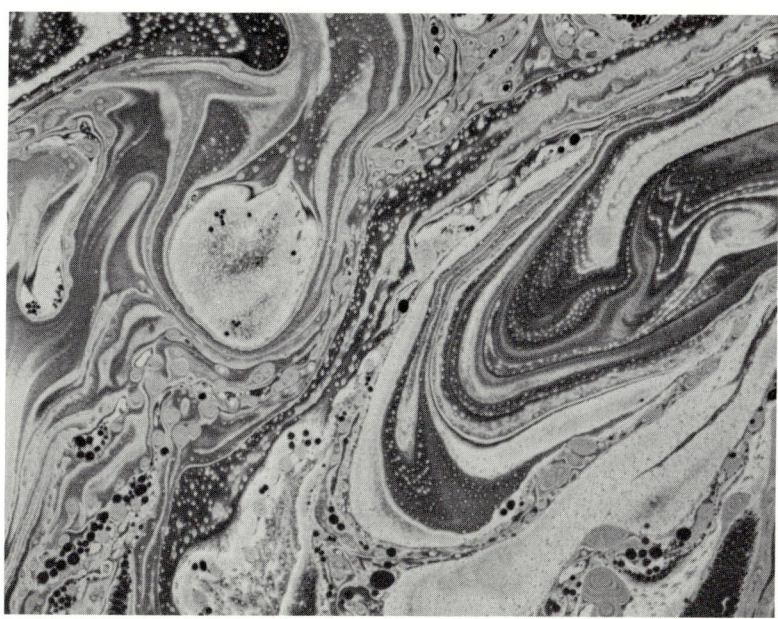

Fig. 5.43. A colored pattern occurs when an uneven thin film of oil floats on water because light of each wavelength undergoes constructive interference at different oil thicknesses. (From Fundamentals of Physical Science by Krauskopf and Beiser, copyright 1966. Used by permission of McGraw-Hill Book Co. Inc.)

those shown. As the number of slits is increased, this effect becomes more pronounced. Thus, if a screen were erected at right angles to the beam after passage through such a multiple slit, it would exhibit a continuous pattern (spectrum) of colors, running from violet at the center through blue, green, yellow, and finally to red at the extremes of the screen.†

† Students who perform the spectrometer experiment will have ample opportunity to study these phenomena in detail and to observe them firsthand.

Refraction of light waves was studied in great detail by Newton, especially in connection with his famous proof, using two glass prisms, that white light does indeed consist of a mixture of all the known colors. Figure 5.45 provides a brief explanation of how, in a similar manner, falling raindrops may produce a rainbow as a result of interference from refracted and internally reflected light rays. Each drop separates the colors as a result of varying refraction effects for different wavelengths. However, an

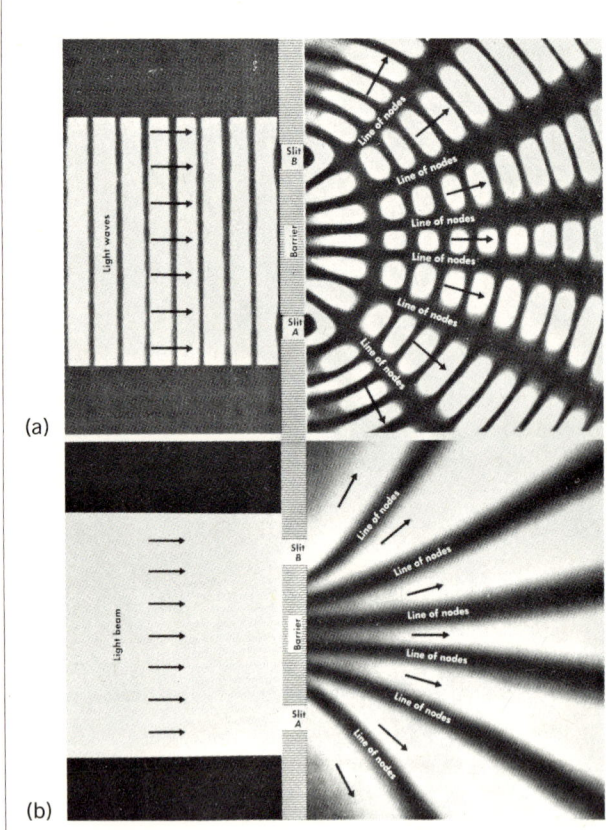

Fig. 5.44. Patterns created by monochromatic light. (a) Light intensity on both sides of double slit at a given instant of time. Waves and interference pattern are propagating to the right. (b) The double slit intensity pattern averaged over time. This is exactly the same as (a) except that it is viewed over a time interval that is long compared with the period of oscillation. (From Fundamental Physics *by Jay Orear, copyright 1964. Used by permission of John Wiley & Sons, Inc.)*

observer sees only the color which happens to come directly to his eye from each drop. Thus drops that are higher up refract and reflect red light to his eyes, those at intermediate levels pass on only the yellow, and so on throughout the complete spectrum.

In these examples we have seen that light exhibits certain wave properties. However, as noted before, Newton was reluctant to accept the wave theory to explain the nature of light. One of the most obvious reasons why we must, even today, recognize that light is more than an ordinary wave phenomenon lies in the fact that it propagates best through the vacuum of empty space. Any pure-wave phenomenon must, it would seem, have a conducting medium to carry the disturbance from point to point. In the complete absence of any conducting medium, what is it that carries the light "wave" from a distant star to the earth? This question and the dualistic wave-particle properties of both radiant energy and matter have constituted main interests of science to the present time, and will be discussed in some detail in Chapters 11 and 19.

5.16 Elastic vibrations in solids, liquids, and gases

It is possible to deform most solid bodies by applying an external force to them. Three types of deformations are possible —bending, stretching, and twisting. When the external force is removed, the internal forces within the solid tend to make it regain its original or undeformed shape. As we saw for the spring, these forces are proportional to the amount of deformation. In attempting to regain its equilibrium shape, the body may overshoot, setting up oscillations or vibra-

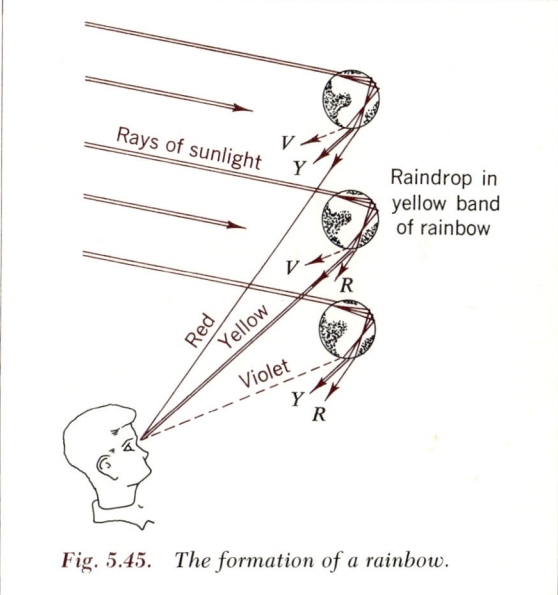

Fig. 5.45. The formation of a rainbow.

tions called *elastic vibrations*. The coefficient of proportionality between applied force and amount of deformation is called the *coefficient of elasticity*. The oscillations set up in an elastic solid generally die out rapidly, owing to internal friction.

Elastic disturbances that propagate through solids can be of two forms—*transverse waves* and *longitudinal waves*. In Fig. 5.46(a) the impact causes a bending of the bar, which propagates as a *transverse* wave (that is, the wave is perpendicular to the bar) along the length of the bar. In Fig. 5.46(b) the impact causes a compression of the bar, which propagates as a *longitudinal* wave (that is, the wave is within the bar) along the length of the bar. Both types of waves are transmitted through the earth as a result of earthquakes, in which case they are called *seismic* waves. The difference in propagation speed of the two types of waves in various media gives evidence of the internal structure of the earth (see Chapter 14).

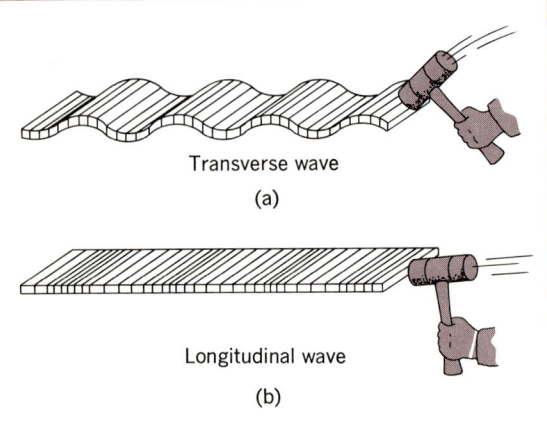

Fig. 5.46. Longitudinal and transverse waves in a solid bar.

5.17 Sound waves

Sound, or *acoustic*, waves are longitudinal compression waves and may be transmitted in any compressible medium—solid, liquid, or gas. These waves play a vital role in every phase of our life, for they carry the vibrations of the vocal chords of a speaker to the ears of a listener, are used to sound the depths of the oceans (sonar), transmit information via radio and telephone, and provide a medium for art form (music).

The speed with which a sound wave propagates depends on the medium through which it is transmitted. In any medium, this varies with temperature and density. In normal air (sea level at a fixed temperature and pressure) the speed of sound is 330 m/sec (or about 1100 fps, or 1 mile every 5 sec). Compared with the speed of light (which we shall see in Chapter 11 to be about 3×10^8 m/sec = 186,000 miles/sec), sound travels very slowly. Thus, although a lightning flash and its accompanying thunder are produced simultaneously, we see the flash long before we hear the thunder. In water the sound speed is about 4.5 times greater than that in air, while in steel it is 15 times greater.

Like all waves, acoustic waves are characterized by a wavelength, λ, and frequency, f. For a sound wave of frequency 440 cps, the wavelength is 0.75 m. The normal human ear can detect sound waves of frequency between 20 cps and about 15,000 cps (or even as high as 20,000 cps in some cases). Frequencies above 20,000 cps are referred to as *ultrasonic*.

The musical notes that we perceive with our ears may be analyzed mathematically. This was first done by Pythagoras, who noted the relation between the "pitch" of the sound emitted and the length of a vibrating string securely fastened at each end and plucked at its center. A wave pulse travels symmetrically in both directions from the center, is reflected at the ends of the string, and returns to the opposite end, where it is again reflected. These two oppositely traveling waves interfere to form a *standing wave* (as in Figs. 5.47 and 5.48) in which the end points remain fixed and the entire string vibrates as a unit. We call this the fundamental tone of the string, and it is the note that the violinist reads in his music. But a string has many other modes of vibration (as shown in Fig. 5.48), called harmonics. It is the presence (or absence) of harmonics that determines the quality of the sound.

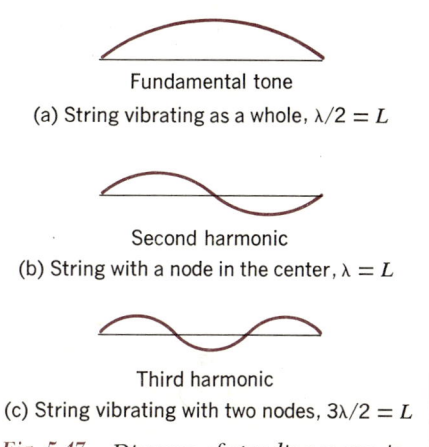

Fundamental tone
(a) String vibrating as a whole, $\lambda/2 = L$

Second harmonic
(b) String with a node in the center, $\lambda = L$

Third harmonic
(c) String vibrating with two nodes, $3\lambda/2 = L$

Fig. 5.47. Diagram of standing waves in a string. All models of vibration may be present at the same time if properly plucked or struck.

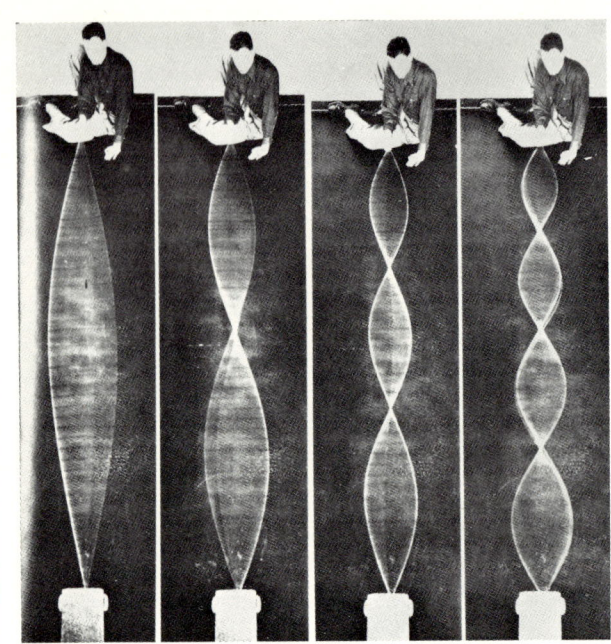

Fig. 5.48. Photograph of standing waves on a vibrating string. The number of standing waves is increased by decreasing the tension of the string. (From Fundamental Physics by Jay Orear, copyright 1964. Used by permission of John Wiley & Sons, Inc.)

The notes of the musical scale are arbitrarily fixed in frequency (or wavelength) by choosing 440 cps (note A) as the standard. The other notes of our present-day major scale are then given as shown below.

Notes that sound most harmonious to our ears are those which bear the simplest numerical ratios; for example, C-C' (2:1, octave), G-C (3:2, fifth), E-C (5:4, major third).

The superposition of many tones produced by different musical instruments yields a large number of possible patterns. Some instruments emphasize certain harmonics that are absent in others. The composer, when orchestrating his work, must select those combinations of instruments (and voices) that are most pleasing to the ear or which best create the image or effect he seeks (or, hopefully, both).

	C	D	E	F	G	A	B	C'
f in cps	264	296	330	352	396	440	495	528
Ratios of f		9/8	10/9	16/15	9/8	10/9	9/8	16/15

5.18 Supersonic and shock waves

Consider the wave produced by the bow of a boat traveling through the water. When the boat is traveling at a speed V and is producing waves that travel outward from the bow with a velocity v, which is greater than V (that is, $v > V$), the wave advances outward in front of the boat [see Fig. 5.49(a)]. When, however, $V > v$ (that is, the boat moves faster than the propagation speed of its bow wave), then this wave forms a "vee" at the bow and the faster the boat, the sharper the "vee."

For an object traveling through the air, however, the speed of the bow wave in air (which is actually a sound wave) is approximately 330 m/sec (or about 750 mph at sea level). When the object is traveling at a rate lower than the speed of sound, its bow wave precedes it. Hence the sound of a falling bomb reaches the ground before the bomb strikes. Behind the object we find a turbulent wake (see Fig. 5.50).

In recent years we have developed supersonic planes and missiles. The speed† of these objects is greater than the speed of their bow waves. A sharp discontinuity in pressure develops along the advancing front edge of the vehicle and remains stationary with respect to it. This discontinuity is called a *standing shock wave* (see Fig. 5.51).

† The speed of a supersonic vehicle is generally expressed in terms of the *Mach number*, or ratio of speed of vehicle to speed of sound in the air around the vehicle. Hence Mach 2 means two times the speed of sound; at sea level this would be 1500 mph.

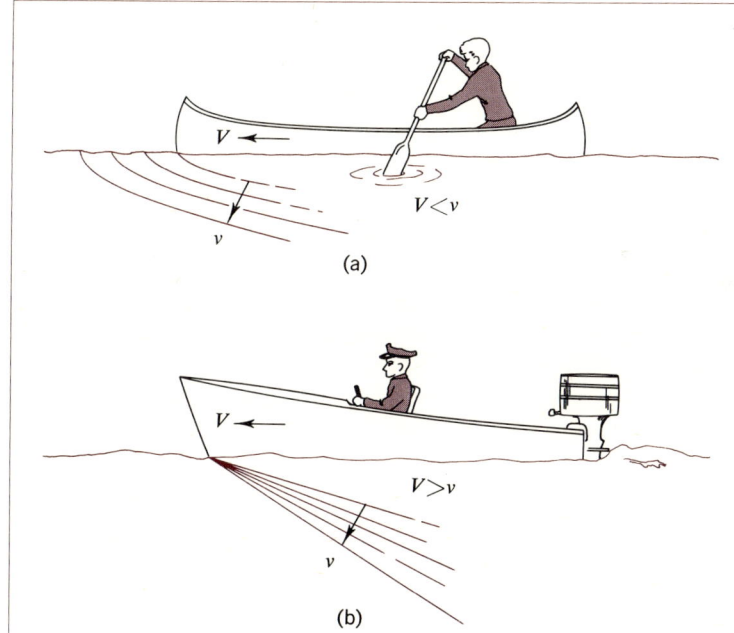

Fig. 5.49. The shape of bow waves (a) when the speed of the wave exceeds that of the boat, and (b) when the speed of the boat exceeds that of the waves.

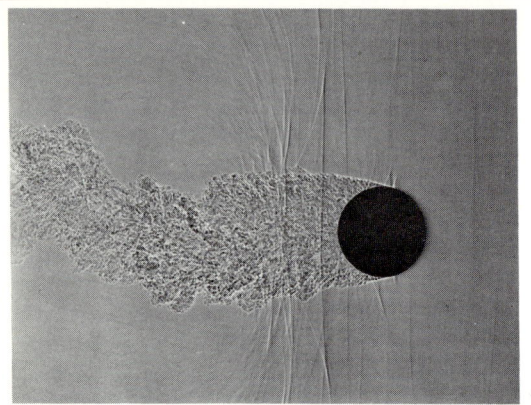

Fig. 5.50. *A metal sphere traveling with 84 percent of the speed of sound (Mach number 0.84) produces a turbulent wake in the air.* (From Matter, Earth and Sky *by G. Gamow, copyright 1965. Aberdeen Proving Grounds Ballistic Laboratory photograph. Reprinted by permission of Prentice-Hall, Inc.)*

Fig. 5.51. *A metal sphere traveling through the air with about twice the velocity of sound (Mach number 2) produces a turbulent wake and a set of shock waves.* (From Matter, Earth and Sky *by G. Gamow, copyright 1965. Aberdeen Proving Grounds Ballistic Laboratory photograph. Reprinted by permission of Prentice-Hall, Inc.)*

When an airplane is traveling at subsonic speeds, the resistance of the air to its motion (called *drag*) increases slowly with increasing speed. As the plane approaches Mach 1, it needs considerable additional energy to overcome the standing shock waves it sets up. The air resistance to further acceleration increases sharply. This resistance, encountered as the plane reaches Mach 1, is called the *sonic barrier* (or sound barrier). Further increases in speed above Mach 1 slows the rate of increase of drag. To a person on the ground, the sharp rise in pressure caused by the shock wave is evidenced as a sound wave, or *sonic boom*.

5.19 Summary

The laws that were systematized and defined by Newton, and the concepts from which they grew, make up almost the full content of the science of the eighteenth and nineteenth centuries. The period that preceded Newton and that continued during his life was notable for its greater daring into new philosophy and into questions of what constitutes truth, validity and reality. The Renaissance, which began about 1450 A.D., had witnessed new investigations in the works of Leonardo da Vinci, Copernicus, and in the studies of the human body by the Belgian anatomist Andreas Vesalius (1514–1564).

There were great movements in social areas as well. The Protestant Reformation came along with the Ninety-Five Theses of Martin Luther (1517). There was new daring and probing into men's thoughts in the art of Michelangelo and in the writings of William Shakespeare and John Milton.

In 1615 William Harvey began to look more closely into the workings of the human heart, which he and a few others believed to be the source of life and the seat of all emotions. He studied the living heart in cold-blooded animals, traced the course of the blood vessels, calculated the quantities of blood that flowed in a day, examined the heart valves and the relationship of the heart to the lungs. His work, published in 1628, finally established what had been speculated on, among others, by Michael Servetus, Realdus Columbus, Cesalpino, Giordano Bruno—the theory of the circulation of blood in the body in which the heart played the role of a pump.

Francis Bacon, the English philosopher and statesman, rejected the utility of Aristotelian logic to get at useful knowledge. Greek philosophy was to him "the talk of idle old men to raw young fellows." He stressed observation and collection of information through systematic observation. Through careful analysis, the information observed would yield general scientific laws. His counterpart on the Continent, the mathematician René Descartes, felt less need for *evidence* to learn the principles of nature. Rather he stressed beginning with intellectually, or reasonably clear, and indubitable ideas to serve as axioms and principles and then from these to deduce true knowledge. He stressed mathematical reasoning. In his *Discours de la Methode* he outlined what he believed to be a thorough and rigorous approach. Although both Bacon and Descartes had great influence in the early days of the growth of science, neither of their methods—the inductive approach of Bacon or the deductive approach of Descartes—was generally used to the exclusion of the other. Does this mean that working scientists do not follow the philosophers of science?

Whether through experiment (experience) or by thought processes (reason), this was a period of active analysis that probed and questioned nearly every accepted notion of nature. Tycho Brahe's observations and Kepler's analyses had established a systematic picture for the outer world. Galileo had brought the phenomena, and analysis of related phenomena through experimentation, to be within the reach and interest of man. There were men like the English physicist and chemist Robert Boyle who conceived small units in the form of atoms to be subject to the laws of nature and of nature's God. There was a feeling of exhilaration that the laws of nature were being revealed and that nature seemed to have the characteristics of a finely designed watch, with each element assigned to its role. God was the great watchmaker. But over all, there had to be the author of nature, the "First Cause" of Descartes, to give origin and direction to the machine-universe and to strengthen the foundations for the "Age of Enlightenment and Reason," that was soon to follow. These men claimed that man had reached mental maturity and no longer needed an authority to tell him what was correct. He could, if he used his reason and the fruits of science, discover it himself. Man had progressed and would progress even further if he was unfettered by the false beliefs of the ancient regime. (See the writings of Condorcet and Godwin, who stressed the perfectability of mankind.)

There was, however, much more to progress than simply the identification and worship of the parts of the machine-universe. Mathematical analysis became a necessary and powerful tool for bringing order to the accumulated information. The new mathematical methods of the calculus, invented independently by

TABLE 5.1
CULTURAL ACHIEVEMENTS OF THE SIXTEENTH–EIGHTEENTH CENTURIES. CHRONOLOGY OF IMPORTANT PEOPLE AND EVENTS.

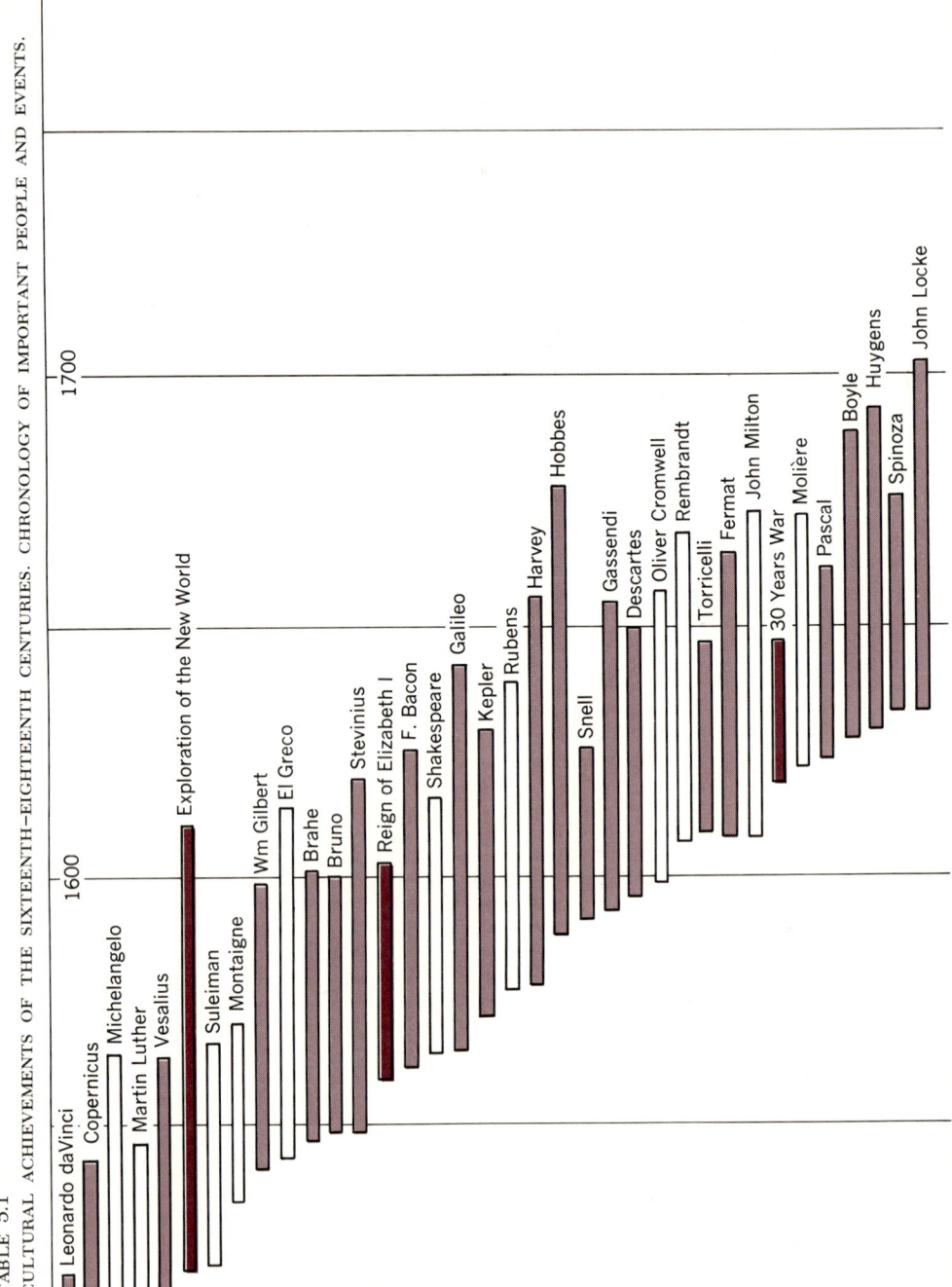

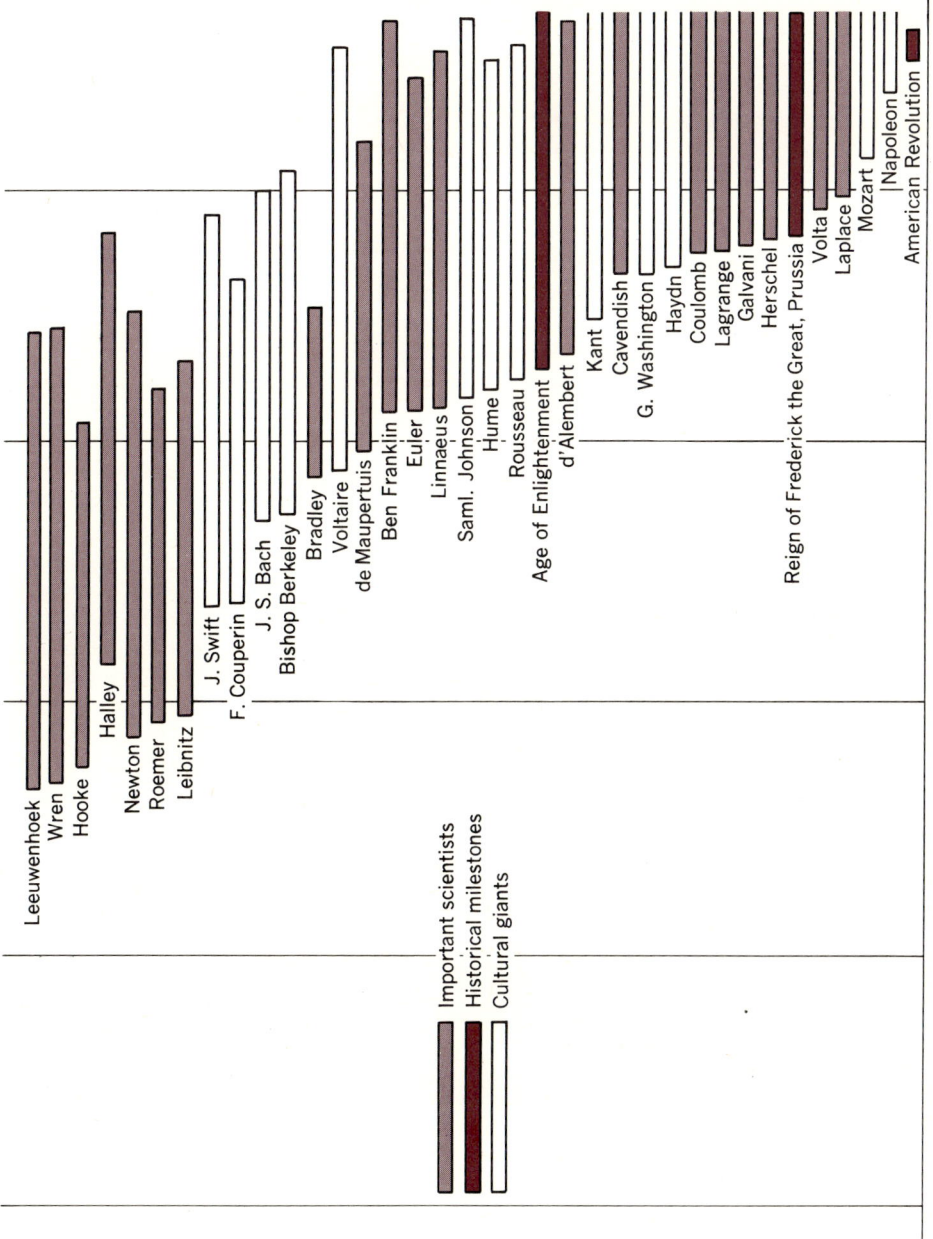

Newton and Leibnitz, had become especially useful for coping with the problems of the moving, changing, dynamic features of nature. But along with these there developed new sensitivity with respect to what constituted absolute knowledge and ultimate science. Every method for exploring nature was encumbered with the fact that objective analysis was impossible because the explorer, man, was himself intimately involved in the environment he sought to study. A mass of matter is understood only through the properties it presents to man, and such properties as wetness, coldness, and color are, it is believed by some, subjective reactions that are not easy to dissociate from the mind of the observer. A contemporary of Newton, the Englishman John Locke (1632–1704), in his *Essay Concerning Human Understanding,* saw experience and observation to be essentially a mental experience, confined to the effects of sensory experience. This raised the serious impasse for which no solution has been yet found. For if all observation and experience of the natural world of man is only the effect of some experience in the mind of man, how can man climb out of his confines to reach the real physical world?† Despite these uncertainties, this was the age when nature was rational, and was governed by laws that were reasonable because God was assumed to be reasonable.

Galileo, using precise definitions of velocity and acceleration, successfully applied these concepts to analyze terrestrial motion, in particular the laws governing objects falling under the influence of gravity. Newton then extended these ideas systematically in two ways: He applied them to celestial objects, introducing the inverse square law of gravitation, and he added the law of action and reaction, which governs all interactions between masses.

By means of the calculus, Newton could show that it follows from the law of gravitation, which applied to all masses, that the mass of a spherical object such as the sun, the earth, or the moon may be regarded as being concentrated at its center. Point-masses became idealized abstract elements in terms of which the dynamics of motion could be analyzed.

Newton's *Principia* stated the three laws of motion and the law of gravitation as postulates, on the basis of which he systematically worked out a vast array of consequences that could be confirmed by observation. Foremost among the consequences were the three laws of Kepler. These empirical laws were thus given an "explanation," in the sense that they were shown to follow from more basic principles. Regarding the postulates (laws) themselves, Newton gave no further "explanation" or proof. Specifically with respect to his idea of gravity, which was set forth "as action at a distance" (in contrast to suggestions that had been advanced by Kepler and Descartes), Newton disclaimed any attempt at explanation: "*Hypothesis non fingo*" (I make no hypothesis). He sought simplicity in nature: "For nature is pleased with simplicity and affects not the pomp of superfluous causes."

On the basis of Newton's systematic treatise, the foundations of the whole of the branch of physics called *mechanics* was now established. In the hands of his successors, Newton's work was rapidly

---

† The doctrine that Locke popularized was a reflection of the sensationalism of Greek Sophists, and of the doctrine propounded before him by Thomas Hobbes. For David Hume this doctrine led to agnosticism regarding our ability to ever achieve real knowledge. See Beth in references for discussion of this topic in relation to ancient doctrines.

extended and the basis for the scientific and industrial revolution was completed. Probably no other single book, with the exception of the Bible, ever had such an impact upon civilization and upon the way men think.

Newton established the law of conservation of momentum. Closely associated with this conservation law, the law of conservation of energy was soon developed. It was shown by Leibniz and Huygens, contemporaries of Newton, that energy of motion (kinetic energy) is given by the formula $\frac{1}{2}mv^2$, and that work is given by the formula "force times distance." The law of conservation of mechanical energy holds for all interacting objects, under the condition that *inelastic collisions are excluded,* and states that the sum of the kinetic and potential energy remains constant throughout any changes or motions of the masses of the system. Later, as we shall see, the law of conservation of energy was generalized to include heat energy, so that the case of inelastic collisions could then be included.

In this chapter, two forms of motion other than rectilinear are studied: rotary motion and simple harmonic motion. If a mass moves in a circle, it is continuously being pulled out of a straight-line path. Therefore it is being continuously accelerated. The acceleration in this case is towards the center of the circle and is given by the formula $a = v^2/r$, where v is the constant speed and r is the radius of the circle. This equation, combined with the law of gravitation, establishes the premise for the discussion of planetary orbits. In a later chapter this discussion will be extended to orbits of artificial satellites.

Simple harmonic motion can be thought of as a paradigm for all periodic motions. Nature abounds in objects which, when stretched or pulled away from an equilibrium or neutral point, are subjected to a restoring force that is proportional to the amount they are displaced from the neutral point. This in turn implies that the acceleration at any time is negatively proportional to the displacement of the object. Whenever this occurs, the object will oscillate about the central point with simple harmonic motion. Such a motion may be best analyzed mathematically by means of sine and cosine functions (which are often for this reason referred to as periodic or circular functions).

When an object oscillates in a medium, a *wave* is usually generated. Energy can be transmitted over a distance either by the motion of a mass or by waves. The transmission of energy by means of waves therefore becomes one of the central topics to which we shall find ourselves returning many times during the course of our discussions.

Questions/Discussions

1. Why are the passengers in a car "thrown forward" when the driver applies the brakes suddenly?

2. A person awakes and finds that he has been placed in a closed, windowless elevator car while he was asleep. Assuming that suitable physics laboratory equipment has also been placed in the elevator car, describe the experiments he might perform to determine the state of motion of the car.

3. If a small sports car collides with a

standard sized (and much more massive) automobile, in which one will the passengers be most likely to suffer serious injuries? Explain your answer in terms of the conservation of momentum and of Newton's laws of motion.

4. Newton's third law asserts that action and reaction forces are *always* equal in magnitude and opposite in direction. How, then, is it ever possible to obtain an unbalanced force as required for applications of Newton's second law?

5. Galileo is alleged to have solved the following problem (can you?): Describe a method for determining the length of a wire that is suspended in a tall tower in such a fashion that the upper end is not visible or accessible but the lower end is.

6. According to the relation $F = G(M_1 M_2/r^2)$, the gravitational force of attraction between any two bodies increases as the distance r decreases, and this force approaches infinity as r approaches zero. When a young man asks a girl to dance and they come very close to each other, why doesn't this gravitational force prevent them from ever breaking away again?

7. Would you expect the total angular momentum of the solar system to remain constant? The total energy? Explain your answers.

8. A prospector carrying a heavy bag of nuggets finds himself motionless in the center of an absolutely smooth and frictionless frozen-lake surface in northern Alaska. Is there any way that he can escape from the lake?

9. An open cart is coasting at constant speed on a frictionless horizontal track. A sudden shower of rain starts to fall and increases the mass of the cart. Would this have any effect on the velocity of the cart?

10. A 4000-lb car is traveling at 30 mph and a 2000-lb car is moving at 60 mph; which has the greater amount of momentum? Of kinetic energy? Explain.

11. What will be the acceleration of a 10-kg mass to which a 25-N force to the right is applied and which is also subject to a 5-N friction drag?

12. An accurate pendulum clock is taken from a location near the equator, where the value of g is 32.088 ft/sec^2, to the North Pole where $g = 32.258$ ft/sec^2. Will it gain or lose time in the new location and by approximately how many minutes per day?

13. A 2-kg ball moving to the right with an initial horizontal velocity of 10 m/sec suffers an elastic head-on collision with a stationary 10-kg ball. What will be the velocity of each ball after the collision?

14. A 3200-lb car is subjected to a sustained, uniform forward thrust of 500 lb for a period of 10 sec while simultaneously being retarded by a friction drag of 100 lb. How much energy has been expended and how much kinetic energy does the car have at the end of this 10-sec period?

15. A 64-lb boy swings back and forth at the end of a 24-ft rope, which is attached to the ceiling in a school gymnasium. If he comes to within 15 ft of the ceiling at the highest point of his swing, what will be his linear speed at the lowest point of the swing? At this same point, what will be his angular speed and what will be the tension in the rope?

16. A 160-lb swimmer dives from one side of a 300-lb life raft with a horizontal

velocity of 10 fps. With what speed does the life raft move in the opposite direction?

17. A 10-kg mass is suspended from a steel spiral spring and its weight extends the spring 0.20 m. The mass is then displaced an additional 0.20 m and released. Find

(a) The spring constant k;
(b) The period of the resulting SHM;
(c) The maximum potential energy (PE) of the system; and
(d) The maximum velocity of the oscillating mass.

Hint: The maximum PE may be found by multiplying the average restoring force (which is one-half of the maximum restoring force) by the amplitude of the motion.

References

The subject matter of this chapter forms the basis of all classical physics. Needless to say, a huge volume of literature on these subjects exists. This Bibliography is intended only to indicate a few additional sources of information.

Andrade, E. N. de C., *Sir Isaac Newton—His Life and Work*. New York: Doubleday (Anchor Books, Science Study Series S-42), 1954. *An excellent treatment of the life of Newton.*

Atkins, K. R., *Physics*. New York: Wiley, 1965. *Chapters 6–10, pp. 69–140 cover Newtonian physics in considerable detail.*

Benade, A. H., *Horns, Strings and Harmony*. New York: Doubleday (Anchor Books, Science Study Series S-11), 1960. *Treats wave phenomena, particularly as applied to music.*

Beth, E. W., *The Foundations of Mathematics*. New York: Harper & Row, 1959.

Dampier, W. C., *A History of Science*. New York: Cambridge Univ. Press, 1929. *Reprinted in paperback in CAM 366 in 1966. Pages 134–177 deal with the contributions of Galileo and Newton.*

Gamow, G., *Biography of Physics*. New York: Harper & Row (Torchbook TB 567), 1961. *Chapter III provides a useful treatment of Newton's accomplishments.*

Holton, G., and Roller, D., *Foundations of Modern Physical Science*. Reading, Mass: Addison-Wesley, 1958. *Chapters 4 and 5 treat Newtonian mechanics; Chapter 11, universal gravitation; Chapter 12, consequences of Newtonian physics; and Chapters 16–18, conservation laws.*

Kock, W. E., *Sound Waves and Light Waves*. New York: Doubleday (Anchor Books, Science Study Series S-40), 1965.

Mason, S. F., *A History of the Sciences*. New York: Crowell-Collier (Collier Books), 1962. *Pages 165–268 are especially recommended.*

Newton, I., *Principia*, 1686. Translated by Motte, revised by Cajori. 2 vols. Berkeley, Calif.: Univ. of California Press, 1962.

Resnick, R., and Halliday, D., *Physics*. New York: Wiley, 1966. *Chapters 15 and 16, Simple Harmonic Motion and Gravitation, are especially recommended.*

Rogers, E. M., *Physics for the Inquiring Mind*. Princeton, N.J.: Princeton Univ. Press, 1960. *Chapters 1, 7, 8, and 10 deal primar-*

ily with Newtonian mechanics, whereas Chapters 19, 20, and 22–24 cover much of the history of these developments.

Singer, C., *Studies in the History of Method of Science,* London and New York: Oxford Univ. Press, 1917, 1921.

Taylor, L. W., *Physics, The Pioneer Science,* Vol. 1. New York: Dover, 1959. *Chapters 1 and 9–18 contain a wealth of supple-mentary material with many historical notes.*

van Bergeÿk, W. A., Pierce, J. R., and David, E. E., *Waves and the Ear.* New York: Doubleday (Anchor Books, Science Study Series S-9), 1960.

Waldron, R. A., *Waves and Oscillations.* Princeton, N.J.: Van Nostrand (Momentum Book No. 4), 1964.

CHAPTER SIX

Systems, Feedback, Cybernetics

As no man is an island complete of himself, and no event emerges unrelated to other events, so every cause is itself an effect, and every effect a cause that moves sometimes forward but oftener backward to alter the source of change.

V. L. P.

THE READER will recall that following the quotation from Teilhard de Chardin in Chapter 1, we proposed extending the scope of our interests to include analysis of *relationship* and *interrelationship of natural phenomena* to each other. We have come to a point that requires a more formal development of such interrelationships.

6.1 Extension of "systems"

One of the accomplishments of the Newtonian period was the strengthening of the concept that in material or physical situations at least, things do not happen without a causing force. A stone does not begin to move or come to a stop of its own volition. In this chapter we shall utilize that concept, but with three extensions.

The first extension takes into account the fact that in most situations surrounding an event (such as the hurling of a stone), the immediate event is itself part of a larger situation or *system* that includes various other articulating parts or related events. (That is, there is a person who throws the stone, and the throwing has relation to some cause or purpose.)

The second extension may perhaps be thought of as related to the action-reaction principle, namely, that within the context of the system involving an event (a stone is thrown) there is often a *feedback* effect (for example, the one at whom the stone is thrown may hurl it back).

The third extension includes in the system both material things (stones) and human beings along with biological processes and the less tangible thought processes.

What do we mean by the term *system*? We might refer to the weight suspended from a spring as a system that executes simple harmonic motion. The governor that controls the speed of an engine is a control system. We also speak of a *system of highways,* the *economic system* of a nation, a *system of thought,* and of many others. The combination locks that protect the vault of a savings bank make up a protective system, but this can also be said to be only a subsystem of the banking institution. The banking institution is itself only a subsystem within the larger community economics, and the latter is a subsystem of national economics. The chain of larger and larger subsystems, or the nesting of subsystems within larger subsystems, may lead to very complex assemblies and relationships.

While an accurate, all-encompassing definition for the term is not easy to give, we can note a few of the characteristics that are usually present in what we call a system:

(1) A system is likely to have two or more parts, elements, or aspects, which tend to have some functional relation to each other (like the bolt and key of the lock, or the president and staff of the bank).

(2) Because systems are usually subsystems of larger units it is usually helpful (and often necessary) to confine one's study to the smallest unit that encompasses the particular functional elements and interrelationships that are under study. (For example, the locksmith can quite properly repair a fault in the lock system of the bank vaults without considering the question of the merits of socialism for the nation's banking system.)

(3) A *control system* has within itself regulatory functions for control of variables such as speed of a motor, the temperature of a room, the price of commodities, or international trade in narcotics.

(4) It is usually possible to identify an "input" and an "output" portion (or aspect) of a system. For example, a key placed in a lock and turned (input) will cause the bolts to move (output); or an order from a president of an industrial firm (input) can double the selling price of its commercial products (output). We shall find, however, that most systems have more than one form of input, as well as a variety of functional relationships that produce quite varied output.

(5) Usually (nearly always in systems that include regulatory functions) there is some form of *feedback* from the output to the input, which may greatly modify the net output of the system. [For example, when the selling prices of the commercial products of paragraph (4) were doubled, the consumers could have initiated strong feedback by refusing to buy the products; and the industry's board of directors could have exerted even stronger feedback by firing the president and hiring another who would hold the prices at a more acceptable level.] The role of feedback will be given considerable attention in the discussion that follows.

We shall now turn to a more detailed introduction to systems, feedback, and control.

6.2 Cyclic character of natural phenomena

In Chapter 5 we learned that a mass suspended from a spring executes simple harmonic motion when displaced slightly from its equilibrium position. When the motion was recorded on a moving sheet of paper (to illustrate the motion as a function of time), the oscillations were recorded as sine or cosine waveforms. It was shown that the motion was initiated when *potential energy was added to the system* of weight and spring (by manually raising the weight from its rest position, against the pull of gravity, or by pulling it down and extending the spring). In either case, the pull of gravity or the pull of the spring alternately introduced a *restoring force,* which tended to return the displaced *mass* to its original position (Fig. 5.24). But since force applied to mass accelerates the mass and thereby increases its velocity (Eq. 5-1), by the time the mass reached the "zero" or initial position it had acquired so much *velocity* (because the *potential energy* we added manually had become kinetic energy at that point) that the mass moved past the zero point to the other extreme. There would have been few or no oscillations at all, on the other hand, if the weight had been subjected to so much frictional drag that the added (potential) energy was lost as heat.† (This might have been the case if the weight moved in a viscous liquid.)

What about cyclic behavior in other phenomena of nature? A very common form can be demonstrated in electric circuits in which the electric energy rapidly passes back and forth between parts of an oscillating circuit until the electric energy dissipates as heat or radiates away from the circuit (as in the transmission of radio waves).

We shall find that there can be many forms of oscillatory behavior when a

---

† We shall learn in Chapter 10 that the kinetic energy of the system goes into faster, random motion of the molecules that make up the parts of the system. The increased molecular motion raises the temperature of the parts of the system, as though it were heated by a flame. There is therefore a correspondence or equivalence between the energy in a flame and mechanical motion of the system.

"disturbance" changes the energy level of a system and introduces a restoring force that causes the energy to convert to another form rather than completely dissipate into the heat energy of the environment. The term *energy* may apply not only to mechanical, electrical, or chemical characters in physical systems, but also to institutional and personal pressures in social situations.

Let us now go to other phenomena that show cyclic or periodic variation. (See Figs. 6.1(a) through 6.1(d), for graphical examples of such cyclic variations.) We might utilize various sensing devices to record changes in the temperature of an air-conditioned room as a function of time, the height of the tides of the sea, wind velocity, the automobile traffic on a road, rainfall, the movements of a tall building or of the long span of a bridge, or the temperature of the earth. We might also look up past statistics on wheat production, the stock market, attendance at church, tourist travel, populations of animals, or the length of women's skirts, and plot these in graph form as function of time. We would find that many phenomena in nature and in animal or social activity have variations of an oscillating character (Fig. 6.1). It can be demonstrated that in all these situations which show oscillations about some average point, there is present a *restoring force* that comes into play whenever there is energy change in a system. To be sure, the magnitude and shapes of these oscillations and waves vary considerably from the sine waves we observe with a weight on a spring. The periods may vary from 10^{-15} sec in the case of light waves, to several hours for the period of the tides, and to many years in the case of other cycles of nature and of some social customs. Nevertheless, all are subject to some common influence, not the least of

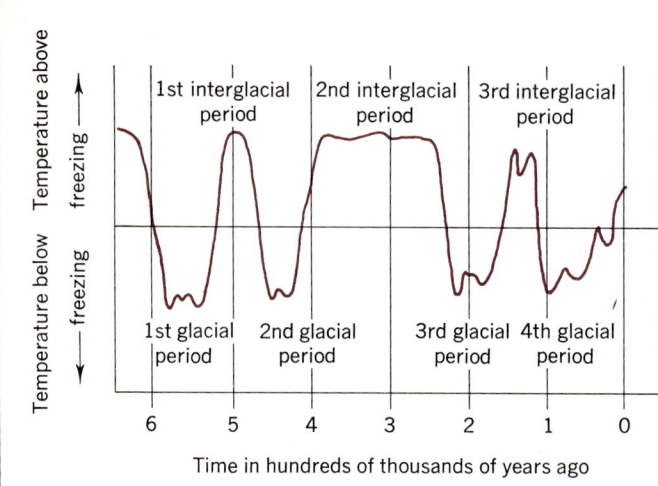

Fig. 6.1. (a) Cyclic temperature variations during the ice ages. Current theory attributes these long, slow temperature variations to relatively minor changes in the atmospheric carbon dioxide content (see Chapter 15, Sec. 2). (Adapted from graph in G. H. Drury, The Face of the Earth, Penguin (Pelican book), pg. 157.)

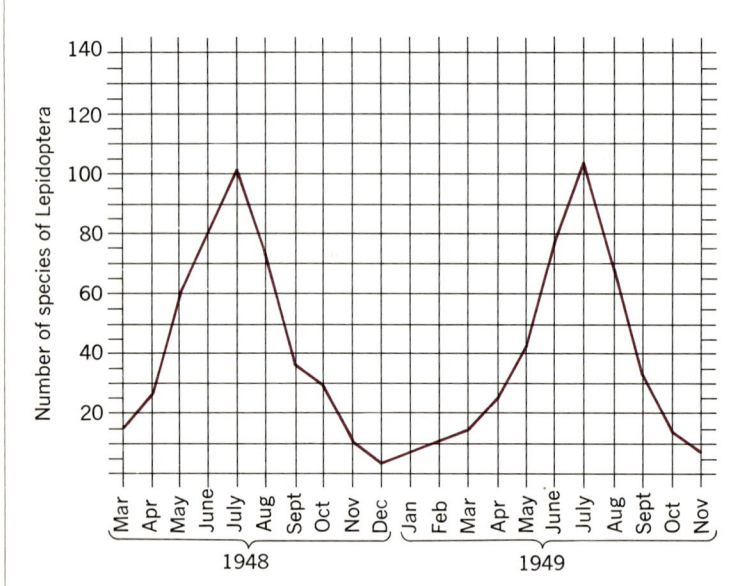

Fig. 6.1. (b) Cyclic variations in numbers of species of Lepidoptera (butterflies and moths) captured in light traps at Woking, Surrey in 1948–49. The number of different species of captured reveals seasonal cyclic variations that are obviously related to weather conditions. Note peaks in successive Julys, when Lepidoptera conditions are ideal, and low values in winter when conditions are poor. (From C. B. Williams, Patterns in the Balance of Nature, Academic Press, 1964, pg. 159.)

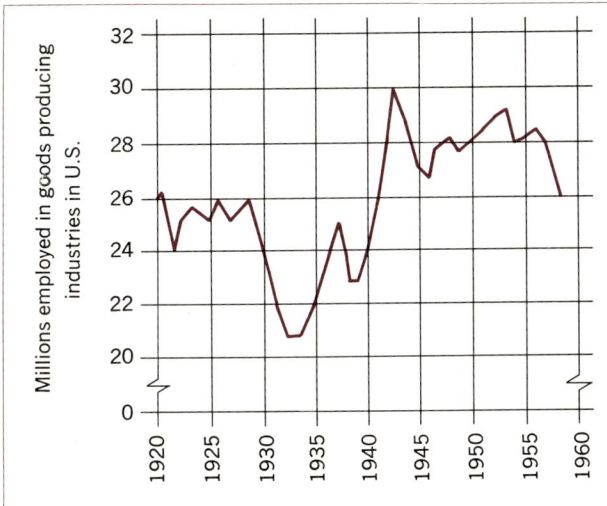

Fig. 6.1. (c) Cyclic character of employment levels in U.S. goods producing industries, 1920–1960. Note large amplitude cycles superimposed on more normal fluctuations as a result of the depression in the early 1930's and of World War II during the early 1940's.

Fig. 6.1. (d) Cyclic variation in value of new construction of religious buildings.

which is the fact that nature is dynamic and in a state of continuous change, and indeed that static situations represent special and almost trivial aspects of nature and of man.

Is the presence of some restoring force sufficient assurance that a system will experience only moderate oscillations without going to extremes? Indeed it is not, as we can learn from the dramatic example of the failure of the Tacoma Narrows suspension bridge of Tacoma, Washington. When the bridge was opened to traffic on July 1, 1940, there were observed, in addition to the ordinary oscillations of the bridge, some unexpected

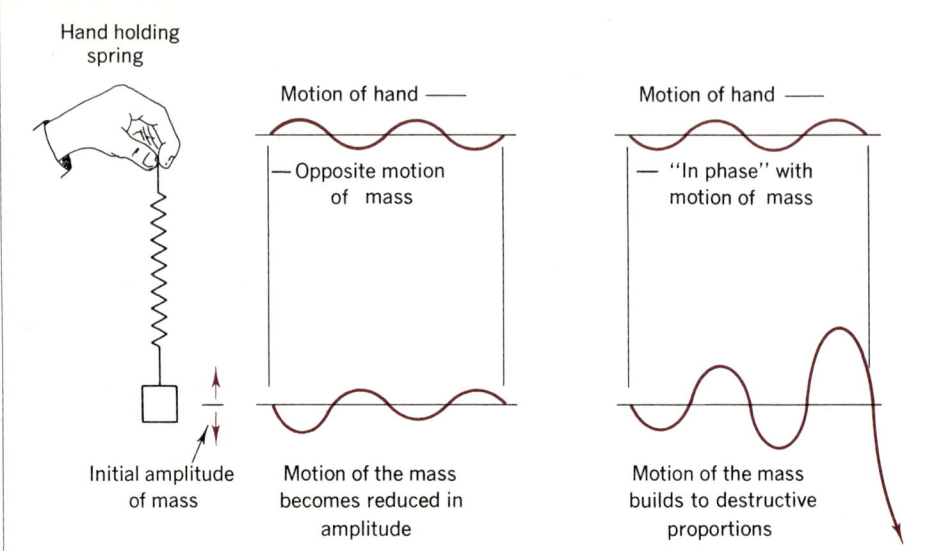

Fig. 6.2. What happens when the mass of a spring is given some additional energy by movement of the hand in two different phase relationships? In the center figure the hand is moved upward when the mass is moving downward. At the right, the hand is moved upward when the mass is also moving upward, causing the mass to take large swings.

transverse (vertical) modes of vibration. On November 7 a wind velocity of 40 to 45 mph made the vibrations so severe that the bridge was closed to traffic, and by 11:00 A.M. the main span collapsed.[†]

[†] A 4-minute film produced by the Ohio State University and distributed by The Ealing Corporation of Cambridge, Massachusetts, gives the very dramatic story of the final oscillation of the bridge prior to its collapse. Every reader should see this film and the variation it offers of "simple harmonic motion" involving the twisting and turning of this huge span of steel and concrete. The new bridge that was built on the original anchorages and tower foundations included deep stiffening trusses instead of girders, and has been entirely successful.

6.3 How oscillations increase despite restoring forces

It is not necessary to resort to the complex behavior of the original Tacoma Narrows bridge to see how a system may have within it strong restoring forces while yet experiencing oscillations that increase in amplitude to the point of destruction. The reader can duplicate the phenomenon with the simple weight on a spring as follows (Fig. 6.2):

Choose a weight and spring combination that gives an oscillatory period between $\frac{1}{5}$ and $\frac{1}{3}$ sec. Hold the spring firmly and steady in your hand, and observe that the weight executes the usual simple harmonic motion, eventually coming to a stop. Now prepare to move your holding hand up or down in synchronism with the motion of the weight and with two alternative movements.

First, *raise* your hand (about a half-inch will do) whenever the weight is moving *downward,* and lower it an equal distance whenever the weight is *moving up.* With a little analysis you can see that the weight tends to *reduce* amplitude because the movements of your hand *increase the restoring force* on the weight. Note that the movement of your hand is *180 deg out of phase* with the motion of the weight.

Next, repeat the experiment with the same up-and-down motion of your hand, but now change the timing to be *in phase* with the motion of the weight. That is, move your hand upward when the weight is moving upward, and downward when the weight is moving downward. There still is restoring force, and the weight continues to oscillate up and down; but now the amplitude of oscillations *increases* until it becomes dangerous to continue the experiment.

Why did the same *amount* of motion of your hand have such opposite effects, depending only on its phase relationship to the motion of the weight? The reason is that in the second case the increments of energy that were introduced by each *in-phase* motion of your hand tended to *add to and increase* the energy of the system represented by the spring and weight. Conversely, the hand motion that was completely out of phase with the motion of the weight detracted from the energy of the system.†

† The reader is urged to perform this experiment and to attempt a careful analysis of the various factors (energies and forces) that become involved in the two cases. For example,

We can now extend this experiment to apply to the early Tacoma Narrows bridge experience. Obviously, the energy of the wind became converted to energy of oscillation of the bridge. Why did the wind energy not become absorbed in the concrete and steel of the bridge? Undoubtedly much of it did become absorbed and changed to heat energy, but not all of it. Apparently when the wind blew to produce a movement of the span at some point along the bridge, the conditions were just right to cause this movement to act as a traveling wave, which on backward reflection returned to the same point in just the right phase to support (rather than oppose) a new movement at that point, caused by the continued blowing of the wind. Had the physical structure of the bridge been different in length or mass, the returning wave could have opposed (out of phase with) any new movement at A, and thus would have added to the stability of the system.

We see, therefore, that for a system to be *stable,* the relationship of the forces and time characteristics must be such that *the amplitude and energy of the system will not increase.* This calls for special attention with respect to the *phase relationships* that obtain between feedback of energy from one part of the system to another part. When the *feedback opposes* the direction of the initial change that produced the feedback, the system tends to be *stable.* In contrast, when the returning *feedback* of energy *supports* the direction of initial change, the system tends to add to the initial energy gain and to be *unstable.* This means we must delve into the theory of system control.

in the second case the increments of energy are added to the spring-weight system. Where does the hand energy go in the first case?

6.4 Modifying cyclic changes: controls

While most fluctuations of nature go their own way without inviting human concern, there are some important cases in which it becomes necessary to interfere, that is, to modify the natural pattern or to control or hold the fluctuations to smaller changes. For example, the farmer may not want to depend entirely on natural rainfall to assure a good crop, so he intervenes by irrigating the fields when there is not enough rainfall. Because in the course of the year there are wide fluctuations in the temperature of the earth, he installs a control system in his home to keep the temperature within comfortable limits.

Many types of controls are involved in our daily life. We shall learn that the human body has a remarkable control system to maintain its own temperature within very close limits. The body's motor functions, by which we move our arms and legs in an accurate and determined manner, are possible only because of the operation of fine control systems. Industrial production relies heavily on control of temperature, pressure, chemical composition, and similar factors. The application of control principles extends to community and national life. Despite their variety, we shall find that there are some common characteristics among them. Also, within a specific control system there can be intermixed a wide variety of elements of widely different types. Take, for example, the very common experience of driving an automobile. Here, the steering control allows the driver to follow the curvature of the road effectively, and many other electromechanical parts as well in the motor and transmission systems affect the driving operation. But we shall learn before long that nearly every aspect of the driver's being—his metabolism, muscle and nerve action, his thinking process—and the life of his community are all parts of the system that encompasses the simple driving experience.

6.5 Introduction to on-off control

We return to the harmonic motion of the weight suspended from a spring and note that, so far, we have neither tried to restrict the amplitude of the motion nor put the movement to some useful application. In each assembly the added energy is converted and reconverted from kinetic energy to potential energy and then back again to kinetic energy. (If there were no frictional losses, the motion would continue forever, since the system would then be self-contained, that is, a closed or isolated system that neither receives energy from nor gives energy to the outside.) Such systems have limited value except as one may use them in a clock or metronome to tell time from the oscillations.†

If there were no frictional or other loss of energy from the system, the motion would have a periodicity of T seconds. Since friction is present, the oscillations become continually smaller in magnitude, and the period of each cycle becomes slightly longer $(T + \Delta T)$ until the mechanical energy dissipates as heat energy and the movement ceases altogether (Fig. 6.3). In general, friction or damping is likely to make a system more stable.

We can design an oscillator to do some-

---

† Of course, as any such device requires periodic additions of energy to the driving springs, and therefore the person who winds the spring becomes part of the system.

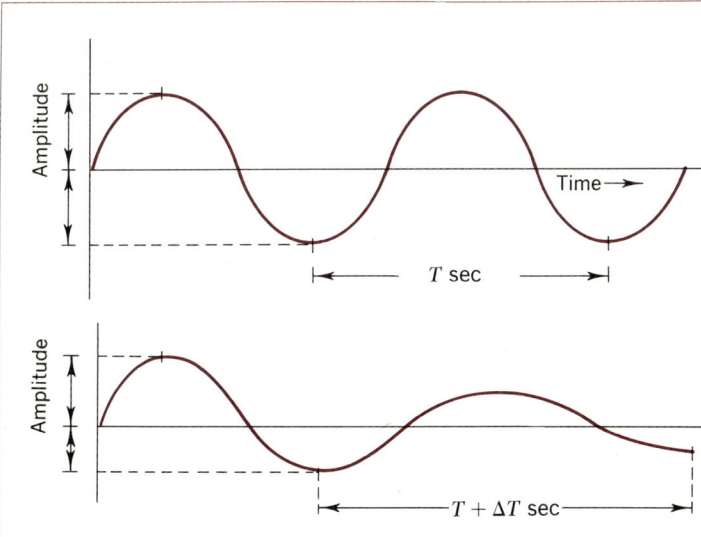

Fig. 6.3. *How the period of simple harmonic motion changes when there is friction in the system. (The period of seconds increases to $T + \Delta T$ sec, while the amplitude of motion decreases.)*

thing more by adding an electric switch so that the dropping weight sends an electrical signal to some device. As we know from common experience, the simple operation of an electric switch can initiate (or trigger) many motor or relay functions that bring into play the vast energy resources of electric power-generating stations. Figure 6.4 illustrates the relationship between input and output, with a transform function that relates the two along with a source of energy.

Suppose that we incorporate such an electric switch as part of a control system for automatically filling a bucket with water. Figure 6.5 illustrates how the dropping pail signals that the pail is full and also turns off the stream of water. This becomes a simple on-off control system in which the electrical signal provides a *feedback* function as part of the control system. (Later we shall introduce the idea that the feedback also represents *information*.)

We examine this process of filling the bucket in a little more detail. When the water flows into the bucket at a very slow rate, the bucket settles slowly and the signal switch has time to stop the flow of water and bring the bucket to a gentle stop. This is shown as curve *A* of Fig. 6.6, which shows very little dropping of the bucket below the desired level (that is, there is very little overshoot beyond the desired control point). The behavior becomes quite different when the water flows into the bucket at a rapid rate, however. The switch operates as it did before, but the rapid dropping of the bucket develops enough momentum to overshoot the desired final position by a substantial amount. The bucket will oscillate violently above and below the desired control height for some time and the switch will open and close erratically (curve *B*). In fact, if the response rates and delays in the switching and valve devices should turn out to be particularly

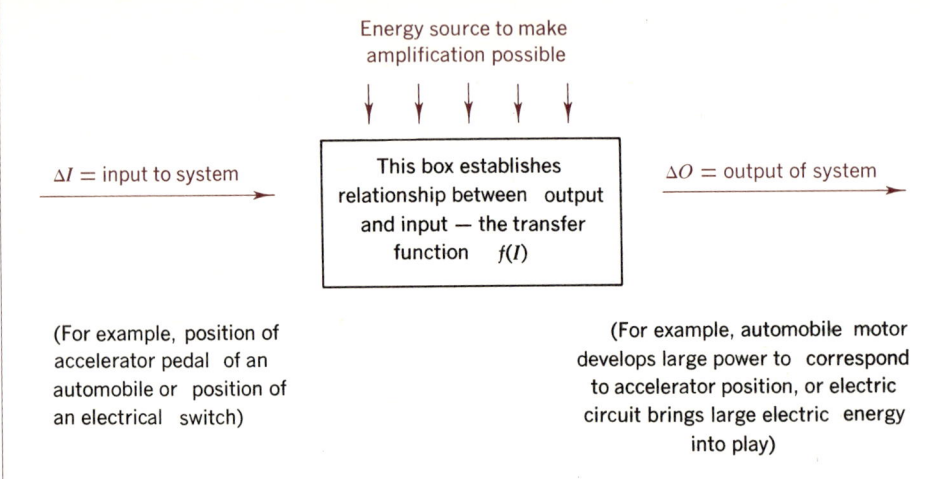

Fig. 6.4. *How a small* input *change (such as the operation of an electric switch) can bring into play sources of energy and thereby produce an* output *that may be quite different in form and magnitude from the input. Each such conversion can be referred to as involving a transformation (transfer function or transform function).*

unsuitable, the water would be turned on and off in such erratic manner as to recall the sad fate of the Tacoma Narrows bridge; see curve B, dotted line, Fig. 6.6.

In the case of room-temperature control, the thermostat is likely to be kept at one temperature (for comfort), say, around 72°F. In the case of the baking oven, the temperature setting will vary with the requirements for baking a cake or roasting meat. In either case, the temperature will vary (or hunt) around the set control point. The hunting or oscillations can be decreased if the rate of heat input is slow. But this would increase the time needed to bring the room or oven to the desired temperature. With on-off control, the heating unit becomes fully hot whenever the control switch turns it on. By the time the temperature at the thermostat reaches the desired temperature to turn off the heat, the region of the heater units becomes much hotter than necessary, and this excess heat drives the temperature well above the desired temperature. A similar delay in reactivating the heating unit as the temperature drops below the desired level causes continual hunting above and below the desired temperature.

We shall appreciate more and more, as we examine more cases, that the "control" of a variable rarely results in an exact holding of the variable to the desired control value. *Nearly always, the variable will hunt or vary about that control value.* Therefore, the function of a successful control system is to hold the variable *within acceptable departures from the desired control value.*

6.6 Negative versus positive feedback

In all the examples given above, while it is clear that control at a point usually ends up as hunting around that point, even this

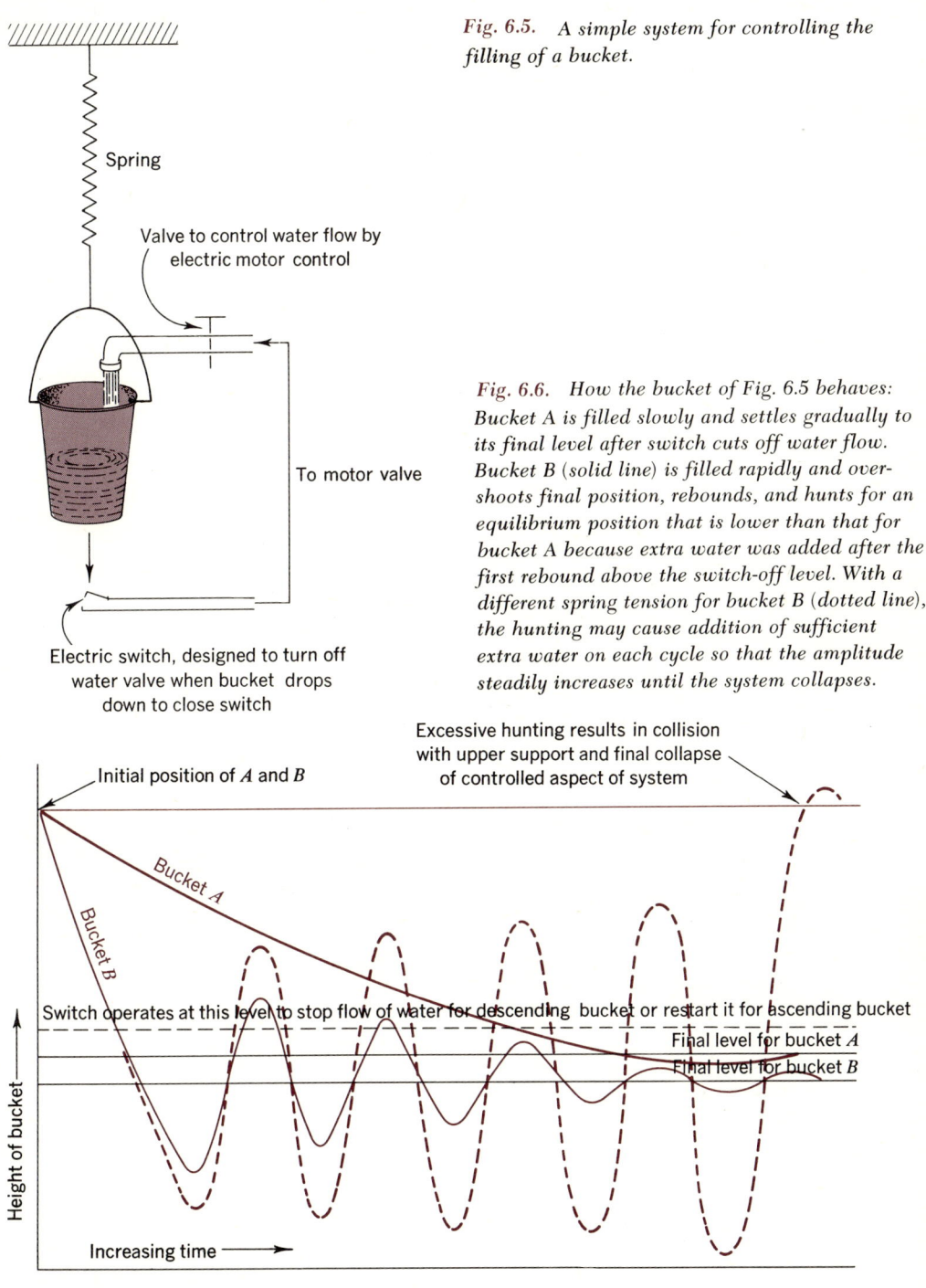

Fig. 6.5. *A simple system for controlling the filling of a bucket.*

Fig. 6.6. *How the bucket of Fig. 6.5 behaves: Bucket A is filled slowly and settles gradually to its final level after switch cuts off water flow. Bucket B (solid line) is filled rapidly and overshoots final position, rebounds, and hunts for an equilibrium position that is lower than that for bucket A because extra water was added after the first rebound above the switch-off level. With a different spring tension for bucket B (dotted line), the hunting may cause addition of sufficient extra water on each cycle so that the amplitude steadily increases until the system collapses.*

degree of control is achieved only when *negative feedback* is present. Thus, in the case of the full bucket, the switch turns off the water (since it was the "water-on" condition that filled the bucket). In the case of room-temperature control (which we shall discuss presently in detail), the heaters must be turned *on* when the room temperature is too low, and *off* when the temperature is too high.

The examples of feedback, as well as the limitations of on-off (sometimes called bang-bang) control can be illustrated further by the example of a blind person walking down a street with his cane. As he progresses along the sidewalk the tapping of his cane tells him when he is too close to the buildings on the right. This *information,* when processed through his brain and muscle system, serves as *feedback* to change his direction. Since his movements have taken him too far to the right, now he must move to the left and therefore the *feedback must be negative.* If the influence of feedback *were positive,* it would support or add to the original direction that took him to the right and would take *him even farther to the right and directly into the wall.* He now continues to the left until his cane warns that he is too close to the curb at the left. This information again converts to become negative feedback, which will oppose the move that carried him too far to the left and thereby will restore his direction until a new signal calls for new action.

Our blind person can negotiate the walk fairly well as long as his movements are slow enough to give him time to receive the signal from his tapping, to interpret these, and to translate them into suitable feedback influence. But now suppose he tries to run down the same sidewalk. Very soon his rate of receiving and responding to signals would be inadequate, and he would be running in a zigzag or colliding with obstacles.

Such an experience, which the reader can himself check rather dramatically, illustrates several features of control that apply fairly generally, namely:

(1) Stable control requires the presence of negative feedback influences.

(2) Stable control of a variable to a "fixed" point usually means maintaining the variable so that is does not hunt around the point beyond acceptable limits.

(3) To be effective for the control of any variable, the control system must be designed to have response rates that are suited for the specific application.

These and other characteristics of control systems will be illustrated in the following sections.

6.7 *Driving an automobile*

To illustrate further the limitations of on-off control, let us apply the technique to driving an automobile in a lane of the road that is marked with white lines. We know from experience that an auto tends to go from side to side (to hunt), and requires continuous steering control. Let us assume an unreal situation in which we turn the steering wheel a small, fixed amount to make the correction, and do this only when a front wheel touches a white line. The experiment would then be like the walk of a blind person. When crawling along at a very slow speed we would find that the car does not go very much outside the lane, but when driving at a moderate speed we would find that this type of correction (applying a fixed amount of adjustment as on-off control) causes the car to weave substantially in and out of the lane. If we were to drive even faster, the car would be likely to leave the road altogether. The amount of overshoot would depend on how slowly we respond to visual signals and take action (see Fig. 6.7).

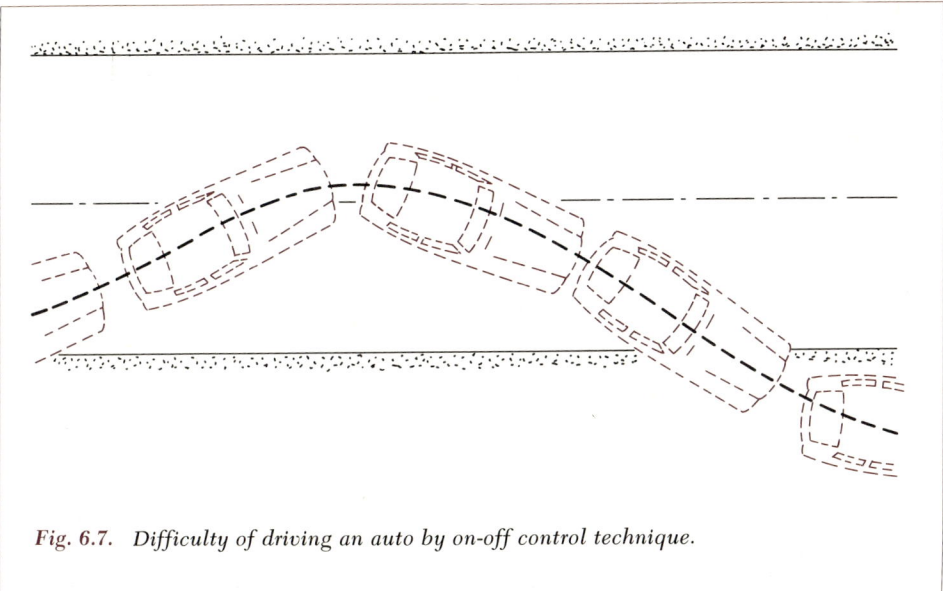

Fig. 6.7. *Difficulty of driving an auto by on-off control technique.*

Fortunately not many people drive in this manner because control of an automobile utilizes a much more sophisticated system of elements than is possible with on-off control. In fact, not many automatic industrial processes can compare with the sophistication and effectiveness of good auto driving, since human judgment enters this operation to a remarkable degree. To begin with, as the auto moves to a new position or direction, the driver is kept continually informed of the nature of each new situation through his sense of sight and general physical awareness. That is, there is continuous feedback, or information, reaching him to guide his next move. The element of judgment or experience also enters. He can vary the sharpness of turn of the steering wheel to conform to the sharp right turn. This is called *proportional control*. In addition, he can see a curve in the road ahead long before the auto has reached the curve. He can therefore anticipate the move (*anticipatory control*) and thus reduce delay in his action (Fig. 6.8).

The driver of an automobile is aware of several elements that make control more difficult. If the steering wheel has looseness or "play" in the shaft or gear system, the steering wheel must be turned several degrees of angle before there is any effect on the front wheel directions. This play, or region of no response, is sometimes called the *dead zone* of the system. The driver himself may be a little slow in judging the situation and taking action. This "lag or slowness of response together with looseness in the steering system, can make for wider overshoot in the movement of the car. If the throttle sticks, the motor hesitates, or the brakes seize, the driver will not be able to assure smooth "feel" and ride. Finally, roughness of the road can introduce random fluctuations that add uncertainty to the normal small feedback of information. A driver is not likely to give delicate

198 Systems, Feedback, Cybernetics

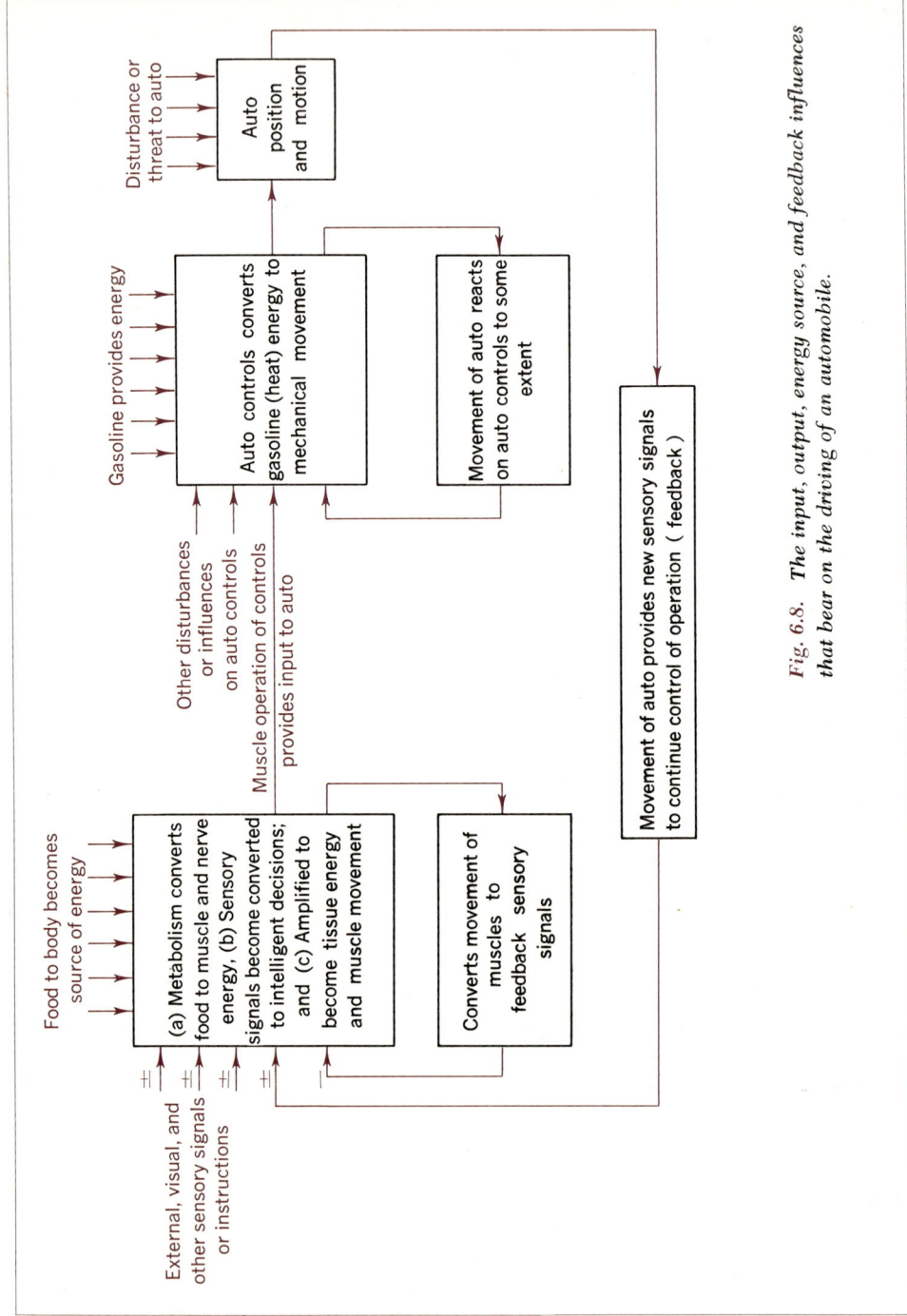

Fig. 6.8. The input, output, energy source, and feedback influences that bear on the driving of an automobile.

guidance to the auto when his whole body is being shaken. This background confusion is often called *noise* or *static* when one is referring to transmission of signal or of information. It exists in almost every type of control circuit, sometimes in the form of vibration of an automobile or plant equipment. It occurs in the normal radioactivity background of the environment, which disturbs radiation measurements. In very sensitive electronic circuitry it shows up in the random movements of the electrons. Similar phenomena are present in social situations and in biological organisms that maintain balance in their internal functions and with their environment.

Because cyclic and control aspects of nature are exceedingly important, we must consider control principles and nomenclature in a little more detail before looking at the several types of systems that are common.

6.8 Some control principles— nomenclature

Before beginning detailed discussion we need some convenient terminology and symbols for representing the elements and functions that make up systems. When the driver of an automobile engages the gears and steps on the accelerator pedal, the motor races and the car moves forward with an expenditure of energy that is vastly greater than the energy applied to the pedal. The power is amplified. We may represent this by a diagram such as Fig. 6.4. The input, ΔI in this case, appears to be simply the change in position of the pedal and the small energy required to make the change. The box in Fig. 6.4 represents the change or transformation (of function) that the input ΔI initiates or experiences; in this case the function produces motor power at a level that is related (and perhaps proportional) to the position of the throttle or accelerator pedal. We can refer to the box as representing a transfer function $f(I)$, or converter, which produces ΔO.

What is the source of the energy that makes this conversion possible? In this case, the energy source is the chemical energy in the tank of gasoline. The pedal, therefore, is nothing more than a lever device for controlling the use of this chemical energy. When one includes the tank of gasoline and the driver along with the automobile, the system becomes a *closed* (or *conservative*) *system*.† Without either one, the system would be incomplete. (It is common practice to omit the *sources* of energy from block diagrams of control systems and to indicate only the energy input and output for a system.)

The operator has freedom to depress the accelerator pedal quickly or slowly, as he wishes. A question that is frequently important for analyzing the behavior of control systems is the following: What is the nature of the output response when the input is given a quick change? A quick increment of input change, which we may represent by ΔI, is usually called an input *step function*.‡ Figure 6.9 illustrates what might happen. Usually there is some lag in the rise of motor power, as shown by the curved rise and fall of the output. This lag will not be serious in the case of the automobile, since the input is not likely to be reversed rapidly very often. In general it is desirable to have as much lag as one can tolerate, consistent with adequate control; otherwise the system will be too ready to "jump" and probably to overshoot the

---

† We neglect the fact that as gasoline is used up it must be replaced, bringing the entire petroleum industry into our system. Likewise, food for the driver is neglected.

‡ The term *step function* is often associated with *on-off* changes because the change of power or direction assumes the form of a sudden change. This is illustrated by the shape of the heat input as it is turned on and off in Figs. 6.9 and 6.10.

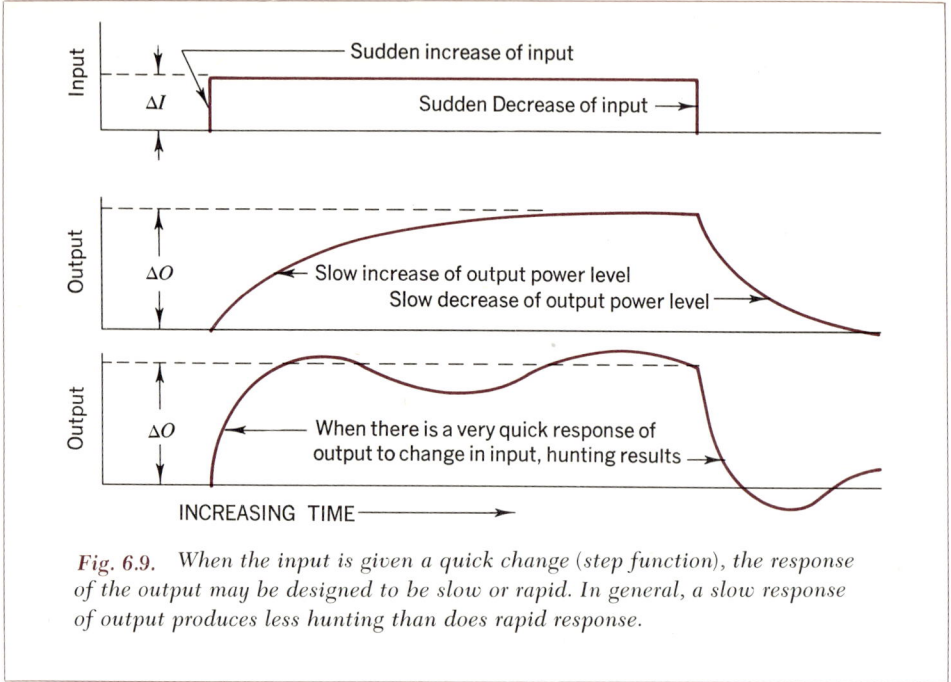

Fig. 6.9. When the input is given a quick change (step function), the response of the output may be designed to be slow or rapid. In general, a slow response of output produces less hunting than does rapid response.

mark and hunt badly before settling down. The lower curve of Fig. 6.9 illustrates the nature of the "hunting" that results when the system is made to respond too quickly to change in input. How the output will respond depends on the characteristics of the system and on the features incorporated in the transform function box of Fig. 6.4.

If the system of moving parts includes large, heavy components such as the flywheel and other parts of the automobile motor, we can appreciate that quicker response is possible only if the motor is designed to have adequate extra power to give the desired acceleration. But excessive power can make control less smooth and more "jumpy," not to mention excessively costly in gasoline and in the complexity of the motor itself. The goal for design of most systems is to find a happy compromise that makes the system *adequately responsive and yet stable against excessive hunting, and which is not too expensive in dollars or in use of energy.*

The system we have been discussing has the features of *proportional control.* That is, the accelerator pedal may be depressed to give large or small change, and the motor power level will respond with some proportional relationship. We backtrack a little to discuss *on-off control* before proceeding further.

6.9 More on on-off control

Earlier we discussed how difficult it would be to hold an automobile within the lane of the road if we applied on-off control principles to adjust the steering. Despite certain limitations, on-off control devices are used very commonly in homes and in industry because of their simplicity. It is very easy to design an electric iron, an oven, or room-temperature control, to operate an electric switch to turn on (or off) the electric power whenever the temperature falls below (or rises above) set values. Figure 6.10 illustrates how this might apply to the thermostat controls for heating a room in the winter time.

As shown in Fig. 6.10(a), 72°F is the temperature desired for this room. But all thermostats and switching devices require a differential zone of temperature change in which to go on and off; otherwise they would act too frequently and probably erratically because of vibration conditions and momentary temperature fluctuations in the immediate neighborhood of the thermostat. We start with the temperature dropping in the upper curve of this figure. When the temperature reaches the lower edge of the differential

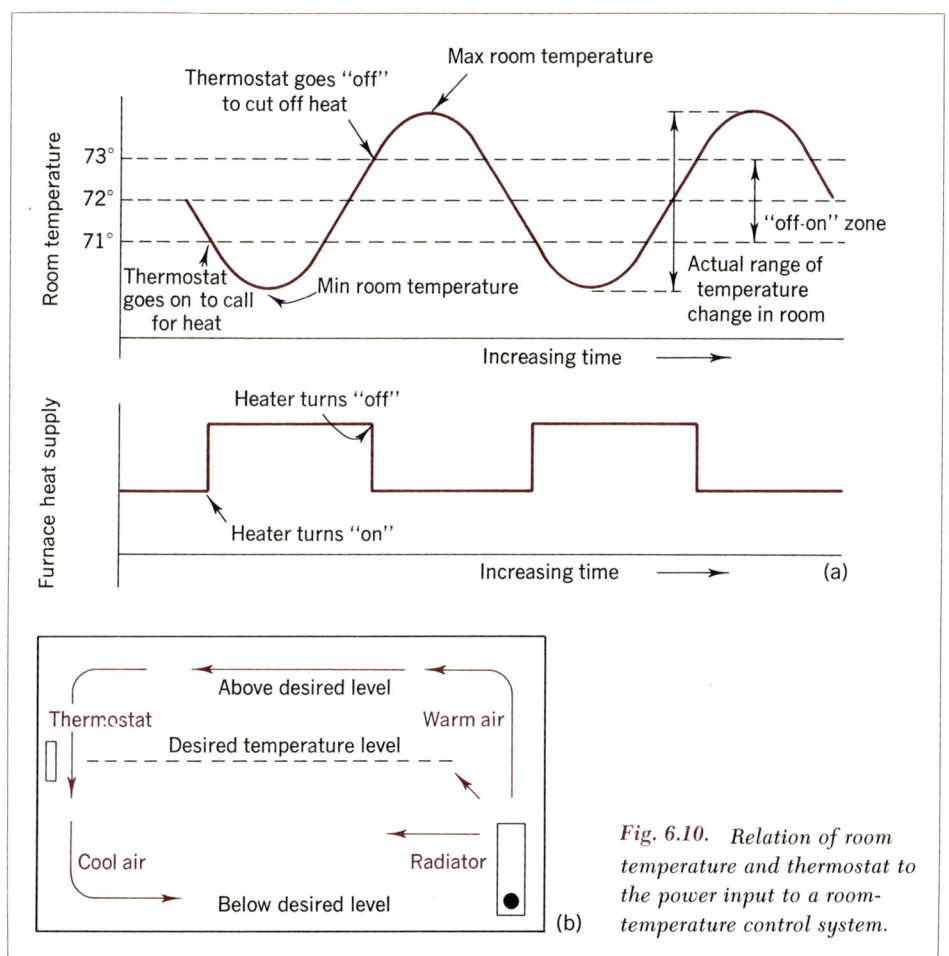

Fig. 6.10. Relation of room temperature and thermostat to the power input to a room-temperature control system.

temperature zone (71°F in Fig. 6.10(a)), the thermostat switch turns on the heater. This assumes that there are no significant time delays in the response of the thermostat or the heater controls. (In actual experience there are always some delays.) The radiators around the room take much more time to heat up, and the temperature of the air in the room continues to drop until it reaches some point which is well below the lower limit of the control range (about 70° in Fig. 6.10(a)).

As the hot radiators heat the air in the room, the temperature at the thermostat starts to climb again, and at the 73°F level the heaters are turned off. But at that point the radiators are fully hot, and the air in the room continues to receive heat and to rise to a maximum temperature which is well above 73°F. The net result is that the room temperature may vary by as much as four or more degrees Fahrenheit. In an actual system there will be a little time lag between the temperature at 71°F or 73°F and the response of the thermostat and heater controls, which can make the overshoot and hunting more severe. Nevertheless, the simplicity and relatively low cost of on-off systems makes them very attractive for use in such operations as controlling temperatures, maintaining water level in tanks, and many other operations. Biological and some social systems, as well as many industrial, mechanical, and chemical processes, usually require the more accurate control that can be achieved through proportional-type systems.

How much power can an on-off system control? It is fairly clear that the switch that turns the heater on and off can be designed to handle any amount of electric or other form of energy. The amount depends on the power requirements to keep the variable that is being controlled as close to the desired value as possible.

A general rule might be to design the power level so that the controller calls for heat about half the time, and the heater remains off half the total time. Sometimes the control is improved by supplying a portion of the power continuously at a low, fixed level, and allowing the control system to add or subtract a smaller increment of power as needed.

6.10 Characteristics of proportional control

The on-off type of temperature control, in which the power is usually turned full-on or full-off, is inadequate for many applications that cannot tolerate the wide surges around the desired control point that often accompany on-off systems. The undesirable surges can be reduced if the power is moderated in proportion to the need. This is exactly what is achieved in proportional control systems, in which the heat input continues at some intermediate level when the temperature is near the desired control point. As the temperature rises somewhat, the controller reduces the heat input *in proportion to the departure from the set control point*. Similarly, the heat input is increased in proportion to a fall in temperature below the set control point. Of course the system becomes more complicated because now the temperature detector must measure the *magnitude of departure from the control point*. (In on-off control, all that the detector has to do is to note that the temperature is above or below the set point.) Also, there must be somewhat more complex interconnection so that the proportionate (or step-by-step) changes in the temperature detector can be translated into proportionate (or step-by-step) action on the part of the valve or motor that controls the fuel or power input.

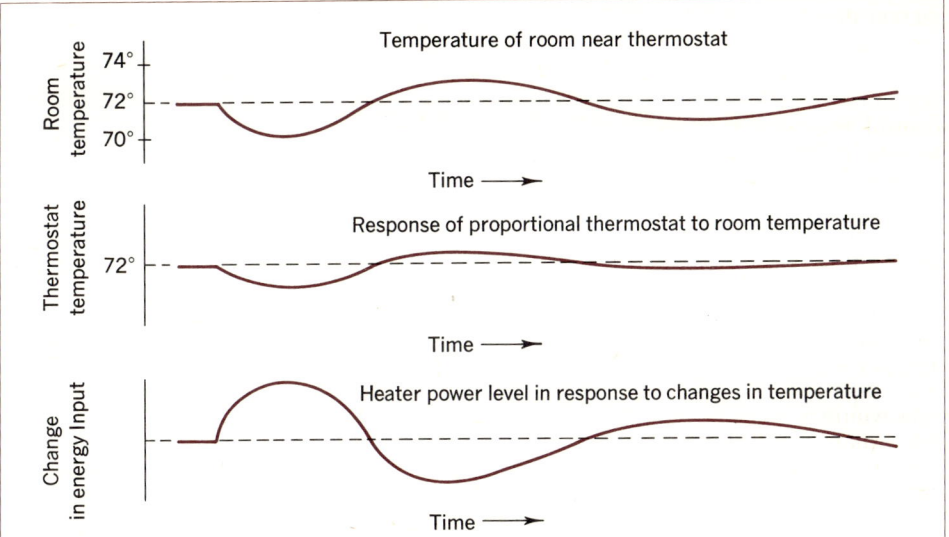

Fig. 6.11. *In a proportional control system, the response of the thermostat is proportional to the departure of room temperature from the desired control point and the change in power input to the boiler is proportional to the response of the thermostat.*

Let us analyze the action of such a system designed to control the temperature in a room.

When the door of the room opens and lets in a draft of cold air, the thermostat responds as shown by the drop in the upper curve of Fig. 6.11. As shown by the middle curve, at that same time the thermostat control calls for a proportional increase of heat, and the heaters respond as shown by the lower curve. As the draft of cold air becomes warmed somewhat by mixing with the warmer air, the proportional thermostat correspondingly reduces its demand for heat. The net result is that the room temperature is maintained much more closely to the desired 72°F than is possible with on-off control. But the proportional control instruments and equipment tend to be more expensive, and for that reason they are not used where on-off control is adequate.

A serious limitation develops in proportional control systems when the load demand changes so that a different average power level must be applied to hold the variable at the desired control value. To understand this, we note that in proportional control, the output ΔO (Fig. 6.4) has a fixed ratio to the input ΔI. This proportionality ratio, or gain, may be represented by $G = \Delta O / \Delta I$. Assume that the room-temperature control we have been discussing is set to control at 72°F *when the outdoor temperature is around 50°F*. We may assume that this requires an average heat input of 10,000 Btu per hour. Suppose that the outdoor temperature drops to 0°F. Obviously, the heater system must provide a great deal more heat to hold the temperature at

72°F, *say*, 30,000 Btu per hour. We therefore need an additional $30{,}000 - 10{,}000 = 20{,}000$ Btu per hour to hold the temperature at 72°F. But since in proportional control more heat is provided only in proportion to the temperature drop from the control setting, how can the additional heat be provided without the actual temperature remaining well below the desired control value?

Let us analyze the situation a little more quantitatively. Suppose that the gain of our control is set so that, for each degree that the temperature drops, the controller permits an additional 2000 Btu per hour to be supplied to the boiler. This represents a gain or proportionality ratio of 2000 Btu per hour per degree fahrenheit. To get the additional 20,000 Btu would require that the temperature of the room go down to 62°F. Or, alternatively, the thermostat setting would have to be moved arbitrarily to about 80°F in order to supply enough heat to hold the room temperature at 72°F as long as the outdoor temperature remained at zero.

This discrepancy could be reduced if the gain were made higher (that is, 1°F could turn on much more than an additional 2000 Btu per hour). But making the gain higher also makes the system more unstable. Other devices can be introduced to change the responsiveness of the controller, such as incorporating into the system an outdoor thermostat that introduces this equivalent of the arbitrary shift of a thermostat setting of 80°F. We need not go into more detail beyond recognizing this severe limitation of proportional control systems.

6.11 Feedback

We must give a little more attention to the important feedback function. When the thermostat of a temperature-control system demands more heat, the additional heat energy continues to pour into the heater boilers and radiators until feedback information (in the form of rising air temperature in its neighborhood) reverses the thermostat demand. In the case of the driver of the automobile, although his foot on the accelerator finds good proportional power response on the part of the motor, only feedback in the form of vision (and the transformation of that information into suitable muscle action) makes driving successful. Without the presence of feedback, the driver could not function as part of the system.

The *kind* as well as the timing (or *phase relationship;* see Sect. 6.3) of feedback are rather important. In the case of the temperature controller, the electrical thermostat reactions must become transformed into heat energy and transfer of this energy to the room if there is to be control of *temperature*. In the case of the driver, the feedback which arrives in the form of sensory information must become interpreted and converted into suitable muscle action on the accelerator pedal to be effective.

In the case of temperature control the feedback must always be *negative*. That is, the rising room temperature causes the thermostat to demand less heat, while a dropping temperature causes it to demand more heat. In the case of driving an automobile, the feedback may be negative (say, when the traffic light turns red and the driver has to let up on the accelerator) or positive (say, when the way is clear for higher speed). When a politician confronts his voting constituents on an important issue, he watches their reactions as he talks, to get some form of feedback, When the response (or feedback) from the audience is "positive," he believes that his statements have been received favorably, whereas a "negative" feedback is likely to make him cautious.

Feedback may take many forms and many types of coupling. Figure 6.12 illustrates a simple modification of an earlier graph. In this illustration some of the output energy is fed back to the input. The box marked "feedback transfer" determines how much of and in what form the output will be fed back. The input is represented by a long arrow with positive increment. The feedback is shown as a small arrow with negative value. In such a setup the net input is *reduced* by the amount of the negative feedback. The effect is to restrain or to limit the output. If the sign of the feedback were positive, the input and the feedback would add and the output would increase continually and build up to destruction, or to the limit of the energy input. A system with feedback is often referred to as a *closed-loop system*. Since such systems incorporate a measure of self-correction, the exact value that the input is permitted to have becomes less critical. This self-correction factor also applies to the automobile driver, who does not have to have a gauge on the foot pedal because the "feedback" of his eyes and ears is enough to guide and restrain his push on the foot pedal.

High values for gain in amplifiers or control circuits tend to make a system unstable, and time lags produce wider oscillations. Negative feedback, on the other hand, tends to stabilize the systems.

6.12 The elements of control systems

Now that we have developed some familiarity with control systems, we can identify the functional elements that make up most systems.

THE VARIABLE TO BE CONTROLLED

First there is the variable that the system is expected to cope with or to control within prescribed limits. Actually it is rare that only one variable is present in a system. In the case of room-temperature control the changes in the outdoor temperature constitute an *independent variable*, while the internal temperature represents the controlled variable. Other independent variables may be introduced, such as children running in and

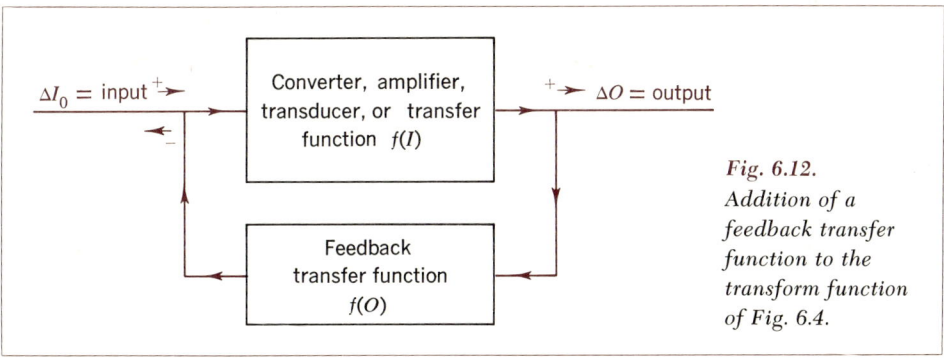

Fig. 6.12. Addition of a feedback transfer function to the transform function of Fig. 6.4.

out of open doors, to cause variable demands for more or less heating.

Similarly, the driver of the automobile has control devices by which he steers and starts and stops the car in relation to the road. But all along the way he is forced to comply with independent demands, such as changing road and traffic conditions, stop signs, and traffic lights, all of which constitute independent variables.

SENSOR DEVICES

Usually there must be some sensor device by which the variable can be measured or gauged. For example, in the case of the temperature measurement we shall learn in Chapter 10 that the temperature of air is actually determined by the velocity of the molecules that make up air. But we cannot gauge the temperature by measuring the velocity of molecules directly. What we can do is to utilize some *effect* that changes with changing molecule velocity. For example: At high air temperature, molecules in a material become more active and bump each other harder and more frequently, causing objects such as fluid in a thermometer or a piece of metal in a thermostat to warm up. A thermostat usually includes a bimetal† that carries an electrical contact; the bimetal changes its position when the air temperature changes and thus makes or breaks an electric circuit.

Similarly, while the position and behavior of the automobile are the variables to be controlled, we gauge these by the use of sensory information (vision, hearing) and the interpretive processes of the brain. The economist also looks for meaningful indices by which to gauge the larger features of national product, industrial trends, and public attitudes. The public utilizes quality and creativity as gauges to evaluate intrinsic or extrinsic return on investment.

ENERGY SOURCE

Whether one deals with a temperature-control system, driving an automobile, or any other situation that involves variables and controls, there must be a *source of energy* by which the job is performed.

MOTOR TRANSFORM DEVICE

In most instances the sensor function must utilize the services of a motor device to restore a variable to its proper value. To do this the motor device, or motor function, utilizes energy from an energy source. In the temperature-control system, the blowers and burners (which are triggered into action by the thermostat) begin to utilize fuel energy to heat the boilers. In the automobile a number of mechanisms come into play to burn the gasoline, to power the steering, and to perform other nondriver functions.

FEEDBACK

Finally there is a feedback device, or feedback function, which in one way or another relates the output to the input and thus controls the net output.

The functional elements of a system cannot always be identified individually or even as subsystems, but they are present in one form or another. One characteristic that will be evident more and more is the wide variety of transformations (transform functions) that are

---

† A bimetal strip is made up of two different metals bonded together. Because the two metals have different temperature expansion coefficients, the bimetal will bend when heated, thus causing the contact to switch on the system. As it cools, it straightens and contact is terminated.

possible in systems. Molecular speeds are transformed into mechanical bending of a bimetal, which completes an electric circuit and utilizes electric energy. This in turn starts a motor and pump device to feed and burn oil in a boiler, which produces heat that is transported or transferred by various means to another area by other motor devices. Similarly, all the tangible and intangible features of human physical energy and human brain processes become involved with electromechanical and chemical systems in driving an automobile. (For each of these transformations we can apply the more elegant phrase *transform function,* and illustrate the nature of the transformation by means of a mathematical equation, a graph, a listing of data, or a simple picture.)

6.13 *Control concepts: cybernetics*

The art and science of control theory has had a long and slow history. In the early days it found application in the sailing and steering of ships. With the coming of the steam engine a mechanical governor was needed to keep the speed of the engine constant. In more recent decades a wide variety of instruments, valves, and other equipment have been developed to maintain uniformity in chemical production processes. Servomechanisms were introduced during World War I for control of gunfire. Electric circuitry and electromechanical systems were given intensive study to improve their responsiveness and stability for purposes of controlling high-speed operations. By the 1940's the pace of automation had quickened as the concepts of control theory and of feedback received wider application in the electrical, mechanical, and processing industries. The term *high-fidelity* became a byword in amplifier design as a result of the introduction of negative feedback.

But the concept of feedback seemed to be basic and useful for a much wider range of applications. In 1947 the mathematician Norbert Wiener and Arturo Rosenbleuth compared the phenomena of control and of feedback, as used in technology, to the nervous system and muscle behavior of the human body. They postulated a close coordination of communication relationships between the brain, the sensory organs, and the muscles, and concluded that this resulted from the extensive use of feedback principles. It seemed that a feedback function is responsible for one's ability to reach down and pick up an object and to know how much farther the hand must move to complete the act. Moreover, they found an identity between *feedback* and *information* and the information content of a signal above the noise level. They gave the name *cybernetics* (from the Greek *kybernes* for *steersman*) to the entire field of control and communication theory, whether in the machine or in the animal.†

The concepts of *feedback* and *information* encompassed by cybernetics permit very extensive applications to the biological and social world. Just as the driver of the automobile performs functions in response to the information he derives

† The broad concepts that make up the science of cybernetics as developed by Wiener and his associates were new. The word itself had much older origin, however. It appears that Plato often employed the word "cybernetics" to mean "the steerman's art." His comment in *"Cleitophon,"* "the cybernetics of men, as you, Socrates, often call politics," suggests a wider implication. In 1834 the French physicist Ampère used the word as "means of governing" people.

from seeing and hearing and evaluating the driving situation, so his reactions under other situations are the result of his relationship or interaction with each new environment. Information and feedback are essential to his every move, every decision, almost every thought and learning process.

We shall have many opportunities to refer to the principles that have just been introduced. There will be applications to strictly technical systems, to systems that involve nature's resources, to biological systems, and to social situations.

The importance of the subject suggests that we summarize a few of the ideas that are most pertinent to our purposes.

1. Nature's processes are characterized by continuous dynamic transformations of energy, which may range from the vast magnitudes of astrophysics to the metabolic adaptation of the smallest living organism to its environment.

2. Much of man's own activities also involves the development of processes for conversion and utilization of nature's energy resources for purposes of assuring his survival and comfort. Indeed, the design of systems that integrate physical and chemical variables into cooperative, controlled systems constitutes a main interest of science and industry to bring about modern civilization and the current standard of living of advanced nations.

3. It is now recognized that the elements that make up a controlled system have common characteristics, whether accomplished by machine components, biological elements, thought processes, or social situations.

4. In such systems, the element of *feedback*, or *information*, which interrelates the output (or behavior) of the system and the input variables, constitutes a major factor for the effective operation and stability of systems.

5. The design of every control system requires careful analysis (and usually compromise) to meet the needs of the process. A prime requisite for most control systems is that they be *adequately responsive to changes*, and that they be *stable*. Also needed is an adequate *source of energy* to perform all the functions that are required of the system. The *input* to the system may be some *variable* such as temperature, liquid level, or pressure. Or it may be information that is itself the product of other operations, such as in *computer systems*.

6. The system performs its function by transforming the input to produce an *output* whose energy content is usually amplified, the added energy being derived from the source of energy of the system. The character of the transformation is designed into the system and is identified by its *transform function* to give the change or *amplification gain* to the output.

7. When a feedback (or information) loop permits some of the energy of the output to be fed back to the input, there can be considerable influence on the nature of the net output and on the stability of the system. In general, feedback that opposes changes (negative) in the input will improve the stability of the system, while feedback that arrives at the input in a manner that increases its changes (positive feedback) tends to reduce the stability of a system.

8. The stability of the system suffers and the system "hunts" more violently when the amplification or gain between output and input is too high or when the system responds too quickly to changes in the input variable. The design must include enough damping to reduce excessive overshoot (or violent hunting) of

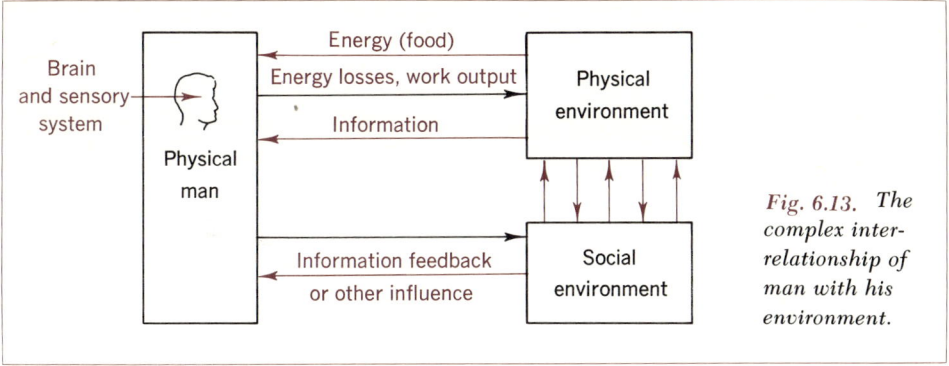

Fig. 6.13. The complex interrelationship of man with his environment.

the system while still providing adequate response. On-off controls offer cost advantages and simplicity, but the need for better control may dictate the use of *proportional* control or of other controls that have more sophisticated design. There can be more than one input to a system, more than one output, and a wide variety of interrelated combinations. In fact, the input may be the statistical output of many interrelated elements or variables.

6.14 Some examples of systems

In Fig. 6.13, which illustrates the relationship of man to his environment, we have identified two aspects of man (his brain and sensory motor system as distinguished from his physical being) and two aspects of his enviornment (the physical and social environments). There is very intimate and extensive interchange between the two aspects of man and between the two aspects of environment, as shown by the proximity and multiple arrows connecting them. Man draws energy and material from the physical environment and returns information and other materials to both.†

In the case of the driving of an automobile, it is difficult to identify all the elements that make up the input to this system. The desire to drive, the sensory activity that provides data to the brain, and the muscle behavior that operates the controls of the car, each is a complex that includes and combines the product of some other part of the system. The energy involved in the seeing, hearing, and judgement operations is negligibly small, but these become greatly magnified by the body's metabolic processes. This transform function of the body is most complex, and is itself made up of innumerable subsystems.

The specific control principles and systems we have discussed thus far are given broader significance by the principles of cybernetics. Cybernetics deals

† It is not easy to distinguish work from information and learning. Physical acts are not readily distinguishable as being separate from sensory response and interpretation that leads to learning, judgment, and decision. Certainly we cannot say that the throwing of a ball, intake of food, reading of newsprint, and a walk around the block are not so much mental processes that lead to future decision or action as they are physical acts.

with elements or variables that are related to each other so intimately that a change in one variable is likely to affect other variables in the system. The elements may be parts of a machine or those of a chemical process. Cybernetics can deal with the very specific behavior of a single molecule among vast numbers of gas molecules or with the behavior of a single cell of the vast numbers that make up an organism. It can as readily (and in general more usefully) consider *the statistical-behavior character of all the gas molecules together, or all the cells of the human body*. It can provide a method for analyzing the economic relationship of a grocer and his customer, or as readily attack questions pertaining to the economics of a whole nation. It establishes functional relationships in the course of changes, emphasizing their coordination, regulation, and control within a systems concept.

From the point of view of cybernetics, the aspect of systems behavior that is of greatest interest is the system's response to a disturbance. This disturbance may be a normal change or a momentary departure (transient) of the input, say as a result of the dropping of temperature of an industrial oven below its control setting when cold material is poured into it. One or more of the input variables or signals may experience changes that sum up to a signal sufficiently large to initiate a major change. For example, many chemical processes go on within the body, such as food intake, digestion, blood cell production, and oxygen utilization. They are not unrelated and all must be considered contributory to whether a person feels well or feels ill. Each process experiences its own daily or hourly variations, which nevertheless may constitute normal operation and good health. There can be occasions, however, when the individual variations in the processes add up to produce sickness of a sort that represents serious imbalance or disturbance of the total system.†

In general,' systems are designed to accept and to cope with very specific variables and to effect reasonably quick restoration whenever some change in those variables upsets equilibrium. The system is considered to be *responsive* when it reacts with *adequate* speed to the upset. A system that responds too quickly or introduces corrective steps that are too large is likely to produce instability around the equilibrium point. A system may also be too sensitive to small fluctuations that are of the order of magnitude of background "noise," and for that reason will be unstable.

We might consider the design of an electrical amplifier system such as that used for a quality phonograph system. Figure 6.12 represents a fairly simple circuit for transforming an input through some form of *transducer* to produce an *output*. The *feedback* to the input in this case was designed to counteract or oppose the input, tending to reduce undesirable excursions in the output due to variations other than the sound signals to be amplified. The system constitutes a channel for transmitting and transposing signals, the input signals being *information*. To be effective, the design must usually incorporate suitable capabilities in such terms as capacity, watts, voltage, range, and frequency. These in turn provide the basis for designing suitable *constraints* into the system.

However, a control system is not likely

---

† As a simple example, the experience of sitting in an awkward position can introduce a combination of neural signals and mental process that suggests the need for a new position and thereby requires a complete readjustment of nerve and muscle systems.

to be designed to control every variable against every change. For example, the body's control of the iris openings of the eyes (to permit only adequate light to reach the retina) has a very specific, limited function and purpose, which excludes sensitivity to other variations of body conditions. The purpose of constraints is to reduce the response of the system to variables that are not considered to be part of the information to be transmitted. There are also natural restraints or constraints on the information and on the variety of information that a channel may transmit. Among these are the limitations and directions imposed by the conservation of energy and the laws of thermodynamics. When a system combines several elements into an integrated organizational and functional interdependence, the interdependence automatically imposes constraints, since the elements are now no longer independent of each other. An amplifier system may have to contend with constraints in the form of costs, against which the designer must balance extra quality or fidelity or amplification.

With only minor modifications the diagram of Fig. 6.12 can represent a quite different system for communication of information. Figure 6.14 illustrates some

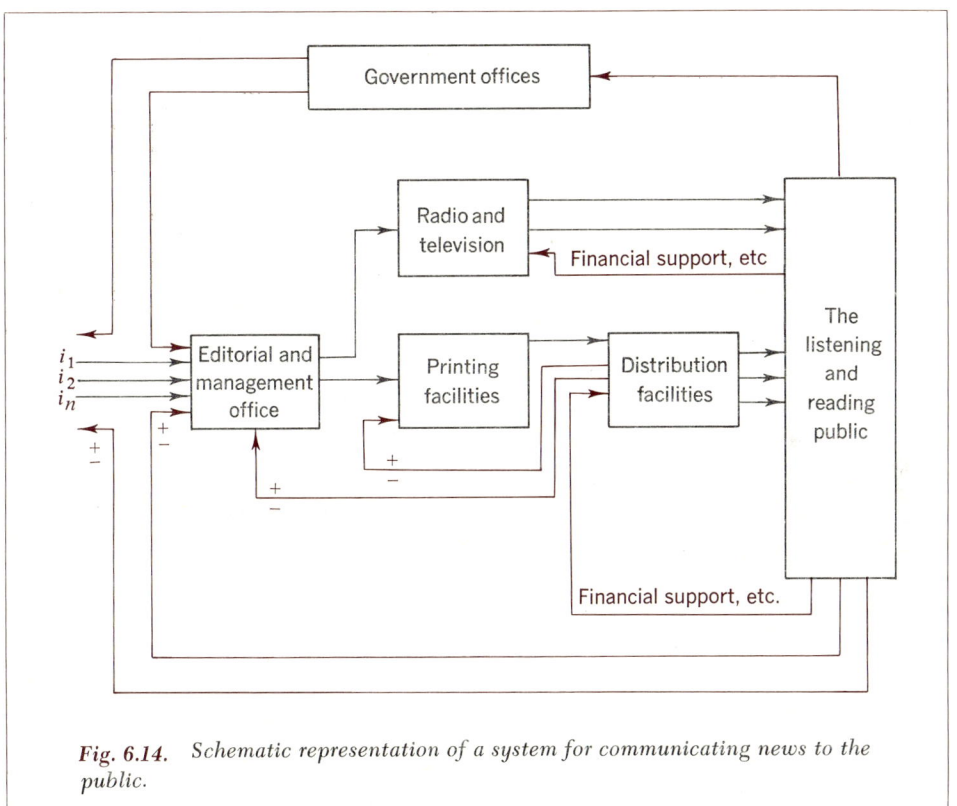

Fig. 6.14. Schematic representation of a system for communicating news to the public.

of the elements that enter into a system for communicating news to the general public. News may be collected from many areas and reported; this news becomes input $(i_1, i_2, \ldots i_n)$ to the editorial offices of a news agency. At the editorial offices this information undergoes modification and shaping, and is put into printed form or given electrical broadcast. There will be close liaison among the several blocks that make up the channels for this communication. There will be government influences as well as government sources of information bearing on the editorial and management offices, much of it in the form of feedback reaction to the communication. The listening and reading public applies "feedback" influence through financial support (or lack of support) of the broadcast and publishing services, through the editorial offices, and through government offices to the sources of information. The constraints in such a system are many. They arise from national and local government policies; from electrical, chemical, mechanical restraints; from the cultural habits and educational level of communities; and from financial considerations. As a total result, such a communication system becomes not a simple amplifier and distributor of simple news information but also a combination system for receiving, modifying, transmitting, and generating of news with built-in restraints and objectives.

One interesting characteristic of a system of this sort arises from the fact that any one of the multiple input signals $(i_1, i_2, i_3 \ldots i_n)$ can suddenly introduce a major disturbance that overshadows all other input signals and that can bring about violent response in either the forward channels or the feedback channels. Such a disturbance might be an act of war, a strike, a catastrophe, or an event that is especially disliked or especially desirable. There may be quite a few surges of output beyond the desired limits of control before the system settles down again.

One may also conceive that the input $(i_1 \ldots i_n)$ can be made up of very many items and elements so that the overall significance of the input is determined by the *statistical character* of the input rather than being overly influenced by any one item.

6.15 Functional relationship: notations

In its simplest form, a cause-and-effect relationship is stated as a simple function such as $y = f(x)$ (meaning y is a function f of x, or $O = f(I)$ (meaning output O is a function of input I). In diagram form this might be written† as representing a transition of I into an output O. Relating

this to our earlier example of the automobile, the power of the motor, O, is some function of the position of the accelerator pedal, I. If we include the driver as well, we have a more complete system *with feedback* and our equation would have to provide a different function for output,

$$O = f(I)F(O)$$

In its simplest form the diagram would be changed to become

The more interesting examples are not likely to be so simple as to comprise

† The approach in this section is considerably influenced by the treatment given by W. Ross Ashby.

only a single input and single output. Within the total system that includes the auto and driver, there are innumerable smaller systems such as cells, neurons, muscles, organs, machine parts, and electrical controls. A study of such assemblies must set out clear objectives before its approach or results can be made significant. For example, are the functions to be studied primarily those that pertain to keeping the auto on the road, or are they functions that determine the state of the driver's gall bladder, heart beat, or temperature? These subsidiary systems are certainly part of the total system of man and auto, but their details are independent of the specific functions that go with driving the automobile. The situation would be different, of course, if part of a study had to do with the effect of heart or temperature function on the driving, for which purpose a new set of elements would be involved when making up the system to be studied.

The situation is illustrated by Ashby in the following diagram, in which one may trace twenty different† circuits.

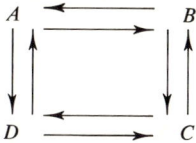

Each subsidiary circuit may have its own mode of feedback and control, and may be either strongly or weakly linked with neighboring circuits. In the case of our driver, vision plays the dominant role in telling him where the auto is going, while his knowledge of the situation is helped by the senses of hearing

† It is suggested that the reader list the twenty different ways in which a signal may travel through the system, starting at A and returning ultimately to A in each case. For example, $ABCDA$ or $ABCBCDA$.

and by the sensations of his body as the car sways. All the input stimuli have some relation to each other. One may picture a strong relationship (or strong coupling) between vision and hearing and a weaker coupling between heartbeat and vision, as far as driving the car is concerned.

Any study must therefore seek first to identify the functionally significant relationships that are the subject of the study, to identify the elements that bear directly on the functions under study, and to eliminate from consideration those elements that are independent of the selected functions.† This is not easy to do in most cases because there are many varieties of influences and "couplings" that come into play. Often the study must assume a series of situations and obtain results and estimates for a wide variety of combinations of systems.

MODELS

Often it becomes necessary to simplify a system or make it understandable by use of a model or models. This becomes imperative for nearly all biological systems, which are enormously complex. But models can very quickly become detrimental to progress when one loses sight of the simplifications and limitations that are inherent to each model.

6.16 "Black box" approach

Ashby considers the interesting case of an experimenter approaching a "black box" that is unknown to him with respect to contents and functions. How should he proceed to investigate and determine the contents and character of the box?

† The importance of such an approach in the study of human behavior is seen in the Gestalt psychology view of phenomena.

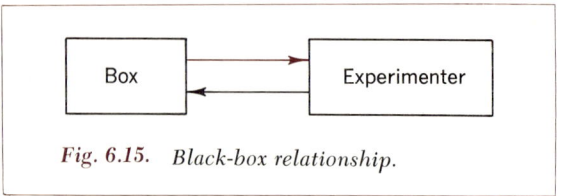

Fig. 6.15. *Black-box relationship.*

(The procedure can be especially important if one imagines that the "black box" could be an explosive bomb.) There is, immediately, a relationship between the experimenter and the box, of the type shown in Fig. 6.15. The diagram illustrates the nature of exchanges and feedback that take place as the experimenter explores the problem by various means. The "means" presumably might include such things as pushing and pulling of the box and levers. For a systematic search, each move would be recorded along with the "state" of the box that accompanied each move. In time the experimenter would presumably be able to identify the "state" of the box for each type of input, and possibly also the function for each type of input. Many of the systems with which we deal are actually made up of "black boxes," and the functional characteristics of the total assembly may be determined by the characteristics of each box and by the nature of their coupling together.† But we may fail to characterize the system because a combination of black boxes may produce an unexpected function that is quite unrelated to the characteristics of any one box. An example given by Ashby is that the approximately twenty amino acids in a bacterium do not individually have the property of being self-replicating, but their combination does introduce this property.

Real-life problems tend to have many "black boxes," often interconnected in such manner as to obscure the specific role of each box, each subsystem. One may make progress in the analysis of the total system and its parts by systematic analysis of "responses" or "states" to questions and input stimuli. One may seek to discover factors that produce

† The reader may picture the similar situation that exists when he first meets a person who is to become his associate on some project.

certain extreme "responses" of "states." The use of computers helps handle large quantities of data and identify common elements or contrasts. But progress in attacking complex problems depends more often on good use of judgment, experience, intuition or insight, persistence, and some luck. It is not always easy to identify and isolate the specific functions that are of importance for the system's functioning. There may be multitudinous other elements within the total system that do not bear on the specific functions under study.

6.17 The closed-loop amplifier system†

It will be helpful to look a little more closely at the quantitative aspects of a system that has feedback characteristics. Figure 6.16 illustrates a system in which an input signal E_i (which may be in volts

† The sections printed in color represent optional reading material.

and related to temperature, pressure, blood count, or other variable) constitutes the control variable. The system may be designed to do something that is proportional to or determined by this control variable i. If the system is a servomechanism, input E_i may represent the angle of rotation of a small motor and output E_o the angle of rotation of a larger motor, the objective being to keep the two motors in step with each other. Or E_i may be the input voltage from a measuring circuit that has high resistance and low power and which is to be converted to an identical voltage in a low-resistance circuit to operate a loudspeaker or solenoid or some other device that requires more power than is available at the input end of the circuit. (Throughout this discussion keep in mind that there must be a source of energy to make this conversion possible, as is illustrated in Fig. 6.16.)

The signal E_i may have a fixed value or may vary with time. It feeds into a comparator element, where E_i is compared (added) to the signal coming as feedback. It both the input signal and

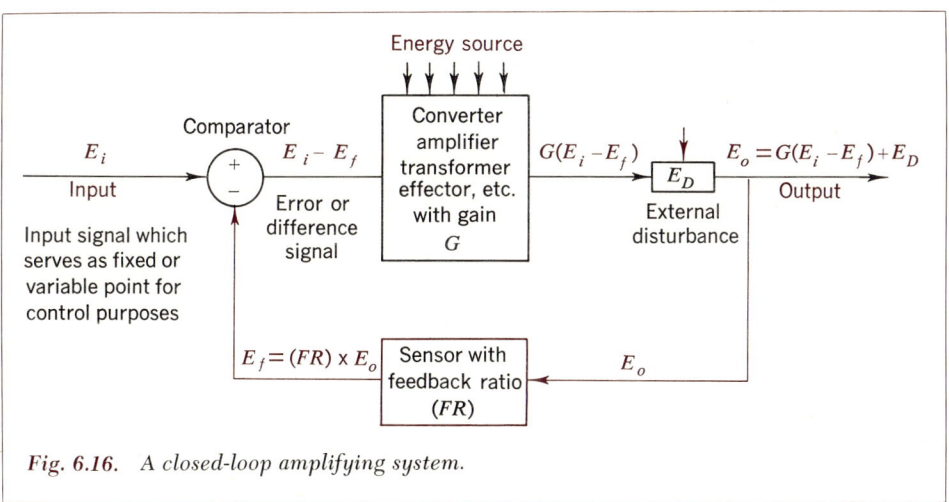

Fig. 6.16. A closed-loop amplifying system.

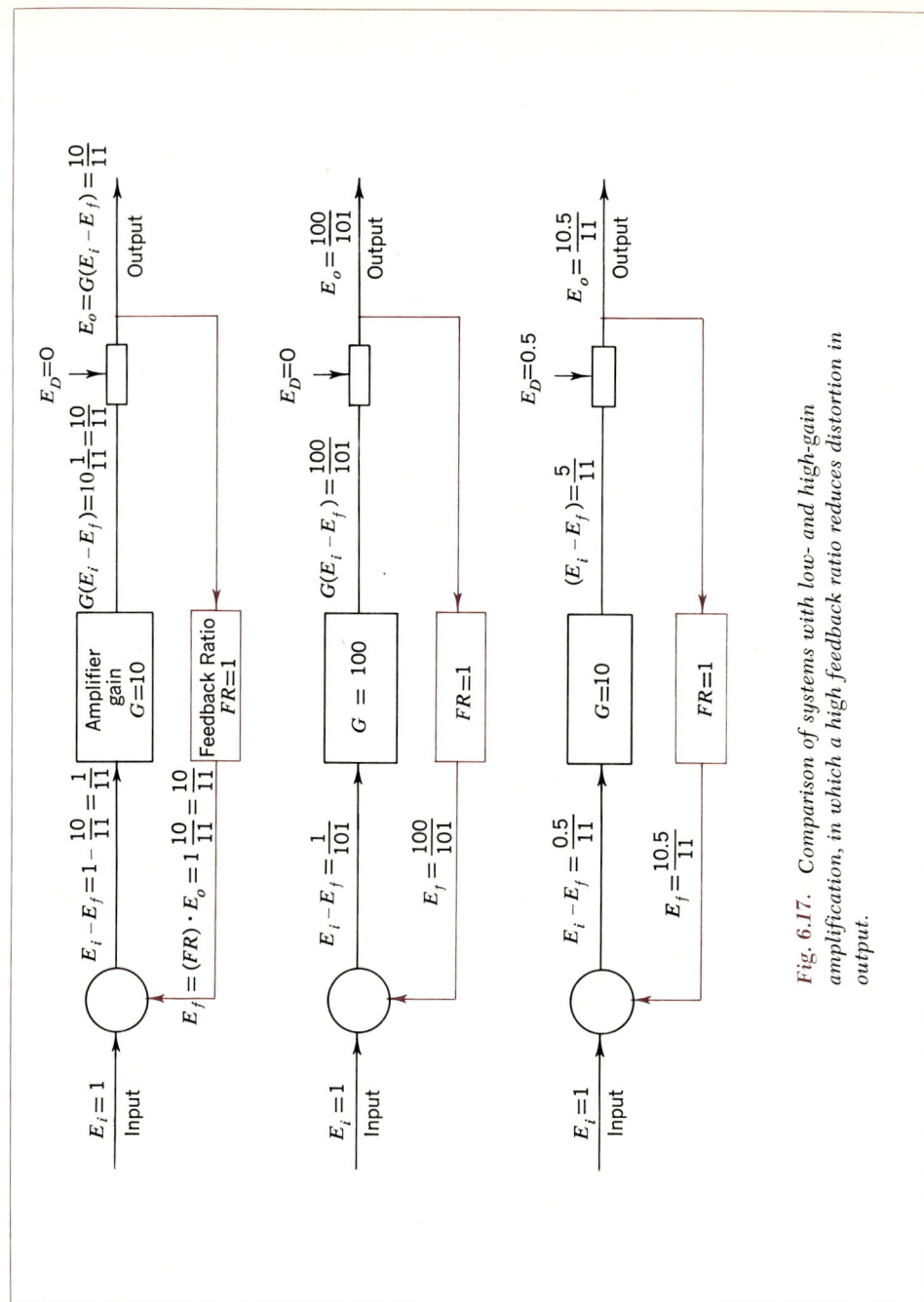

Fig. 6.17. *Comparison of systems with low- and high-gain amplification, in which a high feedback ratio reduces distortion in output.*

the feedback are equal, (equal amounts of positive input and of negative feedback cancel each other), there remains zero *error; conversely, if they are not equal, a difference signal* will be fed to the amplifier or transducer. The output of the amplifier goes through a unit that represents all the undesirable disturbances (E_D) that can upset the balance. These may be a sudden twist of the output shaft in the case of the servomotor, a sudden drop in temperature when a new load of cold metal is injected into an industrial melting furnace, or a sudden bend of the body which changes blood pressure requirements. As the disturbance changes the motor position, temperature, or blood pressure, there automatically develops also a change in the feedback loop. The feedback and input E_i are then no longer in balance, and the sudden error signal which results from the imbalance causes the amplifier to go into action to restore balance.

The amplifier is shown as having a gain G. (This means that if a signal of 0.1 volt enters the amplifier box from the left side, the voltage output of the amplifier itself will be $G \times 0.1$ volt.) In Fig. 6.16 a difference or error voltage of $(E_i - E_f)$ enters the amplifier and then $G(E_i - E_f)$ is available directly from the amplifier to the output. But if a disturbance E_D is introduced the output from the amplifier becomes the sum of the two, or $G(E_i - E_f) + E_D = E_O$. However, the feedback sensor in the output line also receives this voltage E_O, and permits a fraction (feedback ratio) of it to be sent back to the comparator, which adjusts the input voltage E_i as previously explained. This immediately tends to reduce the excursions sent by the error or difference signal to the amplifier, and consequently also reduces the excursions that E_O may experience.

The gain G of an amplifier determines in part how strongly the system will act to restore a balance from a small error signal E_i. The higher the gain, the more sensitive the system will be. But the gain may fluctuate widely as a result of changes in line voltage, age of electronic circuits, and similar variables. A certain input voltage, say 0.1 volt, could give quite different output voltages from day to day or even hour to hour. It is in counteracting such fluctuations that the feedback principle becomes so important, as we shall see now.

Suppose we introduce an input voltage $E_i = 1$ volt into an amplifier that has a gain $G = 10$ (Fig. 6.17). (In the present case let us assume that we amplify voltages, and so G becomes a pure number.) If we do not have feedback, the 1 volt is amplified to 10 volts, directly following the amplification gain factor. If we assume no other disturbance, or $E_D = 0$, the output of the system would be $E_O = 10$ volts. But if we should have a feedback ratio of 1, the full impact of the output voltage is sent back to oppose E_i. The net effect is, of course, that E_O never becomes 10 volts but is held down to a value closer to the input $E_i = 1$ volt. (We can determine what will happen by a few simple relationships.†)

The amplifier gain G is given by

$$G = \frac{\text{(amplifier output)}}{\text{(error or difference signal)}}$$

$$= \frac{E_O - E_D}{E_i - (FR)(E_O)} \quad (6\text{-}1)$$

where FR is the feedback ratio.

Transposing gives

$$GE_i - G(FR)E_O = E_O - E_D,$$

or

$$E_O[1 + (FR)G] = GE_i + E_D,$$

† See James E. Randall.

or

$$E_O = E_i \left(\frac{G}{1+(FR)G}\right) + E_D \left(\frac{1}{1+(FR)G}\right)$$

(6-2)

For the above example,

$$E_O = 1\left(\frac{10}{1+(1\times 10)}\right) + 0\left(\frac{1}{1+1\times 10}\right)$$

$$= \frac{10}{11} \text{ volt}$$

The error or difference voltage therefore becomes $E_i - E_f = 1 - (10/11) = 1/11$ volt. The amplifier must be capable of acting on this voltage if the system is designed to work on such magnitudes of error. One way to improve this is to increase the gain G of the amplifier. If, in this case, the gain G is increased from 10 to 100 while keeping the feedback ratio and E_D the same, the error voltage reduces to 1/101 volt from 1/11 volt. It can be seen that the gain can change markedly without introducing serious error in such a system.

Finally suppose there develops a disturbance E_D amounting to 0.5 volt, with the gain $G = 10$ and $FR = 1$. From the last term of Eq. (6-2), the effect of the disturbance reduces to

$$0.5 \frac{1}{1+1\times 10} = \frac{0.5}{11}$$

as a result of the feedback. Figure 6.17 presents these figures applied directly to the diagram of Fig. 6.16 (following the example of James E. Randall, *Elements of Biophysics*).

The examples thus far apply to static systems. The behavior of systems varies considerably when the input voltage changes too rapidly for the system to follow the changes of E_i or E_D. The subject of controller stability has received a great deal of attention in connection with servomechanism design (for remote control of airplane movements, and similar applications) and electric circuit design for communication, but we cannot delve into that aspect of controller theory. However, one related consideration is "noise," mentioned briefly in an earlier discussion. This also has had considerable study because of the important effect on the capacity of circuitry to convey "information."

6.18 The nature of "living" systems

In the discussions of control systems thus far we have not distinguished between systems involving machine components and those involving living systems. Nor is it our intention to do so now. The fact is that, except for varying complexity, the very same concepts may be applied to living as well as nonliving or machine systems. Each of the sensory organs through which we communicate with the much vaster system of nature is itself designed, oriented, and functionally controlled to achieve certain specific goals or "purposes." It does not matter whether we discuss a nerve cell or an electric wire connection. Both are motor-sensors. Information may be transmitted through the medium of voice, teletype, wireless, visual signals, or the raising of an eyebrow. Each may be an element of a system, and a composite system may include many elements or subsystems. The science of information theory must cope with vast complications to determine the maximum and minimum informational content that an actual system can transmit, even when the role and nature of each link of the chain can be fairly understood.

The acts of stretching the body or of reaching to pick up an object entail the function of a fairly complex system of

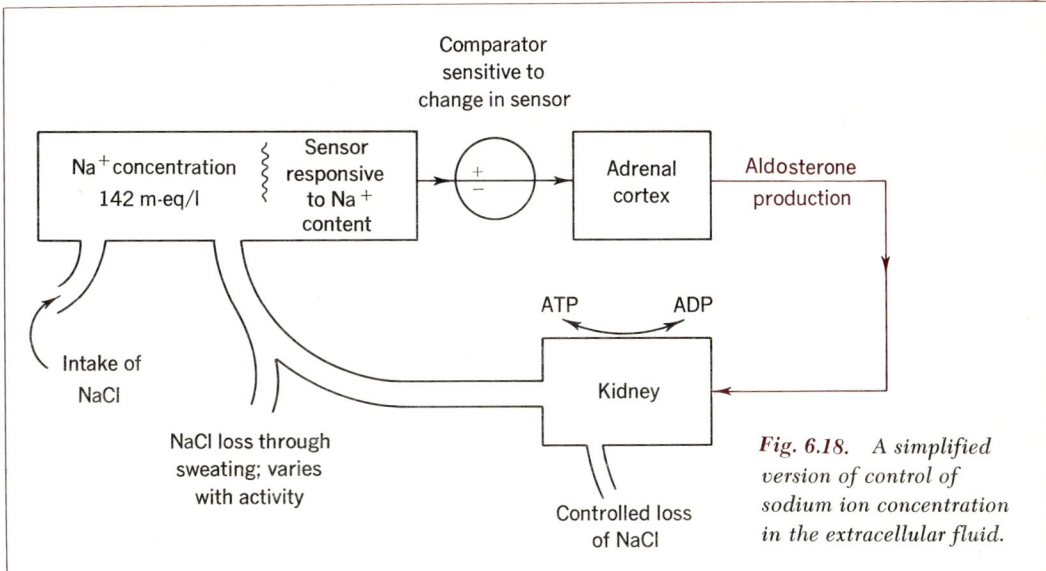

Fig. 6.18. A simplified version of control of sodium ion concentration in the extracellular fluid.

regulation and control. This has been demonstrated by Karl Smith's† experiments with delayed visual feedback in visual motor behavior, which showed that what a person sees is delayed in reaching him when he is performing various other tasks such as writing.

The *intention* to stretch or to pick up an object is itself a complex function, developing in the mind as a product of other activities and influences. The *command signal*, in the form of nerve impulses, originate in the motor cortex of the brain and initiates action in the muscle contractile proteins. There is an amplification, G, which may be expressed as change in muscle length for a unit change in the motor neurons that initiate the discharges. The muscle spindle acts as a *sensor-transducer* to produce nerve impulses, in proportion to muscle length extension, to send back to the brain as

† See K. V. Smith.

feedback on the extension. The original impulses and the feedback impulses are integrated in the spinal cord and give indication of the *error* or *difference* from completion of the intended act. The spindle proprioceptors serve to provide constant information on the state and tone of the muscle system, and assure smooth action of the body. When an individual is deprived of their help, muscle activity tends to be abrupt rather than smooth, requiring dependence on visual sense of position to the extent that he cannot stand when blindfolded.

A person suffering from Parkinson's disease retains some benefit from proprioceptive information, but tends to overshoot when reaching for an object—a motion that recalls damped oscillation.

In later studies of biological systems we shall have occasion to study in some detail a few of the regulatory systems on which life depends. Figure 6.18 illustrates how the sodium ion concentration is maintained constant in the fluid that surrounds the individual cells. There is an elaborate system for maintaining uniform pressure in the circulation of blood. Pressure-sensitive transducers, located in the aorta and carotid arteries, send information about the magnitude of the pressure to an integrating center within the medullary portion of the brain. This results in action that lowers blood pressure by slowing the heart rate and also by producing vascular dilatation.

For respiration there is needed a minimal value for blood carbon dioxide and an adequate supply of blood oxygen. When carbon dioxide concentration in the blood increases, the medullary respiratory center stimulates respiration to eliminate carbon dioxide. The transit time for the flow of blood between the lungs and the respiratory center is only a few seconds under normal conditions (see Randall, p. 108). Body temperature is maintained by a delicate balance between heat loss (from warm-blooded or homeothermic animals) and heat production within the animal through metabolism. The "thermostat" that controls this balance is located in the hypothalamus of the brain and receives information from various temperature transducers of the body to guide its own function.

The regulatory system can extend beyond the body to include the interactions involving climate, geography, geology, agriculture, theology, government, disease, or any other influences. The elements of determinate function, disturbances, control variables, amplification, feedback, informational content, are all three, but they may take the forms of imposed law, self-imposed law, self-imposed restraints, religious restraints, moral obligations, and many other forms that are even less tangible.

The regulatory principles apply to commercial production plants where orders for goods become converted to products for sale, with often quick and direct feedback from consumer to producer. The economist must be aware of the relationship of the key elements of a nation's economy in terms that are identical to those discussed, if he is to succeed in regulating the ups and downs of business within manageable proportions. The problem becomes especially severe when each of the elements of the system is a result of statistical variations, and the statistics lack the assurance of experience or of numbers. The difficulties too often savor of the uncertainties of "black boxes," and yet one must select a suitable model, suitably simple to be manageable and not too far removed from the realities of the situation.

The student is urged to study carefully all the details that have been included in this section on controls. In time he will find that many of the topics that are to come in later chapters will fall more easily into place. For nature and man exist and continue as a result of a balance of forces and utilization of energy, the whole constituting a system that is in a state of reasonable balance and regulation and yet continually changing toward wholly new forms.

In conclusion, we hope that this brief introduction to systems and cybernetics will encourage each reader to view the events of his life with keener appreciation for the *interrelationship* of the factors that bear on the events, and especially for *feedback* influences. A word of caution is in order, however, with respect to overextended use of the term *cybernetics* to situations wherein the relationships are too complex or too obscure, and wherein there are not present the control systems elements which we have discussed.

Questions/Discussions

The assignments for this chapter are intended to give the reader opportunity to discover for himself how broadly the concepts and techniques involving *systems, feedback, control, stability-instability,* and *cybernetics* apply to phenomena in nature and to all aspects of human social relations. It is suggested that from two to four weeks be allowed for completion of this work.

1. For purposes of review, tabulate the five elements of control systems (described in Sec. 6.12) that apply in the following personal situations. Explain also whether the feedback is positive or negative in each case.

(a) The control of temperature of your home.
(b) The factors that control your waking up on a weekday morning.
(c) The factors that control your breakfasting.
(d) One situation or experience of your day that includes strong positive feedback.
(e) A situation or experience of your day that includes strong negative feedback.

2. Select three phenomena or situations, taken from any three of the following categories, and analyze their "systems" aspects in the following terms:

(a) The dependent and independent variables that are involved in each, either as "input" to the system or as disturbances.
(b) The sensor devices or transform functions required at the input end for each variable.
(c) The energy sources.
(d) The motor devices or processes, and the related transform functions.
(e) The gain or amplification between output and input.
(f) The nature of feedback influences (distinguishing between positive and negative feedback and phase relationships) related to each input and each output.
(g) The nature of subsystems that are included.
(h) The factors that make for stability and instability in the total system or subsystems.
(i) The graphical representation of the above elements and processes, with indication of polarity (direction) of feedback between each output and input.

The phenomena or situations are to be drawn from any three of the following seven categories:

I Electromechanical, pneumatic systems, chemical or production processes
II Geophysical or meteorological processes
III Biological processes (plants, animals), ecological relationships
IV Medical, pathological experiences
V Economics (international, national, or personal), business operations
VI Behavioral, cultural, ethical, moral, theological, and psychological aspects of social experiences

Note: It is suggested that each "case" be given adequate discussion and one to two pages of graphical representation. Because of the importance of the subject of "systems," it is suggested that these analyses be given time for class discussion. Group effort on the part of the students is encouraged, although each must present his own final case study.

References

Textbooks

Abbott, L., *Economics and the Modern World.* New York: Harcourt, 1960, pp. 403–405. *On acceleration effect and the national income.*

Abrahams, A., "The Regulation of Heat," in *The Human Machine.* Baltimore: Penguin (Pelican books), 1956, pp. 31–33.

Ashby, W. Ross, *An Introduction to Cybernetics.* London: Chapman & Hall, 1961.

Baker, J. J. W. and Allen, G. E., *Matter, Energy, and Life.* Reading, Mass.: Addison-Wesley, 1965, chap. 5.

Broad, C., *The Mind and Its Place in Nature.* Paterson, N.J.: Littlefield, Adams, 1960, pp. 3–133.

Casey, E. J., *Biophysics.* New York: Reinhold, 1962, pp. 6–25, 295–314. (Also see Index.)

Colburn, R. (ed.), *Modern Science and Technology.* Princeton, N.J.: Van Nostrand, 1965; pp. 633–739.

Ebert, J., *Interacting Systems in Development.* New York: Holt, Rinehart and Winston, 1965, pp. 128–139, 188–195.

Finnigan, R. E., Uthe, P. M., and Lee, A. E., "Modern Press Control," *Modern Science and Technology.* Princeton, N.J.: Van Nostrand, 1965.

Grabbe, E. M. (ed.), *Automation in Business and Industry.* New York: Wiley, 1957, pp. 25–32, 41–75.

Hardin, G., *Biology.* San Francisco: Freeman, 1961, pp. 74–88, 156–164.

Harriott, Peter, *Process Control.* New York: McGraw-Hill, 1964.

Hayek, J., *The Sensory Order.* Chicago: Univ. of Chicago Press, 1963. (*See Index.*)

Hutchings, E., *Frontiers in Science.* New York: Basic Books, 1958, pp. 80–87 (*food problems*), pp. 100–121. *Some relation to feedback relative to man's future.*

Kaplan, A., *The Conduct of Inquiry.* San Francisco: Chandler, 1964, pp. 258–293.

Kapp, K. W., *Toward a Science of Man in Society.* The Hague, Netherlands, 1961, pp. 103–109, pp. 123–137.

Koenig, S., *Sociology.* New York: Barnes & Noble, 1957, pp. 28–41, 42–52, 53–68, 69–106.

Lindsay, R. B., *The Role of Science in Civilization.* New York: Harper & Row, 1963, pp. 133–196.

Oparin, A. I., *Life, Its Nature, Origin and Development.* New York: Academic Press, 1962, pp. 1–30.

Optner, Stanford L., *Systems Analysis for Business and Industrial Problem Solving.* Englewood Cliffs, N.J.: Prentice-Hall, 1965.

Paul, J., *Cell Biology.* Stanford, Calif.: Stanford Univ. Press, 1964, pp. 10–13. *On cell theory.*

Randall, J. E., *Elements of Biophysics*, 2d. ed. Chicago: Yearbook Medical Publishers, 1962, pp. 70–90, 91–110.

Samuelson, P. A., *Economics.* New York: McGraw-Hill, 1964, pp. 35–38, 396–398.

Setlow and Pollard, *Molecular Biophysics.* Reading, Mass.: Addison-Wesley, 1962, pp. 66–76. *On information theory.*

Simpson and Beck, *Life.* New York: Harcourt, 1965, pp. 646–701, 333.

Smith, K. V., *Delayed Sensory Feedback Behavior.* Philadelphia: Saunders, 1967.

Tiryakian, E., *Sociologism and Existentialism.* Englewood Cliffs, N.J.: Prentice-Hall, 1962, pp. 22–68.

Ubbelohde, A., *Man and Energy.* Baltimore: Penguin, 1963, pp. 166–195. *Discusses growth of knowledge, thermodynamics.*

Watt, Kenneth E. F., *Systems Analysis in Ecology.* New York: Academic Press, 1966.

Weisz, P., *The Science of Biology.* New York: McGraw-Hill, 1967, pp. 303–328, 354–381, 466–485, 524–552.

Wiener, N., *Cybernetics,* 2d ed. New York: The MIT Press & Wiley, 1961, pp. 95–115.

Winchester, A. M., *Modern Biological Principles.* Princeton, N.J.: Van Nostrand, 1965, p. 236, fig. 179.

Articles

Angrist, S. W., "Fluid Control Devices," *Scientific American* (December 1964), pp. 81–88.

Bellman, R., "Control Theory," *Scientific American* (March 1964), pp. 182–200.

Bolt et al., "Doctoral Feedback into Higher Education," *Science* (May 14, 1965), pp. 918–928.

Brooks, N., "Scientific Concepts and Cultural Change," *Daedalus* (Winter 1965), pp. 66–83.

Changeux, J. P., "The Control of Biochemical Reactions," *Scientific American* (April 1965), pp. 36–45.

Deevey, E. S., "The Human Population," *Scientific American* (September 1960).

Dorn, H. F., "War Population Growth: An International Dilemma," *Science* (January 26, 1962), pp. 283–290.

Fender, D., "Control Mechanisms of the Eye," *Scientific American* (January 1964), pp. 24–33.

Hoagland, H., "Mechanisms of Population Control," *Proc. Am. Acad. Arts Sci.* (March 1964).

Malkus, J. S., "The Origin of Hurricanes," *Scientific American* (August 1957).

Moshor, R., "Industrial Manipulators," *Scientific American* (October 1964), pp. 88–96.

Opik, E. J., "Climate and the Changing Sun," *Scientific American* (June 1958).

Starr, V., "The General Circulation of the Atmosphere," *Scientific American* (December 1956).

Stone, R., "Mathematics in the Social Sciences," *Scientific American* (September 1964). *Illustrates several models.*

Tepper, M., "Tornadoes," *Scientific American* (May 1958).

Wexlor, H., "Volcanoes and World Climate," *Scientific American* (April 1952).

Woodcock, A., "Salt and Rain," *Scientific American* (October 1957).

Wynne-Edwards, V. C., "Self-regulating Systems in Populations of Animals," *Science* (Mar. 26, 1965). *Contains letters from readers replying to this article.*

CHAPTER SEVEN

Introduction to the Space Sciences

Earth is the cradle of the mind, but one cannot live in the cradle forever.

K. E. TSIOLKOVSKY

THE CONCEPTS AND TECHNIQUES of mechanics and of systems analysis become very useful for discussion of exploration of the space of our solar system, to which we now turn. The subject becomes important in part because of the new science and technology that space exploration has produced thus far, in part because a great deal of international competition is now involved in space activities, and in part because many billions of dollars of the national budget are being allocated for an attempt to reach and study the moon and planets.

7.1 History of space exploration

What do we mean by the term *space?* The word has been given many definitions in recent literature. For our present purposes we can say that space is that part of the universe beyond the earth and its immediate atmosphere which is most conveniently explored by means of rocket-propelled, manned, or unmanned vehicles.

Although the actual exploration of space on a large scale is a recent development in human history, the subject has fascinated and challenged the thoughts of men for over 18 centuries. One of the earliest known fictional accounts of a space venture was written about 160 A.D. by the Greek satirist Lucian in his work entitled, *True History.* Another work of fiction, *Sleep,* published in 1634, which gives vivid description of an imaginary trip to the moon, is worthy of mention because its authorship is attributed to none other than the pioneer German astronomer Johannes Kepler. In 1687, as a direct consequence of his theory of universal gravitation, Sir Isaac Newton mentioned the theoretical possibility of establishing an artificial satellite to earth. Many novelists and theoreticians have written about satellites, rockets, and space travel since the time of Newton.

The beginnings of space technology can be traced back to 1890 when Hermann Ganswindt, a German inventor, prepared a semiscientific design for a spaceship with a rocket motor to be propelled by dynamite or some other type of explosive fuel. In a series of technical articles issued during the period 1911–1913, Konstantin Eduardovitch Tsiolkovsky, a Russian school teacher, developed the idea of large, piloted rockets. In 1919, Dr. Robert H. Goddard, then professor at Clark University, Worcester, Massachusetts, published *A Method of Reaching Extreme Altitudes,* in which he presented theoretical aspects of rocket propulsion and formulas for calculating rocket power necessary to reach the moon. His earlier experiments dealt with powder rockets. Later, he built and tested liquid-fuel rocket motors and rocket stabilizers. In 1923, Hermann Oberth, a German mathematician, published a small book entitled, *The Rocket into Interplanetary Space,* which was destined by circumstances to play an important role in the history of rocketry. As a result of the interest aroused by Oberth's work, the German government established the Peenemünde Research Institute in 1937, for the purpose of developing rocket weapon systems. It was here that the V-2 rockets (the infamous "Buzz-bombs" of World War II) were developed.

A new phase in the development of rockets started immediately after the Second World War. The captured parts of the V-2 rockets were shipped to the United States of America and the Union of Soviet Socialist Republics (Soviet Union), where German engineers and scientists continued their rocket research. Project Bumper in the United States succeeded in firing the first sophisticated two-stage, liquid-fuel rocket in 1949.

Sputnik I, the first artificial satellite of earth, was successfully launched by the Soviet Union on October 4, 1957. *Sputnik I* was followed by *Sputnik II* (November 3, 1957), which had a payload six times heavier than that of *Sputnik I.* In addition to various scientific instruments to measure radiation from the sun, cosmic rays, and the temperature and pressure of the upper atmosphere, it carried a live dog for biomedical experiments.

It was not until January 31, 1958, that the United States of America successfully launched its first earth satellite, *Explorer I.* Cylindrical in shape, 80 inches long, and 6 inches in diameter, it weighed 31 pounds, including 11 pounds of instruments. The instruments within the satellite were designed to measure the intensity of cosmic rays, the impact of tiny meteorites, and the internal temperature of the satellite.

Since 1958, numerous satellites and space probes have been launched, principally by the United States and the Soviet Union, to carry out various missions. Manned space flight was first achieved by the Russians when Yuri Gagarin orbited the earth once on April 12, 1961, in the spacecraft *Vostok I.* John H. Glenn, Jr., the first American in orbit, orbited the earth three times in a *Mercury* capsule on February 20, 1962. Since 1962, many other astronauts have orbited the earth, singly and in groups of two and three.

Table 7.1 (page 226) lists a few significant space achievements during the first ten years of the space age.

7.2 Why explore space?

Billions of research and engineering dollars are being expended each year on space research by the United States, the Soviet Union, and to a lesser extent by several other nations. It is therefore

TABLE 7.1 SOME SPACE MILESTONES

Date	Spacecraft	National Origin	Remarks
Oct. 4, 1957	Sputnik I	USSR	First artificial satellite of earth
Nov. 3, 1957	Sputnik II	USSR	Carried dog into orbit
Jan. 31, 1958	Explorer I	USA	Discovered Van Allen belts
Jan. 2, 1959	Lunik I	USSR	First spacecraft to escape earth
Sept. 12, 1959	Lunik II	USSR	First spacecraft to impact on moon
Oct. 4, 1959	Lunik III	USSR	Photographed far side of moon
Apr. 1, 1960	Tiros I	USA	First meteorological satellite
Apr. 12, 1961	Vostok I	USSR	First manned space flight
Feb. 20, 1962	Friendship 7	USA	First U.S. manned space flight
Aug. 26, 1962	Mariner 2	USA	Close-up study of Venus
July 28, 1964	Ranger 7	USA	Close-up photos of moon
Nov. 28, 1964	Mariner 4	USA	Close-up photos of Mars
Mar. 18, 1965	Vostok II	USSR	First "space-walk"
Apr. 6, 1965	Early Bird	USA	First commercial communications satellite
Nov. 26, 1965	A-1	France	First French satellite
Jan. 31, 1966	Luna 9	USSR	First soft landing on moon
Mar. 31, 1966	Luna 10	USSR	Orbited moon
Aug. 10, 1966	Lunar Orbiter	USA	Photo of earth from moon
Apr. 17, 1967	Surveyor 3	USA	Dug hole in lunar surface
Oct. 18, 1967	Venus 4	USSR	Sent instrument package into Venus atmosphere

not inappropriate to ask: "Why explore space?"

There is no simple answer to this question. Certainly, much of the expenditures made by the various governments for this purpose are made in the hope of realizing military or political advantages. The most expensive part of space research, the development of the rocket boosters, contributes to the development of weapons systems and would likely be carried out in any case. We shall not dwell on these aspects, important though they may be for some purposes. Rather, we see space exploration to be a source of new information about the earth, the moon, and the planets. Among the new information obtained during the first decade of the space age we list the following:

(1) The existence and nature of trapped high-energy particles that surround the earth (the Van Allen radiation belts).

(2) The nature of the lunar surface.

(3) The surface conditions on the planets Mars and Venus.

(4) Quantitative measure of particulate (meteoric) materials in the neighborhood of the earth.

(5) Quantitative measure of the magnetic fields of the earth, the moon, Mars, and Venus as well as the background field of the sun.

The use of space vehicles may therefore be considered to be a natural extension of the astronomical techniques that began with the use of the telescope by Galileo.

Another reason given for exploring space lies in the technological "spin-off" that may accrue. The term refers to the fact that the new materials and techniques that evolve out of the space effort sometimes find application in more conventional fields. For example, there are now new broadcast and communication systems that utilize satellites, as well as new aids to meteorology and navigation.

Indeed the desire to explore space is akin to the motivation that opened the way to the New World in the fifteenth and sixteenth centuries, and which centuries earlier had caused men to probe uncharted seas, deserts, jungles, and mountains. Possibly the most important reason for exploring space is given by Arthur C. Clarke, who wrote

Perhaps one day men will no longer be interested in the unknown, no longer tantalized by mystery. This is possible, but when Man loses his curiosity one feels he will have lost most of the other things that make him human . . . if there were not a single good "scientific" reason for going to the planets, he would still want to go there, just the same.†

7.3 The mechanics of space travel

Let us recall the simple experience of a small stone that is released from the top of a building. It starts out with zero speed. The acceleration due to gravity is about 32 ft per second per second, which means that the stone will pick up a velocity of 32 ft per second (fps) in the first second of its fall, another 32 ft per second in the second second of its fall, and so on. Therefore the speed with which it strikes the ground, or the impact speed, depends on the duration of fall. The greater the duration of fall, the greater is its impact speed. Correspondingly, the longer the distance of fall of the stone, the greater is the duration of fall. It follows that the greater the height from which the stone is released, the greater is its impact speed. If the height is increased by an infinitely large amount, will the impact speed correspondingly increase to an infinitely large value? The answer is "no."† The impact speed of the stone resulting from a fall through infinite distance is only about 7 miles per second, which is equal to 25,200 miles per hour (mph). If, instead of a small stone, we use a larger object such as a space vehicle, will it have a greater impact speed? Since its acceleration due to gravity is identical to that experienced by the stone, its impact speed must be the same. In other words, 25,200 mph is the maximum impact speed that the earth's gravity can produce on any object. Actually, the practical maximum impact speed is less than 25,200 mph when air resistance is taken into consideration.

Conversely, if we consider the reverse situation of an object going up into space against gravity with an initial velocity of about 25,200 mph, the logical inference is that it will go an infinite distance into space. This speed is termed the *escape velocity*, but because this value does not take air resistance into account, in practice we need a velocity of about 26,000 mph. We should note that, contrary to popular notion, the escape velocity does not remove an object entirely from the influence of gravitational attraction. One can only say that as a spaceship travels deeper and deeper into space, the gravitational attraction due to the earth de-

---

† *The Exploration of Space.* Greenwich, Conn.: Fawcett, 1960 (p. 176).

† See next optional section for reason for this statement.

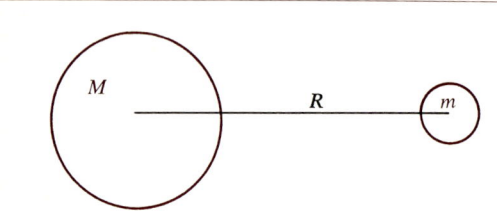

Fig. 7.1. Gravitational attraction between two masses.

creases but never becomes equal to zero. (Of course there will also be gravitational attraction from all the other heavenly bodies.)

NEWTON'S LAW OF GRAVITATION

Newton's law of universal gravitation states that two bodies of masses M and m, whose centers are a distance R apart, experience an attractive force given by the equation

$$F = G \frac{Mm}{R^2} \qquad (7\text{-}1)$$

where G is a constant, known as the gravitational constant (see Sec. 5.8 and Fig. 7.1). If force is measured in newtons, distance in meters (m), and masses in kilograms (kg), G has a value of 6.67×10^{-11} newton m²/kg².

We apply this law to the case of a particle with mass m resting on the surface of the earth. The force acting on m and due to gravitational attraction of the earth is†

$$\frac{GM_e m}{R_e^2}$$

where M_e and R_e are respectively the mass and radius of the earth. According to Newton's second law, this force must equal mg_e, the product of mass and acceleration due to gravity. Thus

$$mg_e = \frac{GM_e m}{R_e^2} \qquad (7\text{-}2)$$

Dividing both sides of Eq. (7-2) by m, we obtain

$$g_e = \frac{GM_e}{R_e^2} \qquad (7\text{-}3)$$

---

† The equation given here assumes that the gravitational attraction of the earth is the same as that of a point mass M_e located at the earth's center. It can be shown, using the calculus, that this assumption is justified for a spherical mass.

The mass of the earth is 5.98×10^{24} kg and its average radius is 6.37×10^6 m. Substituting these values in Eq. (7-3), we obtain† a computed value for the acceleration due to gravity on the surface of the earth:

$$g_e = \frac{6.67 \times 10^{-11} \times 5.98 \times 10^{24}}{(6.37 \times 10^6)^2} \quad (7\text{-}4)$$
$$= 9.83 \text{ m/sec}^2$$

Since 1 m is 3.28 ft, the equivalent value of g in feet per second^2 is $(9.83)(3.28) = 32.25$ ft/sec^2.

The acceleration due to gravity at a height h above the surface of the earth is

$$g_h = \frac{GM_e}{(R_e + h)^2} \quad (7\text{-}5)$$

As h increases, g_h will decrease, approaching the value zero as h approaches infinity. However, for any finite h, g_h will never be equal to zero.

The total mechanical energy of a particle that is moving under the influence of the earth's gravity consists of kinetic energy and potential energy, which always add up to a constant total value. The kinetic energy of a particle of mass m, moving with velocity v, is $\frac{1}{2} mv^2$. The gravitational potential energy is the work done against the force of gravity to displace the particle from a reference position to its given position. The reference position is chosen to be at infinite distance from the center of attraction, at which position we consider the potential energy to be zero. The work done in bringing a body from infinite distance to a point where the distance from the center of attraction of the earth is R is equal to the change in gravitational potential energy U of the body $[U(\infty) - U(R)]$ and is

† Actual best average value is 9.81 m/sec^2 as determined by direct measurement.

given by the integral $\int_\infty^R F\, dr$.

$$\int_\infty^R F\, dr = \int_\infty^R \frac{RGM_e m\, dr}{R^2}$$
$$= -GM_e m \left[\frac{1}{r}\right]_\infty^R$$
$$= \frac{-GM_e m}{R} \quad (7\text{-}6)$$

Thus, if $U(\infty) = 0$, $U(R) = -GM_e(m/R)$. The gravitational force between a body and the earth is attractive, and therefore work is required to counter this force of attraction and separate a body from the earth.

Applying the principle of conservation of energy to a particle of mass m moving with velocity v, in the gravitational field of the earth at a certain height h above the surface of the earth, we obtain

$$\frac{1}{2} mv^2 + \left[\frac{-GM_e m}{R_e + h}\right] = C \quad (7\text{-}7)$$

where C is constant.

If the speed v is to be the minimum escape velocity, v_{esc}, the particle would theoretically arrive at an infinite distance from earth with zero velocity and zero potential energy. Therefore, when $v = v_{esc}$, $C = 0$, Eq. (7-7) reduces to

$$\frac{1}{2} mv_{esc}^2 = \frac{GM_e m}{R_e + h} \quad (7\text{-}8)$$

From this we obtain an expression for the escape velocity:

$$v_{esc} = \left[\frac{2GM_e}{R_e + h}\right]^{1/2} \quad (7\text{-}9)$$

By substituting proper values in Eq. (7-9), we obtain the escape speed of a body with respect to the surface of the earth:

$$v_{esc} = \left(\frac{2 \times 6.67 \times 10^{-11} \times 5.98 \times 10^{24}}{6.37 \times 10^6}\right)^{1/2}$$
$$= 1.12 \times 10^4 \text{ m/sec}$$
$$= 6.96 \text{ miles/sec} \quad (7\text{-}10)$$

In Eq. (7-10), h was set equal to zero.

However, Eq. (7-9) is such that the escape speed with respect to a surface above or below that of the earth can be calculated by substituting correspondingly positive or negative values for h.

Let us consider the problem of launching an artificial satellite into a circular orbit around the earth. The satellite has to be given a certain orbital velocity so that it will stay in the orbit and not fall back to the earth. In our preceding discussion we saw that a body dropped from a certain height above the surface of the earth gains a speed of 32 fps during every second. In other words, at the instant the fall begins, body speed is zero and after 1 sec its speed is 32 fps. During this 1 sec, therefore, its average speed is

$$\frac{0 + 32}{2} = 16 \text{ fps}$$

Furthermore, it is known that the earth's surface, as a result of its curvature, drops about 16 ft in a distance of 5 miles. This is represented schematically, but not to scale, in Fig. 7.2. Therefore an object projected horizontally from the surface of the earth with a speed of about 5 miles/sec, or 18,000 mph, would just reach a circular orbit around the earth.† Since such an orbit would be seriously affected by resistance of the air in the earth's atmosphere, it is better to project the satellite in a horizontal direction from a height where there is practically no air, say, 200 or 300 miles. At such heights, the acceleration due to gravity is also smaller

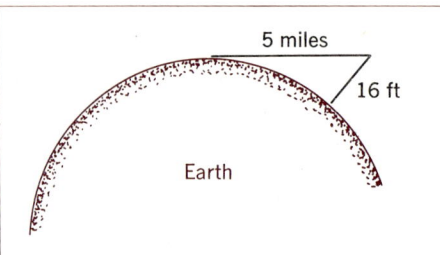

Fig. 7.2. *Effect of curvature of the earth.*

---

† It may be noted that the speed v_h needed to maintain a circular orbit is given by

$$v_h = \left(\frac{GM_e}{R_e + h}\right)^{1/2}$$

See Eq. (7-14). Thus the circular speed and escape speed are related by

$$v_{esc} = \sqrt{2}\, v_h$$

and as a result the needed projection speed will be smaller. The speed needed for a circular satellite orbit at any height up to 1000 miles above the surface of the earth can be obtained from the graph in Fig. 7.3. The numbers marked on the curve indicate the *period* of the satellite—the time (in minutes) needed for the satellite to go round the orbit once.

Example. Suppose a satellite of mass m is to be placed in a circular orbit at a height h above the surface of the earth. Its needed orbital velocity can be found by equating the force of gravity and the centrifugal force. Thus

$$\frac{mv_h^2}{R_e + h} = mg_h \qquad (7\text{-}11)$$

Dividing both sides of Eq. (7-11) by m and taking the square root, we obtain

$$v_h = [g_h(R_e + h)]^{1/2} \qquad (7\text{-}12)$$

From Eqs. (7-5) and (7-3),

$$g_h = \left[\frac{g_e R_e^2}{(R_e + h)^2}\right]^{1/2} \qquad (7\text{-}13)$$

By substituting the value of g_h from Eq. (7-13) into Eq. (7-12), we obtain

$$v_h = \left(\frac{g_e R_e^2}{R_e + h}\right)^{1/2} \qquad (7\text{-}14)$$

Equation (7-14) shows that the orbital speed of a satellite decreases as the height of its orbit above the surface of the earth is increased. Equation (7-14) has been used to plot the graph in Fig. 7.3.

The circular orbital speed serves as a basis for finding the characteristics of any orbit. If the projection speed is very much less than that required for a circular orbit, the satellite fails to go into orbit and falls back to earth (see orbit A in Fig. 7.4). If

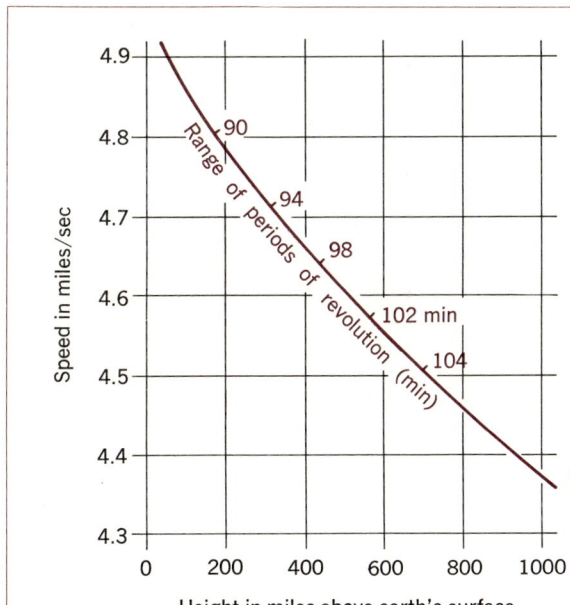

Fig. 7.3. The speed needed for a circular orbit at any height up to 1000 miles. The numbers marked on the curve indicate the period of revolution in minutes.

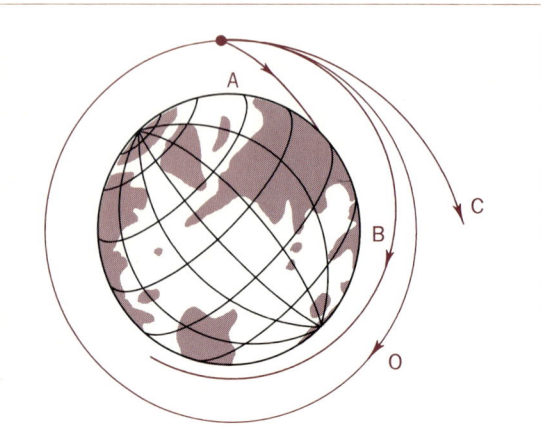

Fig. 7.4. Possible trajectories for satellite launched with various projected speeds.

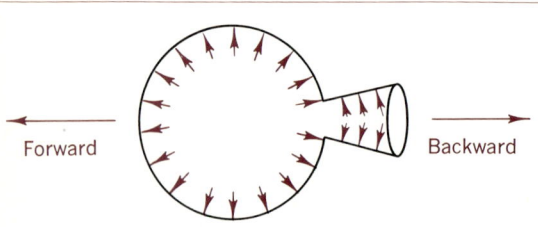

Fig. 7.5. Direction of forces inside combustion chamber of a rocket.

the speed is only slightly less than the circular orbital speed, the satellite drops nearer to the earth, moving in the ellipse that comes closest to the earth at the point opposite to the point of projection (see orbit B in Fig. 7.4). If the speed is greater than the circular orbit speed, the satellite moves out to a greater height and the orbit is an ellipse with minimum height at the point of projection (see orbit C in Fig. 7.4). Then O is the required circular orbit in Fig. 7.4.

For a satellite in an elliptical orbit, the maximum height it attains is called its *apogee* distance; the minimum height its *perigee* distance. Table 7.2 lists a few satellite orbits.

7.4 Rockets

On the basis of the foregoing discussion it should be evident that we must attain speeds of the order of 20,000 mph in order to place a satellite in orbit about the earth. Even greater speeds are needed to send a space probe to the moon or to Mars. A considerable force is needed to accelerate a satellite of reasonable mass, initially at rest on the earth's surface, to the desired orbital velocity at, say, a height of several hundred miles. At the present time, the only way to obtain the needed force is by using rockets.

A rocket consists basically of a chamber in which the chemical energy of the fuel is transformed into kinetic energy of the material (products of combustion) in the chamber. The resulting pressure in the chamber is equal in all directions. Therefore the force that pushes up against the upper chamber wall (as in Fig. 7.5) is counterbalanced by an equal downward push against the lower chamber wall. However, that which pushes forward against the forward chamber wall is not counterbalanced, since the back part of

TABLE 7.2 TYPICAL SATELLITE ORBITS

Satellite	Apogee Height, miles	Perigee Height, miles	Average Height, miles	Period, min
Sputnik I	588	144	365	96
Explorer I	1573	224	899	114
Tiros	468	435	452	99
Friendship 7	141	87	114	88

the chamber is open to allow the burning gases to escape. This results in a net forward *thrust* on the body of the rocket, of a magnitude that depends on the momentum of the escaping gas.

To determine the *thrust*, we need only recall Newton's laws of motion and the conservation principles discussed in Chapter 5. By the third law of motion, the forward thrust (force) on the rocket is exactly equal to the backward thrust on the escaping gas. However, we should note that in the case of a rocket pointed upward, unless this forward thrust is at least a little larger than the weight of the entire rocket assembly, the engine will not rise from the ground. We recall Eq. (5-5) and are reminded that $F \Delta t = \Delta mv$, where $F \Delta t$ is the impulse applied to the rocket and Δmv is its change of momentum. But, by the momentum conservation principle,

$$M_R v_R = \Delta m_g V_g \quad (7\text{-}15)$$

where M_R and Δm_g represent the masses of the rocket and escaping gas, respectively, and the v_R and V_g represent the velocities of these masses. Thus the change of momentum of the rocket must equal the momentum of the escaping gas, and we may equate

$$F \Delta t = \Delta m_g V_g \quad (7\text{-}16)$$

or

$$F = \frac{\Delta m_g \, V_g}{\Delta t} \quad (7\text{-}17)$$

This equation shows that the rocket fuel must be burned at such a rate that the instantaneous rate of change of the mass times the velocity of the burned gas must equal the huge force necessary to lift a heavy rocket. For a given rocket engine, the gas escape velocity is relatively constant, so that Eq. (7-17) may be written as

$$F = V_g \frac{\Delta m_g}{\Delta t} \quad (7\text{-}18)$$

where the $\Delta m_g / \Delta t$ is the rate of fuel consumption.

Two aspects of Eq. (7-18) should be noted. First, to achieve the necessary thrust, fuel must be consumed at a very rapid rate. For example (see Fig. 7.6), in the *Saturn V booster*† $\Delta m_g / \Delta t \cong 15$ tons/sec! Second, it is desirable to have the velocity of the escaping gas as high as possible. Near the earth's surface, ordinary air pressure outside the rocket tends to obstruct the escaping gas and to reduce this V_g somewhat. Thus a rocket engine is

---

† This rocket develops a thrust of 7.5 million pounds and consumes 4.4 million pounds of lox (liquid oxygen) and kerosene in only 2½ min.

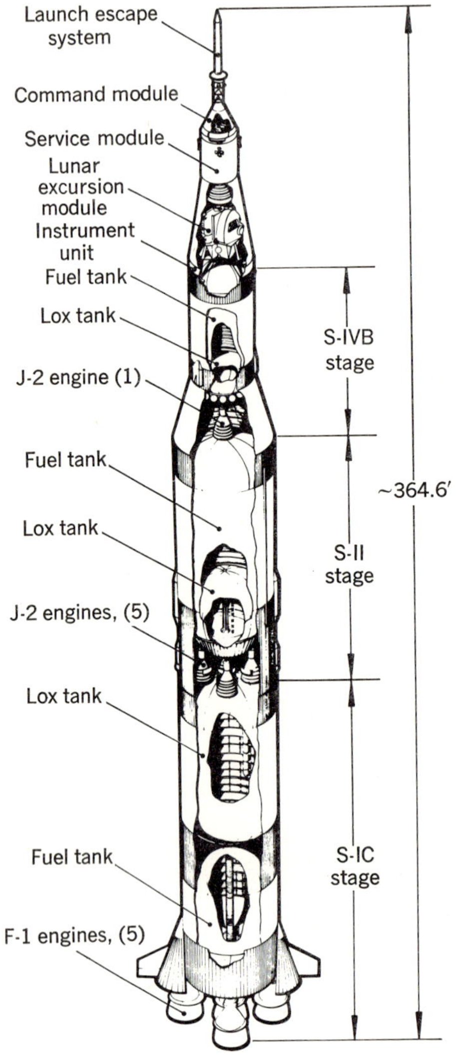

Fig. 7.6. *Schematic diagram of the Saturn V rocket. (NASA photograph.)*

Labels on diagram:
- Launch escape system
- Command module
- Service module
- Lunar excursion module
- Instrument unit
- Fuel tank
- Lox tank
- J-2 engine (1)
- Fuel tank
- Lox tank
- J-2 engines, (5)
- Lox tank
- Fuel tank
- F-1 engines, (5)
- S-IVB stage
- S-II stage
- S-IC stage
- ~364.6′

more efficient when operating above the earth's atmosphere.

A single rocket, or a collection of rocket engines to form one stage, may not produce the thrust needed to accelerate a space vehicle to the high speeds required for orbiting in space. In such cases, multistage rockets are employed. A multistage rocket consists of two or more independent rocket stages arranged in tandem, one on top of the other. Each stage has its own motor and fuel supply. The lowest, or the first stage booster, has the highest thrust because it has to lift the largest mass, whereas the thrust can be reduced for subsequent stages. To launch a space vehicle, the first stage is ignited and the vehicle rises to some height. When all the fuel for this stage is consumed, the exhausted stage is disengaged from the main body of the vehicle. At this instant or very soon thereafter the second stage is ignited. Since at that point the overall mass is considerably decreased, an engine of lower thrust (and lower rate of propellant consumption) is adequate to accelerate the vehicle to new heights. When this second stage attains the required speed or position for the particular mission, the payload is usually detached by means of an explosive charge. Whereas the remains of the early rocket stages usually fall back to earth, the last stage is likely to achieve the final speed of the spaceship and therefore also remain in orbit (Fig. 7.7).

7.5 Guidance, control, and communication systems

A space vehicle has to be guided and controlled not only when it is being launched but also while it is in space. The guidance operation requires:

(1) Measuring the position and velocity of the vehicle.

(2) Computing the control actions that are necessary to make the vehicle follow a prescribed path or to fulfill a particular mission.

(3) Delivering suitable adjustment commands to the vehicle's control system. This amounts to steering the vehicle along a selected path or to maintaining its attitude in a specific orientation in space.

Guidance operations can be classified as three types: radio guidance, inertial guidance, and radio-inertial guidance. In radio guidance, the characteristics of the vehicle's motion are determined by tracking from the earth, whereas in inertial guidance systems this information must be determined from within the spacecraft itself. In radio guidance, the vehicle is in communication with the earth, whereas in a completely inertial guidance system there is no information link with the earth after lift-off. Radio-inertial guidance systems combine features of both radio and inertial guidance systems.

Guidance operations may occur in all or any of the three phases of space flight, the three being termed *initial, midcourse,* and *terminal* phases. There is no single set of initial conditions that will direct a spaceship toward a specific goal. In fact there is an infinite number of possible flight paths that may originate at a point and terminate at the desired destination. It is the task of the guidance system to cause the rocket to follow one of these paths and to assume the proper velocity that will fulfill the mission. The path for a given flight is determined or selected by the guidance system as a continuous operation during the powered portion of the flight, the selections being based on a complex set of criteria involving structural loading, propellant economy, and other factors. This approach is preferred because of the extreme difficulty of

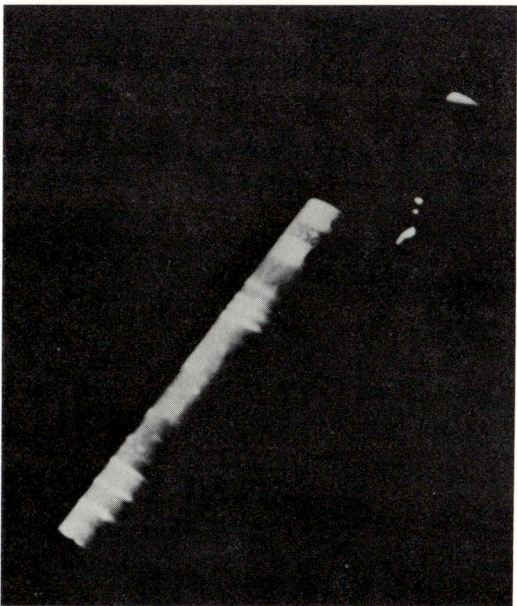

Fig. 7.7. Second-stage rocket as seen from Gemini capsule in orbit. Astronauts Frank Borman and James A. Lovell, Jr. photographed the second stage of the Gemini launch vehicle VII tumbling slowly through space behind them. After the spacecraft and second-stage separation, the spacecraft turned around with nose facing the booster. The GLV VII second stage reentered the earth's atmosphere over the Indian Ocean three days later. (NASA photograph.)

guiding a rocket along a single preplanned trajectory when subject to the uncertainties in engine performance, environmental changes, and unforeseen operating difficulties. When a suitable combination of position and velocity is about to be achieved, the guidance system must immediately signal cut-off of the propulsion system. Of course there must be included something resembling a prediction process to compensate for time lags in the operation of engine controls. In fact, all the elements and problems of control systems that were discussed in Chapter 6 (systems) must be taken into account.

The initial guidance process is similar in all cases, whether a rocket is an intercontinental ballistic missile, a satellite launcher, or a space vehicle carrier. Guidance during midcourse and at the terminal point may or may not be necessary, depending on the mission involved. With suitable instrumentation, the guidance process may be continuous or nearly continuous throughout the flight. If a human navigator were involved, observations would be made only periodically. The computation process depends on the kinds of instruments used, but it has to be prompt, reliable, and fast. Terminal guidance systems require some form of information from the target, such as infrared radiation or radar echo. This information is then used to steer the vehicle to its destination, whether the destination be a planet, a satellite, or a military target.

To illustrate the guidance and control systems of a space probe, let us take the example of a "typical" Ranger mission to the moon. The mission starts when an Atlas rocket (booster) lifts off the launch pad at Cape Kennedy. The thrust developed by the Atlas rocket and the acceleration it produces while the booster is firing are known. The position and velocity of the system are thus known, in theory at least, at the moment of burn-out when the Atlas rocket has used up all its fuel. The Atlas rocket—now an empty shell devoid of fuel—is jettisoned and falls back to earth. The second stage, an Agena rocket carrying the Ranger space probe, is then ignited and its thrust is utilized to place the entire package in a prescribed circular orbit (parking orbit). The length of time the Agena must fire to attain the parking orbit is also determined by earlier computation and programmed into the mission computers. However, since small errors between the actual and theoretical thrusts and firing times can cause significant errors in the path that the spaceship follows, the position and velocity of the system must be carefully observed (by optical and radar techniques) from time of launch to orbit. These observations constitute a feedback of information to mission control, which allows the Agena firing to be lengthened or shortened from its prescribed value to correct for errors.

Once the parking orbit has been achieved, the next step is to fire the Agena again when it is in proper position to send the Ranger to the moon. The time for initiating this phase is also determined from previous computation so as to put the Ranger on the correct course with the correct speed to intercept the moon. Should the Ranger be pointed away from its proper course by as little as 1 deg, it can miss the moon by several thousand miles! Thus the space probe must leave its parking orbit so as to pass through a specified "window" area in space about 10 miles in diameter and with a velocity within 16 mph of the calculated value (around 24,500 mph). (See Fig. 7.8). With such tight requirements, it is clear that the space probe must be observed continuously (either

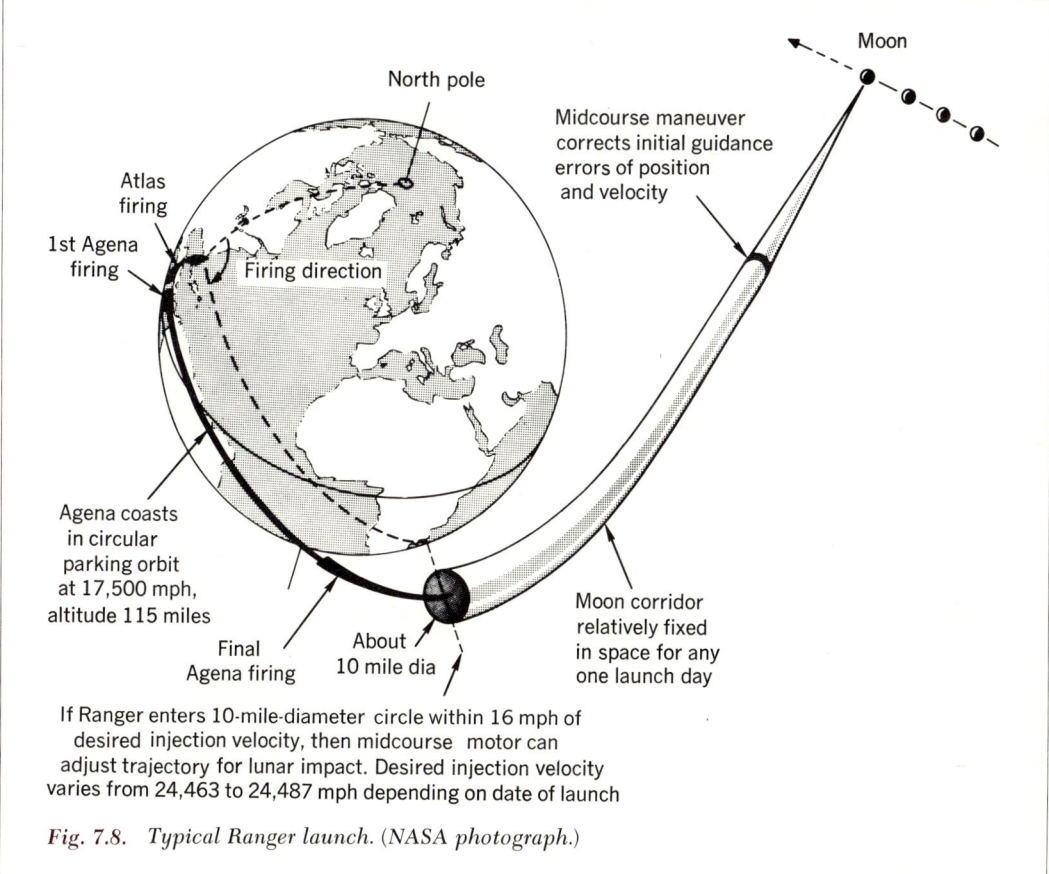

Fig. 7.8. Typical Ranger launch. (NASA photograph.)

by radar or by radio signals transmitted from the probe itself) to compare the actual trajectory with the preplanned one. About 16 hr after launch, a radio signal from mission control causes the Ranger to modify its flight path to better correspond to the desired one. This is accomplished by firing small rockets mounted on the Ranger itself (the Agena was detached after ejecting the Ranger from its parking orbit). If this midcourse maneuver is successful, the instrument package is likely to reach its desired destination. One final correction is possible (if needed) when the Ranger is close to the moon. This allows it to select a specific spot for impact. Figure 7.9 illustrates the elements of the control system just described.

One additional aspect of guidance and control in the Ranger flight is concerned with rotational motions (rolling, tumbling) of the space probe. Since it is important to have the cameras pointed in a specific

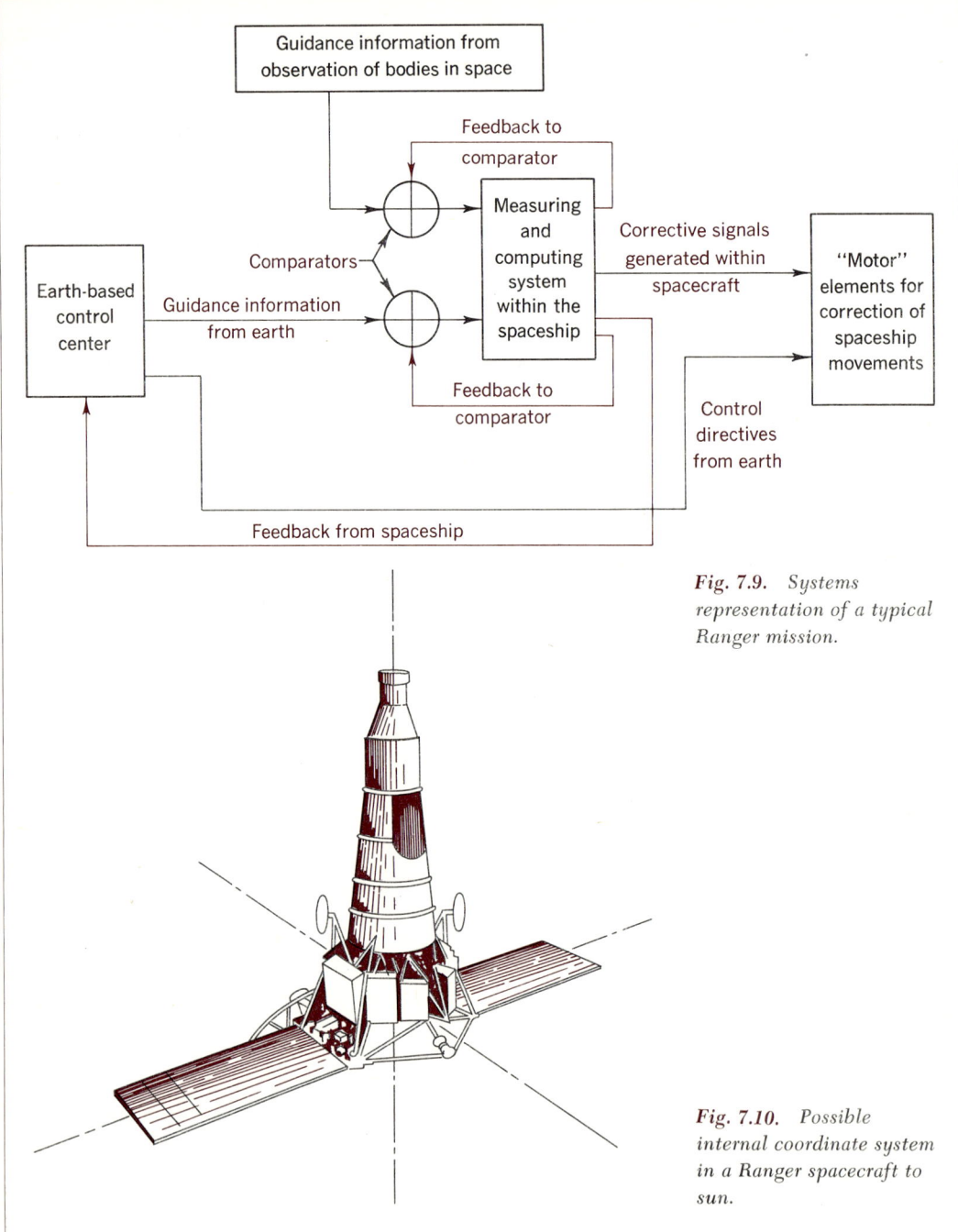

Fig. 7.9. *Systems representation of a typical Ranger mission.*

Fig. 7.10. *Possible internal coordinate system in a Ranger spacecraft to sun.*

direction as the Ranger approaches the lunar surface, the aspect (that is, position of space with respect to a fixed set of coordinate axes) must be controlled. For this purpose, inertial guidance is used. The instrument package contains a number of sensing devices, usually a radio receiver to locate earth, and photocells to locate the sun or possibly also some bright star (usually Canopus). These define two (or three) fixed spatial directions for the satellite. The satellite's own internal coordinate system (see Fig. 7.10) is also known, and the actual angles between the external and internal axes can be found and compared to pre-computed values stored in the satellite's computer memory. Any significant deviation of the observed angle from the computed one causes a specified rocket to fire and correct the aspect.

Satellites and space vehicles contain measuring instruments that gather a wide variety of information. These may include the pulse rate of the astronaut, the intensity of microwave radiation from a distant planet, and other measurements. The process whereby these measurements are conveyed to earth receiving stations is called *telemetry*. Usually, the data (coded or uncoded) are recorded on magnetic tapes as they are observed by the instruments and are "read out" at a rapid rate upon receipt of a command from a ground station. In principle, telemetry is similar to radio broadcasting and reception.

7.6 Life support in space

If a spacecraft has to carry human beings, it must be designed to provide for their needs. Oxygen must be supplied for breathing, and the products of respiration (carbon and water vapor) must be removed. One obvious way is to reproduce man's normal atmospheric environment: 21% oxygen and 79% nitrogen by volume at a total pressure of 14.7 pounds per square inch (psi). This was the method used in some Russian manned spacecrafts. An alternative approach used in U.S. Mercury and Gemini spacecraft was to provide an atmosphere containing only oxygen at about 3.1 psi. Preliminary experiments indicate that such an atmosphere should be suitable for breathing for at least two weeks.

The feasibility of using this low-pressure atmosphere containing oxygen alone came into serious question on January 27, 1967, when three astronauts were killed in a flash fire inside the *Apollo I* capsule as they were conducting a "dry run" rehearsal for a space mission. This tragedy seriously delayed the U.S. space exploration effort because it necessitated a complete redesign of that portion of the life-support system.

For reason of further safety, the astronauts of Projects Mercury and Gemini wore airtight pressure suits. The atmosphere within the suit, which the astronaut breathed, was maintained at required pressure, temperature, and purity by means of an environmental control system. The suit remained closed, although the visor could be opened for eating and drinking. Such suits allow the astronaut to leave his space capsule in flight for extravehicular activity, as was done during several Gemini flights (see Fig. 7.11).

Food in dehydrated and frozen form was used in the Mercury and Gemini flights. Such food can be kept for long periods of time without deterioration and can be reconstituted to approximately its original appearance and taste by addition of water. About 1.4 lb of such food, providing about 3000 calories, are sufficient for a moderately active man per day. The problem of food storage for missions of long duration has yet to be resolved.

Fig. 7.11. White's space walk. Astronaut Edward H. White II is shown performing his space walk during the third orbit of the Gemini-Titan 4 flight. White wears a specially designed space suit for his extravehicular activity. He wears an emergency oxygen supply chest pack. He is holding a hand-held self-maneuvering unit, which he used to move about in the weightless environment. (NASA photograph.)

Pressure suits have facilities for elimination and storage of body wastes. The need for defecation is reduced by the utilization of low-residue foods. For missions of long duration other arrangements will have to be provided.

The vehicle is subjected to acceleration during the launch phase, and deceleration in the reentry phase, of the order of 10 g. The astronaut experiences similar acceleration forces during the launch phase. (During reentry his seating position can be reversed to make him experience acceleration in the same direction.) Astronauts must therefore be tested individually to determine if they can withstand the forces imposed by such acceleration. This is done by generating forces of that magnitude in a "human centrifuge," which consists of a beam about 50 ft long that is pivoted near one end. At the other end of the beam there is a gondola in which the astronaut is seated. By rotating the beam at high speeds, the man can be subjected to accelerations of the order of 10 g. It is found that sustained acceleration may affect the circulation of blood, the respiratory system, and visual clarity. The undesirable effects are found to be least when the accelerating force acts from the back to the front of the body. A form-fitting couch that keeps the knees bent, as in a seated position, decreases the stress. By orienting the space traveler in the proper manner during lift-off and reentry periods, the effects of acceleration are minimized (see Fig. 7.12).

When a spaceship in flight is not acted upon by its own propulsive force or by resistive forces, spacemen experience the phenomenon called (incorrectly) *weightlessness*. During weightlessness, only the subjective sensation of weight is missing. In this state, spacemen experience no restraint as they attempt to stand up, sit down, or bend down. Hence, up-and-down directional feelings are absent. Loose objects tend to float about the space capsule. Liquids break into droplets rather than pour. The experiences of astronauts and animals suggest that weightlessness does not have any significant harmful effect, at least for periods up to two weeks. Normal functions of eating, drinking, and sleeping can be carried out in this state. It may be that prolonged exposure to weightlessness will prove intolerable, in which case some additional force may have to be introduced. One conceivable way of avoiding weightlessness is to cause the spacecraft to spin and thus introduce a force of about 0.1 g.

There are various high-energy radiations in space that are harmful to living organisms. Because they have the property of exciting and ionizing† atoms and molecules, exposure to these radiations can lead to decomposition of such vital compounds as enzymes and nucleic acids in the body and can also affect the genes. The degree of injury depends on the energy and type of radiation and on the period of exposure, and the effect varies considerably among the organs of the body. The main sources of radiation in space are the charged particles of the Van Allen belts and galactic and solar cosmic rays. By lining the spacecraft with a sufficient mass of shielding material, the intensity of the radiation that penetrates into the cabin can be reduced to tolerable levels.

7.7 Reentry problems

The spacecraft has to be decelerated and eventually brought to a safe landing. The most obvious way for doing this is to

† We shall learn later that ionizing radiation can break up molecules and disturb the electron components of atoms.

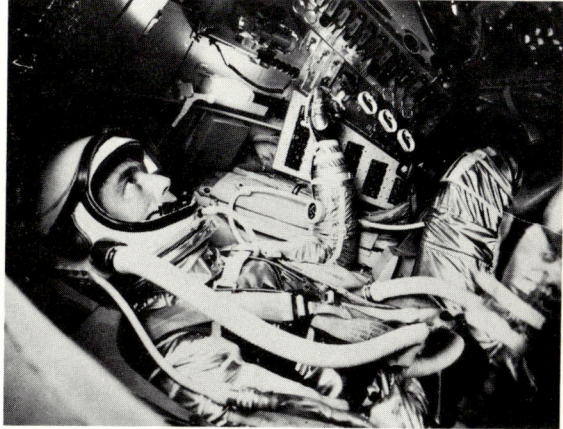

Fig. 7.12. Astronaut (M. Scott Carpenter) in supine position occupied during launch and reentry phases. (NASA photograph.)

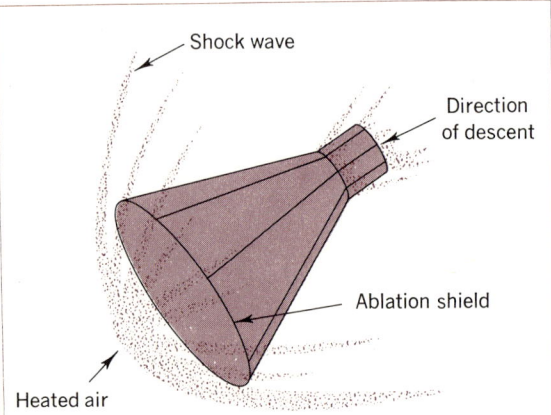

Fig. 7.13. *Formation of shock wave as a spacecraft reenters the atmosphere with the blunt end forward.*

apply reverse thrusts by means of retrograde (retro-) rockets. But since the energy required would be approximately equal to that used in launching the spacecraft, the use of retro-rockets alone is impractical. A practical way is to decelerate the spacecraft by taking advantage of the great resistance offered by the atmosphere to fast-moving objects. This resistance, however, causes considerable heating of the spacecraft, which requires that the interior of the vehicle be protected from excessive heating.

A space vehicle has a better chance of successfully reentering the atmosphere and landing safely it it is blunt rather than sharp, and large rather than small. Part of the kinetic energy of an object moving in air with high speed is spent in producing shock waves and part is spent in overcoming friction. If the body has a slender, streamlined configuration, only a small proportion of the kinetic energy is lost in the shock wave, whereas a blunt or broadnosed shape transfers a large proportion of the kinetic energy to the surrounding air in the form of a bow-shaped shock wave (see Fig. 7.13) with correspondingly reduced frictional heating. The larger the volume of the spacecraft, the greater is the amount of displaced air and therefore the greater is the force acting against the direction of motion and the deceleration.

Despite this improvement in configuration, it is necessary to utilize several other schemes to cope with the overheating of the spacecraft. One is to insulate the walls to reduce transmission of external heat into the cabin. Another is to thicken the outer walls and skin of the spacecraft to absorb more heat. A third is to cool the inner skin surface by absorbing heat in a coolant fluid. Still another resorts to cooling by ablation, a process in which the outer coatings, or surface, of the spacecraft are allowed to vaporize away. Usually a combination of these methods is employed, the choice depending on the structure and design and mission of the spacecraft.

7.8 Some results of space research 1957–1967

Now that we have discussed the techniques for launching space vehicles, let us note the scientific results that have been achieved during the first decade of the space age.

(1) Upper atmosphere of the earth. As noted in Sec. 7.7, the earth's atmosphere serves to decelerate a satellite and cause its orbit to contract toward the earth's surface. From the rate with which the satellite's orbit decays, it is possible to measure the density of the atmosphere as a function of height, especially for heights greater than 100 miles. Relatively good models of the upper atmosphere of the earth have been constructed, based on satellite observations.

(2) Geomagnetic field and trapped particles. The discovery by *Explorer I* of the Van Allen belts (see Fig. 7.14) has led to an extensive study of the earth's magnetic field out to distances of several tens of thousands of miles. Variations in the number and energies of particles trapped in this field (as monitored by a number of satellites, for example, *Injun I*, *Explorer XII* and *XVII*) have led to new theories (or models) of solar-terrestrial phenomena (such as aurorae, radio blackouts, and magnetic storms).

(3) Meteoric matter. Several satellites (for example, *Pegasus*) have been orbited to measure the rate of influx of very small meteoric particles. Also, direct collection of material was attempted on several Gemini flights.

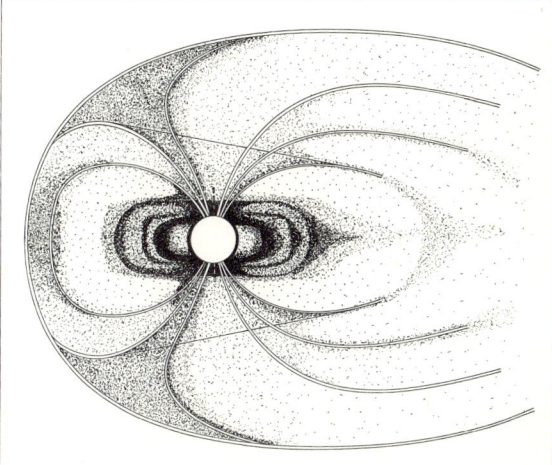

Fig. 7.14. Earth's magnetosphere and Van Allen belts.

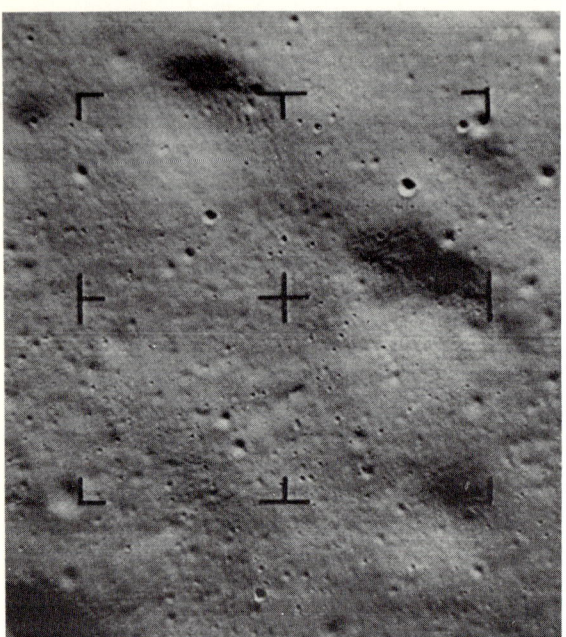

Fig. 7.15. Ranger IX photograph. Television picture by Ranger IX 2.97 sec prior to impact on March 24, 1965 at 06:08:20 P.S.T. View with shadows to the left. North is at the top. Spacecraft altitude is about 4.5 miles above the moon. Dimensions are 2.1 × 2.0 miles. Note the several large, shallow depressions with "tree bark" texture in their walls. Many dimple craters can be seen. Craters down to 40 ft in size may be seen. (NASA photograph.)

Fig. 7.16. Surveyor III photo of lunar surface composition. A scoop from the vehicle had just completed digging a small trench about 6 in. deep and 2 in. wide. Characteristics of the trench suggest that the lunar soil at this point behaves somewhat like coarse, damp beach sand. (NASA photograph.)

(4) Geodesy Satellites have been used as reference points for measuring distances on the earth's surface with higher accuracy than was hitherto possible. The fact that the earth is not a perfect sphere introduces some complications into the orbit of a satellite. By observing these "perturbations" in the orbit, it is possible to determine the shape and/or mass distribution of the earth.

(5) The moon. Since the moon revolves about the earth with the same side always toward us, nearly one-half of the lunar surface has never been seen. Several lunar probes (*Lunik, Zond III, Lunar Orbiter*) have sent back good quality pictures of the unseen hemisphere of the moon. Pictures of the visible side of the moon by the Ranger probes have given a much closer view of the moon's surface than is available through a telescope. Finally, *Luna 9* and the Surveyors have soft-landed on the moon to give information about the nature of the surface material of the moon (see Figs. 7.15 through 7.18).

(6) Mars and Venus. Successful "Flybys" of both planets were achieved by Mariner space probes. In addition, in

Fig. 7.17. This closeup photograph of the lunar crater Copernicus, taken by Lunar Orbiter II *with a telephoto lens, has been called the "Picture of the Century."* Lunar Orbiter II *was 28.4 miles above the surface of the moon and about 150 miles due south of the center of Copernicus when the picture was taken. Looking due north from the crater's southern rim, detail of the central part of Copernicus can be seen. Mountains rising from the flat floor of the crater are 1,000 ft high with slopes up to 30 deg. A ledge of bedrock is visible in the central part of the mountain chain on the floor of the crater. The 3,000-ft mountain on the horizon is the Gay-Lussac promontory in the Carpathian Mountains. Cliffs on the rim of the crater are 1,000 ft high and undergoing continual downslope movement of material. From the horizon to the base of the photograph is about 150 miles. The horizontal distance across the part of the crater shown in this photograph is about 17 miles. (NASA photograph.)*

Fig. 7.18. On November 20th, 1966, the wide angle lens of Lunar Orbiter II *took this picture of the moon's far side. This is viewed with the lunar horizon at the bottom. North is at the top and the moon's equator runs roughly along the top of the photograph. The smallest features that can be detected in the upper portion of the picture are about two-tenths of a mile across.* Lunar Orbiter II *was about 900 miles above the moon when the photograph was made. Distance across the top of the photo is about 670 miles. (NASA photograph.)*

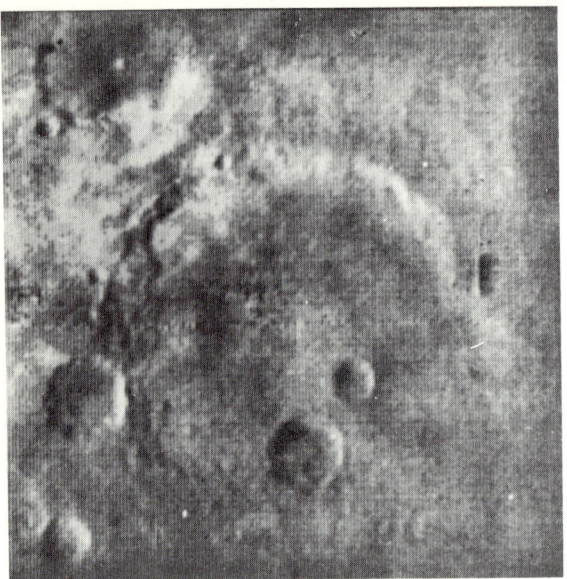

Fig. 7.19. Mariner photo of a portion of the surface of Mars ("one of the most remarkable scientific photographs of this age," in the words of the experimenters) shows a 75-mile diameter crater, with a 3-mile crater in its eastern (upper) rim. Contrast is quadrupled; filter used was green. Sun was 47 deg from zenith, range was 7800 miles, area 170 × 150 miles (west by north) in Atlantis (light region between dark areas; Mare Cimmerium to west; Mare Sirenum to east). (NASA photograph.)

October 1967 the Soviet Union sent an instrument package into the atmosphere of Venus. Data from this mission indicate that the temperature above Venus reaches over 500°F. The atmosphere of Venus is mainly carbon dioxide, with a trace of water vapor but apparently without the nitrogen which makes up the bulk of our atmosphere.

Mars seems to have a surface appearance similar to that of the moon (see Fig. 7.19) and an atmosphere of very low density, composed principally of carbon dioxide.

7.9 Distance scales in the universe

The planning for future space travel requires accurate knowledge of the distances to the various planets of the solar system. The fact that *Surveyor I* landed on the moon within a fraction of a second of the predicted time shows how accurately it has been possible to determine the distance to the moon. These considerations, in turn, lead naturally into more detailed awareness of the dimensions of the universe as a whole. In this section we shall indicate the observational data and the lines of reasoning that led man away from anthropocentric concepts of the universe to his modern ideas about time and space.

According to the third law of Kepler, a planet in an elliptical orbit about the sun with period P and a semimajor† axis a has the relation: P^2 is proportional to a^3. Letting subscripts 1 and 2 refer to any two planets, we may write this relation as

$$\frac{P_1^2}{P_2^2} = \frac{a_1^3}{a_2^3} \qquad (7\text{-}19)$$

† If the term *semimajor axis* in the following discussion proves confusing, you can simplify the orbits from ellipses to circles and substitute radius for semimajor axis, which will not change the validity of the argument.

If, furthermore, we choose earth as planet number 2 and use the period of the earth, $P_2 = 1$ (year), and semimajor axis of the earth's orbit, $a_2 = 1$ (this is the *astronomical unit*), for any planet we have

$$P^2 = a^3 \qquad (7\text{-}20)$$

where P is in years and a is in astronomical units (abbreviated "au"). Table 7.3 gives P and a for all the major planets in the solar system. Columns 4 and 5 give a^3 and P^2, respectively. Note that corresponding values of P^2 and a^3 are almost exactly equal, confirming that the values we have used for a and P are consistent with Kepler's law. In practice, a relative scale of distance in the solar system can be found by observing P for each planet and using Eq. (7-20) to find a for the planet. The small differences between columns 4 and 5 are the result of arithmetic rounding-off errors.

This scale model of the solar system gives the most distant planet of the ancients, Saturn, as being nearly 10 times as far from the sun as is the earth, while the orbit of Pluto, discovered in 1930 as a result of a photographic search of the sky, has a semimajor axis of nearly 40 au.

In order to convert this relative scale of distances into absolute distances, we must convert the distances expressed as astronomical units into more conventional distance units (say, kilometers, km). To do this, we must somehow measure the distance from earth to sun.

According to the method of triangulation (discussed in Sec. 2.13) the height of a tree, h (Fig. 7.20), can be determined if the observer accurately measures off a distance d from the base of the tree and measures the angle θ subtended by the tree at that distance. The trigonometric relation gives

$$h = d \tan \theta \qquad (7\text{-}21)$$

In a similar manner we should, in principle, be able to measure the earth-sun distance. Consider two observers, one on the equator and the other at the North Pole, observing the sun (Fig. 7.21).

To simplify the problem, assume that the observer on the equator sees the sun directly overhead, as in Fig. 7.21, while

TABLE 7.3 PERIODS AND SEMIMAJOR AXES OF THE MAJOR PLANETS

Planet Col. 1	a, in au Col. 2	P, yr Col. 3	a^3 Col. 4	P^2 Col. 5
Mercury	0.39	0.24	0.058	0.058
Venus	0.72	0.61	0.378	0.378
Earth	1.00	1.00	1.00	1.00
Mars	1.52	1.88	3.54	3.54
Jupiter	5.20	11.86	140.7	140.8
Saturn	9.55	29.46	867.9	867.7
Uranus	19.2	84.02	7,012	7,012
Neptune	30.1	164.78	27,543	27,539
Pluto	39.5	248.4	61,350	61,355

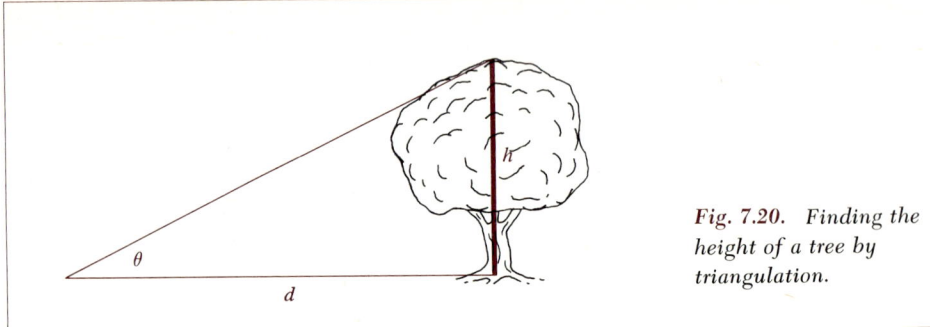

Fig. 7.20. Finding the height of a tree by triangulation.

the observer at the North Pole sees the sun at an angle ϕ from the line joining the center of the earth to the North Pole. Using the trigonometry of a right triangle, we get

$$\tan \phi = \frac{d}{R}$$

or

$$d = R \tan \phi \qquad (7\text{-}22)$$

where $R = 6356.91$ km is the polar radius of the earth.†

† It is assumed that we are able to obtain an accurate measure of the size of the earth. We shall not actually discuss this problem here. See, Payne and Gaposchkin for a brief history of our knowledge of the size and shape of the earth. Also see Sec. 3.9.

In actual measurement it is found that θ is only 8 sec of arc (approximately 1/240 of the apparent angular diameter of the sun) and thus not measurable directly with any accuracy.† Instead we must use an indirect method of finding ϕ (or θ, which is called the *solar parallax*). We resort again to Kepler's third law, which tells us that if the distance of any planet from the sun is known in absolute units,

† Modern instruments are capable of measuring angles much smaller than this, but the difficulty here lies in the fact that the sun is a large disc and both sides of the angle in question must be measured relative to exactly the same specific point on that disc. Since there are no such sharply defined fixed features on the surface of the sun, angles this small cannot be measured with sufficient accuracy for this purpose.

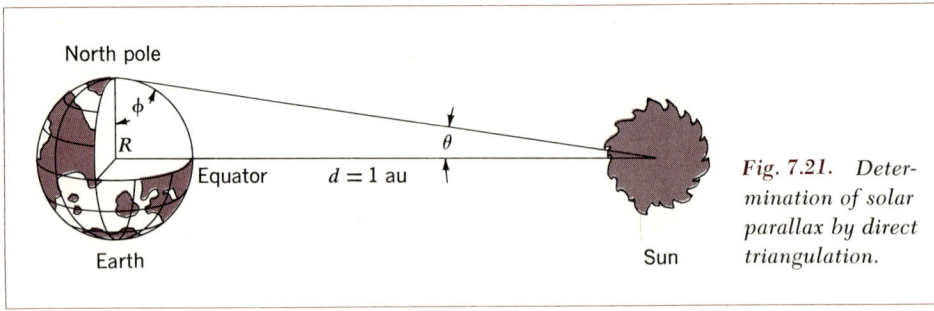

Fig. 7.21. Determination of solar parallax by direct triangulation.

7.9 Distance Scales in the Universe

then the distances of all the planets can be found in these units. Further, if the distance between two different bodies in orbit about the sun is known, then the scale of the entire solar system is again set (the reasoning behind this last statement is straightforward and left as an exercise to the reader). Until very recently, no direct distance measurements could be made to any other planet, however, and we had to search further for a key to the length of the astronomical unit (note that the moon is no help because it is not in direct orbit about the sun). The key was found in the asteroid Eros.

Asteroids (or minor planets) are small stony or metallic objects ranging from a few kilometers up to several hundred kilometers in diameter. They all travel in orbits about the sun, mainly between the orbit of Mars and Jupiter. A few, however, such as Eros, have very eccentric orbits and do at times pass fairly close to earth. In 1931 Eros passed within 25 million kilometers of the earth. Triangulation observations could be accurately made at that time because (1) Eros appears as a point rather than an extended disc, as does the sun, and (2) the angle θ was nearly 1 min of arc for Eros.

With this distance from earth to Eros found accurately, a short calculation gives

$$1 \text{ au} = 150{,}000{,}000 \text{ km}$$

(best value $149{,}598{,}000 \pm 100$ km). Several other methods give nearly the same value for the astronomical unit. Two of these are worth further mention.

In 1958, observers at Lincoln Laboratories of the Massachusetts Institute of Technology succeeded in bouncing a radar pulse off Venus. The transit time between the outward pulse and the returning echo gave a very accurate value of the earth to Venus distance. The experiment has been repeated several times since by other observers, with essentially the same results.

In 1962, the United States successfully launched the *Mariner II* space probe into an orbit which passed within 35,000 km of the planet Venus. Tracking of this probe gave a very accurate measure of the astronomical unit, which was further improved when *Mariner IV* was launched in a Mars orbit in 1964.

These last-named methods, while adding more significant figures to our conversion factor, have not changed the fact that 1 au is approximately 1.50×10^8 km.[†]

Now that we have determined the scale of the solar system, we ask, "How far away are the fixed stars?" As a starting assumption, let us assume that all fixed stars are *not* at the same distance from us, an assumption that seems reasonable, since the stars appear with very different brightness that could be due to differing distances. We again resort to triangulation methods.

Consider an observer in a moving vehicle who notes at some instant a telephone pole (relatively nearby) against a background of pine trees. Looking back at the telephone pole at some later time, he notes that the pole now has a background of maple trees (see Fig. 7.22). He rightly concludes that (1) the pine trees did not magically become maple trees; (2) that he rather than the telephone pole moves; and (3) the pole is closer to him than to either group of trees. We call this phenomenon *parallax*. Similarly, as the earth revolves about the sun we can view the sky from different points in space; hence, nearby stars should appear to

---

[†] The student is referred to the excellent article by McGuire, Spangler, and Wong, in *Scientific American* (April, 1961), p. 64, which deals more exhaustively with the measurement of the astronomical unit.

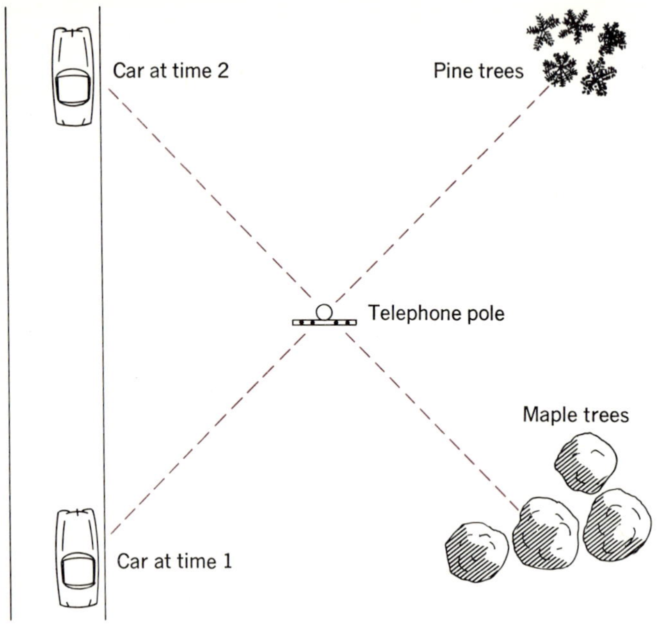

250 Introduction to the Space Sciences

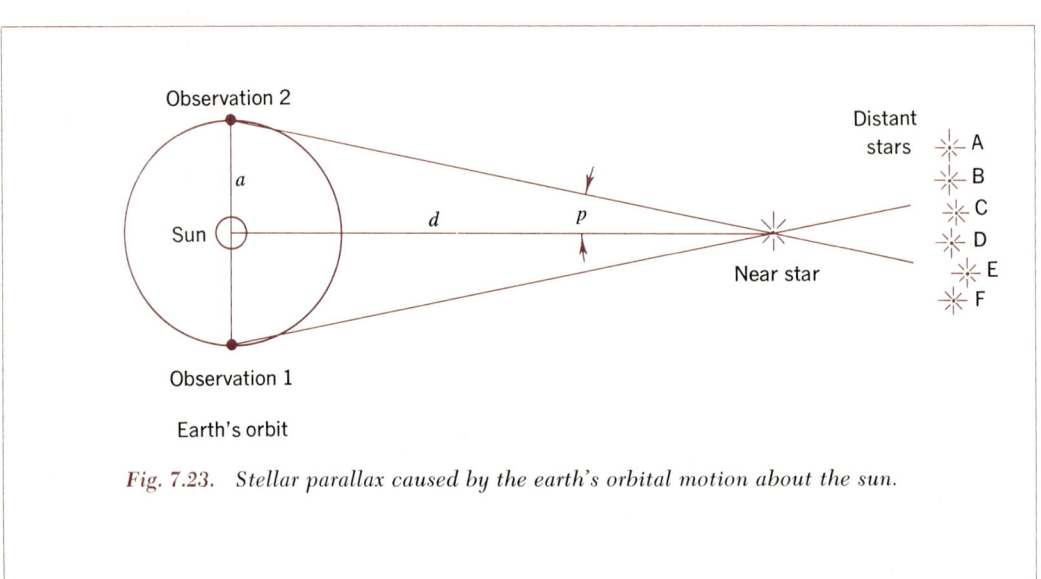

Fig. 7.22. Parallax seen by observer in a moving car.

Fig. 7.23. Stellar parallax caused by the earth's orbital motion about the sun.

change position relative to more distant ones (Fig. 7.23).

At some initial time an observer on earth might note that a nearby star is among stars B and C (observation 1), while 6 months later he would see it to be near star E (observation 2).† This parallax angle, $2p$, is a measure of the star's apparent motion. By trigonometric analysis, similar to that leading to Eq. (7.22), we find

$$d = a \cot p \qquad (7\text{-}23)$$

where d and a are respectively the distance to the star, measured in astronomical units, and a is the earth to sun distance, or 1 au. The angle p is expressed in normal angular measure (degrees, minutes, and seconds).

The first measurements of stellar parallax were made in 1838 when, with the aid of telescopic observations, F. Bessel in Germany succeeded in measuring the parallax of the double star 61 Cygni. During the same year, T. Henderson at the Cape of Good Hope measured the parallax of α Centauri, while F. Struve in Russia observed that of the bright star Vega.

The largest stellar parallax known is that for α Centauri, a very bright star not visible in the United States, owing to its southerly position. It has a parallax of only 0.75 sec of arc! By Eq. (7-23),

$$d = 1 \cot 0\rlap{.}''75 \text{ au}$$
$$= \text{over 250,000 au} \ddagger$$

To attempt to imagine this distance,

---

† This example is grossly exaggerated in order to simplify the explanation of stellar parallax. The term *parallax* thus refers to the angular difference between two directions or lines-of-sight taken from two positions located on a plane or base line.

‡ 1 rad = 206,265″ and the $\cot x \sim \dfrac{1}{x}$ when x is in radians and very small.

consider that we make a scale model of the solar system in which 1 au = 1 ft. Then Mercury would appear about $3\frac{1}{2}$ in. from the sun, Venus about $8\frac{1}{2}$ in. and the earth 1 ft away. Pluto would be almost 40 ft away. On this scale Jupiter (at 5 ft) would be in the living room of a house, Saturn (at 10 ft) might be in the dining room, Pluto would be on the front lawn, while α Centauri would be 50 miles away!

Using present photographic techniques, stellar parallaxes as small as 0.005 sec of arc can be measured. For the distances that become revealed by these measurements, the astronomical unit is much too small a "yardstick" and astronomers prefer to introduce a new unit called the *parsec*. A parsec is the distance at which a star will appear to have a parallax of 1 sec of arc.† Another useful distance measure is the light-hear—the distance traveled by light at a speed of 300,000 km/sec in one year (3×10^7 sec).

The nearest star, α Centauri, is therefore a distance of 1.33 parsecs, or 4.3 light-years away. Thus the light emitted from α Centauri takes over four years to reach the earth!

Trigonometric parallaxes are useful for measuring distances out to about 100 parsecs (325 light-years). The volume of space included in the measurable region contains of the order of 10^6 stars, most of which are invisible to the unaided eye. Using techniques that depend on a quantitative analysis of the light we receive from the stars, we can determine approximate physical parameters (for example, size, mass, temperature, intrinsic bright-

---

† Some conversions:

1 parsec = 206,265 au
= 3×10^{13} km
= 3.26 light-years
1 light-year = 10^{13} km = 6×10^{12} miles

ness)† for many of these stars. From these we have concluded that our sun is a normal star—smaller than some and larger than most.

Thus, in the past 500 years, our earth has been "demoted" from being popularly considered as the center of the universe to just one of nine "satellites" of the sun, while the sun itself is but one of a myriad of stars!

7.10   The galaxies

To consider the expanding picture of man's ideas about the universe, we shall now take a long jump away from the sun and earth and consider the larger aggregates of stars called *galaxies*.

On observing the sky from a dark, clear sky location, one can easily see the faint, whitish band of light that stretches across the sky and which since antiquity has been called "the Milky Way." Astronomers refer to it (as they do to all the visible stars) as "The Galaxy."

When Galileo in 1610 first turned his small telescope on the Milky Way he noted that it was actually made up of millions of stars, too close together to be resolved as individuals by the eye alone.‡ In 1750, Thomas Wright speculated that the sun is a part of an enormous disc-shaped system of stars and that the visible Milky Way is light from the surrounding stars that lie more or less in the plane of the disc. In 1785, William Herschel showed, by counting stars in various directions in the sky, that Wright's suggestion was essentially correct. However, up to the beginning of this century there was no more quantitative information about our Galaxy and the sun's position within it. We had to await new methods to extend our measurements of stellar distances. One such method, discovered by Dr. Harlow Shapley and Miss Henrietta Leavitt of the Harvard College Observatory, and announced by Miss Leavitt in 1912, uses the Cepheid variable stars, which we shall presently describe.

Although to the ancients the stars appeared fixed and unchanging, it was recognized nearly 2000 years ago that occasionally a "new" star appeared in the sky. Often, we now surmise, these new stars were comets, members of our solar system, which occasionally come in close to the sun and are seen by reflected sunlight. Others, such as the new star of 1054 (recorded by the Chinese), Tycho's star (1562), and Kepler's star (1604), are examples of a type of phenomenon known as *supernovae*. From their sudden appearances we simply note that the starry sky is far from unchangeable. Supernovae are examples of variable stars—stars whose brightness may change with time.

Present catalogs list some 15,000 stars that have been observed to fluctuate in brightness. For some, these observations of brightness as a function of time have been plotted to give what astronomers call *light-curves* (see Fig. 7.24). For some of these stars the light-curves show random or unpredictable brightness fluctuations, whereas other variable stars show light-curves that repeat in a periodic fashion. One such type of object is the "Cepheid variable," named after the star in the constellation Cepheus, which was the first such star discovered. It is these stars, the Cepheids, that provide us with a means of measuring distances of millions of parsecs!

---

† See one of the astronomy texts listed in the References for a more detailed account of how this is accomplished.

‡ It might be noted that the Greek atomist-philosopher Democritus suggested about 2000 years earlier that the Milky Way was composed of stars. However, Democritus' view was pure speculation and not based on observation.

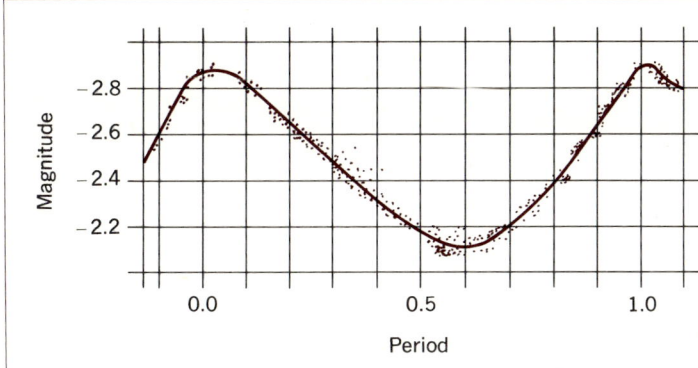

Fig. 7.24. Light curve of δ Cephei.

Just as the earth has the moon as a satellite and the planets are satellites of the *sun*, our galaxy also has a pair of satellites. These satellites, vast aggregates of stars in their own right, are called the *Magellanic Clouds* (see Fig. 7.25). To an observer south of the equator, the Magellanic Clouds appear as faint patches of light, which might have broken off from the Milky Way. Unfortunately these interesting objects are invisible from the United States.

As a part of an extended study of the Magellanic Clouds by Dr. Shapley and other members of the staff of the Harvard College Observatory, Miss Leavitt investigated the light-curves of a large number of Cepheid variables observed

Fig. 7.25. The Lesser Magellanic Cloud. This and the Greater Magellanic Cloud are the nearest neighbors of the Milky Way. (From Bonner, The Mystery of the Expanding Universe, Macmillan, 1963.)

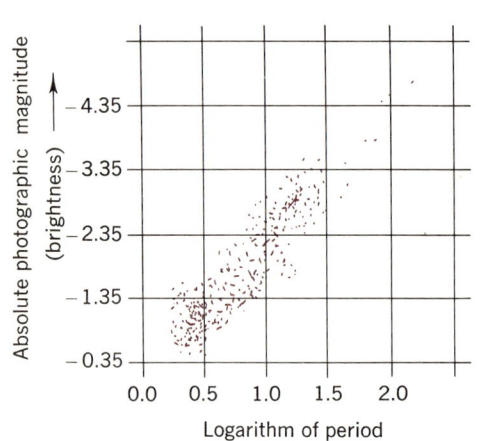

Fig. 7.26. (a) Period-luminosity curve plotted from the observations of Henrietta Leavitt. (After Shapley, in Goldberg and Aller, Atoms, Stars, and Nebulae, Blakiston, Philadelphia, 1943, p. 134.)

photographically in the Magellanic Clouds. She found that when she compared the period of light variation of these stars with their brightness at maximum light, a definite relation appeared between these two quantities in the sense that the longer the period, the brighter the star at maximum. The interpretation that all the long-period stars were at the nearer end of the cloud and the shorter-period stars were at the far side of the cloud and appeared fainter was rejected as improbable. Instead, Miss Leavitt properly concluded that an intrinsic relation exists between the period of light variation and the brightness of the star at maximum light (the *period-luminosity relation*). (See Fig. 7.26(a)).

It can be easily demonstrated in the laboratory that, given a source of light of *known* (intrinsic) brightness, the *apparent* brightness of the source decreases as the inverse square of the distance of the observer from the source. Thus, if we know both the intrinsic and apparent brightness of the source, we can immediately compute the distance between source and observer. If we assume, with Miss Leavitt, that the period-luminosity relation is an intrinsic characteristic of the Cepheids, then observing the period of one of these stars, which can be done with high accuracy, should tell us the intrinsic brightness of the star. However, to calibrate the period-luminosity relation, we need a star whose actual distance is known. Unfortunately, trigonometric parallaxes, the only distance indicator discussed so far, is ineffective in calibrating the relation, and astronomers must use statistical methods on many Cepheids. In the past 20 years, it has been discovered that no single period-luminosity relation is consistent with all observational data and that a multiplicity of such relations must exist, corresponding to the age and evolution of the Cepheids

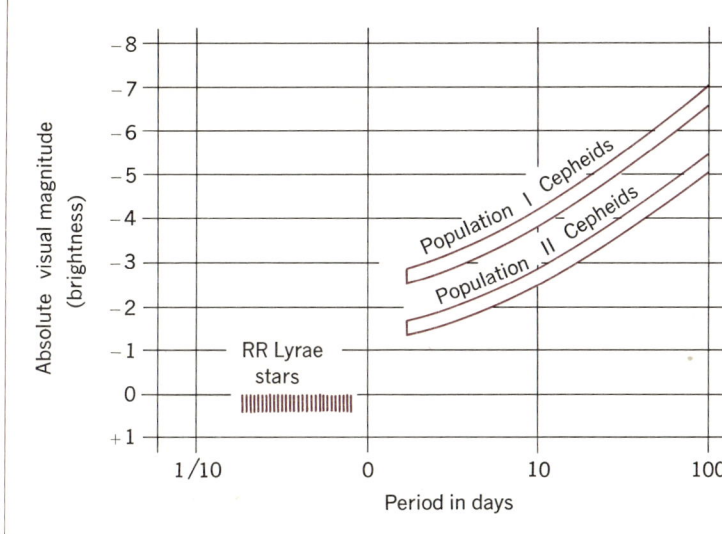

Fig. 7.26. (b) *A schematic diagram showing the period-luminosity laws for Population I and Population II Cepheids. In the lower left-hand corner of the diagram the RR Lyrae variables (short-period variables) are shown. Population I stars are young, Population II stars are old.*

considered (see Fig. 7.26(b)). By measuring several other properties of the Cepheids, such as color and size of the fluctuation between maximum and minimum light, it is possible to distinguish between the several different period-luminosity relations.

By using Cepheid variables, we find the Magellanic Clouds to be between 40,000 and 50,000 parsecs away (125,000–150,000 light-years) and hence to be galaxies in their own right.

The ancients grouped the visible stars into *constellations*. These are artificial groupings, dictated by the apparent positions of the stars in the sky. There do exist, however, many natural groupings of stars into large aggregates. One such type of grouping is the globular cluster (see Fig. 7.27). These vast systems of

Fig. 7.27. Globular star cluster in Hercules (Messier Catalog, No. 13). (A 200-in. photograph from Mount Wilson and Palomar Observatories.)

stars, containing up to one million stars in a compact, spherical cluster (hence the name) are distant members of our own Galaxy. About a hundred such clusters are members of the Galaxy and are generally found away from the plane of the Milky Way. Thus they form sort of a halo around the Galaxy. Harlow Shapley in 1917 used Cepheid variables occurring in globular clusters to obtain their distances and spatial distribution. He found that they formed a spherical system centered not about our sun but about a point about 8000 to 10,000 parsecs away in the direction of the summer constellation of Sagittarius. Shapley correctly concluded that this point represents the center (or nucleus) of our Galaxy. This finding is consistent with the fact that the Milky Way appears denser in the region of Sagittarius than it does in the region on the opposite side of the sky, in the constellation Perseus.

Radio astronomy studies since 1948 have measured the 21-cm wavelength radiation due to hydrogen in the space between the stars. These observations have led to a picture consistent with that of Shapley. Our Galaxy appears to be a vast spiral, centered in the constellation Sagittarius and with a diameter of about 30,000 parsecs (100,000 light-years). The

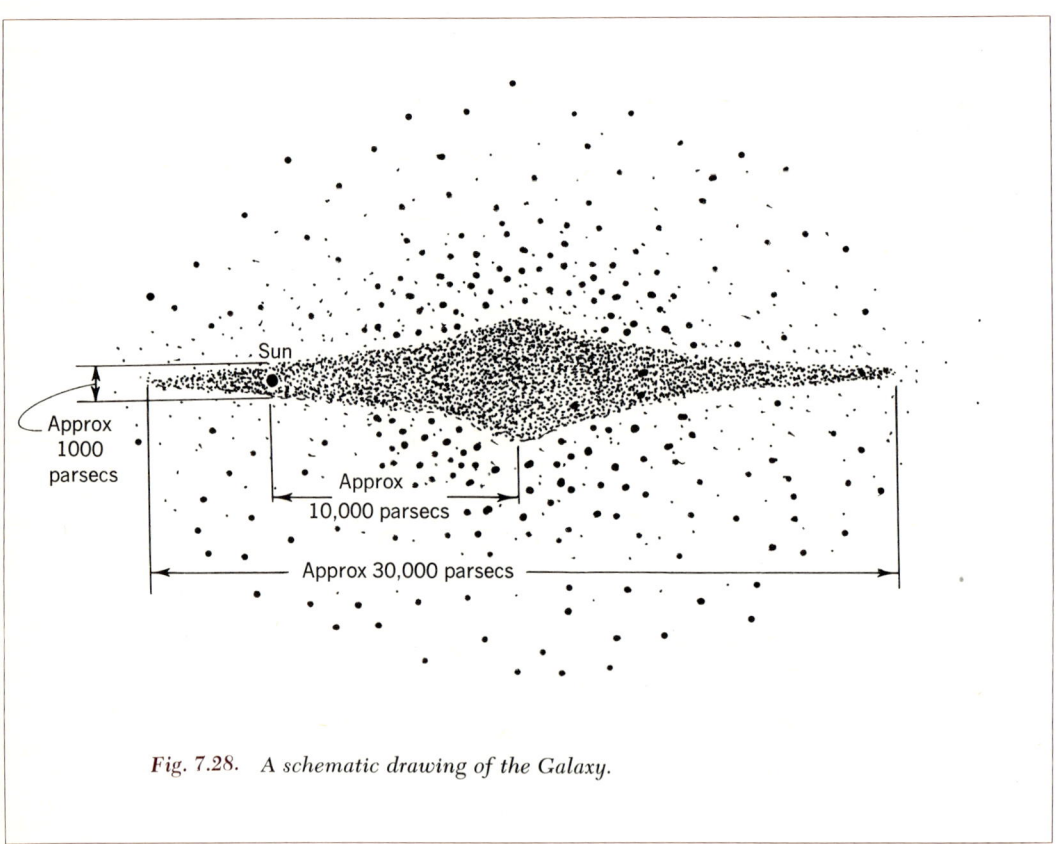

Fig. 7.28. *A schematic drawing of the Galaxy.*

sun is about 10,000 parsecs out from the center. Thus, instead of being the center of the universe, our sun and the earth are now known to be located two-thirds of the way out toward the edge of a large galaxy whose population may exceed 100 billion (10^{11} stars). (See Fig. 7.28.)

In earlier chapters we discussed the effects of the various scientific and technological revolutions on human thought and behavior. It is, however, appropriate that we note at this time the drastic changes in man's ideas about many nonscientific topics, which accompanied this "demotion" of the earth's universal position. It is worth noting how great have been the philosophical, theological, and psychological changes accompanying our growing knowledge about the universe. Drastic changes have also taken place in scientific thought. Whereas astronomers in the sixteenth century spoke circumspectly about a heliocentric universe, modern astronomers extend the Copernican theory to exclude any special position to the earth or the solar system.

Since the beginning of telescopic observation, a number of celestial objects have been noted whose appearance is distinctly nonstarlike. These objects, as a group, were called nebulae (clouds). We now know that the term *Nebulae* embraces many very dissimilar types of objects. One type, however, is of special importance in our discussion of the universe as a whole. These are the spiral nebulae such as the Great Spiral in Andromeda (see Fig. 7.29). Some men considered these spirals to represent solar systems that were being formed; others considered them to be very remote, unresolved objects. In 1755, Immanuel Kant called them "island universes," a connotation that today we know to be very close to the truth.

On April 26, 1920, a historical, astronomical debate took place between

Fig. 7.29. Great Spiral in Andromeda. (Messier Catalog, No. 31.) Satellite nebulae NGC 205 and 221 also are shown. (A 48-in. Schmidt photograph from Mount Wilson and Palomar Observatories.)

Fig. 7.30. Southern edge of Andromeda Nebula resolved into stars. (Photograph from the Mount Wilson-Palomar Observatories.)

Fig. 7.31. The Crab Nebula. This object is about 4000 light-years from the earth. It was formed by a supernova explosion seen on earth in A.D. 1054. (Photograph from the Mount Wilson-Palomar Observatories.)

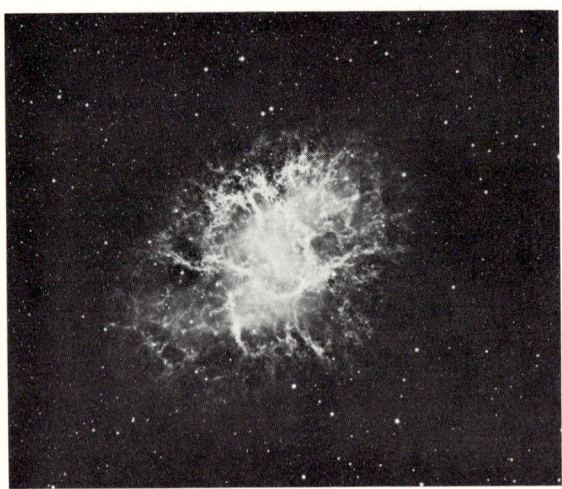

Harlow Shapley of Harvard and H. D. Curtis of the Lick Observatory. Curtis supported the island universe theory of spiral nebulae, while Shapley held that they were within our own Galaxy and that perhaps the Galaxy and universe were synonymous. In writing of this debate in 1961, Allan R. Sandage of the Mt. Wilson and Palomar Observatories said, "Perhaps the fairest statement that can be made is that Shapley used many correct arguments to reach the wrong conclusion, while Curtis, whose intuition was better in this case, gave rather weak and sometimes incorrect arguments from the facts, but reached the correct conclusion."

As a debater, Shapley certainly won. It is ironic that less than five years later, on December 30, 1924, Edwin Hubble, a lawyer turned astronomer, announced to the American Astronomical Society that observations he had made with the 100-in. telescope at Mt. Wilson Observatory had succeeded in resolving part of the Andromeda Nebula into stars (see Fig. 7.30) and that several Cepheid variables were found. We presently conclude that the Andromeda Nebula is nearly 700,000 parsecs away and is a galaxy similar to our Milky Way.

During the period since Hubble's announcement, the distances to galaxies as far away as 10 million parsecs have been measured by using Cepheid variables. For these galaxies, we can determine something of their intrinsic physical properties and can, by using statistical methods, find the "average" properties of a galaxy. If we then observe a more distant galaxy that is (at present) not resolvable into stars, and we assume that it is an "average" system, then we can make an estimate of its distance.

As noted above, stars often occur in large clusters. This also applies to galaxies, many of which belong to clusters containing as many as a thousand members. For these rich clusters, statistical estimates of distance can be very accurate. In this way we have extended our distance scale. In our future discussion of the structure of the atom we shall make use of the data obtained from spectra.† It is by the use of spectra that we

learn about the more distant, observable objects in our universe.

The spectrum of a nearby galaxy shows dark lines characteristic of the chemical elements making up the member stars.

---

† If we pass a beam of "white" light (that is, light containing all colors) through a prism, we break it up into a "rainbow" of colors. Spectra will be discussed in more detail in a later chapter.

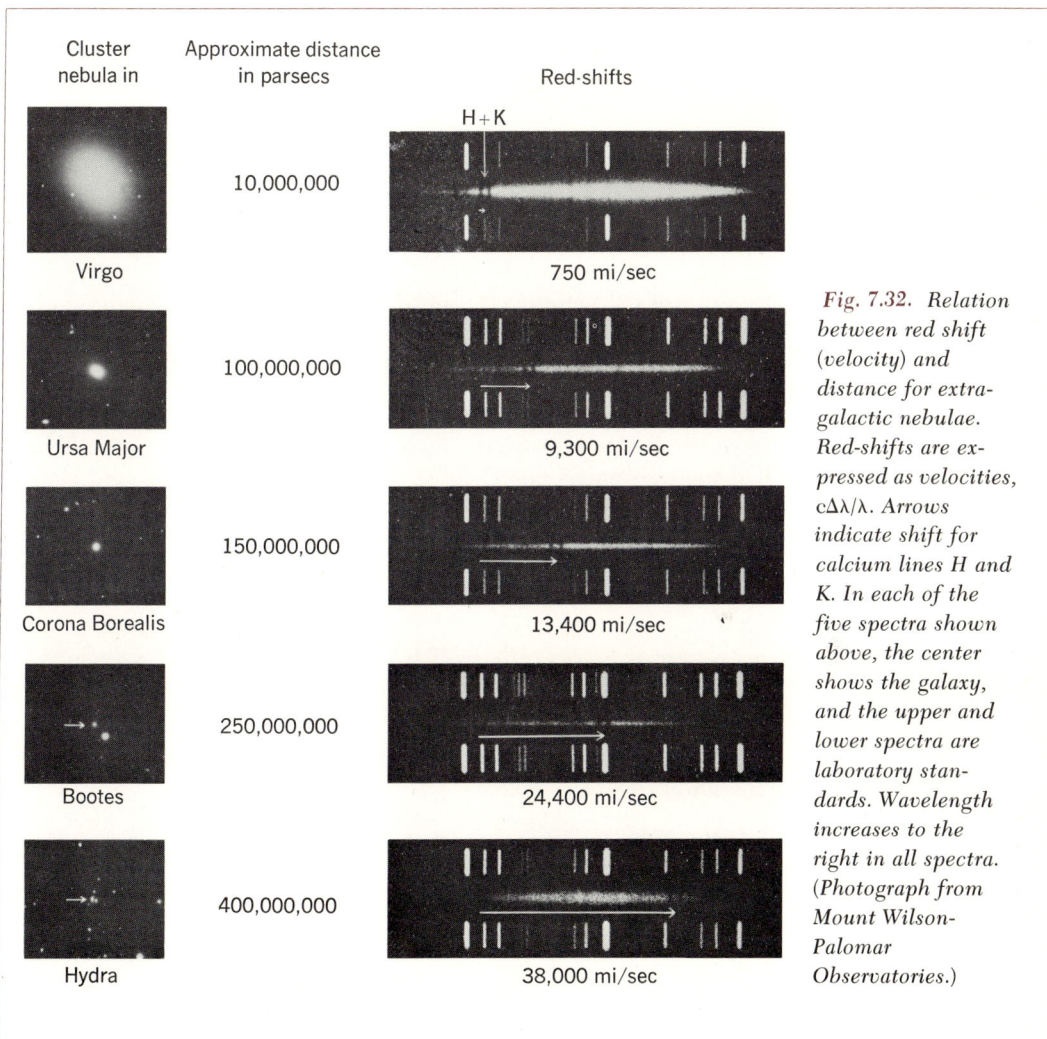

Fig. 7.32. Relation between red shift (velocity) and distance for extragalactic nebulae. Red-shifts are expressed as velocities, $c\Delta\lambda/\lambda$. Arrows indicate shift for calcium lines H and K. In each of the five spectra shown above, the center shows the galaxy, and the upper and lower spectra are laboratory standards. Wavelength increases to the right in all spectra. (Photograph from Mount Wilson-Palomar Observatories.)

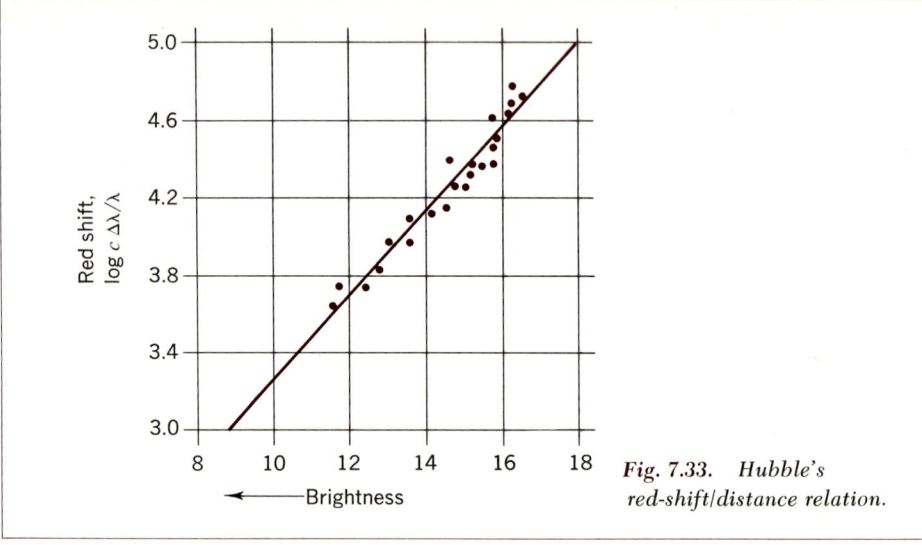

Fig. 7.33. *Hubble's red-shift/distance relation.*

The positions of these lines are approximately the same as those found for the same elements in our laboratory. However, as we observe more distant galaxies, the lines in the spectrum appear shifted toward the red† (see Fig. 7.32). Using galaxies whose distances have been determined by the methods outlined above, Hubble showed that there exists a relation (Fig. 7.33) between the amount of red shift (velocity) and distance to the galaxy. By extrapolating the observed relation to galaxies whose distances cannot be otherwise determined, we can estimate how far away they are.

7.11 Radio astronomy

During the past two decades, astronomers have supplemented the usual optical observations of the sky with radio studies. The sky contains numerous discrete sources of radio radiation, most of which have not as yet been identified as associated with any optically visible objects. Some of the "brightest" sources have been so identified, however; for example, Taurus A with the Crab Nebula, a supernova remnant (see Fig. 7.31). During the past five years, however, a number of these radio sources have been identified with starlike objects called *quasistellar radio sources* (*quasars*). Martin Schmidt and Allan Sandage of Mt. Wilson and Palomar Observatories have pioneered in the study of quasars and find from studies of their red shift that they are among the most distant objects ever "seen."‡ The quasar 3C9 studied by Schmidt is over

---

† The observed red shift of spectral lines is generally interpreted as resulting from the Doppler effect (see Chapter 5) of objects that are receding from us. This gives rise to the picture of an *expanding universe*.

‡ Speculation as to the distance of a quasar depends on identifying its red shift and assuming it to be due to motion following Hubble's red-shift/distance relation. Other interpretations suggested do not place these objects so far away.

3 billion parsecs (10 billion light-years) away!

We may also at this point consider some speculations about the time scale of the universe. We have already noted that using the radioactive decay of uranium to estimate the age of the earth gives values of the order of several billion years. Astrophysicists are able to make models of the internal structure of the sun and of its evolutions and these too give an age for the sun of about 5 billion years. Thus it now seems reasonable to assume that the earth and sun were formed at (or nearly at) the same time some 5 billion years ago.

To get an estimate of the age of our Milky Way, we must consider in detail the motions of the various objects within it—stars, dust, gas, clusters, and so forth. Various models all give ages between 5 and 10 billion years.

Finally, if one interprets the velocity/distance relation as representing the age of the universe since some primordial "big bang," one gets an age of about 13 billion years. Thus, all present data seem to indicate a very old universe.

7.12 Summary

If it were possible to reduce the size of the earth from its present radius of about 4000 miles to 4 in., its surface with all its mountains and valleys and canyons would be almost as smooth as the surface of a crystal ball. Until the present decade, man was restricted in his outward and inward movements to a departure of no more than one-tenth of 1 percent of the earth's radius. But the age of space exploration has now begun. Shortly man will move outward to a distance of some 60 earth radii, and he dreams of penetrating even farther. The spirit of adventure and the possibility of practical discoveries drive man to take up the new challenge which so taxes his scientific and engineering ingenuity.

Two types of difficulties must be overcome in this endeavor: (1) the problem of living in an environment to which man is completely unadapted (the total lack of atmosphere, extremities of heat and cold, dangerous radiations, and weightlessness present rigors of a new order); (2) the engineering problem of attaining velocities great enough to permit man to orbit or to move about freely in space with adequate control of the direction and speed of motion.

By taking his own atmosphere as well as food and fuel with him, and by the use of carefully designed shielding and the avoidance of certain areas of space where dangerous radiations are most concentrated (that is, the Van Allen belts), man has overcome all the problem of weightlessness and this now appears not to be a serious danger.

The problem of *getting into space* presents a different but equally fascinating problem in which scientific theory and engineering techniques of the highest order must be blended. The calculation of orbital speed, once an object is in space, is not difficult and follows simply from Newton's laws, which we have studied. The basic equation used represents the equivalence of the centripetal force of gravitation acting on a satellite and the accelerating force required to keep the satellite continuously accelerated in a curved path about the earth. For circular orbits this equation is

$$G\frac{mM_e}{R^2} = \frac{mV^2}{R}$$

(For elliptical orbits the equation is somewhat more complicated, but the principle remains the same.) If the distance R to the center of the earth is set, the speed V is determined. Note particularly that the mass m of the space ship cancels out.

This implies that two artifical satellites in the same orbit will travel at the same speed regardless of the mass; which is obviously a great advantage in docking maneuvers.

The practical engineering problem, however, of obtaining the necessary speed is immensely difficult. At this point the mass of the space ship to be launched is significant and energy considerations must be taken into account. To escape from the earth altogether, an object would have to be launched at 7 miles/sec. But the more massive the object, the greater the energy requirements (since the kinetic energy is given by $\frac{1}{2} mv^2$). To give a 1-ton mass (roughly 10^3 kg) a velocity of 7 miles/sec (roughly 10^4 m/sec) would require about 50 billion joules of energy. Since a speed of only about 5 miles/sec is required for the presently orbiting satellites, it might be anticipated that the energy would be less. However, the picture is further complicated by two factors: air resistance and the fact that satellites cannot be given too high an initial acceleration, particularly if they carry a man who cannot organically tolerate an acceleration of more than 10 g. Hence, the energy of the fuel cannot be expended all in one burst. A large portion of the fuel itself must be carried aloft and its energy gradually expended. Hence, for a payload of 1 ton, the original mass, including fuel and its container must be as much as 100 times the mass of the payload.

To reach the necessary velocities, the law of momentum (action and reaction) is utilized. The hot gases are given a very high velocity as they are exhausted backward. Thus the backward change in momentum ($\Delta m_g\, V_g$) of the gases, which are light in mass, gives a forward momentum to the more massive rocket. (There is no need for air against which to push.) To exhaust the gas at a high enough speed that will result in an adequate momentum requires extreme heat, and this in itself presents many difficult technical problems (see Sec. 7.4).

Such then are some of the elements and problems involved in this endeavor of man to reach out beyond the limits of his natural environment. But a myriad of other technical developments such as guidance and control, communications, computer science, electronic miniaturization, and medical research all play their part. A few of these have been discussed briefly in this chapter. Inevitably, in addition to the direct practical results that have arisen out of the satellite program (weather forecasting, communication, further astronomical and geophysical knowledge), many unexpected technological and scientific discoveries have been made. As usual, scientific research pursued intensely with one goal in mind has yielded a rich dividend of discoveries that are applicable in other directions.

Along with the exploration of space accomplished by sending up satellite probes, new astronomical probes have been recently developed, including radar, radio telescopes, better spectrometers, and vastly improved instruments. In the twentieth century, for the first time, we have learned to look well beyond the stars of our own Galaxy and at last discover methods of measurement for reaching out into the vast recesses of space. Just as estimates of the time of the existence of our earth have been greatly expanded in our century, so estimates of distance have been expanded.

The simplest method of measuring the distance to an object that cannot be reached is by triangulation. This depends on the observation of the object from two points at different locations on a base line. Given two angles and the length of base line, the distance can be determined. In this way the distance to an object as close as the moon can be determined by observation from two points on earth to any point on the object. But to determine the distance of the sun in a similar fashion is impossible because no base line on earth can be made long enough. The sun is so distant that even the two ends of the longest possible line will form angles with the sun that are too small to be measured with sufficient precision. Section 7.9 outlined a method for calculating such far distances, by combining Kepler's third law with a knowledge of the distance of an asteroid (close enough for triangular determination).

Once the distance of the sun is accurately determined, a new base line is established, namely, the radius of the earth's orbit. Using this base line and parallax angles, the distance to the closer stars (up to about 325 light-years) can be found. Beyond that distance, however, the base line once again is found to be too short for precise determination of parallax angles. But in the past few decades a new method has been developed. The key is that the variable stars known as Cepheids have a periodicity related to their intrinsic brightness by a specific equation. Knowing the intrinsic brightness, then measuring the apparent brightness, and combining these two values with the inverse square law relating the intensity of light to distance, the distance of these stars can be calculated even though the distance is far greater than can be determined by parallax.

Still more dramatic has been the discovery of Hubble's relation, which gives a clue to the distances that separate our own galaxy, the Milky Way, from very distant galaxies—so distant that it takes several billion years for the light to reach the earth. Hubble's law relates the so-called spectral red shift to velocities of recession of galaxies and also relates these velocities of recession to the distances of the galaxies. It is partly on the basis of this law, as well as on the observation of radio stars, that astronomers now have extended the spatial dimensions of the observable universe to a minimum of 5 billion light-years. Both space and time have been expanded by a factor of more than a million since the beginning of the twentieth century.

Questions/Discussions

1. Why are astronauts required to keep themselves in superior physical condition?

2. What are the major factors involved in selecting the proper time to launch a space probe to a planet such as Venus?

3. Observers on earth always see the same side of the moon. Would an observer on the moon always see the same side of the earth?

4. Is there a "dark side" of the moon where the sun never shines?

5. Would the earth exhibit "phases" to an observer on the moon? If so, how long would it take to complete a full cycle from "new earth" to "new earth"? (That is, what would be its synodic period?)

6. Which of the following devices could be used successfully as they are ordinarily used on earth in an earth-orbiting laboratory by the "weightless" astronauts on board? (a) Knife, fork, and spoon, (b) Chopsticks, (c) Broom, (d) Vacuum cleaner, (e) Wrench, (f) Paper weight, (g) Hammer, (h) Soda straw, (i) Aerosol mosquito "bomb," (j) Fire extinguisher of the type that must be inverted to operate, (k) Pendulum clock, (l) The shift mechanism of an ordinary typewriter, (m) A drip-style coffee pot.

7. In your opinion, which of the planets, if any, are capable of supporting either plant or animal life?

8. How does the total rocket fuel required for a trip from the earth to the moon compare with the fuel required for the return trip?

9. What are the difficulties inherent in attempting direct radio communication with suspected planets revolving around some of the closest stars?

10. Assuming the mass of the moon to be 7×10^{22} kg and its radius to be 1.7×10^6 m, calculate the value of escape velocity from its surface.

11. Explain, by means of a specific example, why it is impossible to include in a book such as this a diagram drawn to scale that shows both the relative sizes of the sun and planets and also the "average" radii of their orbits. It is suggested that you start with a $\frac{1}{4}$-in. diameter circle to represent the sun and then compute the scale size of Mercury and also the scale distance from the sun to Pluto.

References

A large number of books contain discussions of space exploration. A few of these are listed below.

Clark, A. C., *Man in Space*; Morristown, N. J.: Silver Burdett, 1966.

Glasstone, S., *Sourcebook on the Space Sciences*. Princeton, N.J.: Van Nostrand, 1965. *An "encyclopedia" of information on astronomy and astronautics.*

King-Hele, D., *Satellites and Scientific Research*. London: Routledge and Paul, 1960. *Covers the early period of the space age (up to 1959) in good detail. Gives clear physical explanations of the techniques used in space research.*

Lundquist, C. A., *Space Science*. New York, McGraw-Hill, 1966. *An up-to-date, highly readable book intended to supplement the usual astronomy texts.*

The astronomical discussion in this chapter is well covered in most elementary astronomy books. Some recent texts are:

Abell, G., *Exploration of the Universe*. New York: Holt, 1965.

Motz, L., and Duveen, A., *Essentials of Astronomy*. Belmont, Calif.: Wadsworth, 1966.

Payne-Gaposchkin, Cecilia, *Introduction to Astronomy*. Englewood Cliffs, N.J.: Prentice-Hall, 1954. pp. 24ff.

Struve, O., *Elementary Astronomy*. New York: Oxford Univ. Press, 1959.

Struve, Otto, and Zelburgs, Velta, *Astronomy of the 20th Century*. New York: Macmillan, 1962. *A clear account of the history of the astronomical theory during the first 60 years of this century.*

CHAPTER EIGHT

Probability and Statistical Concepts in Human Affairs

It is remarkable that a science which began with the consideration of games of chance should have become the most important object of human knowledge.

PIERRE SIMON DE LAPLACE

THUS FAR WE have been concerned with situations in which the behavior of individual objects could be analyzed on the assumption that a series of individual measurements could be made which would be adequate for the analysis. For example, when a body is caused to move because it is subjected to a force, the measurement of such quantities as force, direction, distance, weight, and time permit a fairly complete analysis to be made of its behavior. But what can we do when the number of objects becomes very large, or when the object is subjected to a very large number and variety of forces in a brief period of time? Or how do we analyze phenomena in which chance plays a part, as in the fall of dice? We find that a great many of the phenomena in nature, and indeed in our social and personal lives, either involve elements of chance or include so many numbers or variables that we are forced to resort to statistical methods to understand them. This chapter presents several concepts and techniques of statistics that are useful for such understanding. We shall defer discussion of historical aspects and basic probability concepts, and introduce the subject with an analysis of a testing operation that is very common in industry.

8.1 The importance of statistical reasoning

When a housewife experiments with a new recipe, it becomes easy enough to inquire at the dinner table whether the members of the family would like the dish again at some future meal. The number of people is small and the relationship quite direct. But how does a food manufacturer who caters to the tastes of millions of consumers, whom he never sees, decide that it is profitable and safe to place a new product on the market? Even more important, how can a pharmaceutical firm and the responsible agencies that control the sale of drugs place a new product on the market with assurance that there will not be serious harmful consequences to health and serious risk of lawsuits?

Obviously the testing of new food products and drugs can be attempted only on a limited number of people, through what may be called a *sampling* process. How large should the size of this sample be to assure that the results are meaningful? How should the individuals be selected to make up a sample group that is representative of the population by which the product will be used? In the case of a new drug, how large must be the sample of people and the test, and what tests should be made of the selected sample to determine its effects under diverse conditions such as pregnancy, unusual diet or working conditions, alcoholism, and use of other common drugs? Only recently the world was shocked to learn that a particular drug produced badly malformed infants when given to expectant mothers.

Question 8.1.† An opinion research institute decided to sample adult opinions in a large city concerning an antipoverty program. Suppose it obtained its information by interviewing every one-hundredth person whose name was listed in the current telephone directory. Would this sample be likely to represent adequately the opinions of all segments of the city's population? Why, or why not?

Question 8.2. Would doubling the size of the sample by selecting every fiftieth name in the directory make the results more satisfactory? Explain your answer.

Question 8.3. Suppose a sample of the same size as that in Question 8.1 were drawn from the city's voter registration lists instead of from the telephone directory. Would this make the results more representative than before? Explain your answer.

The fact is that sampling methods have come into common use in practically every aspect of our life. Dresses and other items of clothing are manufactured according to information derived from sampling. Census figures are amplified from samples. The tax income estimates of state and federal governments are obtained the same way. The division of funds among the various regions of the country is based on these methods. The parts that go into the automobile production line, the radio production line, or any other production line are tested by sampling techniques. The durability of parts that go into machines, of gadgets, and of furniture is designed to be good enough to stand up under use, but not so good as to last forever and thus be prohibitively expensive, heavy, or otherwise impractical. Some failures are expected. This requires that there be ade-

---

† In this chapter the reader will note a substantial change from the style used in earlier chapters. While the subject of this chapter is far from difficult to grasp, it does require that the reader pause to think through the logic of each situation. As a general rule it becomes useful for the reader to pursue an approach somewhat like the following analysis:
 (a) What is the thing I am looking for?
 (b) What is given to work with?
 (c) What principal law or axiom applies?
 (d) What is given that is needed?
 (e) What is needed but is not given?
 (f) Can what is needed, but not given, be deduced or obtained elsewhere?

(g) Which mathematical statement or formula is appropriate?
(h) What is the logical or numerical answer?

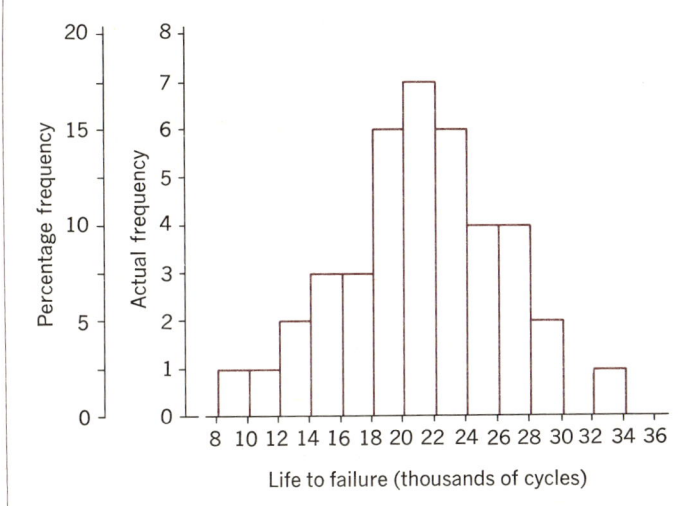

Fig. 8.1. *Histogram for the frequency distribution of Table 8-1. This graphical presentation of the data listed in Table 8-1 is a histogram, or bar diagram. In addition to the actual frequency of failure observations falling within a given interval, the histogram shows the relative frequency (actual frequency divided by the total number in sample).* [From Gen. Motors Eng. J., Second Quarter (1965), p. 15.]

quate testing of samples, exercising care to make sure that the samples are: (1) large enough and (2) representative enough to keep failures down to the limits that are acceptable. This is sometimes called *quality control,* and becomes a regular part of production operations.

For example, a General Motors plant, which manufactures a certain kitchen timer, selected a random sample of 40 units to determine their endurance in terms of cycles of operation until failure (see Boase). Of course the conditions under which the timers were tested had to be comparable to the conditions under which the timers were used by the purchasers. The actual data obtained are shown in Table 8.1 (page 270). One failed after only 9120 cycles. More failed as the cycles were continued. The last one did not fail until nearly 34,000 cycles had been reached. The data are plotted in Fig. 8.1 as a histogram. In this table and graph the data on failures were grouped (for convenience of handling and interpreting the results) into 2000-cycle intervals, although the counting equipment recorded exactly at which cycle each failure occurred. For example, the one that failed after 9120 cycles is simply listed as one failure in the 8000- to 10,000-cycle interval, and the two that failed at 13,245 cycles and 12,203 cycles are listed in the interval 12,000 to 14,000. In this histogram each timer unit corresponds to a certain rectangular area, which we can think of as a "unit" area. The area under the histogram curve is therefore a total of 40 unit areas. One timer thus is represented by one-fortieth of the total area. Each timer failure therefore represents a failure of $1/40 = 0.025$, or 2.5 percent of the total. The ordinates of the histogram can be easily changed to give the failures in terms of (1) the number of timers that fail (or actual frequency of failure), or (2) *relative* frequency (or fraction of the total number that fail), or (3) percent frequency (by treating the 40 as though it were 100).

268 Probability and Statistical Concepts in Human Affairs

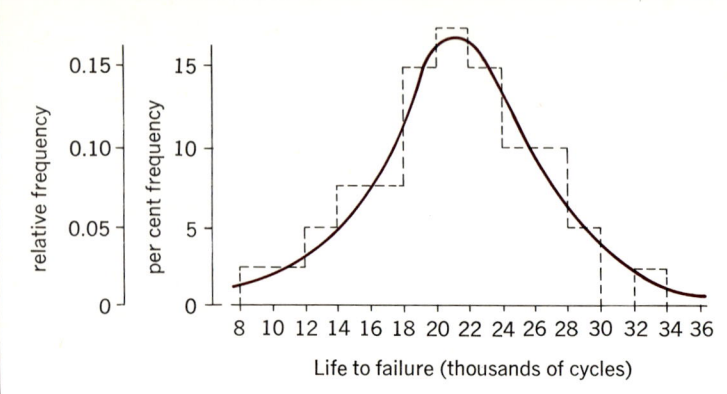

Fig. 8.2. *Population frequency curve.* To obtain the approximated population frequency distribution curve (Table 8-1), a smooth curve is constructed through the outline of the histogram (Fig. 8-1) indicated by the dashed lines. [From *Gen. Motors Eng. J.*, Second Quarter (1965), p. 15.]

A smooth curve can be drawn through this histogram, as shown in Fig. 8.2. The curve is now thought of as a *population frequency curve*.

Question 8.4. Compute the relative frequencies for the first, sixth, and eleventh intervals, expressing them both as decimals and as percentages.

Question 8.5. If the total area of the histogram is 40 units, in terms of each timer representing a unit area, what will be the total area when the vertical scale is converted from frequency of failures to relative frequency? If converted to percent frequency?

Question 8.6. The company could obtain complete information about its timers by testing, in the manner described, all the timers it produced. Why would the company not want to do this?

8.2 Reasoning from sample data to population expectancy

The population frequency curve introduced above is of great practical as well as theoretical importance. If the sample of 40 timers is properly selected and tested, we may assume that the histogram, though portraying only test data for the 40 units, gives information that applies to a much larger number of units—in fact to the whole "population" or total number of the timer units that are made in an identical manner. The smooth curve gives a functional relationship, which can be expressed as a mathematical function $f(x)$. The manufacturer will from time to time select other samples of 40 units and test them in a similar way to determine whether the quality of the product has changed. A new shipment of brass or a fault in soldering operations could shift the whole histogram down toward lower cycles for failure. Sampling checks of this type permit the manufacturer to maintain the desired quality of the product.

Figures 8.1 and 8.2 illustrate how the data of Table 8.1 can be plotted to reveal useful information from the results of the test. In each figure, the abscissa is divided into the cycle intervals of Table 8.1, and the actual number of failures is then plotted for each interval. This gives the ordinate scale, marked *actual frequency of failures*. It is useful to have also the *relative frequency* of failures at each interval; this is obtained by simply dividing each actual frequency ordinate by the total number of 40. (This makes the sum of the ordinates equal to unity instead of 40.) To obtain *percent frequency*, we multiply the relative frequency by 100. (This makes the area under the curve of Fig. 8.2 equal to 100, so that any portion of that area between two cycle values becomes the *percentage* of failures within that range of cycles.)

A unit that lasts too long is generally more expensive to manufacture, and a competitor can take over the market with a product that performs well enough and sells at a lower price. Each wants to make sure, however, that the life expectancy of the timers is not so low that the reputation of the products (and therefore sales and profits) suffers in favor of a competitor. The relationship that identifies the fraction of failures as a function, shown in Fig. 8.2, is of very great importance to a manufacturer.

Before we pursue the manufacturer's use of this functional relationship, let us turn to the sampling of a quite different "product," namely, the height of women. In Fig. 8.3, a histogram is plotted from a total sampling of 1375 women. This gives the number of women who fall within each height interval, which is chosen as 1 in. From this histogram is constructed a smooth curve that gives a functional relationship, a new $f(x)$, for the number of women with various heights.

Question 8.7. If you were to convert the ordinates of Fig. 8.3 from number of women to relative frequency, what would be the relative frequency corresponding to 50 women? To 150 women? To 137.5 women?

The reader will note immediately the remarkable similarity between the curves of Figs. 8.2 and 8.3. A curve of this general shape is of great importance for describing many features of natural phenomena. It is similar to the "normal" or *Gaussian* curve.† We shall find that a curve of this

† The curves of Figs. 8.2 and 8.3 are somewhat unsymmetrical, as contrasted with the symmetrical nature of the Gaussian curve. The curve is named in honor of the German mathematician Karl Friedrich Gauss (1777–1855).

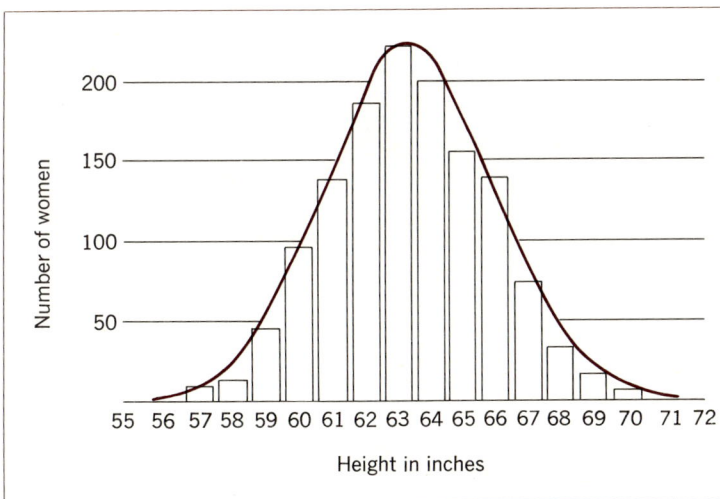

Fig. 8.3. *Heights of women produce a histogram to which the normal-distribution curve can be fitted. The bell-shaped curve conforms to many other empirical distributions found in the physical and biological worlds.*
(From *Probability* by Marc Kac. Copyright © September, 1964 by Scientific American, Inc. All rights reserved.)

TABLE 8.1

FAILURE DATA AND FREQUENCY DISTRIBUTION†

Test results showing life-to-failure in cycles			
19,998	20,098	29,887	18,571
14,032	21,015	13,245	21,827
27,215	15,545	17,231	29,653
25,101	27,728	23,283	9,120
18,673	23,807	24,778	24,051
21,186	17,008	20,005	21,917
23,049	26,114	12,203	22,753
10,053	16,077	32,787	18,003
15,100	21,428	25,753	26,210
18,273	22,237	23,092	19,547

Life to failure, cycles	Frequency, no. of failures
8,000– 9,999	1
10,000–11,999	1
12,000–13,999	2
14,000–15,999	3
16,000–17,999	3
18,000–19,999	6
20,000–21,999	7
22,000–23,999	6
24,000–25,999	4
26,000–27,999	4
28,000–29,999	2
30,000–31,999	0
32,000–33,999	1
Total	40

† A random sample of 40 timers was selected and tested simultaneously under the same specified conditions. Each timer's life-to-failure was recorded in cycles, listed at the top. Failure data were grouped into the number of failures observed within 2000-cycle intervals. This grouping (bottom) is a frequency diagram, which represents the distribution of the frequencies at which the sample failures occurred. This is the sample frequency distribution.

SOURCE: From General Motors Eng. J. Second Quarter (1965), p. 14.

type frequently characterizes phenomena in the physical, biological, and even behavioral or social sciences. Whether we plot the results of coin tossing, the frequency with which radioactive atoms emit radiation, the distribution of height of grain in the field, the velocity of gas molecules of the air, the heights or weights of people, certain intellectual abilities, or other biological, social, or physical phenomena, we seem to find useful representation in the general shape of the normal or "Gaussian" curve. One application, which often leaves a student unhappy, is the use of such curves for grading the class.†

With this more general character of the distribution curve in mind, let us return to Fig. 8.3 to see what other information we can derive from the curve. We shall now talk about the probability of failure instead of numbers or percentages. In effect we are now going to extend our results from the actual experience with the 40 timers to a much larger population of timers. The assumption is that the experience with the 40 timers will *probably* be the experience with the larger number.

† In the treatment of statistical distributions, this chapter will deal exclusively with the characteristics of the Gaussian (or normal) distributions. This distribution is particularly applicable to analyses of phenomena that display continuity or wherein the numbers are so large as to be essentially continuous characteristics. There are other models and forms of distribution, however, which offer advantages under other conditions. One of these is the Poisson distribution. The Poisson probability model is useful when one is interested in the distribution of events within a relatively short interval of time during which the number of events is small, and when the distribution of events during one interval is independent of the events that occur during other intervals of time. Although the Poisson distribution offers some advantages for dealing with many phenomena in which the events are of discrete character, we shall not pursue the method further.

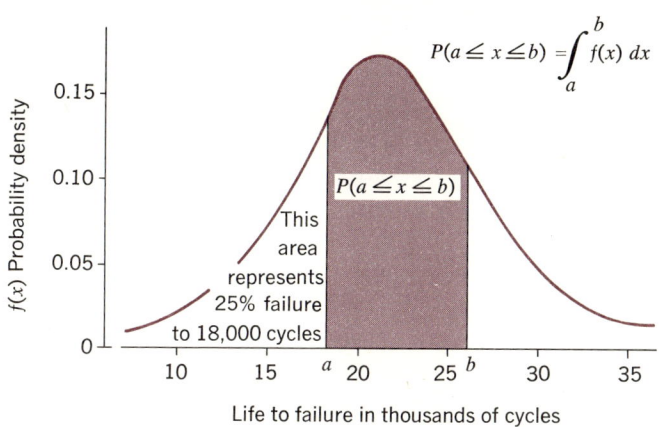

Fig. 8.4. Probability density function: calculating probability from the population frequency curve. The population frequency curve shown here (from Fig. 8.2) depicts the probability density function f(x), where x is the life to failure. From this function, the probability P that a random failure whose life-to-failure x will fall in an interval a-b can be calculated. The probability is the area under the curve f(x) from a to b, which is expressed by the integral shown. For example, to calculate the probability that a given timer from the population would fail after 18,000 cycles but before 26,000 cycles, a and b would be set equal to 18,000 and 26,000 cycles, respectively, and the area under the curve between these points determined. [From Gen. Motors Eng. J., Second Quarter (1965), p. 16.]

Each value of the ordinate of our curve of Fig. 8.2, at each frequency interval, represents a *probable* behavior pattern that is likely (if the sampling was good) to be repeated among larger numbers.

In order to extend our results with 40 timers to a larger population, we may replace the concept of relative frequency of failure with the general one of *probability of failure*. Thus, in Fig. 8.4 we have changed the ordinate from the number of failures—or fractions failing—(Figs. 8.1 and 8.2) to *probability density*. By probability density in this connection we mean the curve for which the ordinate represents the probability of failure at any interval of cycles of test. Note that the ordinates are those we called *relative frequency* in Fig. 8.2. Note also that on this basis the total area under the curve is unity.

The curve now expresses a more general functional relationship, $f(x)$, which is the probability of failure as a function of number of cycles of test (represented by the variable x). But the area under the curve of Fig. 8.4 also continues to represent the total number of units that fail between one value for cycles of test and any other number of cycles of test. That is, the curve not only answers the question "What is the probability for failure of timers (in terms of fraction of the total) at any particular value of test cycles?" but also "What total fraction of timers is likely to have failed by the time the test has reached that number of cycles of test?" In our case, the number of units that failed up to 18,000 cycles of operation is 10, which corresponds to the area to the left of the 18,000 cycle point of Fig. 8.1. We can say that the probability for failure is 10 out of 40 units, or 25 percent up to 18,000 cycles. The cumulative probability for failure was 100 percent, on this same basis, at 34,000 cycles.

By the same token, we can say that at 18,000 cycles, 30 of the units survived (which is simply $40 - 10$) and the *probability* for *survival*, or the *reliability*, of the units is 75 percent up to the 18,000-cycle point. The area under the curve can therefore be designed to tell us, as Fig. 8.4 does, that the area to the right of any frequency point represents the total probability for survival or reliability. Of course the sum of the two, or the number that have failed plus the number that have not failed, must total 100 percent (or, more conveniently, unity). When the first unit failed within only 10,000 cycles the ratio of failures was 1 in 40, and there does not seem to be any other way of stating it. But when the 7 units failed between 20,000 and 22,000 cycles, there are two ways of expressing it. We can say, as we did for the first failure, that there were 7 failures of the same total of 40, or a ratio of $7:40 = 0.17$. But this is not quite the correct picture, since at the beginning of the interval in question only 24 timers remained operable; therefore the failure rate was 7 out of 24, or a much higher ratio of $7:24 = 0.29$. It is more meaningful, therefore, when we want to determine the *failure rate*, to determine that rate in terms of the number of units that have survived up to the interval for which the rate is desired.

The smooth curve in Fig. 8.5 shows how this method is applied for more general use. The failure rate at x (or rate of success or of any other change, depending on what is desired) is shown as a ratio of the area of a thin rectangle to the area lying to the right of x (including the rectangular strip). That is, the area lying under the curve to the right of the point x represents all the units that have survived and are still operating to that point, while the area to the left of x represents the failures. What is the failure rate at the point x? We take this in four steps:

(1) We must assign some interval between x and a neighboring $x + 1$ (just as we assigned 2000-cycle intervals for Table 8.1 and Fig. 8.1).

(2) Then we determine the segment of area represented by that interval (which is simply that interval times the height up to our curve at the midpoint of the interval). This is identified as the Area $= 1 \times f(x)$ in the Fig. 8.5.

(3) We then find by some graphical or other means the area of the shaded portion lying to the right of x, which represents the surviving units and is designated in Fig. 8.5 as $R(x)$.

(4) Dividing one by the other gives the failure rate:

$$x = \frac{f(x)}{R(x)}$$

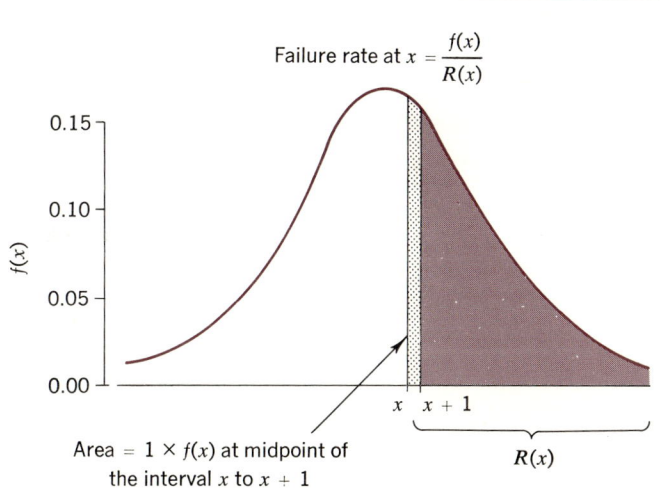

Fig. 8.5. *Determination of failure rate. The failure rate at a given time of a population whose failure frequency is described by a probability density function f(x) (plotted here) would be calculated as being equal to the area under the curve from life=x to x+1 divided by the area under the curve to the right of x. The area under the curve to the right of x is merely the probability of survival, or reliability, at x. The area under the curve from life=x to x+1 is the probability density function evaluated at x. The failure rate, therefore, is expressed by the equation shown, where R(x) denotes the reliability at x. The function describing the failure rate is called the hazard function. [From Gen. Motors Eng. J., Second Quarter (1965), p. 18.]*

Question 8.8. At the start of the interval 28,000–30,000 there remained only three timers, of which two failed in that interval. Compare the failure rate (1) in terms of the three remaining timers and (2) in terms of the total of 40 timers. What conclusions can you draw?

Question 8.9. One of the failure rates of the timers turned out to be 1.000. How should this result be interpreted?

8.3 Errors of sampling

We seem to have drawn a great deal of useful data from the simple numbers listed in Table 8.1. From the test of 40 timers we extrapolated the histogram to give a smooth curve, and from it drew implications affecting perhaps 100,000 timers or more. An analysis of the characteristics of a small group of people could be extrapolated to provide useful information about a large community.

How dependable are these assumptions? Statistical analytic processes are indeed powerful. In fact, they constitute the only practical method available to modern civilization for maintaining an ordered society that involves large numbers of persons and where fluctuations and rapid changes are more characteristic of a large city than stability and status quo. Picture the hopelessness of maintaining an accurate census count for a large city or state in the face of the births, deaths, and travel in and out of the area. The city fathers can, however, make a fairly good estimate by the use of sampling statistics, which can project how many births or deaths there will be and also how many people are likely to move in and out of the area in the course of a day. Of course they would not be able to tell exactly which individual will go out or move in. But a good sampling process applied to each of the variables can provide a fairly valid and reliable picture of the population as a whole.

The entire procedure has value only to the degree that the sampling and selection methods are adequate. When, therefore, is the sampling process adequate? In other words, how does one decide on the adequacy of sample size and select the individual elements that make up a truly representative group? How far may one generalize from the sample to the larger population? How accurately, for example, can one assert that the distribution of height of women as given in Fig. 8.3 applies to women of various nations or times and places? The heights in inches that are given can vary from one country to another. Certainly the figures given do not apply to pygmies, nor to various other races. The inferences or conclusions that are drawn must therefore be limited to the population from which this sample was drawn or to other populations of the same type. Experience has shown, however, that as one goes from one race of women to another, only the numbers of the abscissa will change while the shape of the curve remains essentially the same; that is, it is still a normal curve.

It is not easy to give a firm answer to the question of adequacy of sample. Indeed the choice of the sample group itself must be considered tentative and be subject to check and recheck. For example, some samples of timers may give results that are far removed from the symmetric curves of Figs. 8.2, 8.3, and similar illustrations. The peak of the curve may fall to the left of the center, indicating a much more rapid rate of failure after only 10,000 cycles. Does this indicate that the sample lot is biased, or that the test procedure is faulty, or that the unsymmetric behavior is indeed the characteristic of all the timers that are being produced at the time when the sample of 40 is chosen? Only further sampling and further testing can answer the question. Many useful techniques have been developed for the purpose, but these lie outside the scope of this text.

8.4 Probability: its early beginnings

Suppose that 40 timers of the same type that have been discussed were placed before us prior to the beginning of the test. Could we tell which one would fail first and which one last? Unless there is some obvious fault, the answer would be "no." We cannot know which one will fail first, or last, or in between. And yet we know that there will be a first, and a last, and that the failures will take some order as indicated by the shape of the graphs of Figs. 8.1 and 8.2. There is a probability that a *maximum number* of timers will fail during the interval between 20,000 and 22,000 cycles compared with other equal intervals. The probability for failure among the remaining timers that are still working goes up, as shown in Fig. 8.5 (working up to 1 for the probability of failure for the last remaining timer).

We shall now change our approach and seek some better understanding of the behavior of the timers in terms of probability concepts applied to a few simple situations. We shall consider tossing pennies and dice and note some other phenomena. The subject is far from limited to games of chance, however. In fact, some of the most profound questions of atomic systems involve estimates of probabilities. The philosophy of determinism is also very intimately concerned with probability concepts.

The pioneer in the field of probability was the Italian mathematician Gerolamo Cardano of Milan (1501–1576). His insight into the laws of probability gave birth to a famous gamblers' manual, the *Liber de Ludo Aleae* (Book of the Game of Chance), in which he presented many of the fundamentals of probability theory applied to dice and other games of chance. But neither this work nor a ten-volume edition of his complete works attracted much attention.† (See Newman in References.) Not until a recent publication by Oystein Ore did his full contributions become widely known and appreciated.

The science of probability received greater impetus from the genius of the French mathematical physicist Pierre Simon de LaPlace (1749–1827). After

† There have been many other situations wherein a new idea has failed to take firm root because "the time was not ripe." Mendel's observations on genetic characteristics is another example.

making significant contributions to the fields of celestial mechanics and mathematical analysis, he turned to the study of the laws of probability. He managed to keep his head during the violent turmoil that came with the French Revolution and with Napoleon Bonaparte, his success being due as much to his opportunistic tendencies as to his brilliance as a scientist. His own explanation of the laws of natural processes (Cartesian determinism) followed the thinking of the period, namely, that *the present is the direct product of the past and the determiner of what will happen in the future.*†

"Present events are connected with preceding ones by a tie based upon the evident principle that a thing cannot occur without a cause which produces it."

(See Newman in References.)

There were other opinions on the meaning and significance of probability. The English mathematician and logician De Morgan, following LaPlace, opined "I consider the word probability as meaning the state of mind with respect to an assertion, a coming event, or any other matter on which absolute knowledge does not exist." Since absolute knowledge or complete certainty rarely exists, the degree of certainty of a proposition, its "probability," is the degree of belief with which it is held. This attitude, formulated by these two men, is sometimes referred to as the *classic view*.

Lord J. M. Keynes (1883–1946), the famous economist, saw probability as an essentially unanalyzable, but intuitively understandable, logical relation between propositions. Many situations require the use of intuition to estimate the prob-

† "The most important questions of life are, for the most part, really only problems of probability," said de LaPlace in his *Theorie Analytique des Probabilities*.

able relations between propositions, and this estimate offers a basis for acting on one hypothesis rather than another. In his article (see Newman in References) Keynes states, "probability begins and ends with probability. [There is no certainty in its direction toward truth or success.] The importance of probability can only be derived from the judgment that it is *rational* to be guided by it in action; and a practical dependence on it can only be justified by a judgment that in action we *ought* to act to take some account of it."

Propositions such as "on the evidence as to their moral character, it is more probable that witness A speaks the truth than does witness B" or "relative to experiment D, the Einstein theory is more probable than relative to experiment E" are examples of the Keynes interpretation.† So are "it is not probable that he could have forgotten me," in which there is assertion of belief in the implication that the degree of probability is in some way an index of the intensity of one's convictions.

Of the several interpretations of probability just discussed, that of LaPlace is still much used today. The following quotation‡ gives LaPlace's basic theory of probability:

"The theory of chance consists in reducing all the events of the same kind to a certain number of cases equally possible, that is to say, to such as we may be equally undecided about in regard to their existence, and in determining the number of cases favorable to the event whose probability is sought. The ratio of this number to that of all the cases possible is the measure of this probability, which is thus simply a fraction whose numerator is the number of favorable cases and whose denominator is the number of all the cases possible.

"When all the cases are favorable to an event the probability changes to certainty and its expression becomes equal to unity. Upon this condition, certainty and probability are comparable, although there may be an essential difference between the two states of the mind when a truth is rigorously demonstrated to it, or when it still perceives a small source of error."

For instance, when tossing a symmetrical coin, the possible outcomes are two in number, namely, "heads" and "tails." The probability of obtaining "heads" on a given toss is the ratio 1 (the outcome in question, namely, "heads") to total possible outcomes, namely, 2. Thus, the probability is $\frac{1}{2}$.

The simplicity of LaPlace's a priori approach is at once apparent. One drawback, however, is that complex situations frequently do not analyze into a set of *equally* likely outcomes. As an example, the probability that a fifty-year old man will die within the next year cannot be computed because the two events (life or death) cannot be broken up into equally likely outcomes. Yet life insurance companies must have some means of at least estimating many probabilities of this type.

A *relative frequency* definition of probability, given by various scientists, is much used today. This definition is based, as was that of LaPlace, on a repeatable experiment. Suppose an experiment has been performed n times and the outcome each time has been recorded. The *relative frequency* of any event associated with the experiment is then

---

† These are quoted from the article by Professor Nagel (see Newman). The student who has interest in philosophical implications of probability concepts is urged to read this article. Applying probability concepts to validity of theories is thought by some to be totally wrong, however.

‡ See Newman in References.

the ratio of the number of occurrences of the events to n, the total number of trials of the experiment.

For example, suppose a coin is tossed 1000 times and lands heads 512 times. Then the relative frequency of heads is 512/1000, or 0.512. Now, if for larger and larger values of n (that is, repetitions of the experiment) the relative frequency of the event becomes closer and closer to a certain fixed number, this number is defined to be the *probability* for the event. It may be noted that insurance companies tend to use this concept in accumulating statistics and applying the concepts of probability.

The significance of probabilistic thinking and the theory of probability lies in the help they give for coping with situations that go beyond the tossing of coins; that is, for situations wherein there is not enough knowledge to permit perfectly secure analysis in the sense of rigorous mathematics or classical logic. We have learned that many questions involving our lives, economics, happiness, and safety, and questions of national security and of national deterioration, rarely permit decisions that are unquestionably wrong or right, black or white, wholly successful or wholly unsuccessful.

8.5 The basic notions of probability

Consider the simple act of tossing a coin and the probability that it will show heads or tails on its fall. If the coin is symmetrical in every respect and the toss is not of a kind to prejudice the result, the probability of its landing heads is equal to that of its landing tails. Suppose that the first throw gives heads, H. We proceed to a second toss. What is the probability now for a head to appear? Does the fact that the first toss was heads influence the probabilities for the next toss? For example, is there an influence that has been termed *maturity of the chances* which insists that when heads have come up more often than tails, the next toss is likely to produce tails? The answer is "no." It has been said that the coins have neither memory nor consciousness, and we know nothing that will disprove this. The next toss is as likely to produce heads again as tails. And so it will continue to be for every next toss: The likelihood for heads is exactly $\frac{1}{2}$, as it is for tails, whether there have been a series of 30 heads only or 15 of each.

What are the results when we toss this same coin 100 times, 1000 times, or more? We find that although the result of any single toss of a coin is completely unpredictable, there develops a remarkable tendency toward regularity when the results of many tosses are tabulated, and the tendency becomes more marked as the number of tosses is increased. In the case of the single coin, the more times it is tossed, the more closely the relative frequency of heads approaches $\frac{1}{2}$; similarly, the relative frequency of tails approaches $\frac{1}{2}$. We might express this idea more briefly by writing $P(H) = \frac{1}{2}$, and $P(T) = \frac{1}{2}$, respectively. While this case is a very simple one, the principles apply equally well to more complex cases of scientific and practical importance.

Question 8.10. Of the definitions of probability discussed in Sec. 8.4, which one corresponds to the statement that "the more times it is tossed, the more closely . . . approaches the value $\frac{1}{2}$?"

Let us call the tossing of a coin an experiment. Is there a "correct result" for this experiment? When our experiment consists of measuring a weight, or length, or voltage, we assume that there is a correct result. We recognize also that there is a correct result. We recognize there can be an error in trying to find that correct value. In our coin toss experiment, however, either one of two possible results is a "correct" result, H or T. The notion of *the correct result* for a determinate quantity is here replaced by the notion of the *possible* results for a quantity subject to chance variation. Such a quantity may be called an indeterminate quantity or a *stochastic† variable*. For a symmetric coin, the two possible results are H and T, each with probability $\frac{1}{2}$. For an unsymmetric coin, we have the same two possible results, but they are not equally likely. Perhaps H comes up 60 percent of the time in the long run and T comes up 40 percent. In this case, the probabilities are $P(H) = 0.60$ and $P(T) = 0.40$.

In general, complete information concerning a stochastic variable consists of (1) a listing of the possible results or values, and (2) a listing of the probabilities for the occurrence of each of these results. We note that a single experiment to observe the stochastic variable yields only one of the possible values. In this sense the possible values are mutually exclusive. We note further that the sum of all the probabilities is unity, which is equivalent to the statement that *no values other than those listed are possible*.

We present one more view to illustrate the basic similarity between the tossing of a coin and the more practical problems of sampling timers or a population of voters. We imagine a very large "population" of tiny balls in a box. Assume half of these to be marked H and half T. With the box well stirred, take out one ball at random. This is equivalent to our problem of tossing a symmetric coin. Here the probabilities are also $P(H) = \frac{1}{2}$ and $P(T) = \frac{1}{2}$. But if many withdrawals from such a box gives 60 percent heads and 40 percent tails, we would say the results indicate that the box contains 60 percent balls marked H and 40 percent marked T, which is equivalent to tossing a coin that ultimately gave $P(H) = 0.6$ and $P(T) = 0.4$.

Again we might imagine each tiny ball to be marked with a number of cycles that tells how many times each timer coming off the production line may operate before failing. The fraction of all the balls marked 10,000 is then the probability that that fraction of timers will fail when operated through 10,000 cycles. If this collection of marked balls is truly marked, and if we were to pick at random a representative group of 40 timers from the lot, we would presumably find the 40 to have about the same distribution of failure curve as does the total population. The sum of the fractions, or probabilities, would add up to unity.

Similarly, if each ball were marked with the height of one of the women we mentioned earlier, a proper sampling would give a distribution that truly represents the total distribution.

There is one distinction we may note between the tossing of coins and the populations of balls in the box, namely, that one may toss the coin without limiting the number of tosses, while the number of timers and the number of women is finite.

Question 8.11. Let us imagine that a large homogeneous group of voters is equally divided on a referendum; that is, half in favor and half against. Is the problem of sampling opinions from this group equivalent to the problem of tossing a symmetrical coin? Consider the following situations:

† Stochastic = conjectural.

1. We are taking a poll of a fairly homogeneous group of 100,000 voters to find their views on a referendum. In some random fashion we select 1000 of them and record their positions. We find 550 in favor and 450 against. What can we conclude?

2. We are tossing a coin about which we have no previous information concerning its symmetry. After 1000 tosses, we found 550 H and 450 T. What can we conclude?

3. A certain disease is known to result in 50 percent mortality in laboratory mice. We wish to test the efficacy of a particular drug. In a test of 1000 mice that were treated with the drug, 550 survived and 450 died. What can we conclude?

The statistical aspects of these three problems are identical, and are illustrative of a great many problems about which our major source of information is only statistical. The use of the theory of probability for making intelligent inferences from such statistical information is one of the most important applications of the theory. We shall learn in this chapter that the answers to the question posed above in three equivalent forms involve the use and applications of a few basic principles in the Theory of Probability. These concern mainly the calculation of the probabilities of combinations of events.

8.6 Two axioms concerning combinations of events

In the experiment of tossing a symmetrical coin, let us introduce the quantity

$x =$ *number of heads in one toss*

The possible values are $x = 0$, when the coin comes up tails (T) and $x = 1$ when the coin comes up heads (H). The following table gives the complete information on the stochastic quantity x.

THE TOSS OF ONE COIN

Possible value, x	Probability, $P(x)$
0	$\frac{1}{2}$
1	$\frac{1}{2}$

Let us now consider the experiment of tossing two symmetrical coins. For convenience in the discussion, we label them as coin number 1 and coin number 2. We focus on the stochastic variable:

$x =$ number of heads in toss of two coins

We may also write

$$x = x_1 + x_2$$

where

$x_1 = 0$ when coin 1 is T_1
$x_1 = 1$ when coin 1 is H_1
$x_2 = 0$ when coin 2 is T_2
$x_2 = 1$ when coin 2 is H_2

What are the possible values of x, and their probabilities? The possible values are clearly $x = 0$, corresponding to the outcome T_1T_2 (coin 1 tail and coin 2 tail); $x = 1$, corresponding to H_1T_2 and to T_1H_2; and $x = 2$, corresponding to H_1H_2. To compute the probabilities for these x values, we proceed as follows:

The two possibilities for coin 1 (H_1 and T_1) combined with the two for coin 2 (H_2 and T_2) lead to the four basic possibilities in the two-coin toss:

H_1H_2, H_1T_2, T_1H_2, T_1T_2

The probability or expected frequency of

occurrence of two heads, $x = 2$, will be the same as the probability of H_1H_2. That for $x = 0$ will be that of T_1T_2. The probability of one head, $x = 1$, will be the sum of the probabilities of H_1T_2 and T_1H_2. We shall see later that the four basic possibilities are equally likely, each expected to occur $\frac{1}{4}$ of the time. We then have the solution in the following table.

THE TWO COIN TOSS

Possible value, $x = x_1 + x_2$	Probability, $P(x)$
0	$\frac{1}{4}$
1	$\frac{1}{2}$
2	$\frac{1}{4}$

The table tells that one head is expected to come up $\frac{1}{2}$ the time; and zero or two heads each will come up $\frac{1}{4}$ of the time. The three possible values of x are not equally likely. The most probable value, one head, is what we might have expected in a toss of two symmetrical coins.

When we added the expected frequencies of the occurrence of H_1T_2 to that of T_1H_2 to obtain the expected frequency of either one of these possibilities (and therefore of the occurrence of one head, $x = 1$), we used one of the basic axioms of the theory of probability, namely:

The probability for the occurrence of one or another of a set of mutually exclusive possibilities is given by the sum of the probabilities that are possible in that set.

We now show that the four possibilities —H_1H_2, H_1T_2, T_1H_2, T_1T_2—are equally likely. We know that coin 1 is expected to come up heads $\frac{1}{2}$ of the time, and tails $\frac{1}{2}$ of the time. Consider only that half of all occurrences in which we have H_1, that is, coin 1 coming up heads. We now argue that the coin 2 result is entirely independent of the coin 1 result. There is no reason for coin 2 to favor heads merely because coin 1 came up heads. Thus, in the half of all occurrences in which coin 1 comes up heads, coin 2 will do what comes naturally, namely, come up heads $\frac{1}{2}$ of the time and tails $\frac{1}{2}$ of the time. We then have that H_1H_2 and H_1T_2 each occur $\frac{1}{2}$ of $\frac{1}{2}$, or $\frac{1}{4}$, of the time. The same reasoning process applied to the occurrences T_1 leads to $\frac{1}{4}$ for the probabilities of T_1H_2 and T_1T_2. The reasoning can be pictured in the following diagram.

	H_2	T_2
H_1	H_1H_2	H_1T_2
T_1	T_1H_2	T_1T_2

In this diagram the square represents all possible combinations of heads and tails: the upper half, those for which coin 1 is heads; the left half, those for which coin 2 is heads. The square is then divided into quarters.†

This "natural" reasoning lies behind the second basic axiom of the theory of probability:

The probability that a number of independent events will occur all together is given by the product of their separate probabilities.

† A square array of this type can be useful for many applications.

8.7 Some applications to combinations of events

We illustrate the application of this reasoning by considering the experiment of tossing *three coins*, assumed to be symmetric. The two possibilities for each coin give rise to eight (that is, $2 \times 2 \times 2$) possible outcomes for the three coins (see Fig. 8.6 for our suggested method of arranging the complete set of all possible outcomes):

$H_1H_2H_3$, $H_1H_2T_3$, $H_1T_2H_3$, $H_1T_2T_3$,
$T_1H_2H_3$, $T_1H_2T_3$, $T_1T_2H_3$, $T_1T_2T_3$,

where, for example, $H_1T_2H_3$ represents the outcome that coin 1 comes up heads; coin 2, tails; and coin 3, heads.

The result for any one coin is independent of the outcome for any of the others. Thus the probability of the outcome $H_1T_2H_3$ is given by $\frac{1}{2} \times \frac{1}{2} \times \frac{1}{2} = \frac{1}{8}$ when our second axiom is applied. In words, we consider first the half of all occurrences in which coin 1 comes up heads. Half of these, $\frac{1}{4}$ of all occurrences, will be those in which coin 2 comes up tails. Thus H_1T_2 occurs $\frac{1}{4}$ of the time. Of this, coin 3 comes up heads $\frac{1}{2}$ the time, resulting in a net probability of $\frac{1}{8}$ for $H_1T_2H_3$. The application of the axiom to each of the listed outcomes results in $\frac{1}{8}$.

We introduced the quantity (stochastic variable)

$x =$ number of heads that come up in one three-coin toss

As before, we write

$$x = x_1 + x_2 + x_3$$

where each x is zero when that coin comes up tails and is 1 when it comes up heads. The possible values for x are seen in Fig. 8.6 to be 0, 1, 2, and 3, with 0

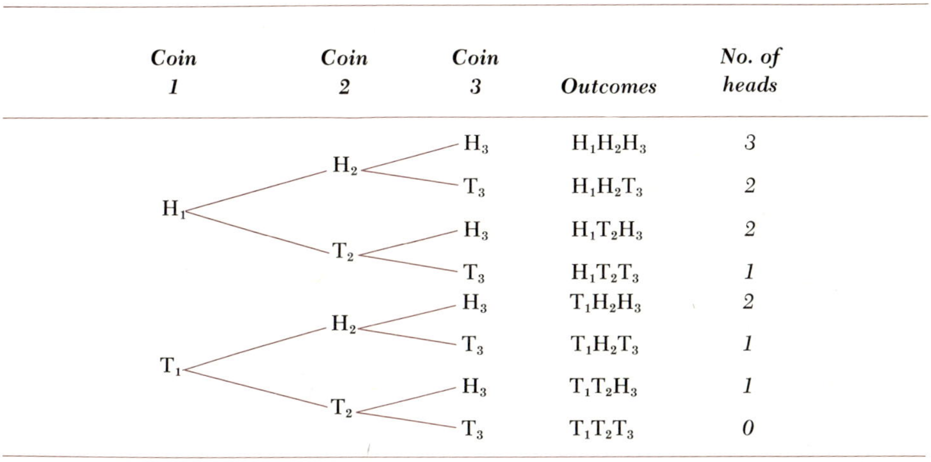

Coin 1	Coin 2	Coin 3	Outcomes	No. of heads
H_1	H_2	H_3	$H_1H_2H_3$	3
		T_3	$H_1H_2T_3$	2
	T_2	H_3	$H_1T_2H_3$	2
		T_3	$H_1T_2T_3$	1
T_1	H_2	H_3	$T_1H_2H_3$	2
		T_3	$T_1H_2T_3$	1
	T_2	H_3	$T_1T_2H_3$	1
		T_3	$T_1T_2T_3$	0

Fig. 8.6. A schematic method of arranging the complete set of all possible outcomes from a toss of three coins.

coming from outcome $T_1T_2T_3$, 1 coming from one of the outcomes $T_1T_2H_3$, $T_1H_2T_3$, and $H_1T_2T_3$. Applying our first axiom, concerning the addition of probabilities of mutually exclusive possibilities, leads to the results in the next table.

Possibility, $x = x_1 + x_2 + x_3$	Probability, $P(x)$
0	$\frac{1}{8}$
1	$\frac{3}{8}$
2	$\frac{3}{8}$
3	$\frac{1}{8}$

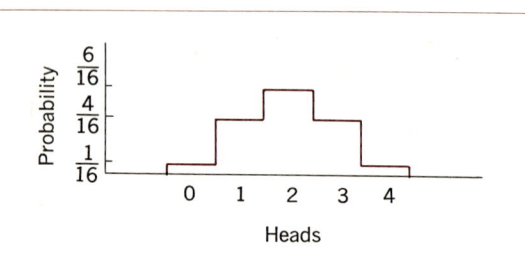

Fig. 8.7. *Probability for occurrence of heads.*

We note that one and two heads are equally likely and are three times more probable than no heads and three heads. The expected average number of heads is $1\frac{1}{2}$, (=6/4) as it should be in repeated tosses of three coins.†

Question 8.12. Apply the same method to the case of a four coin toss. Show that the following table results for the number of heads, $x = x_1 + x_2 + x_3 + x_4$.

Possible values, $x = x_1 + x_2 + x_3 + x_4$	Probability, $P(x)$
0	$\frac{1}{16}$
1	$\frac{4}{16}$
2	$\frac{6}{16}$
3	$\frac{4}{16}$
4	$\frac{1}{16}$

What do these results suggest as to the most likely expectation for the number of heads? Are these results consistent with the expectation that the number of heads should average to 2?

† The average number of heads is found by adding the four different values of x (that is, $0+1+2+3=6$) and dividing by the number (four) of different outcomes.

8.7 Some Applications to Combinations of Events

We might look a little longer at the tabulation for the four coins. If we were to plot the probability fractions in the form of a histogram, we would obtain Fig. 8.7.

Question 8.13. We are given a pair of symmetrical, "unloaded" dice. Let x_1 represent the number that results from the roll of die† 1. The possible x_1 values are the integers 1 through 6. Their probabilities are all equal; $P(x_1) = \frac{1}{6}$. Our "experiment" is the rolling of a pair of dice. Our "observations" (or stochastic variables) are $x = x_1 + x_2$, the sum of the outcomes of each die. Show that the possible x values and their probabilities are those listed in Table 8.2. *Hint:* We can imagine listing the 36 equally likely basic outcomes in the roll of a pair of dice. For example, 4,1; 4,2; 4,3; 4,4; 4,5; 4,6 are the six possibilities for which die 1 comes up as a 4. Each of these basic possibilities has probability $\frac{1}{36}$, by our second axiom. By the application of our first axiom we should then be able to solve the problem.

Question 8.14. In one game of dice, coming up with a 7 or 11 on the first roll *wins* immediately, but coming up with a 2 or 12 on the first roll *loses* immediately. If some number other than 7 or 11, or 2 or 12, comes up, the player continues to roll until either he repeats his first number, in which case he wins, or he rolls a 7 and loses.

(a) What is the probability of winning on the first roll?
(b) What is the probability of losing on the first roll?
(c) On the first roll, what numbers other than 7 and 11 give the player the most favorable chance of sub-

† Presumably every reader is aware of the cubical shape of a die (plural = dice) in which each side has dots, starting with one and ending with six dots on the sixth side.

TABLE 8.2
OUTCOMES IN THE ROLL OF A PAIR OF DICE

Possible value, $x = x_1 + x_2$	Combinations	Probability, $P(x)$
2	$1+1$	$\frac{1}{36}$
3	$1+2, 2+1$	$\frac{2}{36}$
4	$1+3, 3+1, 2+2$	$\frac{3}{36}$
5	$1+4, 4+1, 2+3, 3+2$	$\frac{4}{36}$
6	$1+5, 5+1, 2+4, 4+2, 3+3$	$\frac{5}{36}$
7	$1+6, 6+1, 2+5, 5+2, 3+4, 4+3$	$\frac{6}{36}$
8	$2+6, 6+2, 3+5, 5+3, 4+4$	$\frac{5}{36}$
9	$3+6, 6+3, 4+5, 5+4$	$\frac{4}{36}$
10	$4+6, 6+4, 5+5$	$\frac{3}{36}$
11	$5+6, 6+5$	$\frac{2}{36}$
12	$6+6$	$\frac{1}{36}$

sequently winning? If he rolls one of these numbers on his first roll, is he then more likely to win or to lose?

Question 8.15 An electronic system consists of three components wired in series, so that the failure of any one of them disables the system. The components operate independently, and their separate probabilities for successful operation are 50, 80, and 90 percent (that is, 0.5, 0.8, and 0.9). What is the chance (as a percentage) that the system fails? (*Answer:* 64 percent, or 0.64.)

(a) Answer the question first by finding the probability that the system performs successfully. This may be calculated by a direct application of the second axiom. The probability of failure is then unity minus the probability of success.

(b) The eight basic possibilities are listed in Fig. 8.8, where S and F indicate success and failure. Note that the probabilities for each may be computed from the given information, as in the example shown for $P(S_1F_2F_3) = 0.01$. In this problem, they are *not* equally likely. Note further that the probability for failure of the entire system may be computed by addition of the probabilities of all (seven) the possibilities containing at least one F.

(c) Find the probability for the basic possibility $F_1F_2S_3$. Let us note the simi-

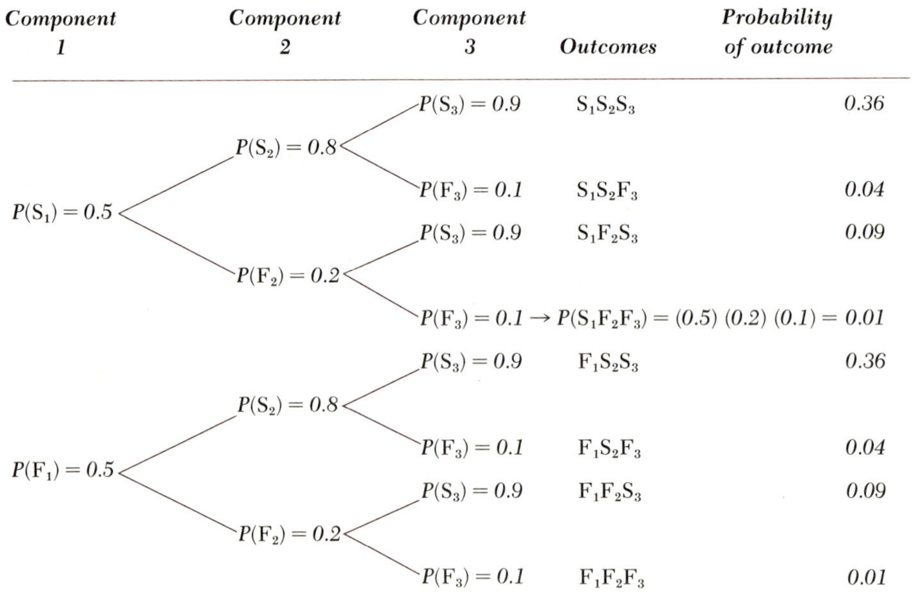

Component 1	Component 2	Component 3	Outcomes	Probability of outcome
$P(S_1) = 0.5$	$P(S_2) = 0.8$	$P(S_3) = 0.9$	$S_1S_2S_3$	0.36
		$P(F_3) = 0.1$	$S_1S_2F_3$	0.04
	$P(F_2) = 0.2$	$P(S_3) = 0.9$	$S_1F_2S_3$	0.09
		$P(F_3) = 0.1 \rightarrow P(S_1F_2F_3) = (0.5)(0.2)(0.1) = 0.01$		
$P(F_1) = 0.5$	$P(S_2) = 0.8$	$P(S_3) = 0.9$	$F_1S_2S_3$	0.36
		$P(F_3) = 0.1$	$F_1S_2F_3$	0.04
	$P(F_2) = 0.2$	$P(S_3) = 0.9$	$F_1F_2S_3$	0.09
		$P(F_3) = 0.1$	$F_1F_2F_3$	0.01

Fig. 8.8. *Diagram for computing the probabilities that series-wired electrical components 1, 2, and 3 (each having different probabilities of success, S) will simultaneously succeed or fail. The probability that component 1 succeeds while components 2 and 3 fail, $P(S_1F_2F_3)$, is computed on the diagram.*

larity between this problem and the three-coin problem. The fact that the probabilities here are not all equal to $\frac{1}{2}$ does not change the basic chance aspects of the problem. The same eight basic possibilities occur, but now with unequal probabilities.

Question 8.16. According to the 1958 Commissioners' Standard Ordinary Mortality Table, currently in use by life insurance companies, the probability that a sixty-year old man will live for the next ten years is 0.73, while that for a fifteen-year old boy is 0.98. Assuming (as insurance companies do) that people live and die independently of each other, what is the probability that the man (age sixty) and the boy (age fifteen) will both live for the next ten years?

Question 8.17. Can you describe a situation in which the assumption that people live and die independently of each other does not hold?

Question 8.18. What is the probability that at least two students in a class of 25 have the same birthday?

DISCUSSION OF SOLUTION
QUESTION 8.18

The surprising result is that the probability is greater than $\frac{1}{2}$ for groups of 23 persons and larger.

The problem is approached as follows. We assume that there are 365 possible birthdays, all equally likely, and therefore each with probability $\frac{1}{365}$. We then play the imaginary game of selecting 25 numbers at random from the equally likely set of 365 possible birthdays. The game can be made real in two ways. In the first, we use 365 tiny balls in a box, each marked with a day of the year (or simply numbered from 1 to 365). A ball is replaced after selection, and the box stirred before the next selection. In the second way, we spin a dial on a circular card divided into 365 equal sectors, each sector representing each day of the year. The game is finished when 25 numbers (birthdays) are chosen. It may then be played many times to determine the expected frequency, or probability, that two or more of the chosen numbers come out the same.

The imaginary game can be solved as follows for the probability that there is *no* duplication of birthdays. The probability of duplication is then the remaining (from unity) probability. Select the first number (birthday—imagine spinning the dial). It is one of the 365. Now select the second number. The probability that we do not obtain duplication is $\frac{364}{365}$. Thus, in a class of two students, the probability of duplication is $1 - (\frac{364}{365}) = \frac{1}{365}$. Now select the third birthday. To be different from the previous two selections, it must come from the remaining 363 birthdays; this results in a probability of $\frac{363}{365}$. To have nonduplication for the three birthdays, we must also have nonduplication for the first two which occur $\frac{364}{365}$ of the time. For 22 students, then, the nonduplication probability is $(\frac{364}{365})(\frac{363}{365})$ . . . down to $(\frac{344}{365})$, and is then slightly larger than $\frac{1}{2}$. The probability of duplication is then slightly less than $\frac{1}{2}$. For 23 students, the additional factor $\frac{343}{364}$ changes the nonduplication probability to less than $\frac{1}{2}$ and the duplication probability to larger than $\frac{1}{2}$.

8.8 The stability of statistical ratios—the law of averages

Let us consider an indeterminate quantity, or stochastic variable, which can take on two equally likely results, each with probability $\frac{1}{2}$, such as a symmetric coin. Let us use a toss (or selection) mechanism. We now make repeated selections. Perhaps the first ten coins tossed (or selections) were

H H T H H T H T T H

After each selection, we record the fraction of the number of coins that were heads up to that time. For the sequence given, the fractions are

$$\frac{1}{1}, \frac{2}{2}, \frac{2}{3}, \frac{3}{4}, \frac{4}{5}, \frac{4}{6}, \frac{5}{7}, \frac{5}{8}, \frac{5}{9}, \frac{6}{10}$$

Imagine now that we continue to toss many many times, recording the fraction of heads following each selection. In what way does the heads ratio stabilize to the value of $\frac{1}{2} = 50$ percent?

We would perhaps like to believe that the ratio of heads stabilizes so as to depart from $\frac{1}{2}$ by amounts that grow smaller and smaller, becoming infinitesimally small as the number of tosses grows larger and larger. After all, in any succession of ten tosses, the possibility of 6T and 4H is as likely as 6H and 4T. Equal numbers of such runs average out to 50 percent heads. Further, in any succession of ten tosses, the most likely result is 5H and 5T, with a very small probability, $(\frac{1}{2})^{10} = \frac{1}{1024}$, for 10H or for 10T.

On the other hand, every selection is independent of the results of the previous selections. Just because the first ten gave 6H and 4T is no reason at all to expect the next ten to tend more toward tails. The independence of each selection prevents this "maturing of the chances." The law of averages does not work this way. Future selections are not expected to cancel the effects of past runs of predominantly heads. All they can do is provide more and more data.

It is simply not true that the deviation of the heads ratio from $\frac{1}{2}$ always becomes infinitesimally small (as small as we please) if only we make a sufficiently large number of selections. The occasional occurrence of improbable, though possible, long runs of predominantly heads or tails results in occasional fluctuations of the heads ratio, both away from and back toward the expected value of $\frac{1}{2}$.†

A more accurate description of the way in which statistical ratios stabilize as the number of selections increases is as follows: We expect the observed ratios generally to grow closer to the expected values (the probabilities) as the number of selections is increased. We also expect to see fluctuations away from and back toward the expected values because of the occasional occurrences of improbable runs. However, large fluctuations, being less probable, occur less frequently than small fluctuations. Further, for a given size of fluctuation, the frequency of occurrence decreases as the number of selections increases.

The described aspects of the stability of the statistical ratios can be illustrated by considering some actual possibilities. Imagine, for example, that we have completed the first 100 tosses of our symmetric coin, obtaining a heads ratio of $\frac{52}{100} = 52$ percent $= 0.52$. We then toss an improbable (probability $\frac{1}{1024}$) run of 10 straight heads. After 110 tosses, the ratio is then $\frac{62}{110} = 0.564 = 56.4$ percent. This improbable run changed the heads ratio from 52 to 56.4 percent. We now show that the same run has far less effect when occurring after 1000 selections. Suppose we have completed 1000 tosses with a ratio of $\frac{520}{1000} = 52$ percent, and that now

† Problem: Take a bag of 1000 pennies. Throw them, and pick up all the tails. Throw the remainder again and pick up all tails. Repeat about ten times. Show that the last penny you pick up will have fallen heads the ten (or so) times. What does this indicate about the probability of one selection being all heads?

our improbable run of 10 heads occurs. The ratio after the run is $\frac{530}{1010} = 52.5$ percent. The run of 10 heads has very little importance when compared to the weight of 1000 selections. To make a large fluctuation after 1000 selections would require a much more improbable run, perhaps 100 heads.

The actual situation is frequently misunderstood by many, who believe and act according to the "doctrine of the maturity of the chances." According to this, a discrepancy in one direction increases the subsequent probability of a compensating discrepancy in the other. From this viewpoint, for example, it is argued that if a symmetric coin comes up heads five times in succession, it is then more likely to come up tails on the sixth toss. In fact it has the same chance of doing that as it had all along.† Charles Dickens' refusal to travel by train late one December because the accident rate for that calendar year had not yet come up to the annual average exemplifies the same misunderstanding.

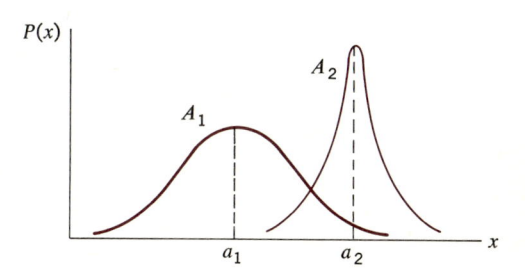

Fig. 8.9. *Two examples of normal curve. The total area under each curve is unity and denotes the sum of the probabilities of all possible x values.*

8.9 The normal distribution

Earlier in the chapter we remarked on the similarity of the bell-shaped frequency distribution curves relating to the failure of the timers and the heights of women. It is found that a standard function agrees remarkably closely with the observed frequencies describing many phenomena in all areas of knowledge. It is called the *normal* curve, or *Gaussian* curve, or distribution, and describes the probability (or expected frequencies),

---

† If we didn't already know that the coin was symmetric, the outcome of five heads in the first five tosses should alert us to the possibility of nonsymmetry. In this case, the prediction of a head on the sixth toss is preferable.

$P(x)$, for the various possible values of a (stochastic) variable subject to chance variation. Figure 8.9 shows two such normal curves. The abscissa is the axis of possible x values and the ordinate is the probability for the value. We note that

(1) The normal curve is symmetrical about its central or average value.

(2) The normal curve is characterized by a central value (a_1 and a_2) and by a spread. In this case, curve A_1 is broader than A_2.

(3) The areas under both curves, the sum of the probabilities of all possible x values, is unity. This is why the broader curve A_1 has a lower peak value.

(4) If we compress curve A_1 horizontally about its center and simultaneously let it push up vertically to maintain a constant area, the shape of curve A_2 would result.

(5) A standard mathematical form describes both curves. It contains two parameters, the central or average value $\bar{x}$ and a parameter σ describing the spread and called the standard *deviation*. This is expressed as

$$P(x) = \left(\frac{1}{\sigma\sqrt{2\pi}}\right) \exp\left[\frac{-(x-\bar{x})^2}{2\sigma^2}\right]$$

(6) Strictly speaking, the normal (or Gaussian) distribution applies only to cases in which the possible values of variable x can range continuously about the average value, but it may also be applied to cases in which the possible x values are finite in number but large. Thus, consider the variable $x =$ the number of heads in the experiment of tossing 100 coins. Here, the possible values of x are all the integers up to 100, including zero, and the probability for these x values follows the normal curve almost exactly (except for the improbable values of x near zero and 100).

As discussed earlier in this chapter, the normal curve also offers an approximation to the results of the test of the timers (Fig. 8.2). Let us now develop some terminology and definitions of the normal curve, using the test results on the timers as our specific example.

8.10 Range, average, standard deviation

Several terms that are useful for describing quantitatively the character of a curve that is similar to the normal curve.

THE RANGE

Figure 8.1 does give some immediate information, namely, that the *range* is from 8000 to 34,000 cycles. The range is simply the difference between the maximum and the minimum.† This information has some value. But it would have less and less value for describing the total performance of this group of timers if one of them should have failed at only 3000 cycles, or only one persisted for as much as 50,000 cycles.

THE AVERAGES: THE ARITHMETIC MEAN

Again looking at the data on the timers, is there some way of averaging the results to give a meaningful average value for the performance of the group? The average weight of ten men, the average wage of a group of factory workers, or the average rainfall for the year can be readily determined and can have considerable value for purposes of comparison or of analysis.

There are several ways of averaging the values. There is a *simple arithmetic mean*, $\bar{x}$, which is obtained by

† Note that we went from the lower limit of the lowest class interval (8000 to 10,000) to the upper limit of the last class interval (32,000 to 34,000).

(1) Adding all the data on all the items to obtain a sum.
(2) Dividing the sum by the number of the items.†

Taking the data of Table 8.1, let us use for the values of x the midpoints of the frequency intervals (9,000, 11,000 and so on) and add the sum of the frequencies times the number of units corresponding to each frequency:

$$(9000 \times 1) + (11{,}000 \times 1) + (13{,}000 \times 2) + \cdots$$

This gives a total of 840,000 for the 40 timers. Dividing the sum by 40 gives an *arithmetic mean* of 21,000 cycles.

Sometimes it is desirable to give more (or less) importance to certain of the values before adding the data. For example, some of the data may have included uncertainties, which suggests that less importance may be assigned to them. Each datum can then be multiplied by a factor, w, that gives it proper "weight" before the summation is attempted. For example, if the first timer failure of Table 8.1 occurred at a time when the test equipment itself was acting up, or if the last timer continued to 34,000 cycles but its behavior was not especially good during that last interval, a weighting factor (of, say, one-half) might be applied to these two units before adding up the totals and dividing to give the *weighted arithmetic mean*.‡

† A simpler way of expressing the arithmetic mean is

$$\bar{x} = \frac{\sum_{i=1}^{n} x_i}{n}$$

where the bar over the x identifies the average x, n is the number of items, and $\sum_{i=1}^{n} x_i$ is the sum of all the values of x that are to be added together, from 1 to n items.

‡ Of course all the other units could have been multiplied by 2 instead. In any case the new average would give the arithmetic mean as

$$\bar{x} = \frac{\sum_{i=1}^{n} (wx)_i}{\Sigma w}$$

where the w represents the weights assigned to the various values.

The arithmetic mean is very useful for general purposes, especially because it is computed so very easily. One must be careful in drawing conclusions from the value alone, however, For example, if nine of ten families have a yearly income of $10,000 each, and the tenth family has an income of $1 million, the arithmetic mean for their incomes would say that the average income for the group of ten families is $109,000 each. If such a case, how useful is the mean? A lack of symmetry in the distribution can make the arithmetic mean quite meaningless.

AVERAGES: THE MEDIAN

The *median* is the value of the middle item in a statistical distribution. It has as many items above it as below it. In the case of Table 8.1, the median value is that lying between the frequency value of the twentieth timer and the twenty-first timer, which lie in the class interval 20,000 to 22,000 cycles. The arithmetic mean and the median therefore lie in the same interval in this case.† The median has less value for most applications than does the mean, but it can have some value for identifying the midpoint for many distributions.

THE MODE

In the distribution of Fig. 8.1, the most populated (most popular, most "fashionable," most "preferred") interval is that

† See Balsley and other texts to learn how the various averages and the median can be identified more accurately.

between 20,000 and 22,000 cycles. This interval therefore represents the mode of the distribution and offers a nonmathematical description of the distribution. In our particular example the mode carries little additional information; it could have had more value if the distribution had been less symmetrical.

The preceding expressions give information on the character of a distribution, but the significance of the information varies enormously with the shape of the distribution curve. When the distribution lacks symmetry, or the range is too wide, the average may have relatively little significance.

AVERAGE DEVIATION

The range noted above gives a rough measure of the spread of a distribution. A better measure is given by the *average deviation* of the distribution from some chosen central or representative value of the set of terms. We can readily use the arithmetic mean $\bar{x}$ for this central value and then see how far the other values are spread from this mean.

To obtain the average deviation† in the case of Table 8.1, we use the mean obtained previously for $\bar{x}$, namely, 21,000 cycles. We list, for convenience, the midvalues of the frequency intervals of Table 8.1 as the first column of Table 8.3, taking a separate line for each term of the set of 40. The second column is obtained simply by subtracting from each value of x of the first column the mean we selected (21,000). Entries in the second column are marked as minus or plus, but our addition will disregard the signs and take into account only the absolute value of the number. The total sum shown when divided by $n = 40$ terms, gives the average deviation from the mean of 4000 cycles.

$$\text{Average deviation} = \frac{160,000}{40}$$
$$= \pm 4000 \text{ cycles}$$

The average deviation (a.d.) tells a little more about the spread of the distribution than is indicated by the range alone. It takes into account every item under consideration, not just the extremes. But an even more useful measure of the spread of a distribution is given by which is called the *standard deviation*.

STANDARD DEVIATION

Like the average deviation, the *standard deviation*, σ, takes into account every item of the set under consideration. It is also called the *root mean square deviation* because it is determined by taking the *squares* of the deviation from a selected mean, adding those squares, dividing by the number of terms, and then taking the square root of the quotient.†

In the example of Table 8.3, the third column lists the squares of the figures of the second column. The sum of the squares is 1,056,000,000. Dividing this by the 40 terms that make up the set, we obtain 26,400,000. The square root of this number is about 5140 cycles, which is the standard deviation (s.d.) for this particular distribution

$$\sigma = \text{standard deviation (s.d.)}$$
$$= \sqrt{\frac{1,056,000,000}{40}}$$
$$= \sqrt{26,400,000}$$
$$= 5140 \text{ cycles}$$

† In symbolic language,

$$\text{a.d.} = \frac{\sum_{i=1}^{n} |d_i|}{n} = \frac{\sum_{i=1}^{n} |x_i - \bar{x}|}{n}$$

where the term $|d_i|$ and $|x_i - \bar{x}|$ are the absolute values of these numbers (that is, the summation disregards the plus and minus signs).

† In symbolic language this becomes

$$\sigma = \sqrt{\frac{\sum |x_i - \bar{x}|^2}{n}}$$

TABLE 8.3
AVERAGE DEVIATION

Value of x	Deviation $d = x - \bar{x}$	The square of the deviation d^2
9,000	−12,000	144,000,000
11,000	−10,000	100,000,000
13,000	− 8,000	64,000,000
13,000	− 8,000	64,000,000
15,000	− 6,000	36,000,000
15,000	− 6,000	36,000,000
15,000	− 6,000	36,000,000
17,000	− 4,000	16,000,000
17,000	− 4,000	16,000,000
17,000	− 4,000	16,000,000
19,000	− 2,000	4,000,000
19,000	− 2,000	4,000,000
19,000	− 2,000	4,000,000
19,000	− 2,000	4,000,000
19,000	− 2,000	4,000,000
19,000	− 2,000	4,000,000
21,000	0	0
21,000	0	0
21,000	0	0
21,000	0	0
21,000	0	0
21,000	0	0
21,000	0	0
23,000	+ 2,000	4,000,000
23,000	+ 2,000	4,000,000
23,000	+ 2,000	4,000,000
23,000	+ 2,000	4,000,000
23,000	+ 2,000	4,000,000
23,000	+ 2,000	4,000,000
25,000	+ 4,000	16,000,000
25,000	+ 4,000	16,000,000
25,000	+ 4,000	16,000,000
25,000	+ 4,000	16,000,000
27,000	+ 6,000	36,000,000
27,000	+ 6,000	36,000,000
27,000	+ 6,000	36,000,000
27,000	+ 6,000	36,000,000
29,000	+ 8,000	64,000,000
29,000	+ 8,000	64,000,000
33,000	+12,000	144,000,000
	$\sum_{i=1}^{n} |d_i| = 160,000$	$\sum_{i=1}^{n} d_i^2 = 1,056,000,000$

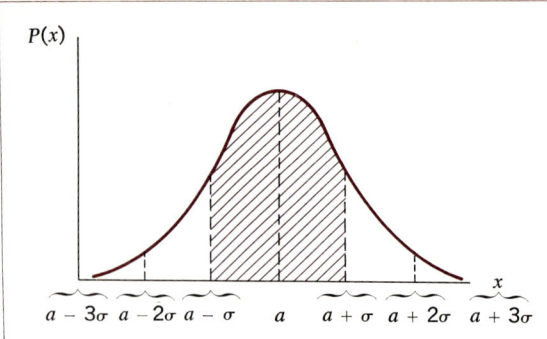

Fig. 8.10. Gaussian (or normal) curve.

Because the standard deviation utilizes the squares of the deviations, it is more indicative of the effects of large deviations from the mean. It is therefore also a more sensitive measure of dispersion.

What is the significance of the standard deviation σ? Qualitatively we may say that x values that deviate from the central or most probable value a by more than a few standard deviations (for example, $x = a + 5\sigma$, or $x = a - 5\sigma$) occur with extremely low probability. Almost all the time our observation of or sampling from a normally distributed stochastic variable is expected to be within a few standard deviations of the central value a.

To make some quantitative remarks, we consider Fig. 8.10, which shows a normal curve sectioned off into units of σ. By measuring or calculating the areas underneath the sections, we obtain the probabilities that the value of x lies in the range corresponding to these sections. Thus the probability to obtain an x value in the range extending from $a - \sigma$ to $a + \sigma$ is given by the center (shaded) areas in the figure, and is close to $\frac{2}{3}$. The probability to fall outside this range, that is, to depart from a by more than $\pm\sigma$ is then $\frac{1}{3}$. In other words approximately two-thirds of all observations of a normally distributed quantity are expected to be within one standard deviation of the average value a, and one-third are expected to depart by more than this. Table 8.4 presents this result along with the corresponding results for some other ranges.

We return now to the question with which this discussion began. What can the manufacturer reply when he is asked about the behavior of his timers in the course of this test? He can now simply report that the test of these timers indicates that they have a life expectancy given by a mean of 21,000 cycles and a standard deviation of 5140 cycles. That is, not more than one-third of the timers in production are likely to last less than 15,860 cycles or longer than 26,140 cycles. Also not more than 5 percent of the timers are likely to fail at less than 10,720 cycles ($=$ mean $- 2\sigma$) or longer than 31,280 cycles. Does this represent satisfactory performance? Possibly he requires a mean for performance at about this value, and the tests indicate that quality production is being maintained. But it may be that 25,000 cycles is the mean for the standard for production, perhaps with a standard deviation of about 5000. In that case the production processes must be analyzed to determine what has gone wrong to prejudice the performance.

8.11 The central limit theorem

Why are so many observed distributions in nature approximated so closely by the normal distribution? The answer is given by what is called the *central limit theorem of the theory of probability*. The idea is that a quantity, or stochastic variable, that is subject to the influences of many small variable effects, each subject to its own probability distribution, is expected to follow a normal distribution rather closely. The proof of the theorem is based on calculations involving the probabilities of combinations of the various possible values for the many variable effects that combine to determine the value of the stochastic variable.

As an example, the number of heads in a toss of n coins can take on many values, from zero to n, depending on the chance occurrence of heads or tails on each of the n coins. The theorem states that the number of heads in an n-coin toss follows the normal distribution more and more closely for larger and larger values of n. Again consider the example of women's heights, where we expect that the variation in height is the result of many effects, some genetic and some environmental. If the group being considered is fairly homogeneous ethnically and in other ways, we expect that each one of the genetic and environmental effects is by itself a small factor. In this case the theorem leads us to expect a normal distribution of heights.

A more precise statement of the central limit theorem is given in the following paragraph.

When a stochastic variable x is itself influenced by n other stochastic variables, y_1 up to y_n, with all variables small in the sense that the contribution of any one of them to the value of x is negligible compared with the contribution of the remaining $n-1$, then the probability distribution for x approaches the normal distribution as n becomes larger. Further, the average value $\bar{x}$ is the sum of the individual average values $\bar{y}_1$ to $\bar{y}_n$, and the square of the standard deviation σ^2 is the sum of the appropriately defined quantities, σ_1^2 up to σ_n^2. The variables y_1 to y_n themselves could have probability distributions of more or less arbitrary shape.

In the example of women's heights, the height of a woman x may be written as

$$x = \bar{x} + \bar{y}_1 + \bar{y}_2 + \cdots + \bar{y}_n$$

where $\bar{x}$ is the height all women in the

TABLE 8.4
NORMAL DISTRIBUTION

Range	Probability of lying within range, %	Probability of lying outside range, %
$a - \sigma$ to $a + \sigma$	$\frac{2}{3} = 67$	$\frac{1}{3} = 33$
$a - 2\sigma$ to $a + 2\sigma$	$\frac{19}{20} = 95$	$\frac{1}{20} = 5$
$a - 2.3\sigma$ to $a + 2.3\sigma$	$0.99 = 99$	$\frac{1}{100} = 1$
$a - 3\sigma$ to $a + 3\sigma$	$\frac{399}{400} = 99\frac{3}{4}$	$\frac{1}{400} = \frac{1}{4}$

group would have if they were identical genetically and environmentally. Each of the variable quantities $\bar{y}_1$ up to $\bar{y}_n$ represents the height deviation caused by one of the genetic or environmental effects.

In the n-coin toss, the number of heads x is given by

$$x = x_1 + x_2 + \cdots + x_n$$

where x_1, for example, is the number of heads (0 or 1) for coin 1. The probability distribution $P(x)$ for the values $n = 1,2,3,$ and 4 were tabulated in Sec. 8.8 and are graphed in Fig. 8.7. If a dotted curve were passed through the points for $n = 4$, it would resemble the normal curve.

The central limit theorem tells us that the possible values of x, zero up to n, follow the normal distribution, with probability given by

$$P(x) = \left(\frac{1}{\sigma\sqrt{2\pi}}\right) e^{-(x-\bar{x})^2/2\sigma^2}$$

approaching more and more closely as n grows large. The average value $\bar{x}$ is given by†

$$\bar{x} = \frac{n}{2}$$

Where n represents the number of cases tested (in this case the number of coins tossed), and the standard deviation squared is given by

$$\sigma^2 = n s_1^2$$

where s_1 is the spread in the toss of a single coin. For a single symmetric coin, the possible results, $x_1 = 0$ and $x_1 = 1$, each deviate from the average value of $\frac{1}{2}$.

We then have

$$s_1 = \tfrac{1}{2}$$

† This formula holds only in cases such as this, where n *also* represents the largest possible value of the stochastic variable x.

the standard deviation σ in the number of heads in the n-coin toss is then

$$\sigma = (\tfrac{1}{2})\sqrt{n}$$

For the case that $n = 1000$, the standard deviation is $(\tfrac{1}{2})\sqrt{1000}$, or approximately 16. Thus, using Table 8.2, we find that a 1000-coin toss is expected to result $\tfrac{2}{3}$ of the time in a number of heads lying in the range 500 ± 16, that is, from 484 to 516. Deviations as large as or larger than $3\sigma = 48$ are expected to occur only once in 400 tosses of 1000 coins.

8.12 Inference—mostly statistical

It is practically impossible to face a new situation with a completely open mind. Past experience, assumptions, attitudes, values, and beliefs of various kinds lead us to entertain certain ideas and mold our way of thinking and viewing new situations.

The new situations or questions may grow out of or have resemblance to others in our past experiences. The way in which we approach them and learn from them is largely conditioned by our past experience. The same phenomena may evoke different reactions from observers with differing experience. Imagine our stone age ancestors faced with an airplane, or auto, or flashlight!

Will the sun rise tomorrow?† David Hume argued that there is no logical way to deduce that because it rose on a previous morning, it will rise again. We can therefore say nothing more, but can only wait and see. There is a difference between the problem in logic about induction and calculation of the probability of its occurrence in a particular instance. Man has always sought to discover the rules and mechanisms of how things

† This is the problem of induction that has yet not been solved!

work. Of course the sun will rise tomorrow! That is the way Apollo rides. Or that is the way the epicycles work. Or that is the way the earth spins. There is no reason at all for the earth to stop spinning in the middle of the night, or for the sun to stop shining. Our laws of physics tell us these things.† Let us spend our time on more interesting and more urgent matters rather than proving logically that the sun will rise tomorrow. Such arguments belong to seminars on the logic of induction and not to the world of daily life.

Draw a potful of water from the kitchen sink and put it on a high-temperature hotplate. What happens? We all know that the water will get hot and eventually boil. But—what would we say if we observed just the opposite: the water became cold and then froze? Would we think of it as an optical illusion? Was the hotplate in reality a refrigerating device that only appeared to be a heater?

Take a quarter from a roll of quarters obtained at the bank and toss it in some random fashion 10 times. What would we say if we observed 10 heads. What would we say if we observed 100 heads in 100 random tosses?

Consider first the 10 tosses of the quarter. Let us assume that we already know from past experience that our tossing mechanism is a fair one. What are some possible conclusions from the observation of 10 heads? Which one would we prefer to believe if we had to make a choice with no further observations? To simplify the discussion, only two possible conclusions will be considered.

(1) The coin is symmetrical. We happen to observe the unlikely event of 10 heads.

(2) The coin is strongly unsymmetrical

† Of course, if the sun doesn't rise tomorrow we had better amend our "laws" of physics.

with a high probability of coming up heads.

Thus, a run of 10 heads is not at all unexpected. What do we know about the symmetry of the coin? If we saw it, felt it, or otherwise measured it, we could form conclusions concerning the symmetry. Further, assume that we know from past experience that it is rare indeed for a quarter obtained as described to be unsymmetric. In fact, let us assume that such quarters are found only once in 10,000. We realize now that we are faced with an improbable event, no matter which of our two conclusions we chose. However, with conclusion (1) the observed event has a likelihood of $(\frac{1}{2})^{10} = \frac{1}{1024}$, while with conclusion (2) the likelihood is only $\frac{1}{10,000}$. We therefore choose conclusion (1).

Applying the same method to the case of 100 heads in 100 tosses, we find that with conclusion (1) the observed event has a likelihood of $(\frac{1}{2})^{100}$, which is less than one-millionth, while with conclusion (2) the likelihood is again $\frac{1}{10,000}$. Here, then, we prefer to believe conclusion (2).

In the case of the water cooling and freezing when placed on the hotplate, we would regard almost any other explanation as preferable to the actual happening. It is much more likely that we were fooled in some way than it is for a potful of water to cool and freeze when heated.†

Four basic notions used in statistical inference from limited observation are illustrated in some of the examples described. Following these notions

(1) We bring to bear on the situation all our related knowledge.

(2) Based on this knowledge (and on our patterns of thinking and facing new

† Another "explanation" is that "a miracle" occurred.

situations), we formulate possible conclusions.

(3) Assuming that each conclusion is true, we calculate or estimate the probability of occurrence of the actual observations.

(4) We then prefer that conclusion for which the likelihood of the actual observation is largest.

We realize that the conclusion arrived at in this way is tentative, especially when there are others that result in almost equal likelihood. However, the less likely an event is, the less frequently it occurs, and the method described will lead more often to correct conclusions. If there is still reason to doubt, further observations eventually determine a conclusion that leaves little statistical room for doubt in the sense that the preferred conclusion gives the observations a far greater likelihood than do the other possible conclusions.

Many new situations in science are not approached by the methods of statistical inference. The problem of explaining observations, understanding them, and making a theory or mechanism is central to these situations. Remember the observations faced by Newton. His "invention" of laws of motion and of gravity caused many observations to be readily "understood" both qualitatively and quantitatively. We shall learn later in the course that applying Newtonian mechanics to the motion of the electrons in an atom gives results that appear in flat contradiction to observation. Here, again, statistical inference is not directly involved. A new theory, a new understanding must be created. Following the work of Planck and Einstein, Bohr made the beginnings of a new quantum theory of mechanics of the atom.

Perhaps the following comment is a correct way of describing those situations in which statistical reasoning plays essentially no role. When one of the possible conclusions in the preceding step 2 seems to unify, explain, and make reasonable a wide variety of our observations (that is, to "make sense," or "look right") and at the same time leads us to make further conclusions and to formulate new questions, we then say we are on the right track and continue as if this conclusion were true. If at some place along the line we find ourselves faced with a contradiction, we try hard to hold on to our theory by making modifications or additions. When all these attempts fail, we search for new ideas.

We close this section by considering the statistical problem of testing a conclusion on the sole basis of statistical evidence. Here, no "theory" or "understanding" is involved. The conclusion to be tested is that the coin we are tossing is symmetric. The observations are simply the results of coin tossing.

Let the first outcome be heads. What can we say at this stage? Assuming our hypothesis of a fair coin and tossing mechanism, the observed outcome is as likely as the only other possible outcome. We certainly can hold on to the hypothesis.

Let the first 10 tosses come out to a total of 6H and 4T. What can we say now? How likely was this outcome as compared with the other possibilities of no heads up to 10 heads? Was this outcome in the likely range of outcomes for a symmetric coin tossed 10 times? Probably, because 5H would not have caused any questioning, and 6H is only somewhat less likely. The actual calculations can proceed, using the method of Sec.

8.7, by listing all the $2^{10} = 1024$ basic and equally likely outcomes and counting† all those that contain various numbers of heads. With only small error, however, we can use the normal distribution for the number of heads, x, in n tosses, with $n = 10$ in the present case. Using the results of the last part of Sec. 8.11, we find that the probability distribution for the number of heads in 10 tosses of a symmetric coin is closely normal, with a central value of $a = \frac{1}{2}(10) = 5$ and a standard deviation of $\sigma = \frac{1}{2}\sqrt{10} = 0.87$. Our observation of $x = 6$ differs from the central value by $6 - 5 = 1$, which is only slightly larger than one standard deviation. A toss of 6H is therefore not at all one of the least likely outcomes. There is at this stage no strong reason to doubt our hypothesis of a symmetric coin.‡

We continue the selections until 100 tosses are completed. Assume the outcome of 55H and 45T. We analyze as before. Here the probability distribution for the number of heads in 100 tosses of a symmetric coin is closely normal, with a central value of $a = \frac{1}{2}(100) = 50$ and a standard deviation of $\sigma = \frac{1}{2}\sqrt{100} = 5$. Our observation of 55H is one standard deviation away from the central value. Again we have no strong reason for disbelieving our hypothesis.

---

† It is not necessary actually to count. The methods of "combinatorial analysis" provide simple formulas for the appropriate counts.

‡ Note that here we are testing one hypothesis and not comparing the credibility of a number of hypotheses. If we were comparing various hypotheses, the preferred one at this stage, assuming no knowledge at all other than 6H in 10 tosses, would be that the coin was not symmetric and had a heads probability of $\frac{6}{10} = 0.6$. However, the hypothesis of a symmetric coin would be only slightly less pref-

What if we observed 60 heads in the first 100 selections? On the basis of the symmetric coin hypothesis, this would be two standard deviations away from the central value. This is already one of the less likely of the possible outcomes. Deviations as large or larger than this are expected to occur only with a probability of 5 percent, according to Table 8.4. We can retain our hypothesis if we like, but our "confidence level" is about 5 percent, according to the statistical data.†

8.13 Summary

Over the recent decades the theory of probability and techniques of statistical analysis have assumed enormous importance both among the mathematical sciences and in connection with the everyday activities of society. Whether analyzing the distribution of radioactive emanations or of mechanical particles, the significance of a set of measurements, planning of forests or farm land or the growth of a city, testing the products of a factory, calculating the economics of a nation, or placing a bet on a game of chance, the theory and techniques are found to be remarkable useful. The usefulness extends to analysis of what has already happened and what is likely to happen in the future. We shall not attempt in this particular summary to repeat the ideas that have been dealt with at some length.

---

erable. On this basis again, we would not have strong reasons to doubt the symmetry of the coin.

† In this case, with no further information, the hypothesis of nonsymmetry with $P(H) = 0.6$ would be significantly more likely by our Step 4 in statistical inference.

Questions/Discussions

1. Joe and Mary are two members of a 20-member club that has decided to select, by means of a random drawing of names, two of its members as delegates to the national convention. What is the probability that both Joe and Mary will be selected?

2. Mary is very hopeful that Joe will take her to the movies on a certain Friday evening. However, she has only recently recovered, she thinks, from a case of mononucleosis. She can go to the movie only if (a) Joe is able to borrow his father's car, which she estimates he has a good chance of doing because he generally gets the car about three out of every four times he asks for it; (b) if her call to the doctor results in his reporting that her last mono test was negative, and she estimates her chance of success here as 50–50; and (c) if Joe decides to invite her to go out this Friday, and she estimates that there are nine chances in ten that he will. What is the estimated probability that she will be able to go to the movies with Joe?

3. A certain radioactive isotope has a half-life of one year, that is, one-half of the atoms will decay during the course of the first year after a pure sample has been prepared, half of the remainder during the second year, and so on. What is the probability that any given atom in the original pure sample will not decay during the first nine years? At the beginning of the tenth year, what is the probability that any one of the undecayed atoms will decay during that year?

4. A man lives in a large city approximately midway between the homes of two girls in whom he is equally interested. Subway trains run at 10-minute intervals in each of the two desired directions. The ones running in the direction of the home of girl A leave on the hour, at 10 past the hour, 20 past, and so forth. Since the man is equally interested in the two girls, he has adopted a practice of going to the subway station at random times, taking the first train that comes in, and visiting whichever of the two girls he reaches by this random selection process. After a few weeks he discovers that he has visited girl A nine times out of every ten. What is the schedule of the train that goes toward the home of girl B?

5. A different kind of dice game has been devised using pyramidal dice with each die having four triangular faces and four numbered vertices. Thus, a roll of two of these dice will result in each one having one of its vertices pointing up, and the total possible score ranges from $1 + 1 = 2$ to $4 + 4 = 8$. The rules of the game are as follows:

   (i) If a player rolls a 5 on his first throw, he wins.
   (ii) If his first roll is not a 5, he rolls again and again until either he again obtains the number that came up on his first roll and wins, or rolls a 5 and loses.

   (a) What is the probability of winning on the first roll?
   (b) Which numbers other than a 5 on the first roll are more likely to result in a subsequent win?
   (c) Assuming the first roll was a 7, what are his chances of winning on the 2nd roll?

(d) How do a player's chances of winning on the first roll in this game compare with his chances of a first-roll win in the game described in Question 8.14?

6. The following table gives the quality point averages of 16 students who are graded on a 4-point scale.

3.0	1.8	3.0	2.6
3.4	2.4	1.9	3.8
3.7	2.2	2.4	2.6
2.5	1.6	2.5	2.2

Find the mean $\bar{x}$, the average deviation a.d., and the standard deviation σ.

(a) What percentage of the class falls within plus or minus one standard deviation of the mean?

(b) How many members of the class have averages that exceed the mean by 2σ or more?

7. A certain auto driver has estimated that there is only 1 chance in 50 that he will suffer an accident if he violates the law and drives through a certain street intersection without stopping when the traffic light is red. Assume that he commits this violation on the average of 5 times a week and does not have an accident there for a full year. Discuss the validity of each of the following statements, with regard to his chances of an accident during the second year at that corner, assuming that he continues this bad habit and that other traffic conditions remain unchanged.

(a) According to the law of averages he is bound to have an accident very early in this second year.

(b) The chances are about 10:1 in favor of his having an accident during the second year.

(c) The chances are at least 10:1 in favor of an accident the next time he commits this violation.

8. You are attempting to predict the result of a closely contested two-party election in a large city, by polling a random sample of the electorate. If, in fact, only 48 percent of the electorate is in favor of the incumbent party, what is the chance that your sample will show a majority in favor of the incumbents (and thus lead you to an erroneous prediction) if the number of people you ask is (a) three, (b) seven?

Under these conditions (that is, 48 percent in favor), how large a sample do you think you should take in order to reduce your chance of making a wrong decision to only $\frac{1}{100}$ (that is, 1 percent probability)?

9. What, if anything, is wrong with the logic expressed here?

A traveler has heard that there is only one chance in a million of there being a bomb on a commercial airplane. Furthermore, the chances of there being *two* bombs on the same plane is only one-millionth of one-millionth. Therefore, he always carries a bomb in his own luggage on the assumption that he is thereby greatly reducing the chance of there being another one on that same plane.

10. Assume that a customer has purchased one of the timers that was manufactured in the same production run as that tested and reported at length in this chapter. What is the probability that he will have gotten one that will last for over 26,140 cycles? If he buys two of them, what are the chances that they will both fail before 15,860 cycles.?

References

Balsley, Howard L., *Introduction to Statistical Method.* Paterson, N.J.: Littlefield, Adams, 1964.

Boase, Ronald L., "The Application of Mathematical Statistics to Industrial Problems," *Gen. Motors Eng. J.,* Second Quarter (1965).

Gamow, George, *One, Two, Three, Infinity* (A Mentor Book). New York: New Amer. Press, 1947.

Huff and Geis, *How to Take a Chance.* New York: Norton, 1959.

Kac, Mark, "Probability," *Sci. American* (September 1964).

Levinson, Horace C., *Chance, Luck and Statistics.* New York: Dover (paperback), 1963.

Moroney, M. J., *Facts from Figures.* Baltimore: Penguin (a Pelican book), 1951.

Mosteller, Rourke, and Thomas, *Probability with Statistical Applications.* Reading, Mass.: Addison-Wesley, 1961.

Newman, James R. (ed.), *The World of Mathematics.* New York: Simon and Schuster, 1956. *Includes several excellent articles and reviews in a section titled Part VII, "The Laws of Chance." The student will find very interesting reading in the commentaries of Newman and in the articles titled:*

"Concerning Probability" by Pierre Simon de Laplace

"The Red and the Black" by Charles Sanders Peirce

"The Probability of Induction" by Charles Sanders Peirce

"The Applications of Probability to Human Conduct" by John Maynard Keynes

"Chance" by Henri Poincare

"The Meaning of Probability" by Ernest Nagel.

Ore, Oystein, *Cardano, the Gambling Scholar,* Princeton, N.J.: Princeton Univ. Press, 1953.

Weaver, Warren, *Lady Luck; The Theory of Probability.* New York: Doubleday (an Anchor book), 1963. *This gives a very interesting and informative introduction to probability concepts.*

Wilks, S. S. *Elementary Statistical Analysis.* Princeton, N.J.: Princeton Univ. Press, 1951.

CHAPTER NINE

Observation, Measurement, Evaluation

When you can measure what you are speaking about, and express it in numbers, you know something about it; but when you cannot measure it, when you cannot express it in numbers, your knowledge is of meager and unsatisfactory kind.

LORD KELVIN

AS WE HAVE seen in previous chapters, the early concepts of the workings of nature were based on qualitative observation and experience. Fortunately, observations did not remain simply qualitative; those who had occasion to think of objects as being tall, or hard, or heavy, often had to note that other things were taller, or harder, or heavier. Comparison led to the ordering or ranking of objects. This was probably accompanied by considerable confusion as the varieties of objects and observations increased, due to the absence of precise units by which to compare or categorize objects. We can picture that the repetitive rising of the sun and moon may have not only provided a method for counting time, but also helped develop early awareness of the value that recognized units can have for the communication of information. The size of the foot and of the hand offered something by which comparison of lengths was made easier. The growth of commercial activities required units of measure for purposes of sale or barter. At any rate, from these experiences and needs there emerged a more systematic process of *measurement*. We have already discussed some of the units by which it has been made possible to measure time, space, motion, force, and other quantities. Before the conclusion of this volume we shall find that it has been possible to develop either basic units or guides for giving quantitative evaluation to most of the variables in natural phenomena that are of significance to our existence.

9.1 The role of measurement in knowledge

The ancients developed language as an effective method for communication from one person to another, and writing made language more effective for communicating from one generation to another. The logical reasoning that framed early Greek philosophy was recorded and made available to people of many lands and many languages. This was done with considerable success, as we have seen. But we have also noted that a greater rate of progress came when logical disputation made room for *measurement* and testing of ideas. These, in turn, demanded a base in *units of measurement*. And, once units became available for general use, the search for the laws of nature became more nearly a universal interest and activity, despite differences of language. Students in Europe, Asia, and the Western Hemisphere have the benefit of observations made in the past and have easier access to those being made at present, simply because measurements and the units of measurement have made it possible to communicate *quantitatively*.

In this chapter we explore the role that observation and measurement play in our own lives and environment. We must begin by noting that our "environment" includes more than the grass and trees and machines that surround us. It includes the influence of society, of schools and churches, and of books through which we have a gateway to the recorded experiences of past generations. It is an environment of greater variety and richness than that enjoyed by any previous generation, and it has a larger variety of instrumental techniques than any other generation for exploring natural phenomena and making measurements of them.

Along with the spirit of the Renaissance came a greater dependence on systematic measurements. It also brought a realization that even the relatively simple phenomena of nature are usually encumbered by complications that detract from accurate knowledge. For example, every measurement of the motion of a body seems to be troubled by the effects of friction, or by several types of forces acting at once. On a few important occasions the discrepancies in measurements have required explorations that eventually led to wholly new concepts. This was the case when careful measurements of the speed of light revealed that, contrary to expectations, the speed remains constant under a selected series of test conditions, which found answer in Einstein's special theory of relativity (see Chapter 12). Another revolution came about when the radiation emitted by a hot body was carefully measured and found to be at variance with the predictions of the then-current theories. In this case also it became necessary to resort to very unusual concepts in order to explain the discrepancies (see Chapter 17).

Measurement, being essentially quantitative, led to the development of new mathematical methods. For example, the need for new techniques with which to analyze the behavior of objects in motion required development of the calculus by Newton and Leibnitz. The latter foresaw the day when scholars would calculate, instead of argue about, even social processes. On occasion there have been innovations in mathematics which in later years became important for the solution of physical problems. In any case, the remarkable progress that came with measurement and mathematics caused Lord Kelvin, one of the major figures of nineteenth century science to make the statement that introduces this chapter. Today, measurement and mathematics are so intertwined that the progress and maturity of any science are often evaluated in terms of the degree of sophistication of its measurement and mathematical techniques. In this respect the physical sciences, because they deal with situations that are easier to express in mathematical formulas, appear to have had a decided advantage over the social or behavioral sciences. One may question, in fact, whether the Kelvin statement can have significance in areas outside the physical sciences. Certainly the experiences of human passions and human hopes convey "knowledge" by their intensity that reaches beyond possibilities for measurement at this time.

But numbers alone do not give the whole story. Along with increased dependence on measurements there has been need for greater discrimination as to the purposes and context for each set of measurements. Measurements of intelligence, height, and weight of people have quite different significance to the anthropologist and anatomist than they do to the flight engineer who must fuel the airplane in proportion to total weight. As a consequence, a considerable specialization of interests has developed with respect to measurement.

The progress of science and technology over the past two centuries has had close ties with the progress of the art and science of measurement, for they are interdependent. Many standards have been established for the units of length, weight, hardness, and similar quantities. Even more important, it has been possible to measure unseen phenomena, such as the field of an electric charge, radio waves, or the properties of atoms. It has been necessary to seek basic information in these areas by techniques that are quite indirect, in contrast with such direct approaches as placing a yardstick against a man in order to measure his height.

Despite this great progress, we are increasingly aware that there is no such thing as an exact measurement, just as we are aware that there are no exact sciences.

We shall have more to say about these aspects.

Finally we must note that while our generation enjoys a much larger storehouse of knowledge from which to draw than was available to previous generations, we as individuals continue to suffer many of the same limitations that encumbered their efforts. As they did, we learn through the intermediary of the senses of sight, hearing, smell, taste, touch, and through the same experiences of pain, pleasure, hunger, satisfaction, questioning, and contemplation. We experience all the hazards of misinformation that troubled them, due to the limitations of sensory experience. More often neither the process of observing with our own eyes nor of deriving benefits from the wisdom and experience of past generations offers information that is particularly dependable. Fortunately, instrumental techniques help to reduce the unreliability of our senses, but even they cannot be completely freed from error.

Before concluding this section we should note that most experimental programs are also subject to boundary conditions with respect to cost in dollars and cost in time. When planning an experiment, the investigator must limit his choice of instruments and of techniques to those which are within the limitations of his budget and which can be completed within the time that is stipulated. Needless to say, this is often a major factor in determining the degree of accuracy with which the work can be done.

9.2 Effect of errors in the planning of experiments

Measurement in and of itself does not convert an observation into a scientific study. "To be a good butcher," said Plato, "one must know the parts of the beast if one is to cut it properly."

Errors of measurement fall into various categories. To begin with, every instrument has its specific capabilities and limitations. Every measurement taken with an instrument, whether it is a yardstick, a micrometer, or an electrical meter, is accurate only within finite and (hopefully) determinable limits. We shall discuss this presently in more detail.

To the limitations of the instrument itself, we must add the inadequacies of the person who selects and uses the instrument. The measurements wanted in scientific research tend to demand the full capabilities of instrumental techniques. For example, it is very common to require an electrical meter to be so sensitive that the normal movements of the electrons within the wires of the instrument itself may generate a larger signal than is given by the phenomenon under study. Sometimes a mechanical or pneumatic system, or a chemical detector, can be more suitable for a particular measurement than the electrical one can be. In any case, to avoid unnecessary errors, the experimenter must have a rather good understanding of the measurement he is after, as well as of the techniques that are suitable for that measurement. Often the design of an experiment entails the careful design of measuring instruments and techniques that are specific to the purposes of that experiment. It is not unusual for the instrument portion of an experiment to be 20 or more times more intricate and expensive and temperamental than is the equipment that houses the phenomenon itself. The analysis of the limitations and capabilities of such instrument systems can engage the experimenter from several months to several years before the data can be considered dependable. This has been especially true of nuclear research.

One of the most serious problems with which the researcher must grapple is the risk that the instruments measuring a phenomenon will change it in the process. We can measure the dimensions of a man simply by placing a yardstick against his body or by wrapping a tape measure around his biceps; the method does not work very well when applied to the study of the size of a mosquito in flight, however. (The analogy is not out of place when it comes to measuring the behavior of atoms) The devices that are useful for measuring the presence of an atom are likely to move that atom entirely out of its normal place. This leads, as we shall see before long, to a declaration of the indeterminacy principle, which says that our knowledge of both the position and momentum of small particles such as electrons cannot be more precise than a certain limit of accuracy.† But, of course, ordinary measurement techniques give faulty results that far exceed the theoretical limits noted above.

We should note that errors introduced by the measurement technique itself are not unique to physical research. For example, suppose we ask a test question that has emotional significance to a student, on an issue or subject in which he has strong personal interest. Will he be likely to think and to comment on that question in the same way that he might have *before* the question was asked? In other words, will not the very asking of the question influence his attitude and his answer? Similar problems arise in nearly every situation where accurate measurements are needed.

9.3 Errors of observation and of judgment

Whenever possible, a research worker is likely to want the data of his experiment to be recorded mechanically, to avoid

† Heisenberg proposed the *uncertainty principle* (see Chapter 19).

human errors. This is not always possible or feasible, however. There can be many occasions when the eye and ear and judgment of the experimenter enter into the observation of data. This means that there can be human errors of observation in addition to instrumental errors.

Since the turn of this century there have been many studies in experimental psychology to develop better understanding of human errors of observation and judgment. It has been confirmed that personal attitudes, interests, and assumptions tend to encourage an observer to take notice of some features of a phenomenon while neglecting other features that are also important. The selection of instrumental techniques, procedures, and measurements to be taken is determined according to the attitude of the experimenter. Previous experience (mental set) plays a role in determining what he will see thereafter.

For example, it has been determined that what a person sees at one instant is likely to influence what he sees in succeeding instants of time. Suppose that a person is shown only a vertical line for an instant, and in 0.09 sec he is shown only a horizontal line for another instant [Fig. 9.1(a)]. The subject may say that the line fell to the right, or he may say it fell to the left, to get to the B position. But now let the test be repeated several times, beginning with line A located not in the vertical position but at an angle θ to the horizontal, and repeating with increasing angle until A is at an angle of as much as 115 degrees. Although this now locates A at *25 degrees on the other side* of the original, vertical position, the subject is likely to say that in every case the line A falls *toward the right* from its various positions, even from the 115 degree position.†

In a second example of an *objective set* that prejudices observations, a subject is first shown a completely ambiguous drawing. He may give it some name, or he may respond that the drawing is meaning-

† See Wertheimer in References.

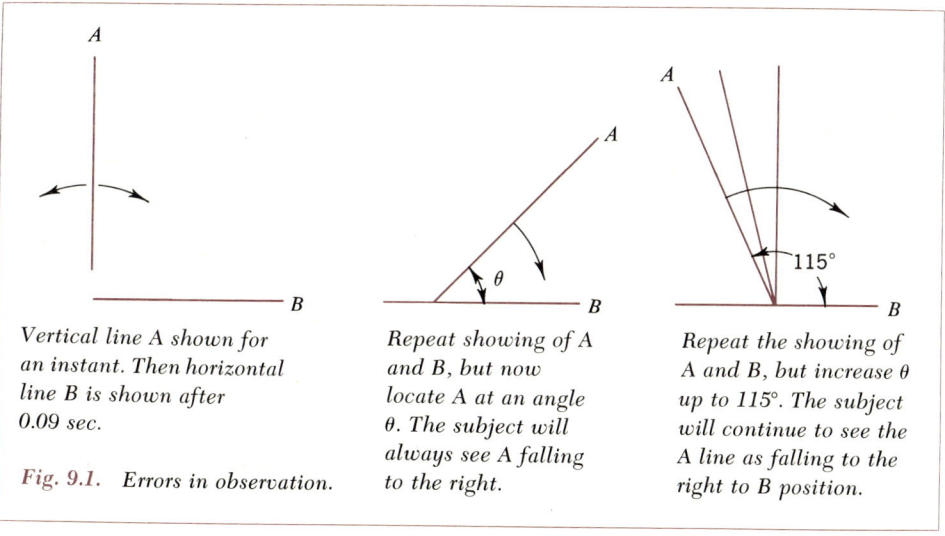

Vertical line A shown for an instant. Then horizontal line B is shown after 0.09 sec.

Repeat showing of A and B, but now locate A at an angle θ. The subject will always see A falling to the right.

Repeat the showing of A and B, but increase θ up to 115°. The subject will continue to see the A line as falling to the right to B position.

Fig. 9.1. Errors in observation.

less. He is then shown a series of 21 drawings, one at a time in quick sequence, with the same ambiguous drawing being the middle one (the eleventh) of the series. The first drawing of the series of 21 is a clear profile of a head, and the last (the twenty-first) is a clear drawing of a milk bottle. Starting with the clear profile, the drawings become more and more indeterminate on the way to the middle one (which is completely ambiguous) and then change to become more and more like the final drawing of the bottle. The subject is likely to continue to "see" the profile long after the series has passed the middle "meaningless" drawing. In fact, he may continue to see the profile until almost to the end, when the bottle suddenly becomes clearly defined. Similarly, if he were to start from the bottle end of the series, he might continue to see only a bottle until almost to the profile end of the series.†

Errors of this type tend to be reduced as one resorts to measuring techniques that utilize electromechanical devices and computers, but they are difficult to eliminate altogether.

There are very many other examples of possible error in what the observer sees. Some of the sources of error lie in the characteristics of vision, which we shall presently discuss. The color of an object will vary, depending on the color of its background; a grey thread may appear white on a dark background but quite dark when placed on a sheet of white paper. A line of a particular length appears short when next to longer lines, but appears to be substantially longer when placed near lines that are much shorter. It is easier to respond to *differences* and to *contrasts* than to absolute values. It is easy to be quite accurate when determining which is the taller of two people, providing they are standing back to back, close to each other. Every woman who wants to match the color of yarn or of dress material realizes that she must place the two next to each other to be accurate in judgment.

9.4 Measuring lengths

With these general introductions, let us now describe a few situations that require quantitative information on some variable or characteristic of a system.

We might begin with a situation that requires we find the diameter of a metal rod. We can get this diameter by placing a rule against the rod and sighting (as illustrated in Fig. 9.2) with the eye. This will tell us that the rod is a trifle over 1-in. diameter. With care we can see that it is about 1 in. and less than $1\frac{1}{16}$ in. But the errors of sighting (parallax, an uncertainty as to the exact coincidence of the edges of the rod and the divisions of the scale) prevent doing much better with the rule. The rule itself is not likely to be precisely graduated, and there will be some errors due to temperature effects; but these errors are meaningless in the face of the larger errors that accompany the use of the eye under these conditions. It is likely that the best we can do with this approach is to state that the diameter is between 1 and $1\frac{1}{16}$ in. This would be adequate information if we plan to use the metal rod to support a shelf or to serve as a clotheshanger support; but it is not likely to be adequate if the rod is to be used as part of a machine tool.

We can improve on this by using a micrometer (Fig. 9.3). In this case the spindle and the anvil of the micrometer make firm contact with the rod and allow measurements that may go to four figures. Let us say that the first measurement

† See Luchins in References.

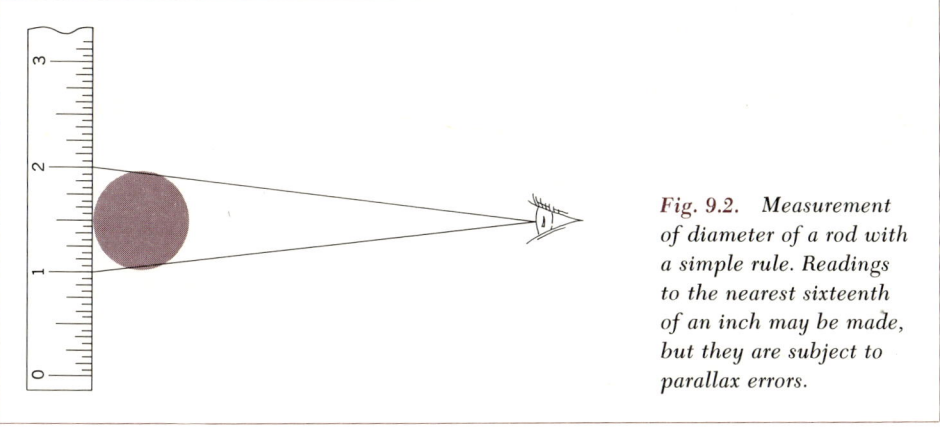

Fig. 9.2. Measurement of diameter of a rod with a simple rule. Readings to the nearest sixteenth of an inch may be made, but they are subject to parallax errors.

gives 1.025 in. Is this one reading enough? It might be, if all that we need is accuracy to within ±0.01 in. (±1 one-hundredth of an inch). But many machine dimensions must be known to accuracies of ±0.001 in. (±1 one-thousandth of an inch) and sometimes to within even closer limits. Let us assume that an accuracy within ±0.001 in. is our goal, or as near to that as we can get with our micrometer.

We find, however, that a second try with the micrometer gives 1.028 in., which is larger than the first reading by three times the tolerance level we said we wanted. If we continue to make measurements, we might find that ten tries give the listings of Table 9.1. We may ask: Why so many variations? The answer is simply that a number of variables or fluctuations enter into every measurement, and what we are recording is the different ways that these fluctuations can add and subtract to give the observed values. For example, one may tighten the screw of the micrometer more tightly with some measurements than with others. Or the spot on the rod that was measured may have varied from one reading to another. The temperature of the micrometer or of the rod may have changed with the handling, causing the

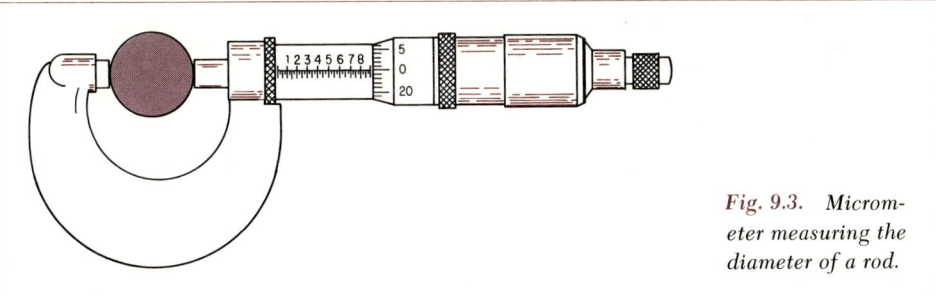

Fig. 9.3. Micrometer measuring the diameter of a rod.

TABLE 9.1
VARIATIONS IN MICROMETER MEASUREMENTS

Measured value x, in.	Deviation, $d = x - \bar{x}$	Square of deviation, d^2
1.025	0.000	0.000,000
1.028	+0.003	0.000,009
1.019	−0.006	0.000,036
1.020	−0.005	0.000,025
1.030	+0.005	0.000,025
1.031	+0.006	0.000,036
1.026	+0.001	0.000,001
1.024	−0.001	0.000,001
1.027	+0.002	0.000,004
1.020	−0.005	0.000,025
10|10.250 1.025	$\Sigma\|d\| = 0.034$ $a.d. = \dfrac{0.034}{10} = 0.003$	$\Sigma d^2 = 0.000,162$

$\bar{x}$ = mean = 1.025 in.
a.d. = average deviation (or average variation).
σ = standard deviation = $\sqrt{\dfrac{0.000162}{10}} = \sqrt{0.0000162} = 0.004$.
Measurement = 1.025 ± 0.004 in.

metals to expand or contract. Or a dust particle may have been caught between the rod and the jaw of the micrometer. We call these *chance errors* or *statistical variations*, and *these are present with every form of measurement*.

In view of these variations, what can we record as the diameter of the rod? From Chapter 8 we learned to evaluate measurements in terms of the *mean* and of the *standard deviation* σ.

Let us find these for the ten measurements. Adding the ten readings, we obtain the sum 10.250; when divided by the number of readings, this gives a mean of 1.025 in. We can now obtain the standard deviation as shown in Table 9.1 (see also Chapter 8), and find it to be ±0.004. The value we can record is then 1.025 ± 0.004 in.† This says that if we were to take additional measurement, not more than one-third of them would have departures from our mean by values equal to or greater than ±0.004.

Obviously we have not, with the results

---

† In practice, when a measured quantity is given along with an evaluation of error, the error is likely to be given in terms of probable error instead of standard deviation. The probable error equals 0.6745 times the standard deviation, and indicates the limits within which half the additional data are likely to fall.

of these ten tries, reached the accuracy within ±0.001 in., which we said we wanted. What can we do about it? We can improve the picture by taking many more measurements, say 100, to improve the value for standard deviation. It may be, however, that even 1000 readings will not improve the picture if the rod diameter itself is not uniform or if the micrometer is faulty. The decision must then be made whether to settle for the figure that is now obtained or to try to improve on it by using better instruments and more careful techniques. For example, the effect of pressure differences on the micrometer screw may be reduced by using an electric circuit to indicate when contact is just made.

Thus far we have dealt only with chance errors. There could also be large systematic errors. For example, the "zero" reading of the micrometer could be off its mark (not calibrated). Or the person reading the micrometer might have a heavy hand and actually cause the micrometer jaws to bend under pressure and compress the bar, and give generally low readings.

Question. Suppose that the "zero" of the micrometer reads 0.10 in. instead of 0.000 in. What will be the standard deviation? If, instead, the operator is careless and varies the pressure on the micrometer screw, how would the standard deviation change?

We must note the difference between *accuracy* and *precision*. Accuracy refers to the "true" value, whereas precision refers to the *consistency* with which one obtains the same value for a particular measurement. If the "zero" of the micrometer were badly off, one could repeat a particular measurement without much variation in the results. The measurements would then have good precision, but poor accuracy. To be accurate, the measurement must reduce both systematic and random errors as far as possible. For many applications in industry it is more important to assure precision than accuracy. For example, a chemical process is often designed to require constancy of temperature conditions at a particular part of the process. The temperature control schemes, then, are designed tó assure this constancy, without too much concern as to whether the temperature-measuring device is off by a few degrees from absolute standards.

The statistical techniques introduced in Chapter 8 and in the preceding example can be very useful for identifying and evaluating various types of random or chance errors. But it should be clear to the reader that it offers less help for eliminating systematic errors.

9.5 Measuring velocity and time

In Chapter 4 we discussed the meanings of velocity and acceleration without indicating how measurements could be made. Suppose we wish to develop complete information on the performance of a runner over a track course of 100-yd length. We can measure the track length to be that length to as high an accuracy as we please, to within a small fraction of an inch if we want such accuracy. However, the larger errors of timing and of the runner's relation to the starting and finishing lines are such that it would be a waste of effort to determine the length too accurately.

Timing offers more difficulty. A good hand-operated stopwatch can be fairly accurate over the 10 sec or so a runner might require for the dash, but the problem of starting and stopping the stopwatch at the correct instants of time presents possibilities for error which far exceed the errors of track length determination. For this reason it is useful to use electrical mechanisms. For example, the noise of the starting shot can trigger an electrical mechanism to begin recording time, and to do this much more accurately than manual operation can assure. The finish line can be a cord across the path of the track, designed to stop the timing mechanism as soon as the cord is touched by the runner. Of course there still will be errors that vary with the arm length of the runner, errors due to wind and weather, and variations in the way the cord is attached. There will be some variations also in the types of electromechanical system, although their rapid rate of operation reduces their internal timing variations to negligible magnitudes.

This example suggests two principles that are important for our purposes:

(1) An electrical or electromechanical system can be used to advantage to reduce human errors by factors of from 10 to 10,000 where timing is concerned. This is simply the result of their faster rate of response (100 to 100,000 times faster, roughly) than the human sensory system.

(2) When considering the overall accuracy of a measuring system that involves several parameters or variables, it is necessary to determine the range of error variations to which the measurement of each variable becomes subject. It is often the case that the measurements of one parameter (for example, time, dimension, voltage, or temperature) are subject to such large errors as to dominate the overall accuracy capabilities of the entire system. By the same token, if such a dominating source of error is good for accuracies of only ± 5 percent, it would be foolish to spend a lot of effort and money to make the accuracies of other measurements of the system to better than ± 1 percent. For example, there are measuring methods by which the track length can be marked for 100 yd to within 0.0001 in., at some cost, but this would be quite foolish and wasteful if other sources of error, such as in timing, amount to an equivalent error in distance of ± 2 in.

9.6 Temperature measurement

Although biological sensory systems are responsive to hot and cold, and the human hand can be very comforting on a fevered brow, our bodies are not equipped to gauge temperatures very effectively. The body temperature remains at a "normal" of about 98.6°F, but the reaction of the body to external temperatures varies from individual to individual, and can vary in the same person from one hour to the next. In any case, there are very few applications where we can gauge temperatures simply by physically sensing them.[†]

Temperature, and the measurement of temperature, have an important influence on our social and economic life. Within a household the room temperature must be held to the neighborhood of 70°F. To heat the house in winter, the interior of the boiler or furnace must be heated to temperatures that exceed 1300°F. Similarly, cooking and baking operations require heater temperatures that far exceed the temperatures of the food being

---

† The student can appreciate how poorly our bodies judge temperature conditions by trying a simple experiment. Take three beakers. Fill one with hot water, another with ice water, and let the third have water at room temperature. Ask an associate to dip one index finger in the cold water, the second in the hot water. After a few moments have him place both fingers in the water at room temperature, and ask his reactions.

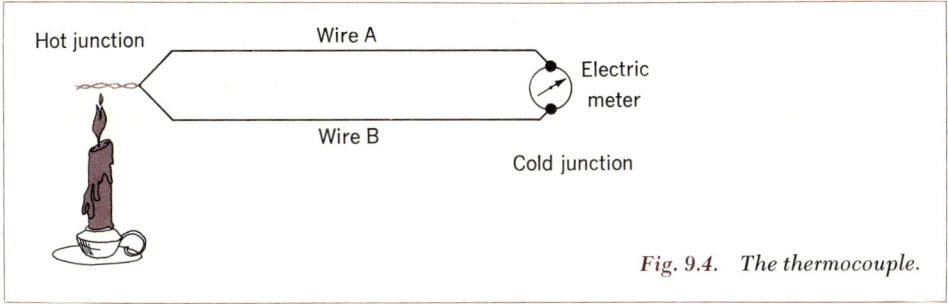

Fig. 9.4. *The thermocouple.*

cooked. The proper measurement and control of temperatures constitutes an important activity in the chemical, petroleum, and metals industries. As we shall see, a rise of only 18°F can double the rate of some chemical reactions or change the direction of other reactions altogether.

For this reason it has been necessary to develop a variety of techniques by which to measure temperature, to meet the various requirements (high temperature range, high accuracy, high sensitivity, small size, and so forth) that are needed by the large variety of applications.

What can we use for measuring temperature? It turns out that metals—and in fact nearly all solids, liquids, and gases —expand when heated. A metal rod becomes longer, and since length is relatively easy to measure, the expansion can be calibrated to give temperature readings directly. Liquids and gases expand in volume, but this again is as easy to measure as changes in length. In fact, the expansion of a gas such as hydrogen has become one of the important standards for determining the temperature scale. The expansion of mercury or of other liquids in thermometers is the most common technique for use in household and clinical thermometers and for many industrial applications that have ranges roughly from −30°F to +1000°F.

Thus, by utilizing the thermal expansion properties of materials, it becomes possible to measure *temperature as a function of length or volume.* It is relatively easy also to utilize a bimetal† or a mechanical linkage that converts *linear* expansion into *rotation* of a meter indicator, so that one can measure *temperature as a function of angle of rotation.*

Other temperature-sensitive properties of materials have been found useful. Most materials become soft and finally melt when heated. The metals and ceramic industries use rodlike cones of alloyed materials, which droop at known temperatures.

A more common device, the thermocouple, produces an *electric voltage* that varies with temperature. The principle is based on the Seebeck effect, discovered in 1821, and is illustrated in Fig. 9.4.

† A bimetal is made up of two strips of metal joined along their length. The two metals have differing expansion coefficients, so that as one of them expands more than the other the strip tends to bend. They can be designed to twist and untwist in the form of a helix. The expansion of a metal to length L would be described by an equation such as $L = L_0(1 + \alpha \Delta T)$, where L_0 is the initial length at some reference temperature T_0, ΔT is the difference in temperature between T and T_0, and α is the expansion coefficient for that particular metal expressed as per-degree of temperature.

When two wires made of different metals are joined at one end and heated (hot junction), and the other ends are connected to a suitable voltmeter that is left at room temperature, an electric current will flow and the meter will give a reading proportional in some way to the *difference of temperature between the hot and cold junctions*. Many special alloys have been developed for use as thermocouple wire. Still another useful property is the increase of *electrical resistivity* of metals when heated. In this case the resistance wire that becomes heated is connected to a battery source and an electric ammeter, and the ammeter reading goes down as the temperature goes up.

The instruments that utilize the thermoelectric properties or resistance change for measuring temperature vary greatly in price, depending on the accuracy that is needed. For measurements that do not require better than ±5 percent accuracy, a system involving a simple thermocouple and meter may not cost more than $20. For industrial uses that require accuracies of ±0.2 percent (or roughly 25 times higher accuracy), there will be needed expensive potentiometric recorder systems that cost over $400 (roughly 20 times more expensive). Laboratory measurements that require ten times better accuracies also require equipment that is ten times more expensive, not to mention the very much higher costs in time and skilled technicians to perform good measurements.

The preceding discussion may suggest that since an increase of temperature produces so many changes in properties of materials, we have wide choice as to which of the changes is utilized as a measure of temperature. There are, however, some severe requirements that must be met before changes in properties can be made useful. For example:

1. The change in property must be sufficiently large to be measurable. That is, the coefficient of expansion of the metal rod must be high enough to produce a *measurable increase* of length for each degree of temperature rise. In the case of the thermocouple wires, while all pairs of metals show some thermoelectric effect, only a relatively few pairs produce enough voltage to be measured easily and dependably.

2. Each of the materials and techniques is useful over a limited *range* of temperatures. The metals can melt or suffer some other change when subjected to temperatures beyond the useful range.

3. Over the useful range, the effect of temperature change must be fairly *uniform*. That is, there cannot be discontinuities. For example, some pairs of metals give very irregular voltage change with uniformly increasing temperatures, which would require a very nonuniform temperature scale for the meter. Also, although materials expand with temperature whether they are in solid, liquid, or gaseous state, there are discontinuities as the material goes from solid to liquid state and from liquid to gaseous state. For this reason the temperature range of such materials must usually be limited to the solid state, or to the liquid state, or to the gaseous phase without crossing from one to the other in the same instrument.†

Still another method for measuring temperature utilizes the fact that at high temperatures (400°F and above) metals and materials radiate heat energy at a rate that is some function of increasing temperature. Measuring the emitted radiation with *radiation pyrometers* can give a fairly good measure of the temperature when properly calibrated and properly

† This may not be altogether so if the instrument is designed to work on vapor pressure.

used. Also, at still higher temperatures (1200°F and above) the hot metals and materials will have a color which again is a function of temperature and can be measured with an *optical pyrometer.*

9.7 Calibration of instruments

Whether one is measuring with a yardstick, a micrometer, a thermocouple, optical pyrometer, pressure gauge, stopwatch, or any other device, the instrument or instrument system must be calibrated against some standard. We have already discussed the standards for measurement of time and of length. It will be instructive to discuss the calibration of temperature gauges.

Some time was required to establish temperature standards, since we lacked such obvious cycles as the day and night and the year to influence our thinking. But as the study of materials progressed, it was discovered that nature has provided some fairly useful temperature check points in the form of the melting (or freezing) point of solids and the boiling point of liquids. For example, when ice approaches its melting point from a lower temperature, the temperature will hold to one value until all the ice has melted, after which the temperature will rise again as heat is added. During that transition, a great deal of extra heat (latent heat of fusion) is absorbed by the ice to melt it. If the heating is continued to the boiling point, the temperature of the water-vapor mixture will again hold steady until all the water has become vapor. Again a great deal of heat energy (heat of vaporization) is needed to convert the water to vapor.

Because of these well-defined characteristics, and because water is so important and common on earth, it was natural to adopt the melting point of ice as 0°C, and the boiling point of water as 100°C, to establish the centigrade temperature scale. The 0°C point corresponds to 32°F of the Fahrenheit scale and to 273.16°K on the Kelvin scale. The latter has strong foundations also in the thermodynamics of an ideal gas. The divisions between 0°C and 100°C are made uniform.†

This scale of 100°C between freezing and boiling points of water can be extended to much lower and much higher temperatures by using the melting and boiling points of metals, gases, and various chemical compounds. In general, these are useful because they also exhibit well-defined temperatures and require substantial latent heat of fusion and vaporization at the transition points, so that many laboratories around the world can check each other's work. For example, the boiling point of nitrogen is one of the fixed points at low temperatures (-195.8°C), while the melting points of silver (960.8°C), chromium (1890°C), tungsten (3370°C) are useful for the higher temperatures.‡

Considerable care must be exercised when temperature measurements with an accuracy of ±0.1 percent are required. We shall outline the procedure involved in checking the melting point of ice, in order to give some idea of the precautions that have to be taken.

---

† The history of the thermometer provides an interesting aspect of the development of temperature measuring techniques. See Wolf in References.

‡ To convert °C to °F, multiply the centigrade degree number by 9/5 and add 32. The two scales are equal at -40°, the freezing point of mercury. Temperature scales have historically been named after their originators (Reaumur, Fahrenheit, Kelvin). The temperature scale commonly called *Centigrade* is now called *Celsius* after its inventor. Fortunately, there is no need to change the abbreviation C.

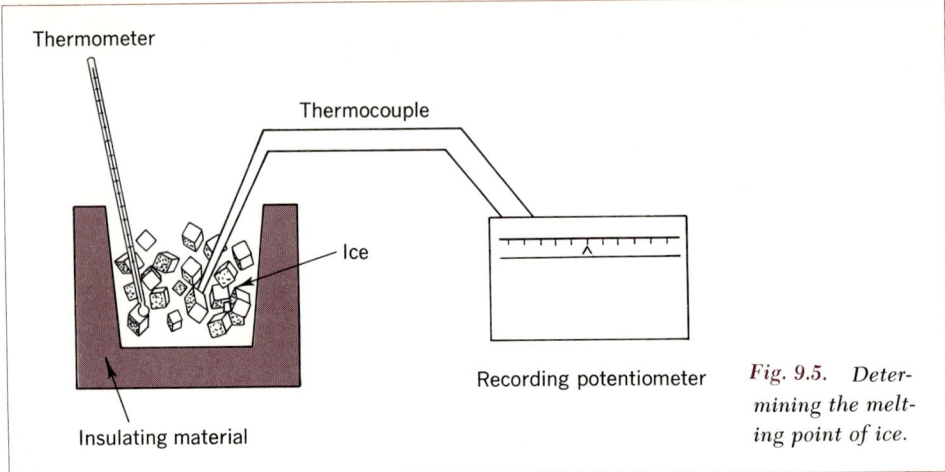

Fig. 9.5. Determining the melting point of ice.

Figure 9.5 shows an ice container placed inside a large container made of insulating material to reduce the rate of melting of the ice. If the ice melts too rapidly, differences of temperature will develop within the ice itself. In the illustration shown, the thermometer and the thermocouple conduct some heat from the atmosphere to the ice; this can cause errors by introducing a temperature gradient (local variation of temperature) just where we least want the gradient (at the point being measured). This requires that the thermometer be eliminated, if possible, and that the thermocouple wires be made of fine gauge (thin) wire. Stirring the ice and water a little helps to make the temperature uniform. If all goes well, the recorder will record a temperature change with time, as shown in Fig. 9.6. The long plateau of the curve is the latent heat of fusion absorption period and

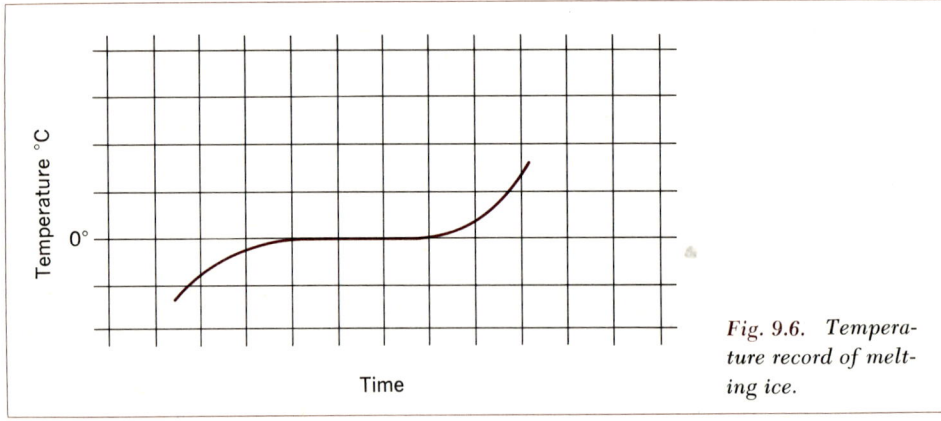

Fig. 9.6. Temperature record of melting ice.

defines the zero melting point. It will be necessary to add a correction for barometric pressure, however, when the atmospheric pressure is higher or lower than the standard atmosphere.

9.8 Altering the length being measured

In Sec. 9.5 we assumed that the metal rod we were examining was firm enough to resist being squeezed to a smaller diameter by the micrometer when measured carefully. Suppose that instead of the metal rod, we measure the thickness of a very soft rubber eraser. We find that as the micrometer spindle approaches the eraser, we have difficulty in knowing when it just touches the eraser, which, being soft, compresses very easily under the pressure between the spindle and the anvil. This is a case where it is difficult to measure the thickness of the eraser without disturbing the very dimension we are trying to measure. It is a common problem in measurement. When we insert a thermometer into a hot solution, we reduce the temperature of the solution by the amount of heat required to bring the thermometer up to the temperature of the solution. When we connect a meter to a thermocouple to measure its voltage, the voltage drops because the meter draws current, and the flow of that electric current through the thermocouple wires causes a voltage drop at the terminals of the wire.†

Since so many measuring processes cause changes in the variable being measured, it becomes necessary for every technician, physicist, biologist, census taker, psychologist, economist, government official, politician, theologian, or whoever attempts to measure or evaluate a phenomenon to ask himself, "Is my measurement (or evaluation) of the situation a true measure of what was there before I entered the picture, or did I change the very parameter I was trying to measure?" He should also proceed to determine, if possible, just how accurate his measurement is.

9.9 Significant characteristics of measurement schemes

Whether we are dealing with the sense of vision, or smell, or with a device that measures the wear of tires, certain characteristics become important to know for each measurement scheme. Among them we can generalize with the following:

1. Variables to be measured. What are the specific variables that the system is designed to measure or to detect? These may be spatial dimension, weight, velocity, charge, odor, and so forth.

2. Measurement versus detection. Is the system designed to give a quantitative measure of the variable, or simply to detect differences between variables or between a variable and a standard? Judging the height of a man or the quality of a voice might be in the first category, while comparing the height of two men of nearly the same height could be in the second category.

3. Range of variable. Over what range can the variable change without going beyond the capacity of the instruments to measure the variable?

4. Linearity of range. How uniformly effective is the measurement system over the range within which the variable may be found? For example, are the dial

† When we study electricity we shall learn that wires have resistance to the flow of electric current, and the product of this resistance and the current represents a voltage difference (or loss) which subtracts from the voltage developed by the thermocouple.

readings of its meter or gauge crowded at one end of the range? Sometimes the crowding at one end is desirable; sometimes the linear calibration is preferred.

5. Sensitivity. How small a change in the variable can the system detect or measure? How does this sensitivity vary over the range of the variable? How large is the *dead zone* within which the instrument does not respond?

In the case of some physical measurement systems, there can be response to changes that are of the order of 10 down to 0.1 percent of the value of the variable. In the 1830's, the German scientist E. H. Weber considered this subject in connection with the sensory system. What is the *just noticeable difference* that a person can detect through the sensory mechanism? For example, if someone is holding a weight of 1 lb, what addition or subtraction of weight will be just noticeably detectable? It is found that this difference is one-thirtieth of the amount being carried. That is, a change in weight of about a half-ounce can be detected if the original weight is 1 lb, but the just detectable change would have to be 3.33 lb if the original weight were 100 lb. This may be expressed as $\Delta W/W = $ constant, the constant being somewhat different for each of the sensory mechanisms. This equation, with different constants, holds for most of our sensory experiences, but not when the stimulus is very small or very intense.†

6. Reproducibility or precision. What are the errors in reading when the variable returns to a particular value at different times or from different directions?

7. Speed of response. How rapid are the changes in the measured variable likely to be? Is the measuring system equal to this in speed of response? What are the likely time delays and errors?

8. Fatigue effects. What are the errors likely to be due to continuous use and fatigue?

9. Errors from ambient changes. What are the errors that may develop from effects of changes in humidity, ambient temperature changes, power voltage fluctuations, and similar variables?

10. Errors from effects of other variables. What other variables are likely to affect the accuracy for measuring the selected variable? For example, if the measured variable is the intensity of light, will the use of different light sources change the frequency spectrum of the light and thereby vary the response and the calibration of the system?

11. Effect of measurement of the variable. Does the process of measuring the variable have an effect on the variable? For example, does the process of measuring a small signal by connecting a meter to the system introduce a change in the measured variable? This becomes more and more serious as the measured variable becomes smaller and smaller in energy content.

12. Background level or "noise." What is the magnitude of the background "noise," or static, that will interfere with the measurement of small signals or small changes of variable? Every system intended to convey information or to measure a variable has within it a certain level of erratic fluctuation called *noise*, which causes troublesome interference with low-level energy signals.

9.10 Psychological measurements

Thus far we have considered only measurements that involve physical quantities and mechanical or electromechanical instruments. These are not the only

† See Graham in References.

measurements that are required, however. For example, there are wide variations in the skills, aptitudes, and interests found among professional specialists, businessmen, mechanics, writers, artists, and unskilled workers. There are comparable variations in the mental and physical capabilities of students. We have found that a nation's economic, social, and technological progress requires that each individual be helped to achieve the measure of education and training that will help him to perform with highest skill the job for which he has interest and ability. But this requires that there be some means for measuring, or at least estimating, the degree of intelligence, interest, aptitude, and special skills. This has been difficult to do, but some progress has been made, beginning with the work of Alfred Binet in 1905.

Binet was asked by the Paris School Commission to devise a method for determining children's scholastic ability and aptitude. He, with his collaborator Simon, devised a series of tasks, such as giving definitions of words and performing computations, each of which increased in complexity. He thus constructed tasks that could be completed by two-thirds to three-fourths of a given sample of children chosen according to chronological age. A child of chronological age six who could complete all the tasks that were prepared for a child of age six would be considered to have a mental age of six. If he passed the seven-year old test, the child would be said to have a mental age of seven. William Stern, a German, developed the idea of taking the ratio of the mental age achieved on the test to the child's chronological age at the time of the test. This ratio was called the *intelligence quotient* (IQ), which, until recently, was thought to remain constant. Should a six-year old perform only the tasks developed for a three-year old, his IQ would be $\frac{3}{6}$, or 50; but a three-year old child who performed the tasks that were developed for the six-year group would have an IQ of $\frac{6}{3} = 200$. [The terms *mental age* (MA) and *chronological age* (CA) are frequently used in this connection.]

Binet tested many children when developing these tests and then compared the results with the performance of these children at school work and with the opinions of their teachers. The results seemed to be satisfactory in that the tests seemed to be a valid index of the mental ability to do school work. Since Binet did his work, the tests have been developed further by the Americans L. M. Terman and D. Wechsler. Despite improvements, the predictability of scholastic and other achievements from these test results is accurate in only about six out of ten cases over the entire population. The value of intelligence testing lies in its application to a large population and not to any particular individual.

During World War I there was need to place millions of draftees of the United States Army in various schools. Psychologists developed a series of paper and pencil tests to measure the mental ability of the soldiers, called the *Army Alpha* (verbal) and *Army Beta* (nonverbal) tests. As a result of this experience it became clear that socioeconomic status, linguistic ability, familiarity with the material and questions of the test, attitude toward being tested, fatigue, and other influences controlled the results of the test.

During World War II specific tests were devised to predict *aptitude* for specialized jobs such as radio operators. This turned out to be more successful. That is, instead of a general intelligence test, tests designed with specific situations in mind tend to have higher validity and reliability than more general tests.

Psychologists and sociologists have learned that each test must be standardized for the particular people to be tested if it is to be suitable and productive. A test that is standardized for children of a high socioeconomic level is not likely to be suitable for children who come from a substantially lower economic level. A test designed for selection of employees for a department store will not reveal the capabilities of employees for totally different work situations.

9.11 Vision

We noted earlier that our sensory systems provide the medium through which we have awareness of our surroundings. It will be useful to consider one of these in a little detail in the light of the earlier discussion.

While all our senses are constantly "telling" us something about our immediate environment and all perform important functions in our daily life, most people would agree that sight and hearing are the two senses that provide the greatest amount of communication with our surroundings. Of the two, sight (or vision) is probably the sense on which we depend most heavily.

There was an ancient Greek idea that the eye, like a searchlight, sends out a beam which touches objects and thereby reveals the object to the observer. We know now that it is light which emanates from outside sources that reaches the eye, and what we "see" is the result of the effects of that light on the eye. The eye is a very complex organ which not only senses the nature of the light that falls on it, but also provides its own movements to enhance its ability to respond to the variability of the light environment as well as to detect motion more effectively.

The eye has been compared (rather erroneously) with a camera because the camera also has a lens that focuses an image on a light-sensitive detector (a photographic film); in both the eye and the camera lens the image is inverted. But in the camera the lens is usually of fixed focal length, whereas the lens of the eye can thicken to change focal length. There are, of course, many more differences than similarities between the two.

PHYSICAL CHARACTERISTICS OF THE STIMULUS

We shall have frequent occasion to refer to light as a moving stream of particles, or photons. That is, certain experiments show light to behave as though its energy were concentrated in small particle form or in small bundles. In Chapter 11, we shall see that light is an electromagnetic radiation which shows interference and diffraction phenomena as do other waves. This wave-particle duality in the nature of light will be discussed in Chapter 17.

One property of any wave is its wavelength, λ, and frequency, f. On the basis of the measured values of λ for light, we can say that its *spectrum*, or *range*, of wavelengths is visible to the eye between 0.39μ and 0.70μ.† For most people the sensitivity of the eye varies substantially over this range, however, with the higher sensitivity in the yellow-green (0.55μ) region of the spectrum.

We ascribe "colors" to the various wavelengths of light. The sensations we receive from light at 0.4μ is known as violet; 0.48μ, blue; 0.55μ, green-yellow; and 0.7μ, red. However, it must be remembered that a 0.7μ wavelength is not in itself red; rather, when the eye is stimulated with light of this wavelength

† The symbol μ represents the micron; $1\mu = 10^4$ Angstrom (Å) units (Abbrev. 10^4 Å).

we experience the sensation to which we assign the color red. In addition to wavelength, the other property that we are concerned with is the intensity of the light, for the intensity of the stimulus correlates with the brightness of the visual experience. The usual way of measuring the intensity of a light source is to compare it with the intensity of a "standard candle," which is defined as having a *candlepower* of 1 unit. An average 100-watt electric light bulb has a rated candlepower of 1630.

How much light is needed to provide a minimum detectable stimulus to the eye? For this question, we may consider light as formed of tiny particles called *photons*, or *quanta*. The photon represents to the physicist the amount of energy carried by the light. Psychological studies of human perception indicate that as few as ten of these photons in the visible spectrum can produce a sensation in the human eye.†

PHYSIOLOGY OF THE EYE

A brief review of the structure of the eye will be useful for discussing its functioning. Figures 9.7 and 9.8 give some details of the eye structure. Light enters the eye through the cornea, the transparent outer surface of the eye. The iris is similar in function to the diaphragm in a camera, regulating the amount of light that enters the inner eye. When the intensity of light incident on the cornea is high, the iris closes, allowing only a portion of the incident light to pass through the pupil into the inner eye. When the light level is low, the iris

† See Graham in References. More recent work suggests that perhaps a single photon may be detectable.

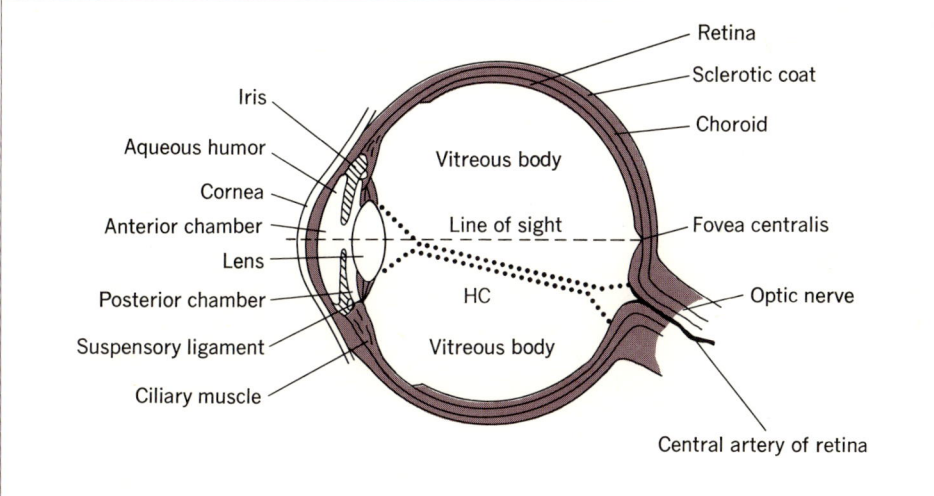

Fig. 9.7. Horizontal section through the eyeball. The dotted lines HC mark a narrow canal in the vitreous body (hyaloid canal) which, in the embryo, lodged the hyaloid artery. (Adapted from Best & Taylor, 1965.)

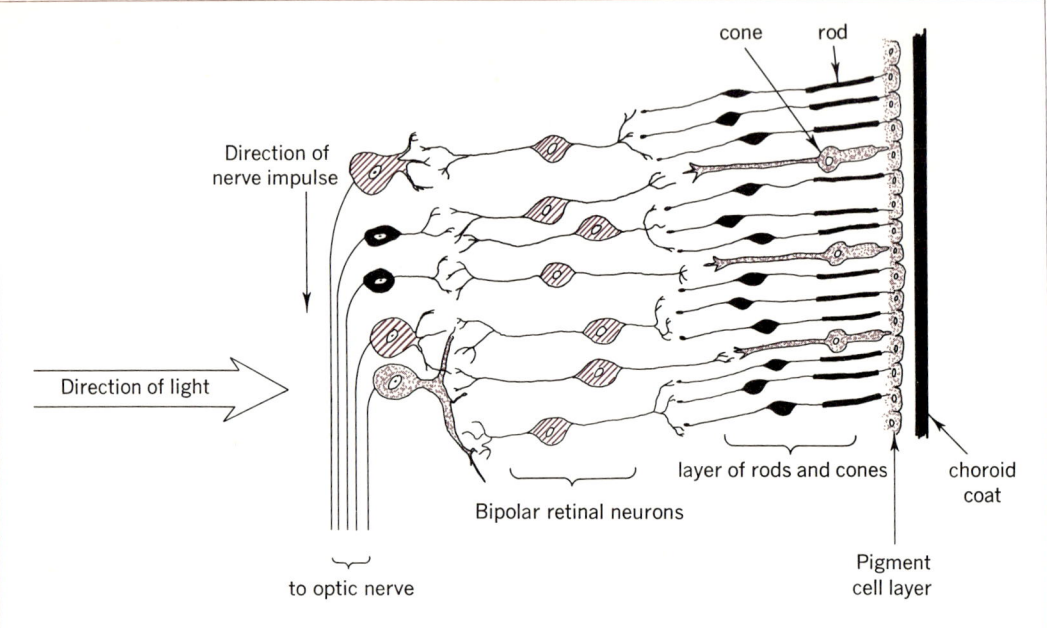

Fig. 9.8. The layers of the retina. Recent evidence suggests that there are many more interconnections between the cells of the retina than is indicated in this illustration. These interconnections help to organize the stimuli to produce some sense of form and motion in the message that is sent to the brain. That is, the eye itself undertakes some prior interpretation of the stimuli.

dilates to allow a much larger fraction of the incident light to reach the inner eye.

After passing through the pupil, the light is brought to focus on the *retina* by a lens. The focal length of this lens is varied by changing its shape through the action of the ciliary muscles. Thus we are enabled to adjust quickly to see either distant or nearby objects as required.

The retina is the photosensitive element of the eye. The astronomer Kepler was one of the first (1604) to note this. In 1619, the German ecclesiastical scientist Christoph Scheiner removed the outer coating from the back of the eye of an ox to demonstrate that the image on the retina is inverted and not erect. This raised the question of how it is possible to see, for example, a man standing upright when his image on the retina is upside down. There was also the question of how it was possible to perceive depth (to see that one object is farther away than another object) when the images of all objects fall on a flat two-dimensional surface. This led to controversy and experimentation in visual perception and became one of the first topics studied by the science of psychology.

Gottfried R. Treviranus, a German naturalist, reported (about 1835) that the retina is not composed of homogeneous layers of identical cells but that it contains specialized cells which constitute the photoreceptors, of which there are two types. One type, the cone, receives its name from its conical shape. There are nearly $6\frac{1}{2}$ million cones in the eye, most of which are clumped together in a small region of the eye called the *fovea* (region of clearest vision), although a few are scattered throughout the retina. The long slender *rods*, nearly 100 million, are spread over the periphery of the retina.

The light arrives at the retina first and then passes through layers of nerve fibers and ganglia and other cells to the rods and cones; any residual light is absorbed by a final layer of pigment cells. The dark brown pigment serves to absorb the light that otherwise might be reflected back and forth, much as the black interior of a camera prevents scattering of light.

There is a rose-colored pigment in the outer segments of the rods, called *rhodopsin* (from the Greek *rhodon* for a rose and *ops* for the eye), and also inaccurately called *visual purple*, which is of protein nature. A pigment of the carotene group is linked to the protein. On receiving light, the rhodopsin undergoes a photochemical (bleaching) reaction that divides it into a protein and an orange-yellow pigment called *retinene*. The reaction initiates nerve discharges, which travel by way of the optic nerve and other "switching" centers to the visual centers of the cortex. The products of the photochemical action then are resynthesized to the original rhodopsin (in the absence of light) by way of a reaction involving vitamin A and the protein.

The photochemical reactions are such that the rods detect light without being able to distinguish color, while the cones are able to distinguish color through more complex chemical processes. For this reason the cones are the primary receptors during daylight to provide sensitivity to color, while the rods give perception of shades of gray in low illumination. The vision offered by the fovea is much clearer, while the rods have much greater sensitivity for low-illumination conditions and for detection of motion. The eyes therefore do not represent an instrument with only a single range, but a remarkably versatile combination of several ranges for varying illumination conditions. There is the range with respect to distance: the eye can adapt its focal length to view objects from distances of a few inches to many miles. There is another range with respect to color spectrum, which we have already noted. There is also a range and adaptation with respect to intensity of light, which can vary from very low to very high intensities.

Some psychologists have theorized that if the eyes transmit not the direct image of an object but discrete nerve discharge signals to the brain, there must exist suitable intermediate devices that code and decode the signals to make them equivalent to the image of the object. More recent evidence suggests, however, that the eye itself organizes the light that impinges on it into groupings of stimulation in such a manner as to give an appearance of form and motion perception. The message that is sent to the brain is therefore an organized message. We have noted that the eye is wonderfully adaptive with respect to different situations of light, color, and distance. How faithful is the coding of information from object to image on the retina, and finally to the image that emerges from all this through the brain and interpretive processes?

In this connection we must emphasize the important role of experience and the supporting role of other sensory experience that we enjoy along with sight. The percept one gets via visual experiences is not merely a function of the reception of visual stimuli but also a function of the action that the person indulges in. The "feedback" from acting on what one "sees" helps to shape the percept. For example, the interpretive mechanisms and experience through the other senses seem to provide the necessary reversal to get the inverted image on the retina to be "seen" as right-side-up objects. Experiments with prisms placed before the eyes to reverse yet again the image on the eye (to make the image erect) have shown that people can become adapted, in time, to either situation.

9.12 Visual acuity, accommodation, intensity discrimination, depth perception

The smallest separation that two points may have and still be recognized as separate points by the eye is termed its *visual acuity*. For a normal eye, this limiting angular separation is about 1 minute of arc, corresponding to a separation of about 0.0045 mm on the retina, if a focal length of 15 mm is assumed for the eye. When an object is moved away from the eye and its visual angle becomes progressively smaller, details of form and structure that could be discerned when the object subtends an angle of a minute or more at the near point become gradually imperceptible. The usual way of expressing visual acuity is to compare the vision of the tested person with that of the normal (average) person. If at a 20-ft distance, a person can barely read letters of exactly the same size that the normal person can barely read, that person is said to have $\frac{20}{20}$ vision, which is normal. If the letters must be as large as those that the normal person can read at a distance of 40 ft, then that person's vision is said to be $\frac{20}{40}$. Occasionally a person may show better than normal vision so that he can read at 20 ft what the normal person can read only at 15 ft, in which case his vision is said to be $\frac{20}{15}$.

The lens of the eye is doubly convex. The radius of curvature of the lens surfaces may be changed by muscular contraction for the purposes of focusing. This power of adjustment is termed *accommodation* and the least distance of distinct vision of a normal eye or near point is around 25 cm. The accommodation of the eye varies greatly from one individual to the other, and varies in everyone with age.

The eye, as do other sense organs, not only gives information about the intensity of stimulation which in general we recognize as levels of brightness, but also discerns differences in stimulus intensity and identifies the brighter of two objects. Since the eye can adapt to various ranges of illumination, it detects stimulus differences over a remarkable range of light intensities. The incremental ratio for noticeable differences in intensity, $\Delta I/I$, is referred to as the *Weber fraction*. This fraction is applicable to vision only in an approximate way. At lowest intensities the Weber fraction is around 1, while at highest intensities it may be as small as $\frac{1}{167}$.

Question. Note the similarity of $\Delta I/I$ to $\Delta W/W$ of Sec. 9.9. What are the implications?

Brightness is not directly related to stimulus intensity. Equal intensities of light of different wavelengths are not equally bright. The assessment of the brightness of an object is dependent on the content of the visual field and the background. As noted earlier, a black background makes a white object brighter, and gray is darker against a white rather than a black background. The immediate past experiences of the eye can also influence the perception of brightness. With colored lights, brightness contrast is often accompanied by color contrasts. A gray on a colored background becomes tinged with the hue complementary to the background color, so that against a green background it looks quite different from gray against a red background.

The human eyes estimate the distance of an object by such factors as the size of the image on the retina and by the phenomenon called *parallax*. To determine the distance of an object by the image size, one should have previous knowledge about the actual size of the object. If the object is distant, its image is very small, and if it is nearby, the image is considerably larger. It is a matter of experience that the images of a single object viewed by the two eyes are slightly different. The parallax method of estimating distance depends on this fact, and its working is illustrated in Fig. 9.9.

The eyes can follow the discontinuous movement of objects at a fairly rapid rate until the displacements take place in such short intervals of time that the motion becomes blurred and resembles continuous motion. In the case of cinema presentations, the eye retains and blends the images when the moving picture frames are changed at a rate faster than 24 frames per second. There are similar limits to the perception of separated sounds and to separate touch sensations.

Although we now conclude our discussion of vision, the student should realize that these few paragraphs hardly do justice to this subject.

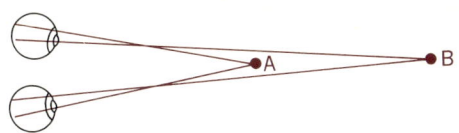

Fig. 9.9. *The two eyes are observing an object A of known size and distance near the eyes and an object B of unknown size and distance farther from the eyes. In the left eye, the image of B lies to the right of the image of A, but in the right eye the image of B lies to the left of the image of A. In other words, the images of A and B are actually shifted on the retinae because of the angles from which the two eyes observe them, and the brain interprets their distances by the amount to which they are shifted. Only the relative distances of the objects and not the actual distances are interpreted this way. However, since the distance of A is known, the distance of B can be estimated.*

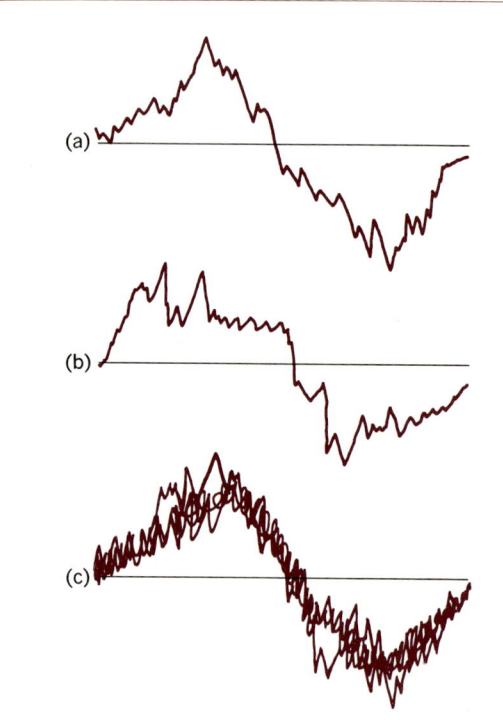

Fig. 9.10. Determining the original signal in the presence of excessive noise.

9.13 Extension of the senses through instruments

We have seen that man's senses are limited in sensitivity, frequency range, and analyzing ability. Instruments have been constructed by man to supplement his perceptions in four ways:

(1) To observe in frequency ranges not available to the senses.
(2) To observe with greater sensitivity than is possible with our senses.
(3) To observe more quantitatively than is possible with our senses.
(4) To detect quantities such as magnetic fields, to which we do not have any sensitivity.

Among these instruments are telescopes, cameras, electron microscopes, galvanometers, geiger counters, microphones, amplifiers, photometers, and spectrometers. With the aid of these devices we can detect quantitatively and make reproducible a much wider range of phenomena than are accessible to human senses alone.

Most instruments have visual outputs: printed data, the position of a needle on a meter, photographs from cameras or telescopes, or electron microscopes. These may be thought of as extensions of vision, so that we often say loosely that we "see" when we use eyeglasses, a microscope, or a telescope, when we use an electron or X-ray microscope, or even when we take an X-ray diffraction pattern and carry out calculations to obtain atomic positions in solids. We use other senses, too, as when we drive a billiard ball and hear the sound of that ball striking another ball. The striking of one ball with another illustrates an important technique of modern physics, namely, scattering technique. In scattering, we shoot particles (which usually are invisible entities) at other particles (which may be individually invisible but may compose a solid substance), and study the angular distribution of scattering of these first particles as a function of incident energy. By such scattering studies we are able to determine the electron distribution of scatterers in atoms, the distribution of nuclear matter, and even the structure of the neutron and the proton.

It was mentioned that many measurements tend to push the capabilities of instruments to such a degree that the statistical fluctuations of atomic particles generate an interfering "noise." Electrical noise generate acoustical noise in our loudspeaker. This random variation becomes important when we try to measure exceedingly small quantities of the same order of smallness as the noise. This noise makes it difficult to interpret information from a signal. We are familiar with the difficulty of understanding a weak radio signal when there is a great deal of noise, or of perceiving an intelligible photograph when film has been greatly overexposed. The reader may have read of the great difficulty in detecting the radar echo from Venus. The scientists who carried out that experiment spent many hours trying to identify the signal of the radar, which seemed to be lost in the noise that their receivers picked up. Complex mathematical means were employed; simply stated, they looked for correlations in signals as a function of time. As an example, assume some signal to vary with angle, as shown in Fig. 9.10(a). Let us run through it again and assume the result is shown in (b). Repeat this over and over again and finally superimpose all these patterns. The result might be something like that shown in (c), in which the repeated superimposition of unclear signals such as those of (a) and (b) tends to produce meaningful information nevertheless.

9.14 Microscope and telescope

As an example of extending the capabilities of human vision, we shall mention briefly the characteristics of the microscope and of the telescope. The apparent size of any object as seen with the unaided eye depends on the angle subtended by the object (see Fig. 9.11).

Fig. 9.11. The angle subtended by the object determines the size of the retinal image. As the object is brought progressively closer to the eye, from P to Q to R, accommodation permits the eye to change its magnifying power and to form a larger and larger retinal image. This is accomplished by the muscles controlling the lens of the eye which either flattens or bulges at its center as the muscles relax or contract.

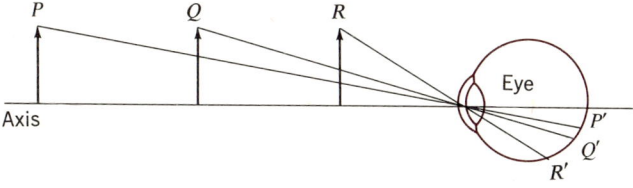

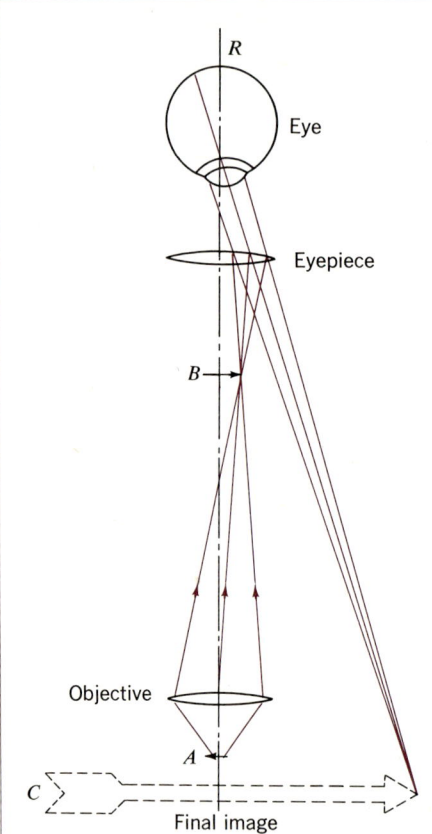

Fig. 9.12. *Principle of the microscope, shown with the eyepiece adjusted to produce the final image at the distance of most distinct vision. The object is located at A just outside the focal point of the objective so that a real magnified image is formed at B. This image serves as the object for the eyepiece. The eyepiece forms a large virtual image at C. This image serves as the object for the eye itself which forms the final real image on the retina R.*

There is a limit to how close an object may be brought to the eye if the latter is still to have sufficient accommodation to produce a sharp image. This limitation can be circumvented by use of a microscope. In its simplest form, a modern microscope consists of two lenses, one of very short focal length called the *objective,* and the other of somewhat larger focal length called the *eyepiece.* Figure 9.12 illustrates how an inverted and greatly enlarged but distinct image is formed by the lenses in a typical microscope.

The telescope is an optical system with which an enlarged image of a distant object may be obtained. The earliest telescope of which we have definite knowledge is that of Galileo, constructed in 1609.[†] The elements of this telescope are still in existence and may be observed on exhibit in Florence, Italy. The principle of the astronomical telescopes of the present day is the same as that of the seventeenth century telescopes. The principle of an astronomical telescope is illustrated in Fig. 9.13.

In general, terrestrial telescopes are similar to astronomical telescopes except that an erect image is obtained by the help of either prisms (as in a binocular) or lens systems.

Microscopes and telescopes[‡] extend the range of vision into regions where the unaided eye cannot probe by itself. However, using lens systems such as in a telescope or microscope introduces certain "side effects." By proper design and

---

[†] Galileo constructed his telescope after hearing about the invention of such a device by a lens grinder in Holland. Galileo is reported to have been first to examine the heavens with this instrument.

[‡] See the experiment on optical systems in the Laboratory and Mathematics Supplement.

construction, many of these are reduced or remedied in costly microscopes and telescopes.

9.15 Concepts of time

Vision and hearing, combined with the other senses and with experience, give us a fair sense for space and for changes in spatial organization. There are within us life processes that are time-dependent, such as breathing, the beating of the heart, and the cycle that demands more food. We often awaken at the same hour. The story is told that in one area a cannon was fired each morning at 8:00 A.M. The inhabitants had become accustomed to the firing, and slept through it. But when one morning the cannon failed to fire, one man jumped from sleep promptly at 8:00 A.M. with the cry, "What was that?" However, we lack a *sense* of time that is in any way comparable to our other senses.

The universe is similarly replete with cyclic phenomena that are time-dependent. The daily cycles of the sun that bring night and day with regularity, the longer cycles that spell the periods of nature's death and rebirth with each fall

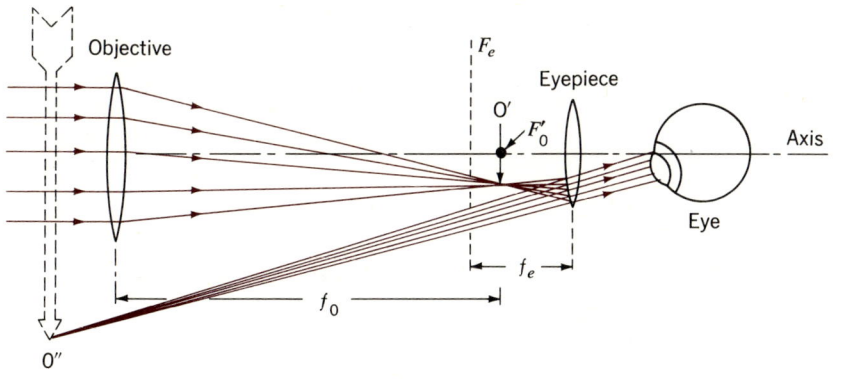

Fig. 9.13. Principle of the astronomical telescope, shown with the eyepiece adjusted to give the image at the distance of most distinct vision. Light from a distant object enters an objective with a long focal length and is focused to form a point image at O'. If the distant object is assumed to be an upright arrow, the image is inverted and real. If the eyepiece is adjusted so that the real image lies just inside its primary focal plane, F_e, a magnified virtual image at O'' may be seen by the eye at the near-point. Usually the real image is made to coincide with the focal points of both lenses with the result that the virtual image is at infinity. The final image is always the one formed on the retina by the rays which appear to come from O''.

and spring—each has had dramatic effect on the habits, hopes, and fears of man since time immemorial. But in the past there seemed to be small connection between the regularity of heavenly bodies and the processes of living and survival. During the day, the movements of the sun were related to changes in the position of the shadow of a tree or of an erect stick, which ultimately produced the sundial. But the movements of the sun did not conform to the uniform subdivisions that the Egyptians drew for their sundial. Time did not flow evenly, but seemed instead to demand a different length of scale division of the sundial for each hour in the course of the day. The Egyptians tried to find devices that would conform to the different lengths of each passing hour, and which would solve the problem of variation of the total length of the day and of each hour with the change of seasons. Their examples were adopted by the Greeks, including also the uneven subdivision of the sundial and uneven duration for each hour. The Babylonians and the Chinese used the sundial, but they measured time also through the uniform flow of water. This gave them two kinds of hours, one of uniform duration and another that changed through the day. The pendulum and the clock that maintain uniformity are relatively new inventions, and so is the concept of time as we now accept it.

Newton thought of time as a stream that ". . . of itself, and from its own nature, flows equably without regard to anything external, and by another name is called duration."† This view of time, as flowing evenly forward and providing a common coordinate to which all events can be referred to make them functions

† Isaac Newton, *Principia; Scholium to the Definitions*. The interested reader is urged to consult Schlegel in References.

of time, appears to have existed from the Greek period. It carried the notion of *absolute time*, which exists and persists independently of any other change or absence of change in nature. It implied the existence of a great master clock that marked its own undeviating course with complete indifference to the events that came and passed away. The view was quite generally held until a few decades ago.

We understand now, however, that the existence of time is related to *change*. That is, time "moves on" only because there are changes that occur in nature as nature passes from one state to another. It is observation of the succession of different states of nature that causes us to assign differences of time, and thereby flow of time, in close association with the changes.†

Physical changes establish our time base, especially changes that are of cyclical or periodic nature, such as the rotation of the earth on its axis and the changes of the seasons. The pendulum clock, electric clock, and vibrational characteristics of materials and systems offer other gauges of time change. The vibrational frequencies of electron motion within atoms, or atomic clocks, give us our most accurate measure, which is more dependable than the earth's rotation for accuracy.

If time is determined by a succession of states, does it also have direction? Do we think of the world and the events as having *forward* direction? What constitutes movement in a forward direction

† Question: If the flow of time is determined by changes of state or of phenomena, what do we mean when we say that a phenomenon is *time-dependent*? Also, discuss the circularity of argument when we say that cyclic or periodic changes (such as the movements of a clock) both measure and are measured by time.

versus backward direction? We say that the states that have occurred are in the past, while those that have yet to occur are in the future. But what distinguishable differences are there between a state and succeeding states?

A little later we shall learn that changes in entropy (see Sec. 10.16) identify the direction, and according to the theory, *Successive later states of the system are states with successively greater total entropy values.*

While absolute time and space may have little or no meaning, the measurement of increments of time, and of motion or change of some sort within that increment of time, are very important for our daily living and for learning the laws of nature. Our senses are moderately sensitive for detecting changes. We are subject to many possibilities for misinterpreting phenomena that take place more rapidly or much more slowly than those to which the senses can respond. For this reason we are forced to depend largely on instruments and instrument systems to detect changes. The intervals of time that we must distinguish may be discrepancies of a few seconds in planetary cycles that are centuries long. In the more routine processes we are satisfied with measurement that involves miles per hour or centimeters per second. In atomic work it sometimes is necessary to measure movements that take place in 10^{-9} sec.

9.16 Summary

The earlier chapters should have made clear how important it has been for the human race to achieve the ability to make careful observations and measurements of natural phenomena. In this chapter we have given a picture of the difficulties that beset our efforts to make accurate, dependable observations and measurements, in a manner that can make them useful to other people as well.

Every measurement program must look first to the capabilities, limitations, and overall suitability of the instrument techniques that are available for making the measurement. What accuracy do I require? What instrument types are available? Can I afford to pay for the best? What are the alternatives? Will the instrument technique I would like to use cause a change in the parameter I am trying to measure? Does the instrument technique permit calibration of a type that will make my data meaningful to my colleagues in other nations? These are questions that the investigator must face.

We have noted that even in the so-called "exact sciences" (physics, for example), human judgment plays a major role and the science is therefore subject to major errors. On beginning any investigation, each of us is subject to a bias that comes from the nature of earlier training. This reveals itself in the selection of approach to the study. When the taking of data involves the human sensory system, there is often present a "mental set" that can bias the observations.

We have given simple examples of how the significance of the data one takes can be given some evaluation in the presence of "chance errors;" but the presence of "systematic errors" is even more serious and has no easy identification. It is clear, however, that every experiment requires analysis and evaluation of the individual errors that have prejudiced the result.

We found that many effects of temperature change can be utilized to construct systems for measuring temperature itself. The successful use of such effects requires that the sensitivity, uniformity, and range characteristics be suitable. In this connection the ability to calibrate accurately is important.

Since sensory observations are ever present in measurement work, we have introduced some details of vision and the physiology of the eye. We have shown how, through the microscope and the telescope, it is possible to extend the capabilities of the eye by many magnitudes. But with each gain of one type, there is a corresponding loss of some other aspect of measurement.

Finally, we have interjected some philosophical questions with respect to time, the direction of time, and absolute time. In the later chapters, we shall look for some explanation of the theory: "Successive later states of the system are states with successively greater entropy values." The topics introduced in this chapter will continue throughout the remainder of the course.

Questions/Discussions

1. A field has the dimensions of 410 × 320 ft. A man estimates the dimensions to be 400 × 300 ft.

    (a) What is the percentage error of his estimate for each dimension?
    (b) What is the actual area of the field?
    (c) What is the area of the field on the basis of his estimate?
    (d) What is the percentage error in the estimated area?
    (e) Show that the percentage error in the estimated area could have been approximated by adding the percentage errors of the factors needed to compute the area without ever actually finding the area.

2. What are the "gains" and the "losses" when one uses a telescope, as compared with the capabilities of the eye?

3. Answer Question 2 with respect to the microscope.

4. Consider the experiment of Fig. 9.5. If the record shows a continuous rise in temperature, without the horizontal plateau of Fig. 9.6, what could have been at fault?

5. Although the Celsius (centigrade) scale and the Kelvin scale are most often used in scientific work, the Fahrenheit scale is more popular in everyday use. Convert +10°C, +200°C, −100°C, −273°C to Fahrenheit and Kelvin temperatures.

6. Suppose that ten micrometer measures of the thickness of a soft eraser give the following readings: 1.020, 1.032, 1.040, 1.050, 1.035, 1.005, 1.002, 1.028, 1.044, 1.058. Determine the mean; compare the magnitude of the standard deviation to that of the metal-rod measurements and explain the similarities and the differences.

7. How does the measurement of intelligence differ from the measurement of a person's height and weight in terms of

    (a) The standards used. Are they direct measures or indirect measures?
    (b) The operations by the measurer?
    (c) What the measured person does?
    (d) Possible sources of error?

8. What are the scientific justifications for grading school children on a normal curve?

    (a) What are its good features?
    (b) What are its bad features?

(c) Has "science" really proved that the only valid and reliable way of measuring school children is on a curve?

(d) Are the implications for the use of the curve different when applied to school grades than when applied to natural phenomena?

References

Chauncey, H., and Dobbin, J. E., *Testing: Its Place in Education Today*. New York: Harper & Row, 1963. *A relatively nontechnical survey of achievement testing. The appendix includes a defense of multiple choice tests by Eductional Testing Service.*

Ebel, R. R., and Damrin, D. E., "Tests and Examinations," in *Encyclopedia of Educational Research*, C. W. Harris (ed.). New York: Macmillan, 1960. *A good summary of achievement testing. Concludes in favor of multiple choice test but gives reasons why the essay test is here to stay. Bibliography of 133 references included.*

Graham, C. H., *Vision*. New York: Wiley, 1965.

Luchins, A. S., *Examination of Rigidity and Flexibility*. New York Regional Office of the Veterans Administration, 1949.

Mueller, Conrad G., *Light and Vision*. New York: Time Inc. (Life Science Library), 1966.

Parsegian, V. L., "Pyrometry for the Ceramic Industry," *Am. Ceram. Soc. Bull.*, **27** No. 1 (January 1948), pp. 1–24.

Ripley, J. A., *The Elements and Structure of the Physical Sciences*. New York: Wiley, 1964. *Chapter 2, "On Size, Shape and Measurement," is particularly recommended.*

Schlegel, Richard, *Time and the Physical World*. East Lansing: Michigan State Univ. Press, 1961.

Temperature, Its Measurement and Control in Science and Industry. New York: Reinhold (AIP), 1941.

Wertheimer, Max, Translation by Willis D. Ellis, *Source Book of Gestalt Psychology*. New York: Harcourt Brace, 1938.

Wilson, E. Bright, Jr., *An Introduction to Scientific Research*. New York: McGraw-Hill (paperback), 1952.

Wolf, A., *A History of Science, Technology and Philosophy in the 16th and 17th Centuries*. New York: Macmillan, 1935. See Chapter XII, "Meteorological Instruments."

Young, Hugh D., *Statistical Treatment of Experimental Data*. New York: McGraw-Hill (paperback), 1962.

CHAPTER TEN

Heat and Thermodynamics

A theory is the more impressive the greater the simplicity of its premises is, the more different kinds of things it relates, and the more extended is its area of applicability. Therefore the deep impression which classical thermodynamics made upon me. It is the only physical theory of universal content concerning which I am convinced that, within the framework of the applicability of its basic concepts, it will never be overthrown (for the special attention of those who are skeptics on principle).

ALBERT EINSTEIN†

IN CHAPTER 5 we noted that as a consequence of friction, the kinetic energy associated with a moving body becomes transformed into *heat*—a term we did not define, but whose intuitive meaning was assumed to be clear. In Chapter 7 we again made note of a transformation of kinetic energy to heat, in reference to the deceleration of a space capsule, due to atmospheric drag. In this chapter we consider the phenomena of "heat" and temperature more carefully. We shall see that they are closely related to the kinetic energy of moving particles—the atoms and molecules that make up all matter. We shall see further that the number of these moving particles is so large that the mechanics we have introduced so far are inadequate to treat their motions. Instead we shall have recourse to the statistical techniques of Chapter 8.

10.1 Early concepts of heat and temperature

Judging by the effort and expense involved in protecting ourselves against cold and heat, and by the interest the state of the weather commands in conversations, we are exceedingly conscious of temperature. In a later chapter we shall see the reasons for this concern in terms of our body needs. It was not different for people of the early civilizations; for them, too, procurement of shelter and clothing was a major interest. Fire was needed for warming the body against the cold of the night and of winter. Fire was needed also for cooking of foods. It was no accident that even at the risk of eternal torture, a Prometheus was needed to steal fire from the boundless resources of

† From Vol. 1, p. 33, of Autobiographical Notes, *Albert Einstein, Philosopher-Scientist;* edited by Paul A. Schilpp, Harper Torchbooks, 1959.

10.1 Early Concepts of Heat and Temperature

the heavens and to bring it to earth for man's use. Fire has a mystical, attractive quality, which made it an object of veneration or near-worship in religions such as Zoroastrianism. The household hearth's fire was sacred to the Greeks and Romans. To some philosophers (for example, Heraclitus) the fundamental element, basic to all matter, was fire. To other Greeks, fire was one of the four basic elements (earth, water, air, fire). The four basic qualities of the elements (hot, cold, dry, wet), which they identified with the four elements, seem to be particularly related to temperature sensibilities. Picture what they learned (and what we may learn) from observation alone when watching a campfire on a cold night. Fire represents a state of high temperature from which smoke and heat spread in all directions. The heat warms us as we watch. Other objects exposed to fire also become hot.† It appears that something comes out of the fire and passes into the surrounding objects. In fact an object like wood may itself become hot enough to burst into flame and become fire. It is no wonder that confusion about the significance of heat persisted until long after Newton.

To understand the nature of heat we must understand the nature of matter. The concept that matter is made up of small, indivisible and eternal "atoms" was put forth by the atomists of ancient Greece—most notably Leucippus and Democritus.‡ The view was taken over by the Epicureans,§ but was rejected by Plato and Aristotle, partly because common experience found no evidence of atoms. Nearly two millenia were to pass before this concept was revived in the seventeenth century by such men as Francis Bacon, Descartes, Gassendi, and Newton.

One barrier to the acceptance of an atomic picture of matter lay in the fact that many persons associated it with the doctrines of the hedonists Epicurus and Lucretius, who used the cosmological approach of the atomists because it fitted in with their rejection of the notion of creation and divine intervention in human affairs. Newton, a deeply religious man, was quite concerned about this aspect of the atomic model. Newton's solution to this problem was that God had created the "world machine," but from then on He let it run according to the "laws" of nature, without continual intervention. The mechanistic philosophers of the eighteenth century believed that if we only knew all the laws of nature, we could determine the future of the world. Some dared to suggest that if Newtonian laws of mechanics were applied to analyze the behavior of atoms, it would be possible to predict the whole future course of the atoms, and hence of the universe.

The modern view of the atomic structure of matter is the product of the past 150 or so years. In 1808 John Dalton published *A New System of Chemical Philosophy,* in which he put forth a *model* of the structure of matter to satisfy the experimental chemistry of his time. He proposed that all matter is made up of a very large number of very small particles, called *atoms,* incapable of being subdivided. According to Dalton, atoms are of different types and weights, and atoms of one type and weight constitute one of the *elements* of nature. The atoms (elements) differ in atomic weight and in chemical properties. Elements can combine in various groups to form "com-

---

† The reader is reminded of the discussion of the sensations of hot and cold in Sec. 9.6.
‡ Democritus taught: "The only existing things are atoms and empty space; all else is mere opinion."
§ The Roman Lucretius (First Century B.C.) expounded the view of atoms in his poem "De Rerum Natura" (On the Nature of Things).

pound atoms," the groups usually involving small numbers of atoms and simple ratios of atom types. An atom, he felt, never wholly loses its identity.

Experimental work by Gay-Lussac had shown that gases always combine in such a way that the ratios of the volumes of reacting gases are small whole numbers.† Amadeo Avogadro explained this experiment in 1811 by saying that equal volumes of different gases under the same temperature and pressure contain equal numbers of molecules. This statement is known as *Avogadro's law*. Note that Avogadro's law combined with Dalton's theory can immediately explain Gay-Lussac's experiments.

Hence, we have a model that explains the structure of matter in terms of atoms. The model does not purport to describe what an atom is or what it looks like, but rather describes quantitatively the manner in which they may react with one another.

10.2 Early theories: the caloric fluid theory of heat

Assuming matter to be made up of individual atoms, there was undoubtedly space between the atoms. Perhaps the heat that went from a hot object to a cold object was actually a form of "fluid" that flowed into the spaces between the atoms of the colder body. Such reasoning led to the "caloric" or "caloric fluid" theory. It had been observed that some materials (such as water) seemed to take up much more heat than did others (for example, the metals) to rise to the same temperature; but this seemed to be acceptable, since the theory allowed for the different materials' having much different spacing between the atoms to accommodate more or less "caloric."

It had been observed that a metal rod becomes somewhat longer when heated; apparently the "caloric" was pushing the atoms outward as it filled the interstices. The caloric theory seemed to give a fairly good explanation for the observations, at least to satisfy most people who were thinking on the subject, up to the end of the eighteenth century. The fact that observations required the caloric fluid to be weightless, to have self-repulsive tendencies, and to be invisible did not seem to be unreasonable. Besides, the "fluid" theory of heat was in the spirit of the Aristotelian concept that there is in each object an "essence" that gives to the object some of its properties. This is an example of the influence which Aristotle continued to exert on the natural sciences, even into the nineteenth century.

Although the caloric theory of heat was popular among most scientists of the late eighteenth and early nineteenth centuries, there were other theories. One of these, proposed by the English philosopher John Locke, declared that: "Heat is a very brisk agitation of the insensible parts of the object, ... What in our sensation is heat, in the object is nothing but motion." Another proponent of this theory, the Dutch physician Hermann Boerhaave, published a book entitled *Treatise on Fire*, which was to have a profound influence on the man who was to become one of the founders of our present-day theory of heat, namely, Benjamin Thompson (Count Rumford).

Thompson (1753–1814), who was later to become a Count of the Holy Roman Empire, was born in Woburn, Massachu-

---

† For example, 1 volume of oxygen gas plus 2 volumes of hydrogen gas (at the same temperature and pressure) give 2 volumes of water vapor. The hydrogen gas is made up not of individual atoms of hydrogen (H), but of the hydrogen molecule (H_2), meaning that two atoms combine and act as one unit. Similarly, oxygen gas exists as the oxygen molecule (O_2), made up of two oxygen (O) atoms.

setts Colony, but remained loyal to the crown during the American Revolution to the extent of commanding a loyalist regiment. In 1784, after three years in London, he became a military advisor to the Elector of Bavaria in Munich. During the next 14 years, during which he served as Minister of War and Minister of Police, and held various other posts in the Bavarian government, Rumford performed many important experiments that involved heat. While investigating the warmth of the clothing issued to his soldiers, he discovered *convection currents*. While he was overseeing the boring of cannons, he noted that both the cannon and the scrap removed from the bore became hot. He devised a scheme for measuring the amount of heat liberated in the drilling process. This consisted of immersing the cannon in water and measuring the temperature of the water as a function of time. The caloric theory stated that an isolated object (such as the cannon) should yield heat to the surroundings (water) under the action of friction, as long as any "caloric" remained. When its "caloric" was exhausted, continued friction would not produce further heating. *Rumford found, contrary to the caloric theory, however, that the longer the drilling continued, the hotter the water became.*†

10.3 Mechanical equivalent of heat

Although Rumford made quantitative measures of the heat produced in the drilling process and of the work done in producing this heat, he did not quantitatively relate these two terms. The earliest measurements of the relation of heat and mechanical energy were due to James Watt (1736–1819) of Edinburgh, who is better known today for his improvement

† For a fascinating account of the life of this remarkable man, see Brown in References.

of the steam engine. The more accurate measurement of the mechanical equivalent of heat was pursued by J. P. Joule in England from 1839–1872. By the end of that period, others had also undertaken similar investigations. There were experiments by which a descending weight was geared to turn a paddle in water, thus converting mechanical energy into heat and producing a rise in the temperature of the water. Mechanical pressure on a piston pushed water through fine tubes, raising the temperature of the water. Others used compressed air systems, electric battery systems, electric motors and generators, metal rubbing metal, and many other devices to accomplish the conversion of energy to heat. *It seemed as though every kind of energy could be converted to heat energy to increase the temperature of objects.* Out of these efforts there emerged a fairly clear relationship between mechanical energy and heat energy.†

The *combustion process* was studied by chemists and physicists. In 1780 it was even suggested by the French chemist Lavoisier (1743–1794) and the French astronomer-mathematician de Laplace (1749–1827) that heat resulted from the "insensible movements of the molecules of a body." They also proposed that combustion in the fireplace was not basically different from the utilization of

† The relationships are

(a) 1 hp = 33,000 ft-lb/min. (This represents a *rate* of doing work);
(b) 778 ft-lb of work will raise the temperature of 1 lb of water by 1°F; or
(c) 4185 joules is equivalent to 1 Cal of heat (where a joule is 1 newton-meter, and 1 Cal is the heat required to raise the temperature of 1 kg of water by 1°C from 14.5°C to 15.5°C (1 Cal = 1000 cal = 1 kilocal. Note that the use of the capital letter in Calorie designates the kilocalorie). Another unit of energy, the erg, is equal to the dyne centimeter.

food in the body of the animal by a form of combustion, in that both required oxygen and both processes gave out carbon dioxide. They performed several experiments that substantiated their statement on animal heat. A similar experiment was performed independently in 1777 by the Englishman Adair Crawford. A guinea pig confined to an insulated box, and given a measured amount of food and oxygen, seemed to produce about the same temperature rise as did burning the same amount of oxygen in a coal or wax fire placed in the same box.

This quantitative work seemed to dispel the caloric fluid concepts and to address more attention to the ideas proposed by Boerhaave, by Lavoisier and Laplace, and by John Locke, namely, that heat was a manifestation of atomic motions.

We must turn now to look more carefully to this "very brisk agitation of the insensible parts . . . ."

10.4 Brownian motion

We have now arrived at the idea that molecules are perpetually in motion and that the quantity of heat (and therefore temperature) possessed by matter is determined by the energy represented by that molecular motion. Is it possible to observe such motion? Robert Brown, a Scottish botanist, in 1827 did observe a phenomenon that seemed to find later explanation in the kinetic motion of molecules. The phenomenon has been called *Brownian motion* and may be observed easily by the reader as well in the following experiment.

The smoke produced by fire (or by a cigarette) is made up of very tiny particles. When we focus on one of these particles with a microscope, the particle can be seen to execute random movements, in zigzag fashion, going from one position to another, somewhat as illustrated in Fig. 10.1. The nature of the motion suggests that the particle is subject to occasional forces (or kicks) that come randomly and cause it to move from place to place in zigzag fashion. The smaller the particle, the more violent are its movements. The same phenomenon can be observed if one watches a colloidal† particle suspended in a liquid such as water.

What explanation can we give to these phenomena? One explanation seems to confirm the molecular, kinetic motion theory of heat and of temperature which has just been introduced, according to the following argument:

Let us assume that atoms and molecules are so small as to be invisible by any means available to us.‡ By the same token, the number of atoms in a unit volume of air or liquid must be enormously large, so large that at any instant of time there would be many atoms colliding with each other and with the walls of the container, and moving in all directions. Suppose that we now place a tiny smoke particle among the molecules of air. The *particle*, although very small and barely visible under the microscope, would in fact be *thousands of times larger than a molecule of oxygen or nitrogen of the air*. The particle would be bumped by the molecules, and there would be a transfer of momentum with each collision. But since it is so (relatively) large, the particle would *on the average* receive as many bumps from one

---

† The colloidal particle could be any particle substantially larger than the molecules that make up the liquid itself, and which remains suspended in the liquid (for example, gelatin).

‡ Only the large organic molecules are visible through the powerful electron microscopes.

side as from the opposite direction. Normally, therefore, the forces and momentum transfers generated by the bumps would balance out and the particle would remain unmoved.

But since the molecular collisions are random, occasionally there can be many more collisions coming from one direction than from all other directions.† When that happens, there is a net, unbalanced force, which in fact becomes a "kick" that moves the particle in the direction of the net force. A particle suspended in a liquid and subject to the random movements of the molecules of the liquid is subjected to similar forces.

This, then, both explains the phenomenon of Brownian motion and gives observational support to the kinetic molecular theory of heat.

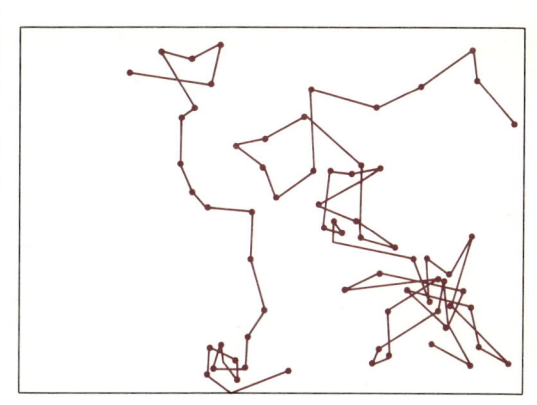

Fig. 10.1. Brownian motion.

10.5 Microscopic versus macroscopic analysis

Until now our interests and measurements have usually involved only large movements that could be analyzed through use of such instruments as metersticks, clocks, and balances. These constituted operations in the *macroscopic* realm. We now suddenly find ourselves dealing with phenomena that involve atoms and molecules, which are invisible to our instruments and whose numbers are so large as to make any effort at their individual measurements utterly hopeless. We use the term *microscopic* in referring to these, realizing that the atoms and molecules cannot be seen even with the microscope.

---

† The probability that ten consecutive collisions out of, say, 1000 collisions are in one direction is exactly the same as the probability of tossing a run of 10 consecutive heads in 1000 tosses of a true coin. As we say in Chapter 8, this is not negligible.

Since the analysis of individual particles in the microscopic realm is impossible for us, the question then becomes: "Can we utilize observations and measurements that are of macroscopic dimensions to give answers that pertain to microscopic phenomena?" That is, can we, by study of the properties of a *container* of gas or liquid, develop an understanding of the characteristics of the atoms and molecules that are in the container? The answer is that we can do so by taking advantage of the statistical concepts and techniques that were developed in Chapter 8.

Therefore, by study of the properties of a *container* of gas or liquid or the properties of a block of solid material (such properties as weight, pressure, volume, hardness, chemical composition, temperature, heat content, and absorption of radiation), we work in the *macroscopic* realm and thereby develop information on *microscopic* phenomena. In this sense, therefore, the observation of Brownian motion is in the macroscopic realm in that what we observe is the averaged or statistically integrated result of very many interactions at the "microscopic" level.†

10.6 Atomic numbers and dimensions

In our discussion so far we have frequently referred to "large numbers of particles." How large is a *large number?* In Sec. 10.1 we mentioned Avogadro's law (equal volumes of gases, under the same temperature and pressure conditions, contain equal numbers of molecules). The number of molecules in

† The reader must try to avoid confusion in this instance where, by common use, a microscope is used to study an effect that is said to be of macroscopic dimensions.

1 gram molecular weight† is called *Avogadro's number*.

Brownian motion can be used to determine Avogadro's number. If the number of collisions in a unit time is very low, each collision will cause the particle to move in the direction of the momentum of the incoming particle. Since the time between collisions will be relatively long, the particle will travel a considerable distance before another collision alters its course. If, on the other hand, an infinite number of collisions occurs in a unit time, no motion of the particle will result. (Why?) From the average distance traveled by the particle in a unit time, we can determine the average number of collisions in this time and (since the number of collisions per unit time is closely related to the number and speed of the colliding molecules) we can, with appropriate assumptions about the speeds, determine the number of gas or liquid molecules. Perrin, in France, published in 1908 the results of a determination of Avogadro's number by such a method. His value was approximately 6×10^{23} molecules! This is *indeed* a large number.

Using Avogadro's number we can obtain some interesting results about atoms. Eighteen grams of liquid water occupy a volume of 18 cc‡ and contain 6×10^{23} molecules. Thus, each molecule weighs

$$\frac{18 \text{ g}}{6 \times 10^{23}} = 3 \times 10^{-23} \text{ g}$$

and a volume of 3×10^{-23} cc (that is, it is

† Each element consists of atoms, all of which have essentially the same weight; for example, hydrogen is 1 unit, oxygen is 16. Thus, a molecule of water weighs 18 units (whatever a unit may be in grams!). One gram molecular weight of water is 18 grams of water (1 gram for each unit of molecular weight). A gram molecular weight of any gas under standard temperature and pressure occupies a volume of 22.4 liters = 22,400 cc).

‡ Recall that the density of water is 1 g/cm³.

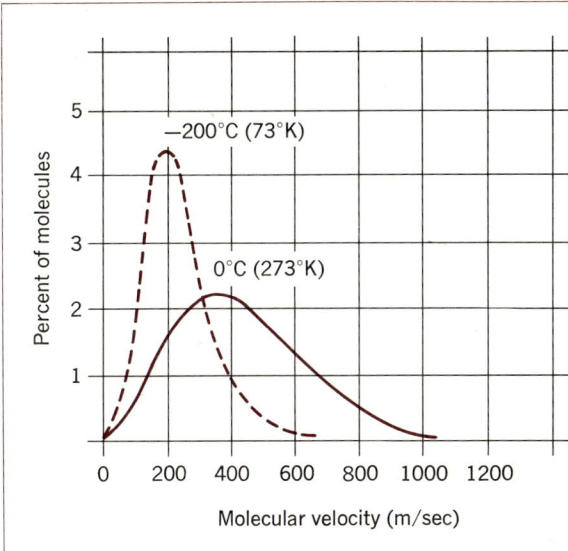

Fig. 10.2. Maxwell-Boltzmann velocity distribution.

assumed to occupy a sphere of about 10^{-8} cm in diameter!).†

From these numbers and dimensions we can readily understand why (1) an atom or molecule is invisible, and (2) with such large numbers, analysis by the statistical techniques that were outlined in Chapter 8 should give very accurate and useful results.

10.7 Kinetic theory of heat

We can now explore the properties of matter using the assumption that *the property we call the heat of a body is related to the kinetic activity of atoms and molecules that make up the body.* This involves the mass, velocity distribution, number of collisions per unit time of the particles. (We must presently

† From the relationship

$$\text{volume of a sphere} = \tfrac{4}{3}\pi r^3$$

where r is the radius of the sphere.

also relate the heat content of a body to its temperature in a more formal way.)

Using these concepts we may develop some better understanding of the kinetic energy of the molecules. This requires that we know something about the distribution of velocities in a gas assembly. That is, we know that under the random conditions that exist in a container of molecules, a certain fraction of them will (at any instant of time) have high velocity while other fractions have lower velocities. Knowledge of the distribution of velocities will give the information we need concerning the total kinetic energy of the molecules, which we can then relate to heat content and temperature. Ludwig Boltzmann and Clerk Maxwell studied this subject and derived the distribution indicated in Fig. 10.2.

Using the kinetic energy model outlined above, the *average* kinetic energy of a molecule is considered to be proportional to the temperature of the body.

If m is the mass of a molecule (considered here to be the same for all molecules in our body) and v its (random) velocity, then

$$\tfrac{1}{2}m\overline{v^2} = \tfrac{3}{2}kT \qquad (10\text{-}1)$$

where k is the constant of proportionality (called Boltzmann's constant; $k = 1.38 \times 10^{-23}$ J/°K). The factor $\tfrac{3}{2}$ on the right-hand side is included for basic reasons of classical physics.† Equation (10-1) may be considered to be a definition of temperature‡ in this theory.

---

† According to classical physics each "degree of freedom" contributes $\tfrac{1}{2}$ kT to the energy. The degrees of freedom represent the number of independent ways a system (in this case a particle) may change. A degree of freedom may be caused by rotation, vibration, or transitional movement. The particles considered above do not rotate or vibrate, they do, however, move in three dimensions (the x, y, z axes), thus resulting in the $\tfrac{3}{2}$ in our formula.

‡ It should be carefully noted that the temperature defined by Eq. (10-1) is the Kelvin (or absolute) temperature. The temperature is zero when $\overline{v^2}$ is zero (that is, no random motions of the molecules). That is referred to verbally as "degrees absolute" or "degrees Kelvin" in honor of the physicist Lord Kelvin who was responsible for clarifying the concept of temperature during the middle years of the nineteenth century. The absolute scale of temperature is much more commonly used in physics than is the centigrade or Fahrenheit scales because the zero of the absolute scale defines the lowest possible temperature that can be attained. The relationship between the temperature (T°K) as measured on the absolute scale and the same temperature (t°C) as measured on the more familiar centigrade scale is given by

$$T°K = (t + 273.15)°C$$

It follows that the absolute zero of temperature is

$$0°K = -273.15°C$$

The temperature at which ice melts under a pressure of 1 standard atmosphere (the ice point) is

According to the above definition, the temperature of a body increases as its heat content, or internal kinetic energy, increases. Conversely, as the temperature increases, the average kinetic energy of the molecules composing any macroscopic body increases and the average speed of the molecules increases. We shall return to this point later in this chapter.

Using Eq. (10-1), we find that at 300°K (=27°C = 81°F), the average speed of a water molecule ($m = 3 \times 10^{-23}$ g $= 3 \times 10^{-26}$ kg) is

$$v_{\text{rms}} = \sqrt{\frac{3 \times 1.38 \times 10^{-23} \times 300}{3 \times 10^{-26}}} \approx 650 \text{ m/sec}$$

This means that if no collisions occurred, the water molecule would travel 650 m in 1 sec. But we say that there is, on the average, one water molecule every 10^{-8} cm $= 10^{-10}$ m. Thus, in traveling 650 m, our original molecule would collide with nearly 10^{13} other molecules. Even in air, where the distances between molecules is large compared with those in water, the number of collisions suffered by any one molecule (in 1 sec) is very large. This large number of collisions experienced by each molecule tends to randomize the direction of its motion.

$$0°C = 273.15°K$$

The temperature at which water boils under a pressure of 1 standard atmosphere (the steam point) is

$$100°C = 373.15°K$$

The most familiar scale of temperature in the United States and Great Britain is the Fahrenheit scale, in which the ice point is 32°F, the steam point is 212°F, and the absolute zero is −459.67°F. The absolute scale is much to be preferred to the Fahrenheit or the centigrade scales for scientific work.

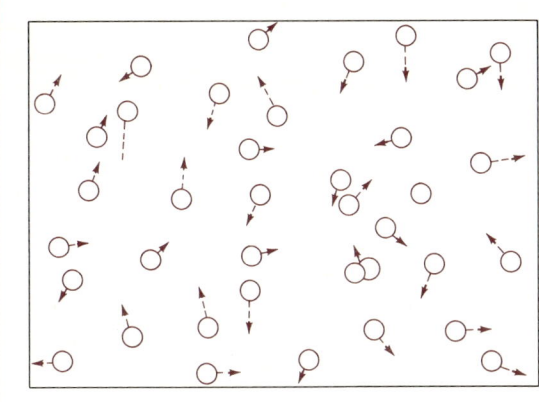

Fig. 10.3. *The kinetic theory picture of an ideal monatomic gas. The tails are intended to create an impression of the motion of the atoms and do not imply that the atom leaves any sort of wake behind it.*

The collisions also serve to distribute the kinetic energy of the molecules and to randomize the molecular speeds. This idea of *equipartition* (literally "equal distribution") of energy is fundamental to the Maxwell-Boltzmann statistical theory of molecular motions.

10.8 Distinction between heat and temperature

Before proceeding further it is important to point out again the distinction between heat and temperature. We have defined *heat* to be a form of energy that manifests itself in matter in the form of kinetic energy[†] of the constituent particles—the *internal activity* of the body. We have used Eq. (10-1) to relate kinetic energy and absolute temperature. However, the temperature of a body as measured by a thermometer is not related to the internal

[†] Throughout this chapter we have ignored all other forms of energy. In practice this is not strictly correct except for an "ideal" gas. The arguments of this section remain valid even when we substitute the actual internal energy of the body for the total kinetic energy of its component parts.

energy of the body by a universal constant, but rather depends on certain physical properties of the material of which the body consists. One of the most important of these, in macroscopic terms, is the *specific heat* of the material, that is, the amount of heat energy that must be supplied to a unit mass of the material to raise its temperature a given amount (say, by 1°C). The total mass of the body also is needed to relate its *total* internal energy to temperature.

10.9 The pressure of an ideal monatomic gas

Assume that the atoms of a gas are in rapid motion (Fig. 10.3) and, because they continually strike the walls of the container, the resulting time rate of change of momentum from the steady bombardment becomes a force pushing the wall outward. Because there are so many atoms, every unit area of the containing vessel experiences the same outward force, even when the container has a very complicated shape (Fig. 10.4).

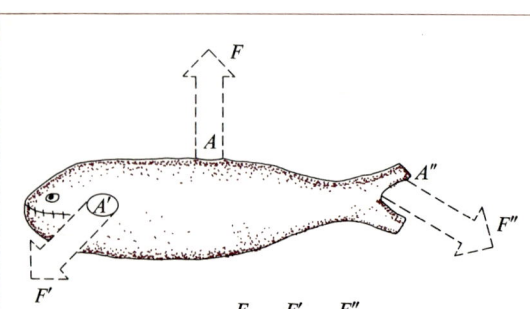

$$P = \frac{F}{A} = \frac{F'}{A'} = \frac{F''}{A''}$$

Fig. 10.4. *The pressure inside a balloon with a complicated shape.*

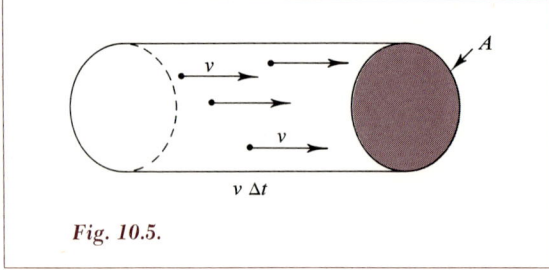

Fig. 10.5.

Consider a small area A of the wall of the container. If the total normal outward force experienced by A is F, then the pressure of the gas is F/A. In words, pressure = normal (perpendicular) force per unit area, or

$$P = \frac{F}{A} \qquad (10\text{-}2)$$

Assuming that the temperature is constant throughout the whole of the gas, and ignoring some very small effects due to the earth's gravitational field, the pressure has the same value on all parts of the wall. Notice that pressure is measured in dynes per square centimeter or newtons per square meter.†

To calculate the pressure, we need only calculate the momentum transferred in 1 sec to a unit area of the wall by the gas atoms. We do not need to know anything about the forces between the gas atoms and the atoms of the wall except to recall that the collisions obey Newton's third law of motion and that the law of conservation of momentum is therefore applicable.

Consider a stream of gas atoms, all moving with the same velocity v. If we take an area of surface A normal (perpendicular) to the direction of the motion of this stream, all the atoms that, at some time t, are contained in a cylinder with this area as a base and a length, $v\,\Delta t$, will strike the surface during the time interval Δt (Fig. 10.5). The total momentum transferred, Δp, to the surface in the time interval Δt is equal to the number N_0 of atoms originally present in the volume times the mass m of each atom times the velocity v, or

$$\Delta p = N_0 m v$$

† 1 cm of mercury = 1.33×10^4 dyne cm^{-2}. 1 standard atmosphere = 76 cm of mercury = 1.0×10^6 dyne cm^{-2}. The reader should prove that 1 newton m^{-2} = 10 dyne cm^{-2}.

10.9 The Pressure of an Ideal Monatomic Gas

If N' is the total number of atoms having the specified velocity in our volume V, then $N_0 = (N'/V)v\,\Delta t A$. (Why?) The total force (according to Newton's second law) is the time rate of change of the momentum $\Delta p/\Delta t$. By Eq. (10-2),

$$P = \frac{mv^2 N'}{V} \qquad (10\text{-}3)$$

If the gas atoms are moving in random directions, then the number N' moving in any specified direction is equal to one-third of the total number of atoms,† or

$$N' = \tfrac{1}{3} N$$

Thus Eq. (10-3) becomes

$$P = \frac{1}{3} mv^2 \frac{N}{V} \qquad (10\text{-}4)$$

---

† To obtain this result, recall that velocity is a vector quantity. Any vector can be considered as the (vector) sum of three component vectors: one parallel to each of three mutually perpendicular directions (axes). Thus the actual velocities of the gas atoms can be thought of as consisting of three parts: one perpendicular to A and the other two in mutually perpendicular directions in the plane parallel to A. Although, for any single atom, its velocity components along these three axes at any moment will not in general be equal, when considering a large number of atoms, the *average* velocities will be equal along each axis. Indeed, since (on the average) as many atoms move to the right as to the left, for example, the average value along *any* specified direction is *zero*! However, the root mean square velocity along any axis is *not* zero in general. Furthermore, the root mean square velocity is the *same* along each of our axes for a random velocity distribution. Thus, if we let v be the root mean square velocity along one direction, one-third of all atoms may be considered to be traveling in this direction with this velocity.

It should be noted further that since Eq. (10-1) contains only terms in v^2, we can associate a kinetic energy of $\tfrac{1}{2}(kT)$ to the motion along *each* axis or $\tfrac{3}{2}(kT)$ for the total kinetic energy. The use of $\tfrac{1}{2}(kT)$ to denote the kinetic energy in any direction is taken here to be by definition. (See footnote † in Sect. 10.7.)

Finally, we have indicated that, in practice, the molecular speeds will be distributed according to Fig. 10.2, and hence our initial assumption that all gas atoms are traveling at the same speed is unrealistic. However, if we choose v to be equal to the root mean square speed† of the ensemble v_a, our relation will be correct:

$$P = \frac{1}{3} \frac{N}{V} mv_a^2 \qquad (10\text{-}5)$$

Using Eq. (10-1), we may note that $(\tfrac{1}{2})mv_a^2 = \tfrac{3}{2}(kT)$. Thus

$$P = k\frac{NT}{V} \qquad \text{or} \qquad PV = NkT \qquad (10\text{-}6)$$

where k is Boltzmann's constant.‡ This is the so-called *equation of state* of an ideal gas.

Note the following macroscopic consequences of Eq. (10-6):

(1) At constant pressure P, the temperature T is proportional to the volume V (Charles' law; also known as Gay-Lussac's law). Thus, if the temperature of a quantity of gas is raised by giving it heat, the volume must also be increased, since otherwise the pressure goes up. (*Question:* What would happen to the temperature if the gas were to expand while holding the pressure constant?)

(2) At constant volume V, the temperature T is proportional to the pressure P (Amonton's law). That is, if a container of fixed volume is heated, the pressure goes up in direct proportion to increase of temperature.

(3) At constant temperature T, the pressure P is inversely proportional to the volume V (Boyle's law). A volume of

---

† See previous footnote.
‡ Note that Boltzmann's constant (k) has the dimension of energy per unit degree. Thus, from Eq. (10-6), the student should also confirm that PV has the dimension of energy.

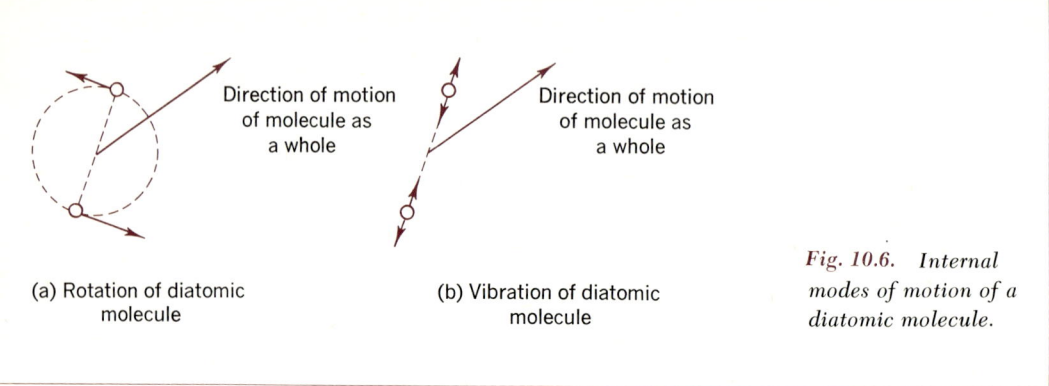

Fig. 10.6. *Internal modes of motion of a diatomic molecule.*

(a) Rotation of diatomic molecule
(b) Vibration of diatomic molecule

gas experiences increase in pressure when it is reduced to smaller volume.

The three "laws" were known before the development of kinetic theory and were deduced from experiments performed on macroscopic volumes of gas. Although the model of an "ideal" or perfect gas makes two unrealistic assumptions—(namely, (1) the gas atoms do not interact with one another, and (2) the gas atoms have zero size)†—the equation of state derived above is found to hold fairly well in a large number of actual physical systems; for example, the center of a star may be considered, with high accuracy, to be such a gas.

10.10 Actual gases

All gases approach the ideal state when expanded to a very large volume V because the molecules are then very far apart and the forces between them are negligibly small. Also, most gases are polyatomic; that is, each molecule contains more than one atom. The calculation of the pressure (by considering the momentum transferred to the wall per second) proceeds as before, and Eq. (10-6) applies here also, where N is now the number of molecules contained in the sample of gas. But a real difference shows up between an ideal polyatomic gas and an ideal monatomic gas in that the total internal energy of a polyatomic gas is not given by its kinetic energy alone.† It is still true that the kinetic energy associated with the motion of the molecules is related to the absolute temperature, as in Eq. (10-1). But the linear motion of the molecule is not the only source of internal energy because the individual atoms are able to move inside the molecule. In Fig. 10.6 the molecule as a whole can move with a velocity v, but in addition the two atoms *rotate* about one another and also *vibrate* along the line joining their centers. These modes of motion of the atoms inside the molecule represent extra kinetic energy and an extra contribution to the internal energy. Therefore

† Note the similarity of these assumptions to the atomic theories of Democritus.

† Additional degrees of freedom are now present.

the total energy of a polyatomic gas is

$$U = N\tfrac{3}{2}kT + \text{rotational energy} + \text{vibrational energy} \quad (10\text{-}7)$$

The rotational and vibrational energies increase as the temperature is raised. When the volume is decreased and the molecules (or atoms) come close enough together, we can no longer ignore interatomic forces.

The air around us is a little too dense to be a good ideal gas, but for many purposes we can utilize the equations that are applicable to an ideal gas.

The rotational frequencies and vibration frequencies of molecules will vary considerably, depending on the mass of the individual atoms and the nature of the bonding forces. In fact, *each molecule has a different set of frequencies, and this uniqueness is the basis for some very powerful techniques for identifying the molecules and their structure.* This is done by subjecting the material that is being studied to an alternating magnetic or electric field that has a variable frequency. As we increase or decrease this frequency, we reach a frequency corresponding to the natural frequency of rotation or the frequency of vibration of the molecule. At that point the molecules absorb more of the energy from the imposed field, and this shows up in the indicators of the apparatus. From this method it becomes possible to *identify* the presence of each type of molecule in the container and to calculate many of the structural and behavioral characteristics of the molecule.

10.11 The phases of water

Let us now consider the various states or phases of water: solid (ice), liquid (water), and gas (steam or water vapor).

A water molecule consists of two atoms of hydrogen and one of oxygen. If two volumes of molecular hydrogen (H_2) are combined with one volume of molecular oxygen (O_2), an exothermic (heat-producing) chemical reaction takes place to form two volumes of water molecules (H_2O) (see Fig. 10.7).

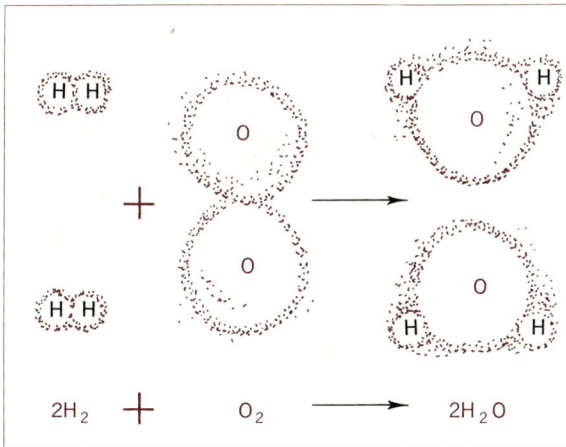

Fig. 10.7. *Two molecules of hydrogen combine with one molecule of oxygen to form two molecules of water.*

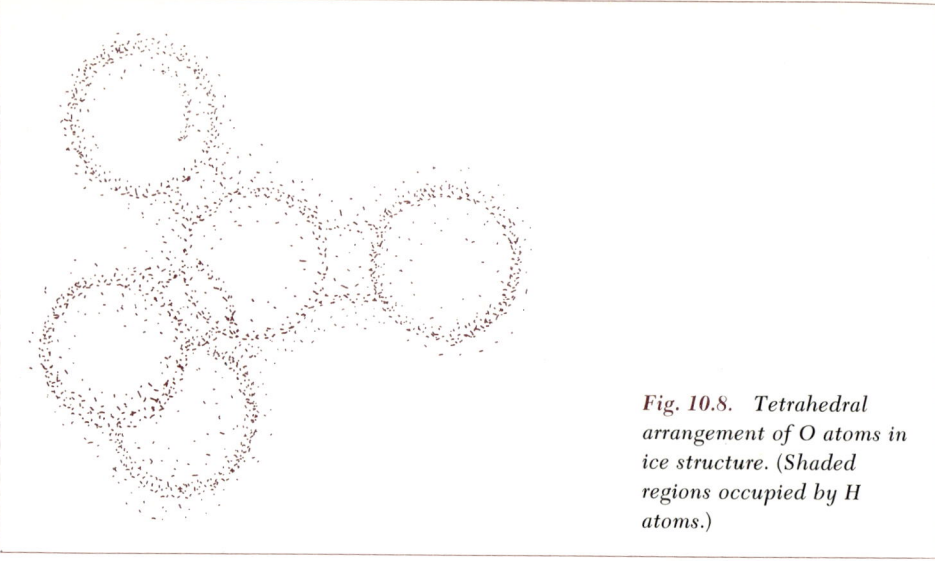

Fig. 10.8. Tetrahedral arrangement of O atoms in ice structure. (Shaded regions occupied by H atoms.)

The study of the arrangement of atoms or molecules within a solid is called *crystallography*. We saw that an atom is of the order of 10^{-10} m (1 angstrom unit) in size. To detect the presence of any object by using electromagnetic radiation, we must use radiation whose wavelength is not much larger than the size of the object we are studying.† X-rays are radiation of wavelength of the order of 10^{-10} m. X-ray studies of ice show that the oxygen atoms in neighboring H_2O molecules form a tetrahedron, with the hydrogen atoms between them (see Fig. 10.8). This configuration is continued in three dimensions throughout the ice crystal (Fig. 10.9). In Fig. 10.9 we note the "honeycomb" arrangement of hexagonal empty spaces. These empty spaces give ice its relatively low density.

As heat is added slowly to a block of ice at 0°C, the input of heat energy is used to free the constituent H_2O molecules from the orderly *lattice* structure of the ice. Under ideal conditions all of the heat energy is used in this way and none goes into raising the temperature of the system, until all H_2O molecules are free, or as we observe it, the ice is entirely melted. We call the heat required to melt the ice without raising the temperature of the system the heat of fusion.†

The disordering of the ice crystal lattice converts solid ice into liquid water.

---

† That is, we can measure large things with short wavelengths, but when we try to measure an object that is much smaller than the wavelength of light or other radiation, the long wavelengths behave like the large water wave that flows around a stick and scarcely feels its presence. The waves must be stopped or deflected in order to have it indicate: "There is something here."

† About 80 Cal of heat per kilogram of ice are required to convert ice at its melting point (0°C) to water at the same temperature. See footnotes on pages 335 and 348.

At 0°C the water is not totally without structure. Studies show that the tetrahedral pattern of oxygen atoms (Fig. 10.10) persists, although the separate tetrahedrons are now more mobile (that is, able to move with respect to their neighbors and to dissociate). Because of the peculiar structure of ice, containing as it does the aforementioned hexagonal empty spaces, the increased mobility of the water tetrahedrons results in a *more compact* structure and hence higher density (see Table 10.1). Since the total mass of H_2O molecules does not change during the melting process, the volume must decrease to account for the higher density. Conversely, on freezing water at 0°C, the volume must increase. H_2O is one of the few common substances that expand from freezing. This effect is quite noticeable when we attempt to freeze water in a confined volume such as a pipe. The

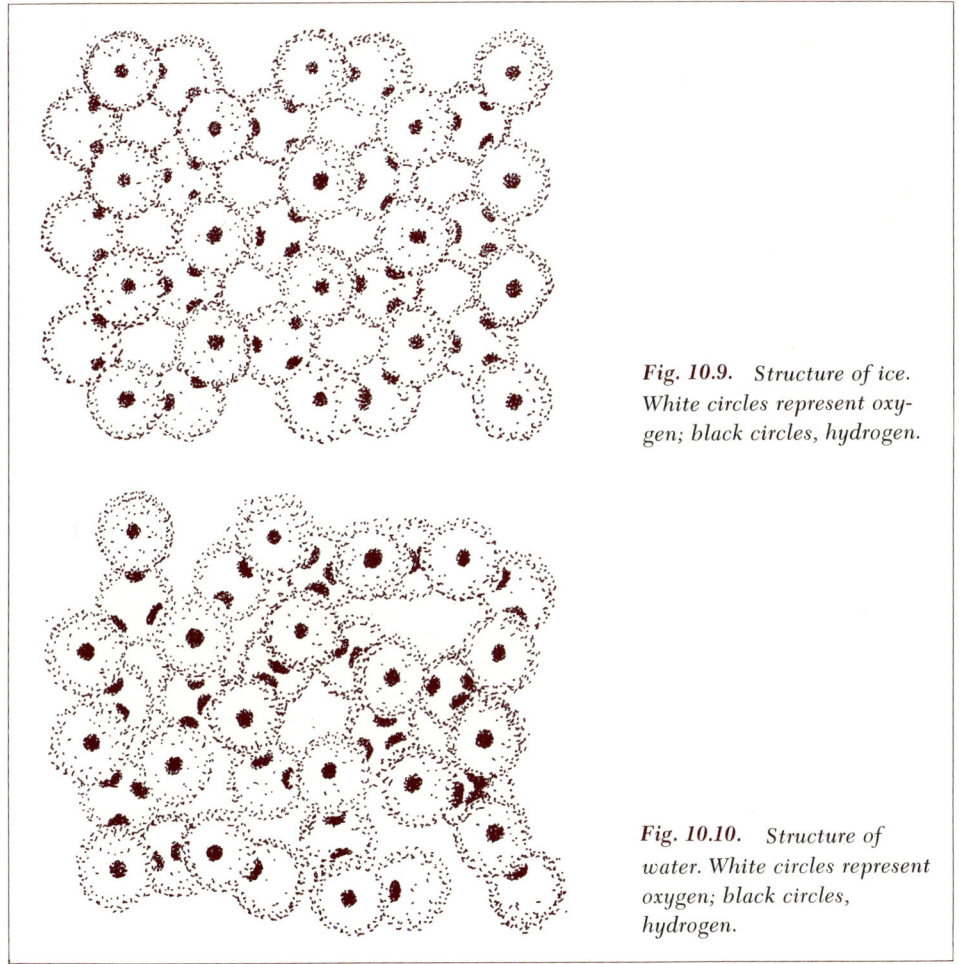

Fig. 10.9. Structure of ice. White circles represent oxygen; black circles, hydrogen.

Fig. 10.10. Structure of water. White circles represent oxygen; black circles, hydrogen.

TABLE 10.1
DENSITY OF WATER AT VARIOUS TEMPERATURES

Temperature, °C	State	Density, g/ml
0	Solid	0.917
0	Liquid	0.9998
3.98	Liquid	1.0000
10	Liquid	0.9997
25	Liquid	0.9971
100	Liquid	0.9584

forces resulting from the expansion are often strong enough to burst a metal pipe.

As we continue to add heat to the water, its temperature rises above 0°C. This is a result of the heat being now converted into kinetic energy of the structures within the liquid. The tetrahedral structures are further broken down and the average space between H_2O molecules becomes greater. This once again reduces the density.

When the water reaches a temperature of 100°C, another change of phase occurs. Additional heat does not change the temperature, but rather supplies the energy to the H_2O molecules needed to free them from the liquid. The quantity of heat needed to do this is called the *heat of vaporization*.† After all the liquid water has been converted to gaseous steam, additional heat will again cause a rise in temperature.

In the preceding paragraphs we have considered each of the various phases. In reality, two (or even all three) phases can coexist. We shall consider the coexistence of the gas phase with both solid and liquid.

The molecules of a gas are not held together, but move freely in a volume that is rather large compared with the volume of the molecules themselves. This means that there is very little attractive force between the molecules.‡ Because of this freedom of molecular motion, a quantity of gas does not have either def-

---

† Approximately 540 Cal of heat per kilogram of water are required to convert it at a temperature of 100°C to steam at the same temperature and sea-level pressure.

‡ This is due to the fact that the magnitude of the attractive force between molecules—the so-called van der Waals force—decreases rapidly with increasing separation of the molecules. Thus it is strong in solids but very weak, on the average, in gases.

inite shape or definite size. The gas simply shapes itself to its container. The molecules of a gas do exert pressure on the walls of the container by striking the walls and rebounding from them. We can understand, therefore, that when the volume occupied by a specified quantity of gas is decreased, the molecules strike the walls more often, and the pressure accordingly increases. This is why we find it increasingly difficult to push down on the bicycle air pump on the downstroke.

10.12 The evaporation of a molecular crystal

At a very low temperature the molecules in a crystal tend to be bound and to have limited motion. As the temperature increases, the molecules become more and more agitated; each one bounds back and forth more and more vigorously in the little space left for it by its neighbors, and each one strikes its neighbors and rebounds more and more strongly. This increase in molecular motion causes the crystal to expand somewhat, giving each molecule slightly larger space in which to move.

What holds a molecule when it happens to be on the surface of the crystal? Why does it not fly away? A certain number of molecules do fly away, despite an attractive force that operates between all molecules when they are close together, called *van der Waals attractive forces*.† Since these attractive forces are quite weak, occasionally molecules will become so agitated as to break loose from their neighbors. If the crystal is in a

† When the molecules are *very* close together, the van der Waals force becomes repulsive. We shall consider only the attractive force between molecules at this time.

vessel, these molecules will accumulate in the space within the vessel, each moving in a straight-line path and occasionally colliding with another molecule or with the walls of the vessel, thus changing the direction of their motion. These give rise to the *vapor pressure* of the crystal (or liquid). In the case of the ice crystal, these free molecules constitute water vapor. Of course the free molecules may also collide with the crystal itself and stick to it again. In time a balance will be reached whereby as many molecules return to the crystal as escape from it, if the vessel is closed long enough. The molecules in the vapor are very much like the molecules in the crystal, with essentially the same interatomic distances *within* the molecule but with much larger distances *between* molecules.

From the preceding discussion we can understand that molecules on the surface of a crystal can evaporate directly into a gas, just as from a liquid. We know, for example, that solid pieces of camphor or of naphthalene (moth balls) left out in the air slowly decrease in size because of the evaporation of molecules from the surface of the solid. Snow may disappear from the ground without apparent melting, by *evaporation* of the ice crystals at a temperature below that of their melting point. The process of going directly from the solid to the gas phase is called *sublimation*. Sublimation becomes accelerated when air currents help remove the water vapor from the immediate neighborhood of the snow crystals and prevent the vapor from condensing again on the crystals. The process by which molecules escape from a solid, then strike and stick to the surface of another solid, is also called *sublimation* of the gas molecules.

How fast do molecules evaporate from

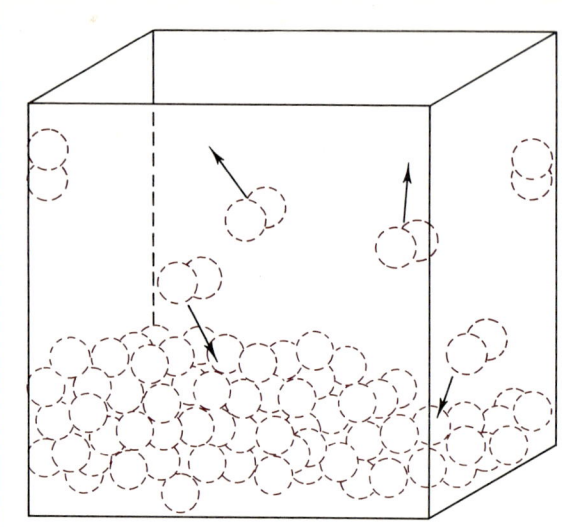

Fig. 10.11. Equilibrium between molecules evaporating from a crystal and gas molecules depositing on the crystal.

the surface of a crystal? The rate is proportional to the area of the surface and is essentially independent of the pressure of the different surrounding gas. But the rate at which the free gas molecules strike and return to the crystal surface is proportional both to the area of the crystal surface and to the concentration of the molecules in the vapor (the number of gas molecules in unit volume). As noted earlier, after a while the process reaches an equilibrium state. The rate at which gas molecules strike the crystal surface and stay there is just equal to the rate at which molecules leave the crystal surface. The gas pressure that corresponds to this equilibrium condition is called the *vapor pressure of the crystal*.

This type of steady state of equilibrium, which results not from things being static but rather from a dynamic or numerical balance of things that are in motion, is representative of many physical, chemical, and biological processes of nature.

Equilibrium is not a situation in which nothing is happening, but rather is a situation in which opposing processes are taking place at the same rate, so that a person who observes only the gross results observes no overall change. This is indicated in Fig. 10.11.

10.13 Average properties of a large number of atoms

When we realize how vast are the numbers of atoms that we take into our bodies with each breath, we become quite convinced that a complete description of the universe in terms of the motion of each atom, electron, proton, or neutron represents a quite impossible task. In fact 1 cc of air contains over 10^{19} molecules.

Fortunately, the fact that a drop of water or a breath of air contains such an enormous number of molecules introduces a simplification since many interesting properties of matter depend only upon the statistical or average behavior of molecules. In the case of the ideal monatomic gas, we saw that pressure depended only on the number of atoms per unit volume and upon the average kinetic energy of these atoms. It was neither necessary to follow the path of each atom in detail nor to know in detail the nature of the forces between the gas atoms and the atoms of the wall. In our discussion we assumed that the temperature of the gas is the same everywhere; that is, the atoms are in *thermal* and *statistical equilibrium* with one another.

What do we mean by statistical equilibrium? We can easily explain what it is *not* by applying heat rapidly to one corner of a box that contains air. That corner will be hotter than any other part of the box and the air molecules that find their way to that corner will gain additional energy. Even if we use a fan to stir the air, we would fail to achieve statistical equilibrium [Fig. 10.12(a)]. If the fan is taken away and the hot flame is removed, the gas then settles down into a state of statistical equilibrium. The hotter atoms move into the bulk of the gas, collide with the colder atoms, and eventually distribute their excess kinetic energy among all the atoms. When statistical equilibrium is reached [Fig. 10.12(b)], there is no net velocity of flow in any particular direction and the average velocity of atoms will be the same for samples taken from any part of the box. This is equivalent to saying that the temperature is now the same everywhere and there is no tendency for energy to flow from one part of the system into another part.

In the next section we shall introduce the subject called *thermodynamics*. Properties such as pressure, temperature, and internal energy depend only on the *average* behavior of the atoms, and lend themselves to analysis by statistical techniques and the laws of thermodynamics. Though the laws of thermodynamics are few and simple, they constitute important expressions for the behavior of large aggregates, just as Newton's laws are important for dealing with individual particles.

Statistical methods as used in ther-

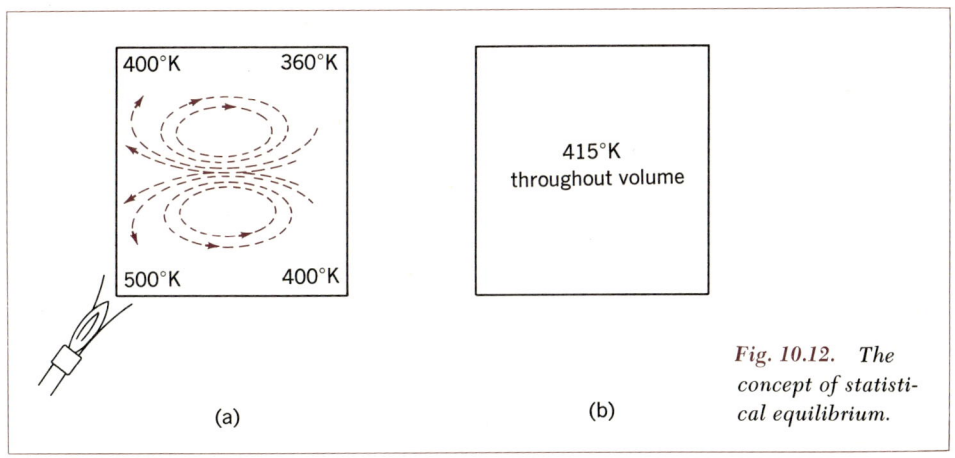

Fig. 10.12. The concept of statistical equilibrium.

modynamics have become exceedingly important for modern physics because of the impossibility of dealing with individual particles and individual events of natural phenomena.

10.14 Zeroth law of thermodynamics

The "zeroth law" of thermodynamics asserts that there is no net transfer of energy between bodies that are in *statistical equilibrium*. The law could also be taken as a definition of statistical equilibrium.

The converse of the zeroth law—if an isolated system† is *not* in statistical equilibrium, then heat will flow from the hotter regions of the cooler ones—allows the measurement of the temperature of a body to be meaningful. If we put a thermometer, initially at room temperature, say, into a hot liquid, we expect some of the heat of the liquid to be transferred to the thermometer. We allow the thermometer to remain in contact with the liquid until the mercury column in the thermometer stops rising, (that is, the thermometer has reached statistical equilibrium with the liquid and hence both are at the same "temperature").‡

10.15 The first law of thermodynamics

The first law of thermodynamics is a very specific statement that energy is conserved. That is, any energy given up by

---

† An isolated system is one that has no heat exchange with its surroundings.

‡ This does not mean that there is no molecular exchange of energy. It does mean that energy that goes *into* the thermometer by molecular interaction is balanced by an equivalent amount of molecular energy going *out* of the thermometer. This assumes that there is no chemical reaction between the thermometer and the liquid.

---

a body is accounted for by an equivalent gain in energy somewhere else, the energy being in any of several forms. More formally the first law says:

The heat added to a system is equal to the increase in the internal energy of the system plus the work done by the system against its surroundings.

Let us interpret this in terms of what happens when we apply heat to a bowl of ice. The heat we add melts the ice and converts it to steam, which can exert considerable pressure. The pressure can be allowed to do work, as when it pushes against a piston that moves. The first law simply says that the total heat applied to the bowl containing the ice is exactly equal to the kinetic energy that the steam molecules acquire, plus the heats of fusion and vaporization, plus the heat dissipated by the hot bowl to the surrounding air, plus whatever work the steam may do on expansion. (*Question:* What do we mean by "work done?") In summing up the total we must include not only the kinetic energy of the water molecules, but also the rotational and vibrational energy *within* each molecule of water.

It is traditional to measure heat in calories. One small† calorie (cal) is the heat required to raise the temperature of 1 g of pure water from 14.5°C to 15.5°C. Since heat is really energy and we have already defined the units of energy, the erg and joule (J) (Sec. 10.3), it is more satisfying to measure heat in ergs or joules. The conversion factors are

$$1 \text{ cal} = 4.185 \text{ J}$$
$$= 4.185 \times 10^7 \text{ ergs}$$

The first law may be stated in equation form as follows: Let ΔQ represent the heat added to the system, ΔW the work

---

† Compare with large Calorie (or kilocalorie) defined in Sec. 10.3.

done by the system against its surroundings, and ΔU the change in internal energy. Then

$$\Delta Q = \Delta U + \Delta W \qquad (10\text{-}8a)$$

This equation says, in symbols, exactly the same thing that our original statement said. Let us rearrange the terms of the equation:

$$\Delta U = \Delta Q - \Delta W \qquad (10\text{-}8b)$$

The right-hand side of Eq (10-8b) is the difference between the heat energy added to the system and the work done *by* the system. If the system was initially in some thermodynamic *state,* defined by its *state variables* (temperature, pressure, and so on), the addition of heat ΔQ will cause a change in the state variables. The work done by the system will again change the state variables; that is, when the steam is allowed to move a piston, the pressure and temperature drop while the volume increases. In general, the values of these variables will be different from their initial values after the processes denoted by the right-hand side of Eq. (10-8b) have been completed. The change in internal energy, ΔU, must therefore be related to the change in the state variables of the system, implying that the internal energy itself is a function of these variables. Indeed, the first law implies that the internal energy is a function of the state variables *only*. In other words, the internal energy can itself be used as a state variable of the system. That is, if we could know the amount of internal energy (vibrational, rotational, electromagnetic) possessed by a molecule, we could then know what pressure and temperature conditions gave that internal energy.

To realize the significance of the preceding statement, note that both terms on the right-hand side of Eq. (10-8) can be defined in macroscopic terms: ΔQ as the *thermal* energy added to our system and ΔW as the *mechanical* (*dynamical*) energy extracted from it. Thus ΔU is defined in macroscopic terms. We have already defined the internal energy from the point of view of kinetic theory (microscopic), but the first law is a macroscopic relation. The fact that the internal energy is a state variable, which is consistent with the kinetic theory approach, is independent of the choice of any microscopic model.†

10.16 Order and disorder

When a baseball of mass m is thrown through space with a velocity v, its kinetic energy $(\frac{1}{2})mv^2$ would not normally be called heat. However, when it strikes the ground, both the ball and the portion of the ground it strikes become warm; the kinetic energy of its bodily motion becomes partially converted into kinetic energy of the individual molecules or atoms of the ball and the ground, which is called *heat*. But since the motion of these molecules is random, *we can think of heat as disorganized energy*.

This concept of disorder is very important in thermodynamics. In this example of throwing the ball, the kinetic energy of the ball represents an *orderly, directed, organized* form of energy; but when the ball strikes the ground, some of this directed energy becomes transformed into disorganized, random motion of the atoms and molecules. Can it go in the reverse direction also, with the disorganized molecular motion becoming "ordered" energy to hurl the ball through space? We cannot say that such an ordered arrangement can never occur in an infinitely long period of time, but experience tells us that it is quite unlikely to

† For more details on thermodynamics see Atkins in References.

happen, since ordinary collisions tend to reduce the assembly to an even more *disordered* arrangement.

The tendency toward a disordered array is the basis for the second and third laws of thermodynamics, as we shall presently see. We now introduce a quantity S, called *entropy*, which is simply a measure of the *probability* that a particular type of motion will occur. We assume that it is *more probable* that a system will become *more disordered than ordered* in its array of motions and behavior, and *S becomes a measure of the probability of a state of motion.* (*Question:* Trace the steps by which the ball was hurled in the first place; that is, the steps by which the random motion of atoms in the soil and sun become an organism that can hurl the ball. Is this not contrary to the "normal" direction of entropy change? What makes it possible? This idea will be very important to us from now on.)

10.17 Second law of thermodynamics

The second law of thermodynamics is stated as

When a closed system containing a large number of particles is left to itself, it assumes a state of maximum entropy; that is, it becomes as disordered as possible.

This second law of thermodynamics says that an ordered situation is unlikely to occur or persist. The more disordered the state of a system, the larger its entropy S. *The states with larger S are more likely to occur.* In other words, a larger value for entropy corresponds to a greater probability for an event to occur, and "what is the more likely to occur will occur." The assumption is that the probability of what is likely to occur is so overwhelmingly large that any other possibility can be completely discounted.

Consider a gas consisting of a large number N of atoms in a very large container (so that we can neglect the collisions with the walls for the moment). Suppose, at some time t_0, these atoms are neatly arranged in space and all moving at the same velocity, like a platoon of soldiers on parade. They would be acting like a rigid solid body in this respect. But the solid body depends on the existence of forces between the atoms to hold them in line or to restore individual atoms when they stray (much as the drill sergeant does for the platoon of soldiers). Our gas has no such restoring forces, since the interatomic forces are very small. If an atom strays, it is lost from the orderly array we have pictured. In fact, since N is large, there is great probability that other atoms will collide with it even before it leaves the array. Very soon there will be very many collisions until a condition of utter confusion is reached.†

In the absence of restoring forces between atoms such as exist in a solid, any event that causes the change in motion of even one atom has a great probability of starting a "chain reaction" whose ultimate effect is to completely change the original orderly motion into random

---

† To continue the analogy with the soldiers on parade, the situation is similar to what happens if one soldier turns left while all the others turn right. He soon collides with a companion and both go off in arbitrary directions, colliding with still more paraders whose direction and speed of march become altered. Soon our orderly parade becomes a disordered jumble of random motions. Order can be restored only if the drill sergeant (or other soldiers) exert sufficient force on the platoon to reorganize it. A similar restoring force is necessary in our gas to return it to an ordered state of motion.

motion. It can be shown analytically that, although orderly motion is possible in a gas, random motion is overwhelmingly probable. Thus it appears that disordered or random motion has a higher probability of occurring than does ordered motion.

Let us now try to approach this second law more quantitatively by choosing a system that is simple, such as a gas consisting of only four atoms in random motion confined in some container. If we assume this container to be divided into two equal parts, we may ask: "In how many ways can the four atoms be placed into two equal volumes?" Table 10.2 enumerates the possible numbers in each region.

Here we represent case A as having all four atoms in one of the two halves of the volume, and case E as having all four in the other half.

For cases A and E there is only *one* way to arrange atoms. For cases B and D there are four ways (why?) and for case C there are six ways (why?). Thus case C represents the *most probable* situation; that is, four atoms will most probably be arranged in two equal volumes so that there are two atoms in each volume.

If we let P represent the probability of finding an ensemble of particles in a given arrangement of motions and positions, we can *define* a second quantity S, called the *entropy*, as

$$S = k \ln P \qquad (10\text{-}9)$$

where k is Boltzmann's constant, which was introduced earlier. According to this definition *the entropy is large for states with large probability of occurrence* and low for those with low probability. Since, as we have seen, isolated systems will most probably assume a state of disordered motion, Eq. (10-9) indicates that the entropy of such a system will increase. According to Eq. (10-9), the state of maxi-

TABLE 10.2

No. of Atoms	A	B	C	D	E
Volume 1	0	1	2	3	4
Volume 2	4	3	2	1	0

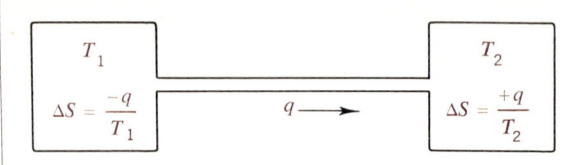

Fig. 10.13. *When heat flows from a hot body to a cold body, entropy of the system increases.*

mum entropy is also the state of maximum probability, which, as we have seen, is the state of greatest disorder. This is in agreement with the earlier statement of the second law.

There is a condition in the second law: the system is to be "left to itself." This means that no heat must be added to or subtracted from the system and no work must be done on or by the system that might change its internal energy. Under these circumstances, the system will come to a state of statistical equilibrium, which has the maximum possible entropy (or disorder) that can be achieved for this particular fixed value of the internal energy.

We can seek a method for determining this quantity called entropy by considering what will happen when we add a little heat to a system. We shall start with a system in which statistical equilibrium has been reached at temperature T with an entropy S. Now let us add a small quantity of heat ΔQ so slowly that the system passes through a sequence of equilibrium states. The system will eventually arrive at a new statistical equilibrium with temperature $T + \Delta T$ and entropy $S + \Delta S$. We shall *define* the change in entropy, ΔS, as:

$$\Delta S = \frac{\text{heat added}}{\text{initial temperature}} = \frac{\Delta Q}{T} \quad (10\text{-}10)$$

We simply state here, without any attempt to prove it, that Eq. (10-10) is consistent with the second law of thermodynamics as stated above and as embodied in Eq. (10-9). We can now restate the second law as

The entropy of a closed, isolated system can never decrease due to internal processes.

This means that all internal processes (such as collisions) either leave the entropy of the system the same or tend to raise it. Left long enough, the system will settle down to an equilibrium situation where the entropy is a maximum, as described by our earlier statement of the second law.

Since both temperature and amount of heat are quantities we can easily measure in the laboratory, this relationship gives us a means for measuring the entropy change, ΔS. This is fortunate, for up to this point this seemingly very important quantity called entropy has been no more than a mathematical concept that is related to probability.

Let us test the second law of thermodynamics by connecting a body with temperature T_1 to a body with a lower temperature T_2, using a rod to conduct the heat (Fig. 10.13). We know that heat will flow from the hot body to the cold body until their two temperatures are equal. (Of course this process takes place fairly rapidly and not through a series of slowly changing equilibrium states such as we have assumed.) If a small amount of heat, q, leaves the hot body and enters the cold body, the entropy of the hot body will have decreased by q/T_1, and the entropy of the cold body increased by q/T_2. The change in the total entropy of the system is therefore

$$\Delta S = \frac{-q}{T_1} + \frac{q}{T_2}$$
$$= q\left(\frac{1}{T_2} - \frac{1}{T_1}\right)$$

Since q itself is positive, T_2 being less than T_1 makes $1/T_2$ larger than $1/T_1$. Therefore this entropy change is *positive*, or increasing, as predicted by the second law. Heat flows from a body at a high temperature to a body at a lower temperature. This increases the total entropy of the system. If the heat q were to flow from the cold body to the hot body, the change in entropy would be negative, contrary to the second law.†

In the example of Fig. 10.13, if we had considered only the one body and not the other, we would have gotten wrong answers. One part of a system may show a decrease in entropy while another part shows an increase, but the entropy of the *entire* system never shows a net decrease. ‡

There are many examples of the influence of the second law. When a group of men push in one direction (which is an example of an ordered state) they can produce more useful work in moving a stationary automobile than is possible when some of them push in one direction

† It is for this reason that we can never hope to recover the vast amount of heat that is in the oceans (without the expenditure of energy from other sources) and make useful application of this heat; for while the ocean temperatures are decidedly higher than absolute zero, they are lower than the temperatures of our bodies and of our environment.
‡ Entropy remains constant only in a reversible process; in all others, it increases.

and some in other directions. Clearly a disordered state tends to decrease the system's total *capacity to perform work*. That is, a state of random motion has less energy to do work than an ordered state which corresponds to a higher potential energy. We know this from the experience that water runs downhill. Light, electricity, chemical energy all tend to become molecular kinetic energy or heat, and thereafter are capable of less work. Ultimately all the potential energy, or energy from ordered systems will become *kinetic energy or heat*. George Gamow† reminds us that by the same token, all a housewife needs to do to promote disorder in the house is to sit pretty and do nothing. Also, roads automatically become impassable if the highway commissioner leans back in his chair and discontinues maintenance work.

10.18 Systems and conservation of energy

But while water normally runs downhill and nature seems to prefer disorder, we have the amazing process by which plants convert the disordered elements of soil and air into ordered compounds. We call the process *photosynthesis* and plant growth. Certainly the gathering of the minerals from the soil and the organizing of these with the carbon dioxide (CO_2) of the air and water (H_2O) to produce complex but beautifully symmetrical and organized molecules is the reverse of disorder. Similarly, the organs of the body cannot be said to perform their functions haphazardly; yet they must utilize matter (oxygen, blood corpuscles, food particles) that follows disordered phenomena not unlike those we have

† See References.

attributed to gases. Do these phenomena disprove the second law?

Indeed they do not. But to see that they do not we must take into account the whole system of which these are a part. The growing plant requires the energy of the sun. The entropy changes that we take into account *must include all the changes of the system.* We will find that the photosynthetic process again demonstrates an increase of entropy if we take into account the energy of sunlight and the transformations that sunlight experiences from the time it leaves the sun. The photosynthetic process *at the site of the plant* is simply given as

$H_2O + CO_2 +$ (light energy
$\qquad$ + entropy decrease) $\rightarrow$ wood + O_2

The reverse reaction, or combustion of wood, is given as

Wood + O_2 $\rightarrow$ (energy
$\qquad$ + entropy increase) + $H_2O + CO_2$

The example of Fig. 10.13 illustrates the error that can be made if one judges the changes in only one of the two bodies for total change in entropy (or in one part of a large system) and not the whole system.

10.19 *Entropy and time*

At the end of Chapter 9 we noted that the direction of time can be related to the change in entropy of any closed system in the sense that "successive later states of a system are states of successively greater entropy." This means that if you view a closed system at two different times, the time when the system shows greater disorder (or higher entropy) is the latest state.

We now note some reservations about the preceding definition. If only reversible processes take place, there will be no change in the entropy and hence no way of distinguishing which of two states of the system occurred earlier. This is like running a motion picture film backward. If the events depicted are completely reversible, we cannot tell which way the film is going. Fortunately, nearly all events in nature are irreversible, and hence we shall adopt the statement for the *increase of entropy as being indicative of the direction of time.*

To justify further this use of increase of entropy as an indicator of the direction of time, we note that all the equations introduced in mechanics, as well as those we shall introduce in Chapter 11 on electricity and magnetism, have the following property: If we substitute $-t$ for $+t$ whenever time occurs, the equations remain the same, that is, they are *invariant to time reversal.* This is a property of equations dealing with macroscopic bodies. In contrast, our statistical considerations that led to formulation of the entropy concept involved microscopic levels. However, when we consider macroscopic equations in terms of the microscopic phenomena, we find (for example, wherever friction occurs in a mechanical system) that time reversibility becomes impossible because of the increased random motion of the constituent particles of our system. It must be noted, however, that in several instances (to be encountered in Chapter 20), time reversal plays an important role. It was from such a consideration of time reversal that Dirac predicted the existence of a positively charged electron.

We can thus conclude that although our definition of the direction of time in terms of entropy change will hold in most, if not all, situations, it must be applied with care.

10.20 *Summary*

In this introduction to heat and to the laws of thermodynamics we have been forced to resort to some new assumptions and techniques. We can begin with a piece of ice, which we can see and move around and to which we can readily attribute the properties of dimension, mass, and well-defined behavior when subjected to an external force. We also have a good sense of cold versus hot when we touch ice and other objects. These observations are sometimes referred to as *operational* information, or knowledge obtained from experience which is in the *macroscopic* realm. By this we mean that the data deal with phenomena that are large enough to be seen with the unaided eyes and measured by conscious human effort. We then introduced a model of atoms in motion because there seemed to be no other choice; *but in the process we jumped from what we could see to a model that was unseen and only a conjecture.*

This was not all. As we proceeded to explore the nature of heat transfer, experience told us that hot bodies always tend to lose heat to their colder surroundings, and never the reverse. From such common experiences we proceeded to construct laws of thermodynamics with respect to heat transfer. Again it required a jump from what was *experience of the past* to assumptions about realms that were still unknown. The concept of entropy especially posed a level of abstraction that required all the skill at our command, utilizing also a molecular model which is clearly limited by the observations and measurements we can make.

To extend and generalize the law of conservation of energy required passing from the dynamical laws of mechanics to statistical laws that reach out of the macroscopic world of observed phenomena to the microscopic world of unseen, postulated entities, called atoms and molecules. These rapidly moving particles have properties differing from the properties of the bodies they compose. Our dynamical observations become observations, then, of statistical averages.

Many of the concepts used before—kinetic energy, force, momentum, and mass—appear again. But new and powerful concepts also emerge, including the idea of heat as the average kinetic energy of a large assemblage of particles in *statistical equilibrium,* temperature as a measure of the direction in which the average kinetic energy flows between such ensembles, and finally *entropy* as a measure of the spontaneous internal process by which isolated systems finally attain statistical equilibrium. These concepts are used to restate the law of conservation of energy (hereafter referred to as the first law of thermodynamics) and to establish a new law of equally wide scope and power, the second law of thermodynamics.

Because we are so accustomed today to thinking of heat as energy or as a mode of motion of small particles, it is hard to realize that it was only a little over one hundred years ago that the alternative theory of heat as a fluid (formerly called "caloric") finally died out. Even now we speak of "heat flowing" or "piping heat" from one place to another, as if it were a liquid. Indeed most observable phenomena involving heat can be explained in terms of a fluid flow. The kinetic theory of heat, as developed in this chapter, depends for its full development upon a set of ideas that goes far beyond what can be observed.

It is not, however, enough to show merely that mechanical energy can be turned into heat energy. Another big idea is needed and this is the idea of *statistical equilibrium,* which "explains" why heat flows always in one direction from hot to cold and never in the opposite direction in any system of particles which is isolated in terms of energy from the surrounding environment.

The first law of thermodynamics refers to the conservation of energy, including heat. The second law states that heat flows from hot to cold regions because in nature there is a tendency for systems to move from organized, ordered regularities to states of greater disorganization. This is a truly remarkable law of greatest generality. It has never been observed to be contravened. It tells us among other things that heat differs from all other forms of energy; any other kind of energy —mechanical, electrical, chemical—can be turned into heat with 100 percent efficiency, but heat cannot be turned into any other form of energy with 100 percent efficiency.

In order to illuminate, although hardly to explain the operation of the second law of dynamics, the concept of *statistical* equilibrium and the concept of *entropy* were introduced in this chapter. In previous chapters there has been discussed the stability of closed systems in dynamic equilibrium (for example, the solar system) and the stability of systems with feedback controls (for example, living organisms). The statistical equilibrium of a system is of a somewhat different nature. It is attained spontaneously, internally, inexorably whenever any large assemblage of molecules or atoms is completely isolated from any outside source of energy.

The simplest model for illustration of these points is the kinetic model of a gas. A cubic centimeter of a gas at normal pressure and temperature contains more than a billion times a billion molecules, all moving with a wide variety of velocities in a wide variety of directions. No discernible rules can be formulated as to the distribution of position, velocities, momenta of the molecules. Each one of an "almost infinite" number of distributions is equally likely. But under these circumstances, an orderly arrangement— such as the collection of all the molecules in one-half the volume—is so highly improbable that such a given distribution can be said never to occur in practice. There is an enormously (by a ratio of billions upon billions) greater chance that the molecules will be distributed randomly throughout the volume. Therefore, left to itself, such an assemblage of molecules, even if arranged initially in a somewhat orderly configuration (for example, a specific concentration of gas molecules in one region or a higher kinetic energy in one side of the volume than in the other), will move toward a randomly ordered or disorderly condition. This tendency for an isolated ensemble to move from order to disorder until statistical equilibrium is attained is one way of expressing the second law of thermodynamics. The measure of disorder present is given the name *entropy*. Entropy always increases spontaneously in an isolated system. It follows that if two systems, each in individual but differing statistical equilibrium (say, with differing temperatures or average kinetic energy), are brought into contact, heat will flow from the system at the higher temperature to the system at the lower temperature until a new overall statistical equilibrium (with a single temperature) is attained.

Now, in order to give these two laws of thermodynamics a mathematically precise formulation, it was necessary to postulate the existence of elementary particles, which themselves either have no internal parts or are already in a state of internal statistical equilibrium. Otherwise, the particles themselves, atoms or molecules, could absorb heat from the surroundings. Originally the atoms were conceived of as having no internal parts. They were thought of as perfectly elastic little spheres. Much later, as we shall see, it was discovered that atoms have internal parts and can be disrupted under the right conditions. But when they are the constituent parts of a gas at normal temperatures and pressures, they do indeed act as elementary and perfectly elastic particles.

Questions/Discussions

1. How would you describe the difference between changing motion into heat and heat into motion?

2. What do we mean by absolute zero of temperature?

3. Does evaporation from a surface increase or lower the temperature of the surface? Explain on the basis of kinetic theory of heat.

4. Occasionally one hears of temperatures being expressed as millions of degrees. Since all matter is in gaseous state at such high temperatures, how can one measure such temperatures?

5. Throughout this chapter we have disregarded forces of gravitational attraction on the atoms. Can you explain why the moon does not have an atmosphere?

6. Describe Brownian motion and its significance to kinetic theory of gases.

7. Suppose a dust particle of mass 10^{-8} g suspended in air experiences Brownian motion. We observe that in a period of 10 min it travels a total of 0.2 cm in staggering fashion. If in the course of this travel it experiences 10^{10} collisions, what would you say is the *mean free path* between collisions?

8. If a gas mixture contains atoms of different masses, would you expect the velocities to be the same for all the atoms?

9. Discuss the meaning of entropy. Explain its significance for explaining the experiment of Fig. 10.13.

10. Discuss the significance of the entropy concept applied to a container of air molecules. Having in mind that the universe is occupied by very many billions of stars and other bodies, would you expect there to be a parallel application of thermodynamic (entropy) principles to these star bodies?

11. Explain what we mean by pressure exerted by a gas.

12. When clothes are hung to dry at a temperature of $-5°C$ (that is, below the freezing point of water), they will eventually dry. Explain.

13. Distinguish between heat and temperature.

14. Referring to Question 12, how would the rate of drying compare with that of clothes hung out to dry in a temperature of $+5°C$? Why?

15. Define the boiling point of a liquid.

16. It is said that dried beans cannot be cooked in an open saucepan on top of Pike's Peak. Explain. How would you resolve this cooking problem?

References

Angrist, Stanley W., and Hepler, Loren G., *Order and Chaos; Laws of Energy and Entropy*. New York: Basic Books, 1967. *A highly readable account of the discovery of the laws of thermodynamics and their application to technology, culture, and life processes.*

Atkins, K. R., *Physics*. New York: Wiley, 1965.

Brown, S. C., *Count Rumford*. New York: Doubleday (an Anchor book), 1962.

Cowling, T. G., *Molecules in Motion*. New York: Harper & Row (Harper Torchbook), 1950. *An easily read introduction to the kinetic theory of gases.*

Gamow, G., *Matter, Earth and Sky*, 2d ed. Englewood, N.J.: Prentice-Hall, 1965. *Chapter 4 treats heat and thermodynamics.*

Gamow, G., *Mr. Tompkins Explores the Atom.* New York: Cambridge Univ. Press, 1945. Also available in *Mr. Tompkins in Paperback* (same publisher). *Chapter entitled "Maxwell's Demon" treats entropy. A very delightful presentation.*

Holton, G., and Roller, D. H. D., *Foundations of Modern Physical Science*. Reading, Mass.: Addison-Wesley, 1958. *Chapters 19 and 20 give an excellent historical account of the development of our ideas about heat. Chapter 25 is devoted to the kinetic theory of matter.*

Ripley, J. A., *The Elements and Structure of the Physical Sciences*. New York: Wiley, 1964. *Chapters 10 and 11 cover the material of this chapter.*

Schlegel, R., *Time and the Physical World*. East Lansing: Michigan State Univ. Press, 1961. *Several chapters are devoted to time and thermodynamics.*

CHAPTER ELEVEN

Electricity and Magnetism

But nature has in reserve one great surprise—electricity

SIR ARTHUR S. EDDINGTON

FOR THE MOST PART, our studies thus far have been concerned with mechanical aspects of the natural phenomena that surround us. We have studied the effects of interactions between objects making contact, such as the bat and ball, weight and string, and in Chapter 10 we made statistical studies of the reactions between atoms and molecules. We now turn to a field that is not directly related to the mass of bodies, namely electricity and magnetism. One cannot see an electric charge or a magnetic field, but the effects are everywhere about us in communications, transportation, and in every aspect of our daily lives. Electricity has become a most versatile servant of modern mankind.

11.1 The discovery of electric charge

There is evidence that ancient man (as early as 3000 B.C.) knew of the strange properties of amber (the Greeks called it *elektron*). But to Thales goes the credit for first describing how, when rubbed with fur, amber would attract small bits of matter. Amber was a prized material to those who were interested in demonstrating their powers of "magic" and to those who associated its qualities with the supernatural. Plutarch (First Century A.D.) wrote of the "flammeous and spirituous nature" of amber and how, when rubbed, it would attract objects as the lodestone does. There gradually grew from these observations the concept that the object that possessed this property of attraction was charged with *electricity*. Much later, in the seventeenth century, it was discovered that when a charged body touches another object, this object may take on some of the same attraction for other particles of matter, but that it is then *repelled* by the *body* from which it ac-

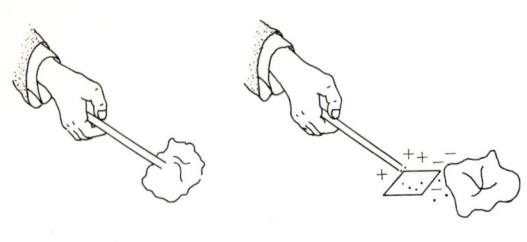

When glass is rubbed with silk it attracts small particles. The silk also attracts particles.

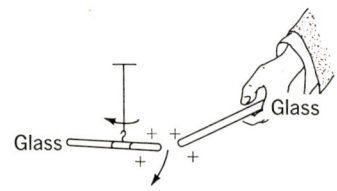

When two pieces of glass are rubbed with silk the two pieces repel each other.

But a glass rod and a plastic or hard rubber rod, rubbed in the same way, attract each other.

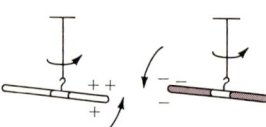

Like charges repel

and unlike charges attract

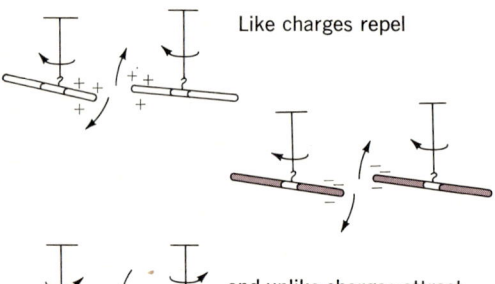

We conclude that the charge that develops on the glass rod is different from that which develops on the plastic rod.

Fig. 11.1. Attraction and repulsion of charged rods.

quired its own *charge*. Also, there seemed to be two kinds of electric charges; the one produced when glass was rubbed (vitreous); the other developed when amber was rubbed (resinous). These two kinds would attract each other, but repel their own kind. Many believed there were two kinds of electrical "fluids" that gave rise to these phenomena. Figure 11.1 shows these effects. Figure 11.2 shows additional effects on conductors and detectors of the two "types" of charges. Here, however, where no contact is made between the two bodies, it will be noted that an opposite charge from the charging body is *induced* onto the object being charged. Figure 11.3 shows the electroscope to be a sensitive device for identifying electric charges, and gives the conditions under which it is charged by contact procedure. But how could one charge the electroscope by *induction?*

Benjamin Franklin† (1706–1790) doubted the two-fluid explanation; he proposed instead that every nonelectrified body contains a "normal" amount of a single electrical "fluid." This fluid, he said, is made up of electric particles, and a body becomes electrified when it has either too much of the fluid and is positive, or has lost some of it and is left with a negative charge.

Meanwhile, the French physicist Charles Coulomb (1736–1806) quantitatively measured the attractive and repulsive forces between charged bodies. From his experiments he found that the repulsive force between two small

---

† Benjamin Franklin played an important role in the American Revolution, and was, like many intellectuals at that time, interested in science. Franklin is known for bringing lightning from the sky with a kite and key in a bottle, thereby "demonstrating completely the sameness of electric matter with that of lightning." His experiments led him to invent the lightning rod.

11.1 The Discovery of Electric Charge

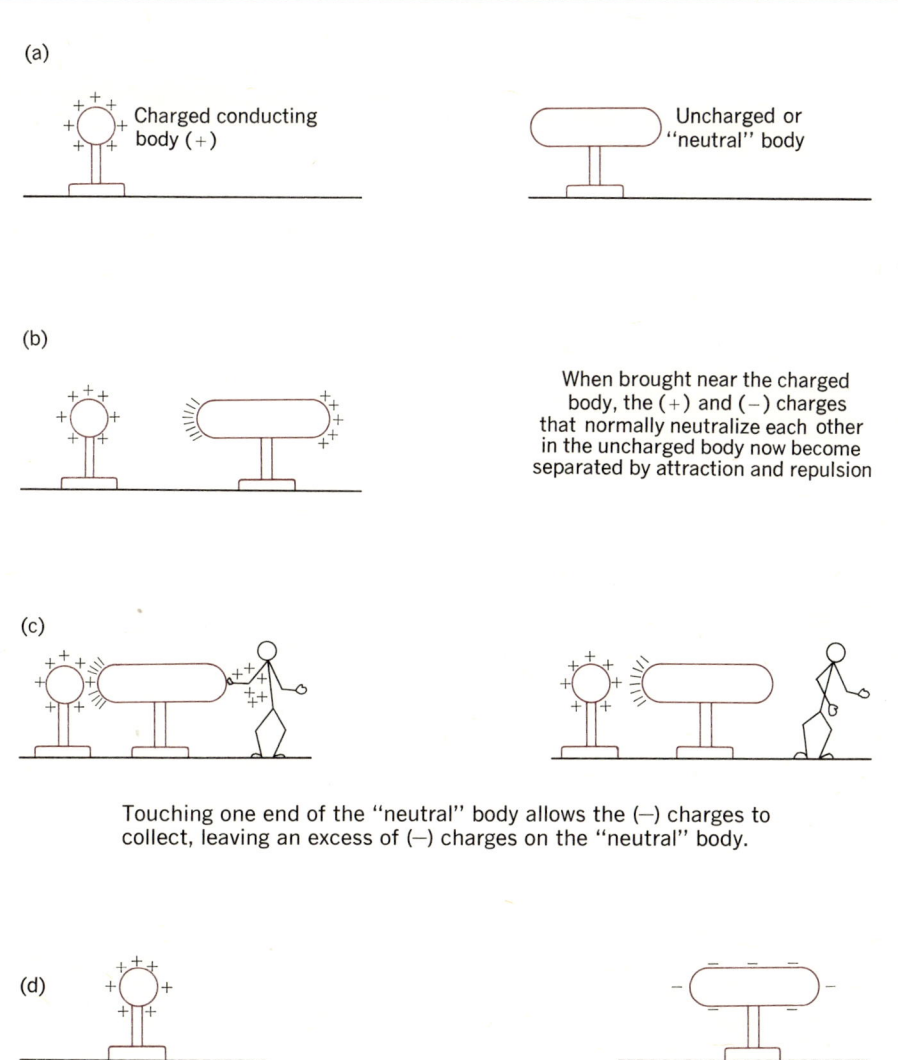

Fig. 11.2. Charging by induction. A conducting body that acquires an electric charge tends to distribute it around its surface because of the mutual repulsion of the individual units of charge. An uncharged body can be made to acquire a charge by the steps indicated in (b), (c), and (d).

spheres charged with the same type of electricity is inversely proportional to the square of the distance between the centers of the two spheres but directly related to the quantity of charge on each sphere. This statement was later summed up in the following equation as Coulomb's law:

$$F = k \frac{Q_1 Q_2}{r^2} \quad (11\text{-}1)$$

where k is a proportionality constant and r is the separation between the two charges Q_1 and Q_2.† When the charges

---

† The unit used in the mks system for the quantity of charge, Q, is the coulomb. The coulomb (C) is a charge of such magnitude that when two like charges of a coulomb each are placed 1 meter apart, they repel each other with a force of 9×10^9 newtons (a newton is equal to 0.225 lb of force). It can be seen that the coulomb is a sizeable charge, and no one is likely to place two like charges of this magnitude 1 meter apart.

Coulomb, in 1784, used the torsion balance and uniformly charged pith balls to determine

have the same polarity the force is repulsive, and is attractive when the two have different polar charges. The reader should be impressed by the resemblance of Eq. (11-1) to the one for Newton's law of universal gravitation.

The public took an interest in electrostatic charges and their effects, and by the eighteenth century, large generating machines had been devised which mounted lumps of amber on a rotating wheel to rub against fixed brushes of cat's fur. It is reported that at the court of Louis XV, Carthusian monks of the Paris convent were arranged in a 900-ft circle

---

Q. He electrified two pith balls of equal size by contact with a charged needle, and measured the force F between them. He then selected a third uncharged pith ball of the *same diameter* and touched it to one of the charged balls. Charge flowed by conduction to the new ball until it was equally shared between them. Coulomb then removed the third ball, and brought the original two to their former positions and measured the force between them. What result would you anticipate?

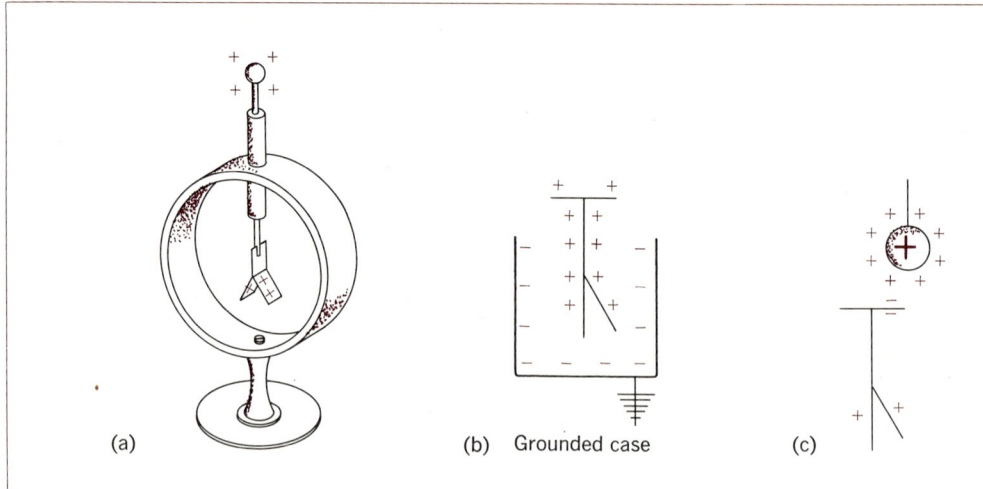

(a) (b) Grounded case (c)

to receive the discharge from a Leyden jar† on which charges had been accumulated. As the story goes, at the moment of discharge the entire line of monks hopped into the air, and found that the experience could "enliven the dull, loosen the tongue, and impart fiery spirits to even the mildest men."

11.2 Magnets and magnetic fields

The Greeks also discovered that lodestone (an oxide of iron), which they found in the province of Magnesia near Miletus in Asia Minor, had the property to attract iron. In fact, when iron was in contact with lodestone, they found it also had the properties of attraction and could influence other pieces of iron. It was not until the time of the Englishman William Gilbert (1540–1603) that a clear distinction was made between the nature of charged particles and magnets. Gilbert found that a magnet always has two poles that are inseparable, while a charged body has a single polarity (positive or negative). When a magnet is broken into two or more pieces, *every part takes on a north and a south pole* (Fig. 11.5). From a long period of observations, it has been found that one pole cannot exist without the other.†

When iron filings are scattered around a magnet they assume the characteristic pattern shown in Fig. 11.6(a). The filings appear to orient themselves along definite lines, as sketched in Fig. 11.6(b). Using terminology that we shall introduce later in this chapter, we call these

† The Leyden jar was the forerunner of the modern condenser or capacitor. It consists of a glass jar coated inside [Fig. 11.4(a)] and outside with a thin metal. The glass, acting as an insulator (dielectric) prevents the two charges from canceling each other. Today's condensers use mica, plastics, paraffin, and other materials as dielectrics between closely spaced sheets of metal to increase the capacity. See Fig. 11.4(b).

† Although the search for the magnetic monopole still goes on, it has not been located at the time of this writing.

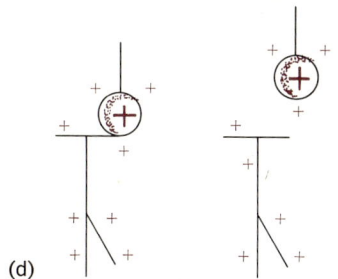

Fig. 11.3. (a) The principle of the electroscope. (b) In practice, the leaves are mounted in a metal container which can be grounded. (c) The electroscope will respond to a charged body that is brought into its neighborhood, but (d) it will not itself acquire a charged state until there is contact with a charged body.

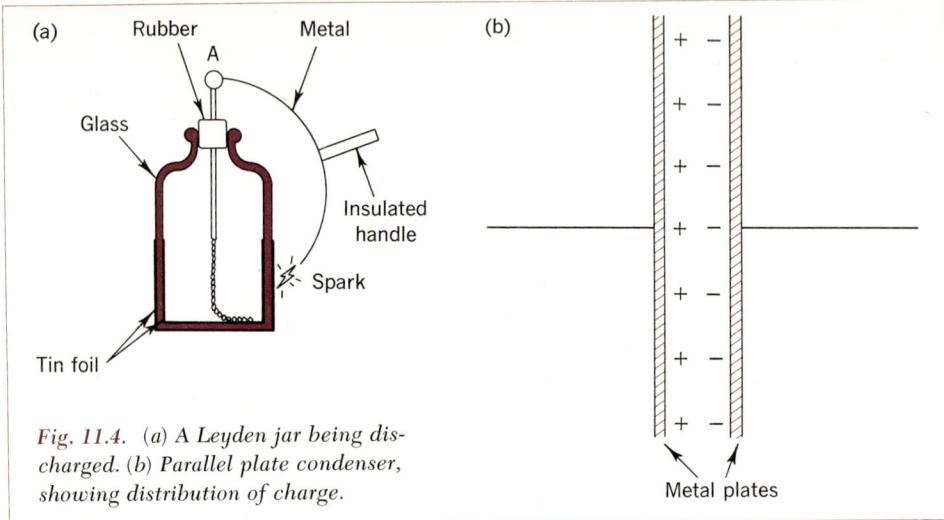

Fig. 11.4. (a) A Leyden jar being discharged. (b) Parallel plate condenser, showing distribution of charge.

lines *lines of force*; the patterns shown in Figs. 11.6(a) and 11.6(b) are representations of the *field* of the magnet in the plane of the paper. A magnetic compass can also be used to explore the field around a magnet.

Gilbert also found that a magnetized needle suspended by a thread will align itself in a north-south direction, with one end of the magnet always seeking north. He concluded, correctly, that the earth is a huge magnet with lines of force running from the north to south magnetic poles, and that the needle was simply aligning itself with the earth's field. Similar to the ways by which the induction of charges takes place, as shown in Fig. 11.2, it is found that induced magnetism occurs in easily magnetized substances (such as iron and nickel) when placed near a strong magnet, as indicated in Fig. 11.6(b). It should be observed that

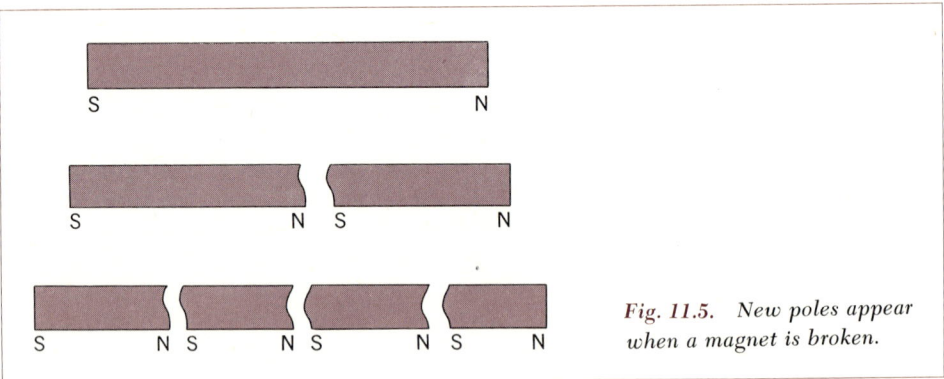

Fig. 11.5. New poles appear when a magnet is broken.

the south pole of the "parent" magnet induces a polarity of the north pole in the induced magnet. The process of induction is an important physical phenomenon in the investigations of electric charges and studies of magnetic fields.

11.3 Electric currents

CHARGES IN MOTION

The leap from electrostatics to electricity in the form of moving charges was due to observations of the frog. In Luigi Galvani's (1737–1798) laboratory in Italy a freshly severed frog's leg was accidentally touched with a wire from a charged electrostatic machine. The leg twitched and thus became the first *galvanometer*.†

Alessandro Volta (1745–1827), another Italian scientist, repeated Galvani's experiment and found that it was not "animal electricity" that made the frog's leg jump, but that the moisture in the leg conducted the electricity from the electrostatic machine to the nerves. Volta continued his studies with a variety of materials in moist media, and finally found that strips of copper and zinc in an acid solution would produce a current in an outside connecting wire. This electrochemical (voltaic) cell produced a

† A galvanometer is an instrument used to detect the presence of a current and its direction of flow.

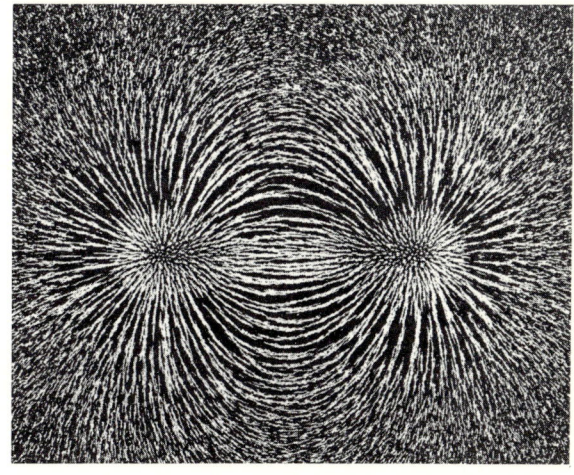

Fig. 11.6. (a) *The magnetic field of a bar magnet as revealed by the iron filings technique. (Courtesy of B. Abbott, Physical Science Study Committee.)*

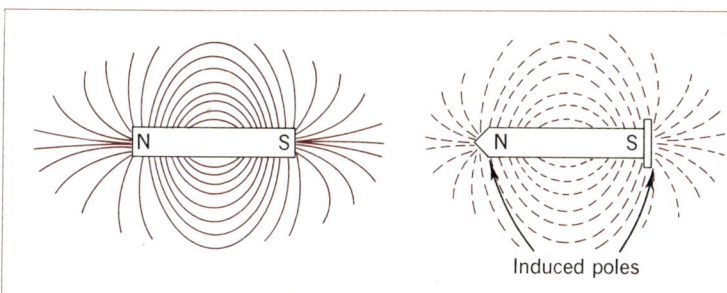

Induced poles

Fig. 11.6. (b) *Sketch of magnetic field about bar magnet at left and about induced magnet at right.*

steady flow of charges and it was much more dependable than the electrostatic generator. Also, by connecting several cells so as to form a battery, Volta found that a much larger charge could be produced in a steady flow.

MAGNETIC EFFECTS OF ELECTRIC CURRENTS

The Danish physicist Hans Christian Oersted (1777–1851), during a lecture demonstration in 1820, caused a charge to flow in a wire above a magnetic compass needle. He was, according to his students, "quite struck with perplexity" to see the needle deviate until it stood at right angles to the wire. Oersted had the presence of mind to reverse the direction of flow of charge in the wire and observed that the needle deviated in the opposite direction, but that it again came to rest at right angles to the wire. This accidental discovery (Pasteur has said, "In the field of experimentation, accidents favor the prepared mind.") was one of the great moments in the history of science. It was to initiate the fusion of electricity and magnetism into the branch of physics known as *electromagnetism*.

Earlier we observed that an electric charge always exerts a force on another charge. Now, from Oersted's experiment, we see that a *moving electric charge*† also exerts a force on a magnet. The amazing and unpredictable element in Oersted's observation was that the magnet set itself perpendicular to the wire carrying the charge (Fig. 11.7). This was not expected because, in dealing with forces between two masses, charges or magnets at rest, we have found the force to act along the line connecting the units, not perpendicular to it.

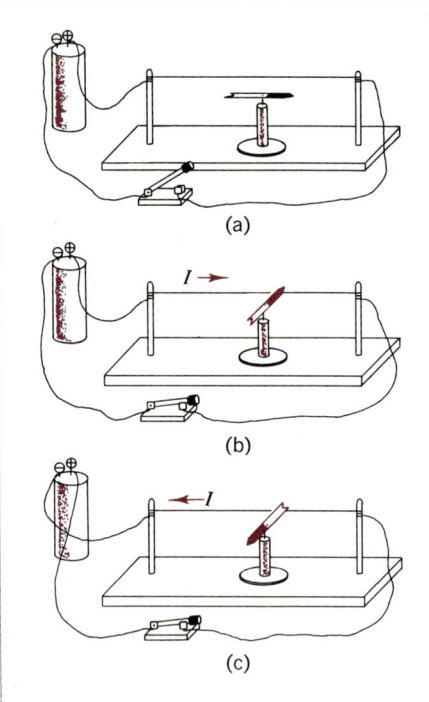

Fig. 11.7. Compass orientations. (a) With no current in the wire; (b) with current in wire; (c) with reversed current in wire.

† The moving charges are the negative electrons that flow through the wire, as we shall see in the next section.

EFFECT OF A MAGNET ON A CURRENT

We have just observed the effect produced by charges moving in a wire near a magnet. We may now ask, will a magnet have a similar effect on a wire that is carrying a current? Michael Faraday (1791–1867) devised an apparatus in 1822 to find answer to this question (Fig. 11.8). A current is passed into the vessel on the left through a wire in the bottom (both vessels contain a conducting medium such as mercury). The current is conducted to the right-hand vessel through a wire that touches the surface of the mercury in each vessel. The bar magnet in the left vessel is tilted so that it is free to swing around the wire, while the wire that touches the mercury in the right-hand vessel is pivoted so it can rotate around the fixed magnet. *When a charge is sent through the system, the pivoted magnet and wire do rotate around the fixed components.* In this apparatus, Faraday had a primitive motor capable of converting electric energy into mechanical work. Faraday was not much concerned with the applications of his discovery, but when asked by Prime Minister Gladstone of its possible use, he replied, "Sir, you may be able to tax it."

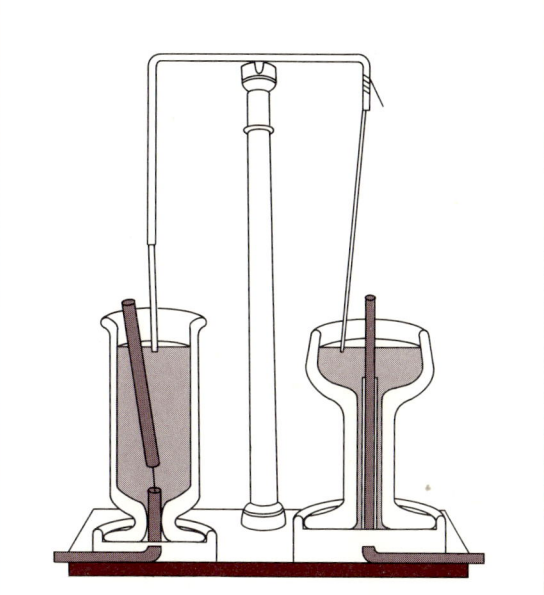

Fig. 11.8. Sketch of Faraday's apparatus for identifying a magnetic field.

INDUCED CURRENTS

From the simple electric motor, let us turn to an experiment conducted in 1831, in which Faraday moved a conductor (such as a coil of copper wire) in a magnetic field. A current was induced in the coil *as long as it was moving across the magnetic field.* It works just as well when the magnetic field moves relative to a conductor; see Fig. 11.9.

It turns out that this principle is more general, for a *changing* current in one fixed circuit will *induce* a current in another fixed circuit (Fig. 11.10).

The discovery of electromagnetic induction is a good example of simultaneous

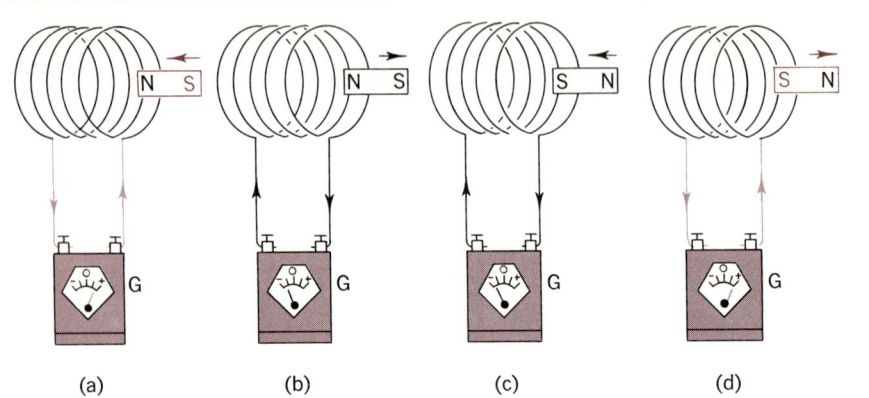

Fig. 11.9. Induced electric currents. (a) When magnet moves through coil, galvanometer deflection indicates current flow. (b) When the motion of the magnet is in the opposite direction, galvanometer deflection is reversed, indicating that the current flow is reversed. (c and d) When polarity of magnet is reversed, current flow is reversed from that of (a) and (b) respectively.

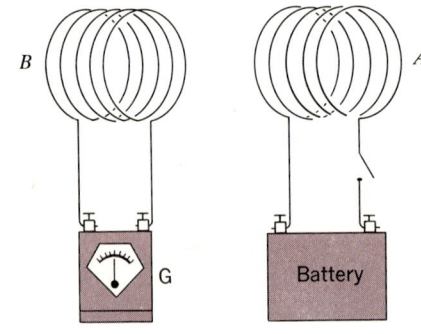

Fig. 11.10. Current is induced in the coil of circuit B while the current supplied by the battery in circuit A is increasing or decreasing; the currents induced in B in the two cases are in opposite directions.

developments in science. Chronologically, Joseph Henry, a teacher at the Albany Academy (New York), was ahead of Faraday in performing induction experiments, but the pressure of his school duties delayed the publication of his findings. At about the same time in Russia, H. F. E. Lenz was engaged in similar studies. He discovered a principle, now known as Lenz's law, which predicts the direction of an induced current. Briefly, the principle is that an induced current must be in such a direction that its own magnetic field will oppose the original action that produced the induced current.

11.4 The electron

From these early discoveries we have learned that charged bodies may attract or repel each other, and that there can be considerable interaction between *moving* charges and *moving* magnets. Let us jump ahead a little and immediately identify the *electron* as a principal actor in electrical phenomena. The electron is a tiny particle with a mass of 9.1×10^{-31} kg and with a negative charge of 1.6×10^{-19} C.†

The particle model of electricity was very tentatively suggested by Faraday in 1832 as a result of his work on electrolysis (see Chapter 16). Faraday found that 96,500 coulombs (C) of charge passing through a solution of common salt (sodium chloride) would deposit 1 gram-atom of sodium (23 g) at the cathode (negative terminal), at the same time liberating about 35.5 g of chlorine gas (1 gram atomic weight) at the positive terminal (anode). According to Faraday's suggested model, each sodium ion in the solution requires the same amount of charge to neutralize and deposit it at the cathode, and since (as we saw in Chapter 10) 1 gram molecular weight of any substance contains Avogadro's number of molecules (atoms in our case), then 6.02×10^{23} atoms require 96,500 coulombs of charge. One atom, therefore, requires

$$\frac{96{,}500}{6.02 \times 10^{23}} = 1.6 \times 10^{-19} \text{ C}$$

Faraday did not actually calculate this value because the numerical value of Avogadro's number had not been accurately determined. Faraday's work was discussed in a series of papers by G. Johnstone Stoney in 1891, who gave the name *electron* to the unit of negative electric charge.† Further investigations of the electron were made by J. J. Thomson, who at Cambridge University, in 1897 used what would be considered today a very primitive cathode ray tube (Fig. 11.11). (Note that the anode, or positive terminal, has a hole through A and B, while the cathode, or negative

---

† There are two major systems of units used in physics: The meter-kilogram-second system (mks) and the centimeter-gram-second system (cgs). In dealing with electricity and magnetism, the mks is more adaptable, so we shall use it. In this system the coulomb is a unit of charge equivalent to the charge of 6.3×10^{18} electrons.

† Stoney actually used the word "electron" to refer to both positive and negative charges. We shall apply it only to the negative charge, in conformity with present-day usage.

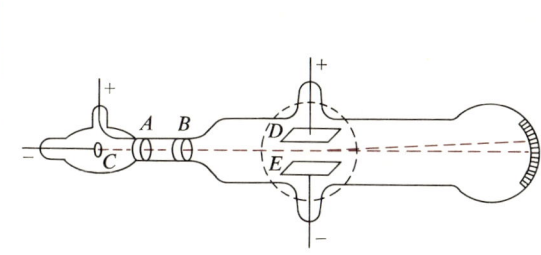

Fig. 11.11. Sketch of cathode ray tube used by J. J. Thompson. A voltage V applied between A and B will accelerate electrons emitted by the cathode C, to an energy of eV electron volts.

terminal, has a thin metal disc at C.) A beam of rays may pass through the holes at A and B into the body of the chamber, which is evacuated, and between plates D and E. When a difference of potential exists between D and E, the beam will be deflected as shown in Fig. 11.11. Thomson made many experiments with this device and found that the "negative rays" were particles with mass. He was able to determine that there was an unique value for the charge-to-mass ratio of these particles. The value thus determined turned out to be more than 1000 times larger than that determined for the electrolysis† of water, and Thomson concluded that his cathode rays were the hitherto undetected electrons.

The first independent measurement of the charge on the electron was made by Millikan in his famous oil drop experiment.‡ (See Experiment 11 in the Laboratory and Mathematics Supplement.) His value for the charge was very close to the 1.6×10^{-19} C given in the preceding equation.§

It was mentioned in an earlier chapter that all atoms have a nucleus made up of protons and neutrons and that surrounding the nucleus are electrons. The neutrons do not have an electric charge, but each proton has a *positive* charge, which is exactly equal in magnitude to the charge of an electron, but of opposite polarity. There are as many negative electrons in a normal atom as there are positive protons, thus making the atom neutral in charge.

We shall find that while the nucleus of an atom is not easily disturbed, the electrons are easily removed from atoms. For example, if a piece of zinc metal and a piece of copper metal are placed in an acid solution, and the two metals are joined by a wire, a current of electrons will flow through the wire as a result of the voltage or potential difference generated at the surface of the two pieces of metal.† We shall find that a flow of current also takes place when a wire circuit is whirled in a magnetic field.

Let us examine electron behavior more closely. To begin with, the electron has mass, and so all the laws that govern the behavior of a billiard ball (Newton's laws of motion) apply to it. It can be accelerated by a force ($F = ma$). It develops momentum and kinetic energy ($p = mv$, $K = \frac{1}{2}mv^2$). It should experience gravitational attraction.

The electron has all these properties, as do other particles, but in addition it also has electric charge. This property gives it a number of additional characteristics to which we shall now turn.

11.5 Conservation of electric charge

There are several instances in which man has camouflaged his ignorance by the

---

† In the electrolysis of water, the hydrogen ion (proton) mass is used in determining the number of electrons per unit mass. The proton mass is nearly 2000 times that of the electron.
‡ See Anderson and Millikan in References for descriptions of the work of Thomson and of Millikan.
§ See Millikan (1960).

† The *volt* is defined as being a potential difference between two points such that 1 joule of work is required to bring a coulomb of charge (6.3×10^{18}e) from one point to another. The unit of current is the *ampere* (amp or I) and it is the flow of 1 C/sec. The unit of resistance in a circuit is the *ohm* (Ω or R), and it is such that a potential difference of 1 volt (V) will cause a current of 1 amp to flow through a circuit of 1-ohm resistance ($V = IR$, Ohm's law). The *watt* (W) is the basic unit of power (P) for electricity and is equal to 1 volt times 1 ampere. The watt-second is a unit of energy equal to a *joule* (J), or volt times ampere times second (or V × C).

use of invisible, massless, and undetectable exchanges of fluid flow. Recall the belief in "caloric" heat flow. Also, in this chapter, we have made reference to both one-fluid and two-fluid mechanisms for explaining electrification processes. As more and more knowledge concerning atomic structure developed during the past half-century a more precise understanding of electric-charge distribution also evolved.

Basic to the accounting system whereby electric charges are interchanged is the concept of *charge conservation*. This assumes that all matter reaches charge equilibrium, whether at the individual atomic level or in macroscopic aggregates; in either case the number of positive charges tends to be neutralized by an equal number of negative charges. It has now been ascertained that negative charges (electrons) are the medium of exchange, whereas positive charges (protons) are found to be less mobile. Why is this so? In Figs. 11.1 and 11.2, the reader will note that a dislocation of a certain number of negative charges from an object A results in its becoming charged positive and *another* object B acquires the surplus of negative charges, becoming charged negatively. In each case the return to charge neutrality occurs when object A receives the same number of negative charges previously removed and the number of surplus negative charges is removed from object B.

It is not surprising, then, that early scientists constructed a scheme of "unseeable fluid flow" to explain charge transfer, since at that time there were no instruments to detect this process. Today although electrons and protons are still "unseeable," we do have reliable detecting instruments for measuring charge transfer. We shall return to the concept of charge conservation in later chapters dealing with chemical and nuclear reactions. Similar to the concepts concerning momentum and energy conservation learned in Chapter 5, we have the concept of conservation of charges which says that the net electric charge in the universe is constant and is not changed by any physical process. This principle of conservation is fundamental to our understandings of electricity in all its phases.

11.6 The electronic properties of matter

We have noted in previous sections of this book that our present view of matter is that it consists of microscopic particles called *atoms*. In Sec. 11.5 we noted that atoms themselves are possessed of charges, although on the whole they are neutral. In this section we briefly note a few of the implications of the charged-particle model of the structure of matter.

If we take a copper wire of a given diameter and length, and establish a potential difference or voltage V across it, a current I will be caused to flow in the wire. The relationship between V and I is (for moderate voltages and currents) given by Ohm's law:

$$V = IR \qquad (11\text{-}2)$$

where R is called the *resistance* of the wire. In general, R depends on the shape and size of the wire used as well as on its temperature and chemical composition. Ohm's law can also be expressed in terms of the electric field E† and *current density J* (where $J = I/A$, A being the cross-sectional area of the wire) in the form

$$E = J\rho \qquad (11\text{-}3)$$

where ρ is called the *resistivity* of the

† See Section 11.7.

material. The resistivity is related to the resistance R, but is independent of geometrical considerations; it is a property of the kind of material of which the wire consists and of its physical state.† Equation 11-3 can be rewritten as

$$E\sigma = J \qquad (11\text{-}4)$$

where $\sigma(=1/\rho)$ is called the *electrical conductivity* of the material. For most metals, σ is large and the material is a good *conductor* of electricity. For glass, porcelain, and most plastics, σ is low, and these materials are poor conductors of electricity. We call them *insulators*.

Conductivity may be described with reference to the kinetic theory model of matter introduced in Chapter 10. In a metal the outer (sometimes called valence) electrons of each atom are not bound to an individual atom, but are free to move around in the lattice structure of the metal. These electrons are termed *conduction electrons*. In the absence of any external electric field, the conduction electrons engage in random motion inside the metal, and collide with each other many times in each second. The average distance, λ, traveled by such an electron between collisions is called the *mean free path*. When a moderate electric field E is applied to the wire, the electrons acquire a second motion, a slow drift in the opposite direction to the field, which is superimposed on the random motion. This drift speed—which is of the order of 10^{-10} of the average (root mean square) random speed of the electrons—is observed as the resulting electric current. A relation can be observed‡ between the resistivity of the metal, the mean speed of the conduction electrons, and the mean free path in the metal.

Although we previously noted that both conductors (high conductivity) and insulators (low conductivity) exist, there also exists a third group of substances with intermediate values of conductivity. These substances are termed *semiconductors*, examples of which are silicon and germanium. These materials will conduct electricity only when the electric field E is very large.

In recent years it has become possible to grow very pure crystals of silicon and germanium, which can be impregnated with minute quantities of certain impurities to produce different effects. If arsenic (having 5 valence electrons) is used as the impurity, the crystal lattice has extra electrons, which increases the number of conduction electrons. Such crystals are called *N-type*. If, on the other hand, aluminum (valence 3) is used as the impurity, the resulting crystal lattice has too few electrons—we say that it contains *holes* through which free electrons may pass and be captured. Such a crystal is termed *P-type*. By joining various N- and P-type crystals, we can make a very useful device known as a *transistor*, a device that has played a large role in the growth of the electronics industry during the past two decades.

In an ordinary conductor we know that a voltage V across a wire will cause a current I to flow in the wire. We have noted that the current is due to the slow drift of electrons in the electric field and that this drift is superimposed on the much larger random (or thermal) motions of the electrons in the wire. The drifting electrons lose some of their kinetic energy, owing to collisions with other electrons, thus increasing the random motions. The change from organized

---

† Different metals have different resistivity. The same metals alloyed or processed in different ways vary in resistivity. Copper or silver are standards against which resistivities of other metals are compared.

‡ See Halliday and Resnick in References.

(drift) motion to disorganized thermal motion produces heating of the wire. Hence, in any normal conductor, some of the energy carried by the current is lost as heat. The amount of energy loss is given by the relation

$$P = IV = I^2 R \qquad (11\text{-}5)$$

where R is the resistance of the wire and P is the power† lost.

Certain substances have the property that when brought to extremely low temperatures, their resistance (and hence power loss) becomes zero. For lead, the temperature needed is about 7°K. Thus, if a current is induced in such a *superconducting* wire and the source is removed, the current will continue to flow indefinitely.‡ Superconductivity is now used to construct electromagnets having very large magnetic fields where the power loss in conventional wires is prohibitively large.

11.7 The concept of field

When we considered gravitational attraction in Chapter 5, we learned that the force of attraction between two masses M and m, is expressed as

$$F = \frac{GMm}{r^2} \qquad (11\text{-}6)$$

where G is the gravitational constant and r is the distance between the centers of the two masses. In his investigation,

---

† Power is the time rate of change of energy and is measured in watts (W) or kilowatts (kW).

‡ Within the limits of our present-day experimental techniques, the resistance and power loss are strictly zero. However, other effects, such as radiation from accelerating electrons in the current loop, will limit the life of this current. Currents have been held in superconducting wires for nearly a year with no measurable loss.

Newton offered no hypothesis to explain the attraction between widely separated masses, but he did reject the then-current concept of "action at a distance."†

Let us consider the force of gravity in a little more detail. We have seen that a body of mass m, in the presence of a second body of mass M, experiences a force F in the direction of the line connecting the centers of the two bodies and directed toward body M of magnitude

$$F = \frac{GMm}{r^2}$$

where r is the distance between the centers of the two bodies. We can interpret this equation as saying that *any* body (of mass m) will experience an acceleration of magnitude $a = GM/r^2$ when placed at a distance r from a body of mass M. Indeed, we could make a "map" of the space around M and label each point in space according to the acceleration (relative to M) that *any* other body would

---

† The concept of action at a distance considered that the force between two separated bodies could act without the presence of any intervening medium. Newton's rejection of this concept is indicated in an excerpt of a letter he wrote to the Rev. Richard Bentley:

"That gravity should be innate, inherent, and essential to matter so that one body may act upon another at a distance through a vacuum and without the mediation of anything else . . . is to me so great an absurdity that I believe that no man who has in philosophical matters a competent faculty of thinking can ever fall into it."

But the arguments for action at a distance were not so easily answered, and at the end of Book III of the *Principia*, Newton declared: "But hitherto I have not been able to discover the cause of those properties of gravity from phenomena, and I frame no hypotheses . . . . To us it is enough that gravity does really exist, and act according to the laws which we have explained, and abundantly serves to account for all the motions of the celestial bodies and of our sea."

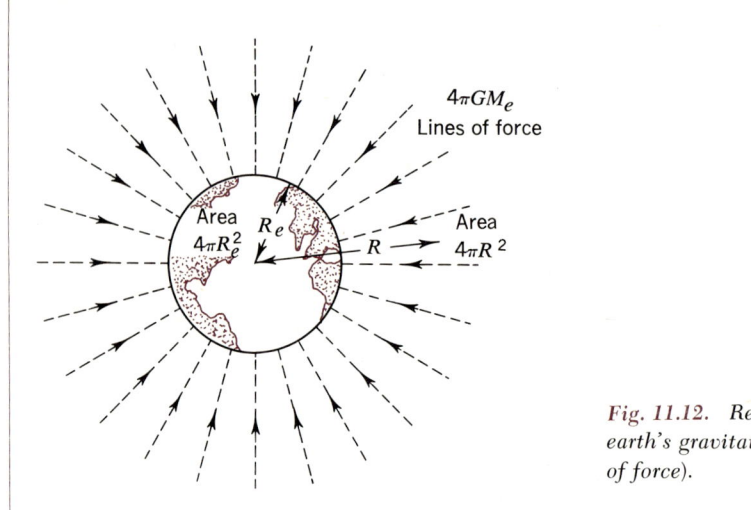

Fig. 11.12. Representation of earth's gravitational field (lines of force).

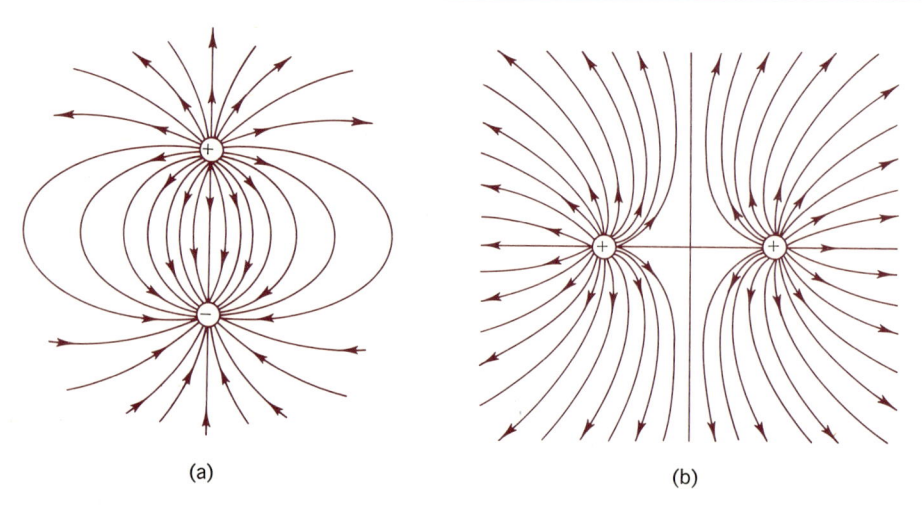

Fig. 11.13. Electrostatic fields. (a) Region of unlike-charged bodies (dipole). Note similarity of pattern to that of magnetic field in Fig. 6. (b) Region of similarly charged (positive) bodies.

have if it were located there. Thus, *every point in space has assigned to it a value of a* due to the gravitation of the mass *M*. Any other body of mass *m* will experience a force $F = ma$ when placed at a point in space where the acceleration has the appropriate value *a*. This array of values *a* is called the gravitational *field* of the mass *M*. Figure 11.12 shows the gravitational field of the earth.

Although never mentioned explicitly, the early developers of the field concept assumed the presence of some medium to transmit the force to an external body. In this respect, the field concept stood in opposition to the concept of action at a distance. We shall see in Chapter 12, however, that no medium is needed, nor indeed is one implied, by the field concept. Instead, we may consider the field itself to be the medium we seek.

Faraday originally introduced the field concept to account for the forces between electric charges at a distance from each other. His admirer, James Clerk Maxwell (1831–1879), the great English theoretical physicist, expanded on the idea and pictured lines of force that emanated from a charge of one polarity and extended through space until they found terminals on an opposite charge. According to this picture a positive and a negative charge are connected in the field by lines of force, as shown in Fig. 11.13(a). Note in Fig. 11.13(b) how the lines of the field are repelled in the vicinity of like charges.

In a similar manner we picture an electric charge $+Q$ with lines of force emanating from it, while an equal number of lines of force come *into* the point charge $-Q$. See Figs. 11.14(a) and (b). At any given point *P* in an electric field, the magnitude of the field intensity *E* is taken as the force *F*, in newtons, that would be exerted on a charge *Q* placed at

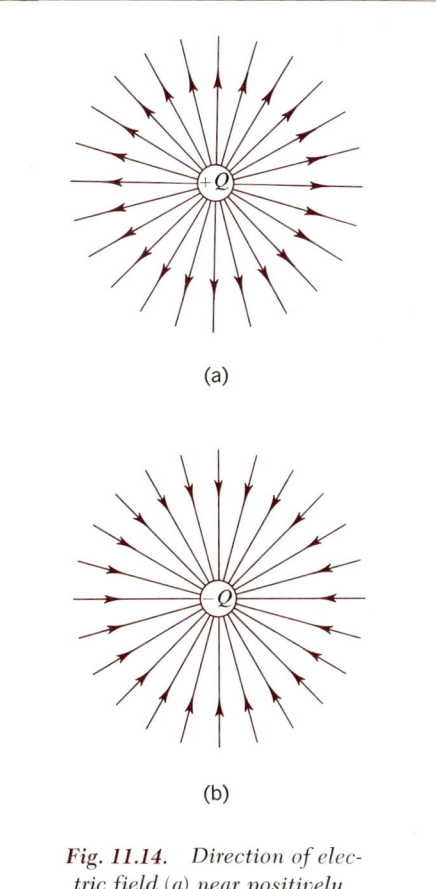

Fig. 11.14. *Direction of electric field (a) near positively charged object; (b) near negatively charged object.*

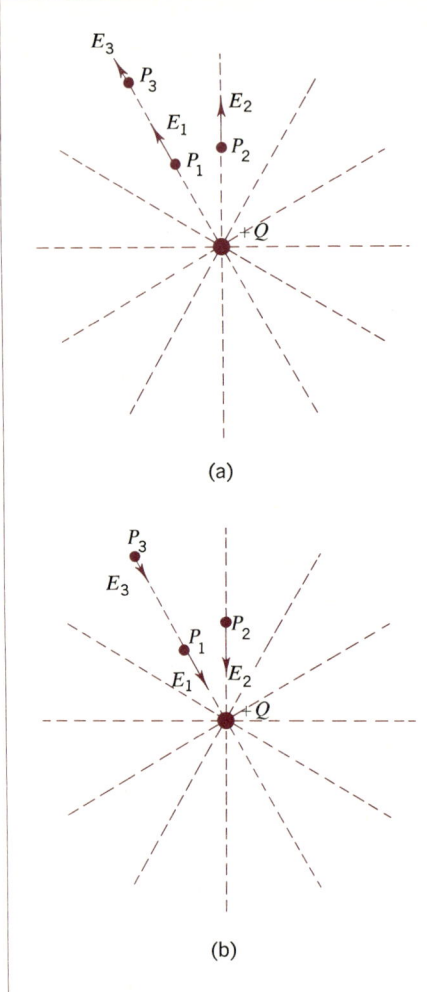

Fig. 11.15. *Force relationships in region of positively charged object. (a) Unit positive test charge; (b) unit negative test charge.*

that point. Thus

$$E = \frac{F}{Q} \quad (11\text{-}7)$$

and the force on any given charge may be computed by using $F = EQ$. When the influence of more than one line of force is present at a point, the net effect is the vector sum of the lines, as in Fig. 11.15(a). The electric field at a point decreases inversely as the square of the distance of that point from $+Q$. A positive unit test charge† located at point P_1, distance r_1, would experience a *repulsive* force along the radius numerically equal to the field E_1 at P_1. A negative unit charge located at that same distance from $+Q$, say at P_2, would experience an *attractive* force equal in magnitude but opposite in direction to the field E_2 at P_2. The same negative unit charge located at a more distant point P_3 would experience a reduced attractive force equal in magnitude to E_3, the field at P_3. These cases are illustrated in Fig. 11.15(b).

A summary of the similarities and differences between gravitational and electrostatic fields reveals that they are unlike in more ways than they resemble each other. It is true that they both give rise to vector forces, and their field intensities are reduced according to the square of the distance between masses or charges. However, the magnitude of the gravitational force between two charges is infinitesimal when compared with the electrostatic force. Furthermore, electric charges may experience either attraction or repulsion, while gravitational forces are always attractive. And, when a wire or magnet or charged particle is in motion in the field of a charge, some very different phenomena

† It is convenient to speak of a *unit* test charge when discussing field strength.

come into play—quite different from anything observed in the gravitational field. In the following sections we shall study some of the effects of charges in motion.

11.8 Moving charges and magnetic field

Up to this point we have been pursuing a reasonably satisfactory analogy between gravitational forces and fields and electric forces and fields. As far as we know, there is no difference in the force of gravitational attraction for a mass that is in motion and that for a stationary one. The situation is quite different, however, for charged particles in motion and for moving magnetic fields.

We shall first consider the case of electric charges flowing in a wire at a constant speed. An electric current is made up of many charged particles (electrons) that pass through a wire circuit under the influence of a battery or of some other source of voltage. Figure 11.16 shows a wire that is part of a circuit in which electrons are flowing. The direction of the electrons is indicated by the arrow.† The motion of the electrons generates a magnetic field with lines of force in a circular pattern around the wire, as shown. If the electrons were to flow in the reverse direction, the direction of the lines of force of the magnetic field would also reverse. As soon as the electron motion ceases, the magnetic field collapses to zero.

This remarkable phenomenon that accompanies moving charges constitutes the basis for the performance of every electric motor, transformer, and almost

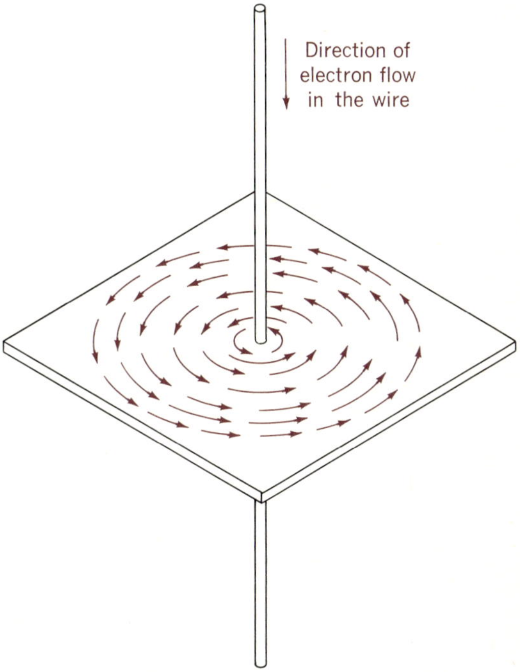

Fig. 11.16. *Orientation of magnetic field about a current-bearing wire.*

---

† It has become customary to identify "current" flow as being opposite to the direction of electron flow.

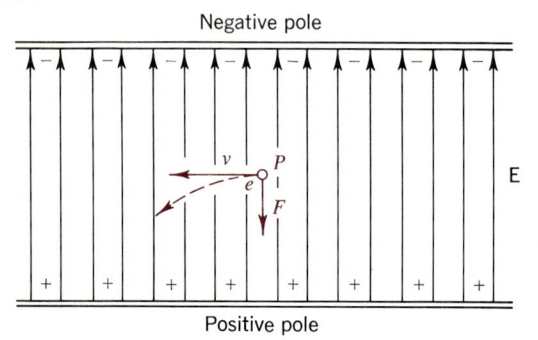

Fig. 11.17. Graphic representation of moving electron in uniform electric field. Note the parabolic-shaped trajectory.

every electromechanical device that we have. We should note that this phenomenon accompanies the *uniform motion* of charged particles. (A little later we shall note that still another phenomenon comes into play when a charged particle is *accelerated.*)

But this is not all. A test electron located near the wire of Fig. 11.16 experiences a repulsive force due to the coulomb forces between that electron and the electrons that are in the wire. The test electron also experiences a force due to the magnetic field in a different direction. The test electron moves under the influence of these forces, and, while in motion, that electron also generates its own magnetic field. These all combine to give a complex interaction pattern of electrostatic and electromagnetic forces.

11.9 Motion of charged particles in electric and magnetic fields

Of the many charged particles with which nature provides us, the electron is of greatest importance. We shall use electrons as the charged particles in our examples. Suppose then, that we have an electron in a uniform electric field, $E = 10{,}000$ newtons/coulomb (or volts/meter) between two charged plates as shown in Fig. 11.17. There will be exerted on this electron a force

$$F = EQ$$
$$= \left(10^4 \frac{\text{newtons}}{\text{coulomb}}\right)(-1.6 \times 10^{-19} \text{ coulomb})$$
$$= -1.6 \times 10^{-15} \text{ newton}$$

The minus sign indicates that the force is opposite in direction to that of the field, that is, the force is downward (see Fig. 11.17). If the electron is originally at rest at P, it will accelerate downward. The force on it is constant, so it falls with a uniform acceleration but *not* the acceleration of gravity, g! We find that by using

Newton's second law, we can calculate the magnitude of this acceleration a as

$$a = \frac{F}{m}$$
$$= \frac{1.6 \times 10^{-15} \text{ newton}}{9.1 \times 10^{-31} \text{ kg}}$$
$$= -1.8 \times 10^{15} \text{ m/sec}^2$$

Compare this acceleration a with the value of g due to gravity (9.80 m/sec²) at the earth's surface. It is quite apparent that the electric field has a much greater influence on the electron's motion than does the gravitational field. However, it is important to remember that Newton's laws of motion still hold (see Chapter 5). If the electron had a velocity v at point P, it would follow a trajectory through the field just like those of the projectile trajectories Galileo studied. One point is very important. The effect of the air upon the electron motion is very great. In an ordinary TV tube, the electrons must be controlled by fields so as to fall on different points of the screen at different instants. If the tube were not evacuated, most of the electrons would never reach the screen at all because of their collisions with air molecules. In a vacuum, then, the motion of electrons in a uniform electric field is analogous to the projectile motion of a body in the earth's gravitational field.

When the desired motion of an electron is linear, or like that of a projectile, we shall most often find electric fields being used. On the other hand, if one introduces an electron or a stream of electrons into a uniform magnetic field, such as the ones shown in Figs. 11.18 and 11.19, the resulting motion will have a circular component due to the force that the electron experiences when moving through such a field.

The force on an electric charge depends not only on where it is, but also on its velocity.

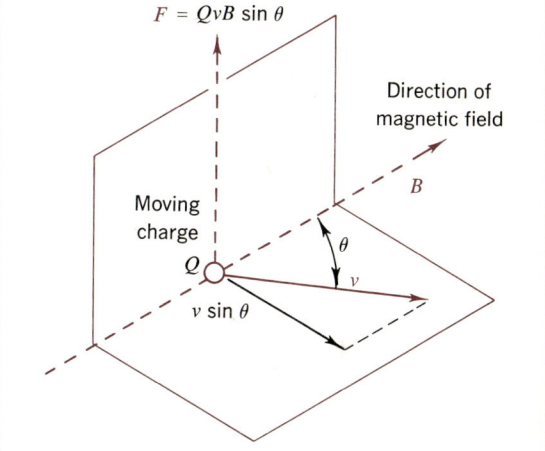

Fig. 11.18. Vector representation of electron moving in the direction of an angle $-\theta$ with respect to a magnetic field **B**.

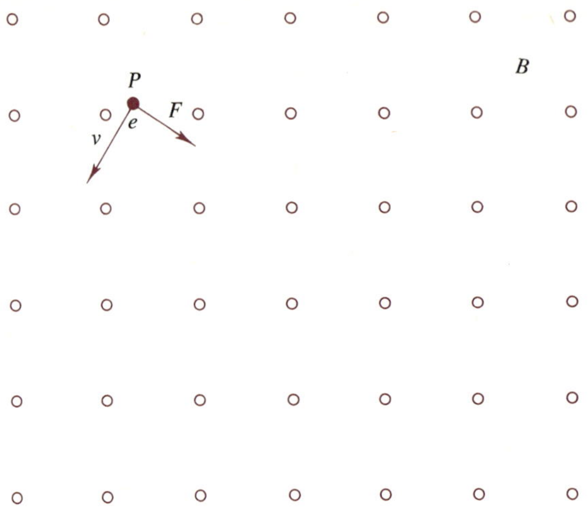

384 Electricity and Magnetism

Fig. 11.19. Centrally directed force acting on electron moving in uniform magnetic field B, with direction into the page.

Every point in space is characterized by two quantities which determine the resultant force on any charge. First, there is the electric field *E*, which gives a force component independent of the motion of the charge. Second, there is an additional force component, due to the *magnetic field*, which depends on the *velocity v* of the charge. This magnetic force has a strange directional character. At any particular point in space, both the *direction* of this force and its *magnitude* depend on the direction of motion of the particle; at every instant the force is always at right angles to the velocity vector. Also, at any particular point, the force is always at right angles to the direction of the so-called *magnetic field vector* **B**, and the magnitude of the force is proportional to the *component* of the velocity at right angles to this specified direction.† These relationships are shown

† The commonly used unit for **B** is given the special name of the weber/meter². An earlier

by the following equation:

$$\mathbf{F} = Q\mathbf{v} \times \mathbf{B} \qquad (11\text{-}8)$$

where the space relationship of the vectors **F**, **B**, **v** is shown in Fig. 11.18.

Since we have already chosen the units in terms of which **F**, *Q*, and **v** are to be expressed, we can say immediately that **B** will be expressed in the units of newton second‡/coulomb meter. (Why?) A pictorial representation of the magnetic field involves *lines of induction*, related to the vector **B** in the same general way as lines of force are related to the electric field vector **E**. But there is as striking a difference between the patterns of electric and magnetic forces. The lines of

―――――

unit for **B**, also commonly used, is the *gauss*, expressed as:

$$1 \frac{\text{weber}}{\text{meter}^2} = 10^4 \text{ gauss}$$

‡ At this point we must identify the symbols that are vectors.

force in an electric field emerge from positive charges and end on negative charges. On the other hand, there are *no* magnetic charges from which lines of **B** can emerge. If we think in terms of "lines" of the vector field **B**, they can never start and they never stop. Then where do they come from? Magnetic fields "appear" *in the presence* of electric currents: Wherever there are currents, there are lines of magnetic field making loops around the currents. Since lines of **B** do not begin or end, they will always link back on themselves, making closed loops. But there can also be complicated situations in which the lines are not simple closed loops. Whatever they do, however, they never diverge from a point. No magnetic "charges" have ever been discovered.

Returning now to the motion that results when an electron with some initial velocity v is subjected to a uniform magnetic field **B**, consider Fig. 11.19.

An electron at P is moving with velocity **v**. The force† on it must be perpendicular to *both* lines of field **B** and velocity vector **v**, and is in the direction shown. This results in uniform circular motion of the electron.

---

† The force **F**, has the magnitude

$$F = QvB \sin \theta \quad (11\text{-}8a)$$

and in this case $Q = e$ and $\theta = 90$ deg, so $\sin \theta = 1$, and therefore

$$F = evB$$

where e is the charge on the electron. Since the electron is in motion on a circle of radius R with speed v, we can set

$$evB = \frac{mv^2}{R} \quad \text{or} \quad R = \frac{mv}{Be} \quad (11\text{-}9)$$

Another way of writing this equation is

$$\frac{e}{m} = \frac{v}{BR} \quad (11\text{-}10)$$

By the proper choice of field one can produce linearly accelerated or projectile motion on the one hand and circular motion on the other. The whole problem of charged particle dynamics is of great importance in such diverse situations as TV and oscilloscope tubes, particle accelerators, electron microscopes, and mass spectrometers. By using electric and magnetic fields simultaneously, almost any desired motion can be imparted to a stream of electrons.

Moving charges thus exhibit properties that static charges do not seem to have. In fact, moving charges are always accompanied by magnetic fields, and other charges moving through these fields experience magnetic forces. In this respect the behavior of charges is quite different from the behavior of masses, for no phenomenon analogous to magnetic fields is known to occur for moving masses. The magnetic field produced by a moving charge and the force exerted on a moving charge are both indicative of a possible unification of electricity and magnetism in the more comprehensive concept of the electromagnetic field.

11.10 *Accelerating charges*

INDUCTION

Both Faraday and Henry discovered, almost at the same time, that a changing magnetic field produces an electric current in a conductor. In Fig. 11.20 we have a method of illustrating this phenomenon as follows: Loop I contains a source of current, the battery, and a switch to turn the current in the loop on and off, as desired. Loop II contains only the current-detecting device, the galvanometer. When a steady current is flowing in loop I, there is no deflection of the needle in the galvanometer of loop II, but when

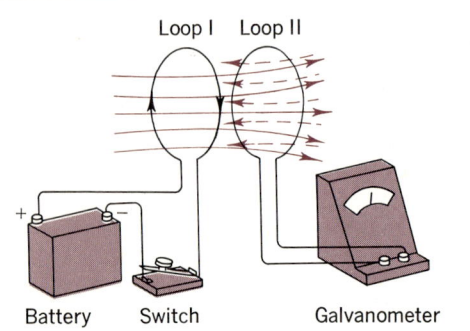

Fig. 11.20. Arrangements for showing induced current production (Note that there are no material connections between loop I and loop II.) The solid arrows represent the magnetic field set up by the current in loop I. The dotted arrows in the reverse direction represent the magnetic field generated by the induced current that flows in loop II.

the current in loop I is switched on and off, the galvanometer is activated in the second circuit. Thus a current is induced in loop II when the current in loop I is changing. This phenomenon is known as *electromagnetic induction.*

An explanation of this event can be made in terms of accelerating electrons. The moving electrons in loop I produce a magnetic field extending to loop II, and when the current in loop I is changing, the magnetic field in loop II is changing. A changing magnetic field is accompanied by an electric field, some of the lines of force of which lie inside loop II, setting some of its electrons in motion.

As far as the electromagnetic induction is concerned, the same result is obtained if loop II is moved relative to loop I when a steady current is flowing. This has the same effect as rotating a conducting coil in a magnetic field to produce a current (the principle of the electric generator).

In addition to the electric generator, a great number of physical and technical applications of electricity and magnetism depend on the law of induction. The dynamo, the transformer, the induction coil, and innumerable other kinds of apparatus and machines are appliances for inducing electric currents by means of changing magnetic fields.

MAXWELL'S THEORY

Maxwell undertook, from 1864–1873, to combine all known electromagnetic phenomena into a unified theory. He was anxious to fuse the field ideas of Faraday with the newly discovered phenomena of electromagnetic induction. He postulated the idea that a changing electric field is always accompanied by an electric *displacement current,* not only in matter, but also in the ether.† He

---

† The ether concept was used extensively in the nineteenth century to explain how light

visualized this current as the separation and flowing of electric fluids in "empty" space or in the insulating space between the plates of a condenser, for example. The displacement current was not pictured as a made up of physical charge, but rather as a kind of "strain" or "displacement" in the medium, which had the same effect as if true charges were moving (or as if a true current were flowing).

In 1864, Maxwell presented to the Royal Society a paper entitled "On a Dynamical Theory of the Electromagnetic Field," which contained his now famous four equations.

Without trying to express these equations in their elegant mathematical form, let us look at the ideas implied:

(1) If electric lines of force, E, diverge from a point or converge on a point, there is a charge located at that point.

___

and other types of waves could travel in space. The ether was a very subtle medium, thought to exist everywhere in space and even between the atoms of matter. To do all required of it, the ether had to be a frictionless fluid, weightless, but still firm and sufficiently elastic to propagate waves.

(2) Magnetic lines of force, B, cannot start at a point or end at a point; they always form closed loops.

(3) A changing magnetic field produces an electric field.

(4) A changing electric field produces a magnetic field.

From statements (3) and (4) it follows that when a charge is accelerating, the electric and magnetic fields it produces will change with time. The mathematical equations of Maxwell's theory can be used to show that an oscillation of an electric field is coupled with an oscillation of a magnetic field at right angles to it. Furthermore, the directions of these oscillating field vectors are at right angles to the direction in which the energy is traveling as a wave (Fig. 11.21).

To illustrate the production and character of such an electromagnetic wave, let us consider the transmission and reception of a radio wave. The transmitting aerial is a long metal rod or wire carrying an electric current that changes its direction of flow with a frequency which (in a typical case) may be a million times per second. We have shown that an electric current is due to the motion of the free

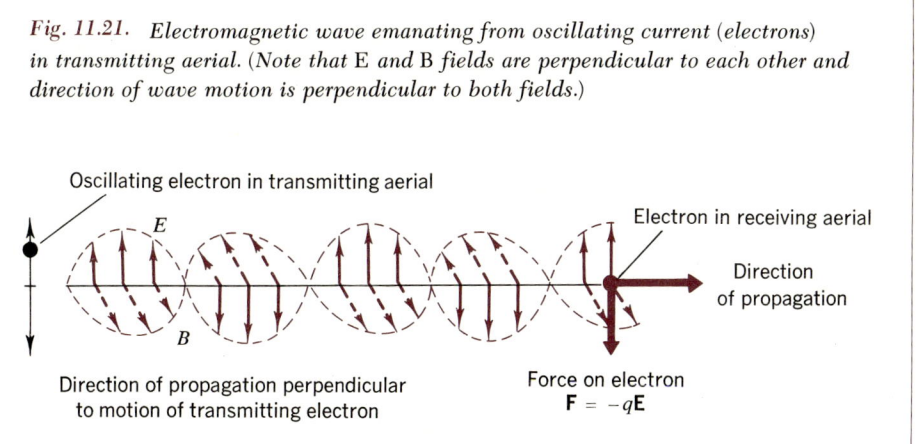

Fig. 11.21. Electromagnetic wave emanating from oscillating current (electrons) in transmitting aerial. (Note that E and B fields are perpendicular to each other and direction of wave motion is perpendicular to both fields.)

electrons in a conductor. In the present instance, each free electron performs a simple harmonic motion with a frequency of 1 million cycles per second (10^6 sec^{-1}). During its simple harmonic motion the oscillating electron is accelerated; as it continually changes its position and velocity, the electric and magnetic fields it produces in its immediate vicinity are continually changing. In accordance with Maxwell's equations, these changing fields produce other fields in their vicinity and the disturbance travels outward with the speed $c = 3 \times 10^8$ m/sec.†

When Maxwell's equations are applied to this situation and solved, it is found that (except in the immediate vicinity of the oscillating electron) the outgoing electromagnetic wave has the form shown in Fig. 11.21. In this diagram we are considering the electric and magnetic fields at a single instant of time at *points along a line*, which for convenience is taken to be in a direction perpendicular to the motion of the electron. At each point the electric and magnetic fields are each represented by a vector. The electric field *E* is seen to be parallel to the motion of the electron, but its direction and magnitude vary along the line in such a way that the tip of the vector traces out a sine wave function. The magnetic field *B* is perpendicular to the electric field and to the direction of travel, and it traces out a similar sine wave function having the same phase as that of the electric vector.

† $c = \sqrt{\dfrac{k}{k^L}} = \sqrt{\dfrac{9.0 \times 10^9 \text{ newtons m}^2/\text{C}^2}{1.0 \times 10^{-7} \text{ newtons/amp}^2}}$

$= \sqrt{9 \times 10^{16} \text{ m}^2 \text{ amp}^2/\text{C}^2} = 3 \times 10^8$ m/sec

k = constant in Coulomb's law of charges
k^L = constant in law of magnetic force (force exerted between two long wires a meter apart and each carrying a current of 1 amp)

The diagram in Fig. 11.21 represents the electric and magnetic fields at points along the direction of radiation *at a fixed instant of time*. The whole pattern should now be imagined to move along the line away from the electron with the speed c. As various parts of the pattern pass over a fixed point, the vector representing the electric field oscillates between a maximum value in an upward direction and an equal value in the downward direction. Similarly, the magnetic field oscillates in and out of the page. It is important to realize, however, that we are describing the electric and magnetic fields in free space at *points* along a straight line and that *nothing* is moving in a direction perpendicular to this line. The significance of the picture is that if an electron *were* present at any point in space, the oscillating electric field at that point would exert a force on it which would oscillate at the same frequency as the field. The electron would be made to perform simple harmonic motion, and hence radiate, just as the electrons in the transmitting antenna did. This is what actually happens when the electromagnetic wave encounters the receiving antenna in which free electrons are present. Thus the oscillations in the transmitting antenna are reproduced in the receiving antenna.

Figure 11.21 shows that both the electric and magnetic fields are perpendicular to each other and to the direction of propagation of the radiation.

Normal alternating currents in our home circuits have frequencies of the order of 60 cps. In a laboratory experiment in 1888, Heinrich Hertz of Germany succeeded in verifying Maxwell's theory for frequencies near 100 million cps (expressed as 100 *megahertz*, MHz, shortwave radio radiation). We shall see later that at much higher frequencies

TABLE 11.1
TYPES OF ELECTROMAGNETIC RADIATION

Frequency, cps	Wavelength, cm	Method of Production and Use
60	5×10^8	Electric Power Utility
0.5×10^6 to 2×10^7	6×10^4 to 1.5×10^3	Radio Broadcast Range
4×10^7 to 2×10^8	7.5×10^2 to 1.5×10^2	Television and FM Radio
10^9 to 3×10^{11}	30 to 0.1	Microwaves
3×10^{11} to 4.3×10^{14}	0.1 to 7×10^{-5}	Invisible Heat Rays from Hot Bodies (radiator)
4.3×10^{14} (red) to 7.5×10^{14} (violet)	7×10^{-5} to 4×10^{-5}	Visible Range (lamps, sun)
7.5×10^{14} to 10^{16}	4×10^{-5} to 3×10^{-6}	Ultraviolet Light
10^{16} to 3×10^{20}	3×10^{-6} to 10^{-10}	X-rays
10^{18} to at least 3×10^{20}	3×10^{-8} to 10^{-13} (or lower)	γ-rays and Bremsstrahlung, Nuclear Reactions Using Accelerators

(say, 10^{14} to 10^{15} cps) we get visible light. We refer to the radiation obtained at different frequencies as the *electromagnetic spectrum* (see Table 11.1).

We may now ask: "Although at a distance from the accelerated charges we measure time-varying E and B fields, do the variations come simultaneously with the changes in the velocity of the charged particles?" Earlier in this section we answered this question in the negative (without proof) and gave a value for this finite speed c of electromagnetic radiation.

Now that we have identified the Maxwellian electromagnetic radiation in part with visible light, we shall henceforth talk about c as the speed of light. Several experiments and observations have shown that this same velocity applies to *all* electromagnetic radiation.

The question as to whether the speed of light is finite or infinite continued unanswered for many years.† Galileo attempted to measure it by using a lantern and shutter arrangement and concluded that the propagation of light "If not instantaneous, it is extraordinarily rapid." Ole Roemer in 1675 gave the first good approximation when he concluded that the reason the eclipses of Jupiter's moons were observed to occur later, when the earth was on the opposite side of the sun from Jupiter (as compared to when we were on the same side), was due to the finite time needed for the light to cross the earth's orbit. Since 1675 several different experiments have been conducted to measure the speed of light. The best value at present is $299,729,900 \pm 200$ m/sec.

Thus we see that light—or more generally the electromagnetic radiation emitted by accelerated charged particles

† See J. H. Rush in References.

—is propagated outward at a speed c, which is nearly 3×10^8 m/sec, or 186,000 miles/sec. Although this very large speed is considered as infinite in most everyday activities (that is, we do not have to consider any time delay when we strike a match and see its light), the finite value of c has very great significance in modern physics, as will be evident in later chapters. This finite speed also imposes some limitations on long-distance communications. For example, the *Early Bird* communications satellite is in a nearly circular orbit about 3.6×10^7 m above the earth's surface. The time needed for a signal to go from the United States via *Early Bird* to, say, England (neglecting any delay times in the satellite equipmeant) would be greater than

$$\frac{2 \times 3.6 \times 10^7 \text{ m}}{3 \times 10^8 \text{ m/sec}} = 0.24 \text{ sec}$$

The delay time for a round-trip signal from earth to *Mariner IV* space probe as it neared Mars was approximately

$$\frac{2 \times \text{distance}}{c} = \frac{2 \times 235 \times 10^9 \text{ m}}{3 \times 10^8 \text{ m/sec}}$$
$$= 1567 \text{ sec}$$
$$= 26 \text{ min}$$

Thus, if a button on earth were pushed at 10 A.M., the signal would reach *Mariner IV* nearly 13 min later, and the signal from the space probe that the command had been received would return to earth at about 10:26!

To extend the idea of the finite speed of light, we noted in an earlier chapter that a standard unit of astronomical distance is the light-year (10^{16} m) and that the *nearest* star to our solar system is 4.3 light-years away; that is, it takes 4.3 years for the light emitted by this star to reach the earth.

We now summarize the properties of electromagnetic radiation.

(1) Radiation is emitted by accelerated charged particles.

(2) It behaves like a wave and therefore is subject to interference and diffraction phenomena.

(3) It behaves like a transverse wave and can be polarized.

(4) Its energy is proportional to the square of the amplitude of either the **E** or **B** field which comprise it.

(5) Its intensity decreases as the square of the distance traveled in free space.

(6) It has a spectral range from very long waves (low-frequency $f \approx 0$) to very short waves (high-frequency, $f \to \infty$).

(7) It is propagated with a finite velocity, $c = 3 \times 10^8$ m/sec, in free space.

Although we have stressed the wave nature of light in detail, it must be noted that in later chapters we shall encounter several phenomena explainable only if light behaves like a particle. This apparent paradox has led to the picture of a wave-particle duality, which plays an important role in modern physics.

11.11 Summary

Early in this chapter we considered some observations of the Greeks that were included under the term *electrical phenomena*. These early observations were surrounded by mystical ideas and explanations. However, in 1600, a new development began with Gilbert's publication of his systematic observations of electricity and magnetism. A century and a half later, Franklin became excited over the experiments he could perform with electricity using simple equipment and with the help of nature. The outcome of Galvani's experiment with frogs' legs was Volta's electric cell. The work of these men and others stimulated widespread interest in electricity in the eighteenth century.

Meanwhile, knowledge of the laws describing electrical and magnetic phenomena was developing, and it became important to consider the space surrounding electrified or magnetized bodies. Although Coulomb's law had a formal similarity to Newton's law of gravitation, significant differences were found between the transmission of the several types of forces. One difference, simple to handle mathematically, is the presence of repulsive as well as attractive forces. (Up to the present no repulsive gravitational force has been detected.) Another difference is the direction of the forces involved. Whereas a gravitational force is transmitted in a straight line between two masses, the direction of a magnetic force curves about the magnetized body. Furthermore, in the interaction between moving electrically charged particles (that is, a current) and magnetized bodies in the neighborhood, the force between them is not in the straight-line path between the two, but at right angles to it.

For these reasons the concept of a *field* was developed. This concept describes how the magnetic and electric forces vary in both strength and direction at every point in space in the neighborhood of electrically charged or magnetized bodies, whether at rest or in motion. Such a formulation of what happens in space was begun by Faraday and refined by Maxwell in the nineteenth century.

Space itself is now endowed, so to speak, with physical properties, handled best through the field concept. The strength of the force field at each point is designated by a vector quantity called the *field intensity* which varies continuously from point to point in direction and magnitude. Furthermore, it has been shown that if within any domain of space there is a variation in the intensity of the electric field, there is a corresponding variation in the magnetic field, with the variations, however, being at right angles to each other. The complementary consequence holds for variations in the magnetic field. Therefore, if an oscillation occurs in either the magnetic or the electric field in any one portion of space, the oscillation will spread out into neighboring portions of space. It was Maxwell's bold speculation that such electromagnetic vibrations or waves must exist and spread out with constant velocity in all directions whenever a current is changing. He further speculated that since the speed of propagation of such a wave would be equal to the speed of light, light itself must be an electromagnetic vibration. His speculations were soon confirmed. Another beautiful generalization had been found, ultimately bringing together into one category the apparently dissimilar phenomena: radio waves, radiant heat waves, visible light, ultraviolet light, and X-rays.

A few of the highlights of this development have been traced in this chapter. Some points need recapitulation and emphasis. The electrostatic forces between "elementary" charges (electrons or protons) are incomparably greater than gravitational forces between elementary masses (atoms, electrons, protons). The apparent magnitude of gravitational effects arises because of the tremendously large accumulation of atoms in bodies such as the earth. This accumulation is possible because there is no gravitational repulsive force. A similar accumulation of one type of charge, positive or negative, is impossible.

Because of the very large accumulation of mass in the earth, the intensity of the earth's gravitational field appears very nearly uniform at the distances from the surface of the earth where man usually operates. But the intensity of electric fields varies with great rapidity over very short ranges (that is, distances less than a centimeter). The intensity of electric and magnetic fields had to be taken into account in the earliest stages of investigation of these fields.

It has been shown that a moving charge gives rise to a magnetic field whose direction is at right angles to the direction of motion. If the charge is accelerating, that is, changing either its direction or speed, the resulting magnetic field must also be changing. If the charge is vibrating with simple harmonic motion, the resulting magnetic field will oscillate correspondingly, and this in turn will produce an electric field, similarly vibrating, in the surrounding neighborhood. Since a current is constituted of electrons (charges) in motion, an alternating current can be used to propagate the electromagnetic radio waves predicted by Maxwell.

Questions/Discussions

1. How could you verify that an insulated rod is electrically charged? If the insulated rod is electrically charged, how could you determine the nature (+ or −) of the charge?

2. Given two iron bars which are similar except that one is magnetized. How would you determine which one is magnetized?

3. How many electrons flow through a cross section of a wire each minute if it is carrying 10 amp of current?

4. Prove that the product of a volt and an ampere is equal to a watt of power.

5. The electron volt (eV) is a unit of energy like the joule (J) or the erg (1 J =

10^7 ergs). Find the equivalent energy of 10 billion eV in joules.

6. A charged particle enters a uniform magnetic field that is at right angles to its direction of motion. How will the magnetic field affect the particle's mass, charge, speed, momentum, and kinetic energy?†

7. An automobile battery produces a potential difference of 12 V between its terminals. A head light bulb is connected directly across the terminals and dis-

† Your answers will be different after we study relativity theory in Chapter 12.

sipates 20 W of power (joule heat). What current will it draw and what is the resistance of the lamp?

8. Two electrons in a vacuum are 6 cm apart. Calculate the gravitational and electrostatic forces between them.

9. What is the wavelength of a radio wave of 1 MHz? Compare this wavelength with that of X-rays (10^{18} cps).

10. Find the speed of an electron that has "fallen" through a potential difference of 2000 V. What is its energy in joules? What is this energy in electron volts?

References

Anderson, David W., *Discovery of the Electron.* Princeton, N.J.: Van Nostrand, 1964.

Atkins, K. R., *Physics.* New York: Wiley, 1965. *A well-illustrated modern approach.*

Beiser, A. (ed), *The World of Physics.* New York: McGraw-Hill, 1960. *A collection of some readings in the history and nature of physics.*

Bitter, F., *Magnets.* New York: Doubleday (Anchor books), 1959. *An elementary treatment of magnetism.*

Bonner, F. T., and Phillips, M., *Principles of Physical Science.* Reading, Mass.: Addison-Wesley, 1957. *Chapters 15, 17, and 18.*

Boorse, H. A., and Motz, Lloyd, *The World of Atoms,* 2 vols. New York: Basic Books, 1966. *An excellent source.*

Halliday, D., and Resnick, R., *Physics,* 2d ed. New York: Wiley, 1964.

Holton, G., and Roller, D., *Foundations of Modern Physical Science.* Reading, Mass.: Addison-Wesley, 1958. *Chapters 26–29.*

Kaempffer, F. A., *The Elements of Physics.* New York: Blaisdell, 1967. *A new approach to the major topics in physics.*

Magie, W. F., *A Source Book in Physics.* New York: McGraw-Hill, 1935. *Extracts of important contributions of Oersted, Faraday, Maxwell, and others.*

Maxwell, J. C., *A Treatise on Electricity and Magnetism.* New York: Dover, 1954. *Maxwell's Preface is good.*

Miller, Franklin, *College Physics.* New York: Harcourt, Brace, 1954. *An excellent reference for problems and exercises (with solutions).*

Millikan, Robert A., *The Electron.* Chicago: Univ. of Chicago Press, 1960. *Describes some of Millikan's early work on the electron.*

Ripley, J. A., *The Elements and Structure of the Physical Sciences.* New York: Wiley, 1964. *Chapters 13 and 14 especially recommended.*

Rush, J. H., "The Speed of Light," *Scientific American* (August 1955), p. 67.

CHAPTER TWELVE

The Theory of Relativity

Henceforth space by itself, and time by itself, are doomed to fade away into mere shadows, and only a kind of union of the two will preserve an independent reality.

HERMANN MINKOWSKI

THE YEAR 1905 was made notable by a number of milestones in the history of science. These included the first mathematical analysis of Brownian motion, which for the first time seemed to give mathematical support to the atomistic hypothesis that we studied in some detail in Chapter 10. The concept of the photon was introduced along with an explanation of the photoelectric effect. We shall study this also as one of the major steps in developing atomic science. There was introduced also the special theory of relativity, which is the subject of this chapter. But the more remarkable feature was that all three accomplishments were achieved by a single individual, a young man by the name of Albert Einstein who was a minor consultant at the Swiss Patent Office at Berne.

12.1 Introduction

Several centuries of research had given Newtonian mechanics an unquestioned position in the conceptual and practical realms of science. A whole industrial revolution followed its development. Maxwell's theory of electromagnetic radiation seemed destined to find an equally firm place in science. Yet, with a paper that carried the unpretentious title of "On the Electrodynamics of Moving Bodies," Einstein dealt a body blow to the science of the day when he published his theory of relativity.

The theory of relativity was not immediately or generally accepted by Einstein's colleagues. Before a decade had passed, however, it was successfully applied by Sommerfeld to the extension of Bohr's theory of the atom. By the 1930's, abundant experimental verification of the new theory was evident in a wide range of applications. Until quite recently those who were not acquainted with the physical sciences considered the theory to be among the most difficult concepts to comprehend. We shall attempt to prove in the succeeding sections of this chapter that relativity theory in itself is neither complex nor difficult to understand. Why then the belief about its great difficulty? The difficulty lies not in its own specific concepts, but rather in the fact that it is at variance with some of our most strongly held ideas, especially ideas that have firm roots in everyday personal experience and in the experience of a sophisticated, technological civilization. Perhaps the greater lesson to be derived from this chapter is that personal experiences that have not been analyzed are very inadequate as guides to understanding the basic laws and features of natural phenomena. To some extent, common sense is actually the seepage down to the population of the accepted philosophical and scientific theories. Galileo's ideas were against "common sense" in the seventeenth century. Newtonian physics actually replaced the old common sense. It is always difficult to renounce ideas that have given one a meaningful world and which seemed to be, in technology, so useful. The people had begun to see the world of technology as concrete proof of the Newtonian world machine.† But a change in such ideas was exactly what Einstein asked of theoretical physicists and of all other men too.

Let us look first at some of the physical concepts of the prerelativity period, and prepare the way for the Einstein revolution.

† The experience is not uncommon. It was not easy to accept the idea of negative numbers, for how can one subtract 8 from 3? It is said that even Descartes, who gave us Cartesian coordinates, hesitated to extend the abscissa and ordinates to the left and below zero.

12.2 The concepts of absolute space and absolute time

In the Aristotelian picture of the universe all motions were considered to be referred to some fixed and absolute *frame of reference*,† in this case the center of earth (which was also the center of the universe). All objects either had a fixed position or were in a state of motion with respect to this earth frame of reference.

Galileo, and later Newton, established the law of inertia, according to which a body in uniform rectilinear motion would, in the absence of external forces, continue in this state of motion forever. The concepts of motion and position demand a frame of reference relative to which the motion or position can be measured. The center of the earth had earlier served Aristotle as a unique reference point. But the Copernican theory, which was accepted by both Galileo and Newton, had removed the earth from any "preferential" position in the universe, and in its stead Newton introduced the concept of *absolute space* with the declaration:

Absolute Space in its own nature and without regard to anything external, always remains similar and unmoving.

Absolute space was conceived by Newton to be an infinitely large three-dimensional container that holds all the objects of our universe and obeys the rules of Euclidean geometry. Within it, objects

† The concept of a frame of reference was expressed analytically by Descartes many centuries after Aristotle. By a frame of reference we mean a system of "guideposts" relative to which all positions and motions may be measured. We generally have used Cartesian frames, which consist of three infinitely extended, mutually perpendicular, straight lines with reference to which every point in space may be located. In two-dimensional graphs we call them *ordinate* and *abscissa*.

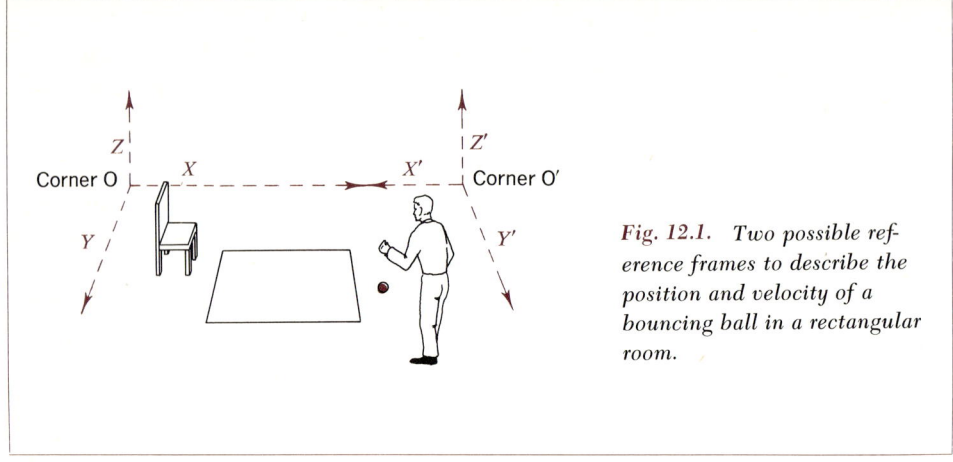

Fig. 12.1. Two possible reference frames to describe the position and velocity of a bouncing ball in a rectangular room.

may be located by utilizing Cartesian coordinates, and their motions may be described by the three laws of motion.

In a similar way, Newton introduced the notion of *absolute time*—the "true and mathematical time" that flows uniformly on without regard to anything external. Therefore, to Newton, absolute space and absolute time were unrelated and independent of one another. All motions in the universe could be referred to them, and there was considerable freedom as to what frames of reference could be used in solving common problems involving motion. For example, consider the drawing of Fig. 12.1, which shows the two corners of an ordinary room, some furniture, and a bouncing ball. If we were asked to tell someone exactly what position the chair, the rug, or the ball occupied, we could utilize the corner O of the room and the three edges of the walls and floor identified as X, Y, Z as a frame of reference, or as a Cartesian coordinate system. Any point in the room can be specified quite accurately by giving the distance along the chosen X-axis, Y-axis, and Z-axis. Even the changing position of the bouncing ball can be given in terms of these reference axes and a new variable, namely, time. But, we ask, does the use of corner O have any advantages over the use of another corner of the room, say, Corner O'? Unless there exists some unusual reason, the one reference system is as good as the other. (Of course one corner may be occupied by a large sofa, which prevents easy access to that corner for purposes of measurement.)

Although we do have a choice as to which corner of the room we use as a reference axis, it is assumed that the room, and any of its corners, could be given a definite relationship to an absolute space. The room could even be a room on a railroad train that moves at constant velocity.

Again, suppose that we are standing on a dock, watching a boat that is moving at a rate of u meters per second and on which a man is playing with a ball. Suppose that he throws the ball vertically upward with an initial velocity of v_0 meters per second. How would we see and describe the motion of the ball as compared to the way the man sees the motion?

The man on the boat would probably choose a frame of reference that is attached to the boat, as being the most convenient, as in Fig. 12.2(a). In this frame the ball would rise to a height y_m (the maximum height) directly above the thrower and return to the thrower at the same height at which it was released, in a time t. In an actual case the values for y_m and t can be found for each numerical value of v_0 from the relations given in Sec. 4.7.

But how would we see the motion of the ball from the shore? In our case the most convenient frame of reference would be the shore or the dock. We would see that the boat, the man, and the ball all have a forward velocity of u meters per second. When the man throws the ball directly upward, within his (moving) frame of reference he sees only an upward motion of the ball, as shown in Fig. 12.2(a). From the shore, however,

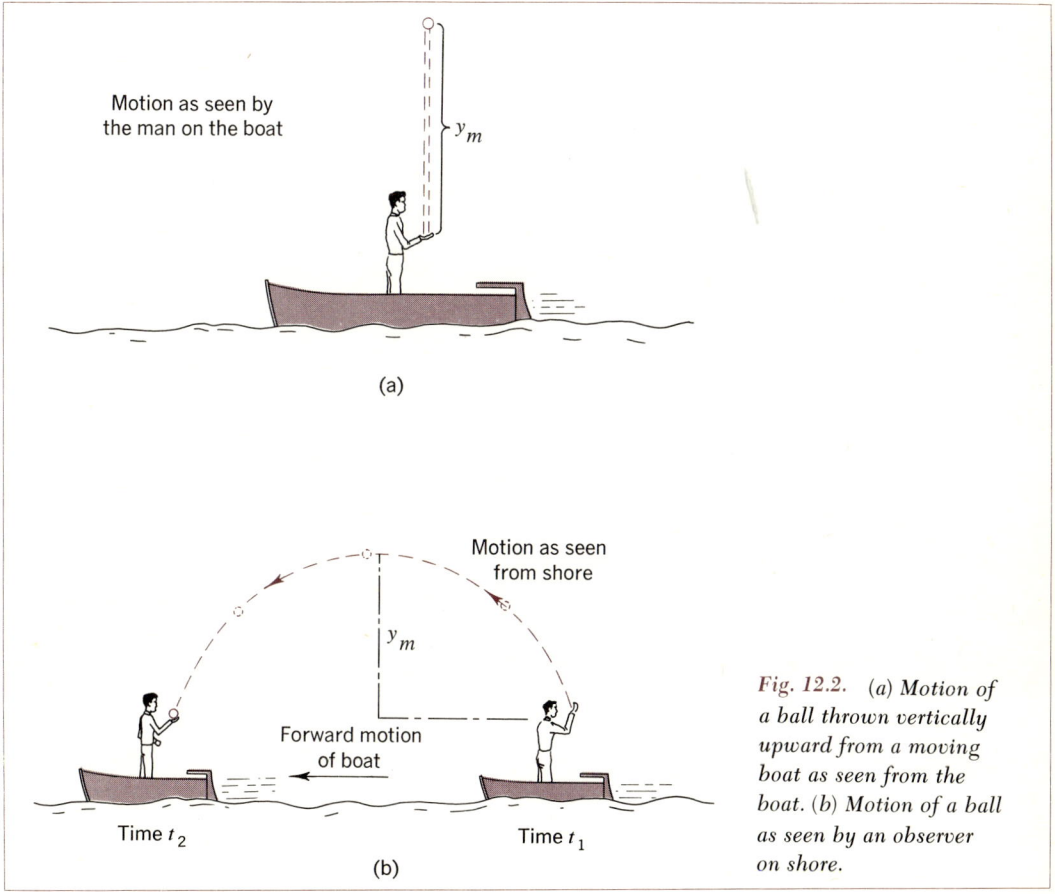

Fig. 12.2. (a) Motion of a ball thrown vertically upward from a moving boat as seen from the boat. (b) Motion of a ball as seen by an observer on shore.

we see that the ball not only executes the upward motion but also a forward motion which is equal to that of the boat. (Or we can say that the ball always has the same forward motion as the boat, while the man gives it an *additional* vertical motion.)

From our frame of reference the initial velocity of the ball would therefore have two components—an x-component of magnitude u in the direction of the boat's motion and a vertical y-component of the magnitude v_0. We may treat these separately.†

Since the ball's x-component is in the same direction and of the same magnitude as the boat's motion, the man and the ball will *always* travel as one with the boat. They will not sense any difference of *horizontal* motion relative to each other unless the man purposely moves or throws the ball sideways. But from the shore we would see the man, the ball, and the boat moving forward at a uniform velocity of u meters per second. In other words, the boat's frame of reference moves with a velocity of u meters per second relative to our own frame of reference.

What about the vertical motion of the ball; how does our view of it compare with the view as seen by the man on the boat? The answer is given by another question we may ask, namely: Does the boat move in the vertical direction with respect to our frame of reference? The answer is "no"; the boat remains at the same vertical level. Therefore our view of the vertical height to which the ball rises is identical to that seen by the man on the boat. That is, we arrive at values of y_m

---

† The horizontal or forward motion was given to the ball when the ship started to move, and the ball would continue in that motion until a reverse force stops its motion.

and t that are identical to those observed by the man on the ship.

The answer to these similarities and differences seems to lie in the relationship of the two frames of reference. The one is moving along one coordinate in relation to the other, but the motion is uniform. That is, they are *unaccelerated* with respect to each other, and for this reason they are suitable for the application of the laws of mechanics developed by Galileo and by Newton. They are known as *inertial frames*. For these, Newton proposed that there exists no "privileged" frame of reference. That is, one frame of reference may be as useful as another as long as they are not accelerated relative to each other. We call this the *principle of Newtonian relativity*.

Although all inertial frames must be considered equivalent to one another as far as the validity of Newton's laws is concerned, the quantities measured in different frames can be different. In the preceding example the observer on the ship would see no x-component of velocity (that is, for him, $v_x = 0$), while the observer on the shore would find $v_x = u$. Is there a simple way of going from one inertial frame to another? The answer is given by the *Galilean transformations*.

Consider two inertial frames: one with axes X, Y, and Z, and the other with axes parallel to the first, denoted by X', Y', and Z'; let the primed frame be moving with a uniform velocity u along the positive X-axis as seen by an observer in the unprimed frame. Consider an object with coordinates x', y', and z' in the primed frame (Fig. 12.3). What will be its position in the unprimed frame?

First we shall assume that the two frames actually coincided at some time $t = 0$ (this doesn't change our arguments but will simplify the results). Then, since all the relative motion is in the x-direction, y- and z-coordinates must remain the same at *all* times (that is, $y = y'$ and $z = z'$). However, the relation between x and x' must involve u, the relative velocity between the frames, and the elapsed time t. We find this relation to be $x = x' + ut$. But are we correct in using the same t for both frames? Of course we are, since both may be referred to absolute time (according to Galileo and Newton, that is, $t = t'$). To repeat

$$x = x' + ut \quad (12\text{-}1a)$$
$$y = y' \quad (12\text{-}1b)$$
$$z = z' \quad (12\text{-}1c)$$
$$t = t' \quad (12\text{-}1d)$$

These four equations, (12-1), are called the *Galilean transformations*.†

12.3 Michelson-Morley experiment

Can we actually observe the absolute space of Newtonian relativity? For several centuries following Newton, physicists attempted without success to do so. However, with the introduction of Maxwell's theory of electromagnetic radiation, the possibility arose of accomplishing this. As is known from our common experience with waves, a wave needs some medium in which to travel. The medium for light was called the *ether* and was considered to be closely related to Newton's absolute space. Many attempts were made to attribute material properties to this ether, but all were without success. Michelson and Morley, just prior to the beginning of the present century, made a final attempt in a celebrated experiment to detect the ether.

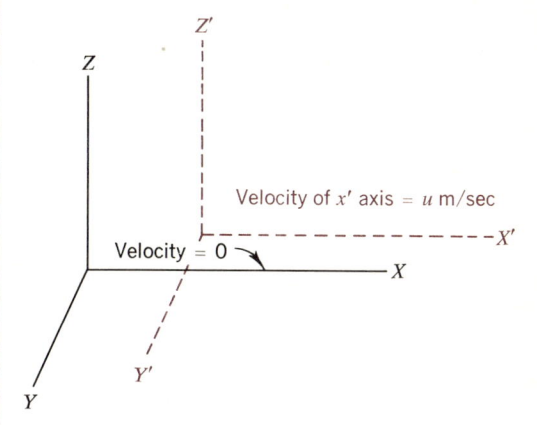

Fig. 12.3. Inertial reference frames for use with Galilean transformations.

---

† Galileo and Newton would not have included $t = t'$ because it is assumed to be obvious.

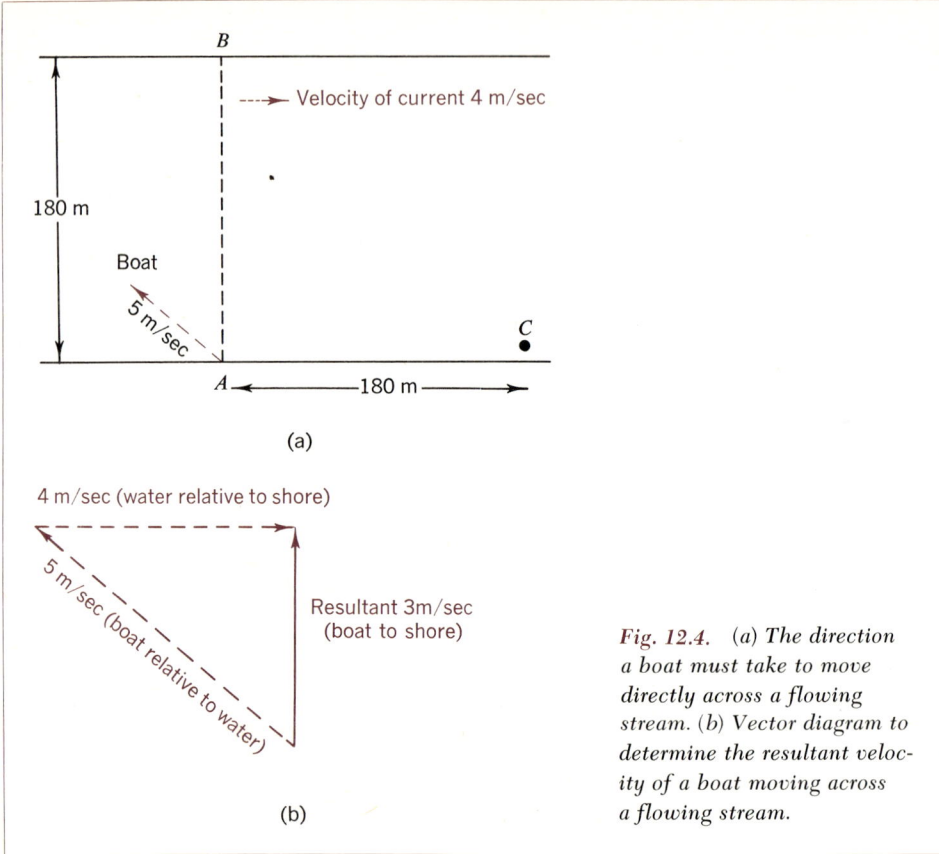

Fig. 12.4. (a) The direction a boat must take to move directly across a flowing stream. (b) Vector diagram to determine the resultant velocity of a boat moving across a flowing stream.

To understand this experiment, let us consider a very simple example of a boat crossing a river. Let the river flow with a velocity of 4 m/sec relative to an observer on the shore, and assume that the boat moves at a velocity of 5 m/sec relative to the water of the river. We set out from point A on the shore, first to go directly across the river (a distance of 180 m) to point B, as shown in Fig. 12.4(a), and then return to A. How long will the total trip require?

Because the flowing river carries the boat downstream, to go from A to B the boat must head *upstream* at such an angle with the shore that it moves upstream 4 m/sec with respect to the water, which is the velocity of the river current. Under these circumstances the boat is carried downstream by the current at the same 4 m/sec, thus the boat travels directly in the line from A to B as seen from the shore. From Fig. 12.4(b) we see that if the boat moves at a rate of 5 m/sec, there is a net or resultant velocity of 3 m/sec from A toward B. It would therefore take 60 sec to traverse the 180 m between A and B. The return trip is similar, the boat again heading upstream at the same angle as before and requiring 60 sec of time. Thus, 120 sec are required for the round trip.

Now let the boat, traveling at 5 m/sec, go downstream from A to C in Fig. 12.4(a), and then return upstream to A. As seen from the shore, the boat travels downstream at a speed of 9 m/sec (the sum of the speeds of the boat relative to the water and of the water relative to the shore), and covers the 180 m in 20 sec. On the return trip, however, the boat's speed as seen from shore is only 1 m/sec (the vector sum of velocities of boat and river, the two being in opposite directions) and hence the trip takes 180 sec. This round trip, A to C and back, takes 200 sec! But the trip of equal total length from A to B took only 120 sec.

Michelson and Morley reasoned as follows: Consider a fixed absolute ether in which the light travels with its characteristic speed of c (3×10^8 m/sec). This is like a boat moving in still water. Now, in its motion around the sun, let us assume that the earth moves through the ether. As seen from a position on earth (which serves as frame of reference similar to the shore in our earlier example), the time required for light to travel the distance between two fixed points should be different, depending on whether the direction of the light's travel is parallel to, or perpendicular to, the direction of the earth's motion. In this case the motion of the earth relative to the ether would be expected to introduce an effect somewhat similar to the effect of the flowing river on our boat. The earth's speed is only 3×10^4 m/sec, however, which is quite small compared with c. To make the very precise measurements required for this experiment, a device called an *interferometer*, invented by Michelson† in 1881, was used. The apparatus is shown schematically in Fig. 12.5.

A beam of light from the source is allowed to fall on a half-silvered mirror so that one-half of the incident light is

† For this invention, Michelson was awarded the Nobel Prize in Physics in 1907, the first American to receive the award.

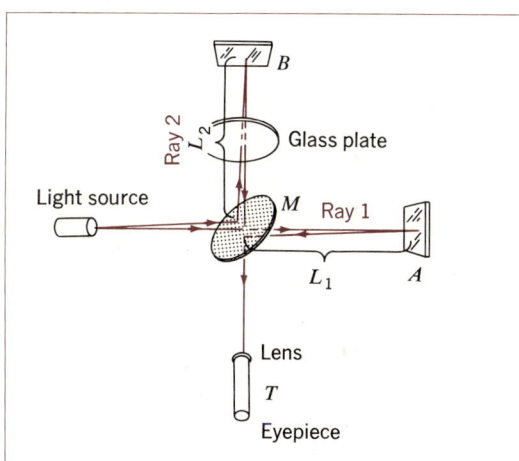

Fig. 12.5. Schematic diagram of a Michelson interferometer.

transmitted through the mirror M to a second mirror A, from which it is reflected back to M and thence to an eyepiece lens system. The other half of the incident beam striking M is reflected to mirror B and then back through M to the same eyepiece that received the first half of the light.

Initially the incident beam can be made to have all its waves in the same phase so that they add constructively. Although the path lengths L_1 and L_2 cannot be made equal within a small fraction of a wavelength, the system can be adjusted so that when ray 1 and ray 2 recombine at the eyepiece, they are in phase (constructive interference).

The Michelson-Morley experiment (1887) worked as follows:

(1) The interferometer described above was adjusted to give a bright fringe in the eyepiece. This was done with the instrument oriented so that the path for, say, ray 1 was perpendicular to the earth's tangential velocity.

(2) The arms L_1 and L_2 were now turned so that L_1 pointed along the direction of the earth's tangential velocity (that is, parallel to the direction of the ether current) and L_2 was perpendicular to this direction.

(3) Comparing this with the experience with the boat, we would expect that it should have now taken the light longer to traverse the path L_1 than to cover the path L_2, and hence the two rays would no longer be in phase. Using the interferometer, very small phase changes could be noted if they existed.

The surprise was that although this experiment was carried out with the greatest precision, *no phase changes were noted!* Michelson and Morley chose to interpret this as meaning that the ether did not exist, but the implications were much greater than might at first appear from such a conclusion. As we shall presently emphasize, the results are consistent with the idea that the conventional rules for addition of velocities do not hold when the speed of light is involved.

12.4 Einstein's special theory of relativity

The failure of the Michelson-Morley experiment to detect any trace of an effect that might be attributed to an ether seemed to portend a crisis to the physicists of the first years of the twentieth century. For the concept of absolute space and absolute time and the concept of an ether had been the foundations of the two greatest achievements of classical physics as represented by Newton and Maxwell. Unless the absence of an effect due to the ether could be explained in purely classical terms, there was danger that the model of the world founded on these theories might crumble.

Many attempts were made to explain the failure of the Michelson-Morley experiment to detect an ether effect. Most noteworthy was that made by G. F. Fitzgerald, a theoretical physicist at the University of Dublin, and H. A. Lorentz, a Dutch physicist. They independently proposed that relative to the absolute frame of the ether, a moving body *contracts* (we shall refer to this as the Lorentz contraction) along the direction of its motion. They went on to apply this concept to the explanation of various problems of electromagnetic theory, with some success. Similar work was pursued by the French mathematician Henri Poincaré. In many ways these men anticipated Einstein's theory, but they were not able to put their results on a firm physical base.

In the years prior to 1905, Albert Einstein considered this same problem in a completely different manner from that of his colleagues. His approach took cognizance of the philosophy of science espoused by the nineteenth century physicist Ernst Mach, who stated that "concepts and statements which are not empirically verifiable should have no place in a physical theory." Einstein agreed strongly with this principle. He reasoned that Maxwell's work on electromagnetic phenomena made successful use of the speed of light, c, as a basic constant of these phenomena. And now the Michelson-Morley experiment seemed to say that the speed of light in a vacuum is that same value of c, whether one measures the velocity while standing still or while running toward or away from the source of light.† That is, the velocities did not add up as had been proposed in Galilean or Newtonian physics (as we assumed for the example of Sec. 12.3 for the velocity of the boat and of the river when we added them to give 9 m/sec). Perhaps there was a basic error in all Newtonian physics because of failure to recognize that since we depend on vision and on signals to measure time and dimensions, we should make allowance for the fact that the signals (and vision) have *finite speed c*. Of course c is very large (3×10^8 m/sec) compared with ordinary mechanical speeds, and so the time required for a signal to reach an observer is usually negligible. But when one deals with objects that move with high speed or are at great distances, the effects may no longer be negligible.

† Einstein, although he appears to have heard of the results of the Michelson-Morley experiment, was apparently unfamiliar with its details. His choice of the constancy of the speed of light was based on other considerations.

Certainly the situation is quite different in the case of light emanating from very distant stars. For example, the light that today reaches our eyes from the Andromeda nebula actually was emitted by that body almost 2 million years ago, and it has taken all that time to reach us. What can we say, then, about the simultaneity of events between what we see now, and what may have happened to change the nebula in the interim of 2 million years? In fact, does simultaneity of events have any absolute significance when so much depends on when the signal of an event reaches an observer, and where the observers are located? Thus, Einstein stated that *the concept of simultaneity has no place in a physical theory*. He then set about to investigate and define the actual processes that become involved when one attempts to make physical measurements.

As a starting postulate, he chose to avoid any use of the notion of absolute space and absolute time, since these could not be verified. Instead, he chose as his basic postulates:

(1) There is no preferential frame of reference, but *all* inertial frames are equivalent. The laws of physics are the same in *all* inertial frames. (The principle of relativity.)

(2) The speed of light (in free space) is the same for *all* observers in *all* inertial frames. (The principle of the constancy of the speed of light.)

The first postulate denies the existence of an absolute space. It further asserts Einstein's belief that the laws of physics —that is, those of mechanics and electricity and magnetism—must appear to be the same to every observer in *any* inertial frame. The second postulate is necessary to preserve Maxwell's theory and is, as we have seen, consistent with the results

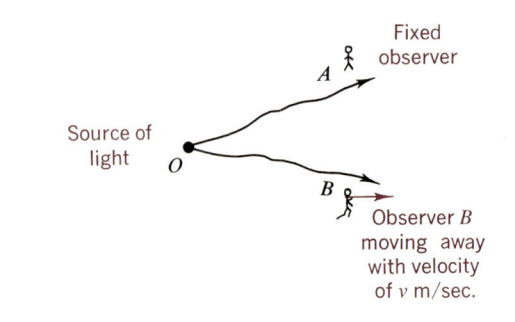

Fig. 12.6. Both A and B measure an identical speed for the light in a vacuum.

of the Michelson-Morley experiment. Consider the situation shown diagrammatically in Fig. 12.6. Here is shown a source of light at point O, with an observer at a fixed point A, and an observer at B who is moving away from the source of light with a velocity v. According to Einstein's second postulate, both observers measure the speed of the light to be c (in vacuum).

Postulate (1) says that there can be no *absolute motion* but only motion relative to the frame of reference of the observer. According to (2), the usual rules for adding velocities must no longer hold. This latter is completely contrary to our experience, which tells us that if an observer on a train sees a ball thrown forward with a velocity of 20 mph, and if a second observer (outside the train) sees the train to be moving at 50 mph in the direction the ball was thrown, this second observer will see the ball to be traveling at 70 mph. That is, in symbol form, if v' is the ball's speed as seen by an observer on the train, and u is the velocity of the train as seen by the second observer, then the velocity v of the ball as seen by the second observer is given by

$$v = v' + u \qquad (12\text{-}2)$$

or

(70 mph = 20 mph + 50 mph)

The result of Eq. (12-2) follows directly from the Galilean transformations of Eq. (12-1).

We see, however, that Einstein's postulate (2) violates Eq. (12-2). According to the postulate (and the Michelson-Morley experiment), when the observer on the train shines a light, he sees it to be traveling at speed c, and the outside observer also sees the light to be traveling with c instead of with speed $c + u$. To make this possible, in place of Eq. (12-2) Einstein found he had to substitute a new relation:

$$v = \frac{v' + u}{1 + \dfrac{u \cdot v'}{c^2}} \qquad (12\text{-}3)\dagger$$

What he did was to divide the right side of the conventional Eq. (12-2) by a factor $1 + uv'/c^2$, where c is the speed of light, u is the speed of travel of the frame of reference (in our case, the train) and v' represents the speed of the traveling object. In our case, the traveling object is the light itself, so we can change the designation v' to c. Now, if $v' = c$, Eq. (12-3) becomes

$$v = \frac{c + u}{1 + \dfrac{c \cdot u}{c^2}} = \frac{c + u}{\dfrac{c + u}{c}} = c$$

and this is the final speed, which is observed to be the same in both frames.

Equation (12-2) follows from common experience and represents the Galilean transformations, but it does not meet the needs of the new postulates and Eq. (12-3). For these we must find a new set of transformations. Einstein found he had to use the following sets, in which the primed coordinate system is moving with respect to the observer while the unprimed coordinates represent the coordinate system of the observer:

Albert Einstein

$$x' = \frac{x - vt}{\sqrt{1 - (v^2/c^2)}} \qquad x = \frac{x' + vt'}{\sqrt{1 - (v^2/c^2)}} \qquad (12\text{-}4a)$$

$$y' = y \qquad y = y' \qquad (12\text{-}4b)$$

$$z' = z \qquad z = z' \qquad (12\text{-}4c)$$

$$t' = \frac{t - (v/c^2)x}{\sqrt{1 - (v^2/c^2)}} \qquad t = \frac{t' + (v/c^2)x'}{\sqrt{1 - (v^2/c^2)}} \qquad (12\text{-}4d)$$

† This equation holds only if v' and u are in the same direction. The result for motions perpendicular to the motion of the frame (u) will not be treated in this text.

406 *The Theory of Relativity*

These equations are the so-called Lorentz transformations of special relativity and relate the space-time coordinates of an event as seen by two observers moving with respect to each other with velocity v along the x-axis. Since there is no motion of the y- and z-coordinates in relation to y', z', and since the properties of space are the same in all directions (isotropic), $y = y'$ and $z = z'$. When v is small compared with c, these equations reduce to the Galilean transformations (12-1). It is possible to show that these equations preserve the constancy of the speed of light as observed by the Michelson-Morley experiments. This requires a simple application of elementary differential calculus and can be found in any book on relativity theory.†

12.5 The relativity of simultaneity

In ordinary discussion we often refer to two events as happening at the same instant of time (that is, simultaneously), although the events may take place under quite separated and different environments. This is partly the result of the assumption that there is such a thing as absolute time, which is itself independent of events, and partly because we disregard some of the details of the *process of seeing* the events. The theory of relativity, on the contrary, says that the idea of simultaneous occurrence of two events is quite meaningless, and offers the following explanation of what happens when we observe two events that are separated in space.

Suppose that we are riding on a very long railroad train that is going along at a constant velocity v in the direction indicated by Fig. 12.7. We, who are riding, can use the train as a reference (coordi-

† See Max Born in References, pp. 263 on.

nate) system (that is, you can say that you will be in compartment D of car number 738 at a particular time). An individual who is standing along the tracks on the embankment will see the train to be moving past him with a velocity v.

Now suppose that this man on the embankment, standing at point M, sees two flashes of lightning occur at the same instant of time, one at point 1 and the other at point 2, the points being equally separated from his position at midpoint M. He claims the flashes to be simultaneous. But now we ask: How do those flashes of lightning appear to us who are riding on the train?

It is easy to see that the man on the embankment sees the flashes to be simultaneous only because the light from the two flashes traveled and arrived at his location at the same time (because he was standing still, and points 1 and 2 were equidistant from his position). But what is the time of appearance of the two flashes to those of us who are on the moving train?

To answer this important question, let us note that with the train standing still, we can mark off a distance between points 3 and 4 on the train that are exactly equal to the distance between points 1 and 2, and that there is also a midpoint M'. With the train now moving, let us assume that at the instant when the lightning flashes occur (before they are seen by anyone), point 3 is exactly alongside point 1, and therefore points 4 and 2, and midpoints M', M are also equivalently placed. Will the two flashes appear to be simultaneous to us as they are to the man outside? They *will not* be, for the reason that the light traveling from the flashes will not reach midpoint M' at the same instant. The reason is, of course, that we (along with point M') will be hastening toward the flash that is in front (point 4) and away

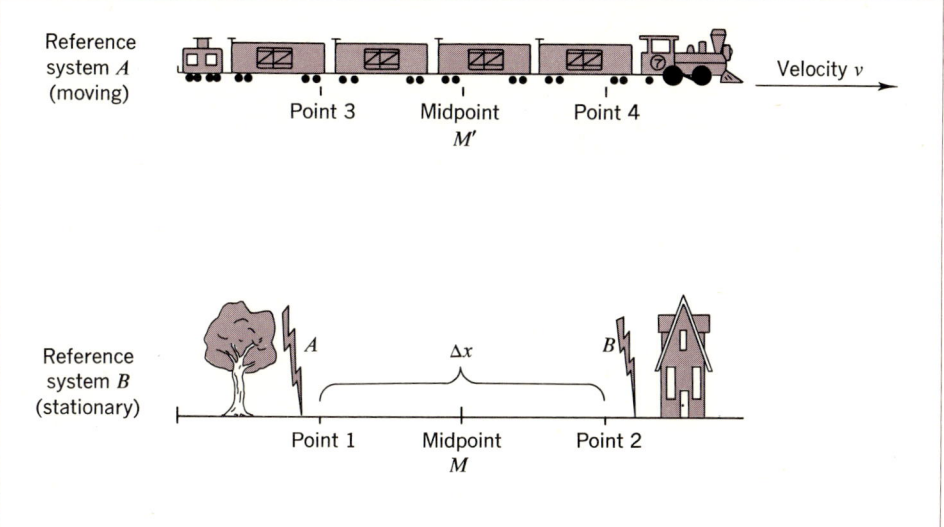

Fig. 12.7. Two frames of reference; one (a railroad train) is moving with velocity v relative to the other, which is stationary.

from the flash to the rear (point 3) with velocity v. The light coming from the forward flash will reach us before that coming from the rear because the distance to the forward flash is constantly decreasing while that from the rear flash is increasing.

We see, therefore, that events may appear to be simultaneous to the man on the embankment and not be at all simultaneous to those who have a frame of reference that is moving with respect to the embankment. In other words, every reference or coordinate system has its own particular time characteristic; therefore, unless we know the reference system to which a statement of time refers, it is not possible to assign any meaning to a statement that gives the time for an event.

Let us use the Lorentz transformations (12-4) to derive the result obtained in Einstein's example. Suppose the observer on the embankment (unprimed frame) sees two flashes of lightning at times t_1 and t_2 and points x_1 and x_2, respectively. By the Eq. (12-4d) we see

$$t'_2 - t'_1 = \frac{t_2 - (v/c^2)x_2}{\sqrt{1-(v^2/c^2)}} - \frac{t_1 - (v/c^2)x_1}{\sqrt{1-(v^2/c^2)}}$$

Rearranging terms we get

$$t'_2 - t'_1 = \frac{t_2 - t_1}{\sqrt{1-(v^2/c^2)}} - \frac{(x_2 - x_1)(v/c^2)}{\sqrt{1-(v^2/c^2)}}$$

or, using the Δ-notation introduced in Chapter 4,

$$\Delta t' = \frac{\Delta t}{\sqrt{1-(v^2/c^2)}} - \frac{\Delta x(v/c^2)}{\sqrt{1-(v^2/c^2)}} \quad (12\text{-}5)$$

There are several important interpretations of Eq. (12-5) with regard to simultaneity:

(1) If the observer in the unprimed frame (that is, on the embankment in our case) sees two events as occurring simultaneously ($\Delta t = 0$), then $\Delta t'$ can be zero only if Δx is zero. That is to say that the moving observer will see these events to be simultaneous also only if both occur at the same point in space.

(2) If two events are observed to occur in the unprimed frame at different times (that is, Δt is not zero) and if the two events take place at the same location (that is, Δx is zero), no observer will see them to be simultaneous.

Indeed, from Eq. (12-5) we can show that if the separation of the events in the unprimed (rest) frame, Δx, is less than the distance that light can traverse in the time interval between the events as viewed in this frame, $c \Delta t$, then there does not exist any velocity v for a moving frame in which the observer would see these events to be simultaneous. Moreover, the *order* of the events will be the same in all inertial frames, although the time interval $\Delta t'$ measured in different moving frames will be different. This important result assures us that if event A *causes* event B to occur, as seen in one frame, the causality is preserved in *all* inertial frames.

12.6 Time dilation

As we just saw, one of the important consequences of the relativity theory is that intervals of time are not the same for two people who are in different frames of reference that move in relation to each other. Suppose that two identical clocks mark off each second by a series of ticks or by light signals. If one of these clocks is at rest with respect to an observer while the second moves with a uniform velocity with respect to the first, the time interval measured by the fixed observer on the *moving* clock will be shorter than that on the clock at rest. That is, there is time dilation, or slowing down of time, with velocity.

Consider a light-clock constructed as follows: Place a source of light at one end of a rod of length L, as in Fig. 12.8(a), and a mirror M at the other end. A beam of light travels from the source to the mirror and is reflected back toward the source. If a detector is placed coincident with the source, the time between the emission of a photon from the source and its detection is Δt_0, given by

$$\Delta t_0 = \frac{2L}{c} \qquad (12\text{-}6)$$

This assumes that the source and detector are at rest with respect to one another and that there is no time lag in detecting the received photon. This is called the *proper time interval*.

Now assume that this same observer watches the same sequence of events in a light-clock located in a frame that moves with speed v with respect to him and the first clock. He would picture it as in Fig. 12.8(b). The second clock would move a distance $v \Delta t$ during the time Δt between emission and detection of the light. The distance traveled by the light is†

$$2 \sqrt{L^2 + \left(\frac{v \Delta t}{2}\right)^2}$$

and the time of travel (again as measured by the outside observer) would be Δt, given by

$$\Delta t = \frac{2}{c} \sqrt{L^2 + \left(\frac{v \Delta t}{2}\right)^2}$$

† Consider this figure as two right triangles back-to-back; then the light travels two hypotenuses.

Squaring both sides and bringing all terms containing Δt to the left-hand side yields

$$\Delta t^2 - \frac{4}{c^2}\frac{v^2}{4}\Delta t^2 = \frac{4}{c^2}L^2$$

or

$$\Delta t^2 = \frac{[2(L/c)]^2}{1-(v/c)^2}$$

Thus,

$$\Delta t = \frac{\Delta t_0}{\sqrt{1-(v/c)^2}} \qquad (12\text{-}7)$$

where t_0 was taken from Eq. (12-6). Equation (12-7) describes the phenomenon of *time dilation*. That is, the time interval measured by an observer in one frame, on a clock in another frame that is moving with respect to him, is *shorter* than that measured on a clock that is at rest with respect to him. For example, if he measures a time interval of 5 min on the clock in his frame, he will read the passage of only 1 min on a clock in a frame traveling at 98 percent the speed of light with respect to him.† Thus the 1 min he reads on the moving clock takes 5 min to pass, according to his rest clock, or we say that the time interval in the moving frame is *dilated*. Thus, if two identical clocks are synchronized at some moment of time, and one is at rest with respect to the observer while the other is moving with a velocity v with respect to him, he will note at any later time that the moving clock runs slow by the factor $\sqrt{1-(v/c)^2}$.

It should be noted that Eq. (12-7) can be easily obtained from Eq. (12-5) by noting that $x = 0$ (that is, in the frame of the clock at rest, the position of the clock when the initial reading is made (x_1) and that when the second reading is made (x_2)

† When $v = 0.98c$, $\dfrac{1}{\sqrt{1-(v/c)^2}} = 5$.

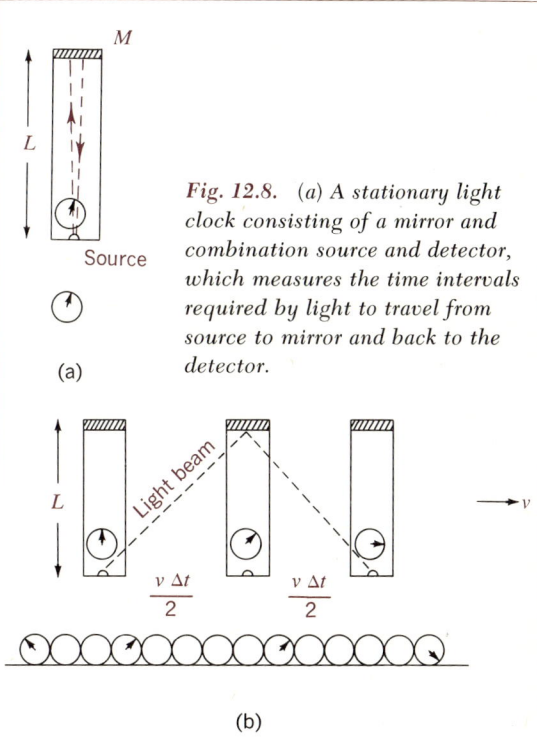

Fig. 12.8. (a) A stationary light clock consisting of a mirror and combination source and detector, which measures the time intervals required by light to travel from source to mirror and back to the detector.

Fig. 12.8. (b) A light clock moving with velocity v measures the longer time interval due to the longer (dotted line) path followed by the light.

are the same). In the notation of Eq. (12-7), the proper time interval is Δt_0, while in that of Eq. (12-5) it is called Δt. Likewise, Δt of Eq. (12-7) is the same as $\Delta t'$ of Eq. (12-5).

Time dilation, or the notion that a moving clock runs slower than a clock at rest, is one of the most important predictions of relativity theory. It has been verified in a number of experiments. In one of these, μ-mesons (radioactive particles present in cosmic rays) have been used as a clock. If these μ-mesons are brought to rest in our laboratory, they decay in a random manner, with a half-life of about 2 microseconds. Only about 5 percent of them survive as long as 6 μsec. In nature (that is, if our equipment wouldn't stop them) the μ-mesons travel at a velocity of 0.995c (that is, 99½ percent of the velocity of light), which means they travel 1000 ft in 1 μsec. If we have two laboratories, one at 6300-ft altitude and the other at sea level, we would expect that for every 1000 mesons passing the upper laboratory, on an average less than 50 (5 percent) would survive the 7-μsec trip (as measured by our lab clocks) to the ground. This is according to classical theory. Instead, experiment shows that nearly 800 survive to reach the ground. That is, many fewer decay than we would have expected from referring to our laboratory clocks. It is as if the mesons had traveled for only 0.7 μsec instead of the 7 μsec indicated by our clock. In other words, if we use as a clock those mesons traveling at 0.995c, they will tell time at one-tenth the rate that a clock at rest would tell it. Equation (12-7) gives this factor to be

$$\sqrt{1 - \frac{v^2}{c}} = \sqrt{1 - (0.995)^2}$$
$$= \sqrt{1 - 0.99}$$
$$= \sqrt{0.01}$$
$$= 0.1 \quad \text{or} \quad \tfrac{1}{10}$$

which is in agreement with our experiment. Another way of saying this is that, when viewed from the earth, the time clock of the traveling mesons runs more slowly; therefore they do not pass through as many half-life decay periods as we expected.

In general, clocks that measure the time between events (proper time) will measure a smaller interval (run slower) than clocks that measure the improper time between the same events.

12.7 Lorentz contraction

In addition to the time dilation, a shortening effect on spatial dimensions is also predicted by relativity theory. Suppose that an object is 1 m long when measured by an observer at rest with respect to that object. Now let this meter stick move in the direction of its length with a velocity v with respect to the observer. How does the 1-m length appear to him now?

The meter stick in the rest frame lying along the x-axis, with its ends at x_1 and x_2 will have the length $x_2 - x_1$ to the stationary observer ($t = 0$). But when the meter stick moves with the relative velocity v, the corresponding ends will be x'_1 and x'_2, and from the first Lorentz equation we have

$$x'_2 = \frac{x_2 - vt_2}{\sqrt{1 - (v^2/c^2)}}$$
$$x'_1 = \frac{x_1 - vt_1}{\sqrt{1 - (v^2/c^2)}}$$

But the length of the moving stick is the distance between the ends measured at the same instant, so $t_1 = t_2$ and

$$x'_2 - x'_1 = \frac{x_2 - x_1}{\sqrt{1 - (v^2/c^2)}}$$

or

$$x_2 - x_1 = (x'_2 - x'_1)\sqrt{1 - v^2/c^2} \quad (12\text{-}8)$$

It follows from these equations that any length L would have its greatest length when at rest; when moving with the velocity v, it appears to an observer at rest to be shortened in the direction of motion by the factor $\sqrt{1-(v^2/c^2)}$.

Now, if both time and space become altered in a moving frame, what is the relationship of the new time and new spatial dimensions? Is it the same as the relationship of conventional time to conventional spatial dimensions? Let's explore this question in connection with the travel of the μ-mesons down the 7000 ft of length between laboratories. Let one observer travel with the μ-mesons. We noted above that the observer and the μ-mesons would cover the 7000-ft distance in only 0.7-μsec time instead of ten times that interval. How would that distance of 7000 ft appear to this traveler? The distance would be $L' = L\sqrt{1-(v^2/c^2)}$, or also only one-tenth as long (=700 ft). That is, both distance and time are one-tenth that seen from the fixed frame. Therefore $700:0.7 = 7000:7$; therefore the *ratio* of distance to time as this fast traveler sees it in his system is the same as the observer on earth sees it.

Again let us repeat that the difference between the conventional Galilean transformations and the relativity or Lorentz transformations lies in the fact that the former assumes the speed of light to be infinitely fast.

To those who have grasped the view of modern physics with respect to the nature of space and time considered to this point, the question of whether lengths will shorten and time will slow becomes irrelevant. The theory of relativity is seen to imply certain properties of space and time. Thus, when considering a rod in a reference system that is moving with a constant velocity with respect to a rod at rest, the observer must assign to it a length *in the moving system* shorter than the one he gives in his own system. He must do this to be consistent with the principle of relativity, and there arises no question of the *real* length of the rod. A similar situation holds with respect to time. The true test of relativity lies in the agreement of its deductions with the results of experiment.

12.8 Relationship of time and space

From the Lorentz transformations themselves or from the previous two sections, we note that time and spatial coordinates are not independent, but are related rather closely. Measurements of the one must entail involvement of the other. Thus, to trace fully the motion of a particle, we must plot not only x, y, z (the three space coordinates), but also t, the time. The reason is simply that we are not conscious of spatial dimensions except as we *see* the dimensions, and the light by which we see requires time in proportion to distance to reach us. Time thus seems to serve the dimensional aspects as do the three space coordinates.

It is assumed in classical physics that any two points located in space have an invariant distance between them, regardless of time. This distance, s, was specified by the Pythagorean theorem as follows:

$$\sqrt{(\Delta x)^2 + (\Delta y)^2 + (\Delta z)^2} = s$$

using the coordinates x, y, z for three-dimensional space. The Δ represent the difference in each coordinate between the two points.

Since light propagates with velocity c, this distance must be equal to $c\,\Delta t$, where t is the time required for light to travel. We can now write the equation as

$$\sqrt{(\Delta x)^2 + (\Delta y)^2 + (\Delta z)^2} = c\,\Delta t$$

or
$$(\Delta x)^2 + (\Delta y)^2 + (\Delta z)^2 = (c\,\Delta t)^2$$
or
$$(\Delta x)^2 + (\Delta y)^2 + (\Delta z)^2 - (c\,\Delta t)^2 = 0\,\dagger$$

The fact that time enters into physical laws as a fourth and necessary coordinate is all that is meant by "time is the fourth dimension." All the vectors previously expressed in terms of three components now become four-component vectors, including a time component, and the laws of physics assume a greater generality than they did before.

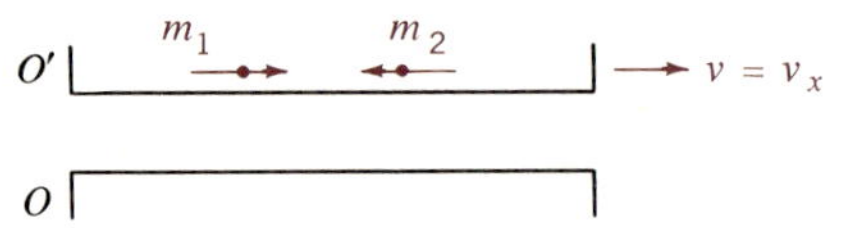

12.9 Mass change with velocity

Even before Einstein developed his theory of relativity, there was some experimental evidence that the mass of electrons increased with velocity. Einstein showed that this effect is in agreement with the theory of relativity. We cannot give the derivation in this text, but shall outline how it may be carried out. Suppose two masses collide in the coordinate system of O' (Fig. 12.9). We know that, as seen by an observer in this frame, the momentum will be conserved in the collision of m_1 and m_2; that is, the net momentum of m_1 and m_2 will be the same after the collision as before. Now suppose this same collision to be observed from the system of O (the O' system has rel-

Fig. 12.9. Relativistic mass is computed from Lorentz-transformed equations for simultaneous conservation of momentum in two reference frames.

† In the notation of the calculus,
$$dx^2 + dy^2 + dz^2 - c^2\,dt^2 = 0$$
where dx, dy, and dz are the incremental distances in the $x, y,$ and z directions, respectively, between our two points, and dt is the increment of time needed for light to traverse the distance between the points. If we let $i = \sqrt{-1}$ or $i^2 = -1$, then we get
$$dx^2 + dy^2 + dz^2 + i^2 c^2\,dt^2 = 0$$

ative velocity of $v = v_x$ with respect to system O). We believe that the conservation of momentum must also be valid, for the collision as observed from O, even though m_1 and m_2 will have velocities for O different from those for O' (we recall that x' and t' variables, which define velocity for O', must be Lorentz-transformed to give the x and t variables that define velocities of m_1 and m_2 for the O observer).

Calculation shows that we have conservation of momentum for O only if the masses m_1 and m_2 are increased from their rest values in the coordinate system of O'. The equation for the increased mass m is

$$m = \frac{m_0}{\sqrt{1-(v^2/c^2)}} \quad (12\text{-}9)$$

That is, m_1 or m_2 (or any other mass) that has the value m_0 to an observer at rest with respect to it must have the value m for an observer who measures the mass when it is moving at a relative speed v with respect to him.

Abundant confirmation of Eq. (12-9) may be found by using accelerators that carry particles to speeds close to c, where the equations for the motions of the particles must take into account the mass increase with velocity, if they are to be accurate.

It should be noted that as the speed of a moving object, v, as viewed from another frame of reference *assumed* to be at rest, approaches the speed of light c, its mass m measured from the rest frame approaches infinity. This must be interpreted as meaning that no material body with a finite rest mass m_0 can ever attain the speed of light as seen from *any* frame of reference.† However, some "particles,"

† There cannot be any frame of reference moving at the speed of light with respect to

which we shall introduce in later chapters (the photon and the neutrinos), do travel at the speed of light. Moreover, they transfer a finite amount of momentum when they are captured by any material body, implying that the mass m is finite [since we classically define momentum as the product of mass times velocity, and Eq. (12-9) results from our preserving this definition]. The only way for Eq. (12-9) to yield a finite value of m when the denominator is zero is to have m_0 (the rest mass) be zero also! Thus these particles are considered to have a zero rest mass; that is, they have no existence unless they are in motion at speed c.

12.10 Mass-energy equivalence

The Cornell University synchrotron has accelerated electrons to $0.99999992c$. Using Eq. (12-9),

$$m = \frac{m_0}{\sqrt{1-(0.99999992)^2}}$$

$$= \frac{m_0}{\sqrt{0.00000016}}$$

$$= \frac{m_0}{4 \times 10^{-4}}$$

$$= 2500 m_0$$

Thus the Cornell electrons have mass characteristics that are 2500 times those for an electron at rest.

Equation (12-9) tells us that as we increase the mass of a particle, we give energy to the particle; for, as we recall,

---

our observer. There can be, therefore, no frame attached to light itself. Note that at $v = c$, the time dilation is infinite (that is, it requires no time as seen by the light to cover a distance that requires a time t_0 in our rest frame) and the length contracts to zero (that is, the distance between *any* two points is zero to light).

the kinetic energy is $\frac{1}{2}mv^2$ (in Newtonian physics), and therefore increased mass for the particle means increased energy. Rigorous use of the relativity theory equations gives the result that any mass is equivalent to energy, in accordance with the relation

$$E = mc^2 \qquad (12\text{-}10)$$

where E is energy and c is the speed of light. In Eq. (12-10), m is the relativistic mass, or

$$\frac{m_0}{\sqrt{1-(v^2/c^2)}}$$

If the particle is at rest, its energy is termed "rest energy," that is,

$$E_0 = m_0 c^2 \qquad (12\text{-}11)$$

The difference between E and E_0 is the energy that has been given to the particle by giving it a velocity v, and is called the relativistic kinetic energy K:

$$\begin{aligned} K &= E - E_0 \\ &= mc^2 - m_0 c^2 \end{aligned} \qquad (12\text{-}12)$$

Using Eq. (12-8), we can write Eq. (12-11) as

$$K = \frac{m_0 c^2}{\sqrt{1-(v^2/c^2)}} - m_0 c^2 \qquad (12\text{-}13)$$

Now, if v is small compared with c, it can be shown that Eq. (12-13) reduces to

$$K = \tfrac{1}{2} m_0 v^2 \qquad (12\text{-}14)$$

Equation (12-14) shows us that relativistic kinetic energy is the ordinary Newtonian kinetic energy, $\frac{1}{2}m_0 v^2$, for velocities that are small compared with the speed of light.

Returning to Eq. (12-11), $E_0 = m_0 c^2$, we see that it indicates that mass has an immense energy equivalent. The transformation between appreciable amounts of mass and energy does occur in nuclear reactions. Conversion of mass to energy in processes of nuclear fission and fusion (atomic energy) will be covered in Chapter 21. An example of nonnuclear conversion of mass to energy, using Eq. (12-11), will give us an idea of the mass-energy ratio.

Since c^2 is a very large number, 9×10^{16} (m/sec)^2, the mass of appreciable amounts of radiant energy is very small if expressed in customary units. Thus a searchlight that emits 6×10^2 joules of light per second becomes lighter by

$$\frac{6 \times 10^2 \text{ kg (m/sec)}^2}{9 \times 10^{16} \text{ (m/sec)}^2} = 7 \times 10^{-15} \text{ kg}$$

The sun, on the other hand, in nuclear reactions loses more than 4×10^9 kg of mass every second by pouring radiation into surrounding space.

We can see from Eq. (12-13) that for $v \to c$, $K \to \infty$. (This follows because $\sqrt{1-(v^2/c^2)} \to 0$ for $v \to c$, and therefore $1/\sqrt{1-(v^2/c^2)} \to \infty$ for $v \to c$.) Since we do not have infinite amounts of energy available, we cannot bring any material particle, for which $m_0 > 0$, to a speed equal to the speed of light. All electromagnetic waves do, of course, move with velocity c in a vacuum. We must therefore regard the rest-mass of light waves as being zero.

12.11 General relativity

Thus far in this chapter we have been concerned with *special* relativity—with the point of view of observers moving relative to each other with uniform speed. Let us now look briefly at *general* relativity, which deals with the laws of physics as they might appear to observers subjected to acceleration. Perhaps you have observed that nature looks and feels quite different to the accelerated observer.

12.11 General Relativity

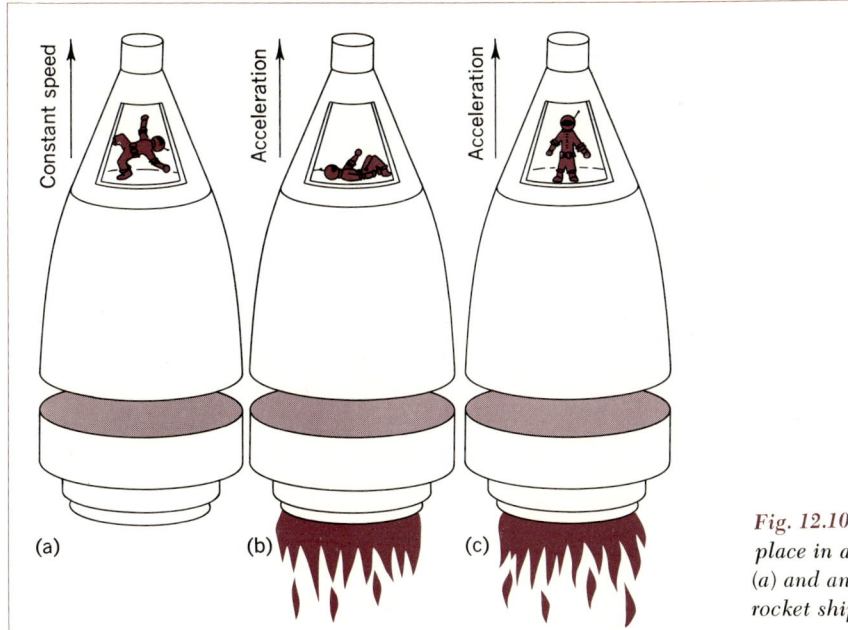

Fig. 12.10. Events taking place in a nonaccelerated (a) and an accelerated (b, c) rocket ship.

ACCELERATION AND GRAVITY

In Newton's theory of gravitation the force $F = Gm_1m_2/r^2$ is one that acts instataneously. This means that a signal or energy could be transmitted instantaneously, and thus would violate one of the basic tenets of relativity, namely, that no energy, not even a signal, can travel faster than the speed of light. So Einstein tackled the problem of a relativistic theory of gravitation. He reasoned that if the laws of physics can be stated so that they are the same regardless of the relative uniform motion of the system of reference, why not attempt to formulate them so that they are independent of accelerated motions? To do this, he postulated what is called the *principle of equivalence* which states that being in a gravitational field is equivalent to being in an accelerated reference system (acceleration produces the same effect as gravity).

For example, imagine that you are in a rocket that blasts off for the moon. In the initial stages you will have the impression that gravity has suddenly increased, and if the ship has an acceleration of $a = 5g$ with respect to the earth, you and everything in the rocket will weigh six times the normal weight. But when your ship settles down to a constant speed, you and everything that is not tied down will float around freely inside the cabin (in the early days of space travel, our astronauts had this trouble with cookie crumbs, and similar small objects). Now suppose you settle into a deep sleep and awake later to find that you feel your weight again and that loose objects have settled to the floor (Fig. 12.10). You could place any one of several interpretations on this: (1) You might say your space ship had started accelerating upward, or (2) that you had landed on the moon during

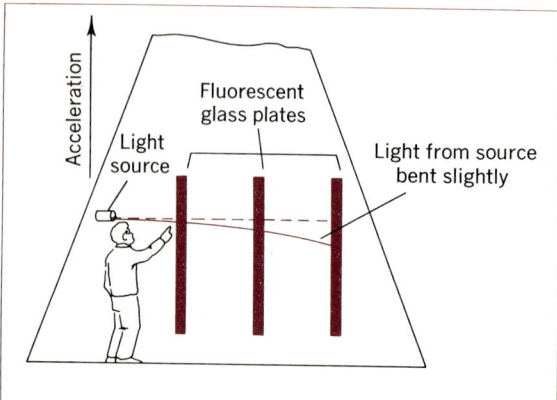

Fig. 12.11. The curvature of light when subjected to inertial or gravitational forces.

your sleep and gravity was acting on the ship and contents, or (3) that you had returned to earth or perhaps had landed on some other planet.

From these observations we can conclude that accelerated motion does not leave the laws of physics unchanged, but produces the same effect as a gravitational field. This *principle of equivalence* is an essential feature of the general theory of relativity, and depends primarily on the fact that the weights of two objects are exactly proportional to their masses so that gravity gives them the same acceleration. Hence, according to the general theory of relativity, the gravitational field and acceleration are essentially one and the same thing and cannot be distinguished in principle! This is equivalent to saying that the mass introduced in Newton's second law ($F = ma$) and in his law of gravity are the same.

LIGHT RAYS IN A GRAVITATIONAL FIELD

Following the line of reasoning of the principle of equivalence, Einstein made an analogous consideration for light rays and predicted that *in general, rays of light are propagated curvilinearly in gravitational fields* (Fig. 12.11). He recognized that the effect would be small in the presence of ordinary gravitational fields. He proposed to test the theory by analyzing star positions when the light coming from a star passes by the sun at grazing incidence. He believed that the position of a star, measured when its light passes close to the sun, should differ from that measured when the sun is not in the neighborhood. Since, in general, we do not see stars close to the sun because their light is weak in relation to the greater brightness of the sun, this measurement is difficult to make. It can be made only at the time when a total eclipse of the sun darkens the sky around the sun sufficiently for stars to be seen. Einstein predicted that the measurements would reveal that the light coming from the star would experience curvature from a straight line corresponding to 1.7 sec of arc. The light would be bent *toward* the sun, which would give to an observer the appearance of the star moving *away from* the sun (Fig. 12.12).

Einstein staked the reputation of his general theory on the correctness of this postulate, and the scientific world waited for a test to be made. A test was made on May 29, 1919, when the sun as seen from Brazil and South Africa was totally eclipsed. The Royal Society of London and the Royal Astronomical Society sent teams of their most celebrated astronomers (including Eddington and Cottingham) to photograph the eclipse. When the pictures were developed and the star positions measured, the deflection of starlight in the gravitational field of the sun was found to average 1.64 sec of arc, remarkably close to Einstein's predicted value. More recent observations have confirmed the existence of the deflection, but there have been departures from the value predicted by Einstein.

As a result of the concepts that developed from the general theory, there had been frequent argument about the validity of the special theory of relativity. The deflection of light in the presence of gravitational fields added to the conflict because it seemed to counter the special relativity requirement that the speed of light in vacuo be constant. (A curvature of rays of light can take place only when the velocity of propagation varies with position.) In this connection Einstein called attention to the need to recognize that the special theory of relativity could not claim unlimited domain of validity;

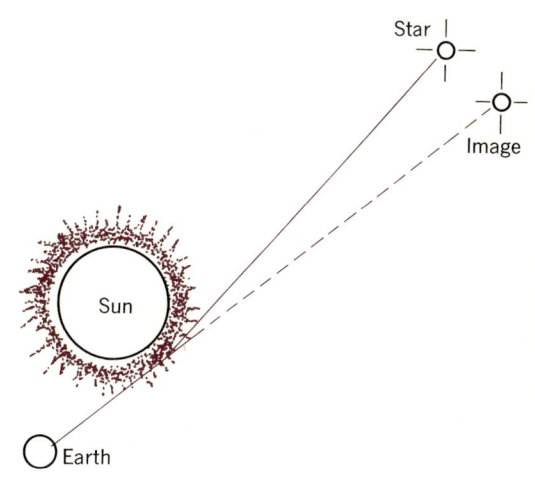

Fig. 12.12. During a total eclipse of the sun, starlight that passes close to the sun has been observed to exhibit a gravitational deflection as shown here and as originally predicted by Einstein.

that, in fact, the results predicted by the special theory were valid only when it was possible to disregard gravitational fields.

The case was not different, argued Einstein, with the course of developments in the field of electricity. The laws of electricity were first associated exclusively with the phenomena involving electrostatics. These were correctly expressed only as long as the charges were at rest. The subject of electrostatic phenomena flourished but finally emerged in the quite different form of electrodynamics when motion of charges and the presence of fields were included. The development into a more comprehensive system of Maxwell's electrodynamics did not in any way deny the correctness of the laws of electrostatics under the special conditions in which these laws hold.

ORBITAL MOTION OF THE PLANET MERCURY

Another example of the deep cosmic significance of Einstein's gravitational laws is the way they account for a seeming anomaly in the orbital motion of Mercury. According to Newton's law of universal gravitation, every planet in the solar system should describe an elliptical orbit, with the sun at one focus. The major axis of this ellipse, which should remain fixed in space (with respect to the distant stars) in the case of a spherical sun and a single planet, in reality precesses (that is, the perihelion, or the point of closest approach of the planet to the sun, rotates about the sun as shown in Fig. 12.13) as a result of the perturbing effect of other planets. For the planet Mercury, the speed of this precession should be 531 sec of arc per century. The observed precession is 574 sec of arc per century. This discrepancy of about 43 sec

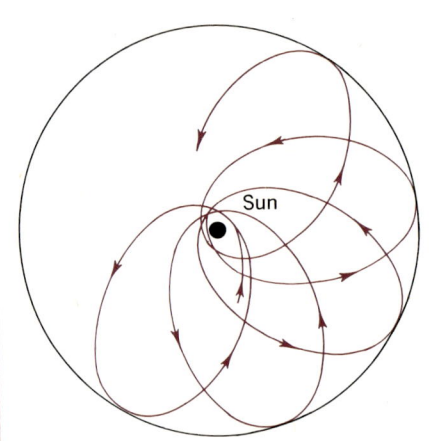

Fig. 12.13. *Mercury's elliptical orbit precesses (rotates) as shown here (greatly exaggerated). Actually, the ellipse advances only 574 sec of arc per century.*

per century can be explained by the relativity theory.† This explanation is based on the fact that (according to relativity theory) the geometrical properties of space are altered near a massive body and the Keplerian ellipse we would expect from Newtonian theory must be similarly altered, resulting in the observed precession.‡

SLOWING DOWN OF CLOCKS AND OF OTHER PHYSICAL PROCESSES IN A GRAVITATIONAL FIELD

Another important consequence of the general theory of relativity is that gravity should influence the rate of all physical processes by slowing them down. For observational purposes a suitable clock is an atom emitting light that exhibits a spectral line of definite frequency. Einstein predicted a displacement (toward the red end of the spectrum) of the spectral lines of light reaching us from very massive stars, as compared with the corresponding lines for light produced in an analogous manner terrestrially. This displacement is called the *gravitational red shift*, and results when light loses energy (which causes a lengthening of its wavelength) in overcoming the gravitational attraction of a massive star. This phenomenon was observed by

---

† Relativity theory predicts an additional precession (over that of 531 sec predicted by Newtonian gravitation) of $43''.03'$. Observation gives $42''.56' \pm 0.94$.

‡ It must be noted that the observed precession could be adequately explained by Newtonian gravitation if the sun were a perfect sphere; rather, it is slightly oblate, as the earth is. This oblateness of the earth causes artificial earth satellites to have precessing orbits. The needed solar oblateness has, however, not yet been confirmed. R. H. Dicke of Princeton has succeeded in measuring the oblateness of the sun, but the value he finds is too small to account for the perihelion motion of Mercury.

Adams in 1924 in the spectral lines of the dwarf companion of Sirius, which were displaced in agreement with the predictions of the general theory of relativity. Thus the tick of an atomic clock on a star (our sun is an example) is running slower than the same atomic clock here on earth.

The same red shift has been observed in a terrestrial experiment that made use of the earth's gravitational field. Such an experiment was performed at Harvard in 1960 by Pound and Rebka. They used a 70-ft (22.5 m) tower and the newly discovered Mössbauer† effect to check the predicted rise in frequency (2.19 parts in 10^{16} per meter) of the photon "falling" the length of the tower. They found that the photons did increase in frequency (2.28 ± 0.23 parts in 10^{16} per meter).

Let us investigate a prediction of the theory of relativity that might affect the future of each reader: This aspect has to do with biological clocks (such as heartbeat, breathing, metabolism) and the slowing-down effect of gravity on living systems. In considering this problem, twins are usually taken as examples, but any combination of individuals—boy and girl, friend and enemy, husband and wife—will serve the purpose.

Let us imagine young twins parting— one to take a rocket journey to the vicinity of Arcturus and the other to continue his life on earth. As the earth twin E watches his brother A traveling at a speed close to that of light, he observes that A's clocks appear to be slow and the biochemical processes of his brother are also proceeding slowly. At the same time,

---

† The German physicist R. L. Mössbauer discovered a very accurate "nuclear clock" in 1959. He used photons from a radioactive nucleus to detect the optical Doppler effect. For more information on this subject, see De Benedetti in References.

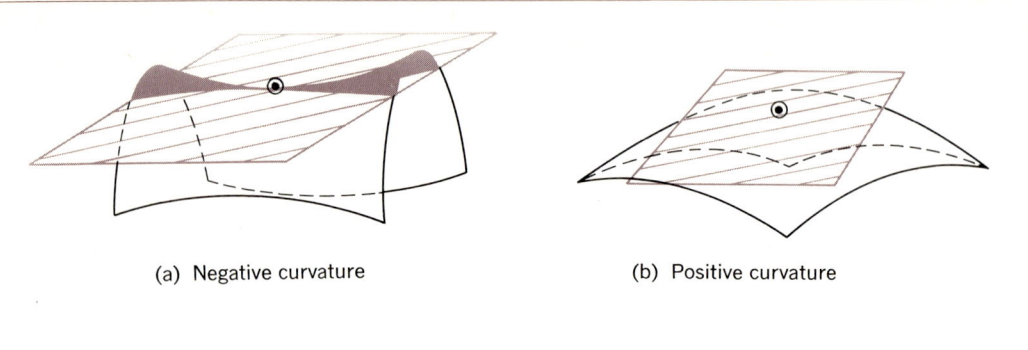

Fig. 12.14. Four-dimensional space exhibits curvature in the vicinity of gravitating masses. In the case of negative curvature, the tangential plane and the curved surface intersect, whereas in the positive case, they do not.

A looks back at his brother E on earth and concludes that E's clocks are slow and that E is aging less rapidly. According to the special theory, there is no reason to prefer either twin's point of view, since motion is relative; that is the earth and the earth twin are moving at a high speed relative to the rocket and the rocket twin! This is the much debated twin "paradox."

But there is an important difference between the two systems! The rotation of the earth and its revolution around the sun produce accelerations that are small, so E's frame of reference may be considered to be an inertial one. But the rocket, in taking off, turning, and landing has a very large acceleration, so A's frame of reference is not an inertial one. And according to the principle of equivalence, an acceleration is equivalent to a gravitational field, so the traveling twin has not aged so fast as his earthbound brother. E had to wait 80 years for brother A to return from Arcturus, and when A returned, he was still young; possibly only ten years older than when he departed! This may seem farfetched, but if we realize that the human organism is regulated by "biological" clocks, there is reason to believe that these biological clocks can slow down just as do physical clocks (Whitrow, 1963).

This is a startling conclusion, but most physicists believe it to be the correct conclusion deduced from the relativity theory.

THE STRUCTURE OF SPACE

According to the general theory of relativity, the geometrical properties of space are not independent, but are determined by matter. Einstein associated gravitational fields with the curvature of four-dimensional space in the neighborhood of gravitating masses. To understand the notion of curved space, perhaps an analogy with curved surfaces that have only two dimensions will help. Mathematicians distinguish between negative curvature, as illustrated in Fig. 12.14(a), and positive curvature, as shown in Fig. 12.14(b).

For curved surfaces like those in Fig.

12.14, the classical plane geometry of Euclid does not hold. We have learned that the sum of the three angles of a triangle is 180 deg. This is true for a triangle on a plane surface such as on the blackboard. But when working with figures on curved surfaces, such as that of a sphere (see Fig. 12.15), the sum of angles of a triangle is greater than 180 deg. On the other hand, if the figure is one of negative curvature, like that of a saddle, then the sum of the angles is less than 180 deg. The lines forming the triangles on curved figures are not "straight" lines, although they are the shortest distance between points on the surface. They are called *geodesic lines* or *geodesics*, and are arcs of the great circle passing through the points in question. Thus, according to the general theory of Einstein, the presence of matter in space makes the space non-Euclidean, and the path of anything moving freely (whether it be a planet or a ray of light) will be along a geodesic in that space.

12.12 Summary

Recurrently in the history of science it is found that older theories, well established, convincing to the human spirit, and apparently successful in relating observations and in predicting future events, must be drastically modified or completely discarded. Scientific theories and beliefs of today become the "superstitions" of tomorrow. The collapse of a theory may sometimes come dramatically and suddenly, but more often by a slow change in man's attitudes and ways of thinking.

Many examples have already been given of the demise of theories: the Pythagorean theory that the world is made of integers and ratios was overturned with the discovery of irrationals; Aristotle's denial of the possibility of a

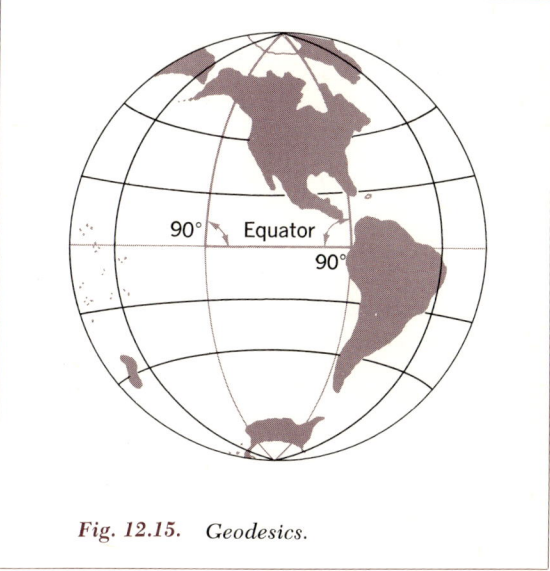

Fig. 12.15. *Geodesics.*

vacuum and his explanation of motion was overturned by Galileo; the grand Ptolemaic system was eventually replaced by the Copernican system. In modern physics, so many apparently well-established theories have been drastically modified that the first half of the twentieth century can be called a period of scientific revolution.

In general, scientific theories go far beyond that which can be directly observed. To explain electromagnetic waves, Maxwell and others hypothesized an "imponderable" frictionless ether. New theories often require the invention of newly postulated entities.

However, scientific theories may not only postulate entities beyond the realm of observation—they may also require postulates that go completely against "common sense" and "tradition." For example, the Copernican revolution violated traditional ways of looking at things and also contradicted common sense. (Do we not still speak of "terra firma"?) Relativity theory violates common sense even more profoundly and contradicts the foundations of classical Newtonian science.

The reader might well pause at this point to consider the way in which authority and classical tradition are challenged by scientists. "Tradition" can never be escaped altogether. It is doubtful whether there could be any religion, art, or philosophy if each generation had to begin *de novo*. Certainly there could be no science. A proper respect for the use of tradition, combined with fearlessness in questioning old-established ways of looking at things is the mark of a creative scientist. Such a man was Einstein—at once a radical innovator and a logically disciplined conservative.

Look at the scientific scene of 1890. Success after success had been piled up by the march of the physical sciences in explaining the world by theories and laws erected upon the foundations laid by Newton; extended and advanced by Maxwell and Hertz in electromagnetics; by Meyer, Joule, and Helmholtz in thermodynamics; by Dalton and Berzelius in chemistry; by Hutton and Lyell in geology; and by Darwin (and his law of the survival of the fittest) in biology.

The basic foundations appeared to be well grounded. Classical physics seemed to be a logically complete and consistent theory. Nothing was left except to fill in the details. All the problems of the universe and even the organization of society could be solved by the rigorous application of the laws of physics. The eighteenth century was dominated by the dream that man could solve all problems rationally. This dream had dimmed because the misery of the people, poverty, illness, death had not disappeared. Man did not seem to have progressed at all in line with the dream. Still, by and large, science promised, with time, to reveal the basic answers following Newtonian methods. It was a question of grubbing after the details to fill in the grand, albeit somewhat pessimistic, design. However, all sorts of strange and dramatic intrusions occurred in the world of science from about 1890 to 1910. All the classical foundations of physics and chemistry collapsed and had to be rewritten.

In this chapter we have undertaken the study of one strand of this rewriting of scientific theory. Closely allied to relativity theory, we shall later find a great variety of other interlocking theoretical innovations: quanta, atomic and nuclear physics, plasma and stellar physics. We have already noted the relatively new way of looking at and analyzing phenomena, termed loosely "cybernetics."

Einstein's special theory of relativity presents no great logical difficulty. It is simpler to understand, far less abstract than the statistical mechanics we treated in Chapter 10. From a few postulates, by simple algebra and analytic geometry, amazing consequences follow. Although the logic is simple, some of the consequences are difficult to accept. However, since all the consequences that can be subjected to test, including the mass-energy equivalence, have been confirmed again and again in laboratory and engineering experience, the special theory of relativity is now fully accepted as the new traditional or classical theory. One point about this new theory, however, needs emphasis; that is, the difference between Newtonian and relativity mechanics becomes of practical significance only at either very high velocities (approaching the speed of light) or in nuclear changes.

In the general theory, Einstein introduced the *principle of equivalence* by which the idea of a force of gravity was replaced by the idea of the curvature of space. This extension of the special theory enabled him to predict facts not previously known, such as the bending of light rays as they pass near the sun, and the change of frequencies in spectral lines in the light from massive stars.

The basic postulates of Einstein should be carefully considered. They well illustrate his conservatism and his boldness. The conservatism is evidenced in his assumption that the laws of physics should be the same regardless of the frame of reference from which they are judged.† If this assumption is to be maintained, and if the results of the Michelson-Morley experiment are to be accepted (which implies the constancy of the speed of light), then Einstein saw that a contradiction arises. His boldness consisted in his abandonment of the three basic postulates of Newtonian physics: (1) the idea of absolute space and time; (2) the principle of addition of velocities; (3) the law of conservatism of mass (now generalizing this into a broader law—the conservation of mass-energy).

The consequences Einstein deduced logically and elegantly from his basic postulates not only are intimately tied up with modern physics, but also have great significance for all scientific philosophies. Only two ideas need to be mentioned here: The emphasis that scientific definitions have to be "operationally" defined, and the understanding (an idea not original with Einstein but given powerful emphasis by him) that the postulates of science are chosen in such a way that the mathematical system results in the most adequate interpretation of the phenomenological world about us.

We have traced the revolution in modern theoretical physics which culminated in the rise of relativity theory. Much of what had been going on in the ordinary pedestrian research still goes on in many ways not so affected by the revolution. This is due to the fact that the theorists have been seeking a logically consistent and complete theory to encompass all that goes on in the scientific study of nature. We often think of the foundations of a science in terms of such a theoretic view. Crises and controversy regarding the foundations do not necessarily hold up the progress in the various empirical pursuits or in research in special areas of a science. We can think of these two trends in science as (1)

---

† This is often overlooked by philosophers, behavioral scientists, and sociologists when they write as if relativity means that there is no invariance. They ignore the problem of the *referent* of the frame of reference in their judgment situations.

Questions/Discussions

1. Discuss what is meant by one frame of reference moving at constant velocity v in relation to one axis of another frame of reference? What would be the Galilean representation for the motion of the moving axis?

2. Suppose a man who is located on the moving frame of problem 1 fires a rifle in the same direction as the motion of his reference frame; the bullet has velocity u. What is the velocity of the bullet in relation to the fixed reference frames, in terms of Galilean transformations?

3. For Question 2, what is wrong with the answer as given by the Galilean transformations? How serious is this error when the velocities are of the ordinary range?

4. Suppose v and u of Questions 1 and 2 are increased to become $0.6c$ each. Can the Galilean transformation be used? What is the velocity of the bullet as seen from the ground? By how much does it exceed the velocity of light, c? Note: *Velocity as seen from the ground becomes*

$$\frac{v+u}{1+(uv/c^2)} = \frac{0.6c + 0.6c}{1+(0.36c^2/c^2)}$$

$$= \frac{1.2c^3}{1.36c^2}$$

$$= 0.88c$$

5. To what extent do the Lorentz transformations depend on the relative positions of objects as contrasted with relative velocity?

6. In judging people's behavior, and in discussions of human values and morality, one often hears it said that the theory of relativity *proves* that all standards are relative; for example, what is moral, true, and good in one time or place may be immoral, false, and bad in another.

   (a) Is this a valid application of the theory of relativity?
   (b) Does relativity prove that there are no values?

7. Graph the equation

$$m = \frac{m_0}{\sqrt{1-(v^2/c^2)}}$$

to show how the mass of a particle (of mass m) increases as its velocity increases from $0.7c$ to $0.9c$. What value will it have at $0.95c$? At $0.99c$? At what v is the mass twice its rest mass?

8. An observer at rest with respect to a rod measures its length and finds it to be 1 m. He also notes that it is oriented so that it makes an angle of 45 deg with the x-axis. A second observer traveling at $0.98c$ in the x direction, as seen by the first observer, also views the same rod.

   (a) How long does the rod appear to the second observer?
   (b) What angle does the rod appear to make with his x-axis?

References

Barnett, Lincoln, *The Universe and Dr. Einstein.* New York: Harper & Row (Torchbook), 1948. *This is completely nontechnical, and students report it to be very helpful.*

Bonnor, William, *The Mystery of the Expanding Universe.* New York: Macmillan, 1964. *A good popular discussion of current cosmology.*

Born, Max, *Einstein's Theory of Relativity.* New York: Dover, 1962. *This is probably the best discussion for the student who wants a little mathematical detail, but not so much as the professional relativist would read.*

DeBenedetti, Sergio, "The Mössbauer Effect," *Scientific American* (April 1960).

Einstein, A., *Relativity.* New York: Doubleday, 1947. *Einstein wrote this account of relativity specifically for the layman.*

Einstein, A., *Relativity: The Special & General Theory.* New York: Doubleday (Masterworks of Science), 1947. *Nontechnical.*

Einstein, A., and Infeld, L., *The Evolution of Physics.* New York: Simon & Schuster, 1942. *This is nontechnical and gives the reader an opportunity to read "the master's" own words.*

Gamow, G., *Matter, Earth and Sky.* Englewood Cliffs, N.J.: Prentice-Hall, 1965. *Gives a lively presentation of special relativity theory (Chap. 7). Then there is both a chapter on general relativity (Chap. 22) and one (Chap. 23) on cosmology.*

Gamow, G., *Mr. Tompkins in Paperback.* New York: Cambridge Univ. Press, 1965.

Katz, Robert, *The Special Theory of Relativity.* Princeton, N.J.: Van Nostrand (Momentum book), 1964. *Written for elementary physics students at high school or freshman college level.*

Ney, E. F., *Electromagnetism & Relativity.* New York: Harper & Row (Torchbook), 1962. *Somewhat more advanced than Born.*

Resnick, R., *Introduction to Special Relativity.* New York: Wiley, 1968. *A paperback that is highly recommended.*

Ripley, Julien A., Jr., *The Elements and Structure of the Physical Sciences.* New York: Wiley, 1964. *Has a good chapter (Chap. 19) on relativity theory, including a discussion of general relativity. The final chapter (Chap. 21) contains a considerable amount of material on relativistic cosmology in section 4.*

Schlegel, R., *Time and the Physical World.* East Lansing: Michigan State Univ. Press, 1961. *Not very mathematical for the most part; the discussions of empirical confirmation of relativity theory (Chap. VI) and the detailed discussion of the clock paradox (Chap. VII) may be useful.*

Smith, James H., *Introduction to Special Relativity.* New York: Benjamin, 1965. *A direct approach with elementary mathematical explanations.*

Taylor, E. F., and Wheeler, J. A., *Spacetime Physics.* San Francisco: Freeman, 1966. *An introduction to relativity and space-time geometry.*

Whitrow, G. J., *The Natural Philosophy of Time,* Harper & Row, 1963 (paperback).

CHAPTER THIRTEEN

Transition from Determinism to Indeterminacy

You believe in a dice-playing god, and I in perfect rule of law. . . .

From a 1944 letter by
ALBERT EINSTEIN to Max Born

THE PREVIOUS CHAPTERS have described some major transformations and periods in the progress of science. It is timely to pause to assess the significance of some of these transformations. In this chapter we take into account some of the conceptual changes that transpired from the time of Newton to the early part of the twentieth century, basing our investigation on the topics that were introduced in previous chapters. Along the way we shall digress to discuss the special role of energy during this transition period.

13.1 The miracle of history

Our discussions have covered vast temporal and spatial dimensions. Their significance lies not especially in the vastness, however, but in the fact that during their development, life and living organisms became ever more complex in structure, functions, and capabilities. To many scientists, this progression of inorganic matter to the many diverse forms of life evokes poetic reaction and such words as the "miracle of nature." In the most advanced form we find one species —namely, man—that has achieved a sense of order out of what often appears to be chaos.

The path of this progress appears to have been tortuous and long, involving many centuries of slow evolutionary changes, years of reasoning, and interspersed with dramatic discoveries of regularity in natural phenomena such as the rising of the sun and the changing of the seasons. It required much time to find causes for these phenomena, for more often than not the causes had no direct relationship to human experience and human comprehension. The concept of gravitational attraction, the laws that describe the courses of planets in the heavens, the brightness of distant stars, and even the more realistic effects of violent winds and storms have very little correlation with the struggles between the human body and its environment.† The laws of physical phenomena became revealed only as the human mind was able to comprehend ideas and concepts that transcended personal, momentary experience. But man did transcend the limitations of the human body, and did achieve the ability to present the laws of mechanics in mathematical terms. In the three laws of Newton, and in the other scientific developments that added sophistication and grace to the period of the seventeenth and eighteenth centuries, man succeeded in describing phenomena and concepts to which his bodily experiences had hitherto given only servile tribute.

While old ideas were reinterpreted and were being replaced by new discoveries and concepts, there developed an interest in identifying the factors that made for progress. Man seemed to have attained a stage of development that permitted progress without depending completely on ancient ideas and old authorities. Could useful knowledge be found at a faster rate by a new approach? Sir Francis Bacon noted that past discoveries had come about largely by chance, and that the progress and span of man's capabilities would be accelerated and increased if he could formulate methods of discovery. Was there a systematic approach that might reveal new knowledge? What constituted a systematic, more effective approach toward discovery? Bacon's general interests were similar to those of Plato, who also sought the betterment of mankind through planned acquisition of knowledge, or rather of *true* knowledge. But their approaches were quite different because of a basic difference in the final objectives for which this true knowledge was sought. For while Plato sought ethical betterment in the *Good Life*, Bacon sought *useful* knowledge. It has been said that Bacon sought to make man comfortable, while Plato sought to make him wise.

13.2 The approaches of Plato and Bacon

As a prerequisite to acquiring true and useful knowledge, both Plato and Bacon stressed that there was need first for man to rid himself of false ideas. To that end Plato devised a plan of education that would free man of false beliefs that had originated from momentary and undependable sensory experience and which would order his thinking processes, to make his mind receptive to the real, causal structure *underlying* sensory experience. He thought that such subjects as mathematics and music† would awaken the mind to bring forth the knowledge that was in the mind. He assumed that these could be learned by the Socratic method, which involved a process by which ideas of the mind were delivered like offspring from a womb.

Bacon also sought liberation from false ideas and from phantoms of the mind that separate man from reality. These were the Idols of the Tribe, of the Cave, of the Market Place, and of the Theatre, which respectively represent human nature, the personality of the individual, linguistic and communication skills, and ideologies

---

† The astrologers and many religions claim considerable correlation, of course.

† Plato considered that branch of music associated specifically with harmonics, with interest primarily in the structure underlying harmonies.

that one accepts uncritically from philosophers and other learned men. Once freed from such mental restraints, it would become possible to collect data in a more unbiased manner.† Bacon devised ways for collecting, classifying, and analyzing data, and for deriving from them the simplest "form" of the phenomena of nature.

But while the common objective of Plato and Bacon was the pursuit of knowledge, the approaches they proposed were substantially different. Plato would awaken the mind by training it in music and mathematics; this would allow it to become *receptive* to the structure of things, the underlying reality, the true knowledge. Bacon sought knowledge that might permit prediction and control of phenomena. Plato stressed reflection and intuitive understanding. Bacon stressed the art of *investigative doing* that would lead to new and useful knowledge. Plato sought ultimate knowledge for its own sake, in the belief that knowledge leads to virtue, but Bacon sought the creation of technologists who would master and control natural phenomena, including man. To Bacon, "knowledge is power." To Plato, knowledge leads to the True, the Beautiful, the Good, and to the *understanding* of nature rather than the mastery of nature.

The centuries that followed Bacon saw tremendous successes, as we shall see, from a pragmatic approach to science, which is credited by some to his influence. The pragmatic influence has not been altogether beneficial, as we shall also see, because a science that largely emphasizes the practical, observable, applied aspects of knowledge is not likely to achieve equal progress at the philosophic, theoretical level.

Bacon, it has been said, pictured the structure of knowledge as having the shape of a pyramid. At the base of the pyramid was a body of general truths and observations that were self-evident to "unbiased" men who had rid themselves of "idols." These would provide the materials which, when assembled and related, would give rise to a natural science, largely of descriptive character. (Bacon seems never to have gone beyond the descriptive phase.) Upon these materials, physics would build a science that would be a product of reasoning and *calculation* applied to the observations. The fourth, and the highest layer, would be *metaphysics*, in which reason attempted to give unity and more basic understanding of scientific concepts than could be achieved by the processes of physics and calculation alone.†

As we shall see, the application of the Baconian pragmatic approach was "in tune" with the discoveries of the periods of Galileo and of Newton, which began the industrial revolution based on mechanical technology. Bacon gave detailed procedures to guide the buildup of the first three layers of his pyramid. He was less helpful in building the fourth layer, the metaphysical, with its philosophical and cultural implications. His failure here was due, in part at least, to the absence of any experience that involved probing of the philosophic and cultural aspects of science. There was

† We might paraphrase this thought as follows: While knowledge obtained from the experiences of other people can have informational value, it also may tend to bias our attitude against achieving new and better information. Of course the same bias can develop from information obtained from personal experiences. But one may argue whether there is such a thing as a "freed" mind.

† See Margenau in References.

some progress in this latter realm, probably exemplified best in the publication of *Critique of Pure Reason* by the German metaphysician Immanuel Kant (1724–1804). But as we shall see presently, the trend in modern science, and in modern physics especially, has been to utilize analytic techniques that savor of a completely different approach from any that Bacon prescribed (or could have prescribed) from the experience of his times.

13.3 Technology, rationalism, and the Industrial Revolution

Over the centuries the search for knowledge alternated between the theoretical and pragmatic approaches proposed by Plato and Francis Bacon. Or perhaps we should say that while both approaches could have existed side by side, the one or the other tended to dominate. Such knowledge as was acquired during the Babylonian period was used for predicting events. During the Greek period, on the other hand, the trend was to divorce "science" from technology and to seek understanding in terms of principles and models from which all knowledge could be deduced.

In the intervening centuries between these developments and the Renaissance, technology continued to be separated from philosophy. To some, such as Thomas Aquinas (1225–1274?) the purpose of man's very existence was the opportunity to live the good life. Truth and knowledge lay inseparable from the church, and were not to be differentiated from church doctrine. Bacon, however, wrote at a time when this outlook was ending. He represented an attitude that had become more prevalent as technology and philosophy (science) began to interact. The technological environment favored this view. The laws of mechanics

that had been clarified and "unshackled" through the work of Galileo and of Newton applied not only to the movements of heavenly bodies but also to the swing of the hammer and to the rotation of a wheel. Meanwhile, improvements in iron and steel technology were making it possible to apply these same principles to the design of machines. When, a little later, the steam engine was invented and could serve as power unit, the machines were ready to take advantage of the new possibilities.

Even before the end of Bacon's own century, the spirit of the times reached for rationalism in all areas, and sought for social reform. The struggles that gave birth to the new knowledge of the seventeenth century could not be confined to the laws of mechanics. It was as if the mind that had suddenly recognized a firm cause-and-effect relationship to exist in the mechanical aspects of its environment could not avoid looking for such relationships in other areas as well. The success that came with the building of machines gave rise to even newer machines and to new applications of the older machines, thus relieving the inventor from burdens that for so many centuries had made servitude a normal way of life. Mechanical progress breeds social progress when mechanisms give shape to the social situations in which they are used.

So it was that early in the eighteenth century there began an *age of reason* (or rationalism) that touched all aspects of life—scientific, technological, religious, political, and social. The science of mechanics had found firm footing,† while the studies of thermodynamics and

---

† Actually the science of mechanics did not have a very solid base in mathematical theory at that time.

chemistry had also begun to find new and extensive applications even before there was more than primitive understanding of the phenomena involved in these areas. The interests of many investigators were directed toward applied science and to technology. Trade and commerce, which had greatly expanded during the seventeenth century, looked to the new machines for new products for trade and for increased productivity. Some historians claim that the new Protestant ethics contributed to the development of technology and science. It is therefore said that a Puritan, nonconformist religious spirit in England had already made technical progress and the pursuit of useful work a prime objective of life. [The preaching of John Wesley (1703–1791), who founded Methodism, reached every man with a message that demanded hard work along with rugged faith.] The concept of a free-enterprise economic system found firm footing in England and gave impetus to the development of inventions that could promote production, trade, and profit. The rise of commerce helped to break down the manorial (economic) system of feudalism, and "free" the peasant from the land. The propertyless peasants flocked to the cities to find "work" through which to earn their daily bread, and thus contributed to the rise of the factory system of production. In time there developed more interest in labor-saving devices to reduce costs. James Watt's steam engine made possible more extensive mining and utilization of iron ore and coal, and by 1787 it became useful for the spinning of cotton as well. The use of coke† in blast furnaces and continuous improvement of metallurgical processes made possible new kinds of steels that permitted design of better machines and machine tools for the textile industry. Production of sulfuric acid and of sodium carbonate on a large scale began in the second half of the eighteenth century and promoted faster progress in chemical industries. By the end of that century, an Industrial Revolution (sometimes referred to as the "second" Industrial Revolution) was well started in England. On the Continent there was greater dependence on the government and on the guild system.

The Puritan spirit also existed in the colonies of England, especially in what was to become the United States of America. Benjamin Franklin was a scientist-statesman (although certainly no Puritan) whose comprehension encompassed the values that lay in a combination of progressive science and technology, shrewd economics, and national progress. His admonition to young men to "waste neither *time* nor *money*" represented the spirit of the Industrial Revolution.

13.4 Energy and modern civilization

The industrial revolution was much more than simply an extension of applied science to produce products for trade. The growth of modern industry took men from the farm and concentrated them in cities and factories. The growth of industry in the hands of industrialists and commercial people brought a revolution among the ruling circles. The factories, which offered new opportunities and new forms of servitude, became a breeding ground for political revolutions that spread from England and the Continent to the new world. There seemed to be in

---

† Charcoal obtained from wood had been used to that time. Coke is produced from coal, which became much more plentiful.

these developments both new hopes and new means for achieving a social revolution that might free the common man from want and virtual slavery. The machines in the factories demanded skilled labor, which in turn demanded a higher level of education for many more people to design and care for the machines and plants. This created not only the need for more people with education, but also a demand for a wider variety of educational preparation. The first institution of higher education devoted to science and technology—the École Polytechnique—was established in Paris in 1794–1795 to meet the military needs of France.† Military needs were then, as they always had been and continue to be to this day, patrons of technological progress.

We cannot dwell long on this vigorous period of history, or probe into many of its exciting facets. There is, however, one feature that deserves special attention; that is, the inventions of machines for utilizing energy sources, and the role of this development in human affairs.

There is something quite fundamental and universal about energy. We have seen that it exists in many forms: kinetic, potential, mechanical, electrical, chemical, thermal, and radiant energy. From the Einstein relationships we have seen (and will pursue in more detail) the mass equivalence of energy. Energy may be transformed from any one form to any other form, with more or less efficiency. We have also learned, however, that energy† is conserved in the process of natural or man-made transformations.

The Industrial Revolution, and all the attending other revolutions that arose during the development of the new sciences, required energy to translate ideas into technological, economic, and social structure. Much more than simple human energy was required, however. As long as man had only another man or an animal to pull the plow, to grind the wheat, or to pull the wagon, his economy could not change too much from that of the Egyptians. But when the steam engine was invented, when coal could be mined with the help of steam power, and the economy could shift from the individual producer or guild system to power production methods, the Industrial Revolution could scarcely be prevented. With steam power came the factories and railroad trains, the latter proving to be a special boon for the expansion of the American colonies into the great West.

There seemed to be no limit to the functions for which machines and factories could be designed, especially as large and small steam engines became available. Before long, experiments with electricity, electric generators, and motors demonstrated that electric energy offered variations and advantages. Filaments of platinum or carbon, and finally of tungsten wire, could be brought to white heat by electricity, to give light. Through the tireless efforts of Thomas Edison in the United States and of various groups in Europe, the electric bulb became a common source of light without which modern civilization would appear drab. Electric motors offered

---

† In the United States, involvement in technology on the part of schools of higher education began in the Hudson River Valley. The military aspects were begun at West Point in the early 1800's, perhaps by 1817, under Sylvanus Thayer. The earliest formal non-military program in civil engineering was established in 1835 at Rensselaer Polytechnic Institute, Troy, New York, the school being founded in 1824 with that end in view.

† Energy is here taken to include the rest energy.

advantages in size and simplicity over steam engines for powering small machinery. The advantages of electric energy gave rise to a vast electrical equipment and power industry, with networks of distribution lines throughout the civilized world.

The new interest in power devices and the new familiarity with the thermodynamics of steam engines had inspired inventors to look for more efficient and higher-temperature energy devices. Before long, experiments were under way to develop internal combustion engines that might replace the steam engine, to serve both as large prime movers and for low-energy applications. The principle was simple: Air mixed with fuel could be compressed and then ignited to develop much higher pressure that moved a piston up and down so that it could perform useful work. The experiments succeeded, and a variety of internal combustion engines and rotary turbines came into being. The new motors were lighter than the steam engines and could power individual vehicles. Meanwhile, the discovery of large oil deposits and the new interest in easily combustible fuel for motor use promoted the development of huge refineries for conversion of crude oil to gasoline. The development of new motors, new fuels, "horseless carriages," and new production methods all followed apace to reach the highly sophisticated stage of the twentieth century.

We shall not pursue these details further, nor take more than passing notice of the coming of the airplane. The development of nuclear energy will be discussed in detail in a later chapter. It will be interesting, however, to compare the energy consumption of modern man with the energy consumption of early civilizations. The primitive societies varied in their use of plant energy, such as wood for fire. The early societies that discovered the use of fire seemed to have carried this use to an extent that may explain the deforestation of large areas of Africa. These early societies were prodigal in their consumption of wood for burning. It appears that even in the hot climates of Africa there was need to keep fires burning for protection from animals. There were other areas, however, where the consumption of energy could not equal that which was possible in forested areas, and the consumption level remained low.

In any case, the utilization of energy resources for purposes other than cooking, warmth, and protection from animals was negligible in very early societies. When mining, smelting, and forging operations became common in the Babylonian and Greek periods, there developed some industrial use of fuel, largely wood. This use increased when coal became more common, such as in Britain in the seventeenth century. From then until now the use of coal, wood, oil, water power, and now nuclear power increases each year. In the United States, for example, the per *capita* use of industrially useful energy amounted to about 13,000 kilowatt-hours (kwh) in the 1950's, whereas only 1000 kwh was used in Italy in the same period. (Unfortunately, nearly all energy sources used in the United States are of the kind that cannot be restored again in our time; namely, coal, oil, and natural gas deposits.) By way of comparison, the annual food needs for an individual amounts to about 930 kwh. Thus, in the most energy-utilizing society, the individual uses about 14 more times the amount of energy of the food equivalent required to support life.† In fact the national growth

† See Thirring in References.

within the United States requires a doubling of the supply of electric energy and other energy forms about every 20 to 30 years. The demands of industrial growth, higher standards of living, and leisure seem to require this rapid rate of increase. An *advanced* society seems to be synonymous with an *energy-consuming* society. How these needs will be met when present fossil sources become depleted presents a challenge, to be discussed in Chapter 21.

We must now return to the main points of the chapter.

13.5 The limitations of Newtonian mechanics

The rationalism that promoted industrial, social, economic and political revolutions had its origin in the age of Descartes and of Newton. The new science gave rise to a view of nature that was highly deterministic. Everything in nature could be understood in terms of its inexorable laws. Nothing was accidental; all was ruled by law. This idea is often called *Cartesian determinism* to distinguish it from the more ancient concept of causality, which resembled a principle of compensation. According to that earlier view, when something happens to disturb nature, there comes into effect a force that tends to make good the damage (or to bring compensation or retribution).

It was part of the Cartesian approach to require that phenomena and concepts be made clear by use of drawings and models. Every object had its form clearly definable in terms of points in space, and space itself was given adequate representation through Cartesian coordinates. Every motion of the object could be traced through this coordinate system. The force required to move things in space offered no mysteries, since the application of force was a common human experience.

It was left to Newtonian mechanics to spell out clear relationship between cause and effect: A body started to move, or was stopped from moving, because a force was applied to it. Nothing happened without a cause. The processes of nature could be determined in advance. Given adequate knowledge of the positions and movements of the particles of the universe, one might predict the entire future course of the universe. Some mechanistic speculation about human behavior would say that, given complete knowledge of the characteristics and movements of the particles that make up the human body, one might chart the whole career of a man. Man being a part of nature, the laws that hold for nature would presumably hold for man as well. Determinism seemed to be built into nature and into man to a degree that seemed to make man a mere automaton equipped (in modern parlance) with a complex computer system that directed his every move. Since the philosophical aspects of Newton's physics were used to justify doctrines about the nature of man, it was said that what appear to be *decision-making activities* in man are nothing other than determined events. Man cannot really choose, and there is no such thing as free will.

Thus, free will seemed to disappear in the face of this determinism.[†] The implications seemed to be that as humans, and hence as part of the universe, we can no more make an independent choice or alter that choice than can the stone along the roadside. Therefore, could man be

---

[†] Or, we should say, the *models* that were used to explain nature seemed to imply that determinism precluded free will, at least within the limited systems that were considered.

held responsible for his actions? The fixedness of nature as outlined by the laws of Newtonian mechanics seemed to offer no alternative, especially since the technological successes of every subsequent decade, through to the twentieth century, seemed to strengthen the validity of these laws. The conservation laws, which at that time were applied to both mass and energy, seemed only to emphasize these concepts and restrictions.

This view of nature did not remain unchanged, however, as we have seen in the previous several chapters. From the discussion of systems and controls we have found that "no man is an island" completely separated from external influences. In fact an organism is intimately a part of its environment, and rarely can be considered to be in isolation from its environment. Neither a stone, organism, nor man himself can be defined or said to have existence wholly divorced from the environment. Neither the changes in the stone nor the life processes of organisms have meaning except in relation to the environment. But in these previous chapters, we noted additional complications that do not lend themselves to comprehension through Newtonian mechanics alone. The pedaling of a bicycle and the mechanisms that move an automobile surely obey Newtonian mechanics, but now the feedback of vision and the interpretive and translational processes of mind and muscle come into play, and these influences make the application of such mechanics alone highly inadequate. For example, informational feedback has consequences on human systems and societies that go beyond the simple concepts of Newtonian mechanics. Clearly, Newtonian mechanics does not offer a complete analytical technique through which to approach many situations that include larger portions of the universe.†

Newtonian mechanics conceived each particle to have a position in space which could be clearly identified through coordinate systems, and each event to have a time of occurrence that is clearly related to the time for every other event. The coordinate system could be a moving frame of reference, but this did not negate either the existence of an absolute frame of reference or of absolute time. But Einstein and the relativity postulate denied the existence of either an absolute frame of reference, absolute time, or absolute motion. The motion of every object and every coordinate system, and time itself, can be considered only *in relation* to other objects and systems.

A feature that seems to be invariant and absolute, regardless of frame of reference as far as experimental measurements can determine, is the velocity of light. The new rules did permit bringing under one roof the laws of mechanics and the newly discovered behavior of electromagnetic radiation; but in the process, Newtonian mechanics was dealt another blow. Nor was this all. The laws of Newton had been derived from study of macroscopic objects—the ball, the wagon, and the planets themselves. For each of these it was possible to assign a mass and a central or specific point where the entire mass could be considered to be concentrated. When two bodies collided, it was possible to analyze the phenomenon in terms of these assigned points and masses.

We have learned, however, that as these objects became smaller and smaller,

† The significance of *determinism* versus *free will* could also change as one goes from a simple to a more complex system.

13.5 The Limitations of Newtonian Mechanics

and their numbers larger and larger (as in volumes of gas), the problems of analyses required a change in tactics. No longer could each particle be treated individually; there were too many particles, and the number of changes in motion experienced by each particle in each second ran to large numbers. Clearly a change of analytical technique was needed.

There was no reason to assume that each tiny particle did not individually obey Newton's laws, but there was difficulty in proving experimentally that it did. Such an experiment required the measurement of the mass and position of the particle at each instant of time. The particles were so tiny that the measuring instruments or the detection principle had to be exceedingly delicate, sensitive, and responsive to the movements of an atom without altering its course. It was found, however, that long before an instrument could be reduced to such delicate sensitivity, the parts of the system became subject to statistical variations of the gas assembly and were hopelessly kicked around in zigzag fashion without stop, as was the case with the tiny dust particles in Brownian motion.

It became necessary, therefore, to resort to a wholly new approach for the analysis of molecular gas assemblies; namely, a statistical approach. The velocity and kinetic energy of each gas particle varied very rapidly because of collisions, but it was possible to assign a distribution curve that might tell how much of the time the particle was likely to spend in each energy state. An average value could also be obtained from this energy distribution curve. Thus the *energy* of the *total* assembly could be determined. The measurement of this total energy proved to be feasible, since it could be performed with the same instruments used for measuring temperature, pressure, volume, mass, and time.

But while the measurement of the energy of a gas assembly seemed to be feasible and not in contradiction with the earlier concepts, a new wrinkle was added. In an elastic collision of two large bodies, the laws of mechanics said that both the total momentum and energy were conserved. Also, the process was reversible (that is, a moving picture film of the collision could be run backward or forward without suggesting contradiction of any rules). The situation was different in the gas system, however, for here although energy and momentum were conserved, a new phenomenon was now observed in that the system was no longer completely reversible. There was an *unidirectional trend*, which was associated with the *entropy* of the system, and which implied that although the total energy was conserved, the system as a whole tended only toward states of increasing entropy.

Finally, we are reminded that the process by which we arrived at a probable energy distribution for any *single* gas molecule is similar to the process involving the tossing of coins and determining the distribution of heads and tails. That is, the statistical treatment of large numbers of similar events, or of large numbers of identical particles, have much in common. In the case of both, chance seems to play a role, for we consider always the *probable* distribution of energies from collisions, and the *probability* for heads or tails. When we speak of the probability of an event, we are saying that certainty is lacking. There can be very great, overwhelming probability that the gas laws will be obeyed, but only when the numbers of molecules are large enough and when a statistical summation of the phenomena assures the high probability. The individual collisions and events do not enjoy this overwhelming probability feature, however. With them the uncertainties are very large, as is true of any individual chance phenomenon, such as whether a coin will fall heads or tails. In fact one is led to question whether in such chance phenomena there is strict adherence to the laws of cause and effect in a sense so rigidly defined by Newtonian mechanics, as interpreted by the followers of Descartes (Cartesians).

We shall see later that for the understanding of the phenomena of atomic physics, we shall be forced to give up this rigid interpretation of cause and effect. It will be necessary to analyze phenomena in terms of their *statistical* characteristics rather than the detailed dynamic features of particle mechanics. When we study the decay of radioactive atoms (already mentioned briefly in Chapter 2 in connection with dating techniques), we shall not be able to say *when* a particular atom will disintegrate, although a large number of identical atoms will follow the characteristic decay rate of that element *very accurately*.†

We are therefore forced to desert the comfort of fixed laws and the "predetermined" course of nature that is given by the Cartesian concept of Newtonian mechanics, and must favor a "science" of nature that has within it large elements of chance. There is some comfort in the thought that through this change we may also have gained some relief from the hopeless concepts of predestination and determinism which accompanied the earlier model. Strict causality postulates that every physical event depends on, and is determined by, other preceding events, and the role of science is to search for the nature of the dependence. But the newer approach finds it necessary to look for the states of a system, without expecting either the dependence of one state on another or the process of transition from one state to another, in the simple manner of Newtonian mechanics.

It is a mistake, of course, to assume that nature is altogether governed by either deterministic laws or chance phenomena. The fact is that in nature there is a composite of what appear to be highly deterministic conditions, while chance seems to be dominant in other phenomena. There are occasions when an individual may "choose" under conditions wherein freedom of choice is apparently possible, while in other situations he may have no choices possible. Nor should one assume that there exists a state of chaos when there is absence of determinism in the classical sense.

13.6 *Philosophy of cause and chance phenomena*

A beautifully presented discussion of the questions and contradictions that are inherent to this subject is that given by Max Born as "Metaphysical Conclusions" to his lectures delivered in England in 1948 and reproduced as *Natural Philosophy of Cause and Chance*,† which we quote through the kind permission of Dover Publications, Inc. Beyond its specific contents, this quotation illustrates in a remarkable way the human aspects of science and the differences of opinion that can exist among leading physicists on the basic questions of this chapter.

---

† That is, the laws that describe the statistical character of such a phenomenon as radioactivity can be stated just as accurately as can the laws associated with the collision of two bodies.

† See References.

METAPHYSICAL CONCLUSIONS

"The statistical interpretation which I have presented in the last section is now generally accepted by physicists all over the world, with a few exceptions, amongst them a most remarkable one. As I have mentioned before, Einstein does not accept it, but still believes in and works on a return to a deterministic theory. To illustrate his opinion, let me quote passages from two letters. The first is dated 7 November 1944, and contains these lines:

"'In unserer wissenschaftlichen Erwartung haben wir uns zu Antipoden entwickelt. Du glaubst an den würfelnden Gott und ich an volle Gesetzlichkeit in einer Welt von etwas objektiv Seiendem, das ich auf wild spekulativem Weg zu erhaschen suche. Ich hoffe, dass einer einen mehr realistischen Weg, bezw. eine mehr greifbare Unterlage für eine solche Auffassung finden wird, als es mir gegeben ist. Der grosse anfängliche Erfolg der Quantentheorie kann mich doch nicht zum Glauben an das fundamentale Würfelspiel bringen.'

"(In our scientific expectations we have progressed towards antipodes. You believe in the dice-playing god, and I in the perfect rule of law in a world of something objectively existing which I try to catch in a wildly speculative way. I hope that somebody will find a more realistic way, or a more tangible foundation for such a conception than that which is given to me. The great initial success of quantum theory cannot convert me to believe in that fundamental game of dice.)

"The second letter, which arrived just when I was writing these pages (dated 3 December 1947), contains this passage:

"'Meine physikalische Haltung kann ich Dir nicht so begründen, dass Du sie irgendwie vernünftig finden würdest. Ich sehe natürlich ein, dass die prinzipiell statistische Behandlungsweise, deren Notwendigkeit im Rahmen des bestehenden Formalismus ja zuerst von Dir klar erkannt wurde, einen bedeutenden Wahrheitsgehalt hat. Ich kann aber deshalb nicht ernsthaft daran glauben, weil die Theorie mit dem Grundsatz unvereinbar ist, dass die Physik eine Wirklichkeit in Zeit und Raum darstellen soll, ohne spukhafte Fernwirkungen. . . . Davon bin ich fest überzeugt, dass man schliesslich bei einer Theorie landen wird, deren gesetzmässig verbundene Dinge nicht Wahrscheinlichkeiten, sondern gedachte Tatbestände sind, wie man es bis vor kurzem als selbstverständlich betrachtet hat. Zur Begründung dieser Überzeugung kann ich aber nicht logische Gründe, sondern nur meinen kleinen Finger als Zeugen beibringen, also keine Autorität, die ausserhalb meiner Haut irgendwelchen Respekt einflössen kann.'

"(I cannot substantiate my attitude to physics in such a manner that you would find it in any way rational. I see of course that the statistical interpretation (the necessity of which in the frame of the existing formalism has been first clearly recognized by yourself) has a considerable content of truth. Yet I cannot seriously believe it because the theory is inconsistent with the principle that physics has to represent a reality in space and time without phantom actions over distances. . . . I am absolutely convinced that one will eventually arrive at a theory in which the objects connected by laws are not probabilities, but conceived facts, as one took for granted only a short time ago. However, I cannot provide logical arguments for my conviction, but can only call on my little finger as a witness, which cannot claim any authority to be respected outside my own skin.)

"I have quoted these letters because I think that the opinion of the greatest living physicist, who has done more than anybody else to establish modern ideas, must not be by-passed. Einstein does not share the opinion held by most of us that there is overwhelming evidence for quantum mechanics.† Yet he concedes 'initial success' and 'a considerable degree of truth.' He obviously agrees that we have at present nothing better, but he hopes that this will be achieved later, for he rejects the 'dice-playing god.' I have discussed the chances of a return to determinism and found them slight. I have tried to show that classical physics is involved in no less formidable conceptional difficulties and have eventually to incorporate chance in its system. We mortals have to play dice anyhow if we wish to deal with atomic systems. Einstein's principle of the existence of an objective real world is therefore rather academic. On the other hand, his contention that quantum theory has given up this principle is not justified, if the conception of reality is properly understood. Of this I shall say more presently.

"Einstein's letters teach us impressively the fact that even an exact science like physics is based on fundamental beliefs. The words *ich glaube* appear repeatedly, and once they are underlined. I shall not further discuss the difference between Einstein's principles and those which I have tried to extract from the history of physics up to the present day. But I wish to collect some of the fundamental assumptions which cannot be further reduced but have to be accepted by an act of faith.

"Causality is such a principle, if it is defined as the belief in the existence of mutual physical dependence of observable situations. However, all specifications of this dependence in regard to space and time (contiguity, antecedence) and to the infinite sharpness of observation (determinism) seem to me not fundamental, but consequences of the actual empirical laws.

† The significance of quantum mechanics will be presented in a later chapter. The term *classical physics* refers to the physics that preceded quantum concepts of atomic structure—THE EDITORS.

"Another metaphysical principle is incorporated in the notion of probability. It is the belief that the predictions of statistical calculations are more than an exercise of the brain, that they can be trusted in the real world. This holds just as well for ordinary probability as for the more refined mixture of probability and mechanics formulated by quantum theory.

"The two metaphysical conceptions of causality and probability have been our main theme. Others, concerning logic, arithmetic, space, and time, are quite beyond the frame of these lectures. But let me add a few more which have occasionally occurred, though I am sure that my list will be quite incomplete. One is the belief in harmony in nature, which is something distinct from causality, as it can be circumscribed by words like beauty, elegance, simplicity applied to certain formulations of natural laws. This belief has played a considerable part in the development of theoretical physics —remember Maxwell's equations of the electromagnetic field, or Einstein's relativity—but how far it is a real guide in the search of the unknown or just the expression of our satisfaction to have discovered a significant relation, I do not venture to say. For I have on occasion made the sad discovery that a theory which seemed to me very lovely nevertheless did not work. And in regard to simplicity, opinions will differ in many cases. Is Einstein's law of gravitation simpler than Newton's? Trained mathematicians will answer Yes, meaning the logical simplicity of the foundations, while others will say emphatically No, because of the horrible complication of the formalism. However this may be, this kind of belief may help some specially gifted men in their research; for the validity of the result it has little importance.

"The last belief I wish to discuss may be called the principle of objectivity. It provides a criterion to distinguish subjective impressions from objective facts, namely by substituting for given sense-data others which can be checked by other individuals. I have spoken about this method when I had to define temperature: the subjective feeling of hot and cold is replaced by the reading of a thermometer, which can be done by any person without a sensation of hot or cold. It is perhaps the most important rule of the code of natural science of which innumerable examples can be given, and it is obviously closely related to the conception of scientific reality. For if reality is understood to mean the sum of observational invariants—and I cannot see any other reasonable interpretation of this word in physics—the elimination of sense qualities is a necessary step to discover them.

"Here I must refer to the previous Waynflete Lectures given by Professor E. D. Adrian, on *The Physical Background of Perception*, because the results of physiological investigations seem to me in perfect agreement with my suggestion about the meaning of reality in physics. The messages which the brain receives have not the least similarity with the stimuli. They consist in pulses of given intensities and frequencies, characteristic for the transmitting nerve-fibre, which ends at a definite place of the cortex. All the brain "learns" (I use here the objectionable language of the "disquieting figure of a little hobgoblin sitting up aloft in the cerebral hemisphere") is a distribution or "map" of pulses. From this information it produces the image of the world by a process which can metaphorically be called a consummate piece of combinatorial mathematics: it sorts out of the maze of indifferent and varying signals invariant shapes and relations which form the world of ordinary experience.

"This unconscious process breaks down for scientific ultraexperience, obtained by magnifying instruments. But then it is continued in the full light of consciousness, by mathematical reasoning. The result is the reality offered by theoretical physics.

"The principle of objectivity can, I think, be applied to every human experience, but is often quite out of place. For instance: what is a fugue by Bach? Is it the invariant cross-section, or the common content of all printed or written copies, gramophone records, sound waves at performances, etc., of this piece of music? As a lover of music I say No! that is not what I mean by a fugue. It is something of another sphere where other notions apply, and the essence of it is not 'notions' at all, but the immediate impact on my soul of its beauty and its greatness.

"In cases like this, the idea of scientific objective reality is obviously inadequate, almost absurd.

"This is trivial, but I have to refer to it if I have to make good my promise to discuss the bearing of modern physical thought on philosophical problems, in particular on the problem of free will. Since ancient times philosophers have been worried how free will can be reconciled with causality, and after the tremendous success of Newton's deterministic theory of nature, this problem seemed to be still more acute. Therefore, the advent of indeterministic quantum theory was welcomed as opening a possibility for the autonomy of the mind without a clash with the laws of nature. Free will is primarily a subjective phenomenon, the interpretation of a sensation which we experience, similar to a sense impression. We can and do, of course, project it into the minds of our fellow beings just as we do in the case of music. We can also correlate it with other phenomena in order to transform it into an objective relation, as the moralists, sociologists, lawyers do—but then it resembles the original sensation no more than an intensity curve in a spectral diagram resembles a colour which I see.

After this transformation into a sociological concept, free will is a symbolic expression to describe the fact that the actions and reactions of human beings are conditioned by their internal mental structure and depend on their whole and unaccountable history. Whether we believe theoretically in strict determinism or not, we can make no use of this theory since a human being is too complicated, and we have to be content with a working hypothesis like that of spontaneity of decision and responsibility of action. If you feel that this clashes with determinism, you have now at your disposal the modern indeterministic philosophy of nature, you can assume a certain "freedom," i.e., deviation from the deterministic laws, because these are only apparent and refer to averages. Yet if you believe in perfect freedom you will get into difficulties again, because you cannot neglect the laws of statistics which are laws of nature.

"I think that the philosophical treatment of the problem of free will suffers often . . . from an insufficient distinction between the subjective and objective aspect. It is doubtless more difficult to keep these apart in the case of such sensations as free will, than in the case of colours, sounds, or temperatures. But the application of scientific conceptions to a subjective experience is an inadequate procedure in all such cases.

"You may call this an evasion of the problem, by means of dividing all experience into two categories, instead of trying to form one all-embracing picture of the world. This division is indeed what I suggest and consider to be unavoidable. If quantum theory has any philosophical importance at all, it lies in the fact that it demonstrates for a single, sharply defined science the necessity of dual aspects and complementary considerations. Niels Bohr has discussed this question with respect to many applications in physiology, psychology, and philosophy in general. According to the rule of indeterminacy, you cannot measure simultaneously position and velocity of particles, but you have to make your choice. The situation is similar if you wish, for instance, to determine the physico-chemical processes in the brain connected with a mental process: it cannot be done because the latter would be decidedly disturbed by the physical investigation. Complete knowledge of the physical situation is only obtainable by a dissection which would mean the death of the living organ or the whole creature, the destruction of the mental situation. This example may suffice; you can find more and subtler ones in Bohr's writings. They illustrate the limits of human understanding and direct the attention to the question of fixing the boundary line, as physics has done in a narrow field by discovering the quantum constant $\hbar$. Much futile controversy could be avoided in this way. To show this by a final example, I wish to refer to these lectures themselves which deal only with one aspect of science, the theoretical one. There is a powerful school of eminent scientists who consider such things to be a futile and snobbish sport, and the people who spend their time on it drones. Science has undoubtedly two aspects: it can be regarded from the social standpoint as a practical collective endeavour for the improvement of human conditions, but it can also be regarded from the individualistic standpoint, as a pursuit of mental desires, the hunger for knowledge and understanding, a sister of art, philosophy, and religion. Both aspects are justified, necessary, and complementary. The collective enterprise of practical science consists in the end of individuals and cannot thrive without their devotion. But devotion does not suffice; nothing great can be achieved without the elementary curiosity of the philosopher. A proper balance is needed. I have chosen the way which seemed to me to harmonize best with the spirit of this ancient place of learning."

Questions/Discussions

This has been an exceedingly brief review of changes in philosophic attitude that have accompanied the transitional periods of the progress of science up to our present century. For readers who are interested in the interrelationships among science, religious and philosophic thought, and the socioeconomics of the western world, the chapter introduces many questions that deserve study, discussion, and class debate.

Among the questions that deserve class time for discussion, definition of terms and meanings, and debate are the following:

1. What are the merits and limitations of the attitudes of Bacon and of Plato with respect to the purposes of science and the purposes of life itself? To what extent were their ideas contradictory? To what extent complementary? To what extent did their views have basis in the socioeconomic characteristics of their own age?

2. Define what is commonly meant by the phrase "The Age of Reason." What were the origins and causes of that period? To what extent does the phrase apply to our present period with respect to science? With respect to religion? With respect to government?

3. Describe what is meant by the presence, or absence, of the exercising of "free will." Discuss its relationship to the Newtonian period, to the Age of Reason, and to the newer ideas on "causes" in science.

4. Referring to the quotation from Einstein that begins this chapter, how would you distinguish the phenomena and results that accompany dice playing from phenomena that follow the "perfect rule of law"?

5. Max Born, in the portion quoted herein, refers to the role of "belief" in scientific matters. How would you distinguish the form of belief to which he refers from religious belief of other individuals? How would you compare, justify, or condemn the Einsteinian clinging to belief in some future confirmation of "perfect rule of law" with religious conviction on the part of individuals?

6. How would you summarize Max Born's approach to the problem of "free will"? To what extent do you agree, and to what extent disagree, with this view?

References

Born, Max, *Natural Philosophy of Cause and Chance.* New York: Dover (paperback), 1964.

Bronowski, J., *Science and Human Values.* New York: Harper & Row (Torchbooks), 1965.

Colodny, Robert G. (ed.), *Beyond the Edge of Certainty.* Englewood Cliffs, N.J.: Prentice-Hall, 1965. *Essays in contemporary science and philosophy. Includes a series of articles by various authors.*

Frank, Philipp, *The Philosophy of Science.* Englewood Cliffs, N.J.: Prentice-Hall, 1957.

Klemm, Friedrich, *A History of Western Technology.* Cambridge, Mass.: M.I.T. Press, 1964.

Luchins, Abraham S., and Luchins, Edith H., *Logical Foundations of Mathematics for Behavioral Scientists.* New York: Holt, Rinehart and Winston, 1965.

Margenau, Henry, "Bacon and Modern Physics: A Confrontation," *Proc. Am. Phil. Soc.,* Vol. 105, No. 5 (October 1961).

Thirring, Hans, *Energy for Man* New York: Harper & Row (Torchbook), *From windmills to nuclear power,* 1958.

Weisskopf, Victor F., *Knowledge and Wonder.* New York: Doubleday (Anchorbook), 1963.

CHAPTER FOURTEEN

The Earth and Its Atmosphere

In rivers, the waters that you touch is the last of what has passed and the first of that which comes: so with time present.

LEONARDO DA VINCI

THE STUDY of natural science may be approached in many ways. An introduction that begins with the concept and role of the atom as the basic, uncuttable (or indivisible) unit of which all matter is composed offers some advantages. We have used this concept to explain certain thermodynamic phenomena, although the details remain to be treated in succeeding chapters. But thus far our approach has been to develop a series of largely macroscopic topics and concepts, which together reveal the historical and changing features of science.

14.1 Man and his environment

We turn now to the subject of earth science, again not from the minuteness of atomic composition but from the point of view of the earth as the environment of man. It seems fairly likely that the directions and patterns that have characterized the development of man (and indeed of all biological organisms) have been determined to a large extent by the nature of the earth environment of these living things. For example, the evolution of organisms seems to have depended on oxygen, since oxygen is present in our atmosphere.† More recently we have become aware of the equally large influence that man exercises in changing the face of the earth to suit his needs.

We shall, in fact, attempt to view man and his planet as an organismic whole. The food he eats, the air he breathes, the materials and means he utilizes to find protection and comfort, all seem to derive directly from his participation in this organismic whole. We shall find the systems concept to be especially useful. There are, within this large whole, very

† But nitrogen is present in even larger proportion!

many systems and subsystems. We are aware of many cyclic changes of large and small dimensions, which recall the oscillatory phenomena of earlier chapters. In the same way we identify feedback influences that tend to restrain cyclic excursions and give a measure of stability to natural processes. But because initial conditions are rarely restored without change, certain trends, which we may call evolutionary changes, can be identified in the history of both geologic phenomena and man.

In this chapter and Chapter 15, we shall attempt to discover order in the composition and structure of the earth and its atmosphere. There is even more remarkable order in the organization of atomic elements that make up living systems. The organization and functions of the one influence those of the other. The changing character of the earth over the long ages has been reflected in even more dramatic changes in the forms and functions of organisms that derive sustenance from the earth.

At this stage of history man seems to have reached a rather unusual turning point in his interrelationship with his environment. He is still dependent on elements that derive from the soil. His state of well-being, mental activity, and comfort is still very much influenced by the daily diet he receives from that soil and by the climate that envelopes him. More than ever before he has learned to control the elements that make up his diet and adjust to the environmental conditions to which his body is exposed. Moreover, his numbers and skills have multiplied to such a degree that the behavioral patterns of his society no longer constitute mere ripples on the course of earth history. His needs for food require skill and large-scale operations in the management of the soil and arable regions of the earth to prevent mass starvation. The forests and coal and oil beds that provide heat energy and raw material for vast industries and for man's comfort are becoming depleted very rapidly. Industrial development and encroachments of his society have greatly changed the features of the lakes and rivers, of flora and fauna. The products of his factories have changed and damaged the life-supporting composition of the atmosphere and of lakes and rivers. And the atomic "devices" that he has learned to produce now have explosive energies that go far beyond the capabilities of the tractor, the bulldozer, or the factory for altering and spoiling the life-supporting capabilities of "mother earth."

We shall explore these matters in some detail, looking in turn to each of the topical areas that make up the total picture. It is important, however, that we achieve some sense of the grand design and function of the total system that comprises earth and living organisms, a system that is complex, dynamic, changing, and which has interminable chains of interrelationships and feedback influences of earth on organisms, organisms on other organisms, reflecting finally back to earth through most intricate and remarkable ways.

Therefore, as we study the details of earth science and the biosphere, three questions tend to emerge from time to time:

(1) How intimately interrelated are the components that make up a dynamic, ever-changing earth system?

(2) How effectively do feedback influences tend to restore equilibrium and to maintain a measure of constancy within this earth system, at least over relatively short periods of time?

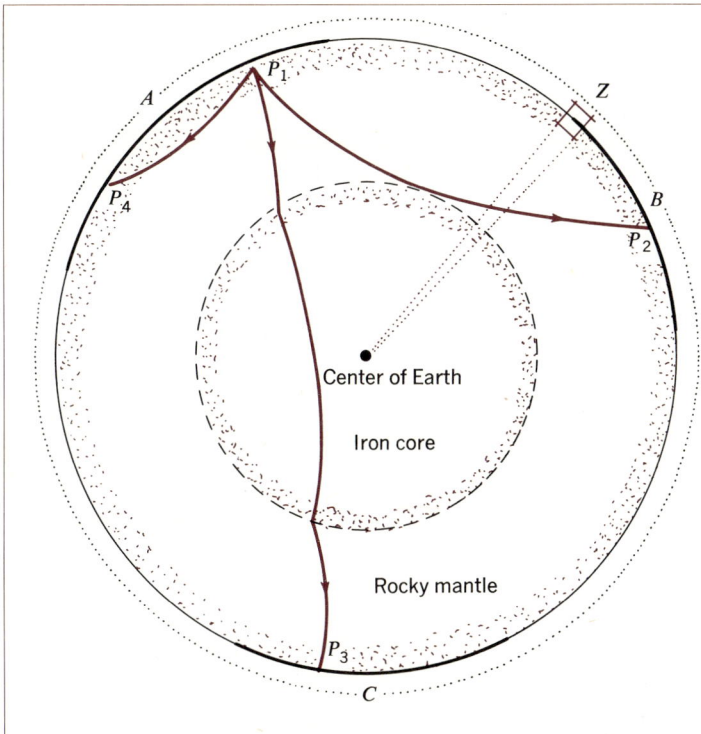

Fig. 14.1. Cross section of the earth. The earth's atmosphere lies almost entirely within 60 miles of the surface (within 1 mm in the figure) and astronauts in "outer space" are not very far out (circular orbit shown by dotted line). About the earth's vast interior, we know very little (see text). A segment such as Z is shown enlarged in Fig. 14.2. The interior lines indicate how an earthquake or nuclear bomb explosion occurring at point P_1 may be detected by vibration effects that reach points P_2, P_3, and P_4. Analysis of the signals that reach these points helps to determine the nature of the materials through which the signals are transmitted.

(3) How, despite this restoration and constancy, does there continue a slow, evolutionary trend toward uncertain ends?

14.2 The earth's interior

Our inanimate environment includes the sun, moon and stars as well as the earth we live on. A little has been said about the sun, moon, and stars in Chapters 3 and 7. It is now appropriate to give our attention to a few facts about the earth.

The component parts of the earth are shown in Figs. 14.1 and 14.2. The surprising thing emphasized by Fig. 14.1 is how little we know about the earth. We know its size. (You can measure the size of the earth by measuring the distance from New York to Montreal and then having your friend in New York measure the altitude of the sun at noon on the same day that you measure it in Montreal. What errors would you have to allow for? Question: Why are New York and Montreal a suitable pair of towns for this purpose? Exactly what quantity would you calculate by this experiment?)

We know its shape, approximately. Experiments of the kind just mentioned have been performed at a sufficient number of different places to give an overall picture. Measurements made by instruments in satellites have improved our knowledge of the shape of the earth.

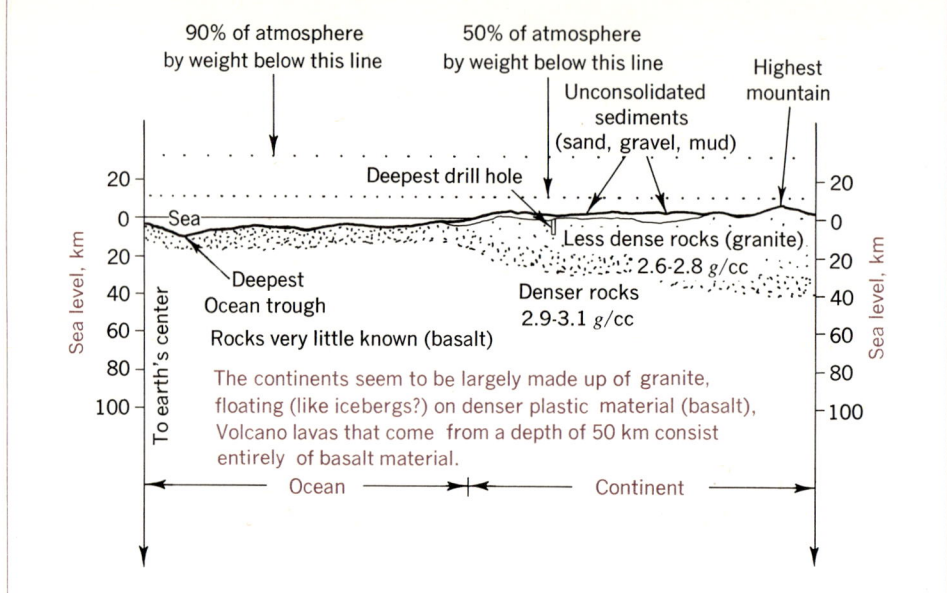

Fig. 14.2. The earth's outer layers. The diagram shows typical dimensions for features of the oceans and continents. It appears that the parts of the earth that we know about, without guessing, form a very thin surface layer. (This segment is shown at Z in Fig. 14.1.)

We know the mass of the earth (recall Cavendish's experiment).

We know also how long it takes for a compression wave that is generated by an earthquake or a nuclear bomb at point P_1 to travel from P_1 to P_2 and P_3 (measured for a very large number of points all over the earth). From these data, using the knowledge of the elastic properties of the various materials that are thought to make up the earth, we can determine the distribution of mass (density) within the earth to a reasonable approximation. We find that:

(1) Probably for most of the earth for most of the time, the density increases continuously toward the center; at least, if there are regions where lower-density rocks lie beneath higher-density rocks, such density "inversions" are probably not marked.

(2) From the travel time for earthquake waves, it is possible to calculate how fast the waves travel at different distances below the surface or, in other words, to construct a "velocity profile" (Fig. 14.3). Although such a profile does not tell us what the density *is* at any depth, it tells us where the density probably changes and where it does not change much with depth, since the density of the earth's material and the seismic wave velocity are rather closely connected.

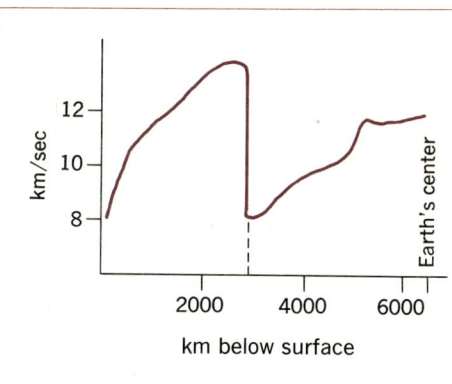

Fig. 14.3. *Change of velocity with depth in earth, for seismic compression waves. Below 2900 km, shear waves are not transmitted, and so the abrupt change at this level can be described as a change from solid to liquid.*

By using guides of this kind, various probable distributions of matter within the earth can be derived; these differ in detail, but not by great amounts, from the arrangement shown in Fig. 14.1.

Seismic observations give additional information about the interior of the earth as follows: In addition to compression waves, earthquakes also generate shear waves; that is to say, they generate waves in which the particles move to and fro across the direction of propagation (in compression waves, particles move parallel to the direction of propagation; see Fig. 14.4). A compression wave can travel through *any* material, however fluid (if air did not transmit compression waves, we would hear no sounds), but air does not transmit shear waves (see Fig. 14.5). If we were to move the transmitter D with our hand so that it made three or four vibrations a second, and if the distance CD were 2 in. and we filled the gap CD with successively stiffer materials, the receiver C would begin to pick up a perceptible signal only after we tried low-viscosity materials such as water and

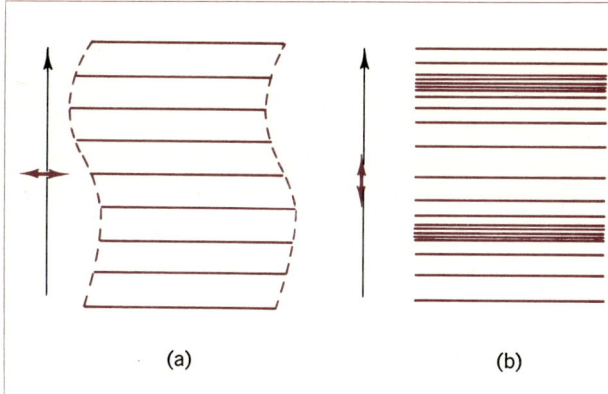

Fig. 14.4. *Shear waves (a) and compression waves (b). In each diagram, the direction in which the wave is advancing is shown by the light arrow, and the direction in which any particle moves to and fro is shown by the heavy arrows.*

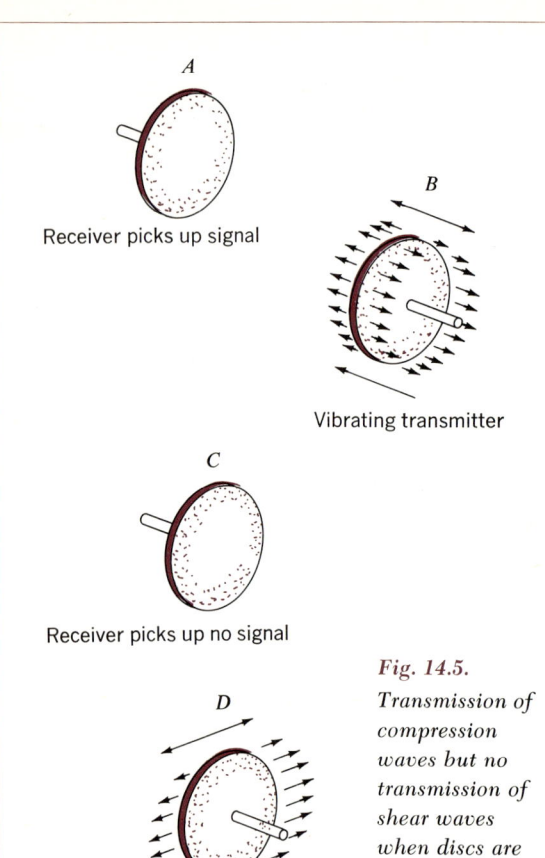

Fig. 14.5. Transmission of compression waves but no transmission of shear waves when discs are embedded in a medium of low viscosity.

corn syrup then changed to materials as stiff as tar or putty. Clearly, the attempt to transmit a shear wave would tell us something about the material between C and D even if this material were inaccessible to every other kind of observation. This is analogous to the situation relating to the interior of the earth: We find that if a shear wave starts along a path that penetrates to 2900 km below the surface (path P_1 to P_2 in Fig. 14.1), it can be received at the far end of its journey, but if it starts along a path that penetrates more deeply than this (path P_1 to P_3 in Fig. 14.1), no effect is felt at the far end of the path. The conclusion is that the core of the earth is too fluid to transmit an earthquake's shear waves, but that the surface regions of the earth have sufficient stiffness to transmit them. (This conclusion is sometimes stated in the form, "The earth is solid except near the center, where it has a liquid core," but the hypothetical experiment described with the vibrator D and receiver C reminds us that it is not easy to say exactly what we mean by the terms *liquid* and *solid*. In science, as in politics, the assigning of labels can be misleading; if our intention is not to mislead, then the more closely our report sticks to what is actually observed, the better.)

If Fig. 14.1 exhausts the physical information we have about the earth's interior (except for a certain amount of information about its temperature, which has not yet been discussed), it is natural to turn to the question, "What is the earth's composition?" Given only an approximate value for the density at a particular depth, we can think of endless materials and mixtures of materials that have densities in that specific range. In addition, it must be remembered that at the pressures that exist inside the earth, all kinds of materials may exist that we shall never become acquainted with on the earth's surface. Out of this wide range of possibilities, the probable composition of the earth (according to any proposed model) has to be judged principally on the following basis: *The material that constitutes the earth is probably similar to the material that constitutes meteorites.* On this basis we could say that the outer parts of the earth are composed of silicates† and sulfides,† and that toward the center the proportion of metals increases, particularly iron and nickel. But if we ask, "Why is it believed that the earth resembles meteorites?" the answer is that the evidences we gather from the chemical analysis of meteorites and from studies of the radiation emitted by the sun and the stars suggest that the composition of the various bodies of the universe seems to be fairly uniform.

14.3 The earth's crust and atmosphere

It might be thought that knowledge of the earth's outer layers would be more extensive and detailed than knowledge of the deep interior, and of course this is so. However, our *direct knowledge* of the material of the earth extends no further than the deepest drill hole, which is about 6 miles deep at present. When we remember that the earth's radius is nearly one-sixth of its circumference (a line the length of the earth's radius would stretch from New York across Alaska into northern Russia), 6 miles is a

---

† Silicates: A silicate is a compound of a metal with silicon and oxygen in the same way that washing soda (sodium carbonate) is a compound of a metal with carbon and oxygen. Compare Na_2SiO_3 (sodium silicate) with Na_2CO_3 (sodium carbonate). Sulfides: A sulfide is a compound of a metal with sulfur but no oxygen; thus, Na_2S is a sodium sulfide.

---

very small fraction of the total. Even in Fig. 14.2, the portion of the earth about which we have direct knowledge is seen to be rather small. However, rivers cut valleys into the earth's crust and it can be argued that some of the rocks now visible at the surface (for example, on the Hudson River around West Point) have been stripped of a cover 10 miles or more thick. This means that we do not rely only on drill holes for our information, and that we can say with some confidence that the outer 20 miles of the earth are composed mostly of silicates, in the tough, compact form seen at the surface. On coasts, in mountain regions, and in some deserts, the rock is separated from the air by loose material (sand, mud, and so on), but the thickness of material loose enough to crumble in the hand is almost everywhere less than 2 miles. The material exposed in highway excavations and such places represents only the outer skin of the earth, which is related to the total bulk approximately as the wallpaper is to the volume of the room it decorates.

Another thing that intervenes between the rock below and the air above is *water*, and in particular sea water, which covers about three-quarters of the earth to a depth of 2 or 3 miles (nearly 7 miles at the deepest point known). The sea is far easier to sample than an equal thickness of rock, so that we know far more about what is 2 miles below the surface of the Pacific than about what is 2 miles below the surface in Kentucky. On the other hand, the sea is an effective screen against most kinds of observation, so that we know much more about what is 7 miles down in Kentucky than we do in the Pacific; the nature of the rocks 1 mile below the ocean floor is almost entirely a matter of conjecture.

The outermost material associated with

the earth, the *atmosphere,* has no precise limit; its density simply diminishes with distance from the earth, as follows:

(1) About 4 miles above sea level, density $= \frac{1}{2} \times$ density at surface (high peaks in the Andes).

(2) About 7 miles above sea level, density $= \frac{1}{3}$ density at surface.

(3) About 20 miles above sea level, density $= \frac{1}{80}$ density at surface.

(4) About 40 miles above sea level, density $= \frac{1}{5000}$ density at surface.

An astronaut orbiting at 200 miles high is outside most of the earth's atmosphere, but the density of particles even 1000 miles from the earth is still well above the interplanetary average. Coming earthward again, we note that 99 percent of the mass of the atmosphere is concentrated within 20 miles of the earth and half of this 99 percent is within 4 miles; if we halve the atmosphere by weight, the denser half forms a sea that is less deep than the Pacific and not much thicker than the skin of sedimentary rocks, which, as noted, is proportionally very thin indeed. In terms of weight, the atmosphere is even less significant. It weighs no more than would a layer of water 34 ft deep, covering the whole earth, almost too little to be shown in Fig. 14.2. (We normally speak of 1 atmosphere (atm) *pressure* being equal to that of 34 ft of water or 760 mm of mercury, but because the reason for their exerting a pressure is their weight, equal pressures imply equal weights. Taking note of the fact that the atmosphere as a whole is a little farther from the earth's center than a 34-ft water layer would be, which would have the greater mass?)

Changes that occur in the atmosphere affect us more *insistently* than changes in any of the other "spheres" (unless we happen to be sailors). It will in fact turn out that in terms of energy, to which we now turn, the atmosphere represents a whirl of activity that has considerable importance, whereas energy conversions within the earth's massive solid bulk are relatively trivial in their effect on man.

14.4 Energy conversions in the earth and atmosphere

The foregoing paragraphs have presented a picture of the component parts of the earth without saying much about what each part is made of. It might be thought logical to describe next the *materials* of the earth and finally the *processes* in which the materials participate. However, there are two objections to this procedure:

(1) It is difficult to describe the earth's constituent materials without giving the impression that they are *inert.* How this impression might mislead can be shown by considering a description of a man as no more than a man consists of about 130 lb water, 8 lb calcium phosphate, and 30 lb of assorted organic molecules. . . . Such a description, while not wrong, is misleading because of its emphasis on the mere *presence* of material, its lack of attention to the incessant motion of the material, and its structure and function. The earth can be misrepresented in the same way: Although, on a human time scale, changes in the earth may be insignificant ("ol' man river just keeps rollin' along"), when viewed on a geological time scale, the earth is seen to be an intensely restless entity. On its surface, mountains rise and fall, ice caps and seas spread and dwindle, and deserts turn into swamps. Although, as noted earlier, these changes are merely skin-deep, the interior presumably undergoes changes of compar-

14.4 Energy Conversions in the Earth and Atmosphere

able frequency. Thus a picture of the earth that is geologically realistic is a picture of interacting processes, just as is a realistic description of a man.

(2) Even if we refrain from taking a geological view and confine our attention to human affairs, it can be argued that what the earth *does* is more important than what it *is*. If a river floods your farm, if an earthquake bursts your gas pipe, if the harbor your ships use cannot be kept clear of silt, the materials involved are of little consequence in comparison with their rearrangement in space.

For both reasons it appears that the earth is best viewed, for our purposes, as a system in action, that is, as a system in which energy conversions are taking place.

If we look around us, what energy conversions do we see? Let us first record what we see in a nonreflective way (like a camera), and secondly, like a thinking man, attempt to distinguish the important from the unimportant in what we have seen. (The advantage of this procedure over that of approaching one's field of inquiry with preconceptions as to what kinds of influence will be dominant is recognized by many people.) Some of the conversions we are likely to notice are:

- Sunlight giving warmth (conversion of radiation energy to heat energy)
- The wind raising water waves (conversion of kinetic energy to potential energy)
- Internal combustion (conversion of chemical energy to kinetic energy)
- Men digging drains (conversion of chemical energy to potential energy)
- The use of an electric shaver (conversion of electric energy to kinetic energy). Into what form is the electric energy ultimately converted?
- A boy pressing snow into a snowball
- A boy blowing up a balloon
- Rain falling
- Fires burning
- Tides rising . . .

(The reader should identify the energy forms involved in the last five items, and think of three more conversions, either by mental imagery or by observation.)

The items in the list can be categorized as important or trivial, depending on one's point of view; but they can also be classed as basic or derivative, which is a less subjective procedure. For example, the conversion of sunlight to chemical energy (photosynthesis), which occurs when plants synthesize atmospheric carbon dioxide and water into more complicated molecules, is a basic process; the subsequent burning of the plant to heat one's home is a derivative process dependent on the photosynthesis just mentioned. A moment's thought reveals that the main basic process is the conversion of sunlight by various mechanisms, and that most of the processes we see (all but two (?) of the examples given above) are derivative in that energy that came ultimately from the sun is converted from one form to another.

The sun's energy is trapped when it causes sea water to evaporate, polar ice to melt, plants to grow, or rocks to get hot; anything that stops *light* can in fact gain energy from the sun (if we distinguish radiant *heat* from radiant *light* on the basis of wavelength, the amount of energy received from the sun as heat is negligible; the sun warms the air principally because the earth's surface absorbs light, gets warmer, and transfers heat back up again to the air). The sun's energy escapes:

(1) When the earth reradiates energy out into space. Part of this reradiated solar energy is scattered sunlight, part is infrared radiation from the earth's surface.

(2) When particles from the upper atmosphere escape completely from the earth's gravity field (the kinetic energy they take with them is derived mainly from the sun by absorption at the earth's surface and transfer back upward through the atmosphere).

There is little reason to suppose that what the earth receives from the sun is *exactly* balanced by what it loses to outer space. Thus there is the possibility that the earth-and-atmosphere system is gradually warming up or cooling down. Before trying to decide whether warming up or cooling down is more probable, we shall remark upon three minor (but not negligible) energy sources. These are

(a) Spontaneous radioactive decay
(b) Kinetic energy of the earth-moon system
(c) Rearrangements of matter within the earth's gravitational field.

Energy from source (a) is continuously generated by the radioactive materials dispersed in small quantities throughout the earth. Energy from source (b) is converted to heat whenever tidal currents "rub" against the ocean floor or against standing water; because of the moon's gravitational attraction, the water of the oceans does not spin around quite so rapidly as the solid earth does, and the drag it consequently exerts on the spinning earth converts some of the earth's kinetic energy into heat. Energy from source (c) appears as heat when the heavier elements such as iron sink into the earth's core and displace lighter elements such as aluminum outward. No one knows at what rate the last process is going on at present, but it appears that if the earth ever consisted of its present materials uniformly dispersed through approximately its present bulk, the mere rearrangement to the form in Fig. 14.1 would heat the earth about 3000°C. That is, the present heat of the earth's interior may have derived largely from this rearrangement at some time in the past and is now slowly seeping away.

The last remark brings us back to the question of whether the earth-and-atmosphere system is warming up or cooling down. It is known that the earth is hotter inside than outside. A consequence is that heat now reaches the surface from the interior in quantity sufficient to melt about $\frac{1}{2}$ in. of ice per year, averaged over the whole earth (in fact, the escape is more rapid at some points than at others, by a factor of 10 or more). It does not appear likely to specialists in this subject that heat is now being liberated inside the earth at a rate equal to the losses at the surface. The favored hypothesis is that heat was generated rapidly at an early stage and is now escaping slowly; that is to say, the earth's interior seems to be cooling. However, it must be remembered that a small fluctuation in the loss or receipt of heat from the sun can cause respectively the growth or melting of a half-inch of ice overnight in some parts of the world. This indicates that the flood of heat coming to the earth's surface and leaving again is great in comparison with the small amount seeping out from within. (See Fig. 14.6.) Could a small percentage of imbalance in the surface flow be as important as the flow from the interior? The earth's average surface temperature has varied, but we ask has any change been due mostly to the cooling of the interior caused by

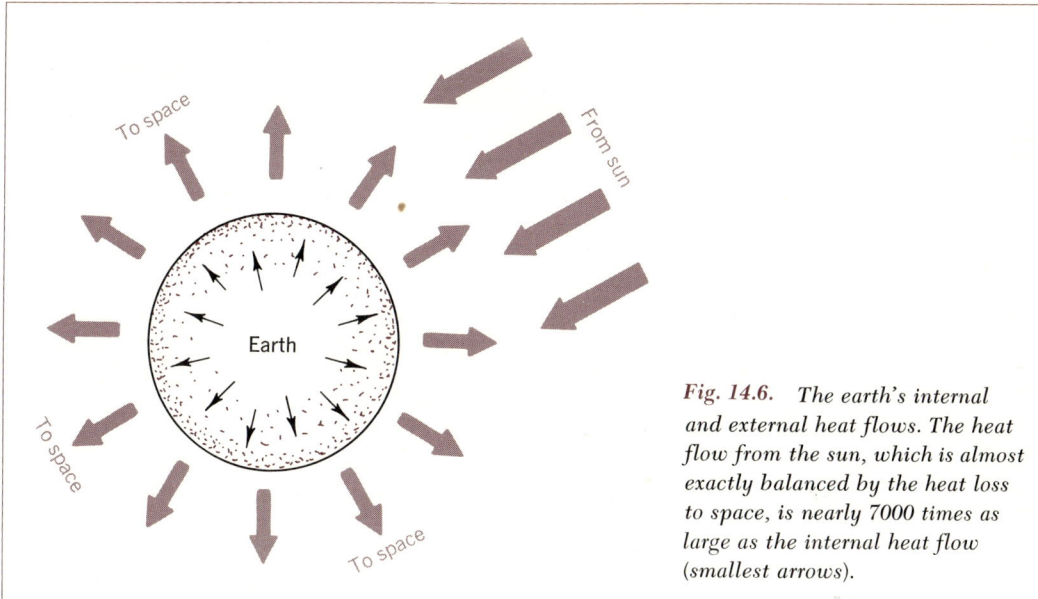

Fig. 14.6. *The earth's internal and external heat flows. The heat flow from the sun, which is almost exactly balanced by the heat loss to space, is nearly 7000 times as large as the internal heat flow (smallest arrows).*

changes in the energy reradiated at the surface or in internal radioactivity? (The tidal effect can be shown to be small in comparison with these.) We cannot, unfortunately, give a definite answer to this important question at this time.

14.5 Self-adjusting systems in the earth and its atmosphere

What kind of system is the sea? Does it have negative feedback, or proportional control? Of course the answer depends on what property of the sea we have in mind: The salinity of the sea is subject to one set of controls, and its level (with respect to some reference point on land) is subject to quite a different set. Let us study the planarity of the sea's surface and let us for convenience think of a small sea so that the earth's curvature can be neglected and the natural condition of the surface will be a plane in the geometrical sense. Let us also study only a vertical cross section of the surface; then an artificial disturbance in the surface can be represented by a kink in a straight line, as in Fig. 14.7.

Suppose such a disturbance of the surface could be generated instantaneously. The effect would be a pressure difference between all points within the dotted lines and all points outside them; in particular, a small pocket of water Q would experience a higher pressure on its left side than on its right; such a pressure difference would cause the pocket of water to move to the right, while a similar pocket P would move to the left. Such movements would permit the original disturbance to subside. The

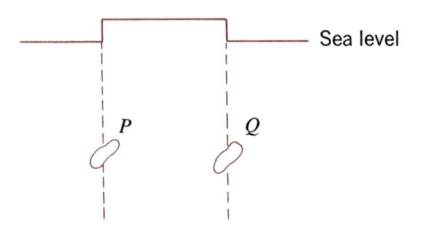

Fig. 14.7. Behavior of a disturbance on the surface of the sea. It has large and very serious aspects when it occurs at the epicenter of a tsunami. A tsunami (from the Japanese for storm wave) is a great sea wave produced by a submarine earth movement or volcanic eruption in the ocean.

water's momentum would cause the system to overshoot, and the result would be a damped oscillation, hunting for the equilibrium condition of planarity.† The system does in fact have negative feedback and, at least in the initial condition, it has proportional control because the restoring force is proportional to the departure from the reference condition; however, the control influences the system's *acceleration* rather than its velocity or displacement, so that the stability often associated with proportional control is lacking.

Does the surface of the land resemble the surface of the sea? In almost every respect, they are markedly different; however, one of the purposes of a study of systems in the abstract is to permit the reader to recognize common features in the behavior of dissimilar arrangements (such as in traffic flow, plumbing, and political institutions). The reader with this facility will notice that the land surface does resemble the sea in at least one respect: To some extent, the higher a hill rises above the general level, the more active are the forces (wind, rain, frost) that tend to reduce it again; the Rocky Mountains are at present being reduced more vigorously than the Great Plains.

If we look, however, at erosion processes on a smaller scale, a different kind of behavior is encountered. The Dutch boy who pushed his finger in the dike realized that a small deviation in waterproofness is not self-correcting but self-magnifying: The effect of a little water penetrating the dike is to remove a few particles and permit more water to pene-

† Question: What is it that damps the oscillation? What happens to the kinetic energy of the water when it passes for the first time through the equilibrium position it finally achieves?

trate. A similar effect, of great economic consequence, is "gully erosion" of pasture. There are many sloping pastures down whose surface rainwater escapes slowly because of the hindrance afforded by the grass stems. As long as the grass cover is intact, it protects the soil from erosion, but many slopes are in such a (marginal, delicate) condition that if the grass cover is accidentally destroyed, even over a small area, the anomaly is self-magnifying. Over the bare earth, rainwater runs faster; some soil is removed, leaving a depression; the depression concentrates the formerly uniform flow and so enhances its erosive power that a channel is formed, whose every increase in size leads to an increase in its destructive ability. This, in fact, is a system with positive feedback, and is dangerous in the way that most such systems are. ("Snowballing" is a common term for the operation of systems with positive feedback, and leads to harm even where moderate operation of the system is beneficial, as was discovered by the Sorcerer's Apprentice. In the buildup of armaments, feedback is positive; the question, "Would a small decrease in armaments have positive feedback?" is probably one of the most important questions now facing the United States.)

The crumbling of a cliff into boulders is another natural example of positive feedback. (See Fig. 14.8.) The overrunning of a lawn by dandelions is another. The more weeds there are, the more rapidly they proliferate (in fact, the exponential increase of populations is as serious a problem as the exponential increase of bombs, and will be discussed further in the next chapter). Combustion of paper, hay, and dry wood has positive feedback; so does the decay of garden rubbish into compost. In fact, while positive feedback is far less common than

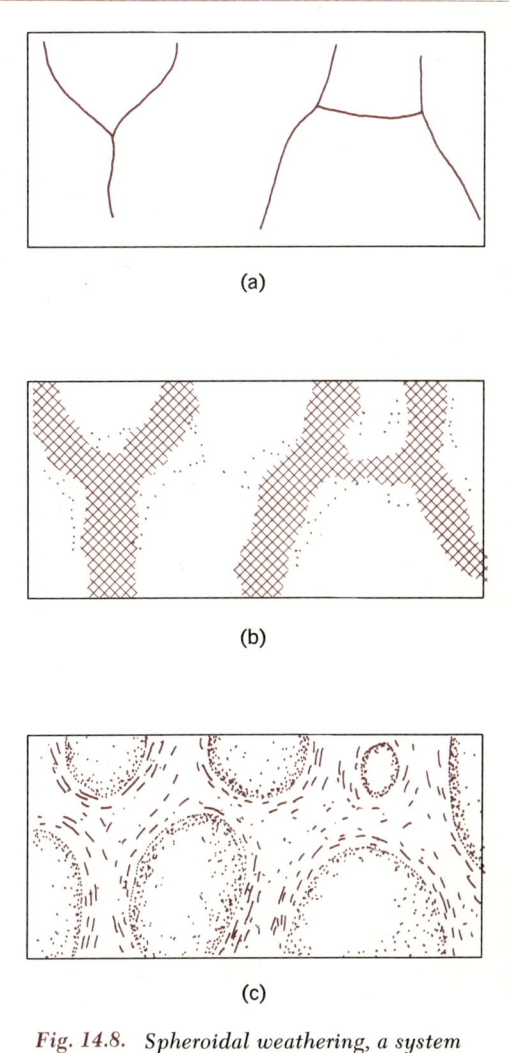

Fig. 14.8. *Spheroidal weathering, a system with positive feedback. (a) Pattern of joints in sound rock; (b) development of rotten rock in neighborhood of joints; (c) reduction of rock to boulders separated by soil.*

negative in man-made systems, in natural systems we find positive feedback processes as common as negative feedback processes.

14.6 Interactions between the inanimate world and human affairs

Exposure to the phenomena of nature evokes many thoughts that intermingle awe and inspiration. For the moment, however, we shall discuss the material aspects having to do with resources, communications, and climate, which are the aspects of human life most directly related to the inanimate world. The importance of the three aspects has changed with time. Primitive man seems to have evolved between latitudes 50° N and 50° S, not because resources were necessarily more abundant but because a temperate climate put less strain on his abilities as hunter, clothier, and homemaker.

We might speculate on some other features of society's interaction with the inanimate world. If climate determined primitive man's *needs* with respect to food, fuel, and shelter, it was the land fertility that determined how easily the needs could be met. Presumably this might also have fixed the fraction of the population that could be spared for non-subsistence activities such as permanent armies, ministering to the religious needs of the community, or participation in creative arts (subsistence-fighting, as practiced by robber bands, is excluded from this discussion). One might expect that in a severe climate, the whole population would concentrate its efforts toward survival, but even then only the more efficient members would survive. It is not surprising that the first great armies, religions, and cultures appeared in fertile valleys of temperate and semitorrid lands.

A modern example of the influence of earth structure on climate, and therefore on demography, is provided by a comparison of the United States and Australia. The two countries are almost exactly equal in area, and similar in dimensions. Australia is somewhat closer to the equator (latitude 15° to 35° compared with 30° to 50° for the United States), but the main difference is that the United States has a large body of water (the Gulf of Mexico) to its south and a great mountain range on the windward side, whereas Australia does not have a comparable combination of mountain and water masses. The consequences of these differences appear to be as follows:

(1) Australia has no inland drainage system (comparable with the Mississippi-Missouri system).

(2) Central Australia is a desert rather than a storehouse of fertility and food.

(3) Communication across Australia is tenuous and expensive. There are no communities to be served en route to share in the costs of installation and maintenance. Fortunately, radio communication offers some relief from this handicap, since intermediate stations are not required.

(4) Although the mineral wealth of the Australian northwest is considerable, its influence on the Australian economy has been negligible until the present decade (cumulative effect on rate of development of wealth).

Until the Industrial Revolution began, the wealth of nations was usually measured in terms of agriculture and in terms of commerce in the products of other nations, the latter including foodstuff, ore, and products of handicraft. During and after the Industrial Revolution, wealth in coal and iron became significant. The size of the effect produced was in part determined by the ease of transportation of ores and fuels. Localities flourished where ores and fuels were juxtaposed, whereas localities where only one or the other was present did not flourish so readily. Countries that achieved dominance for reasons related to their mineral wealth were Great Britain, Germany, and the United States.

The most effective integration of mineral resources into a nation's economy is a continuing challenge. The time when prosperity depended on juxtaposition of ore and fuel has largely passed. What determines a country's prosperity today is the fraction of its manpower it can spare for increasing and utilizing its available resources. Wealth accrues not to the country that *possesses* untapped resources, but to the country that knows how to *exploit* them, sometimes by joint effort such as when French coal is used in Germany, Australian ore enriches Japan, and Venezuelan oil is used in the United States. This suggests that perhaps the poverty of the "underdeveloped nations" is partly a poverty of technologists, and even more a lack of national planning of a kind that promotes cooperative effort among nations and motivation of the people to produce both technologists and products. The point is that the discovery of mineral wealth does not automatically make the possessor-nation rich (how well does this generalization stand up to the example of Kuwait?). Only some special aspects of a nation's life rather than its overall prosperity are now dominated by its mineral resources.

The foregoing summarizes the history of the civilized world roughly into five millenia of *wealth by fertility*, one century of *wealth by ores and fuels*, and the present period of *wealth by know-how*. In a state of wealth by know-how, what factors from the inanimate world remain of importance? It is to this question that we now turn.

14.7 The inanimate world in modern technology

When one asks, "At what point is the technology of a nation limited by the resources of the inanimate world?" one is tempted to answer in terms of metals. It is true that the "hardware" that distinguishes a rich nation from a poor nation is largely of metallic construction, but the availability of metal ores is hardly a limiting factor. If a metal cannot be obtained locally, it can be made available by some more troublesome and expensive method. The time is past when metal that could not be gotten with a pick and shovel could not be gotten at all. It can in fact be said that we are approaching a paradox wherein the more severe limits on technology are imposed by the most abundant materials; for example, we have all the samarium† we need, but are often short of water and stone. The reason is that we have become accustomed to using these materials in great bulk, and bulky materials tend to have high transport costs compared with production costs. The problem is aggravated by the "flight from the land" (that is, increase in the *urban* population) and also by the flight to the suburbs, which increases the area occupied per person in regions of high-density population.

For example, the information given in Table 14.1 may surprise many people of the State of New York. The table lists the production of minerals in the state during 1963 in terms of dollar value at the production site. The largest dollar activity involved the production of gravel and stone, accounting for more than $81 million of the $260 million total. The transportation costs to deliver the ma-

† Samarium, element number 62, is in the "rare earth" category on atomic charts.

TABLE 14.1

MINERAL PRODUCTION IN
NEW YORK STATE—1963

Mineral	Value
Clays	$ 2,186,000
Emery	119,000
Gem stones	10,000
Gypsum	3,339,000
Lead (recoverable content of ores, etc.)	218,000
Natural gas	1,169,000
Peat	178,000
Petroleum	8,854,000
Salt	34,228,000
Sand and gravel	37,274,000
Silver (recoverable content of ores, etc.)	25,000
Stone	44,549,000
Zinc (recoverable content of ores, etc.)	12,304,000
Value of items that cannot be disclosed: Abrasive garnet, cement, iron, lime, talc, titanium concentrate, and wollastonite	115,768,000
Total	$260,221,000

SOURCE: From James R. Dunn and John G. Broughton, *State Government*. Published by The Council of State Government (Summer 1965).

terial to the points of use probably doubled the total cost.

The quantity of stone for the building of cities may be surprising. For example, the Manhattan skyline clearly demonstrates how large quantities of material can be taken out of its natural habitat and projected far up into the air. But where is the "ground" on which the Manhattan skyline stands? It is not at all uncommon to stand in a city street and to be quite unable to determine where the natural ground level was. Underpasses, overpasses, basement parking lots, and elevated shopping plazas remind us that a city dweller occupies not so many *square* feet as he does *cubic* feet of space, and that x cubic feet of space involves y cubic feet of stone, either removed or, more commonly, brought in. It is only rarely that a city can build at one point with the stone it excavates at another point. It is not an uncommon practice to truck out unsuitable rock from an excavation, and truck in more expensive, more suitable rock for building foundations.† Let us examine, as taxpayers, the expenses of building a city, with particular reference to the cost of *stone*.

To build a city at low initial cost, one might choose level land. Rome was not built in a day and one of the obstacles to quick progress was the hummocky terrain; there was plenty of natural rock close to the surface, making each building's foundations a matter of hand sculpturing. This was somewhat expensive for the Romans and would be exceedingly expensive today, when the cost of preparing an individual building's founda-

† Compare the construction of a city block in New York City with the building of the pyramids. What lessons can be learned in terms of use of stone, labor costs, utility with respect of population needs, economy of this nation, productivity, aesthetic values?

tion on rock is high in comparison with the cost of preparing a group of building foundations by routine operations with an excavator. Given flattish surface, what kind of rock would we prefer beneath it?

In the overall view of earth structure in Sec. 14.2 and Fig. 14.2, the bulk of the earth was referred to as *rock*, largely composed of *silicates*, and distinguished from a thin layer of unconsolidated sediments above and a quasiliquid core below; but the distinction between rock and sediments is not so simple as it was made to seem. Sedimentary layers (salt, lime, clay, sand, remains of organisms) accumulate in lowlands and in the sea. To summarize† the processes of rock formation in a sentence, sediments become hard when they are buried in the earth, while hard rocks become soft when exposed to air, rain, and varying temperature. The softening process is familiar to all (though fully understood by none). Upstanding rock formation is cracked by sun, rain, and the activities or organisms, or movement of the earth's crust, and the particles crumble and tumble into the valleys below (positions of lower potential energy). The hardening processes are more varied: When buried, a sediment is both squeezed and baked, but the effects depend on the rates of increase of temperature and pressure and on the initial composition of the sediment. Question: Why does a sediment get warmer when buried? First there is the heat flow from the earth's interior; if heat is flowing into a sedimentary layer from below and the layer is "blanketed" by additional sediments, the heat-flow balance (loss from top equals input from below) will be upset, and the layer temperature will rise until it is so hot that it loses heat through its "blanket" as rapidly as it had done previously when it was exposed to air. Secondly, we know from watching volcanoes erupt that molten rock can move about inside the earth. Sometimes it erupts as lava, and sometimes it solidifies without reaching the surface; either way, as it moves from place to place (in response to the slow drifting and bending and rising and sinking of the "solid" earth's crust) the molten rock heats its new environment, which may include the sediment whose burial we are discussing. The effect is to turn mud and clay into *shale*, sand into *sandstone* and accumulations of limy sediments into *limestone*† or similar sedimentary rock.

In the conversions just mentioned, a conspicuous effect is *compaction*, with expulsion of water or air from the spaces between the grains. A less conspicuous effect is change of grain shape—the grains squeeze together and begin to interlock instead of merely touching at their corners—and there is also change in grain *size*. When deeply buried, increase in grain size is very important: Limestone may turn into *marble* and many other kinds of sediment turn to *gneiss*.‡

† We shall return to this subject in Section 14.11.

† Terminology: If a rock is mainly composed of calcium carbonate, with or without magnesium carbonate, it may be called *limestone*. If a rock is mainly composed of silicate grains more than $\frac{1}{16}$ mm across, it may be called *sandstone*; most sandstones are mainly quartz, but it is possible to have a limestone-sandstone. If a rock is mainly composed of grains less than $\frac{1}{16}$ mm across *and splits easily*, it may be called *shale*. It would be possible to have a limestone-shale, but such rocks are not common, and most shales are composed of silt or clay-sized particles with indeterminate mineral composition.

‡ Terminology: If a rock is highly compacted and *streaky*, it may be called *gneiss*. It is frequently used in monuments, in cemeteries,

There are other kinds of rock upon which cities may be built. Coarse crystalline rocks that are as compact as gneiss, but uniform in appearance rather than streaky, are often called *granite* (although what rocks *should* be called granite is a matter that geologists cannot agree upon). But enough different rocks have been mentioned to permit us to proceed with the selection of city foundations. The requirements are somewhat conflicting: We require foundations that will support buildings at least 20 stories high, which require something substantially firmer than clay and unconsolidated sand within a hundred feet or so of the surface. But we must consider also a city's subterranean requirements: drainage, subways, electric power, gas, and water. These are most conveniently arranged and maintained when the subsurface is *not* too hard—not so hard as gneiss or granite. *Gravel* is a convenient compromise. *Gravel* is an unconsolidated sediment too coarse to be called sand (that is, with most particles over 2 mm in diameter; gravel particles are usually well above this limit). This material can be easily moved without drilling or blasting; yet it can support a building better than finer sediments such as mud. There is a disadvantage to concentrating buildings over gravel beds, however, which is another geological matter that a city planner needs to keep in mind: Once a deposit is built over, it is *lost* to future generations. Suppose a city's first million inhabitants build on all the gravel deposits: When it comes to expanding the city to accommodate the second million (an undertaking that requires more than a doubling of expressways, airports, and downtown services), the very materials most needed for construction are found to be inaccessible.

14.8 Water, earth, and man

Man's dependence on water is far more basic than his dependence on stone. In fact, everything that we call a living thing is water-dependent—man needs oxygen; plants need carbon dioxide; some organisms need nitrogen for survival; but none of these forms could live in the absence of water. Conceivably, some form of "life" could develop in another medium, but whatever is alive on earth is watery. Densities of familiar materials range from less than $0.001 g/cc$ (air) to more than $21 g/cc$ (platinum), but nothing alive has the density far from $1 g/cc$ which is the density of water.

Not only does man as an organism consist largely of water, but he also uses large quantities of it to meet his daily needs, and this may amount to about 100 gallons per day in various societies. From farm through factory to foodstore, water is consumed in very large quantities. Water is not only an industrial convenience for transport, but also a principal requirement for industrial development. (Tons of water are required for the processing of a ton of steel.) Sources of fresh water such as the Great Lakes are national assets, and the prosperity of states like Arizona is increasingly tied to their skill in water-husbandry.

A very large portion of the earth's life forms is found in the waters of the seas and oceans, which also contain considerable amounts of salt and minerals.

---

and on the facings of buildings such as banks; if the ornamental stone is uniform in texture, it is not gneiss, but if you find a streaky ornamental stone, you are probably looking at a sediment that has been intensely squeezed and heated. Many gneisses have been kept red-hot far below the earth's surface for a million years or more. The reader can calculate the pressure they have suffered under, say, 10 miles of overburden, without difficulty.

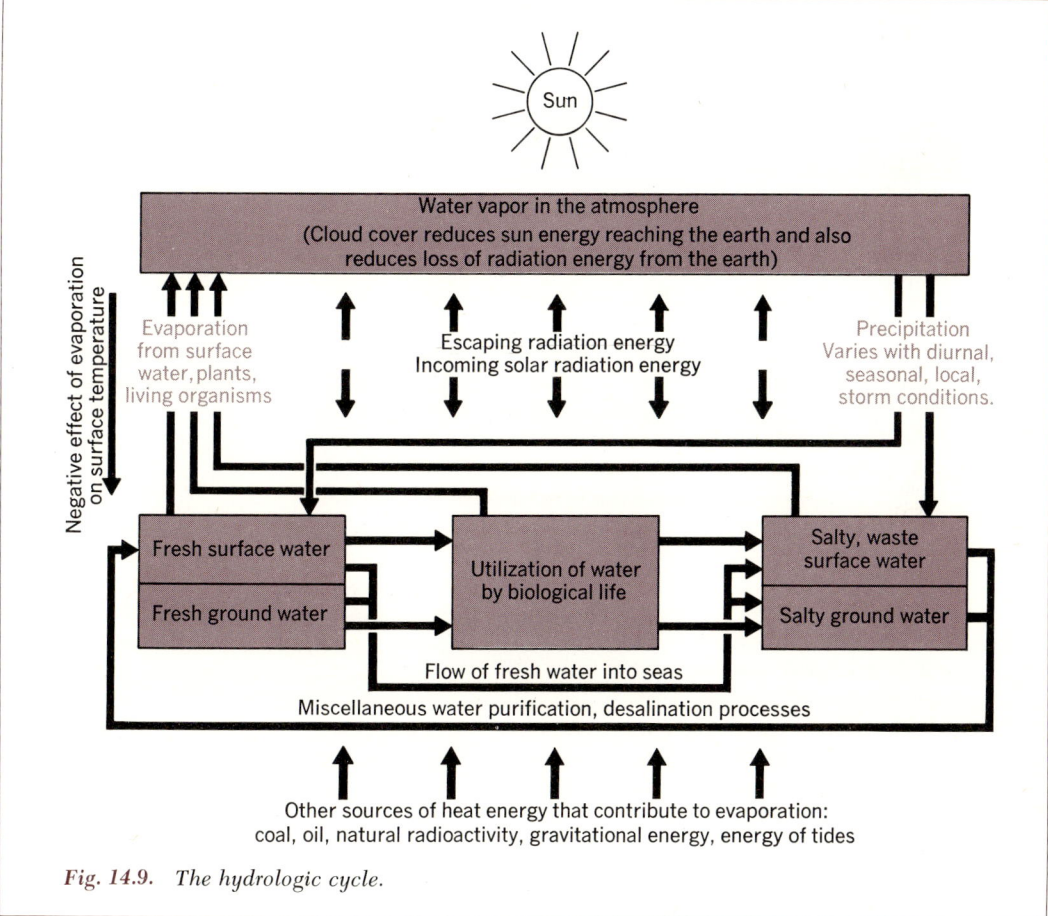

Fig. 14.9. The hydrologic cycle.

But land animals and plants require fresh water or water largely free of salts. The salty oceans nevertheless do provide the land areas with vast amounts of fresh water through the process of evaporation-condensation. The water vapor that escapes from the briny sea is sufficiently free of salt for the purposes of man and plants. In the chapter on heat and thermodynamics we found that considerable heat energy is required to evaporate water. Of course this heat is given up again by the vapor when it condenses. Because the evaporation takes place on the earth's surface and condensation takes place in the atmosphere above the surface, the process causes a vast transfer of energy from the earth's surface to the clouds of water vapor and to the atmosphere.

Let us look into some of the details of the process, keeping in mind the system's interrelationship aspects and especially the feedback elements that come into play in either a positive or negative sense.

The process is illustrated in Fig. 14.9. The bulk of the energy comes from the radiation that reaches us from the sun. The radiation from the sun heats the surface of the earth, increases its temperature, and thereby also the rate with which the surface waters evaporate into the atmosphere. The loss of this heat of vaporization in turn *cools* the surface and thus holds down the *rate* of evaporation from the earth. The vapors condense at higher altitudes to form microscopic droplets that accumulate and form clouds in the sky. The clouds form a partial barrier to reduce the amount of heat that reaches the earth's surface. The presence of the clouds therefore tends to cool the earth still more and the cloud itself heats up. (This last result reduces the likelihood of precipitation until the thickness of cloud builds up to larger proportions.) But we must note also that the same clouds reduce the rate with which radiation escapes from the earth, especially during the night hours. The overall effect of the evaporation process therefore is to reduce the fluctuation of the earth's temperature as well as to provide fresh water.

As illustrated in Fig. 14.9, the largest source of fresh water is due to evaporation from the salty ocean water. But there is large loss of fresh water to sea or to waste as a result of river flow, ground seepage, and careless use by society.

We see that there are many "negative and positive feedback" effects that, depending on the timing or phase relations and magnitudes, tend to suppress evaporation or rainfall, or to bring about severe storms. Eventually the cloud formations experience sufficient lowering of temperature to encourage fusion of the microscopic droplets to larger droplets of water,† which fall to the earth as rain. The cycle involves considerable time, often hours or days. The changes of temperature on the earth and in the atmosphere create large differences of temperature (temperature gradients) between adjacent areas. The warming process expands the gas volume and causes its outward motion, whereas cooling and condensation reduce the atmospheric pressure. This expansion and contraction of the gas mass causes movements of the air and of the cloud formations, which may extend over large distances of terrain from a region of plains and valleys to mountain areas. The effect is to transform the evaporation-condensation cycle, with its moderating "negative feedback" influences, to a cycle that is quite variable and without any control. Sometimes the proper precipitation steps do not follow; sometimes they follow with violence in the form of storms of various types.

When the rains do fall, what happens to the water? Most of it returns to the salty oceans from which it came. Much of it falls on the soil of the earth, which absorbs it in part, and the remainder either accumulates in lakes or runs off as streams and rivers. Continual seepage through the pores and crevices of the earth provides pools of underground water. During this time the evaporation from the earth, and that which is taken up through the roots of plants, will dispose of a good portion of the water. Eventually

---

† The formation of droplets requires the presence of condensation nuclei such as dust particles. In a completely dust-free atmosphere, no drops will form. Fortunately, many natural processes give rise to atmospheric particles that serve as condensation nuclei. Indeed, with the increase of atmospheric pollution by man and his technology, there is a danger of having too many particles in the atmosphere.

some of it returns to the oceans, either through the rivers or drainage from cities, or escapes by underground seepage into open areas such as bogs and swamps.

Water is, of course, "consumed" in a special sense. Although in Arizona even dirty water would be welcome, in many places it is *clean* water that is in short supply. Water is not so much consumed as fouled; even on the banks of the Hudson River, clean water is in demand. The most universal way to clean water is to distill it (that is, to evaporate it and permit it to recondense), so that water-husbandry involves two aspects: minimizing pollution, and low-cost distillation. The former is a social problem, to be discussed later in the course. But cheap distillation is in a way a function of the inanimate world discussed here; in fact it is its main function, since more of the sun's energy is absorbed in evaporating water than in other ways. The amount of energy tied up in the evaporation-precipitation cycle dwarfs all other energy conversions on earth.

It was remarked earlier that the heat escaping from inside the earth in a year is sufficient to melt only a half-inch layer of ice, if distributed uniformly over the earth, whereas heat from the sun *evaporates* and permits precipitation of about 40 in. of water (again averaged over the whole earth; the difference is far more than the factor of 80 in volume between the $\frac{1}{2}$ in. and 40 in.; as the reader will recall, the heat required to evaporate water is about seven times more than that needed to melt an equivalent amount of ice). The vastness of the evaporation cycle is apparent. The smallness of man's power resources can also be quickly shown: It would take about 20,000 power stations, each delivering 1,000,000 kw to melt the same half-inch of ice in a year; at present only a handful of nuclear power stations approach this power. To make the same point another way, if a desert were to be provided with the 40 in. of "rainfall" by nuclear power instead of by the sun, the world's largest power station could supply a plot of about 3 square miles. The idea of using nuclear power to make deserts blossom like roses in a direct way may be rather farfetched.

With these magnitudes of energy in mind, we might ask: "How practical is rainmaking? What about cloud seeding by a single aircraft?" Most people have heard of such attempts, and know their success to be only partial. It seems clear that while *imitating* the sun is beyond our powers, guiding its effect may not be. Operations on all scales, from one's watering can to the TVA or Aswan dam schemes, are in fact examples of man's local interference with the natural operation of the evaporation-precipitation cycle, and cloud seeding is only an attempt at interference by another means. Irrigation systems try to control water flow after the rain has fallen; cloud seeding attempts to direct rain to the site where it is needed.† An aircraft does not perform as a power resource or pumping station in this operation. Its feedback function in making rainfall is its ability to reach the elevation where the energy balance of the cloud formation is receptive to disturbance. Water that has already been evaporated by the sun's energy is simply helped in its condensation by scattering crystals from the airplane.

† The success of an aircraft in precipitating moisture from a supercooled cloud depends upon the energy balance between the form (cumulus) of the cloud and the size of the moisture droplets within it. Silver iodide or dry ice particles dropped into such clouds make the droplets coalesce and form a size (about $\frac{1}{125}$ in. or larger) large enough to fall as rain (or snow).

464 *The Earth and Its Atmosphere*

Another natural process that man might possibly learn to influence by expending less energy than the process itself involves is in retarding the progress of a hurricane or deflecting its path. The air has kinetic energy, thermal energy, and gravitational potential energy, and a hurricane is one of the possible results when by chance a great deal of this energy becomes concentrated in a small space. Man has no method for suddenly absorbing a great deal of energy from the atmosphere, but he does have a method (bombs) for suddenly discharging a great deal; it may therefore be possible to deflect a hurricane's course by adding energy at a point to one side of its line of advance. This type of manipulation is clearly different from triggering precipitation and requires an energy source that need not be equal to that of the hurricane, but at least should approach its force (possibly $\frac{1}{20}$ of the hurricane energy would be sufficient).† There are many arguments against such proposals. Even estimating the amount of energy contained in a natural hurricane involves many uncertainties. Note that we have returned to the topic of energy conversion discussed in Sec. 14.4; even the inanimate world is less a world of *things* than of *processes* that involve *energy conversions*. Study of energy-conversion systems helps us understand the natural world, and study of the natural world helps us understand systems.

14.9 Time and geological processes

The great contributions of geology to our understanding of the processes of nature have to do with the element of *time*, for

---

† Of course the use of currently available nuclear bombs for such a purpose could introduce a "cure" that is worse than the "disease," as we shall find in Chapter 21.

long periods of time are involved in those processes. As stated in Chapter 2, this contribution came toward the end of the eighteenth century from James Hutton, who studied the sequences that seemed to exist in sedimentary rocks. He came to the conclusion that for the interpretation of natural processes, there was need to utilize what one can observe.

". . . no powers [are] to be employed that are not natural to the globe, no action admitted of except those of which we know the principle, and no extraordinary events to be alleged in order to explain a common appearance . . . we are not to make nature act in violation to that order which we actually observe. . . .

"In whatever manner, therefore, we are to employ the great agents, fire and water, for producing those things which appear, it ought to be in such a way as is consistent with the propagation of plants and the life of animals upon the surface of the earth."

Before we can apply the physical principles that we have been studying (specifically, the laws of motion and thermodynamics) to the earth as a whole, it is necessary to spend a little time considering the evidence collected by the geologist over the past 200 years and how this evidence has forced radical changes in our understanding of processes and the time involved in their operation. The awareness of *time* relation in geologic processes grew out of the collection of stratigraphic data, the records of the sequences of the sedimentary rocks that were seen to have formed one on top of another, to have been tilted, folded, and faulted. Hutton insisted that any interpretation of the time and processes involved must be based upon actual observation, that evidence must precede explanations.

The rate of sedimentation has been measured many times and at many places in attempts to find out how long it takes to form a sedimentary rock such as shale or sandstone. Measurements off the coast of Normandy were reported by Buffon in 1807 as 5 inches per year for the rate of deposition. From the total thickness of the sedimentary record which he assumed, a thickness considerably below that we know of now, he estimated an age of 100 thousand years. His approach was valid, but his sample was not.

Later measurements at Memphis in Egypt showed that the Nile is depositing sediment at the rate of 1 ft every 400 to 500 years (Buffon's figures were 1 ft every $2\frac{2}{5}$ years), which, for the assumed total thickness of sedimentary rocks since the beginning of the fossiliferous record, gives an age of some 200 million years before abundant life developed on the earth. But the Nile deposits sediment at an unusually fast rate, so that this figure has always been accepted as the minimum. The search continued for a measure of the rate of sedimentation and a way to recognize annual layers that could be counted.

Such annual layers, called *varves*, were recognized late in the nineteenth century. The counting of varves was first applied to the dating of ice retreat, by de Geer in Sweden. Later they were also recognized in the Green River shales of Wyoming, Colorado, and Utah, rocks completely unrelated to and much older than the Pleistocene continental glaciation. W. H. Bradley showed that the 2000 ft of this formation took $6\frac{1}{2} \pm 1\frac{1}{2}$ million years to form, which is a rate of 1 ft every 3300 years. To date, this is the longest sequence of sedimentary rocks that has yielded directly the time required for deposition. This rate is lower than an average for all types of sedimentary rocks, so it sets a probable maximum age for the fossiliferous record.

Calculations based on what is assumed to be a more nearly average rate of deposition gave 500 to 600 million years of time for the fossiliferous record. These calculations do not take into account the long Precambrian record, a record that is many times longer than that of the fossiliferous record.

Before continuing our discussion of time, let us return to Hutton and see how such considerations agreed with the other problems he studied and which led him to his famous statement: "We find no vestige of a beginning—no prospect of an end." Also, let us see how his statement, "the past history of our globe must be explained by what can be seen to be happening now," was developed and adopted.

At the time that Hutton turned from medicine to geology, the foremost teacher of geology was Abraham Gottlob Werner at the Freiberg School of Mines. Although famed as an inspiring teacher, classifier, and pioneer mineralogist, he has been remembered more for his supporting the position that basalt and granite formed by precipitation from water. A leader of the Neptunist school of thinking, Werner effectively attacked the Vulcanists, who said all basalts were related to volcanic activity and the Plutonists who said that granites, too, had crystallized from a molten state rather than precipitated from water. Hutton supported the Plutonist position.

Another group, which is usually assumed to have agreed with the Neptunists, accepted catastrophic events as the causes of many events that they could not explain. Thus we find that Hutton was, on many points, in opposition to the Catastrophists, although he recognized the importance of temporary and local crises such as earthquakes and volcanic eruptions.

Hutton presented his synthesis of geology to the Royal Society of Edinburgh in 1785, two years after it was chartered. As the result of sharp attacks on his work, he published *The Theory of the Earth, with Proofs and Illustrations* in 1795. John Playfair, a more lucid writer, and a close compatriot of Dr. Hutton, published his *Illustration of the Huttonian Theory of the Earth* in 1802, and through this started the movement toward general acceptance of Hutton's theory.

The theory is often presented in the statement: *The present is the key to the past.* Oversimplified as this is, it succeeds in conveying the essence of the thought that, for determining the history of the earth and of its processes, we shall find answers we seek by consideration of the same principles and processes that operate today. Applied to considerations of time, the recognition that sedimentation required the earth to be much older than generally accepted did two things: It provided time for the operation of geologic and biologic processes at an exceedingly slow rate, a rate difficult to observe directly, and it suggested that experiments to verify geologic interpretations could not be made. Another effect, which will be treated later, was to cause consternation among theologians.

The suggestion to test Huttonian conclusions by laboratory experiments was put forth by Sir James Hall. In deference to Hutton, however, he held off the work until after Hutton's death in 1797. A year later he presented to the Royal Society of Edinburgh the results of his "Experiments on Whinstone and Lava," which involved melting and recrystallizing basalts and lavas to show that either could be chilled to form a glass or a crystalline mass by quick or slow cooling, respectively. Coupled with the clear field evidence gathered by Desmarest in 1793, the volcanic origin of basalt was proved by two independent methods. An ancillary benefit, which has become of increasing significance in twentieth century geology, was to show that laboratory work must agree with and relate to field evidence, not vice versa.†

We have seen that the time necessary for the formation through deposition of sedimentary rocks can be estimated by assuming a rate of deposition based on the observed rates of selected rivers and ocean areas. Now let us reverse the cycle and consider the rate at which the land is being lowered by erosion. This becomes a problem of determining the rate of denudation. Consider the region of the Colorado River and Iowa River basins. These are in areas of different climatic regimes and different geology, but an average of the results is useful for approximation.‡ Let us assume that computation of the average shows 1 ft of erosion in 3000 years. We know that the average elevation of the continents above sea level is 2757 ft, and therefore a steady-state situation with no variation in rate of erosion would require over 8 million years to erode the continents down to sea level.

Examination of our assumptions shows several errors, all of which give us a too short time interval. The first error is that we cannot assume a constant rate of

---

† See the October, 1961, issue of the magazine *Endeavor* for an excellent article on Sir James Hall by V. A. Eyles. The article contains excellent illustrations showing Hall's apparatus and microphotos of thin sections of some of the melts.

‡ There can be large differences in rates of erosion, depending on weather conditions. For example, an Egyptian obelisk, which had stood thousands of years without any particular decay in its original mild, dry climate, suffered serious damage after only a few years of exposure in Central Park of New York City.

erosion as land level is lowered. Comparison of the Iowa River basin and the Colorado River basin shows that when land elevation and slope decrease, the rate of erosion decreases. The second serious error is one we will come to later, when isostasy is discussed. Isostasy involves the elevation or depression of part of crust of the earth to maintain equilibrium in, for example, pressure beneath the surface at some specific depth. Thus, as land is eroded (for comparison, consider snow melting off the top of an iceberg), the underlying land rises to compensate for the reduction in pressure at the specified depth.

But the most serious error in the assumptions is the implication that only static forces are involved, whereas land is continually being actively elevated even while it is being actively degraded. The proof of this statement requires detailed treatment, for which we do not have time here.

We return now to the conflict that arose because the geologists who followed Hutton's lead proposed a great antiquity for the earth. This requirement met two attacks, one from (Christian) theological opponents and the other from physicists. The former argued that the world was created in 4004 B.C. and any suggestion of greater age was irreligious—an attack that sorely offended the deeply religious Hutton. The other attack, led by Kelvin, was based on the existing understanding of the origin of the earth and the application of physical principles to its history. It was generally believed that the earth had been created as a molten mass and was continually cooling down. From measurements of temperatures in bore holes and mines, and the rate of heat loss of the earth, Kelvin calculated the time since the earth's crust had been molten. In 1862 he announced that the crust had an age between 20 and 400 million years. By 1897 he had narrowed the limits to 20 to 40 million years, a value in close agreement with calculations based on modern data and the same assumptions.

Kelvin's basic assumption was wrong, however. He assumed that the earth must be cooling because it is losing heat. Although no source of heat beyond residual heat was known at the time, the geological evidence for a far greater antiquity showed that there was something not considered by Kelvin, which would change his calculation. There were indeed two factors that he could not take into account—the one, radioactivity; and the other related to gravitation. Before 1899, during the period of greatest disagreement between the geologists and Lord Kelvin, there was discovered a large source of heat in the earth in the form of radioactive materials. This evidence became available when laboratory work revealed that heat is given off during radioactive decay. More recently it has been recognized that gravitational attraction, and the resulting disturbances to the oceans and to the land masses, add still more heat to the earth.

14.10 The processes that build mountains

On March 27, 1964, an earthquake devastated several Alaskan towns and sent seismic sea waves 50 ft high into seaports like Seward. Within 4 min a region the size of Maine was lifted an average of 7 ft, while parts of Montague Island rose 33 ft. Other areas of land appear to have sunk to a lesser degree. The water levels in a number of wells in the United States were altered by the earthquake, some raised, some lowered. There have been various theories about the specific cause

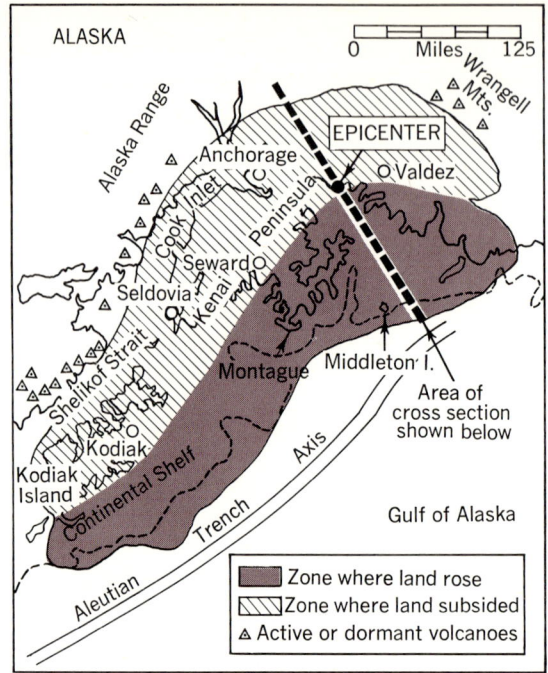

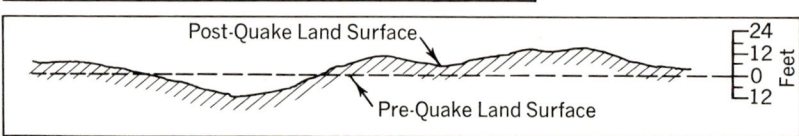

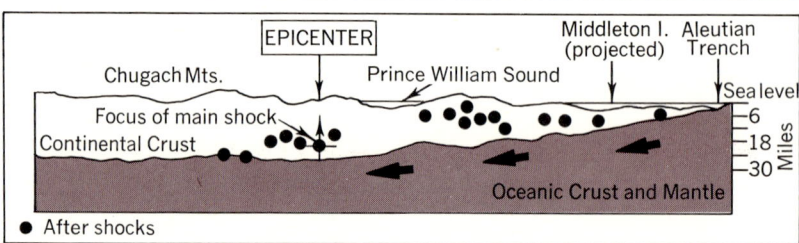

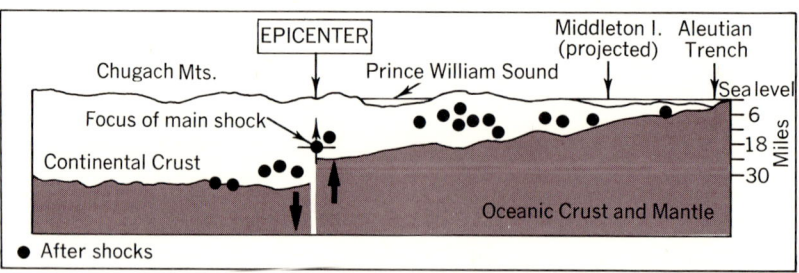

Fig. 14.10. (Left) Land upheavals and subsidence resulting from the Alaskan earthquake. (Below) Theories of Alaskan earthquake causes. (Courtesy of the New York Times.)

of this violent thrust of mother earth; but it did appear that the world was witnessing an example of the processes that build mountains. Figure 14.10, published by the New York Times of July 11, 1965, reports the opinions of the period. It was also opined that the energies involved were the equivalent of 100 nuclear explosions, each of 100-megaton capacity.

Mountain building and the attending deformation of rocks results from the action of compression and tension. Although each of these produces a change in the topographic elevation of the surface of the earth, the subsequent geologic structure is quite different. The Appalachian mountain range and the Sierra Nevada mountain range are prominent topographic features whose manner of formation are decidedly distinct. However, the major mountain ranges of the world—the Andes, Alps, Rocky Mountains, and others—follow more closely the structural pattern seen in the Appalachian mountains, and it was in the study of the latter that one of the major theories in geology was developed: the *geosynclinal theory*. Briefly, this theory proposes that sediments are deposited in slowly subsiding elongated depressions in the earth's crust. The deposits include fossils of shallow-water organisms, many rainprints, mud cracks, salt crystal casts, and even coal deposited as shallow beds. Concurrent with subsidence, lateral forces tend to compress and fold the sedimentary strata. After deposition of several tens of thousands of feet of sediments, the intrusion of molten igneous rock, perhaps from the mantle, produces a general expansion and uplift of the former basin of deposition. It is this uplift that not only accounts for the high elevation of mountains but which also exposes the structurally deformed strata and the intruded igneous rock. All major mountain ranges on the earth, except the undersea mid-Atlantic Ridge, are therefore assumed to be former sites of very thick sedimentary deposits, and all have strikingly similar structural features; for example, an elongated granitic core extending parallel to the general trend of the range. Figure 14.11 illustrates the general features of geosyncline. The great American geologist James Hall (1761–1832) published some significant generalization on their formation.†

† See Mather and Mason in References.

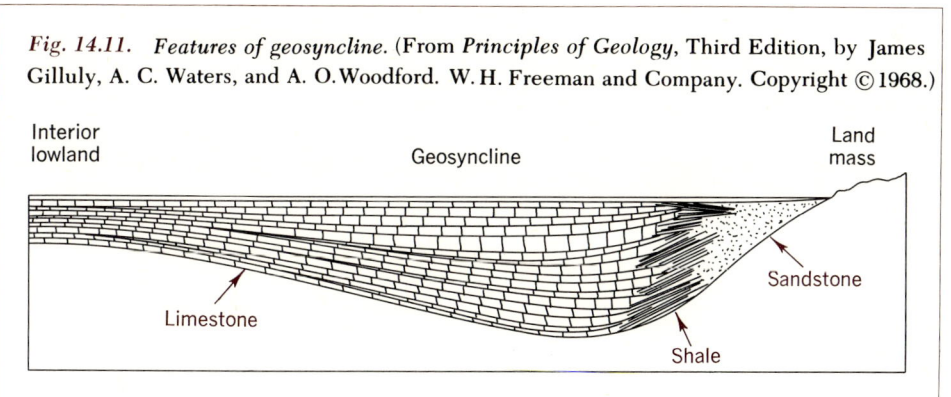

Fig. 14.11. Features of geosyncline. (From *Principles of Geology*, Third Edition, by James Gilluly, A. C. Waters, and A. O. Woodford. W. H. Freeman and Company. Copyright © 1968.)

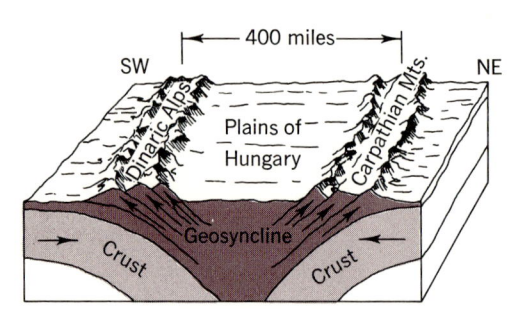

Fig. 14.12. Hypothetical section to illustrate a theory of the connection between geosyncline and the mountains formed from it. (After Leopold Kober in L. Don Leet and Sheldon Judson, Physical Geology, 3d ed., 1965. Reprinted by permission of Prentice Hall.)

The formation of the geosyncline involves a continuous downwarping of the earth's crust along with the deposition of sediments. A trough that begins with a depth of a few hundred feet can sink to much greater depth as the sediment accumulates. The areas that supply the sediments are not easily identified. The beds of sediment sink until there is a balance of buoyancy within the surrounding denser rock. The system thus reaches isostatic† balance.

There are various theories of how mountain formation follows the formation of geosynclines. Perhaps the heat energy inside the earth causes thermal contraction and other phenomena that may account for mountain formation. The heavy weight and pressure that develop at the bottom of the geosyncline apparently are enough to increase temperatures to melt the sedimentary layers. In the molten, plastic state, the layers could experience easier mobility under the pressure of the surrounding denser rocks. Figure 14.12 illustrates a possible process.

The discussion of geosynclines suggests that a type of structural deformation within the crust results in the formation of fold mountains where compression is the primary process. Of secondary but important significance are the mountain ranges produced by tension which, when acting upon the crust, results in block mountains of the type common in the coastal ranges of the western regions of the United States and which are characteristic of the basin and range province of Nevada, Utah, Wyoming, and parts of Montana. Block mountains are formed by large-scale tensional faults which in

---

† Isostatic: Subjected to equal pressure from every side; being in hydrostatic equilibrium, as a body submerged in a liquid at rest.

planview are roughly parallel and whose fault may form as *horsts*, as shown in Fig. 14.13, with intervening valleys, which are structurally referred to as *Grabens*.

The majority of mountain ranges in the western part of the United States are the result of block faulting. The best known example is the Sierra Nevada of California. This 400 mile long block of igneous rock is normally faulted on the east side with a displacement of more than 15,000 ft. Similar structures of Graben and Horst type continue east across Nevada and Utah. It would appear that the entire crust in the southwestern part of the United States has been subjected to a gentle arching and that the tension produced is responsible for the large number of normal faults. The degree of arching is also evident in the general uplift of the area, as seen in the downcutting action of the Colorado River and the resulting erosion of the Grand Canyon.

It should be emphasized that the formation of block mountains in the western United States is essentially a geologically recent process, which is still operative today. Evidence for present activity is clearly shown by the common occurrence of earthquakes in the western states, resulting from the continued movement of blocks along established faults. A recent striking example is the Hebgen Lake earthquake of 1959 in the general area of west Yellowstone. Movement along one of the faults bordering the valley of the Madison River resulted in a large-scale landslide that dammed the river, forming a new lake which was subsequently named Quake Lake. The fault scarp marking the actual zone of movement is quite visible on the ground surface and indicates a vertical displacement of 10 to 15 ft.

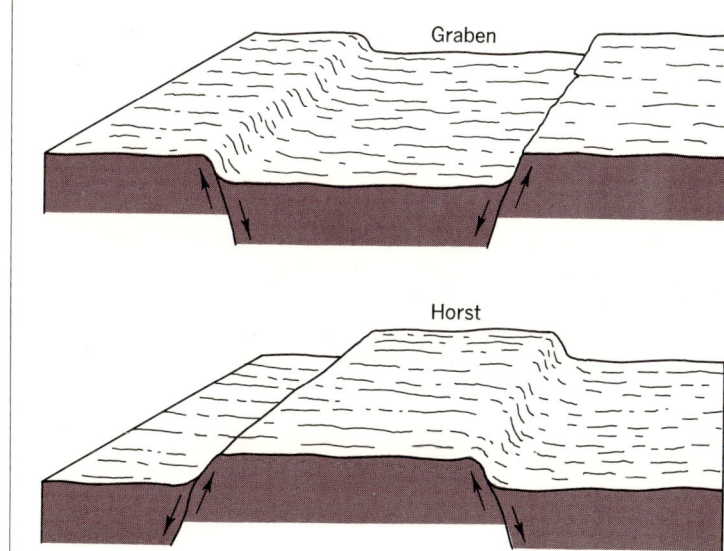

Fig. 14.13. *Graben and horst are simply down-dropped and uplifted fault blocks, respectively.* (From Arthur N. Strahler, Physical Geography. *Courtesy of Wiley.*)

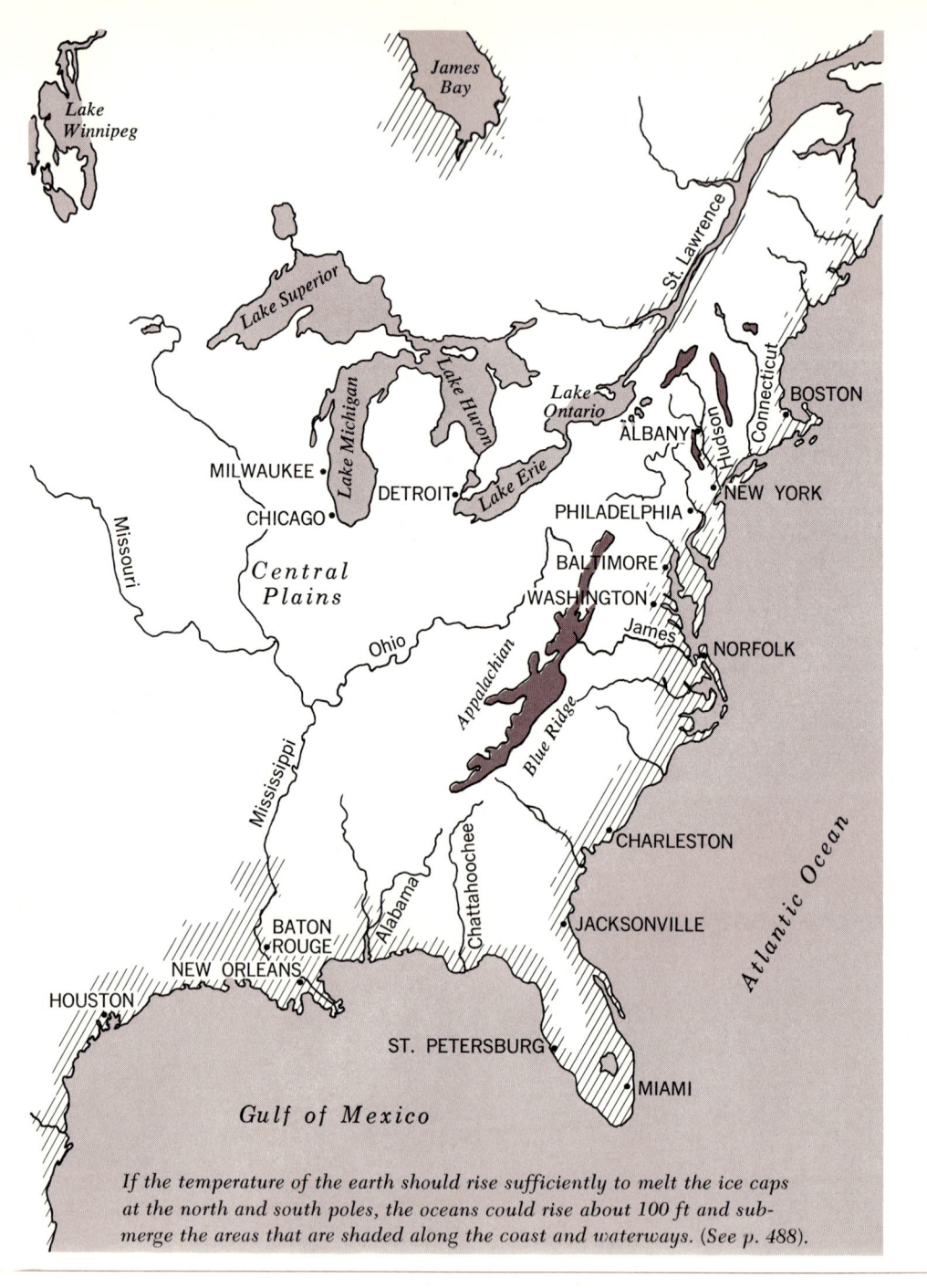

If the temperature of the earth should rise sufficiently to melt the ice caps at the north and south poles, the oceans could rise about 100 ft and submerge the areas that are shaded along the coast and waterways. (See p. 488).

The formation of mountains is a continuing process throughout geologic time, whether deformation of rocks takes place at great depths within the crust, as in geosynclinal-fold mountains, or relatively near the surface, as in block-fault mountains. Although the former type of mountain is produced by compression and the latter by tension, it should be noted that neither is exclusively dependent upon only one type of force.

14.11 Cyclic processes of geology

The buildup and erosion of a mountain represents only one portion of the cyclic character of geologic phenomena. We pursue the subject a little further.

Possibly the best-known cyclic occurrence in geology is related to coal deposits. *Coal* seams are repeated at a series of levels in the ground. The sequences of rocks that lie above and below the coal seam are also repeated, and the total of a repeated unit is called a *cyclotherm*. The coal and associated layers called *Pennsylvania* are about a mile thick in their southern West Virginia portion and include 90 cyclotherms. Each might roughly represent about a half-million years of the total of 50 million years estimated for the formation of these deposits. But they are far from uniform, as a result of shifting stream channels, effects of intervening swamps, variations in crustal warping, and other differences.

The process for accumulation of each cyclotherm might be as follows: An accumulation of sands and clays from the flow of streams might produce a swampy area that supports forest growth, which eventually becomes coal. Waters from the sea might inundate the area, bringing in marine animals whose calcareous shells and fragments accumulate to form limestones. The sea might then withdraw. But repetition of the process requires that the sea levels rise, or that the earlier deposits sink to lower levels, or both. The influences of the cyclic glacial periods could add to the levels of the seas and depress the earth as well.

Other cyclic processes, involving common rocks, have also been recognized. Three types of rocks with which we are familiar are representative of three types of formation. Sandstone, a sedimentary rock, is formed when sand collects in a basin and experiences cementation. Marble, a metamorphic rock, is a limestone that has been subjected to heat under pressure and has recrystallized, sometimes with the addition or subtraction of accessory minerals. Granite, frequently an igneous rock, may form by crystallization from a molten mass, called a *magma*. These are related in Fig. 14.14,† which shows that it is possible for a rock to be altered into another type of rock through geologic processes. Since we realize that the quantity of material available for the formation of rocks is limited, and the time needed for rocks to form is exceedingly great, it becomes obvious that rock material has been recycled many, many times. Hutton expressed it thus in his *Theory of the Earth:*

"We are thus led to see a circulation in the matter of this globe, and a system of beautiful economy in the works of nature. This earth, like the body of an animal, is wasted at the same time that it is repaired. It has a state of growth and augmentation; it has another state, which is that of diminution and decay. This world is thus destroyed in one part, but it is renewed in another; and the operations by which this world is thus constantly renewed are as evident to the scientific eye, as are those in which it is necessarily destroyed."

† See Leet & Johnson in References.

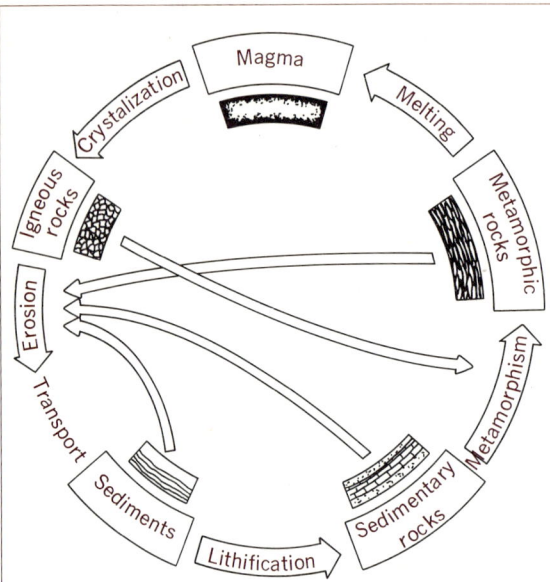

Fig. 14.14. The rock cycle, shown diagrammatically. If uninterrupted, the cycle will continue completely around the outer margin of the diagram from magma through igneous rocks, sediments, sedimentary rocks, metamorphic rocks, and back again to magma. The cycle may be interrupted, however, at various points along its course and follow the path of one of the arrows crossing through the interior of the diagram. (From Leet and Judson, Physical Geology, *3d ed. Courtesy Prentice-Hall.)*

But the recycling involved in the changes from igneous to sedimentary and metamorphic and back to igneous is very complex, and it is this complexity that undoubtedly contributed to the mistaken noncyclic model espoused by Werner and the Neptunists.

Figure 14.15 shows a modification of Fig. 14.14, to indicate the complexities of rock-recycling processes. Hutton's celebrated conclusion that he could find "no vestige of a beginning," has been reconfirmed to the point where no geologist expects to find any direct evidence of the original surface of the earth. His evidence of previous events is found in the fact that true cyclic development does not occur on a worldwide basis and in detail, and that equilibrium is rarely attained throughout the crust.

Figure 14.15 shows only a few of the shortcuts that may have occurred in the cycling of rocks. By placing the atmosphere at the top of the diagram, with sedimentary activity near the top, the diagram emphasizes that this cycling does not involve all the material of the earth but is limited to the crust, a mere skin on the earth. Further, it emphasizes that heat and material are added from the mantle, that is, from the material immediately below the crust of the earth. Not shown is the fact that heat is also locally produced and released within the crust; however, this is a relatively minor process compared to the heat transfer from the mantle.

Now consider the formation of granites. Some granites form by crystallization from a magma at great depth. These may be exposed at the surface, where they are weathered and eroded, with the products being deposited to form sedimentary rocks. We know that the majority of the igneous rocks occurring in the continents are granites (and basalts), and we should expect that the bulk chemical composition of the granites should approximate the bulk chemical composition of the sedimentary rocks. It does. Sedimentary rocks may be metamorphosed and, without becoming a liquid, the material may be reorganized by pressure and temperature to form something that cannot be distinguished from a true igneous granite when seen in a small exposure or a hand specimen. This is also called a granite, and should it be melted and recrystallized under certain conditions, a true igneous granite will be formed. This is what is referred to by the term "granitization" in Fig. 14.15. It is what Barth stresses when he speaks of igneous processes as homogenizing processes, and erosion and sedimentary processes as segregation processes. And it suggests that there is considerable disagreement among geologists on the origin of many large granite masses. The significance of this argument

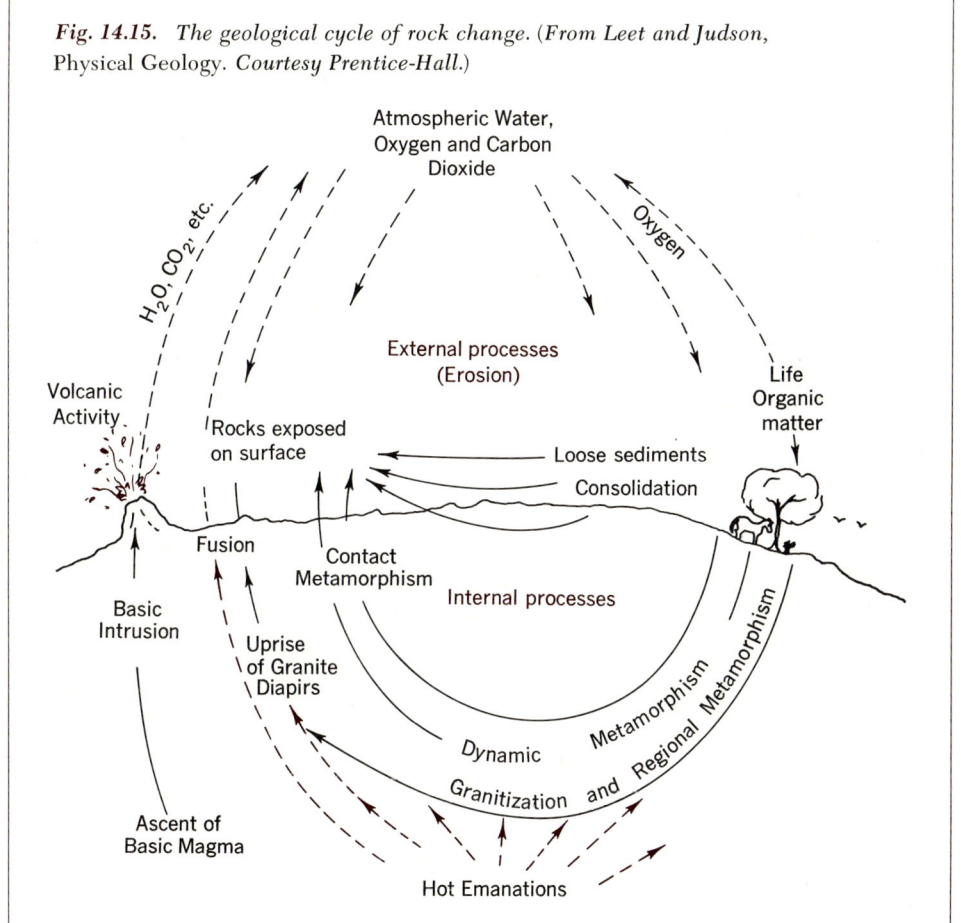

Fig. 14.15. The geological cycle of rock change. (From Leet and Judson, Physical Geology. Courtesy Prentice-Hall.)

appears in connection with analyses on the origin and development of folded mountain systems.

There is, on the other hand, no argument that basalts and related coarse-grained igneous rocks are formed simply from magma.† They differ in an important way from the granitic igneous rocks. As far as we can tell now, they contain material that has been derived from below the granitic and sedimentary crustal material, although there may even be material brought up from the top of the mantle or the interface between the mantle and the crust.

Further consideration of Fig. 14.15 shows that there may be breaks in the cyclic nature of events. From the astronomers we learn that the present atmosphere is probably not composed of the gases that were present as an atmosphere when the earth was first formed. Geologic evidence supports this position. Whence, then, did the present atmosphere, and the present hydrosphere come from? The most valid explanation is that they formed by the escape of gases from the interior of the earth through volcanism. Studies of basaltic lavas that are now escaping show that water vapor is the most abundant gas that is released, with carbon dioxide as next most abundant. Other atmospheric gases are also released. Probably many of the gases arrive near the surface in elemental form (that is, as hydrogen, not water) and react during or just prior to their escape. We shall not go further into the details of atmospheric modification here, but shall state the conclusions that are needed to illustrate the modifications of rock cycles.

The next example of a noncyclic development, which was initially considered cyclic in operation but never in product, comes from paleontology. The fossil record goes back, we know now, to roughly 2 billion years. But the record is very sketchy regarding rocks that are older than about 6 million years. The relative abundance of various animal phyla (groups) has fluctuated through geologic time. The reasons are multiple, including (among others) the habitability of the environment and the accidents of preservation. Cuvier† explained the appearances of successive fossil assemblages as the result of successive episodes of special creation and extinction. After Hutton, the majority of geologists (and, after Darwin, the majority of biologists) explained this cycling in abundance and variety in terms of theories of evolution. For our present discussion, the important point is that cycling has occurred.

The presentation of three types of cycles found in geology serves another purpose; it shows that cycling has been influenced by progressive changes in available material. This, simply, is evolution. Theories on the operation of evolution in the inorganic as well as the organic world require that we recognize a characteristic of time indicated in the earlier discussions, namely, its *unidirectional character.*

We are now faced with a conflict. Modern geology developed as a result of the synthesis of Hutton, a synthesis that delineated nicely the cyclic character of many geologic processes and events. Among astronomers, too, there is a group which believes in a pulsating universe—expanding and contracting, again a cyclic

---

† These are being studied at laboratories in Hawaii.

† Baron L.C.F.D. Cuvier (1769–1832) of Germany, born of Swiss parents. As a comparative anatomist he did much to give a rational basis to paleontology. See Mather and Mason in references.

occurrence. Yet from the same geologic record we find an evolutionary trend that suggests unidirectional time and development. We find ourselves faced with the question: Are cyclic events and unidirectional evolution antithetical, or are they complementary? A return to the development of geologic thinking will help to answer this question.

William "Strata" Smith† was faced in his work with cyclic repetition in rock types. In developing the first modern geologic map, he found the changes in fossil assemblages to be the key to his mapping, and his mapping was the key to his success as the canal builder of early nineteenth century England. Despite his application of the concepts of evolution in his work, he still believed in special creation of species, yet recognized the progression in time.‡ Lyell,§ in his textbooks, developed actualism into uniformitarianism to the exclusion of evolution, nevertheless recording the paleontologic evidence faithfully, and utilized evolutionary progression to set up the subdivisions of the past 65 million years. Darwin (reputedly) said that Lyell's texts had all the evidence necessary for developing the concept of evolution, but was unable to win Lyell over to accepting evolution until he was an elderly gentleman.¶ Once evolutionary progression superimposed on cyclic developments was accepted, the possibility of catastrophic events was excluded. So, by the end of the nineteenth century, a modified form of uniformitarianism was developed.

Still, in the geologic record there were events that defied explanation in these terms. To understand our difficulties, we must recall that the time available for geologic processes is so great that it is difficult to grasp the full significance of such processes. Catastrophic explanations, using the geologist's concept of time, were rejected, but other explanations have not been forthcoming. Today we find that geologically cataclysmic events are often resorted to for explaining the explosive biologic evolution of some 600 million years ago, and the (geologically) sudden extinction of orders and classes a few times since then. Thus, the geologic records provide clear-cut examples of cyclic events and of evolutionary changes that modify them, and provide new possibilities for geologic development. It has been necessary to modify Lyell's development of Huttonian principles in the direction of utilizing the limited cataclysmic explanations which were allowed by Hutton.

14.12 Summary

In an earth environment that is characterized more by the term *changeable* than by *constancy*, living systems have the capacity to maintain their own stability and regularity through "feedback" control in response to a changing environment. But there are distinct limits to this process. The duration and replication of living species requires at least a minimal stability of the environment. (For example, a fluctuation of as much as 5 percent in the surface temperature of the sun could cause the destruction of most living

---

† (1769–1839), an English civil engineer. See Mather and Mason in References.

‡ This is not surprising because commentators on creation had, for at least 200 years before the presentation of Darwin's theory, proposed "progression" or development.

§ See article *Charles Lyell* by L. C. Eiseley, Scientific American Aug. 1959. Also Mather and Mason in References.

¶ Lyell was in good company in his rejection of evolution. Claude Barnard and Karl von Nageli, among others, rejected it because it seemed to be philosophical speculation that had no relevance to the understanding of life.

forms on the earth.) Lacking such complete stability or constancy in either earth environment or living systems, we find changes that appear to be unidirectional, to which the concept and term *evolution* can be applied.

This chapter has concerned itself with the earth-air-water system, and with the deeper regions of the earth. The processes and mechanism of the self-adjustment and stability of this physical system have been analyzed in terms of a few of the major components of the system and of the energy conversions that involve these components. This approach is continued in the next chapter, wherein attention is shifted to the chemical elements and to the interaction between these and the living forms on the earth.

The extent, composition, and some of the properties of the air, of water, and of the earth have been discussed. As important for understanding as the knowledge of these facts is the knowledge of how they have been obtained. Many methods and many theories from a variety of branches of physics and chemistry are used in combination. One of the most striking features of the scientific endeavor is the manner in which a discovery or a successful generalization in one field of study illuminates the problems in another often totally unrelated field. Newton, by "looking up at the moon," could conclude that the earth's center, 4000 miles beneath his feet, was much denser than its surface. (His estimates of 1686 were surprisingly close to present figures.)

The study of the earth's interior, which is so much more inaccessible to man than outer space, has now been charted, using the laws of hydrodynamics and wave analysis, studies of meteorites, and even the study of the earth's magnetic field and its effect on the Van Allen belt. A great mass of information about the interior of the earth has already been accumulated. This, combined with our knowledge and classification of surface rocks, enables us to trace much of the earth's history and perhaps to foretell something of its future. Similarly, the waters over the face of the earth and the air above us—their depth, distribution, composition, temperatures, and currents —have yielded up a great store of information. Some of the more important facts have been discussed.

The purpose of this chapter has been not so much to describe the physical aspects of the environment in which we live as to emphasize the self-adjusting processes that keep the system reasonably stable, yet dynamic and ever changing under the vast energy conversions that are continuously taking place. It is the combination of this measure of stability, local variability, and continuing change that gives to living systems their great variety and changing character.

When we come in a later chapter to the study of nuclear energy, we shall acquire an understanding of how the sun has been able to pour out such a copious flow of radiant energy with such precise constancy over billions of years. In this chapter, however, we have focused on the inanimate energies of the earth. The overall heat balance is primarily maintained by the equality of the radiant heat from the sun reaching the earth and the heat being reradiated from the earth into space. (Radioactivity, tidal and wind friction, and motion of mountains play a very minor role.) The evenness of the climate over the surface of the earth, always fluctuating within narrow limits, is maintained by the atmosphere, ocean, and glaciers in a surprising and often complex manner which meteorologists are only beginning to understand.

Some of the self-adjusting "feedback" mechanisms that keep our physical environment steady have been discussed. In addition, some of the long chain of geological events bringing about the evolution of the world as we know it today have been presented. Many of these changes were significant for living systems, but they took place so slowly over such eons of time that life was able to make its adaptation. Indeed the physical changes undoubtedly aided the tremendous variety of living forms that have arisen. Thus, slow changes, in combination with a generally self-adjusting physical system, have wrought the miracle of evolution in the human and animal societies as we know them today.

Questions/Discussions

1. How would you divide the earth into zones, having in mind geological formations and features that affect life on earth?

2. The density of rocks that are measured in the laboratory rarely exceeds 3.0, yet the average specific gravity of the earth is found to be over 5. How can this be? Describe how the value of "over 5" is arrived at.

3. Describe the nature of the earth's core according to current ideas.

4. Describe the mantle zone as we now envision it.

5. Describe the principle of measurements involving seismographs.

6. How do sedimentary beds become rock?

7. How do sedimentary beds become highlands?

8. Describe a metamorphic process for rock formation.

9. Discuss the influence of "feedback" when considered in relation to erosion of land.

10. Discuss the influence of "feedback" when considered in relation to rainfall.

11. What are the conjectures with respect to changes in the atmosphere of the earth?

12. Describe the techniques by which the (time history) historical sequence of geologic formations can be determined.

13. Give two examples of negative feedback influences with respect to stability of plant life.

14. Discuss the influences of negative and/or positive feedback with respect to concepts of geological evolution.

15. Discuss the nature of the dominant energy conversion processes that characterize the following geographic areas: The North Pole the Sahara Desert, the region of the Nile River, the Alps, Equator, The Great Lakes, the Atlantic Ocean.

16. What are the basic assumptions and basic limitations of the theory of uniformitarianism?

17. Discuss the values, limitations, and risks that accompany interpretation of human history through information obtained from fossils.

18. Analyze the past and future of coal and oil deposits as "system" concepts.

References

Cantzlaar, G. L., *Your Guide to the Weather.* New York: Barnes & Noble, 1964.

Commission on Ground Water, "Ground Water Basin Management." ASCE, *Man. Eng. Practice*, No. 40 (1961), pp. 1–39, 99–118.

Commission on Hydrology, "Hydrology Handbook." ASCE, *Man. Eng. Practice*, No. 28 (1949), pp. 65–73.

Foster, Robert J., *Geology*, Charles E. Merrill, Columbus, Ohio, 1966.

Geikie, A., *The Founders of Geology*, New York: Macmillan, 1905.

Gilluly, Waters, and Woodford, *Principles of Geology.* San Francisco: W. H. Freeman, 1959.

Hodgson, John H., *Earthquakes and Earth Structures.* Englewood Cliffs, N.J.: Prentice-Hall, 1964.

Holmes, A., *Principles of Physical Geology*, 2d ed. New York: Ronald Press, 1965.

Hutton, James, *Theory of the Earth.* (See Mather and Mason for excerpts.)

King, C., *An Introduction to Oceanography.* New York: McGraw-Hill, 1963.

King, Philip B., *The Evolution of North America.* Princeton, N.J.: Princeton Univ. Press, 1959.

Leet and Judson, *Physical Geology*, 3d ed. Englewood Cliffs, N.J.: Prentice-Hall, 1965.

Leopold, L. B., and Langbein, W. B., *A Primer on Water.* U.S. Geological Survey Special Publication, U.S. Govt. Printing Office, 1960.

Mason, Brian, *Principles of Geochemistry*, 2d ed. New York: Wiley, 1960.

Mather, K. F. and Mason, S. L., *A Source Book on Geology.* New York: Hafner, 1964.

Meinzer, O. E., *The Occurrence of Ground Water in the United States* (with a discussion on principles). U.S. Geol. Survey Water Supply Paper 480, pp. 478–485, 507–513.

Meinzer, O. E. (ed.), "Physics of the Earth-IX," in *Hydrology.* New York: Dover, 1942, pp. 1–33.

Panel on Hydrology. *Scientific Hydrology.* Federal Council Sci. Tech., June 1962, 37 pages.

Rankama, K., and Sahama, Th. G., *Geochemistry.* Chicago: Univ. of Chicago Press, 1950, pp. 264–299.

Schwab, C. E., "Pollution, a Growing Problem of a Growing Nation." U.S. Dept. of Agriculture, *Yearbook of Agriculture, 1955*, pp. 636–643.

Williams, H., Turner, F. J., and Gilbert, C. M., *Petrography.* San Francisco: W. H. Freeman, 1954.

CHAPTER FIFTEEN

THE PRECEDING CHAPTER presented some of the *physical*, and especially *geological*, features that characterize our earthly habitat, giving emphasis to the major cyclic transformations and to the systems interrelationships that make the earth habitable. In this chapter our interest is directed more toward the features that bear directly on man and on other life forms.

15.1 What is the biosphere?

The most significant features of earth science and natural cycles have to do with interactions that affect the life of man. Weather, earthquakes, storms, and natural chemical changes have great influence on the life-supporting capabilities that the earth makes available to living organisms. In this chapter we briefly discuss some of the features of the physical *environment*, and the influence it has on the populations of the plant and animal world. The study of the relationships of living organisms with one another and with their physical environment is called *ecology* (from the Greek *oikos*, house). The *ecological system*, or ecosystem, refers to the operation of the ecological interactions in nature. The term *biosphere* usually refers to the portion of the earth in which ecosystems exist. This includes the atmosphere, the soil, and the fresh and salt waters that contain plant and animal life.

A number of natural processes give the biosphere the characteristics that sustain life. Some of them have been mentioned earlier, and a few more are discussed in this chapter. To these natural processes man has added other processes that have assumed such proportions as to greatly modify the natural environment. The extent and magnitude of the modifications that man has imposed seem to have in-

The Biosphere as Environment for Populations

How much longer will the earth that gave birth and nurture to man withstand the assault of man's society?

V. L. P.

creased enormously as the population of man has increased. It was scarcely more than two centuries ago when fossil fuel first became generally used. Now the rate of use of coal and oil has increased to provide a base for huge industrial operations. This increased use has reached such proportions that we are now changing the biosphere in a way that appears to be harmful to man. These changes also threaten dangerous depletion of the natural deposits of fossil fuels in the relatively near future.

The mining and utilization of fossil fuels and the mining for minerals have brought the somewhat deeper regions of the earth nearer the biosphere. Large-scale production of chemical fertilizers to replenish depleted soil nutrients, mechanization of farming, and extensive transportation of food and the products of technology have interwoven the technological processes of man's society with those of the biosphere. The construction of dams and waterways has greatly improved the hydrologic aspects of nature, but this can scarcely counteract the waste that has accumulated from the misuse and pollution of fresh waters by industrial and municipal processes. The careless use of chemicals, insecticides, detergents, fuel oil, coal, and nuclear power create new environmental conditions that introduce substantial unknowns to which man and his societies as well as other living things must adapt. Moreover, because the world's population continues to increase at a rapid rate, we are forced to look to a time when the food resources of the biosphere can no longer maintain the tenuous balance between availability and need. The threat of starvation looms for large numbers of the population.

In this chapter we explore a few of these problem areas that require increasing attention as a result of the transformations that have been wrought by man. Most of the details that have to do with the balance and adjustment of the individual organism with its environment will be reserved for the studies of the second year of the course.

15.2 The influence of geography on man

We are generally aware that there are vast differences in the geologic, thermal, and climatic conditions, and hydrological features, that characterize the various portions of the biosphere. There are deserts and huge oceans, tropical areas, and icy regions that rarely melt even in the summer sun. A person does not have to be a world traveler to realize the extent to which these environmental conditions affect and even determine the habits and thoughts and physical well-being of the inhabitants of each area. The food man eats is often determined by the nature of the plant and animal life that are part of his immediate environment. The clothing he wears is similarly determined. Whether he pauses to think or to indulge in luxuries depends on how much of his time and effort go into providing for the needs of the body. Poetry and philosophy are seldom composed on an empty stomach or while indulging in heavy physical labor. The condition of his health is determined to a large extent by the environment, and sometimes he must change from one climate to another to remain in good health or to recover from illness.

Over many centuries there has been recognition of the great influence that geography can exert on man and on his way of life and thought. One of the early systematic studies of this influence was undertaken by Baron Alexander von Humboldt (1769–1859) who correlated the distribution of plant life with altitude and temperature zones. He studied the density and distribution of people and the relationship of the natural environment with the social and political lives of the people who live in them. The studies seemed to indicate strong geographical and climatic influences on the people. While the various societies of men were able to adapt the environment to their particular needs through agriculture, mining, building of shelters, and similar pursuits, limitations to these adaptations were set by geography; whether the locale was desert sand, tundra of the frozen northlands, river bank, valley, or mountainous terrain. Generally the course of human history parallels closely the course of climatic or geographic history. We can think of aspects of a society's culture as a direct or indirect response to the geographic conditions of the habitat in which the society originated and developed.

Karl Ritter (1779–1859) held the first academic chair in geography, established in 1820 at the University of Berlin. To him the earth was "the preparatory school of the human race," created by the Deity expressly for man. He saw the earth to be a "unitary organism," each portion having its own potentialities for human development. He reasoned that the characteristics and potentialities of each region of this "unitary organism" are reflected in the kinds of plants, in the pattern of agriculture and mining operations, and in the domestication, breeding, and rearing of animals that flourish or are feasible in each region. There was "purpose" in man and nature, a view that was reminiscent of the teleological views of Aristotle, Strabo, and Herder. But the interrelationships among man's economic, social, and political activities and his physical environment were not so close as to deny him the freedom to choose to do or not to do the will of God.

But the spirit of the times did not permit such heavy dependence on teleological views. Friedrich Ratzel (1844–1904), a zoologist who had turned to journalism to finance his zoological studies, led the way to *anthropogeography*, or the study of the dynamics of human culture related to geography. He did not find it necessary to include teleological implications but rather depended on concepts from theories of biological evolution for explaining the origin and differentiation of human culture in relation to environmental conditions.

Ratzel traced the movement of cultural elements to study their diffusion from one region to another. It seemed to him that the Hottentot of the Khalahai Desert, the Arabs of ancient Arabia, the Sumerians of ancient Sumeria, the Chinese of the Yellow River, or the aborigines of Australia became what they were because of pressures imposed by their environment. The way of life that a group of people may develop on moving from moderate to Arctic climates would be quite different from the life and ways they would adopt had they moved to a desert area. In time there would develop pronounced differences in their customs, language, knowledge, skills, personality, religion, physique, morality, social mores, and values. There is no one natural or divine way of life. To understand a people one must understand the "soil" as much as "the blood" from which the people derive their characteristics. Cultural values derive in part from the environmental demands of a region, not merely from a common source that all men share alike. Ratzel became the apostle of *environmentalism*.† It should be noted, however, that in his later writings he gave much more credit

† See Huntington and Cushing in References.

to cultural influences as determining factors in the development of peoples.

But these concepts, which minimized the influences of the exercise of free will on the part of man, did not seem pertinent in a modern world in which technology, commerce, transportation, and communication offered abundant opportunities for transcending the limitations of one's immediate environmental restrictions. The white man who displaced the American Indian did not thereby assume any of the aspects of Indian culture, which had been in existence for hundreds of years. Instead he cut down the forests, dammed the rivers, tilled the soil, killed the wild animals, until the North American Continent became not unlike the Europe from which he came. Obviously man can also change and even shape the environment in which he finds himself. We must note, however, that there are restrictive boundary conditions within which changes are not developed easily. For example, it would be difficult to convert the frozen tundra of the north into European soil.† Also, the background of culture, attitudes, and knowledge that were transplanted from Europe were in part products of the European environment.

15.3 Influence of geography on political and economic development

Some geographers have identified significant influences of geography and environmental pressures on the political and economic life of communities. It has been suggested that in ancient times

† It might also be noted that over 200 years of the white man's influence in Africa has not succeeded in making that continent a counterpart of Europe.

the fertile river valleys of Egypt made possible the strong central governments that developed there, while the mountainous regions of Greece favored the development of loose governmental ties between the various cities. Whittlesey (in his 1939 publication) declared that "a political system is the summation of laws which people make in order to extract a livelihood from their habitat." The laws and practices affecting ethical conduct, morality, adherence to laws, and concepts of religion also derive their pattern largely from environmental conditions, according to this view. But the case for this argument becomes increasingly difficult to defend in the light of modern travel and communication opportunities. Nevertheless there is evidence that even in modern times, governments of nations in warm climates tend to be less stable than are governments in cooler climates.

The economic patterns and the commercial activities of a people seem to be more easily related to the soil than are their political or religious institutions. Agriculture, breeding of animals, mining operations, use of rivers and other water systems for navigation, and availability of harbors are determined in large part by the natural environmental conditions and potentialities of a geographic region. There will be no exportation of raw materials from a region that cannot produce them. There can be, however, very heavy commerce in importation of raw materials and export of manufactured products when the people are capable of converting the one to the other through technological know-how.

With the steady increase of commerce, growth of transportation and communication between regions of the world, and the increasing availability of electricity and other sources of energy to regions that have been deprived of such energy, it becomes more and more difficult to defend the view that the cultural character of a people derives largely from the soil. When adequate sources of energy become available, it is much easier to irrigate land, manufacture goods, control the temperature of living quarters, and to acquire luxuries. Nevertheless when one draws a map of the world to identify the regions of greatest technological and social progress, there is clear indication that man works best where the climate is temperate and favors his working at high physical and mental efficiency. His efficiency, initiative, and physical health are helped by climatic conditions that provide strong but not extreme contrast between summer and winter, with summer temperatures averaging not much above 65°F for day and night, and the average for winter not falling much below 40°F. Variability of weather seems to be very desirable and important, with changes alternating from sun to rain, cold to warm, storm to calm (Huntington). The temperature belts of the world have changed considerably from ancient days and are assumed to be continually changing. For example, the more favorable climatic belt was once nearer the equator; as this belt moved northward, so did the civilization of man. Huntington asserts further that "each type of human culture appears to develop most fully in its own environment and tends to be modified as soon as it is transferred elsewhere." But, as we have indicated earlier, the effects of transfer are minimized in a highly technological culture.

We shall now conclude this brief excursion into the subject of geography and man and turn to the analysis of a few of the biochemical features of the biosphere that are common to all living organisms.

15.4 The carbon and oxygen cycles

We have noted that there are various theories as to how the earth came into being. At the present time it is believed that the earth began as a cold body. (The history of the atmosphere of the earth would, of course, depend very much on the history of the earth itself. When the chemical composition of the earth resembled that of the sun, what might have been called an atmosphere would presumably have been made up largely of hydrogen and helium.) At some point during its evolution, the earth would have become heated as a result of contraction, radiation from the sun, and radioactivity from within the earth itself. At that time the atmosphere presumably was made up of ammonia (NH_3), methane (CH_4), sulfur compounds such as H_2S, and similar compounds. These molecules could have had sufficient kinetic energy to escape from the earth's gravitational field, or they might have recombined when the earth contracted and solidified.

With the coming of solidification, substantial amounts of gases that had been dissolved in the liquid rock material were released to become a new atmosphere. It is thought that this, the second stage of the development of atmosphere, was made up largely of carbon dioxide (CO_2), water vapor (H_2O), nitrogen (N_2), and possibly hydrocarbons. The water vapor could have begun to condense and form water droplets, but probably some time elapsed before the earth was cool enough for these droplets to reach the earth as rain. Once there was rain, the processes of erosion and the formation of rivers and seas could begin.

In time, the high concentration of carbon dioxide was reduced as it became absorbed in the oceans and in the formation of limestone, leaving mostly methane, ammonia, hydrogen, and water vapor. This was possibly the third stage of atmospheric development, occurring about 2 or 3 billion years ago. It was during this phase of violent storms, volcanic action, heavy rains, and earth convulsions that life may have appeared.

We shall defer to later chapters of Part II the questions regarding the origin and progress of life. Here we raise the question of how oxygen originated in the atmosphere. Possibly some of it may have been formed when the energy of sunlight or discharges of lightning decomposed water vapor. It is more likely, however, that the bulk of the oxygen gas was the product of plants, which by their photosynthetic process could release more and more free oxygen as there developed more and more plant life. This, then, brings us to the stage of our present atmosphere, which consists of about 78 percent nitrogen, 21 percent oxygen, and the remainder a mixture of other gases.

A most important chemical element in the body of living organisms is carbon, which in combination with water, nitrogen, oxygen, hydrogen, and some other elements makes up the biological environment and all that feed on that environment. When carbon reacts with oxygen it is converted to carbon dioxide and in the process gives up considerable heat energy. In the simplest case, the burning of the fossil fuels coal and oil ($C + O_2 \rightarrow CO_2$ + energy), the carbon dioxide goes directly into the atmosphere. Light energy and water are utilized by the chlorophyll of plants to convert CO_2 to plant food (carbohydrates) and release oxygen (O_2) to the atmosphere. This process, called *photosynthesis*, is often deceptively summarized as $CO_2 + H_2O \rightarrow$ carbohydrates $+ O_2$. About 674 kilocalories (kcal) of heat must be supplied for every mole of glucose (sugar formed by photosynthesis). The mole amounts to about 180 grams (g), and the sun energy required is enough to raise the temperature of 180 g of water by about 3200°C. The process is roughly 0.1 percent efficient.† (The water will become steam long before that temperature is reached.)

The plant world that makes up a critically important part of our biological environment not only is the ultimate food of the animal world, but also the source of the deposits of coal and oil that were formed in past ages. Animals that feed on plants (the herbivores) convert the carbohydrates to animal tissue and release waste CO_2 to the atmosphere.‡ The plants that are not consumed by animals are subject to decay through bacterial action, which again releases CO_2 to the atmosphere (Fig. 15.1).§

The concentration of CO_2 in the atmosphere (dry air) varies between 0.023 and 0.050 percent by volume. The concentration does not vary too much because an increase tends to drive more CO_2 into

† This means that more than 99 percent of the energy of the sun that falls on plant leaves is lost, as far as the photosynthetic process is concerned.

‡ Organisms that are able to synthesize organic compounds directly from the inorganic form (CO_2) are called *autotrophs* [from the Greek *auto* (self) and *trophe* (nutrition)] or self-feeders; for example, plants. Man, animals, and other organisms that cannot do this and which require intermediate steps are called *heterotrophs* (from the Greek *heteros*, other). Whether autotroph or heterotroph, all organisms rely on the breakdown of organic compounds within the body for the release of necessary energy.

§ The equations may be written as C (carbon or organic carbon) $+ O_2 \rightarrow CO_2$ + energy (water is usually present, and bacteria must also be present for decay. The latter will be mentioned again). The estimates of carbon transfer are those given by Hardin (see References) in his Fig. 9.1 (*Biology—Its Principles and Implications*).

15.4 The Carbon and Oxygen Cycles

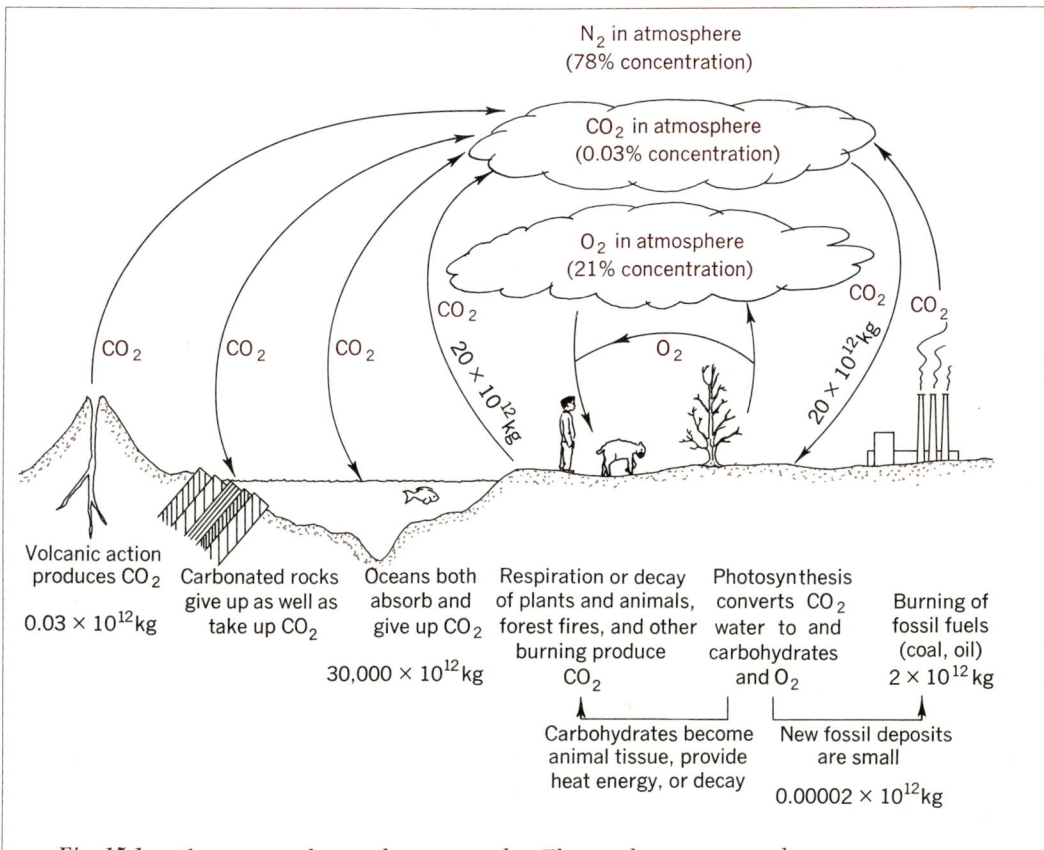

Fig. 15.1. The major carbon and oxygen cycles. The numbers represent the amounts of carbon that are transferred each year by the various processes.

solution in the oceans. There is at present an ever increasing imbalance, which is due to burning fossil fuels at a fantastically higher rate than the rate at which fuel deposits can be restored. Clearly we must eventually reduce or eliminate coal and oil as sources of the growing energy demands of the world. New sources of energy (for example, nuclear energy) must soon take a larger role in supplying heat and energy to meet the needs of this modern technological society. We should note, however, that only a few of these other sources of energy are presently useful and even so are not limitless in productive capacity.

Does the burning of coal introduce other effects? It is known that an increase in the CO_2 content of the atmosphere from this combustion will raise the temperature of the earth: When the sun's rays reach and warm the earth, the earth itself emits heat at (longer) wavelengths that have more difficulty in escaping the earth through the CO_2 blanket.† The rise in earth temperature could change the polar climates to the point of melting large glacier beds. It is estimated that the melting of these glaciers would raise the level of the oceans by as much as 100 ft, thus submerging much land area, including most of our coastal plains (see page 472). But, since the fresh water of the glaciers contains less CO_2 than does the ocean, the rising oceans could absorb more of the CO_2 of the atmosphere. This could reverse the process, cool the earth, and reestablish the glacier beds at the expense of the oceans. In time, these glaciers would move into warmer regions and once again alter the surface of the earth as they also replenish the oceans. The process resembles the cyclic processes with feedback, which we have discussed. The time spans involved may be many thousands of years. The feedback can be slow, and may have a phase relation of the positive kind, which leads to large surges of hunting, rather than of a negative, damping character. Whether the process finds an easy and intermediate balance or develops violent surges around this balance condition depends on other factors such as time lags and momentum of processes.‡

15.5 The nitrogen cycle

The element nitrogen is one of the key components of proteins, nucleic acids, and other compounds that make up organic matter. Nitrogen in the form of the molecule N_2 makes up about 78 percent of the earth's atmosphere, but in this form it cannot be readily used by organisms. In its nitrate, nitrite, or ammonia form, it can be absorbed by plants to become one of the amino groups ($-NH_2$) or other nitrogen-containing living matter.

There are several sources of useful nitrogen-bearing ions or nitrogen compounds. Large deposits of nitrate rock dissolve in water and thus make available the needed nitrate ion. There are also chemical processes for fixation of nitrogen whereby, with addition of considerable energy, the nitrogen molecules of the atmosphere are converted to useful compounds of nitrogen with oxygen or hydrogen. The nitrogen fixation process is used for the production of ammonia (NH_3), which is also a source of nitrogen for plant growth. Lightning discharges convert some of the atmospheric nitrogen to compounds of nitrogen and oxygen, which are then carried by rainfall into the ground and made available to plants.

A major source of nitrogen compounds results from the decay action of bacteria on organic materials. Nitrogen compounds undergo transitions from one form of protein to another, and are converted into amino acids as plants are consumed by animals and the latter are consumed by other animals (carnivores). Death overtakes them all, and bacterial decay reduces the organic nitrogen of both plants and animals to ammonia (NH_3). Through the action of *nitrifying bacteria* the ammonia may be converted to *nitrite* ions, NO_2^-, which are excreted into the environment. The nitrite ions may be absorbed by other bacteria and excreted as nitrate ions NO_3^-. The environment thus receives large quantities of nitrates that become available to supply the plants. Figure 15.2 describes these transitions.

---

† This is sometimes called the "greenhouse effect." A glass greenhouse becomes warm in cold weather simply because the entering sunlight energy becomes converted to longer wavelength (heat) rays, which do not pass through the glass as easily as did the sunlight.

‡ Question: Does this indicate that we should expect another ice age? What should be society's attitude toward finding an answer to this question? If there is clear indication that there is such a trend what should be the attitude, responsibility, and action of society?

15.5 The Nitrogen Cycle 489

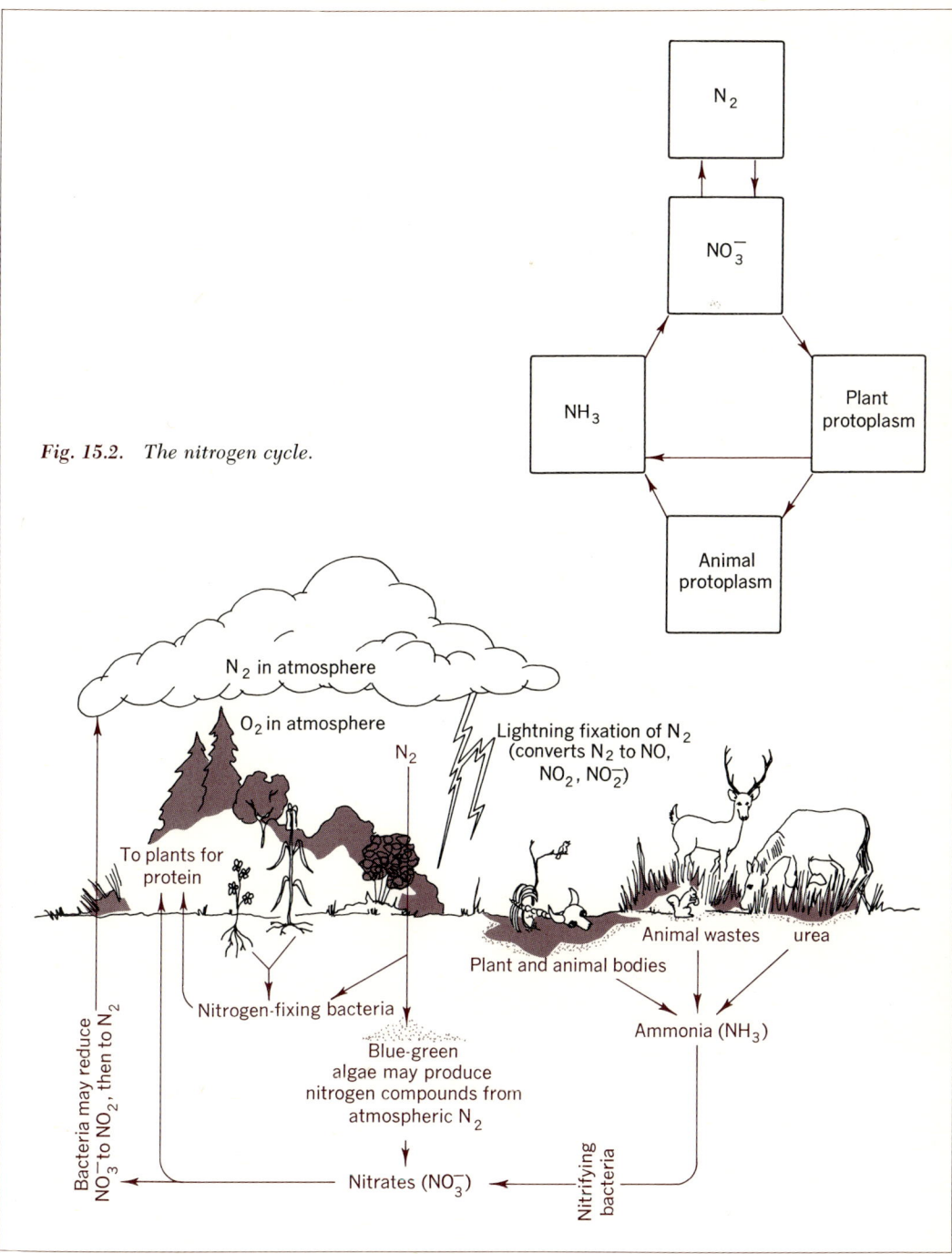

Fig. 15.2. The nitrogen cycle.

Some bacteria and blue-green algae living in water and soil absorb atmospheric nitrogen directly and convert it into amino acids and proteins. Some of these plants yield ammonia to the soil when they decay. In the case of leguminous plants such as alfalfa, clover, and soybeans, bacteria develop in root nodules and make nitrogen available to the legumes.

15.6 The energy requirements for life

The processes we have described for the carbon, oxygen, and nitrogen cycles require a great deal of energy. The principal source of their energy is the sun. The sun's energy is the product of nuclear reactions in which some of its mass is converted to radiation. We shall learn some details of the process a little later. At this point we need only note that this transformation of mass into energy is estimated to reduce the mass of the sun by about 4.7×10^9 kg each second. While this is a considerable mass, conversion at this rate from the time of the origin of the earth, about 4 to 5 billion years ago, would have resulted in 0.03 percent reduction of the sun's mass. Assuming that this rate of conversion continues into the future, we can be reasonably confident that the sun will continue to serve as a major source of energy for the earth for billions of years to come.

It is fortunate that this huge source of solar energy will continue to be available to us, for it will be needed in view of the trend that is represented by the entropy concept of the second law of thermodynamics. That concept assumes that although the total energy content of the universe remains constant (as far as we can extrapolate from experience), every change and every step of nature involving transformation of energy tends to reduce the energy available for future use. In other words, nature tends toward increasing entropy, which means that it tends toward states of higher disorder while preserving the constancy of energy.

Contrary to this interpretation of the increase of entropy, all the biological processes we have described tend to reduce entropy by converting disorder into orderly structures and functions. But since living systems are open systems, there is no contradiction with the second law of thermodynamics because our biosphere and its processes represent only one part of the total system, the other part (with overall increase of entropy) being represented by the activity of the sun as the principal source of energy.†

But while the energy received from the sun is likely to remain constant, there are basic limitations to the capacity of the biosphere to support life. One common although crude representation of both the limitations of food supply and the interdependence of organisms on each other is the following: The photosynthetic process has an efficiency of only about 0.1 to 1 percent. What are the food requirements for a man who subsists on fish? The food chain involved for producing 1 lb of man requires about 10 lbs of bass. These in turn require 100 lb of minnows, 1000 lb of water fleas, and finally 10,000 lb of algae. Similarly, about 10 lb of organic compounds are needed in grass form to produce 1 lb of sheep. There is thus a pyramidal nature in the quantity

---

† We might pursue thoughts on entropy a little further with respect to the question of what makes up a complete system. We preserve the concept of increasing entropy by including the sun in our system. Who is to say that if we included the galaxies that lie beyond our own, the laws of nature would not change drastically within our "system"? We can only report the trends that we can measure.

of food required to build higher and higher levels of the food chain. The example also points up the character of the predatory chain that exists in nature, by which each level lives by consuming the organisms of another lower level. It is not common for the lower forms of life to die of mass starvation, except in local areas of drought or of other unusual, temporary change. The organisms are more likely to be eaten up by higher orders of life. But as one approaches the highest forms of life, and especially the level of man, death by starvation becomes more common, in part because death from other causes becomes less likely. It is for this reason, among others, that the population excess is of growing concern to society. We shall discuss some features of the problem at this time because of its close ties with earth science and the biosphere.

15.7 Some observations on growth of populations

At the present time the threat of hunger and starvation hangs heavy on millions of inhabitants of the earth. Despite this fearful situation, the increase of population continues at a high rate of 2 percent each year, which doubles the total population approximately every 35 years.[†] According to conservative estimates of United Nations agencies, if all factors remain the same, this rate will double the present number of 3.2 billion by the year 2000 and bring a sevenfold increase by the end of the next 100 years. The increase is due only in part to a continuing high birth rate. It is due also to increase of the life span of individuals as food, resources, living conditions, and medical services are improved and extended to underprivileged areas. The average life expectancy increased from about 30 years in 1850 to about 70 in 1950, in the more advanced societies, without large change in the maximum age. But there are vast areas in which the lower life expectancy continues. The birth rate in Asia continues as high as ever, where medical services are especially needed for reducing the effects of disease and extending the life span of hundreds of millions. The situation brings to the foreground once again the fears expressed by the Englishman Thomas Robert Malthus (1766–1834). In *An Essay on the Principle of Population* (1798), Malthus considered the world's population to be increasing in geometric ratio (1, 2, 4, 8 . . .), while food availability seemed to increase by the much slower arithmetic progression (1, 2, 3, 4, 5 . . .). He looked over the course of history and decided that two conditions seemed to be God-given and likely to endure, namely: (1) food is necessary to the existence of man, and (2) cohabitation of the sexes is necessary and will remain near its present state. On these premises he envisioned that the problem of subsistence (hunger) would become a most pressing problem to man.

Nearly every argument advanced by Malthus has been contested and proved to require important qualifications. He did not envision the development of modern industrial capability to produce more food by more efficient methods or the possibilities of obtaining food from the seas or by synthetic production. Nor did he recognize the possibility that as society became technologically more advanced, there would be a reduction of birth rate and a balancing of population against the availability of food. But the facts are: (1) the world's population has increased fourfold over the 160 years

---

[†] A 1 percent rate of population increase doubles the population in about 70 years. A rate of 2 percent doubles it in 35 years, and a rate of 3 percent doubles it in 23 years.

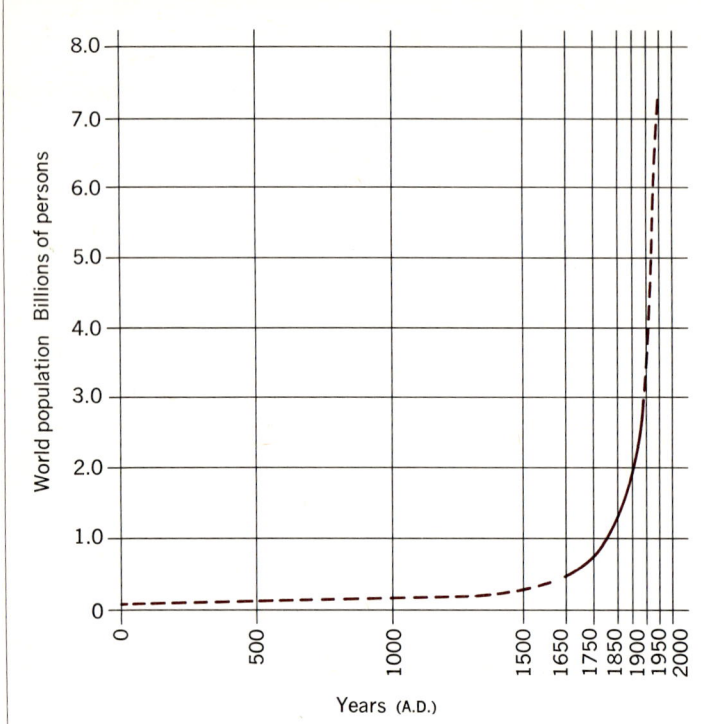

Fig. 15.3. Estimated population of the world, A.D. 1 to A.D. 2000. The solid line represents the portion of the curve about which we have some data.

since his pronouncements, (2) most of the world is not technologically advanced at this time, and (3) there are even today occasions when the number of people who are underfed is comparable to nearly the total world population of his day. Figure 15.3 illustrates the increase in the population of the world since A.D. 1, and the expected increases up to the period A.D. 2000.

Charles Darwin, in his *Origin of Species* (1859) also saw every type of organism "striving to the utmost to increase in numbers." The checks on this increase were of four kinds: Availability of food established the extreme limit to which any species can increase. Predatory animals provided a second check. Physical factors such as are associated with climate and the inroads of ill health and disease were also restraining factors.

The history of population numbers in various parts of the world has been very complex. The assumption that the rate of population increase is determined largely by the availability of food has been proved and disproved many times. In Ireland the introduction of the potato in the late seventeenth century helped to relieve the food problem by increasing the food output per acre. The population also increased from 2 million to about 8 million by 1845. The failure of the crop during that year brought disaster of much larger proportions, when from 1 to 2 million died of starvation and as many emigrated. Since that time the population has remained about stationary, partly because young people defer their marriage to later years.

What are some of the factors that tend to increase breeding rate and population of animals and humans, and what are the influences that stabilize or reduce growth? There seem to be very real differences in the experiences that apply to the nonhuman world and those that apply to human society. For our purposes we shall divide the subject into three categories. The first is a study of the experiences and experiments involving animals. The second is of primitive human societies that depend primarily on collection of wild food rather than on agriculture and technology. The third deals with human societies that utilize agriculture and technology for procurement of food. Within this last group there are wide differences among nations with respect to the efficiency and effectiveness with which technology aids in the production of food.

15.8 Population controls in nonhuman societies

Figure 15.4 illustrates the experiences of a specific animal society. Beginning at the left side, there is increase of population toward the right side. At the top of the diagram the available food supply *per animal* is illustrated as decreasing to the right, corresponding to the greater demands of increasing population. It is interesting to note that while the food level may decrease to a minimum subsistence, general starvation is not common among animal societies within the main population group. There is evidence of natural regulation of numbers in animal societies, which prevents conditions that cause large-scale starvation. For example, studies of the habits of offland sea birds, which depend on the planktonic organisms that float or grow vertically distended in the North Atlantic Ocean, reveal that as the food supply decreases the birds distribute themselves over wider areas. The movement is motivated by a search for food but has elements of competition.

Where laboratory experimentation is possible, as with insects, water fleas, guppies, mice, and rats, a pair will breed up to a predictable total number for a particular confinement situation. There is identifiable an apparent maximum (ceiling) population density (number of animals per unit area) for each situation. The number will increase rapidly until this ceiling population density is reached. The number is not likely to go beyond that ceiling density even with abundance of food and complete absence of predators or disease. A three-year series of experiments on guppies† gave remarkable demonstration of regulatory practices that hold the numbers within the ceiling density. When the numbers were reduced by removal of some of the guppies at regular intervals, the numbers were soon restored as more of the young survived. When the stocks were undisturbed, breeding continued at the usual rate, but the excess of numbers was consistently removed by cannibalism at birth.

Regulatory practices vary with each animal society when overpopulation exists.† In mice, ovulation and reproduction decline and may even cease; thus population of animals tends to be highly dependent on density, the reproduction rate being great when the population is

† See Silliman in References.

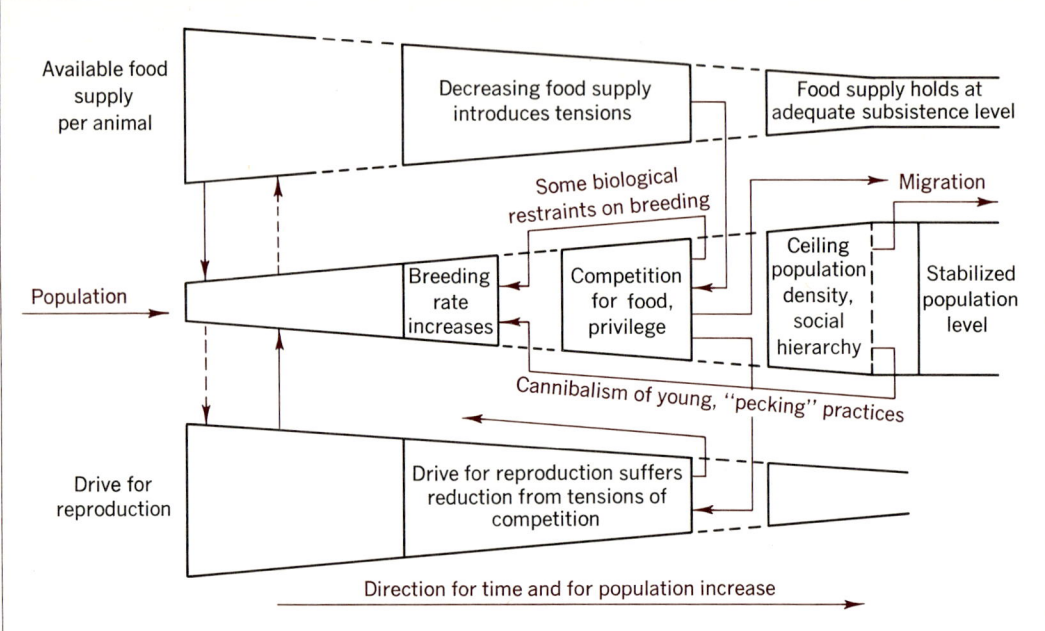

Fig. 15.4. *Some mechanisms for regulation of population density among animals. As the population increases (from left to right of the diagram), the food supply per animal decreases. This introduces competition and tensions that tend to restrain breeding and encourage migration. When a certain maximum (ceiling) density of population is reached, cannibalism and pecking practices hold down further increase, even in the presence of adequate food supply. Arrows pointing toward the left indicate influences that oppose overpopulation.*

well below the ceiling density and becoming reduced when the ceiling is reached.

The importance of a ceiling population density as a determining factor in the adjustment of animal societies to their environment is further emphasized in the case of bird societies, which resort to a pecking order to establish social hierarchy, to reduce breeding of excessive numbers on the same territory, and to encourage migration. Those that migrate may find other food or become victims of predation or lack of food. In experiments conducted at Aberdeen with Scottish red grouse, it was found that every season produced a population surplus, but the aggressive behavior of dominant males drove away the supernumeraries. The establishing of social hierarchy among animals sometimes tends to maintain a strong and healthy superior group at the expense of the subordinates, keeping the total numbers within acceptable ceiling limits. It is interesting that the processes resulting in population limits tend to become conventionalized. As stated by Wynne-Edwards: "Fighting and bloodshed are superseded by mere threats of violence, and threats in their turn are sublimated into displays of magnificence and virtuosity. This is the world of bluff and status symbols. What takes place, in other words, is a contest for conventional prizes conducted under conventional rules. But the contest itself is no fantasy, for the individual losers can forfeit the chance of posterity and the right to survive."†

It is by such limiting devices that animals are maintained in homeostatic‡ equilibrium with their numbers and environmental resources: sometimes by operation of an aggressive social priority; sometimes through hormonal changes that seem to be promoted by excessive population density and that react to discourage breeding; sometimes by cannibalism or inadequate care or desertion of the young. The restraints on increase are not the four listed by Darwin, but are automatic adjustments (feedback) within the society itself to cope with the numerous uncontrollable influences, including food shortage, climate, predation, crowding, and disease.

15.9 Population increase in human societies

The experiences of primitive societies of *Homo sapiens*, as shown in Fig. 15.5, differ in some important respects from those of the nonhuman societies of Fig. 15.4. Food is again largely obtained through hunting and gathering, but with some improvement in that even primitive man can cook and enjoy a larger variety of edibles. He knows how to build shelter for protection and to make clothing for warmth. There are few seasonal limitations on breeding. These all tend to encourage an increase in birth rate that bring about early crowding and early competition for food and privilege. The establishing of a "pecking" order takes violent form in the shape of wars, and disease may spread quickly within the tight assemblies of the tribes. There develop taboos and conventions that often tend to restrain breeding. Inadequate care of the young, desertion of the young, and even cannibalism are practiced on occasion. Despite these restraints, human societies reach states in which large numbers suffer from inadequate food supply, and large numbers suffer starvation in adulthood and childhood. Part of the reason may lie in the long period of development for growth from infancy to adulthood. Birth often occurs at a time when the environment favors breeding and multiplication, but conditions may change severely by the time the child has matured. The effect of this delay in feedback is to increase greatly the tendency to overshoot the control point and induce severe hunting.

Figure 15.6 is probably more representative of the social conditions that prevail at the present time. Here man utilizes agriculture and technology to increase his food supply. The upper portion of the figure shows a general contraction of the food available to each individual as a result of population increase. The reduction is interrupted at two places in the diagram, to represent expansion of available food supply as a result of invention and new technology. It appears that a boost of food supply also often encourages a boost in the birth rate, at least in the early stages of organized society. There also tends to be a boosting

---

† See Wynne-Edwards in References.
‡ That is, there is tendency for populations to maintain a steady state.

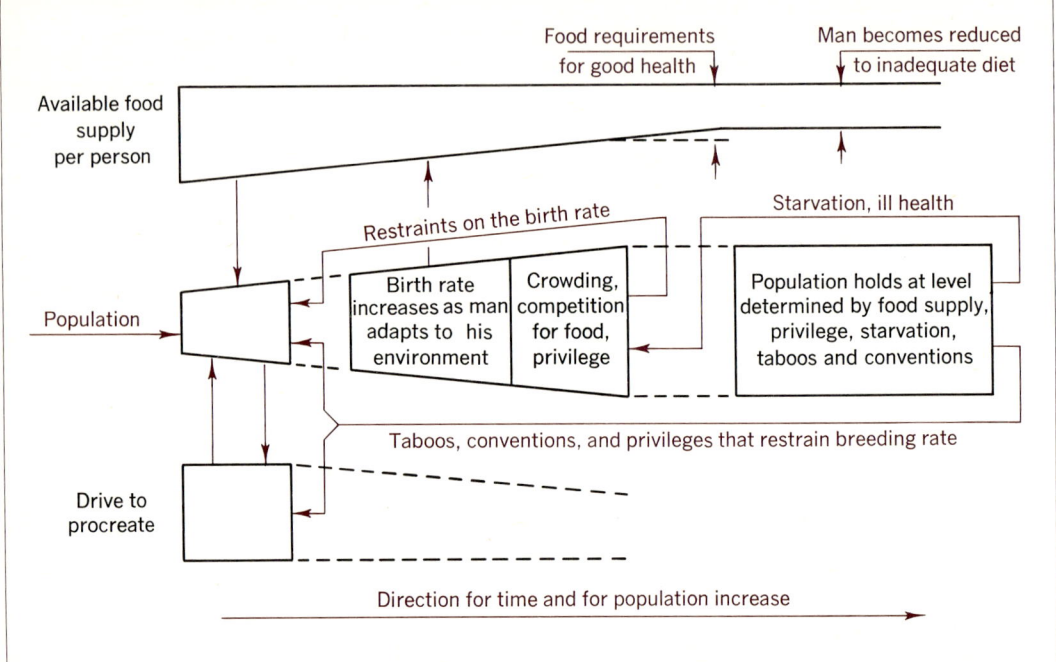

Fig. 15.5. Population practices among those primitive human societies that depend largely on collection of food. Development and adaptation to the environment (shelter, clothing, cooking) encourage increase in population. Crowding results in competition, wars, ill health, repudiation of members, and even cannibalism; all these reduce the population. Taboos and conventions tend to control birth rate. Starvation is common.

of the birth rate due to control of disease, improvement of environment to assure greater safety and longevity, to political influences, and to religious influences that place sanctity on numbers and on all useful "creations of God."

The diagram shows a branching in two general directions as the mass of humanity expands in numbers. One portion finds that the comforts of an opulent society can be achieved and retained only by restricting the size of the family. This portion exercises voluntary restraint to hold down the birth rate, and thus continues with an abundance of food available to its members. There is thereby achieved an equilibrium between population level and food supply. Oddly enough, it is usually the cultures that are highly advanced in technology, and therefore capable of greatly increasing their production of food, that tend to exercise this restraint on population. In societies such as those in the United States, the food needs of the entire population can be provided by engaging only a fraction of the population in agriculture. In the process, gluttony and waste of food have become characteristic of life in the United States, while the larger mass of humanity continues to increase and its food supply falls to starvation levels, much as was the case with the primitive societies of Fig. 15.5. In the present case, competition is tempered as well as defined by conventions that dictate humanitarian and religious restraints, except when there is resort to war. Education brings some awareness of cause and effect, and a sense for the needs of the future; but awareness seems to encompass only short-range goals and avoids looking to the risks of starvation that threaten the more distant future. Every humanitarian effort to increase food supply, reduce disease, improve conditions for healthy birth, assure healthy growth, protect each individual from harm, and assure a long life seems to work in the direction of promoting a greater calamity in the future.

What are the alternatives? Clearly the restraints to be cultivated do not lie in the practices of the animal world to prevent overpopulation. Just as clearly, the masses of people must be helped to ad-

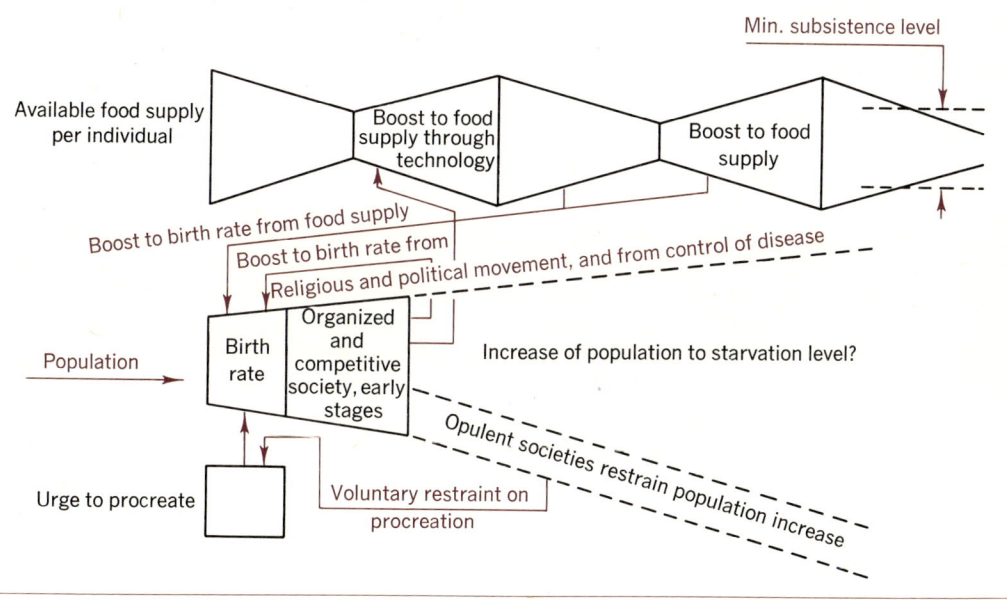

Fig. 15.6. Population trends among societies that produce food through agriculture. The total food production increases periodically through improved technology, but not enough to keep up with population increases. In many social groups the increase of food supply, successful treatment of disease, religious and political influences, all tend to increase population to the point of risking periodic starvation. Although the more opulent societies develop voluntary restraints on the birth rate, there persists an overall threat of overpopulation and large-scale starvation in the world.

vance beyond the present intermediate stage of culture and self-control toward effective and humane restraints on population increase. While the specific reasons offered by Malthus and Darwin as definitive of man's progress do not seem to hold, there is little doubt that the future of our society is uncertain if the population growth is not arrested.

15.10 Summary

Not only are living forms adapted to the environment of the earth, but the biosphere, particularly in the distribution of its chemical elements, is peculiarly well fitted to the support of life. The three elements carbon, hydrogen, and oxygen —which possess the greatest number of compounds of any three elements, as well as the greatest power of combining with other elements—are widely distributed over the face of the earth. This distribution occurs not only because of the presence of oxygen and carbon dioxide in the atmosphere and hydrogen in the oceans, but also because of the unique properties of water and carbon dioxide.

Water is unique among fluids in many of its properties, each of which contributes to its peculiar adaptation to the support of life. To mention a few: its ability to dissolve a wide variety of compounds; its specific heat; its density changes under variation of temperature; its expansion when frozen; its surface tension (so important for ensuring proper soil condition for vegetation); its latent heats of vaporization and fusion. In each one of these respects, water is very nearly unique among all compounds. The properties it has are eminently suitable to ensure its distribution over the land areas of the earth, to contribute to the stability of the environment, and to enter into the many processes of life.

Carbon dioxide also has unique properties that ensure its distribution over the earth at a fairly constant and necessary level in the atmosphere. Neither is it completely washed out of the air and dissolved by the oceans, nor does its relative abundance increase by its coming out of solution beyond an equilibrium level. It is held within narrow limits to perform its function of controlling climate and of feeding plant life.

Three cycles are studied in this chapter —the carbon and oxygen cycles, and the nitrogen cycle. They should be regarded as processes of energy and chemical conversions with built-in controls ensuring, under natural and normal conditions, the long-run possibility of survival of life in a tremendous variety of species.

Because of man's power to affect the environment physically and chemically, and on a tremendous scale, attention has been called in this chapter to the danger of putting into effect irreversible changes that will destroy the present self-adjusting system in which we live. We are apt to take it for granted that nothing can permanently destroy a system that has maintained its stability, with only mild and usually local fluctuations, over a billion years. But there is enough evidence to show that our industrial development is placing us at the threshold of danger. Scientists and statesmen must act on an international basis, for if they do not find a solution, within a relatively short time our environment may be so changed that human life and possibly animal life can no longer survive. An irreversible change with positive feedback might easily be inaugurated.

There is a close interaction not only between living forms and the physical environment but also between living forms themselves in the areas of their biological and social systems. Here also we have noted self-adjusting controls that provide an ecological balance among species. The exponential growth of population of most species in nature is controlled by many curbs, internal as well as external. It is thus rare that an animal or a vegetable species grows in number beyond a certain point. But here again, man, using the scientific technologies (including medicine, agricultural practices, and industrial productivity), has interfered with nature on a grand scale. The exponential growth of the population of mankind has not been slowed down appreciably; misery and starvation faces future generations unless some new controls are provided. In the long run, overpopulation is possibly more of a threat to mankind's future welfare than is the threat of a nuclear holocaust (and note that the two threats are closely related). Again, every scientist, statesman, religious leader, and citizen must cooperate soon if this challenge is to be met.

To alleviate these fears and potential dangers there are those who suggest that, with care, society can provide for the food needs of much larger populations than we now have. The seas hold vast quantities of food possibilities, much more land can be made arable, and the waste of food can be decreased by reducing our standard of living. But those who must cope with the problems of government, of energy sources to power cities and industries, and with problems of waste disposal find little comfort in this reasoning. To them the words of Harrison Brown† bring a fearful threat:

"If we were willing to be crowded together closely enough, to eat foods which would bear little resemblance to the foods we eat today, and to be deprived of simple but satisfying luxuries such as fireplaces, gardens and lawns, a world population of 50 billion persons would not be out of the question. And if we really put our minds to the problem we could construct floating islands where people might live and where algae farms could function, and perhaps 100 billion persons could be provided for. If we set strict limits to physical activities so that caloric requirements could be kept at very low levels, perhaps we could provide for 200 billion persons. . . ."

The problem of population control is not without its moral and teleological uncertainties, however. Whether one regards nature with cold "scientific" analysis, or finds excitement in its immense variety, its awe-inspiring upheavals, its order, its push for life and yet more life, the question is unavoidable as to the "why" of nature. The question takes on even greater force when addressed to the "why?" of man's existence. The reply of earlier generations, which avoided more than answered the question (namely, that man and nature exist for the glory of God) is no longer sufficiently complete or satisfying; and yet no other answer has supplanted it.

Questions/Discussions

1. (a) Suppose that the average temperature of the biosphere were to drop by 5°C. What would be the expected effects on plants and on people living in the present temperate zone and in the torrid zone? (Discuss in terms of variety and quantity of plant life, and in terms of patterns of civilization.)

† See Brown in References.

(b) What effects would be expected if the temperatures were to rise by 5°C?

2. Several correlations have been found between weather conditions and human behavior. Examine each of the following and comment as to whether the effect is due to temperature or to some other aspect of the climate. Note also whether your personal experience confirms or contradicts any of these correlations.

(a) Authoritarian governments are more prevalent in warm climates.
(b) Warm weather tends to make people more excitable.
(c) Crimes such as assault and drunkeness tend to increase in the fall season.
(d) The percentage of people with high I.Q. (intelligence quotient) is greater in the upper than in the lower latitudes.
(e) A greater degree of individualism exists in cold climates than in warm.
(f) Children of cool climates tend to have more respect for parents than do children of warmer climates.

3. The natives of the islands Tierra del Fuego, located near Cape Horn, used to go about naked except for use of some shields against the wind. If geography determines the pattern of living, explain why they did this, and why they now wear clothing and build houses.

4. Against the fears of "prophets of gloom" who see the population explosion to be leading directly to shortage of food and fuel and living space, there are "optimists" who argue that man in the past has always found solutions to problems of survival and will do so again. Discuss the realism of the two positions and the arguments you would give *against* each.

5. Having in mind the problem of depletion of energy resources, discuss what might be alternatives to our present pattern for utilization of oil and coal.

6. Since solar energy seems likely to continue without significant depletion in the several thousand years to come, discuss what might be the pattern for utilization of solar energy in the future. (In this connection we should note that, thus far, the recovery of solar energy by use of reflectors to concentrate the radiation has been impractical because the large reflectors are expensive to build and to maintain.)

7. If you were in a position of adequate authority in the United States, what would you undertake to do with respect to the following:

(a) Population numbers
(b) Energy sources
(c) Waste disposal
(d) Food production

8. Repeat your analysis of Question 7, assuming you are now planning for India.

9. Repeat your analysis of Question 7 for Egypt.

10. Suppose that the earth's human population continues to double every 33 years (approximately) and is now (1968) 4 billion.

(a) What will be the approximate number of human beings on earth in the year 2000?, 2100?, 2500?
(b) If the earth can support 64 billion people, at approximately what date will the population exceed this number?
(c) Suppose we attempt to solve the population problem by shipping

people to distant stars. Let there be assumed to be one habitable planet in every 1000 cubic light-years of space. What total volume of space will be needed to contain mankind (assuming both the earth and distant planets can sustain 64 billion persons in the year 2200? 2500? 2800?

(d) Assuming the volume calculated in (c) is a sphere centered on the earth, how far away from earth (in light-years) is its outer edge in the year 2500? 2600? 2700? 2800? How much did the radius change in the 100 years between 2600 and 2700? What conclusions can you draw?

References

Borgstrom, George, *The Hungry Planet.* New York: Macmillan, 1965.

Bresler, Jack B. *Human Ecology;* Reading, Mass: Addison-Wesley, 1966. (Collected readings.)

Brown, Harrison, *The Challenge of Man's Future.* New York: Viking, 1954.

Carthy, J. D., and Ebling, F. J., *The Natural History of Aggression.* New York: Academic Press, 1964.

Davis, Kingsley, "Population," *Scientific American,* **209** (September 1963), p. 62.

Hardin, G., *Biology, Its Principles and Implications.* San Francisco: W. H. Freeman, 1961.

Hardin, Garrett (ed.), *Population Evolution, Birth Control: A Collection of Controversial Readings.* San Francisco: W. H. Freeman, 1964.

Hoagland, Hudson, "Mechanisms of Population Control," *Daedalus* (Summer 1964).

Huntington, E., and Cushing, S. W., *Principles of Human Geography.* New York: Wiley, 1934.

Loebsack, Theo, *Our Atmosphere.* New York: Pantheon (a Mentor paperback), 1961.

Odum, Eugene P., *Ecology.* New York: Holt, Rinehart and Winston, 1963.

Perpillon, Aimé Vincent, *Human Geography.* New York: Wiley, 1966. (*This volume counters the deterministic view of Huntingon.*)

Sax, Karl, *Standing Room Only* in Hardin, *Population Evolution, Birth Control, ibid.,* p. 246.

Silliman, R. P., and Gutsell, J. S., "Experimental exploitation of fish populations," *U.S. Fish Wildlife Service Bulletin 58,* 1958.

Whittlesey, D., *Earth and the State.* New York: Holt, Rinehart and Winston, 1938.

Wynne-Edwards, V. C., "Self-regulating Systems in Populations of Animals," *Science,* Vol. 147 (Mar. 26, 1965), p. 1543.

CHAPTER SIXTEEN

Transition from the Classical to the Atomic Period

Newton, forgive me; you found the only way which, in your age, was just about possible for a man of highest thought— and creative power. The concepts, which you created, are even today still guiding our thinking in physics, although we now know that they will have to be replaced by others farther removed from the sphere of immediate experience, if we aim at a profounder understanding of relationships.

ALBERT EINSTEIN†

ADOPTION OF THE empirical approach to nature (that is, pursuit of knowledge by experimentation and observation) paid off handsomely in terms of new knowledge. With few exceptions (the theory of relativity being one of them) the experiments were relatively close to the common experiences of man, even though the phenomena were not well understood. For example, although the cause of gravitational and of electrostatic attraction could not be explained, nor could the molecular motions attributed to heat and temperature be visualized, the phenomena involved were nevertheless so intimately a part of daily life that it was not difficult to accept the theories and laws without severe reservations. (In fact, if it were somewhat more difficult to accept the theories as being only theory, there would be less inclination to assume that we know all there is to be known about these phenomena.)

16.1 The microscopic picture thus far

But now we come to a period of history, and to phenomena, wherein experiments become more and more removed from personal daily experiences, namely, the realm of atomic phenomena. That is, the properties we observe and feel in a desk or in a piece of paper require explanations in terms of atomic structure which will seem far removed from what can be seen or felt. We must again note, however, that were it not for habituating effects on our thinking caused by daily experiences, we would recognize that gravitational attraction also must have explanation in atomic phenomena that

† From Paul A. Schilpp (ed.), Albert Einstein, *Philosopher-Scientist* (autobiographical notes), Vol. 1. Harper (Torchbook), New York, 1959, p. 33.

cannot be visualized or felt and for which there are not even satisfactory theories.

With the important exception of the theory of relativity, nearly all scientific theories presented thus far came into being prior to the beginning of the twentieth century. The several decades prior to this century felt the flush of complete success because it seemed at the time that all the important concepts of science had been revealed. The head of the United States Patent Office resigned his position because it seemed to him that even the field of inventions was about to run dry, and before long he would be without any work if he remained in that office. Newtonian mechanics, and the simple relations explaining electricity and light, offered a fairly complete explanation of natural phenomena, all of which had found their proper application. Or so it seemed.

It is true that the invisible phenomena that had to be explained in terms of atoms offered some difficulties. As we saw in Chapter 10, the explanations of temperature and of heat energy depended on statistical analysis and probability concepts involving large numbers of atoms. But this was not too difficult to do because, for all anyone knew, the tiny invisible atoms did not appear to have properties that were much different from those of their visible aggregates. That is, there did not seem to be any objection to applying equations such as: kinetic energy $= \frac{1}{2} mv^2$, whether the symbols represented atoms or aggregates. There was no reason to expect the atoms to be much more complicated entities than dust particles, except for size, although the atoms had been organized in a periodic table.

The time was ripe for a change, however. A great deal of experience and knowledge had been accumulating in two areas, namely, chemical properties of atoms and molecules, and the behavior of charged particles. The interrelationship of these areas, as well as the properties of light, had not yet become revealed.

In this chapter we discuss some of these developments as an introduction to the new age of atomic science, which began early in the twentieth century and which will be discussed in succeeding chapters. The new era was ushered in with concepts that were quite beyond, and seemingly in contradiction to, common experience. Perhaps a little anticipation of the strange and contradictory features that appeared at the time will help prepare the reader to understand those features with reasonable ease when we treat them in greater detail.

16.2 Atomic science in the early nineteenth century

In his *Opticks* (1704) Newton observed: "It Seems Probable To Me, that God in the Beginning form'd Matter in solid, massy, hard, impenetrable, moveable Particles, of such Sizes and Figures, and with such other Properties, and in such Proportion in Space, as most conduced to the End for which he form'd them; and that these primitive Particles being Solids, are incomparably harder than any porous Bodies compounded of them; even so hard, as never to wear or break in pieces; no ordinary Power being able to divide what God himself made one in the first Creation."[†]

This view of the atom, which was not uncommon even in the nineteenth century, made it easy to apply the same rules with respect to collision between atoms

[†] From Sir Isaac Newton, *Opticks* (Dover, New York, 1952, pp. 400–405). Quoted also in Boorse and Motz. The reader will find interesting reading in these volumes.

as were applied more visibly to billiard balls. The kinetic theory of gases emerged from this concept, and the chemists of the period did not find the view altogether unsuitable.

The Newtonian view allowed the atoms to be ". . . in perpetual motion." In his *Principia*, Newton suggested that the existence of a repulsive force, which is inversely proportional to the separation of centers of the atoms, would give a gas the property of having density proportional to pressure. This idea is reflected in Boyle's law for the volume-pressure relationship in gases.

But not all the people of the somewhat later period could accept the same degree of hardness for the atom, since this meant that the collisions between atoms would occur during an infinitely short time of contact (therefore extremely high rate of change of momentum).† Roger Joseph Boscovich (1711-1787), a Yugoslav who helped to promote Newtonian mechanics in Italy, the country of his mother, conceived that atoms experienced a series of forces, now repulsive and then attractive, with increasing separation until at large separations there was only gravitational attraction. These surrounding spheres of attraction and repulsion seemed to offer more flexibility to account for the behavior of atoms in solid, liquid, and vapor states.

A more productive view of atomic phenomena was introduced when, in the person of the Englishman John Dalton (1766-1844), " . . . fates sent an unprepossessing country schoolmaster of silent mien and uncouth manners, who, in a chance flash of revelation, caught a glimpse of an open door and passed through.‡ The son of a poverty-stricken weaver, Dalton's formal education continued into many years of teaching and the study of science and mathematics. He developed an interest in the properties of gases, and especially of "mixed gases" such as the nitrogen, oxygen, and water vapor of the atmosphere. Early in the nineteenth century he began to give fairly authoritative presentations on diffusion phenomena, partial pressures of gases, and the fact that different gases experience chemical combination in definite small number ratios.

Dalton's *New System of Chemical Philosophy*,† published first in 1809, presented the basic principles of chemistry and chemical physics. When atoms are in the *elastic state* (we now call this the *gaseous state*) each atom is separated by large distances from other atoms " . . . and supports its dignity by keeping all the rest, which by their gravity, or otherwise are disposed to encroach upon it, at a respectful distance.

"Chemical analysis and synthesis go no farther than to the separation of particles one from another, and to their reunion. No new creation or destruction of matter is within the reach of chemical agency. . . ." All the changes we can produce, consist in separating particles that are in a state of cohesion or combination, and joining those that were previously at a distance.

Dalton succeeded also in identifying about 20 elements according to relative atomic weights, and showed that simple ratios of numbers of these elements made up the compounds that were familiar, such as water and ammonia. The details of the weights and compounds he gave were not correct, as when he stated water to have one oxygen and one hy-

---

† Question: Explain why this is not a problem with collisions between billiard balls.

‡ Quoted in Boorse and Motz, p. 139. See also Greenway in References.

† Given in part in Boorse and Motz, pp. 147-156.

drogen atom, but the principles were sound and very important for the progress of chemistry and atomic science. Dalton himself, despite serious handicaps in his attitude toward people and society, and a combination of gruffness and awkwardness that was often repulsive, was highly honored by the leading scientific and educational institutions. It is also interesting that it was chemistry, rather than physics which was considered by many to be the more sophisticated science, that strongly promoted the concept of atomism in modern science.

16.3 The Mendeléev periodic table of elements

Dalton's theory stated that chemical compounds were formed by atom-to-atom union of the basic elements in definite weight proportions. The experimental work of the period was not always successful in demonstrating the proportions clearly, however. Nor were Dalton's own experimental techniques very successful. It was Joseph Louis Gay-Lussac (1778–1850)† who demonstrated in 1809 that if gases enter into chemical reaction, they do so in numerically simple volume ratios, and the volume of the gaseous products may be expressed by simple integral numerical ratios to the volume of the original reactants. For example, 2 unit volumes of nitrogen, when reacted with 1 unit volume of oxygen produce 2 volumes of nitrous oxide (N_2O). Or 1 volume of nitrogen reacting with 3 volumes of hydrogen produce 2 volumes of ammonia. The significance of the simple relationships was not clear, in part because Dalton's theory was barely emerging at the time and in part because there was confusion as to whether the reactant gases (hydrogen, nitrogen, and oxygen) existed originally as atoms or as molecules (that is, 2 atoms of hydrogen combined as a hydrogen molecule).

It was the Italian chemist and physicist Amadeo Avogadro (1776–1856) who made it possible to clear up the confusion. He analyzed the experimental results of Gay-Lussac and reasoned that the results could be explained if the volumes contained equal numbers of molecules. He then proposed two postulates: (1) *equal volumes of gases contain the same number of molecules when the gases are at the same temperature and pressure*; and (2) *the elemental particles that make up a gas may be made up of 2 or more atoms*. The hydrogen gas particle is actually made up of 2 hydrogen atoms (H_2), that of oxygen made up of 2 oxygen atoms (O_2), and so on. That is, these particles are *diatomic*. In counting the number of hydrogen atoms that participate in a chemical reaction of the type studied by Gay-Lussac, the 2 volumes of nitrogen represented 2 volumes of N_2 molecules (not of N atoms) combining with 1 volume of oxygen molecules (O_2, not of oxygen atoms) to form 2 volumes of nitrous oxide (N_2O).† Or 1 volume of nitrogen molecules (N_2) reacting with 3 volumes of hydrogen molecules (H_2) produce 2 volumes of ammonia (NH_3).† According to his first postulate, a unit volume of O_2 and a unit volume of N_2 each contains the same number of molecules. The products of these reactions, namely, N_2O and NH_3 molecules, again obey the same rule: A unit volume of N_2O and 1 volume of NH_3 would each have the same number of molecules as the unit volumes of N_2

---

† See Boorse and Motz for a brief account of his career and also that of Avogadro, who will be mentioned presently.

† That is:

$$O_2 + 2N_2 \rightarrow 2N_2O \quad \text{and} \quad N_2 + 3H_2 \rightarrow 2NH_3$$

TABLE 16.1 PERIODIC TABLE

Periods	IA	IIA	IIIB	IVB	VB	VIB	VIIB	VIII			IB	IIB	IIIA	IVA	VA	VIA	VIIA	0
1	H 1																H 1	He 2
2	Li 3	Be 4											B 5	C 6	N 7	O 8	F 9	Ne 10
3	Na 11	Mg 12				Transition Elements							Al 13	Si 14	P 15	S 16	Cl 17	Ar 18
4	K 19	Ca 20	Sc 21	Ti 22	V 23	Cr 24	Mn 25	Fe 26	Co 27	Ni 28	Cu 29	Zn 30	Ga 31	Ge 32	As 33	Se 34	Br 35	Kr 36
5	Rb 37	Sr 38	Y 39	Zr 40	Nb 41	Mo 42	Tc 43	Ru 44	Rh 45	Pd 46	Ag 47	Cd 48	In 49	Sn 50	Sb 51	Te 52	I 53	Xe 54
6	Cs 55	Ba 56	La 57	Hf 72	Ta 73	W 74	Re 75	Os 76	Ir 77	Pt 78	Au 79	Hg 80	Tl 81	Pb 82	Bi 83	Po 84	At 85	Rn 86
7	Fr 87	Ra 88	Ac 89															

Lanthanide Series	Ce 58	Pr 59	Nd 60	Pm 61	Sm 62	Eu 63	Gd 64	Tb 65	Dy 66	Ho 67	Er 68	Tm 69	Yb 70	Lu 71
Actinide Series	Th 90	Pa 91	U 92	Np 93	Pu 94	Am 95	Cm 96	Bk 97	Cf 98	Es 99	Fm 100	Md 101	No 102	Lw 103

[a] The heavy black line separates the metallic from the nonmetallic elements; the distinction, however, is not so sharp as shown. The names of the elements are listed in the Appendix.

or O_2 (assuming all the gases are at identical temperatures and pressures).

Avogadro's explanations and hypotheses were not immediately accepted, however, until they were advocated by the Italian chemist, Stanislao Cannizzaro (1826–1910). In 1860 Cannizzaro pointed out that the Dalton postulates (which proposed atom-to-atom combinations) had brought only confusion and disagreement, whereas the Avogadro emphasis on *molecules* seemed to give a consistent set of atomic weights.

The confusion of proportions of elements seemed finally to be explained, and led to the development in 1865 of a very important constant called *Avogadro's number*. This is the number of molecules that there are in 1 molar weight of any substance, and is 6.02×10^{23},[†] we saw in Chapter 10.

At the time it had been possible to assign relative atomic weights to the elements that had been identified. Dalton listed the weight of 1 for hydrogen, 5 each for "azote" (nitrogen) and for "carbone," 7 for oxygen, 9 for phosphorus, and so forth. These are quite different from the values that were developed later. Nevertheless, the round numbers raised the question: Was there a "primordial" element from which all others are made? Was hydrogen that element? William Prout, an English physician who was also skilled in chemistry, proposed in 1815 that there was such an element and that it was indeed hydrogen. *Prout's hypothesis*, as it was called, inspired a great deal of effort in search of proof of the theory. The effort was to no avail, as we shall see, but it did lead to a great deal of new and useful understanding of atomic properties and reactions.

We shall jump over the intervening period, however, to the time of Dmitri Ivanovich Mendeléev (1834–1907),[†] a Russian chemist. The fourteenth child of the director of the gymnasium at Tobolsk, Siberia, he had the most fortunate influence of a mother whose dying admonition to him was: "Refrain from illusions, insist on work and not on words, search patiently divine and scientific truth." He became interested in physical science, and continued his studies of the physical constants of chemical compounds in Heidelberg.

Mendeléev's studies of the weights and chemical properties of the elements led him to seek a plan or "periodic law" for the relationship of the elements to each other. It had been recognized for some years that certain elements behaved quite alike in chemical reactions, although they differed greatly in their atomic weights. Chlorine, bromine, and iodine were such elements. Lithium, sodium, potassium were another set of "triads." In 1869 Mendeléev published a paper on "The Relations of the Properties to the Atomic Weights of the Elements." In this paper he arranged in vertical columns all the elements that had similarities in chemical properties, beginning with the lightest element at the top of each vertical column. When

---

[†] Recall (Chapter 10) that a molar weight refers to the weight in grams equal in number to the atomic or molecular weight of the substance.

[†] Boorse and Motz, pp. 298–302, give the life story of Mendeléev in the form of the obituary notice that was published in 1910 by the Royal Society of London. The above quotations are from this book. Mendeléev's mother operated a glass factory in which he spent much time. While his father was alive, their home was something of an intellectual and cultural center. There is available a fictionalized but fairly correct biography of Mendeléev with respect to essentials, written by D. Q. Posin, McGraw-Hill, New York, 1948.

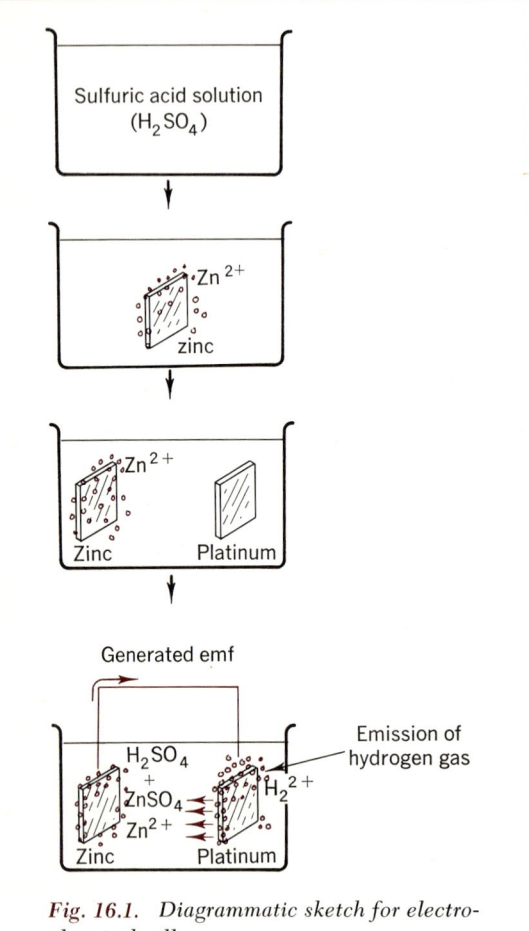

Fig. 16.1. *Diagrammatic sketch for electrochemical cell.*

the columns were arranged next to each other according to the progression of atomic weights, they seemed to place the elements in the order of their valencies, or the number by which atoms combine with other atoms (that is, a valency of 2 permits an atom to join with two atoms that have valency of 1 each). Table 16.1 illustrates a much more advanced form of the Mendeléev table.

What caused the periodicity of the elements? This question could not be answered until the introduction of the newer atomic science of the twentieth century.

16.4 The "fourth state of matter" and the electron

The early studies of atomic structure made very little provision for the electrical characteristics that might be associated with atoms.

While the chemists and others were busily developing the details of the elements and their combinations, a parallel effort was under way to develop more understanding of electrical phenomena. The early developments involving electric charges, magnets, and electromagnetic fields have been presented in Chapter 11. The efforts of two Englishmen, the experimentalist Michael Faraday (1791–1867) and the mathematical physicist James Clerk Maxwell (1831–1879), were especially fruitful because they complemented each other very effectively.

To a considerable degree, Faraday's education was limited to what he could teach himself. His insight into the basic nature of problems and his ingenious design of experiments helped him to become a very productive physicist, although his work included very little mathematical analysis. His experiments on electric currents and the electromagnetic field, and his discovery that electric currents can be induced from one circuit to another circuit through electromagnetic fields, laid the foundation for Maxwell's analyses.

Faraday's work on electrochemistry revealed some remarkable phenomena. When a zinc plate is placed in sulfuric acid, some of the metal is lost into solution. The presence of a second metal such as platinum in contact with the zinc produced even more reaction, including the emission of hydrogen gas at the platinum metal (cathode), accompanied by loss of weight by the zinc plate. When the platinum and zinc were placed separately in the solution and connected by a wire and measuring device, it was observed that the assembly of the two metals in the acid had become a source of electric current—a battery or voltaic cell. It seemed that the hydrogen gas was the result of decomposition of water molecules, but the mechanism of the process was far from clear. Somehow the zinc also went into the solution in proportion to the amount of hydrogen gas that was evolved. (See Fig. 16.1 and Fig. 16.2.)

He repeated the experiments more carefully, collecting the evolved hydrogen gas in a jar and carefully determining the loss of zinc. There seemed to be a very definite quantitative relationship between the two weights. He measured the amount of electric current that passed through the voltaic cell and found that the loss of weight in the zinc plate, the weight of the evolved hydrogen, and the quantity of electric current that flowed in the process were all related. He came to the conclusion that the electricity produced in a voltaic cell is the product of chemical decomposition or of other chemical action. That is, even when the

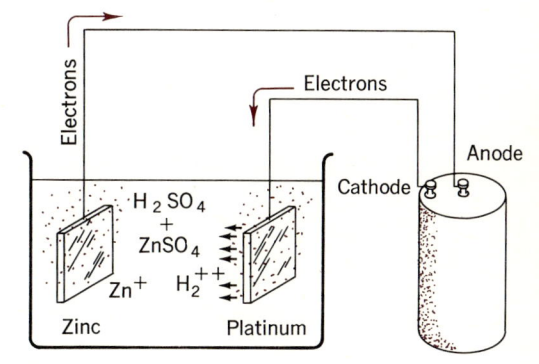

Fig. 16.2. *Faraday's laws of electrolysis. Faraday observed that, for electrolytic processes in which elements are the only products, (1) the weight of an element deposited (released) is proportional to the amount of electricity passed through the solution; (2) the weights of various elements deposited (released) by a given quantity of electricity are proportional to their combining weights; (3) for zinc two units of charge (valence + 2) are required for each atom deposited and thus the combining weight (65 g/$2 = 32.5$ g of zinc) deposited requires one Faraday of charge ($96{,}500$ coulombs).*

solution was varied, the same quantity of material became decomposed when an electric current of a given amount flowed for a fixed period of time. Apparently each ion of a particular kind had exactly the same electric charge. Moreover, he could postulate that electric energy held the atoms together and that they could be separated again if an equivalent electric energy could be applied to separate them.

Faraday also discovered the principle of electromagnetic induction, which is the basis of a device called the spark coil: Consider a "primary" coil of wire to be connected to a source of electric current, and a "secondary" coil consisting of very many more turns of wire to be wrapped around the "primary" coil. When the current in the "primary" coil is interrupted, a high voltage is generated in the "secondary" coil (see Fig. 11.20). When this "secondary" coil has enough turns of wire, the voltage generated becomes high enough to cause a current to jump across an air gap (or spark gap between two terminals). The jump of the spark can be very vivid, colorful, and noisy. Investigators who studied the phenomenon of spark discharge found that very interesting results could be obtained when the spark passed not through air but through glass envelopes in which various gases could be contained at high or low pressures. It was now possible to study electric currents through such "discharge tubes" with more care and control. Professor Julius Plücker of the University of Tubingen studied the discharge phenomenon and discovered that when the gas pressure is low enough, radiation emanates from the cathode (negative pole) of the discharge tube, which on striking the glass walls produces a greenish phosphorescence. The rays came to be known as "cathode rays."

The English scientist William Crookes (1832–1919) studied these cathode rays and ascertained that the rays were electrified particles that were projected at high speed from the cathode of the tube. The particles could be deflected by magnets that were brought close to the tube. The green phosphorescence of the glass was the result of these particles striking the glass wall. (He, like several others, just missed being the discoverer of X-rays, which are produced when charged particles of high energy strike a target).† He found that the glow that appears inside the discharge tube can take many forms as the pressure of the gas and the nature of the gas are varied.

To the solid, liquid, and gaseous forms of matter, there was added a new state, which we today call the *plasma*. It constitutes a very important field of study involving electrically charged particles of all kinds. In the words of Crookes, this new state is:

". . . a world where matter may exist in a fourth state; where the corpuscular theory of light may be true, and where light does not always move in straight lines, but where we can never enter, and with which we must be content to observe and experiment from the outside."‡

---

† William C. Roentgen, while working with a Crookes tube which he covered by a shield of black cardboard, noticed a "peculiar black line" on a piece of barium platinocyanide paper that was on the bench nearby. He guessed that the line was due to something emanating from the tube. These were later determined to be penetrating radiation called X-rays. Other investigators had noted that photographic paper became fogged by nearness to Crookes tube, but gave it no further attention.

‡ From Sir William Crookes, *Philosophical Transactions*, 170 (1879), pp. 135–164, reported in Boorse & Motz, pp. 351–362.

Another Englishman, whose long life and many contributions to science exemplified the life of a productive and very human scientist and leader of young research workers, J. J. Thomson (1856–1940), continued the Crookes experiments on the cathode rays with much more attention to quantitative details. The rays could be deflected easily by a magnet placed close to the tube as we saw above. Thomson introduced parallel metal plates into a glass tube containing a source of cathode rays and found that a voltage impressed across the plates would also deflect the cathode rays, confirming that they were charged particles. The direction of the deflection in both electric and magnetic fields indicated that the charge was negative.

He then undertook a series of experiments in which he filled the tube with various gases such as air, hydrogen, and carbon dioxide. He found that the amount of deflection of the negative particles seemed to indicate that their ratio of mass to charge (m/e ratio) was always the same. Clearly the negative particles were not gas molecules, such as hydrogen, in which the ratio m/e was quite different from that of the ratio for electrons. Possibly the negative particles constituted a subdivision of ordinary atoms that played the role of the primordial substance of Prout's hypothesis. The new particle was about 2000 times less massive than the lightest (hydrogen) atom. It was, in fact, the *electron*, which from now on will be a very important part of our concern.†

---

† Thomson's work, as reported in *Philosophical Magazine*, 44 (1897), pp. 293–311, is reproduced in part in Boorse & Motz, pp. 416–426. Because the measurements of e/m gave the ratio and not the absolute value of e or m, the confirmation of the charge and mass

16.5 Force and energy: concept of causality

For the remainder of this chapter we shall change our approach from discussion of specific topical areas to a discussion of their interrelationships. We shall also jump a little ahead of our story, to introduce some of the major apparent contradictions and unanswered questions that face scientists at the present time. This is being done with the hope that early introduction to the dilemmas of our day will make it somewhat easier and more interesting for the reader to trace the route by which we have arrived at this point.

We noted in Chapter 13 that an important concept in Newtonian physics was that of *causality*. According to this concept, for every observable *effect* there exists a *cause*, and conversely, every causal action produces a definite effect or set of effects. Thus, when a body of mass m is observed to be moving with an acceleration a (effect), it can be concluded that a force f is acting on the body (cause). Conversely, if we impress a force f on a body of mass m (cause), we expect that an acceleration a in its motion (effect) will result.

The concept of causality was not new to Newtonian physics or to the mechanistic philosophy that came with it. It is an embodiment of the idea that nature is "rational," an idea that had existed for more than two millenia and possibly grew out of a human impression that there must be a "reason" behind every observed action. Like any assumption, the existence of causality cannot be proved; nevertheless, it should be noted

---

of the electron did not come until the American physicist Robert A. Millikan published the results of his oil drop experiments in 1911.

that most of the progress in science during the past three centuries has come from a search for causal relationships in nature.

When the phenomena observed and the actions that give rise to them both exist on the macroscopic level, the causal relationships are generally not too difficult to establish. Thus, when a man throws a ball (action or cause) and the ball crashes through a window (effect), we can easily trace the sequence of events and establish the causal relations we seek. Even when we do not observe the man actually throwing the ball, we feel justified in concluding from the observation of the ball crashing through the window that it was indeed thrown (or batted or kicked, all of which, for this example, are to be considered as forms of throwing) by someone. It is a little more difficult to trace the causal relationships in situations where the observable effects on macroscopic bodies are due to causes that are not visible, such as that due to gravitational attraction.†

The search to define the nature and properties of light has also been difficult. Newton thought that light was made up of particles (corpuscular theory). The Dutch scientist Christian Huygens (1629–1695) proposed that light had a wave nature. Later experiments seemed to prove Huygens to be correct, since the wave concept readily explained all the interference and diffraction phenomena (as well as refraction and the slower speed of light when it travels in a dense medium like water) that were observed.

The study of charged bodies and of electricity and the electric current added the concept of electric field. A current flowing in a wire produces a magnetic field around that wire, the magnetic field seeming to always have a direction at right angles to the electric field and to the direction of the moving charges (electrons) that make up the current. A changing magnetic field, in turn, *induces* an electric field. It was Maxwell who gathered together the various laws of electricity and magnetism to produce a single, coherent electromagnetic theory. In Maxwell's theory, an oscillating, charged particle gives rise to both electric and magnetic fields, which also vary with time. The direction of motion (or propagation) of this radiation is at right angles to both the electric and magnetic fields that constitute it. Further, Maxwell's theory showed that both the electric and magnetic fields vary sinusoidally with time to produce a *transverse wave*. Depending on the frequency of the oscillations of the source electrons, the electromagnetic waves are observed as radio waves, light, X-radiation, and other similar phenomena. The energy in an electromagnetic wave could be shown to be proportional to the square of the amplitude of the electric or magnetic field intensities.

Thus, the great achievements of classical physics—Newtonian mechanics and Maxwellian electromagnetic theory—were built upon the concept of causality. Today's vast industrial complex, which sustains our standard of living, is to a great extent the technological offshoot of classical physics. All natural phenomena seem causally related to forces and energy transformations. It was a rationally determined world in which some people of the Victorian era saw assured progress for mankind. Man seemed to be on the verge of a better world through the harnessing of nature's resources.

---

† The reader is referred to Max Born's excellent essay, *Natural Philosophy of Cause and Chance*.

16.6 Energy as matter or motion

What are the similarities and distinctions between energy in the form of a moving ball and electromagnetic radiation such as light? Both can be measured in terms of the energy content. In the case of the ball we know that the kinetic energy is given by the relationship $K = \frac{1}{2}mv^2$, where m is the mass and v is the velocity. For the present we shall disregard the internal energy represented by the mass itself. In the case of light and other electromagnetic radiation, the energy is a constant times the square of either (or of both) the electric or magnetic field intensities E^2 or H^2, respectively. The units for expressing both may be ergs, calories, joules, foot-pounds, or any other appropriate unit.

But there seemed to be a major difference between the two forms of energy, mechanical and electromagnetic. In the first category (as exemplified in the thrown ball), mass or matter is always present. The matter may go up in smoke or in a chemical reaction. It may be moving or stopped, or changed from solid to liquid, but the energy considerations usually include material substance. In contrast with this category, the energy associated with electromagnetic radiation has no mass associated with it;† there is only the electromagnetic field, which has wave characteristics and the speed of light.

In any case, the Industrial Revolution and the related technological work that gave it impetus had little difficulty in utilizing both forms of energy. In some applications the mechanical energy could be converted to electromagnetic energy,

† We are speaking in this section from the point of view of prerelativistic, classical physics.

as when a filament or gas was heated to emit light. Through photosynthesis, light could give its energy to form solid matter. Energy could be transformed from one form to the other. But there was little doubt that the electromagnetic energy contained in light was without mass or that when energy did involve material substance there was total absence of the wave characteristics of electromagnetic radiation.

At this point we must jump a little ahead of our story to note that the coming of the nuclear age and of nuclear energy demonstrated that mass can be converted into radiation energy, and radiation energy into mass. As we shall see in Chapter 21, these nuclear transformations have become important sources of electric energy, to meet the needs of communities and of industries.

16.7 Duality of matter and waves

This separated and well-defined distinction between energy involving mass and energy in the form of electromagnetic waves continued until the beginning of this century. At that time, studies of atomic reactions and atomic structure naturally led to more detailed studies of the behavior of small particles. During the 1890's, J. J. Thomson demonstrated that all chemical substances contain electrons, which carry the negative charge and have a mass that is very much smaller than the lightest atom. These electrons are ejected from the surface of materials when light of the proper *frequency* (not intensity; what is the difference?) is shone on the surface. The light can be reduced to very low intensity, approaching zero, yet *if any* radiation of the *right frequency* reaches the surface, there may be one or more electrons emitted. How is it possible for light waves, which would be expected to spread their energy over a substantial area, to eject an electron at a point? The light acts as though all its energy is concentrated in a tiny mass, which is then able to knock out an electron by a conventional collision process. In 1905, Albert Einstein gave an explanation of this process: *While the light is an electromagnetic wave that exhibits all the interference and diffraction characteristics of waves, somehow in its interaction with tiny masses such as electrons, the full energy of the light wave seems to become available in discrete energy packets.* It is as if the light wave also had a concentration of mass to give it momentum for collision purposes. In other words, radiation (or photon, as he called it) seems to have a dual personality; it has both wave characteristic and the properties of a corpuscle (or particle with mass).†

Some years later it was demonstrated that when electrons pass through a tiny hole they develop most unusual diffraction patterns, similar to those made by electromagnetic waves (X-rays) passing through the same hole. The experiment demonstrated conclusively that matter (or mass) in the form of electrons does indeed behave as though it had wave properties in addition to its normal corpuscular nature. Moreover, if there are two adjacent small holes through which the electrons may pass, each electron that finds its way through one of the holes lands in a different spot on the other side, different from the position it would have found if the second hole were not present. In fact, with the two holes, the electrons will eventually form on a photographic plate an interference pattern similar to the interference pattern that might be produced by a light wave. If we explain the phenomenon in terms of the baseball, we arrive at an amazing result. Suppose we hurl a baseball through a small window to a catcher on the other side. He would catch the ball at the same spot as long as we did not change the direction and speed of throw. Now suppose there is another small window adjacent to the first window, but we continue to throw the ball in identical manner through the first window. We would find that the catcher would have to change his position to catch the ball. Somehow the simple *presence* of the second window introduces an interference phenomenon *that makes the ball seem to have a wave character to its motion.* Of course in real life we see no such deflection in the case of masses of baseball size, but the effect is very real for masses of the atomic dimensions. In the case of the electron, the very *presence* of a second hole or window definitely changes the direction of electron motion. The effect is not dependent on the particle's having electric charge because the same wave property is present in the case of the neutron, which has no charge.

One of the most important new concepts that physicists must investigate in the twentieth century is related to this apparent dual property of matter and light. The "dilemma" is complicated by conflicting evidence that experiments on atomic particles have brought against the whole concept of determinism, the subject we discussed in Chapter 13. These confusing facts will be dealt with in more detail in succeeding chapters, but a little prior discussion will help alert us to what is coming as we pass from the common experiences of the macroscopic world to the unseen phenomena of the

---

† It is interesting to note that Einstein was given the Nobel prize for this work and not for his theory of relativity.

microscopic atomic world. In this new world there will be an even bolder move toward uncertainty and indeterminacy as important aspects of natural phenomena.

16.8 Determinism in macroscopic and microscopic phenomena

Until a few decades ago, the attitude of scientists and philosophers toward the idea of determinism as expressed by LaPlace was as follows: Given the state of the universe at a given time, its state could be calculated at any future or past time. That is, when we know the position and velocity of all the particles of a system and also know the forces acting between the particles, we can (theoretically, at least) determine exactly what the next state of the system will be. This idea is now referred to as *classical determinism*.

The reason for the *classical* qualification is that a distinction is necessary because of more recent studies involving very small particles such as the electron. To determine where a particle will go from one instant to the next requires that we know the position *and* the velocity of that particle. This is not easy to do when the particle becomes very small. We can determine that an electron is located at a particular spot at a particular instant, but in the process of locating it, we tend to change its velocity. Or we can design an experiment to measure its direction and speed, but in the process we lose track of its exact position at any instant of time.

This lack of definiteness arises from the fact that our measuring devices utilize light or particles that are of the same order of magnitude in size and energy content as the particles we are measuring. This contrasts with the more common experience involving measurement of a house or of a baseball: the light we see by, or the meter stick we use, do not disturb the location or motion of the house or the baseball. The results would be drastically different if we had to collide a house with another house in order to determine its mass or other characteristics.

Because we do not have available to us subatomic measuring particles that are much smaller than the objects we are measuring, we face an impasse in the form of an *uncertainty principle,* which states that there is a limit to the amount of information we can have about atomic behavior. We shall discuss this in a little detail. The uncertainty principle seems to say that we can no longer follow in the strict detail relationships between cause and effect. We seem to be forced to reject rigorous determinism as a universal characteristic of all nature. That is, there are realms in which it does not apply. Therefore our world is not altogether determined and logically ordered, as was believed by philosophers such as Descartes and Spinoza who used classical physics as their model.

Uncertainty and indeterminism have very major physical and philosophical implications. If indeed the events that bear on atomic behavior are the result of chance rather than deterministic phenomena, there is no reason to assume that we can predict without serious error the ultimate state of even the larger bodies of the universe. Nor can we assume that chance phenomena do not influence biological processes, since these are at the atomic level. For example, can chance phenomena rather than determined processes play a part in mutation and the function of genes? One scientist has commented that, in view of the complexity of the genetic code (which dictates the characteristics of the embryo), it is a near-miracle that there is so little confusion in the process of reproduction. Another has pointed out, however, that there is some "garbling" of the code during reproduction in that only a relatively few of the fertilized eggs do develop, and of those, no two are exactly alike. This latter result, of course, makes it possible to promote individuality, differences, freedom from exact type, and (in time) evolution.

But now a question: Does indeterminacy also mean that there is anarchy and haphazardness in nature? In answer, some would argue that indeterminacy makes the notion of moral responsibility meaningful. Man has a choice between alternatives. But whether moral responsibility and the indeterminacy concept should be associated has yet to be answered.

These are some of the questions we shall bear in mind as we delve into atomic and nuclear science.

References

Anderson, David L., *The Discovery of the Electron*. Princeton, N.J.: Van Nostrand (Momentum book No. 3), 1964.

Boorse, Henry A., and Motz, Lloyd (eds.), *The World of the Atom*, Vols. 1 and 2. New York: Basic Books, 1966. *These two volumes provide a wealth of material. They include a good deal of biographical material about scientists we mention only casually and include many excerpts from their publications.*

Born, Max, *The Natural Philosophy of Cause and Chance*. New York: Dover, 1964. (Reprint of a 1949 publication.)

Gamow, George, *Thirty Years That Shook Physics*. New York: Doubleday, 1966. *This is a highly readable account of the topics presented in this chapter; available in paperback.*

Greenaway, Frank, *John Dalton and the Atom*. Ithaca: Cornell Univ. Press, 1966.

Ihde, A. J., *The Development of Modern Chemistry*. New York: Harper & Row, 1964.

Millikan, Robert Andrews, *The Electron*. Chicago: Univ. of Chicago Press, 1917. (Available as paperback.)

Nash, Leonard, *History of the Atomic Molecular Theory*. Cambridge, Mass.: Harvard Univ. Press, Harvard Cast Studies, 1950.

CHAPTER SEVENTEEN

The Birth of Modern Physics

MODERN PHYSICS was born early in the twentieth century when, as a result of certain experiments on the radiation given off by hot bodies, a new view of the nature of radiant energy gradually emerged. As we shall see, the "new view" required the adoption of some unexpected concepts of the structure and behavior of atoms, concepts which are the subject of concentrated study and questioning to this day. Before looking at these experiments and their results, let us briefly review the state of knowledge about light and other forms of radiant energy as it existed at the beginning of this century.

17.1 Radiation from charged particles

We begin with a question. How do we see objects? At one time it was thought that we saw objects by means of something sent out by our eyes. Now it is understood that we see a tree because light is scattered in all directions from the tree, some of it entering the eyes and generating the sensations that end up in our seeing that tree. Of course the original light probably came from the sun.

How does the light travel from the sun, through empty space, to the tree, and then bounce again from the tree into the eye? There seem to be two possibilities, identified as the *particle* concept and the *wave* concept. Each is accompanied by its own set of questions and difficulties. Is light like the tiny, discrete particles or molecules that travel swiftly through empty space? Or is light like the waves that travel across a tight rope or over the surface of water? If the latter, what medium performs the role of the rope or

† From his 1919 address before the Nobel Foundation. Quoted in Boorse and Motz.

There is one particular question the answer to which will, in my opinion, lead to an extensive elucidation of the entire problem. What happens to the energy of a light-quantum after its emission? Does it pass outwards in all directions, according to Huygens's wave theory, continually increasing in volume and tending towards infinite dilution? Or does it, as in Newton's emanation theory, fly like a projectile in one direction only? In the former case the quantum would never again be in a position to concentrate its energy at a spot strongly enough to detach an electron from its atom; while in the latter case it would be necessary to sacrifice the chief triumph of Maxwell's theory—the continuity between the static and the dynamic fields—and with it the classical theory of the interference phenomena which accounted for all their details, both alternatives leading to consequences very disagreeable to the modern theoretical physicist.

MAX PLANCK†

the water in what we call "empty" space. How empty is space? These questions have been asked many times over the centuries without finding clear answers.

The two concepts—one that light has the nature of particles or corpuscles, and the second that it is of wave character—have appeared off and on, in one form or another, since the time of the Greeks. Newton supported the corpuscular theory, while the Dutch physicist Huygens proposed the wave theory at the same period of time. When early in the nineteenth century Thomas Young and Augustin J. Fresnel demonstrated that light could experience interference and diffraction phenomena, they seemed to confirm the wave theory as the correct one, although the question of how the waves could travel in empty space remained unanswered.

Meanwhile, as a result of the work of Newton and of others such as Bernoulli, Euler, Lagrange, Cavendish, and Coulomb, the eighteenth century witnessed the development of a firm base of physical laws and mathematical techniques on which to build and test new ideas. The nineteenth century physicists were especially productive in improving understanding of electrical and electromagnetic phenomena.

The Scotsman James Clerk Maxwell (1831–1879) brilliantly demonstrated that light is an electromagnetic phenomenon. The wave nature of the light results from the rapid rise and fall of an electric field, which in turn gives rise to a similarly falling and rising magnetic field. The frequency of light is simply the frequency with which the fields increase and decrease. But what gives rise to the electric field and to light?

We recall from earlier discussions some data about charged particles:

(1) A charged particle that is *standing still* produces an electric field, the intensity of which decreases with the square of the distance from the charge (Coulomb's law).

(2) When that same particle is *moving at constant speed* in a wire, it produces a magnetic field around the wire (or around its path if it is traveling without a wire conductor) in addition to the electric field noted in (1).

(3) Whenever it is *accelerated*, this same charged particle emits a pulse of electromagnetic radiation. The frequency of the emitted pulse is determined by the boundary (starting and stopping) conditions imposed on the charged particle. In all the cases we shall consider, the electromagnetic radiation results from the transition of an atomic electron from one energy level to another, and the frequency of the radiation is determined by the frequency of oscillation of the electron. The intensity of the radiation is proportional to *the square of the product of* magnitude of charge times acceleration, and is inversely proportional to the square of the distance from observer to the source.†

In Chapter 5 we noted that when the oscillation frequency is low, of the order of 10^5 cycles/sec (cps) to 10^8 cps, the radiation is said to be in the radio-frequency range. The wavelength λ in this case ranged from 3000 to 3 meters (m) (from the relation wavelength, $\lambda = c/f$, where c is speed of light and f is frequency). The frequencies from $f = 3 \times 10^{11}$ cps ($\lambda = 0.001$ m) $f = 4.3 \times 10^{14}$ cps ($\lambda = 7 \times 10^{-7}$ m) are referred to as the microwave and infrared regions of the spectrum. Between a frequency of 4.3×10^{14} cps and 7.5×10^{14} cps ($\lambda = 7 \times 10^{-7}$ m to 4×10^{-7} m), the eye becomes sensitive to the radiation, which we then call light.

† The student is urged to review Chapter 11 in which these phenomena were discussed.

Still higher frequencies are unseen by the eye and are called ultraviolet light, X-rays, and gamma rays.

17.2 Radiation from hot bodies

So far we have reviewed only the topics that were discussed in an earlier chapter. Now we turn to a related subject, namely, radiation emitted by a hot body. The discussion will begin with concepts that are deduced from common experience and will then proceed to describe how careful measurements forced scientists to adopt some very radical new concepts, the end conclusions of which are even now not fully understood.

Our first example involves an iron poker and a fireplace. When cold, the poker simply looks and feels cold. When it is heated a little in the fire, it feels warm to the touch. When heated still more, it continues to look as black as ever, but now it sends out radiation which can be felt as heat at a short distance. With continued heating it begins to take on a dull red color, which becomes brighter and whiter as the heating continues. The heat can also be felt from a larger and larger distance.

What has happened to cause the poker to emit this radiation? Since its outside dimensions do not change significantly because of the heating, we must admit that the change must be inside the poker. That is, the atoms that make up the material of the poker must be responsible for the radiation. From our earlier discussion of heat energy, we can safely say that the heating has transferred enough energy to the atoms to raise them to a state of excited motion of some kind that produces radiation.

Next is an interesting phenomenon that we can demonstrate with our poker if we *surround it closely with hot coals*, allowing an opening only large enough to see the poker. The hot poker sends heat out to the hot coals, which in turn return as much heat to the poker, but now the metal will appear hotter than it did when it was not surrounded by the hot coals. Ideally, it would be better to use a short piece of metal that is *completely surrounded* by the hot coals, with only a small opening through which to peek

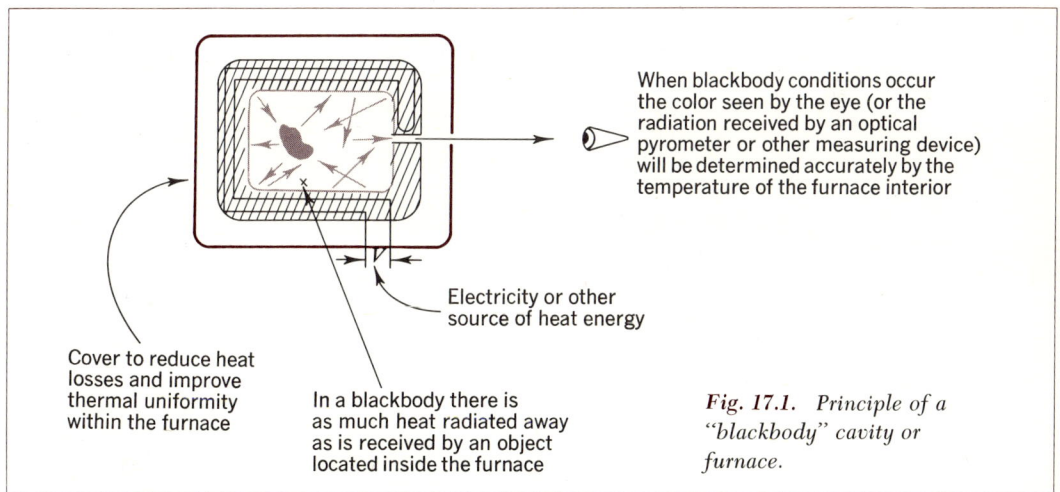

Fig. 17.1. Principle of a "blackbody" cavity or furnace.

(Fig. 17.1). Under these equilibrium conditions the color of the hot interior, the temperature, and the radiation that is emitted through the opening, are said to be *characteristic of a blackbody† of that temperature and completely independent of the material of the metal and the coals*. That is, the amount and the kind of radiation emitted by a blackbody are determined only by the temperature state of the blackbody. As described above, we created blackbody conditions simply by forming a cavity within which the radiation could bounce back and forth (without much loss to the outside) until equilibrium conditions became established. We mention this because the radiation laws as discussed herein assume blackbody conditions to exist at the source of radiation.

According to the kinetic theory of heat presented in Chapter 10, the molecules of matter are in a state of considerable excitement or motion, the energy of the motion being in fact the energy associated with heat of the body. Maxwell calculated that an excited assembly of particles in a blackbody would be expected to have a wide distribution of speeds of the particles (and therefore a wide range of kinetic energies, since the kinetic energy of a particle of mass m is $\frac{1}{2}mv^2$) such as is illustrated in Fig. 10.2. At any particular temperature state of a body, some of the molecules have low kinetic energy (or excited motion of some sort), some have relatively high kinetic energy, and many more have kinetic energy close to an intermediate region, such as is illustrated by the peaks of the curves of Fig. 17.2. As the temperature is raised, the whole *spectrum* of speeds,

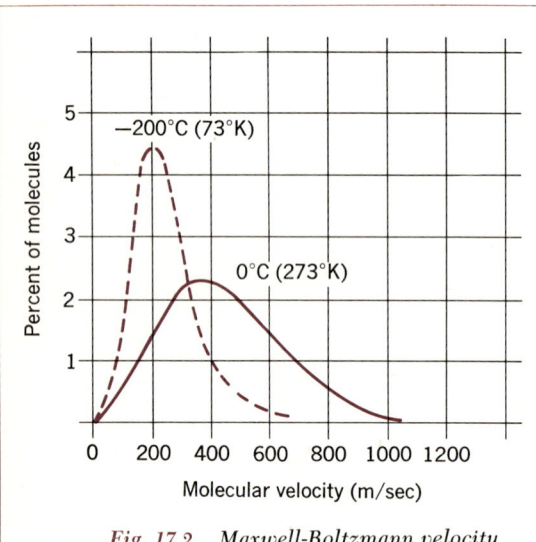

Fig. 17.2. *Maxwell-Boltzmann velocity distribution.*

† If, for all values of the wavelength of the incident radiant energy, all energy is absorbed, the absorber is called a blackbody.

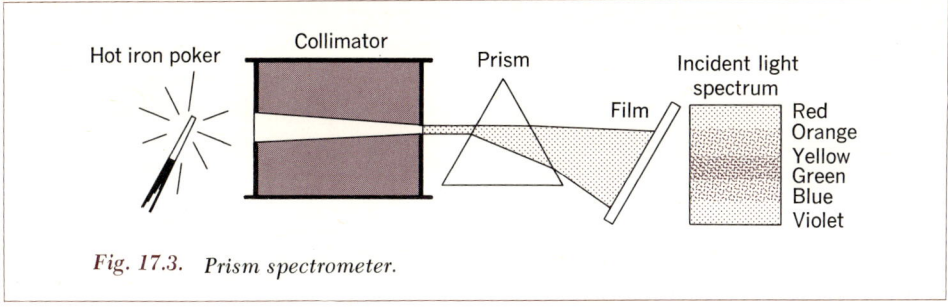

Fig. 17.3. *Prism spectrometer.*

and especially the peaks, shift toward higher speeds. But for each temperature state of the body there is a distribution of speeds that is *characteristic of that temperature*.

How can we know that this is so?

Fortunately there is available an instrumental technique which is just what we need to answer this question. We refer to the spectrometer, which can separate the radiation into its component frequencies. (Spectrometers are studied in laboratory experiments of the course and will not be discussed here.) Figure 17.3 illustrates how such an instrument makes it possible to analyze the radiation that is emitted by our poker. The radiation from the poker (preferably coming from within the cavity so as to maintain blackbody conditions) passes through narrow slits onto a prism that separates the radiation into its spectral components and directs them separately toward a film on which they are recorded. When many atoms emit at the same frequency, the line recorded for that emission is intense. When only a relatively few atoms radiate a particular frequency, the line recorded on the film will be very weak. There is good reason, therefore, to consider that the intensity of each spectral line (that is, its blackness on a film) is proportional to the number of radiation emissions per second, which in turn is a measure of the number of atoms that are in a state of excitement or energy to radiate that particular frequency.

Suppose we were to heat the poker to some moderately high temperature and analyze the radiation spectrum in the manner suggested.† The measurement of intensities versus frequency of radiation, when plotted, would look like that of Fig. 17.4, say, the 1200°K curve. The curve is similar to the curves of Fig. 17.2, indicating some relationship in the radiation process to the numbers of particles versus kinetic energy. Also note that the curve is continuous; that is, all frequencies appear, since atoms of all energies are present in the poker.

If now we raise the poker to a substantially higher temperature and measure the intensity of the radiation as before, the graph might resemble that of the 5000°K curve of Fig. 17.4. From Fig. 17.4 we see that as the temperature is increased, the wavelength at which the radiation curve reaches a maximum value (λ_m) decreases. This result was derived from classical theory by Wien and is

---

† The actual experiment would involve more instruments and experimental problems than this simple statement implies, but the principle is correct.

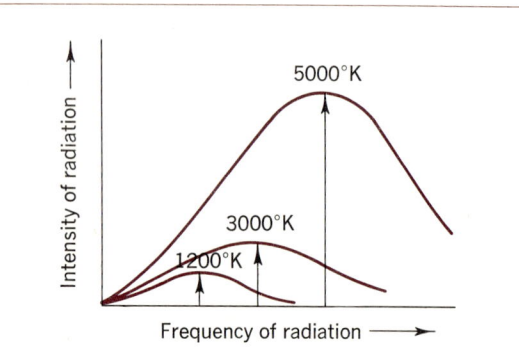

Fig. 17.4. Intensity of radiation versus frequency at different temperatures.

known as Wien's displacement law. Mathematically, it may be written

$$\lambda_m T = \text{constant} \qquad (17\text{-}1)$$

A second relation represented by the same family of curves in Fig. 17.4 is that the total energy radiated per second (represented by the area under the curve, taken from $\lambda = 0$ to $\lambda \to \infty$) increases with increasing T. Independently, Stefan and Boltzmann derived classical relations for this phenomenon. These may be stated as: The total energy radiated per unit of surface area per second, E, is proportional to the fourth power of T, or

$$E = \sigma T^4 \qquad (17\text{-}2)$$

where σ is the constant of proportionality.

The two relations given in Eqs. (17-1) and (17-2) could have been derived on the basis of the classical theories of thermodynamics and electromagnetic radiation. Deriving a single relation to represent the entire family of curves with respect to both T and λ was more difficult, however.

According to Maxwell's theory, electromagnetic radiation results from *oscillating electric charges*. If the walls of our cavity radiator are assumed to be made up of such oscillators, then we might consider that heat energy supplied to raise the temperature may be transformed into the increased kinetic energy of these oscillators. In theory, all one should have to do is to specify how the energy radiated at each wavelength is divided among the various oscillators, and sum up over all oscillators in order to get the total energy radiated at each wavelength.

Many attempts were made to derive the necessary relation. The approach outlined above, with the necessary assumptions as to the energy distribution in the oscillators, led Wien to a relation that could be made to represent the observations very well for a wavelength less than λ_m. But for long wavelengths the relation did not appear to fit. Theoretical work by Lord Rayleigh and Sir James Jeans also led to a relation that represented experiment only at long wavelengths but which failed to account for the observed radiation at short wavelengths.

There was another difficulty from the theoretical side. If one were to assume that all possible frequencies of motion and of radiation emission could take place (and there was no reason for assuming otherwise), the theory predicted that much of the energy given to the body as heat should become converted to radiation of higher and higher frequencies. This can be illustrated by analogy with the behavior of a violin string. If the violin string is plucked at its center (Fig. 17.5), the energy injected into the system would first sound the fundamental tone of the whole string moving. But very quickly we would hear the more complex sound of higher frequencies as the energy goes into a complex form of string vibration containing higher and higher frequencies. If this were similarly true in the case of light waves, the energy contained in the long waves of red light should become transformed into shorter waves of ultraviolet light and even to X-rays. But this does not happen. There is no endless transfer of energy from low frequency to high frequencies of light without limit, as was predicted by the classical theory and the classical equations. And it is fortunate that no such transfer takes place, since we know today that the shorter wavelengths of ultraviolet light and of X-rays are dangerous to the human body.†

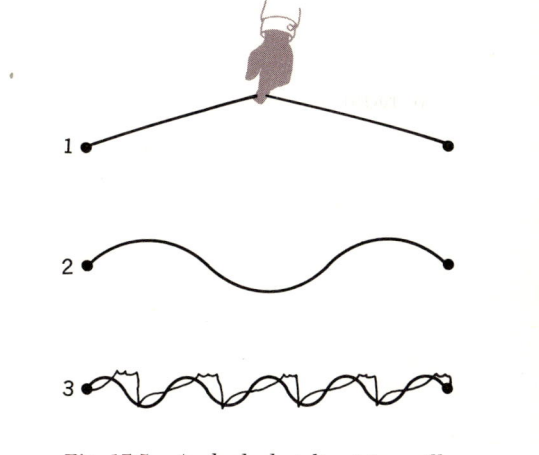

Fig. 17.5. A plucked violin string will presently transfer the energy of vibration to a complex form involving many higher frequencies. There seems to be almost no limit to how the vibration can diffuse to multiples of any whole or fractional number.

† This conversion to higher frequencies was referred to as the "ultraviolet catastrophe."

But why doesn't the energy work into the high-frequency end of the spectrum with disastrous results? This became a question for which an answer had to be found.

17.3 The birth of the quantum hypothesis

The question of high-frequency propagation was still unanswered when the German physicist Max Planck† turned his attention to the problem of blackbody radiation. Starting with the same basic assumptions as his predecessors, Planck applied the energy-entropy relations of thermodynamics to this problem and concluded that Wien's relation was correct. In fact, he was so convinced of this fact for a while that when observations failed to confirm the Wien formulation, he believed the observations to be in error. But by the year 1900, the experimental results had been extended to radio wavelengths (long) and there could be no doubt of the discrepancy between theory and observation. The classical theories of thermodynamics and electromagnetic radiation appeared unable to account for blackbody radiation.

In putting forth a theory to explain a natural phenomenon, a scientist draws upon other successful theories in related fields, on his experience in the field, and sometimes on his intuition. Having evolved the basic assumptions of his theory, he uses these to deduce the expected outcome of experimental or observational tests he intends to make. If his deduced results do not agree with observation (assuming the deductive steps to be correct), he must alter his basic assumptions or add restrictive assumptions to his theory. This was the situation with which Planck was faced in the first years of the twentieth century.

Planck reasoned as follows: Since Wien's formula holds whenever the wavelength is short, and the Rayleigh-Jeans formula when the wavelength is long, the true formula must reduce to each of these when it is applied to the short and long ends of the spectrum, respectively. He then proceeded to derive an empirical formula that would have the desired properties for long and short wavelengths. But the mathematical manipulations that produced this new formula shed no light on the physical processes taking place. Although Planck now had an equation that correctly represented the amount of energy radiated by a blackbody at temperature T per unit-wavelength interval, at any wavelength, he still did not understand the basic processes that gave rise to the radiation.

In his original hypotheses Planck assumed, as did his predecessors, that Maxwell's electromagnetic theory held. The experiments by Heinrich Hertz in producing electromagnetic radiation of very long wavelength (radio waves) by using oscillating electric charges convinced Planck that the troubles did not lie in Maxwell's theory. Planck, again like his predecessors, assumed that cavity radiation was produced by submicroscopic electric oscillators in the cavity walls. Each oscillator has its own fixed frequency f (or wavelength λ) and emits radiation of this frequency only. All frequencies are represented because of the large

---

† Max Planck (1858–1947) was born in Kiel and moved to Munich. He studied physics at the University of Munich and at the University of Berlin. After preliminary appointments as teacher at Munich and at Kiel, he became full professor at the University of Berlin, the highest academic post in Germany, in 1892. His work on the quantum hypothesis earned him the Nobel Prize in 1918.

number of oscillators present, and the sum of all these gives a continuous spectrum—that is, all frequencies are present.

An oscillator that is radiating loses energy; to keep it from cooling, new energy in the form of heat must be added. The heat converts to kinetic energy and is shared among the oscillators by their numerous collisions. As we saw in Chapter 5, the energy in a wave is given by the square of its amplitude; hence, as each oscillator gains or loses energy, only the amplitude of its oscillation changes—not its frequency, which is assumed to be fixed.

So far this is pure classical theory as used to derive the Rayleigh-Jeans law, which was shown not to hold. The classical theories, however, imply two further assumptions: (1) that the energy of any individual oscillator can have any value from zero upward, and (2) that since an accelerated charge radiates, the oscillators must continue to radiate as long as they vibrate. It was these two assumptions Planck was forced to change. In replacing them he was forced to break with classical physics, which started a revolution within the science.

Planck found that he could derive his empirical formula from theory if he replaced the last two assumptions as follows:

(1) Each oscillator can have only certain definite energies associated with it and *each of these* must be an *integer multiple* of hf, where h is a *new constant of nature* (hereafter referred to as Planck's constant) *and f is the oscillator's frequency.*

(2) As long as an oscillator remains in a particular energy level, no energy is radiated or absorbed. The oscillator *radiates* energy only when it changes from one energy level to *any other lower energy level*. The oscillator *absorbs* energy only when it changes from one energy level to *any other higher energy level*. Radiation or absorption of energy always takes place in steps of integer multiples of hf.†

The effect of these radical assumptions was to take electromagnetic radiation, which had hitherto been considered to be continuous, and to require that it be absorbed or emitted only in discrete bundles, or quanta. Although the earlier acceptance in chemistry of the atomic theory had destroyed the notion of the continuity of matter, energy was still believed to be a continuous thing that could be subdivided in any manner desired. For example, a tuning fork maintains one frequency of vibration, once it is energized; but as its energy decreases, it seems to pass slowly through all intervening amplitudes of vibrations (energy states) until it finally stops vibrating altogether. And now Planck's quantum concepts seemed to say that the *loss of energy in atomic oscillators occurred in quanta, or jumps*, rather than in continuous, infinitesimally small increments.

Despite his radical innovations, Planck was basically a conservative man. He strongly believed in the deterministic physics of Newton. As a thermodynamicist he had often taken exception to Boltzmann's statistical interpretations. Yet, by his own work, he was forced to accept the quantization of the energy absorbed and emitted by an oscillating electric charge. He felt certain that in quantization lay the only explanation of the observed radiation laws. Still, he felt that

† Further developments in physics show that the different energy levels of a harmonic oscillator are odd half-integer multiples of hf. Thus, even though Planck's first assumption is not correct, his second assumption remains valid.

perhaps quantization might be really only a mathematical artifice with no physical reality. For well over a decade he refused to embrace some of the extreme consequences of his own innovations, despite their acceptance by some of his colleagues.

But Planck had planted an idea that could not be restrained from blossoming into a flower of an amazingly new variety.

17.4 The photon and the photoelectric effect

For nearly five years after he proposed the quantum hypothesis, Planck himself was ill at ease with his brainchild. In 1905, however, the quantum hypothesis was strengthened, and the revolution started by Planck was given an impetus that continues to this day.

This new interest in the quantum concept stemmed from a paper that appeared in the *Annalen der Physik* under the modest title, "On a heuristic point of view concerning the generation and transformation of light," written by a twenty-five year old clerk in the Swiss patent office in Berne, named Albert Einstein. Earlier in 1905, this same young man had published a paper explaining Brownian motion on the basis of a statistical investigation of the kinetic theory of gases. Later that year he would publish still another paper, this time entitled simply, "On the electrodynamics of moving bodies," enunciating the special theory of relativity. Any one of these three papers would have sufficed to bring fame to its author. It is a monument to Einstein's great genius that he published all three *within a single year!*

Whereas Planck, both in his original paper and in his later writings considered his "quantum" as something to be applied only to the emission and absorption of electromagnetic radiation and not to radiation in free space, Einstein took a much broader view. He proposed the hypothesis that "the energy of light is not distributed evenly over the whole wavefront, as the classical theory assumes, but rather is concentrated, or localized, in small discrete regions." Thus Einstein considered light to be divisible into small "bundles." Once again light was considered to be *corpuscular* in nature!

The root of Einstein's theory lay in Planck's work. Planck, although he demanded that light be absorbed and emitted in little bundles, assumed that it traveled through space as a continuous wave in accordance with Maxwell's theory. Yet, if the radiation incident on the absorber was very weak, how did it manage to wait until the correct amount of energy, hf, was available before absorbing it? "No," said Einstein, "light always exists in these little packets or quanta of energy hf." Later these packets were given the name *photon*.

Einstein's hypothesis, while of great interest philosophically, would not have created such a stir were it not for the second part of the same paper in which Einstein showed that by employing this new hypothesis, one previously unexplained phenomenon became easily understood—namely the *photoelectric effect*.

To understand Einstein's explanation we must backtrack a bit to an accidental observation by Heinrich Hertz in 1887. Hertz had been studying electromagnetic wave phenomena by discharging high voltage across spark gaps. He found that the electromagnetic waves generated by the discharge would produce a discharge also in a nearby wire loop and spark gap. But he did not investigate this phenom-

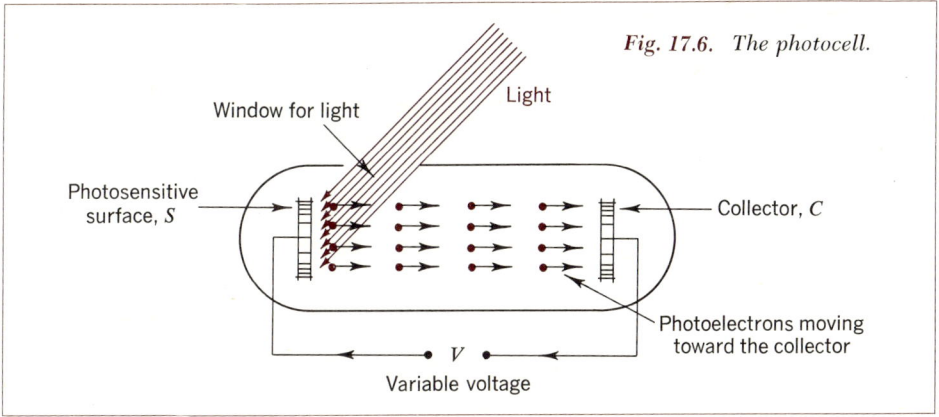

Fig. 17.6. The photocell.

enon and thereby just missed being the discoverer of radio communication.

A spark discharge produces light of some intensity. Hertz noticed that when the light produced by such a spark discharge fell on the spark gap of a second discharge loop, the spark of that second loop could jump across a larger gap. That is, light tended to help the spark to jump larger distances, perhaps by helping to release carriers of electricity from the metal of the spark-gap material. In time it was recognized by others that light, especially ultraviolet light, can cause electrons to be pulled or driven out of a metal surface (the photoelectric effect).

During the final years of the nineteenth century, additional experimental evidence was gathered to show that light can free electrons from the surface of certain metals but not from others, depending on the frequency or nature of the light. Using this phenomenon, a device known as a photocell can be constructed, which is capable of responding to light by passing an electric current when light of the correct frequency falls on its light-sensitive surface (S of Fig. 17.6). Light falling on S releases the negative elec-

trons, which are then carried across the tube to the collector C under the influence of an electric field existing between S and C.

Consider an electron being emitted from S with a certain amount of kinetic energy. If the collector C is kept positive with respect to S, the electron will be subjected to a force field that will accelerate the electron towards C. The electron will reach C with a greater amount of kinetic energy than that with which it started from S. But if C is kept negative with respect to S, the electron will be opposed by a force that will decelerate the electron in its motion towards C. The electron will reach C with a smaller amount of kinetic energy than that with which it started from S. The amount of negative voltage of C with respect to S can be adjusted by varying the voltage V so that the kinetic energy of the electrons just reaching C are made to equal zero. In other words, for some particular value of negative voltage of C with respect to S, the photoelectric current ceases. This cut-off voltage is usually referred to as the *stopping potential*.

Experimental studies of photoelectric effect yield the following *observations* for which classical electromagnetic theory could not provide proper explanation:

(1) The amount of photocurrent that flows decreases as the intensity of light is decreased (keeping the frequency of the light and the voltage V at the same values). However, no matter how weak the intensity of the light, such current as does flow, *flows instantly*. That is, there is apparently no need to accumulate light energy on the surface S to release electrons.

(2) With light of a particular frequency and intensity, changing V to impose a higher negative voltage on C will stop the flow of current at some value of V. It is found that the same value of V will stop the flow of current whether the light intensity is very high or very low, as long as the quality of the light (that is, its frequency) is kept the same.

(3) When the frequency of the light is varied—say, by going from the violet (high frequency, short wave) side toward the red—at some point along the way the photoelectric current disappears. That is, there is no flow when the light has a lower frequency (longer wavelength) than a certain threshold frequency f_{th}.

What were the conclusions to be drawn from these observations? One thing was clear, namely, that increasing the intensity of the light increased the magnitude of the photocurrent, but had no effect whatever on the cut-off voltage V or on the frequency of light f_{th}, at which the current would begin to flow. Clearly the initiation of the photoelectric effect was dependent on the *frequency* of the light rather than on intensity. The explanation of the photoelectric effect as given by Einstein can be stated as follows:

Suppose that an electron is held inside the material S by a binding energy, which we may call a *work function*, W_0. It will take the addition of *at least* that much energy from the light to release the electron from S. If energy higher than the work function W_0 is given to it (say, W_1, where $W_1 > W_0$), the electron will fly out with a maximum kinetic energy equal to the excess energy $W_1 - W_0$ and will require some opposing energy to stop it. The negative voltage V, the stopping potential, provides the required force to oppose this kinetic energy and has the value

$$Ve = W_1 - W_0 = \tfrac{1}{2}mv^2 \qquad (17\text{-}3)$$

where m and v are the mass and *maximum*† velocity of the electron, respectively, and e is the electronic charge.

But what about the mechanism by which energy is transferred from the light to the electron? Surely, if it is a matter of accumulation of energy in the surface S to excite the surface and release the electron, the electromagnetic waves of light of low frequency should accumulate this energy if their intensity is high enough. But it is not simply a matter of total light energy. The first and third observations, (1) and (3), say in fact that accumulation of energy is not a factor at all. Either the light is of a frequency that will release an electron or it is not, and the intensity has nothing to do with this threshold release energy.

In this situation, light behaves as though its energy were concentrated in a tiny bundle, or quantum, or particle. It behaves as though this quantum of

---

† Part of the energy, W_1, acquired by the electron may be lost as a result of interactions within the metal surface before the electron escapes. $W_1 - W_0$ represents the *maximum* energy an electron can carry away.

energy must have enough concentrated energy to strike a particular electron and transfer all its energy to that electron, much as a billiard ball might strike another ball. And since the photoelectric effect is more effective with light of higher frequency, the bundles or quanta of light energy clearly must be proportional to some constant times frequency, or $E = hf$. But this was exactly Planck's postulate stated in his assumption (2).

On this assumption, that $E = hf$ does indeed represent the energy of the incoming light of frequency f, it follows immediately that the ejected photoelectron has *maximum* kinetic energy equal exactly to the energy of the incoming light, hf, less the work function W_0, or

$$(\tfrac{1}{2}mv^2) = hf - W_0 \qquad (17\text{-}4)$$

or, from Eq. (17-3), we obtain

$$Ve = hf - W_0 \qquad (17\text{-}5)$$

This is the celebrated photoelectric equation of Einstein.†

With this new approach the reader is urged to return to the observations (1), (2), (3) and satisfy himself that the explanations are now adequate.

Einstein's photon theory triumphed

---

† Equation (17-5) may be used to evaluate Planck's constant h. W_0 is equal to hf_0, where f_0, the minimum frequency of incident light that gives rise to a photocurrent, can be measured experimentally for any given surface. The electron charge e is also experimentally determined (see Chapter 11). Thus

$$h = \frac{Ve}{f - f_0}$$

The approximate value of h is

$$h = 6.63 \times 10^{-34} \text{ J-sec}$$

It is the extreme smallness of h that tends to mask quantum effects in our everyday life. Indeed, if h were actually zero, we would be back to classical physics!

where classical theory failed, and with this triumph the quantum hypothesis found fresh strength and an extension and broadening of its scope and ramifications.

17.5 Some further thoughts on the "quantum"

Experimental evidence continued to accumulate and to support this radical theory. In 1915, R. A. Millikan experimentally established Einstein's photoelectric equation with great precision. But something more was needed. Light, or the photon, behaves like a particle. But the ordinary particles we dealt with in Chapter 5 also had momentum. Does this photon also have this property of momentum? In other words, can we treat a collision between a photon and any material body in the same way we treated the collision between two billiard balls?

An unexpected answer to this question came in 1923 when A. H. Compton experimented with X-ray bombardment of electrons. He directed the X-ray beam of known frequency onto a graphite target, and studied radiation that came off at right angles to the beam. This latter radiation was "scattered," or reradiated, by the electrons in the graphite atoms. He found that the scattered radiation always had lower frequency (longer wavelength) than did the initial beam. According to the wave theory of light such a shift in frequency (or wavelength) could not be explained. By adopting Einstein's photon theory, Compton interpreted the change in X-ray wavelength as an energy loss. By treating the photon-electron interaction as a two-body collision, he was able to derive a relation which accounted for his experimental results. The momentum of light was known to be E/c. Accord-

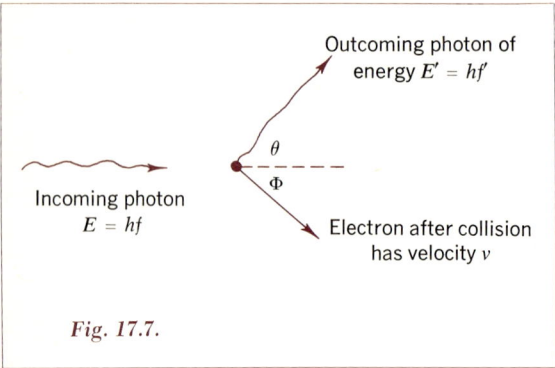

Fig. 17.7.

ing to the photon theory this becomes:

$$\text{Momentum of photon} = \frac{hf}{c} = \frac{h}{\lambda} \quad (17\text{-}6)$$

where c is the speed of light. For this research Compton obtained the Nobel Prize in Physics, and the particle-like properties of light became firmly established.

To understand Compton's experiment, consider an electron of mass m, at rest initially, which is struck by a photon of energy, $E = hf$, as pictured in Fig. 17.7.

Let us apply the laws of conservation of energy and momentum:

(1) Energy conservation: Energy of incoming photon = energy of outgoing photon plus kinetic energy gained by electron. In equation form this may be written† as

$$hf = hf' + m_0 c^2 \left[\frac{1}{\sqrt{1-(v/c)^2}} - 1 \right]$$

where m_0 is the *rest* mass of the electron.

(2) Conservation of linear momentum in the forward direction: Momentum of the incoming photon equals the sum of the component of momentum of the electron (after collision) in this direction, plus the component of momentum of the outgoing photon, again in the forward direction, or

$$\frac{hf}{c} = \frac{hf'}{c} \cos\theta + \frac{m_0 v}{\sqrt{1-(v/c)^2}} \cos\phi$$

(3) Conservation of linear momentum in the vertical direction yields

$$0 = \frac{hf'}{c} \sin\theta + \frac{m_0 v}{\sqrt{1-(v/c)^2}} \sin\phi$$

† We have used the relativistic form of the kinetic energy (see Chapter 12), since the resultant velocity of the electron, v, may be very large.

We thus have *three* equations in *four* unknowns; f', θ, ϕ, and v. Eliminating ϕ and v and recalling that $\lambda f = c$ yield

$$\lambda' - \lambda = \frac{h}{m_0 c} (1 - \cos\theta) \quad (17\text{-}7)$$

which says that the difference between the wavelengths of the outgoing and incoming photons is a constant, $h/m_0 c$, times a function of the scattering angle θ. The term $h/m_0 c$ has the dimension of length and is referred to as the *Compton wavelength* of the electron. Its numerical value is 2.43×10^{-12} m (0.0243 Å.). Experiment confirms this equation and helps to establish the particle character of light.

But despite its successes, the photon theory of light seems contrary to other well-established experimental evidence. Young's double-slit experiment in which light was shown to display the interference patterns expected of a wave, had appeared to spell the doom of the corpuscular theory of light nearly a century earlier. How, then, can we reconcile the two theories? According to the work of Young, Faunhofer, Fresnel, and Maxwell, light must be a wave, but Einstein and Compton have shown that it must act like a particle. Perhaps the best resolution of this seeming paradox was stated by another Nobel Prize winner, Max Born:

"The ultimate origin of the difficulty lies in the fact (or philosophical principle) that we are compelled to use the words of common language when we wish to describe a phenomenon not by logical or mathematical analysis, but by a picture appealing to the imagination. Common language has grown by everyday experience and can never surpass these limits. Classical physics has restricted itself to the use of concepts of this kind; by analyzing visible motions it has developed two ways of representing them by elementary processes: moving particles and waves. There is no other way of giving a pictorial description of motions—we have to apply it even in the region of atomic processes, where classical physics breaks down.

"Every process can be interpreted either in terms of corpuscles or in terms of waves, but on the other hand it is beyond our power to produce proof that it is actually corpuscles or waves with which we are dealing, for we cannot simultaneously determine all the other properties which are distinctive of a corpuscle or of a wave, as the case may be. We can therefore say that the wave and corpuscular descriptions are only to be regarded as complementary ways of viewing one and the same objective process, a process which only in definite limiting cases admits of complete pictorial interpretation. . . ."†

The next great step in quantum physics was already going on while the debate raged over Einstein's broad view of the quantum. When, in 1912, the Danish physicist Niels Bohr proposed a model of atomic structure based on the quantum hypothesis, modern physics began to emerge as a predictive tool of enormous power. We study this in the next chapter.

17.6 Summary

In earlier chapters, radiant energy—its nature and manner of propagation—and the dynamics of matter in motion were treated more or less separately except for a brief reference to the source of electromagnetic radiation. We now have reviewed more thoroughly the interactions of radiation and mass, and the discovery

† From Born, *Atomic Physics*. See References.

of some strange, nonclassical laws governing these interactions, with the additional surprising consequence that Maxwell's theory of electromagnetic radiation must be strongly modified.

That matter itself is "quantized"—that is, it ultimately consists of indivisible "particles," "bundles," or "granules"—is an ancient idea not too far removed from what we observe in such composites as sand. But the idea that energy is also quantized and comes in discontinuous packages caught scientists by surprise. Energy can be measured in terms equivalent to $\frac{1}{2}mv^2$. Although mass may be discontinuous, it is difficult to see how velocity can be other than continuous, that is, able to take on any of an infinite number of values within any given range. If velocity is a continuously varying variable, it would seem that energy should be also. Similar arguments arise out of Maxwell's equations. The energy depends on the square of the intensity of the magnetic or electric fields. If both these fields are continuous variables, how can energy be restricted to quantized packages? It was all very puzzling to the scientists. (Perhaps one day it may be found that space itself, or time, is likewise quantized; this would be hardly regarded by scientists today as more puzzling than was the discovery of the quantization of energy.)

But the evidence we have presented is clear enough, and as more evidence has been uncovered, the theory of the quantization of energy has been more strongly confirmed. We have shown that it was impossible to deduce from classical laws of physics the distribution of energy and frequency given off by a blackbody radiator. Planck developed a purely empirical mathematical formula that fitted the observation. At first he could find "no explanation," that is, no way to account for the form of his equation and especially for one of the constants (h) entering into it. (Compare this situation with Kepler's discovery of his third law, $P^2 = kR^3$, where he could give no explanation of k.) Planck finally, with some hesitation and even distaste, proposed the theory that the particles responsible for radiating the energy did so only in multiples of hf, h being a universal constant and f being the frequency. This was the birth of the concept of the *quantum of energy*, as the product hf has been designated.

Einstein extended this concept and showed that if light or radiant energy could be thought of as composed of quanta (conveniently called photons when speaking of light energy), the *measured* observations connected with the photoelectric effect would be nicely accounted for, whereas they could not be explained by classical theory. (The reader should study this example carefully. It is a simple illustration of how classical theory may be misleading because it implies false consequences, and how another hypothesis may replace it or modify it by leading to conclusions that concur with observations. Note that the first is proved false; the second is shown to be plausible, but is not thereby "proved" logically.)

Soon other confirmations followed. Compton showed that when X-rays strike an electron, the electron recoils with a change in momentum and energy. Whence comes this change in momentum and energy? Compton found that there was a change of frequency in the scattered X-ray and that this change in frequency always could be accounted for by assuming that the X-ray lost energy in an amount equal to $h \, \Delta f$.

Finally, the quantum idea was applied

to the study of atomic structure, as we shall see in the next chapter. Thus the concept of quantization of energy has become an integral part of the body of modern physics. Puzzles remain, such as the problem of the dual nature of matter and energy wherein wave properties and corpuscular properties combine in a photon or particle at one and the same time. We shall return to this puzzle in later discussions, and shall find that the answers provided by physics give rise to some profound philosophical problems, as yet unsettled.

Questions/Discussions

1. Adequate explanation of the photoelectric effect was given by Albert Einstein, but he did not discover this effect. Who did discover it, when, and under what circumstances?

2. As stated in Sec. 17.1, light in the frequency range 4.3×10^{14} cps to 7.5×10^{14} cps is usually referred to as visible light. Find the maximum and minimum energy of visible light photons.

3. Calculate the minimum number of photons, with frequency 7.5×10^{14} cps, striking the eye per second in order to produce the sensation of light. (Minimum power necessary to produce the sensation is 10^{-18} watts.)

4. "Each material seems to have its own critical value for the threshold frequency, f_{th}." Calculate this threshold frequency for aluminum if 4.2 eV of energy are required to just release a photoelectron. Does aluminum show the photoelectric effect for red light of wavelength 7000 Å?

5. Discuss your reaction to the following comment: "Our eyes are not able to see the light waves with frequency lower than 4.3×10^{14} cps, because it is the threshold frequency below which no photoelectric effect can occur within the eyes and generate sensation of vision."

6. The following graph is a plot of the data obtained during a test of the photoelectric properties of a metal surface. The letters V and f have the same significance as given in Sec. 17.4. Discuss the significance of the points V_0, f_{th}, and of the values $\Delta V/\Delta f$, $(h/e)f$.

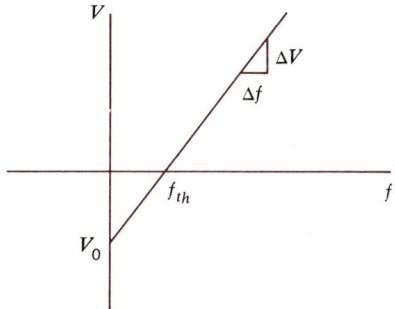

7. Discuss how you might proceed to determine a value for h from an experiment of the type of Problem 6.

8. Referring to Max Born's reference (Sec. 17.5) to the complementarism that seems to exist in the properties of light, how would you react to the thought that complementarism exists in behavioral situations, such as in comparing mercy and justice, or in considering a person to be a unique individual as compared with a social being? Or in comparing war and peace?

9. Referring again to Max Born's comments (Sec. 17.5), how could you relate what one sees and thinks with the expression in language about what is seen and thought? What is this relation? How does the need to communicate help (or interfere with) one to express impressions of objects, people, and events?

References

Boorse, H. A., and Motz, L. (eds.), *The World of the Atom*. New York: Basic Books, 1966. *Contains several papers relevant to the material of this chapter.* (1) Planck: "Origin and Development of Quantum Theory," pp. 462 ff.; (2) Einstein: "Concerning a Heuristic Point of View about the Creation and Transformation of Light," pp. 533 ff.; (3) Compton: "A Quantum Theory of the Scattering of X-rays by Light Elements," pp. 902 ff.

Born, Max, *Atomic Physics*. English translation (by John Dougall) published by Hafner Publishing Co., 1946.

Cline, Barbara Lovett, *The Questioners*. New York: Crowell, 1965.

de Broglie, *Matter and Light*. New York: Dover (reprint), 1937. *A popular account of modern physics by an outstanding physicist and Nobel Laureate.*

Gamow, George, *Thirty Years That Shook Physics*. New York: Doubleday, 1966.

Hoffmann, Banesh, *The Strange Story of the Quantum*. New York: Dover, 1959. *An easily read account of the growth of modern physics. Highly recommended.*

CHAPTER EIGHTEEN

The Bohr Atom

UNTIL ABOUT 1900 the theories proposed for the atomic nature of matter did not consider the internal structure of the atom. It was assumed that there were differences in the structure, since there were large variations in the chemical interactions among atoms. But even the kinetic theory of gases and the laws of thermodynamics found it possible to avoid the problems of internal structure. This could not go on much longer, however, in view of accumulating questions that continued to increase as new experimental data became available.

18.1 Experimental evidence from the spectrometer

In Chapter 17 we saw how studies involving the nature and characteristics of radiation from blackbodies eventually led to the concept of the quantum of radiation. In this chapter we discuss a few simple experiments regarding the absorption and emission aspects of radiation. Such experiments have been enormously fruitful, and still are used for obtaining information about the structure and dynamic characteristics of atoms and molecules.

We are all familiar with the light we receive from the sun or from a filament lamp.† We call these sources *white light* because that is the effect of the sensation they produce in vision. The sensation of white disappears, however, when (by

† The reason for specifying a lamp that has a *filament* (that is, incandescent wire) is that other available light sources, such as gas-filled tubes, give quite different spectra.

‡ From an address delivered in connection with ceremonies at the 1954 Bicentennial of Columbia University. Quoted in *Great Essays by Nobel Prize Winners*, edited by Leo Hamalian and E. L. Volpe, The Noonday Press, New York, 1960.

> *... classical ideas of mechanics and electromagnetism did not suffice to account for the essential stability of atomic structures, as exhibited by the specific properties of the elements. However, ... the discovery of the universal quantum of action to which Planck was led in the first year of our century ... revealed in atomic processes a feature of wholeness quite foreign to the mechanical conception of nature, and made it evident that the classical physical theories are idealizations valid only in the description of phenomena in the analysis of which all actions are sufficiently large to permit the neglect of the quantum. While this condition is amply fulfilled in phenomena on the ordinary scale, we meet in atomic phenomena regularities of quite a new kind, defying deterministic pictorial description.*
>
> NIELS BOHR‡

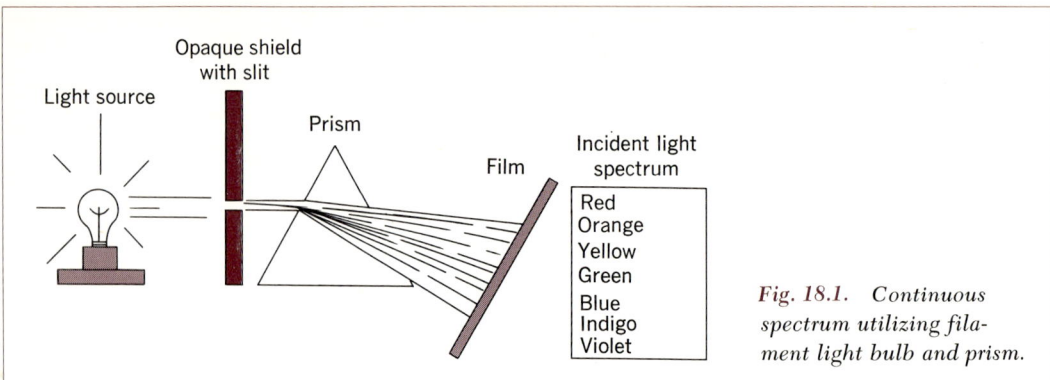

Fig. 18.1. *Continuous spectrum utilizing filament light bulb and prism.*

means of a slit in a diaphragm) we confine the light to a small pencil width and direct it through a glass prism, as shown in Fig. 18.1. If the light that passes through the prism is allowed to fall on a screen, the narrow beam spreads out into a beautiful band of colors, including violet, indigo, blue, green, yellow, orange, and red, as seen in a rainbow. The violet fades and merges into indigo, indigo fades and merges into blue [Fig. 18.2(b)], and so on in a continuous fashion; for this reason we call it a *continuous spectrum*. If we were to utilize a lens in the optical path of the band of colors, we would obtain on the screen a series of *colored images* of the slit in the diaphragm. This indicates that the different colors must have been in the original light. That is, there is good reason to believe that *the prism acts only as an agent of dispersion,* and that it does not introduce these colors into the white light. Therefore, it is correct to conclude that *white light must be a composite of these various colors.*

Instead of using sunlight or a lamp as source of light, let us use hydrogen gas. At ordinary temperatures, hydrogen gas does not emit any radiation as far as our senses can detect. When heated sufficiently, however, it emits visible radiation. One may purchase hydrogen-filled tubes that are designed to admit high voltage to excite the gas. By using one of these tubes in conjunction with a spectroscope, it becomes possible to study the spectrum of the light emitted by hydrogen. In the case of hydrogen we do not observe a continuous spectrum as observed when a filament lamp is used. Rather, there appear a few strong lines: one red, one blue, and several violet ones, separated by dark patches between. This is called a *line spectrum*. Each of the lines has a definite wavelength and therefore a definite frequency. [See Fig. 18.2(a).]

If we add some sodium vapor to the hydrogen gas in the discharge-tube source of light, the lines that were previously observed in the spectrum of hydrogen can still be seen, but other lines also appear, which can be readily shown to be due to sodium (by simply examining the spectrum of sodium alone.) All these observations suggest the idea that *the spectrum of lines one observes is characteristic of the elements that are in the source.*

Suppose that we return to the first experiment in which a filament lamp was

Increasing frequency ⟶
⟵ longer wavelengths

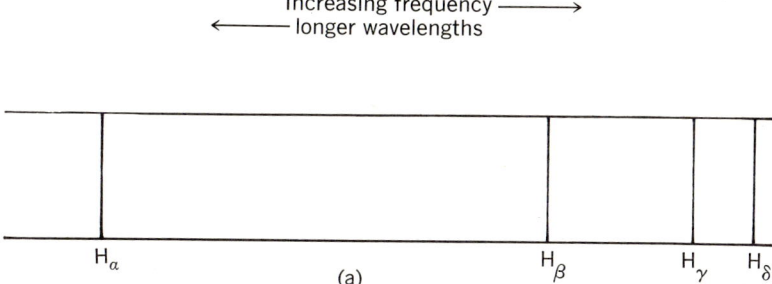

H_α (a) H_β H_γ H_δ

(a) When hydrogen gas is heated by spark discharge (or by other means) and the light it emits is analyzed with a spectrometer, the above four bright lines appear. There are additional, weaker lines that continue to crowd up to a limiting value at the right. Wavelengths: $H_\alpha = 6562.8$ Å, $H_\beta = 4861.3$ Å, $H_\gamma = 4340.5$ Å, $H_\delta = 4101.7$ Å.

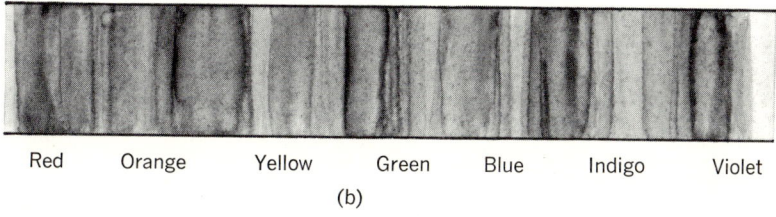

Red   Orange   Yellow   Green   Blue   Indigo   Violet
(b)

(b) When "white" light from a hot filament is analyzed through a spectrometer, there appears a continuous spectrum of lines, going from red at one end through the colors of the rainbow to violet at the other end.

H_α position H_β position H_γ position

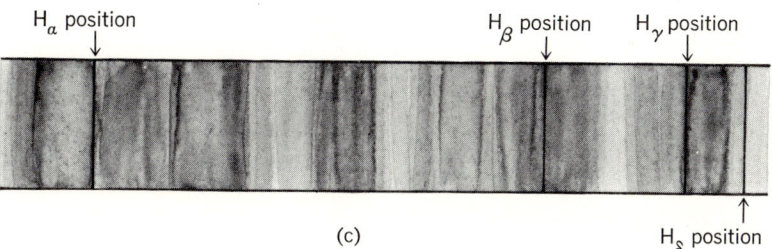

(c) H_δ position

(c) When, while the light from a hot filament is being analyzed, hydrogen gas is interposed in the path of the light beam, four dark lines appear and are superimposed on the rainbow spectrum at exactly the positions where H_α, H_β, H_γ, and H_δ appeared in (a) above.

Fig. 18.2.

the source, again obtaining the continuous spectrum. Now let us interpose hydrogen gas in the path of the radiation that produces the continuous spectrum. If the hydrogen gas is in the proper range of temperature to absorb visible light, we now observe dark lines, superimposed on the continuous spectrum, which are *located at exactly the same frequencies and wavelengths where the bright lines appeared from the hydrogen source* when the white light source was not present [Fig. 18.2(c)]. What does this indicate? It indicates that hydrogen is capable of *absorbing* radiation that has the *same wavelengths* as those of the lines it *emits*. Experiments repeated with gases of other elements give the same results and convince us that *each element absorbs most readily the radiation of the frequencies* which it itself radiates when it is heated.†

Why should a particular element emit and absorb only radiation of definite frequencies? From Planck's formula, presented in Chapter 17, we know that the radiation of a particular frequency represents an energy packet corresponding to that frequency. The higher the frequency f, the higher the energy E from the relationship $E = hf$.

In the late nineteenth century, when scientists first tried these experiments and struggled with the question of absorption and emission of particular energy lines from the elements, neither Planck's formula nor the structure of the atom were known to them. It was therefore quite natural that investigations regarding this question should have proceeded in a direction that tried to find some explanations in the order and relationship of these lines to each other, based on experimental data.

Johann Jacob Balmer, a Swiss school teacher, was interested in investigating the problem of the hydrogen line spectrum. He measured the frequencies and tried to find simple relationships among them. In 1885 he succeeded in identifying an empirical formula that fitted the hydrogen lines remarkably well. A rearranged version of this formula is

$$k_n = \frac{1}{\lambda} = R\left(\frac{1}{2^2} - \frac{1}{n^2}\right) \quad (18\text{-}1)$$

where k_n is the wave number (number of wavelengths per centimeter), and R is now called the *Rydberg constant* in honor of a Swedish spectroscopist. The present value of this constant is

$$R = 1.0968 \times 10^{-3} \text{ Å}^{-1}$$

Substitution of integral values of n, such as 3, 4, 5, and 6, yield wavelengths λ for corresponding lines of the visible spectrum of hydrogen.† These are now known as *Balmer lines*.

Ritz as well as Rydberg made the suggestion that if the numbers 1, 3, 4, or 5 were used in Eq. (18-1) in place of the factor 2 of the denominator of the first term on the right,‡ the formula might express the wavelengths of other lines that possibly might exist in parts of the

---

† When sunlight is made to pass through a prism, the spectrum includes, in addition to the colors of the rainbow, a series of dark lines. Fraunhofer interpreted these dark lines as being due to the presence of gases in the solar atmosphere, which absorbed these particular frequencies. These are referred to as *Fraunhofer lines*. Interestingly, the element helium was observed to be in the sun (through spectral lines that were in sunlight) before it was discovered on earth.

† For example, when $n = 3$, $R(1/2^2 - 1/3^2) = (\frac{5}{36}) R$. The wavelength would be the reciprocal of this, or $\lambda = (36/5)(1/R) = 6564$ Å, which is the H_α line of hydrogen. With $n = 4$, we obtain the H_β line, or $\lambda = 4862$ Å.

‡ That is,

$$\frac{1}{\lambda} = R\left(\frac{1}{2^2} - \frac{1}{n^2}\right) \text{ or } \frac{1}{\lambda} = R\left(\frac{1}{3^2} - \frac{1}{n^2}\right)$$

hydrogen spectrum beyond the visible region. It turned out that this was the case, for Lyman, Paschen, Brackett, and Pfund independently discovered different series of lines in the extreme ultraviolet and infrared regions of the hydrogen spectrum. Each series of lines is named after its discoverer. In essence a spectral line of hydrogen can be obtained from a general formula such as

$$\frac{1}{\lambda} = R\left(\frac{1}{m^2} - \frac{1}{n^2}\right) \qquad (18\text{-}2)$$

where m and n are integers, and with n greater than m (that is, $n > m$).

Empirical formulas, patterned after the Balmer formula, were deduced for other elements and seemed to agree remarkably well with experimental results. While this empirical approach was convenient and useful for grouping lines into different series, and stimulated successful search for missing lines, it offered no understanding whatever of the physical processes that might be responsible for the emission and absorption of radiation. It remained for Ernest Rutherford to pave the way for this breakthrough, and for Niels Bohr to come up with some explanation of the structure of the atom. We now discuss their work.

18.2 The Rutherford scattering experiments

J. J. Thomson visualized the atom as being composed of some positively charged substance, distributed more or less uniformly throughout the entire body of the atom, with negatively charged electrons embedded within it (Fig. 18.3). The number of electrons inside the atom was assumed to be sufficient to balance the positive charges and thus make the entire atom electrically neutral. Inside the atom, the electrons would repel each other, since they carry like charges, while

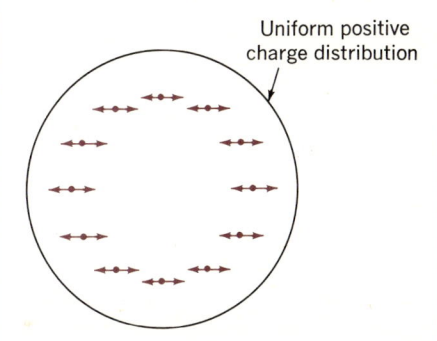

Fig. 18.3. Thompson model of the atom. The dots indicate electrons.

also being attracted in some way to the positive charge distribution. Thomson argued that in the neutral state of the atom, these two opposing sets of forces would be in equilibrium. According to him, when an atom is disturbed from its neutral state (either as a result of collision with another atom or when it passes charged particles), the electrons begin to vibrate around their equilibrium positions. This motion of the electrons involves acceleration, which in turn would cause radiation of energy according to classical electromagnetic theory. By using classical methods, the frequencies of vibration of any atom of a particular element could be calculated. Thomson made some calculations that were based on this model (of negative electrons embedded and vibrating inside a positive atom), with the hope that these calculated frequencies of vibration would turn out to be the same as the frequencies that were measured in the spectrum. To his disappointment, the results were negative. The structural features of the atom were still a mystery.

It was at this time (1911) that the young Danish physicist Niels Bohr (1885–1962)† arrived at the Cavendish laboratory of Cambridge University in England to join the group working with J. J. Thomson. Planck's proposals on the quantized nature of radiation energy had not yet been accepted, although Einstein had already used it successfully to explain the photoelectric phenomenon. Bohr argued that the quantum concept should somehow be introduced into the model of the atom, since Thomson's model based on classical ideas could not explain the radiation emission and absorption phenomena. He wondered whether the mechanical energy of the atomic electrons would have to be restricted to definite values in order to incorporate the quantum idea. It was difficult for Thomson to accept Bohr's ideas. Bohr left the Cavendish laboratory group to work at the University of Manchester, where the chair of physics was then held by Ernest Rutherford, a former student of Thomson. When Bohr arrived in Manchester, Rutherford was already immersed in experiments that unexpectedly gave considerable information about the structure of the atom.

Rutherford proposed to use alpha particles—helium nuclei—as "bullets," shoot them at a thin sheet of gold, and find out why and how they could penetrate the gold foil. His simple experimental setup is shown in Fig. 18.4. The α-particles emanating from a tiny source of radioactive radium are confined to a narrow pencil (by passing through a slit) and allowed to fall on a gold foil. It turns out that as they go through the foil, the α-particles collide with the gold atoms; a fraction of these particles scatter in different directions and can be made to fall on a fluorescent screen, where they produce scintillations. By use of a microscope, these scintillations can be observed and counted, and thus the number of particles that scatter in different directions can also be determined. As expected, Rutherford observed that there were many scintillations (scatterings) in the generally forward direction (through the gold foil to the other side). While not expecting the scattering to be in any other than the forward direction, he nevertheless asked his associates (Marsden and Geiger) to try observations from other directions as well. To their surprise it was observed that some of the α-particles were scattered backward, as though hurled back by a very strong repulsive

---

† At various times he was called the "Great Dane," the "spirit of modern physics," etc. See Boorse & Motz.

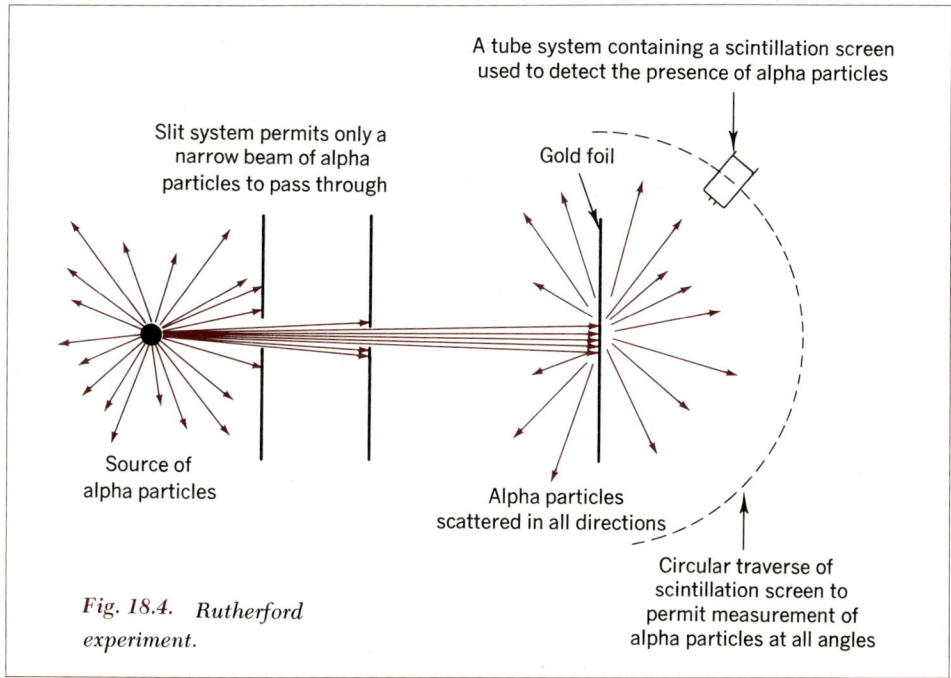

Fig. 18.4. Rutherford experiment.

force. This was a dramatic surprise, and Rutherford compared it to firing a 16-in. gun at a piece of tissue paper and having the shell bounce back to hit the one who fired the cannon.

It seemed reasonable that the repulsive scattering forces resulted from the *positive α-particle's being repelled by the positive charges within the atom*, according to the Coulomb inverse square law.† For the repulsive force to be strong enough to hurl it back, however, it seemed necessary not only to assume that the α-particle does indeed reach very close to the center of the atom, but also that the entire mass of the atom must be very small and located close to the center of the atom. That is, the repulsive forces could not be large enough to explain the observations if the positive portion of the atom was so big that the α-particle could not get very close to its center, or if the α-particle penetrated inside a Thomson model atom.† The Thomson model of the atom, which pictured the individual electrons to be embedded in thinly distributed positive charges throughout the atom, clearly could not permit such large repulsive forces to develop. A new model was needed, one that would recognize that *the repulsive forces required to hurl the α-particles back can be possible only if the positive charge within the atom is very, very small in size*. Rutherford pro-

† See Eq. (11-1).

† This, of course, required that the alpha particle itself also be small.

posed that the atom is very much like our own solar system (the electrons assuming the role of planets), with a *positive nucleus* that has only $\frac{1}{10,000}$ the dimensions of the atom (as the sun is small in relation to the orbits of the planets).†

Rutherford's model of the atom was designed only to be consistent with observations regarding scattering of α-particles from gold foil. Could it also explain the emission and absorption of radiation in terms of definite energies? Here we must recall that electrons that travel in closed orbits around a central positive nucleus must experience continuous radial acceleration toward the center (Fig. 5.12), and according to Maxwell's electromagnetic theory, they should radiate energy continuously. But such a model would require that the radiation from atoms include all frequencies and wavelengths, and not be limited to very specific frequencies such as are observed in the case of atomic spectra. Furthermore, the radiating electron would lose energy, and as a result would move closer and closer to the nucleus, eventually crashing into the nucleus. Such an atom model would not permit the atom to exist for longer than one hundred-millionth of a second, whereas atoms in reality are quite stable.

18.3 The Bohr atom

The picture of the atom now included a small, positive nucleus around which the electrons could revolve; but there did not seem to be any basis for understanding how the electron did (or could) continue in orbit. It remained for Bohr to introduce postulates that would "stabilize" the atom model and at the same time give it such

† See Rutherford's account in *Background to Modern Science*, Macmillan, New York, 1938.

a structure as to explain the discrete energy levels observed in spectral lines. Bohr's approach to the problem was somewhat as follows: He decided to develop a working model for the simplest and lightest of all atoms—namely, the hydrogen atom—by assuming a positive nucleus and a single electron going around it in a circle. The model could assume that this nucleus and its "planetary" electron contained a certain energy, which presumably would be determined by the velocity and radius of the electron in its orbit.

Could the electron have no more than one radius or one velocity? There was no reason to assume that the electron could not move to an orbit of larger radius if it should suddenly gain some additional energy. For example, if a fast and positively charged particle were to pass close to this atom, its electron could be pulled (accelerated) by the Coulomb attraction, and this added energy could force the electron to a higher velocity in an orbit of larger radius. (A fast, negatively charged particle would have a similar effect on the electron by repelling it.)

Each gain of energy would therefore force the electron to find a larger orbit. By the same token, each orbit would represent a *particular velocity and radial distance for the electron and therefore for the energy state of the atom.*

Can an electron that gains energy lose that energy somehow and return to lower velocity and smaller radius? Yes, it can and it does, whenever permitted to do so, according to this model. In fact the early evidence presented by the spectrometer seemed to offer the proof that it did. When hydrogen gas was placed in the beam [Figs. 18.1 and 18.2(c)], there appeared dark lines in the otherwise continuous spectrum. On the Bohr model, the *dark lines appeared because those frequencies of the spectrum became absorbed by the hydrogen atoms,* causing those particular hydrogen atoms to be raised to higher energy states (that is, their electrons moved to larger orbits). But when hydrogen gas atoms became highly disturbed by electric discharge, they emitted four bright lines, as illustrated in Fig. 18.2(a). The explanation was simply that *light is emitted when an electron goes from a higher energy* (larger orbit) *state to a lower energy state* (smaller orbit).

The evidence given by the spectrometer, and as systematized by Balmer and others, suggests that we may be able to go even further. Since the series of lines have very precise, fixed frequencies and relationships, is it proper to assume that the atom itself can take on only certain fixed, precisely determined, energy levels? That is, the electrons cannot take on just *any* orbital path or energy state. Can we say, therefore, that each atom has a number of *energy states* (E_1, E_2, ...) which its electrons may occupy, depending on the energy of the electrons? Indeed each element, beginning with hydrogen and going to all other elements, may conceivably have its own precise system of spectral lines and therefore of energy states.

With these notions and questions in mind, let us attempt to draw a diagram of possible energy states. We begin with a lowest energy state (E_1 of Fig. 18.5) and work upward to higher energy levels, $E_2, E_3 \ldots E_n$. An atom presumably has a certain minimum internal energy (E_1) in its *normal* state (also called the *ground* state). Since we know that atoms are normally stable, we must assume that atoms do not radiate energy when they are in this ground state. However, the atom must be capable of absorbing energy and thus of being raised to a higher energy state. Moreover, it can radiate light only when it is in an energy state from which it may drop to a lower energy state.

We can picture the energy state of an atom to be determined by the combination of electrostatic and electromagnetic interactions of its positive nucleus and its negative electrons, plus the mechanical energy of its internal motions. A higher energy state is therefore representative of more intense activity on the part of its electrons. In fact, we can picture that the change in energy state is largely determined by the electrons occupying different orbits, the higher energies being associated with the larger orbits.

But now we ask: How does the electron go from one orbit to another? The mechanism for its transfer is not known. We do gather from Planck's postulate, however, that the radiation and absorption of energy by the atom *always* takes place in *integral multiples of the quantum of energy, hf*. Therefore, one should be able to relate the frequency f_{mn} of the radiation emitted by the atom, when it changes from the state characterized by energy E_m to a state characterized by a lower energy E_n, by a simple formula such as

$$hf_{mn} = E_m - E_n \qquad (18\text{-}3)$$

Let us see if this picture conforms to the experimental evidence as represented by Eq. (18-2). That equation gave the general formula for different series of the spectral lines of hydrogen developed by Balmer and others. By comparing this formula with Eq. (18-3), it is easily seen that $(E_m - E_n)$ should be equal to $hcR(1/m^2 - 1/n^2)$.† In other words, as-

---

† The student will recall that frequency $f = c/\lambda$. Then $1/\lambda = f/c$, or $hf = h(c/\lambda) = hcR[(1/m^2) - (1/n^2)]$.

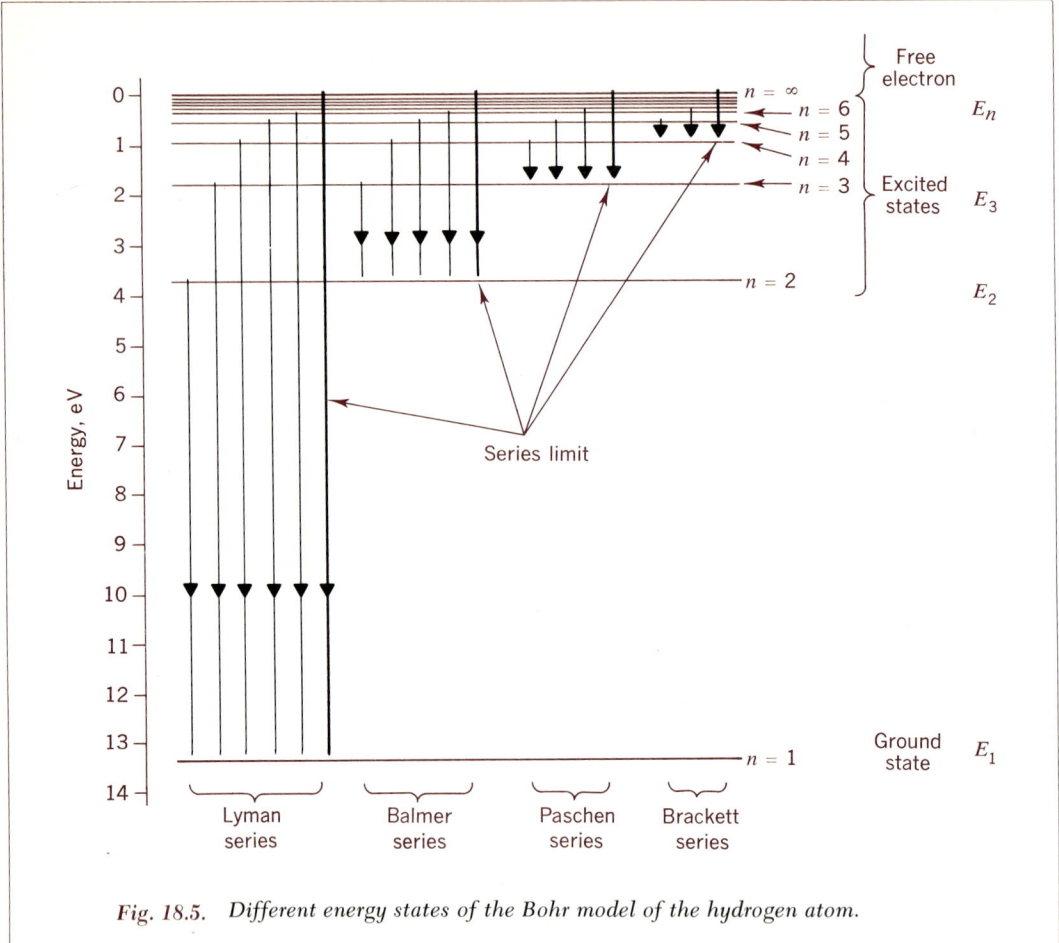

Fig. 18.5. Different energy states of the Bohr model of the hydrogen atom.

suming for E_m the value $hcR(1/m^2)$ and for E_n the value $hcR(1/n^2)$ should provide the means for making the resulting model consistent with observations.

So, in general, any state E_n of the atom of hydrogen should have the form

$$E_n = hcR\left(\frac{1}{n^2}\right) \quad (18\text{-}4)$$

where n is an integer, h is Planck's constant, and R is the Rydberg constant.

Bohr now had the problem of explaining how the energy of the different states of the atom could be of this form, varying inversely with the square of an integer n. We shall merely state that he did solve this problem and in doing so was able to replace the empirical Rydberg constant with terms that are physically meaningful. (See Sec. 18.4.)

Figure 18.6 again illustrates the different energy levels of the hydrogen atom according to Bohr's model. (The size of the nucleus at the center is exaggerated in the figure.) The smallest possible orbit ($n = 1$) is shown to have a radius r_1. The radii of other orbits are expressed in multiples of r_1. The orbits are not drawn to scale in Fig. 18.6, the large radii being much smaller than they should be. An electron would have minimum energy in the orbit with the smallest radius. When the hydrogen atom is in its ground state, the electron revolves in this orbit. The diameter of this orbit is the diameter of the normal hydrogen atom, which is of the order of 10^{-10} m.

The spectral lines of the Lyman series are due to transitions of the electron from orbits with $n > 1$ to orbit $n = 1$. The spectral lines of the Balmer series are due to transitions of the electron from orbits with $n > 2$ to orbit $n = 2$, and the lines of the Paschen series result from transitions of the electron from orbits with $n > 3$ to orbit $n = 3$. The transitions giving rise to the Lyman, Balmer, and Paschen series of spectral lines are shown in the figure.

Let us restate in simplified form the principles that emerged from Bohr's model.

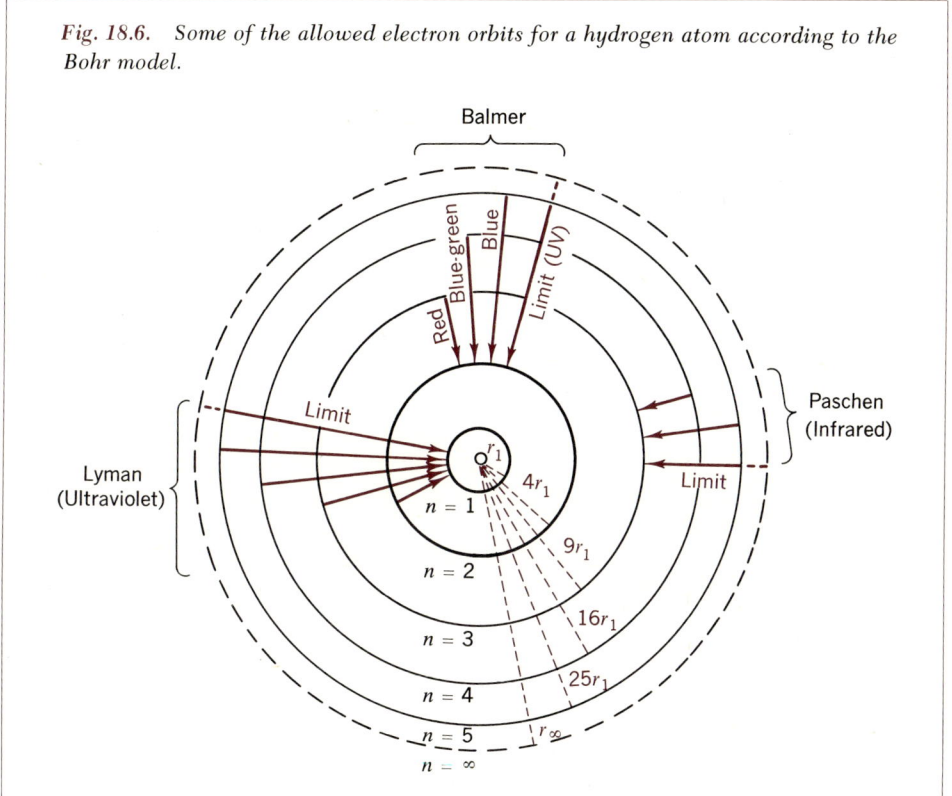

Fig. 18.6. *Some of the allowed electron orbits for a hydrogen atom according to the Bohr model.*

(1) An atom can exist only in certain discrete energy states.

(2) The energy states are quantized in accordance with Planck's postulate.

(3) When an atom undergoes transition from an upper energy state E_m to a lower energy state E_n, a photon of energy hf_{mn} is emitted with the requirement that $hf_{mn} = E_m - E_n$.

(4) If an atom absorbs photon energy hf and undergoes a transition from an energy state E_n to a higher energy state E_m, the same relationship must hold, namely, $hf_{mn} = E_m - E_n$.

(5) An atom does *not* radiate or absorb radiation *unless* a change of energy state occurs.

18.4 Bohr's problem

Bohr's problem essentially consisted of developing a model to express the energy of the hydrogen atom in the form $E_n = hcR/n^2$, where n is an integer and R is the Rydberg constant for hydrogen. The Rydberg constant had to be expressed in terms of other physical constants that have more general physical meaning.

When the electron moves in a circular orbit in the field of the positive nucleus with a specific velocity, it has kinetic energy and electrostatic potential energy corresponding to that velocity and the orbital distances and charges involved. Consider the electron, with mass m, to be in its nth orbit, with radius r_n and velocity v_n. If the charge of the electron is $-e$ for the hydrogen atom (which is the element we shall consider), the charge of the positive nucleus must be $+e$, to account for the electrical neutrality of the atom. Then the potential energy of the electron will be $-9 \times 10^9 (e^2/r_n)$ and it will have a kinetic energy $\frac{1}{2}mv_n^2$. The sum of these $(-9 \times 10^9(e^2/r_n) + \frac{1}{2}mv_n^2)$ is the total energy E_n of the electron:†

$$E_n = -9 \times 10^9 \left(\frac{e^2}{r_n}\right) + \frac{1}{2} mv_n^2 \quad (18\text{-}5)$$

To keep the electron stable in its circular orbit around the nucleus, it is necessary that an inward force of magnitude mv_n^2/r_n act on the electron. The attractive force needed is supplied by the electric force according to Coulomb's law: $9 \times 10^9 [(+e)(-e)/r_n^2]$. See Fig. 18.7. Therefore, $+9 \times 10^9 (e^2/r_n^2) = mv_n^2/r_n$ is the

† The potential energy existing between two charged particles is given by the product of the charges divided by the distance between their centers, all multiplied by a suitable constant. See footnote of Section 11-1 for the constant 9×10^9. See Section 5-7 for forces in rotational motion.

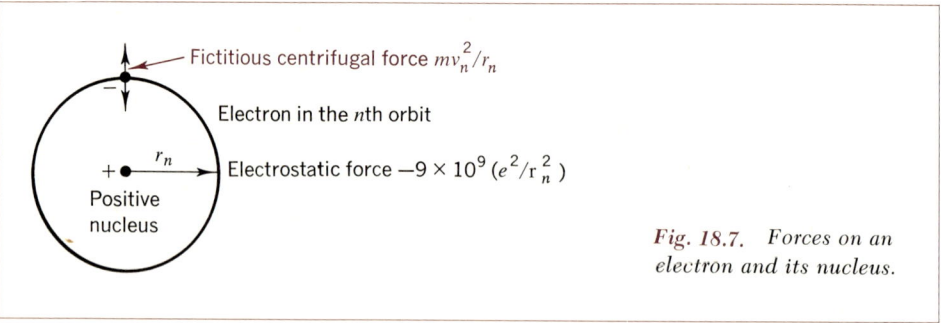

Fig. 18.7. *Forces on an electron and its nucleus.*

necessary condition. This requires that the electron have a velocity given by the relation

$$v_n = \sqrt{9 \times 10^9} \left(\frac{e}{\sqrt{mr_n}} \right) \quad (18\text{-}6)$$

By substituting this into Eq. (18-5), one gets for the total energy of the electron:

$$E_n = -\frac{1}{2}(9 \times 10^9) \frac{e^2}{r_n} \quad (18\text{-}7)$$

How could this be put into the required form given by Eq. (18-4)? Bohr's problem was to develop the model in such a way that the empirical Rydberg constant could be replaced by a term or product of terms that was physically meaningful. Essentially this depended on the significance to be assigned to the radius of the orbit, r_n.

Planck's radiation law can be written in the form of $nh = E/f$. Since n is an integer and f has the dimension of (time)$^{-1}$, it follows that h must have the dimension of: (energy) × (time). This is equivalent to the dimension of: (mass) × (velocity) × (length). The quantity in classical physics that has this dimension is the "action." The electron's action in the nth orbit can be obtained by the product of three quantities: its mass, its velocity, and the length of its circular orbit. By requiring that the *action of the electron be an integer multiple of* h, Bohr decided to incorporate the *quantum idea* into his model of the hydrogen atom.

The condition he introduced is expressed as an equation in the form:

$$m \times v_n \times 2\pi r_n = nh \quad (18\text{-}8)$$

This requires that the radius of the nth orbit, r_n, be given by the relation†

---

† To derive this relation, substitute Eq. (18-6) into Eq. (18-8).

$$r_n = \frac{1}{9 \times 10^9} \cdot \frac{n^2 h^2}{4\pi^2 m e^2}$$

By substituting this into Eq. (18-7), he obtained for the energy E_n:

$$E_n = -(9 \times 10^9)^2 \frac{2\pi^2 m e^4}{h^2} \left(\frac{1}{n^2}\right) \quad (18\text{-}9)$$

which seemed to be the form he wanted to obtain. Equating this to Eq. (18-4), he could resolve the required expression for the Rydberg constant:

$$R = (9 \times 10^9)^2 \frac{2\pi^2 m e^4}{ch^3}$$

By substituting known values for the charge and mass of the electron and for Planck's constant, Bohr found the value for the empirical Rydberg constant to be the same as that assumed in the generalized Balmer formula. Bohr had achieved what he wanted to do.

The reader should have noticed that Bohr accepted Rutherford's model as his base and incorporated just enough new postulates to provide a model that might permit stability of the atom and at the same time explain the spectroscopic data that were available. One of the postulates, Eq. (18-8) was derived from Planck's hypothesis. The assumption that the electron does not radiate as long as it is in a particular quantum orbit was especially bold, since it was contrary to the concept of classical electromagnetic theory, which stated that an accelerated charged particle radiates energy. For developing such a model of the hydrogen atom, Niels Bohr was awarded the Nobel Prize in Physics.

18.5 Success and shortcomings of the Bohr model

The Bohr model "explained" the stability of the atoms, the observed frequency of the emission and absorption spectra of hydrogen-like atoms with one electron, and the measured ionization energies of one-electron atoms.† However, the model was soon found to have limited validity. For instance, it was incapable of explaining the spectra of atoms having more than one electron. It failed to account for the fine details of the hydrogen spectrum (high-resolution spectroscopic measurements show that each "line" predicted by the Bohr theory consists of two or more closely spaced lines or fine structure). It was nonrelativistic and it gave no method of calculating the intensities of spectral lines. Equation (18-8) quantized the orbital angular momentum of the electron; but the quantization of angular momentum actually requires a more complicated rule than that given by the Bohr model.

All these shortcomings were overcome by a relativistic wave-mechanical treatment, which is beyond our scope here because of its mathematical complexity. This newer treatment removed the emphasis of the Bohr model on the *particle nature* of the electron, and substituted for it a *wave* aspect to which we shall turn in the next chapter.

18.6 Extensions of Bohr theory

The Bohr model of the atom focused more attention on the suitability and inadequacies of the model to explain the details of spectral lines. By using spec-

† The ionization energy is the minimum energy that must be given to an atom to remove an electron entirely away from the attraction of its nucleus. An atom is said to be ionized when the electron is removed, and said to be "excited" when the energy of the bound electron is above that of the ground (or lowest) state. When an electron leaves its atom, the remaining electrons become somewhat rearranged.

trometers of higher resolution, and subjecting the emitting atoms to magnetic fields, the spectral lines were shown to consist of multiple lines that are close together. The Bohr model clearly could not explain the phenomena in terms of the simple circular orbits for electrons. It appeared that there were other variables to be considered.

In 1916 the German scientist Arnold Sommerfeld extended the Bohr theory of the atom by replacing the circular orbits with elliptical ones. The reasoning was relatively simple: Since the inverse square force of gravity produces elliptical orbits for the planets, with the sun at one of the foci, is not the inverse square Coulomb force likely to produce elliptical orbits for the electrons, with the nucleus at one of the foci? We must, of course, observe the quantum conditions that restrict the electron to certain allowed elliptical orbits. When we applied the quantum condition [Eq. (18-8)] to circular orbits of the Bohr model, we had only to consider the rotation of the radius connecting the nucleus and the electron, since all other factors remained unchanged. For an elliptical orbit, however, the radius not only rotates but also varies in its length, which calls for the quantization of *two* factors. Sommerfeld extended Bohr's quantum condition to include these factors, each one required to be *an integer times Planck's constant*. Therefore the allowed orbits of Sommerfeld's model are characterized by two quantum numbers instead of just one. One of these is the same as Bohr's original quantum number, n, described more commonly as the *principal quantum number*. The second one is designated by l and is called the *orbital angular momentum quantum number*. The symbol l expresses the eccentricity (or flatness) of the ellipse. The values of l are restricted to integral values starting with 0 and ending with $(n-1)$. Thus, for each principal quantum number, there are n allowed orbits. The most elliptical orbit has $l=0$, and as the value of l increases, the orbits become less elliptical, the final one being a circle and having the value of $l=n-1$. The nucleus is at the center of the circular orbit and at one of the foci of the ellipses. For example, let us look at the orbits for the value of the principal quantum number, $n=3$. Three orbits are possible: one having the value $l=2$, another with the value $l=1$, and the third with $l=0$ (value zero). Of these, the one with $l=0$ is the most elliptical and the one with $l=2$ is a circle. All three orbits are shown in Fig. 18.8.

Sommerfeld also investigated the effect of the electron's relativistic change in mass due to its speed in the orbit. The greatest speed of the electron in the orbit approaches 1/100 the speed of light. This is enough to introduce a very small relativistic correction, which is enough to reveal fine structure for spectral lines. In introducing this correction, Sommerfeld showed that the energies for different orbits for each value of n were slightly different. Transitions that terminate on an energy level represented by $n=2$, for example, could end up on two slightly different energy levels, one for the circle and one for the ellipse. Thus two photons with slightly different energies would be radiated. In this way Sommerfeld could predict the observed elaborate fine structure of H and He$^+$ spectral lines.

The circular and elliptical orbits encountered so far represent motion in only two dimensions. Sommerfeld also considered the problem of movement in all three dimensions. Applying quantum conditions to each of the three coordinates, the orbits turned out to be circles

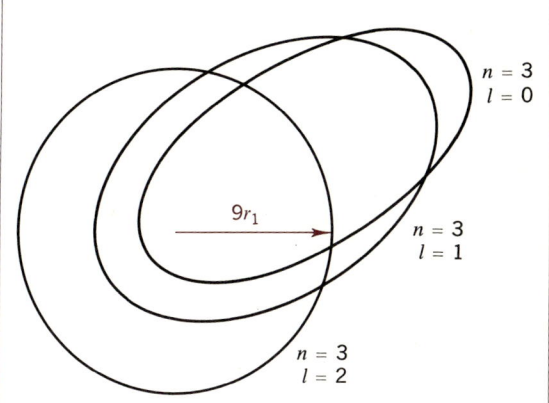

Fig. 18.8. *Sommerfeld's orbits for the principal quantum number, $n=3$.*

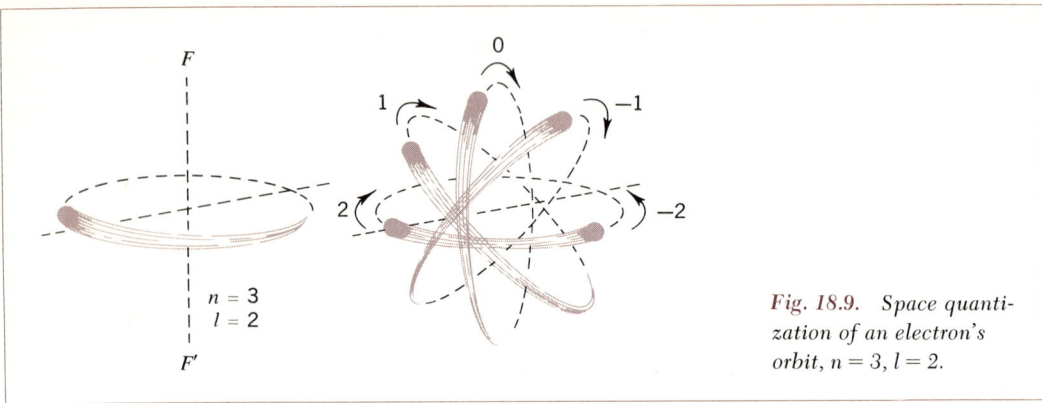

Fig. 18.9. Space quantization of an electron's orbit, $n = 3$, $l = 2$.

and ellipses as described above and specified by quantum numbers n and l. In addition, the plane of the orbit was found to have certain allowable orientations in space. This orientation of the orbit in space was called *space quantization*. Existence of space quantization seemed to explain a phenomenon that had been observed experimentally much earlier in what had been called the Zeeman effect. In 1896, P. Zeeman, who was studying the emission lines from atoms placed in a strong external magnetic field, found that the spectral lines appeared to be broadened. Closer examination of these spectral lines under high resolution showed that what normally appears as a single line becomes split by the application of an external field into two or more closely spaced, sharp component lines. This splitting of the spectral lines into discrete components by a magnetic field is known as the Zeeman effect, for the explanation of which the Sommerfeld hypothesis involving space quantization seemed to offer better possibilities than had existed up to that time.

According to space quantization, the orbital angular momentum vector cannot assume just any direction in an external magnetic field; rather it is restricted to those particular orientations for which the component of the orbital angular momentum vector along the direction of the magnetic field is an integral multiple of $h/2\pi$. These possible values are therefore given by a third quantum number, m_l, the *orbital magnetic quantum number*, which may take on certain integral values determined by l. For a given value of l, these values are $m_l = l, l-1, l-2, \ldots 0, \ldots (-l)$. For example, let us take the case $n = 3$, $l = 2$. The possible values of m_l are five, namely, 2, 1, 0, -1, -2. These five different orbits resulting from the space quantization of an electron orbit, $n = 3$, $l = 2$, when placed in a magnetic field in direction FF' are shown in Fig. 18.9.

It is obvious that for a large value of the principal quantum number n, l can take $(n-1)$ values, and each of these can give rise to a large group of space quantized states.

In the case of the spectral lines of sodium, even in the absence of an external magnetic field, each of the orbital energy levels (except the levels for which $l = 0$) is split into two components. How can the doubling of these states be explained? This fine structure of sodium resembles the Zeeman effect in that both represent a splitting of otherwise single lines, except that unlike the Zeeman effect, this fine structure does not require an external magnetic field for its observation. Could it be that this fine-structure splitting is attributable to an "internal Zeeman effect" within the atom? Such an interpretation requires the presence of an *internal magnetic field* and a new source of magnetic moment and angular momentum within the atom. We have already taken into account the orbital angular momentum of the electron. What other contribution to angular momentum can there be? An answer was not long in coming.

In 1925 two Dutch physicists, S. A. Goudsmit and G. E. Uhlenbeck, suggested that an intrinsic angular momentum might be associated with the electron. They assumed that as the electron moves along an orbit, it also spins on an internal axis, just as our earth rotates on its own axis as it moves around the sun. But since the electron carries an electric charge, would not both its spin and its orbital motion be the equivalent of a moving charge that produces a magnetic field? If so, the spinning charge of the electron creates a magnetic dipole along the axis of the spin, while the orbital motion of the electron creates a magnetic field perpendicular to the plane of the orbit. How is the axis of the spin oriented? On the basis of the motions we have postulated, there would be two directions, both parallel to the magnetic field due to orbital motion, but one in the *same* direction as the field and the other *opposite* to that direction. These two possible orientations are therefore specified by a fourth or *spin quantum number*, m_s, which is assigned two values, $+\frac{1}{2}$ and $-\frac{1}{2}$. (See Fig. 18.10.) The two orientations represent a small difference in the energy levels that are possible to the atom.

By associating different states with allowable orbits for the electrons, we have succeeded at last in explaining the complex spectra of different elements. The model of orbits we have pictured

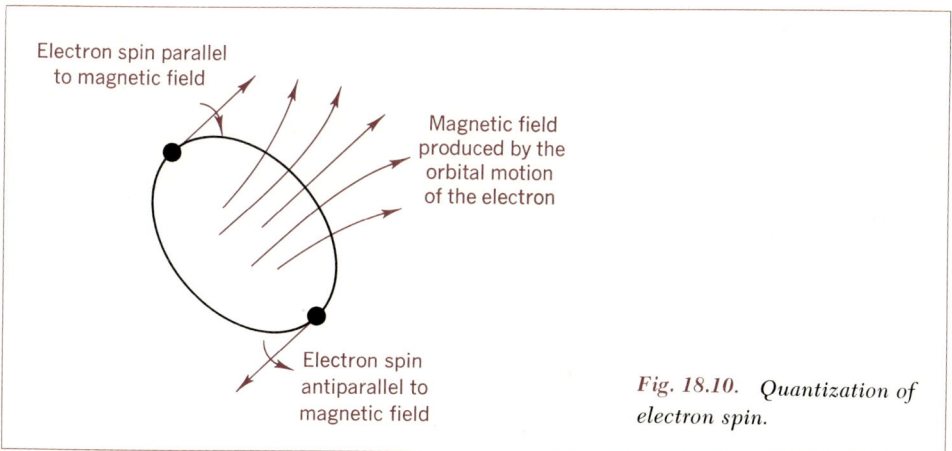

Fig. 18.10. *Quantization of electron spin.*

may not be accurate in the physical sense, but the four quantum numbers n, l, m_l, and m_s nevertheless do describe the discrete energies of the different states.

18.7 Explanation of the periodic table of elements

Thus far we have considered only the mechanical and electromagnetic features that can be conceived to be characteristic of atoms. But the characteristics of atoms as elements of nature have been identified largely with their *chemical* behavior with respect to each other. A major question now must be posed, namely: Can the chemical characteristics of atoms, as listed in the periodic table of the elements, be explained in terms of the structural features we have just discussed? Fortunately, with the aid of the four quantum numbers just proposed, and a principle suggested by Wolfgang Pauli, we can now provide some explanation for the periodic table of elements.

The *Pauli exclusion principle* simply recognized that as far as we can tell from evidence, *within an atom no two electrons can exist together in the same energy state*. Since the four quantum numbers n, l, m_l, and m_s seem to describe the state of an atom unequivocally, what this principle says is that no two electrons can have the same set of quantum numbers. No exception to this rule has been found as yet.

Consider the hydrogen atom with a single electron. This electron may take any set of n, l, m_l, and m_s without violating Pauli's principle. However, $n = 1$ is the lowest possible energy state, and hence both l and m_l are zero. The electron can assume either of two possible substates due to its spin because of the two values $+\frac{1}{2}$ and $-\frac{1}{2}$ that m_s can take, depending on the direction of spin.†

Consider next a helium atom with two electrons. In the ground state with $n = 1$, these two electrons can occupy states characterized by $n = 1$, $l = 0$, $m_l = 0$, $m_s = +\frac{1}{2}$, and by $n = 1$, $l = 0$, $m_l = 0$, $m_s = -\frac{1}{2}$. That is, when the two electrons have identical values for quantum numbers n, l, m_l, they must have different values for m_s. This is simply accomplished when the two electrons are in the same shell but have *opposite spin directions*.

What happens when there is a third electron, as in the element lithium? Can the third electron stay in the same energy state as either one of the other two? According to Pauli's principle, it cannot have the same four quantum numbers of any other electron in that atom, which means that it has to "find" a new energy state, in which it differs by at least one quantum number from any other electron. The third electron, therefore, goes into an outer orbital state. This leaves the first two electrons in what we may call a *closed shell*, since that shell will not accept any more electrons. We shall see that such closed-shell conditions tend to make an atom more stable against change than when there are electrons lacking to complete the shell.

In the case of the lithium atom with three electrons, the third electron moves into a new shell position with the principal quantum number $n = 2$. Within this new shell of $n = 2$, electrons can have $l = 0$ or $l = 1$ ($= n - 1$). Two electrons

---

† The small energy difference between the states $s = +\frac{1}{2}$ and $s = -\frac{1}{2}$ corresponds to a wavelength of 21 cm (frequency of 1530 Mc/sec). It is the transition between these two levels of hydrogen that accounts for the 21-cm radiation in our Galaxy.

TABLE 18.1

ELECTRONIC STRUCTURE OF ATOMS†

Principal Quantum Number n	1	2		3			4		
Orbital Quantum Number l	0	0	1	0	1	2	0	1	
Letter Designation of State	1s	2s	2p	3s	3p	3d	4s	4p	Observations

Z		ELEMENT	V, volts	1s	2s	2p	3s	3p	3d	4s	4p	Observations
1	H	Hydrogen	13.60	1								Highly reactive with other atoms.
2	He	Helium	24.58	2								Inert because the 1s shell is now complete.
3	Li	Lithium	5.39		1							The lone third electron is easily removed.
4	Be	Beryllium	9.32		2							
5	B	Boron	8.30		2	1						
6	C	Carbon	11.26		2	2						C needs 4 electrons to fill the n = 2 shells.
7	N	Nitrogen	14.54		2	3						Could use 3 electrons to fill the n = 2 shell; hence valence 3.
8	O	Oxygen	13.61	*Helium Core*	2	4						Needs two electrons to fill the shell.
9	F	Fluorine	17.42		2	5						Needs one more electron to fill the shell.
10	Ne	Neon	21.56		2	6						Inert because now the n = 2 shells are complete.
11	Na	Sodium	5.14				1					The lone electron can be easily lost.
12	Mg	Magnesium	7.64				2					
13	Al	Aluminum	5.98	*Neon Core*			2	1				
14	Si	Silicon	8.15				2	2				Somewhat like carbon.
15	P	Phosphorus	10.55				2	3				Much like nitrogen in its behavior.
16	S	Sulfur	10.36				2	4				Like oxygen, it needs two electrons to fill a shell.
17	Cl	Chlorine	13.01				2	5				Needs one electron to fill the shell.
18	A	Argon	15.76				2	6				The shell being completed, argon is inert.
19	K	Potassium	4.34							1		The lone electron can be easily lost.
20	Ca	Calcium	6.11							2		
21	Sc	Scandium	6.56						1	2		
22	Ti	Titanium	6.83						2	2		
23	V	Vanadium	6.74						3	2		(Note that the sequence of electrons becomes irregular for Z above 18, the reasons being that the internal energy levels increase in an irregular way with addition of electrons.)
24	Cr	Chromium	6.76	*Argon Core*					5	1		
25	Mn	Manganese	7.43						5	2		
26	Fe	Iron	7.90						6	2		
27	Co	Cobalt	7.86						7	2		
28	Ni	Nickel	7.63						8	2		
29	Cu	Copper	7.72						10	1		
30	Zn	Zinc	9.39						10	2		
31	Ga	Gallium	6.00						10	2	1	
32	Ge	Germanium	7.88						10	2	2	
33	As	Arsenic	9.81						10	2	3	
34	Se	Selenium	9.75						10	2	4	
35	Br	Bromine	11.84						10	2	5	
36	Kr	Krypton	14.00						10	2	6	

† Adapted from Charlotte E. Moore, *Atomic Energy Levels*, Vol. II, National Bureau of Standards, Circular 467, Washington, D. C. 1952.

can have $l = 0$ and differ by spin direction only. Six electrons may have $l = 1$ because they can differ by having $m_l = 0$, $+1$, or -1, and each of these offers space for two electrons if the two have different spin directions. This makes a total of eight electrons that can be accommodated in the shell $n = 2$.

Thus, when the shell for $n = 1$ is complete with two electrons we have the element helium, which is one of the nearly inert, stable, and so-called noble gases. When the shell for $n = 2$ is also complete (with eight more, or a total of ten electrons), we have another noble gas, neon. So it is that completed shells give an inert character to atoms.

The shell structure of some of the elements are given in Table 18.1. In this table the "shells" are customarily identified by the letters $K, L, M, \ldots$, corresponding to the principal quantum numbers $n = 1, 2, 3. \ldots$ There are also subshell designations $s, p, d, f, \ldots$, corresponding to quantum numbers $l = 0, 1, 2, 3 \ldots$ etc. The column under V lists the ionization voltage; that is, the energy required to remove one electron.

18.8 The covalent bonding of atoms

The Pauli exclusion principle helps us to "explain" (if not completely understand) many of the behavioral characteristics of atoms in relation to each other. Although atoms are electrically neutral in that they contain as many negative as positive charges, only a very few of them (the so-called noble gases) are likely to be nonreactive toward other atoms. The concept of "shells," just mentioned, becomes useful for explaining the behavior of atoms. When a "shell" is filled with all the electrons it can hold (according to the Pauli exclusion principle) it is likely to be nonreactive toward other atoms. The remarkable aspect of this is that when the "shells" of atoms do not contain all the electrons they can hold, they combine in such a manner as to share electrons and to give each atom the *appearance* of having a full complement of electrons in simulated shells.

The hydrogen atom is highly reactive. That is, it cannot remain alone if there is opportunity for it to join other atoms. Suppose we locate two hydrogen atoms a substantial distance (say 10 or more atomic diameters) away from each other. Since each is electrically neutral, there will be no forces between them (assuming the gravitational attraction to be insignificant, which is the case). But now let the two approach each other to be only about 1 diam apart (Fig. 18.11). As the electrons move from one side of their nuclei to the other, the distances between the electrons and between the nuclei and electrons vary by large percentages. This in turn causes the local repulsive and attractive forces to change considerably according to the Coulomb inverse square law, so that the "neutral" state of each whole hydrogen atom becomes meaningless in the face of the intense local, changing forces.

Even more important, each hydrogen atom finds a more stable energy state by attracting the electron of the other atom into its own sphere to simulate a helium-like shell. The two atoms, therefore, join together to form a hydrogen molecule (Fig. 18.11). By sharing each other's electron, the two hydrogen atoms find a structure that is more stable and therefore less chemically reactive. This form of uniting by sharing of electrons is called

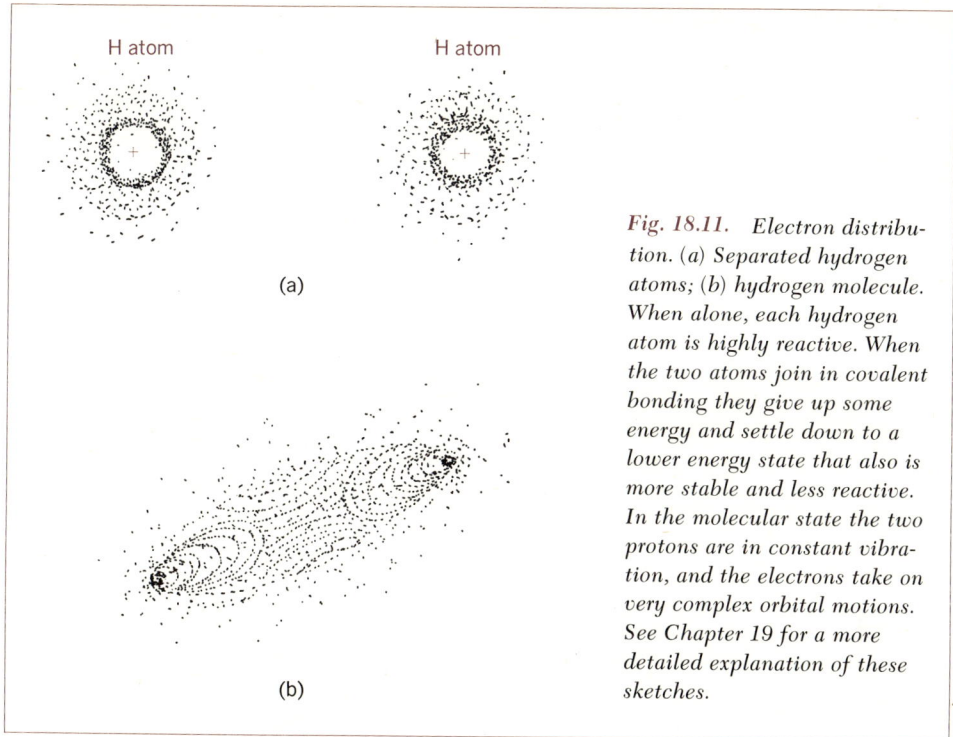

Fig. 18.11. *Electron distribution. (a) Separated hydrogen atoms; (b) hydrogen molecule. When alone, each hydrogen atom is highly reactive. When the two atoms join in covalent bonding they give up some energy and settle down to a lower energy state that also is more stable and less reactive. In the molecular state the two protons are in constant vibration, and the electrons take on very complex orbital motions. See Chapter 19 for a more detailed explanation of these sketches.*

covalent bonding. A covalent bond is one in which two atoms are held together by sharing one or more pairs of electrons that have opposite spins, each electron having orbital motion around both nuclei. That is,

$$H + H \rightarrow H_2 \qquad (18\text{-}10)$$

What is the pattern of electron orbital motion in such a molecule? Figure 18.11(b) illustrates a complex pattern of orbits, and indeed the spectra of molecules are correspondingly complex. The two nuclei maintain an average separation between centers of about 0.74 Å, and vibrate in relation to each other with an amplitude of a few hundredths angstroms.

Each electron now "belongs" to both nuclei. Each nucleus thereby also has its two electrons of opposite spin direction, and thus simulates the more stable helium structure.

What is the nature of the forces that result from covalent bonding? The attractive forces between the positive nuclei and the bonding electrons are present of course, as are also the repulsive forces between the two electrons and between the two nuclei. But these electromagnetic or electrostatic forces are not adequate to explain the magnitude of the bond energy. The Bohr-Sommerfeld model, which we have discussed in this chapter, could not account adequately for covalent bonding. The answer had to await the development of wave mechanics (see Chapter 19). According to the wave model, it requires less energy to make up the hydrogen molecule than it does to make two individual hydrogen atoms in their ground state; therefore the molecular structure will occur whenever possible.

Covalent bonds are very important and common for making up molecular assemblies,† as we shall see in Part II

---

† The covalent bond is commonly encountered in inorganic and biological systems. Covalent bonds, such as the C—C, C—N,

of this text. There exists, however, another important mechanism for molecule formation, the ionic bond.

18.9 The ionic bond: electrolysis

As noted earlier, sodium has a neon-like shell structure of ten electrons plus an eleventh electron that begins the series for the next shell. We noted also that this eleventh electron is easily removed from

---

C—S, S—H, O—H and C—O, are found in fats, carbohydrates, proteins, and nucleic acids, and provide comparatively stable compounds.

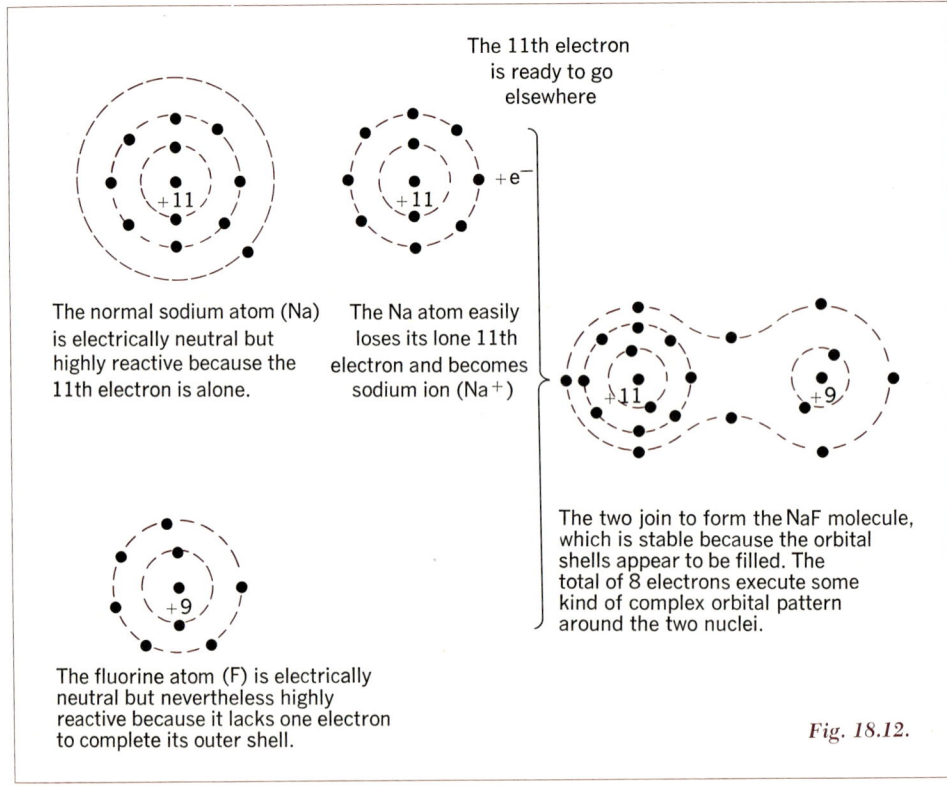

The normal sodium atom (Na) is electrically neutral but highly reactive because the 11th electron is alone.

The Na atom easily loses its lone 11th electron and becomes sodium ion (Na+).

The 11th electron is ready to go elsewhere

The fluorine atom (F) is electrically neutral but nevertheless highly reactive because it lacks one electron to complete its outer shell.

The two join to form the NaF molecule, which is stable because the orbital shells appear to be filled. The total of 8 electrons execute some kind of complex orbital pattern around the two nuclei.

Fig. 18.12.

18.9 The Ionic Bond: Electrolysis

the sodium atom, leaving a sodium *ion*. This ion has an excess of one positive charge; therefore it will experience a force when it is in the presence of another ion or of an electric field. An *ion* is simply an atom, or group of atoms, that bears an electric charge.

In contrast, the element fluorine has nine electrons, but it lacks one electron to form the stable structure of the noble gas neon. Because of this deficiency, we noted that fluorine is highly reactive; that is, it can easily join with other atoms to simulate a neon-like structure. In fact, the fluorine is able to capture an additional electron to become a negative ion (F^-). We can immediately guess that the Na^+ and F^- will strongly attract each other and that the combination of the two elements in the combination NaF will be stable. The bond is referred to as being *ionic* because the sodium atom is able to lose an electron to become a positive ion, and the fluorine is able to pick up an electron to become a negative ion (Fig. 18.12). It should be noted, however, that here also the same forces come into play with respect to the propensity of each atom to simulate stable shell structures for the associated electrons.

We can represent the behavior of Na and F by the following reaction:†

$$Na^+ + F^- \rightarrow NaF \qquad (18\text{-}11)$$

We have seen how fluorine combines readily with sodium through ionic bond. Can fluorine utilize covalent bonds for combination with other atoms? We find that it can. In fact, as is illustrated in Fig. 18.13, two fluorine atoms will combine to share each other's electrons in the fluorine molecule F_2.

† NaF is sometimes used to "fluoridate" municipal water supplies.

A more common example of ionic bonding is illustrated by the elements that make up common table salt, NaCl. The element chlorine (Cl) has 17 electrons, which is one short of the 18 that give the next element, argon, the stable structure of another noble gas. This lack of one electron gives chlorine the propensity to capture an additional electron, which makes it a negative ion (Cl^-). The combination NaCl is neutral and also simulates the stable shell structures.

While we have distinguished ionic bonds from covalent bonds, most molecular assemblies tend to include both forms to a greater or lesser degree. How can we distinguish the compounds that are held together by covalent bonds from compounds that are held together by ionic bonds? The differences show up in the properties of the two forms. For example, covalent compounds can change to molten or gaseous forms at relatively low temperatures, by simply separating the covalent molecules from each other, without having to break interatomic bonds. In contrast, ionic compounds such as NaCl will not become liquid until the temperature is high enough to break the ionic bond. For example, while the covalent molecule F_2 has a melting point of $-233°C$, the ionic compound NaF has a melting point of $980°C$—a vast difference. The fluorine molecule remains as a covalent molecule in the molten and gaseous state, whereas the NaF ionic compound changes its structure considerably.

The ionic compounds also behave quite differently from covalent compounds when subjected to electric fields. When ionic compounds are in the molten stage or are in solution, they become conductors of electricity, whereas covalent substances are rather poor conductors in the molten state. Figure 18.14 illus-

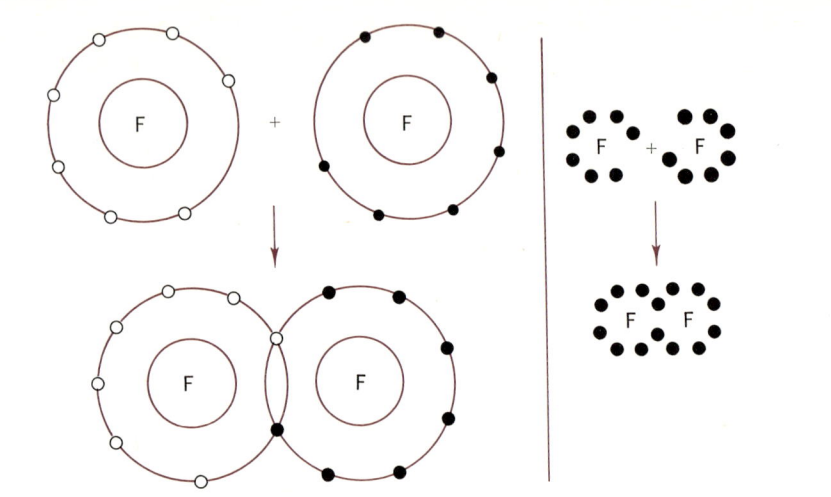

Fig. 18.13. Two fluorine atoms are shown with only the outer electron shells indicated. When apart, each atom has an outer shell that is one electron short of completion. By sharing one pair of electrons, however, each atom acquires a complete octet. The resulting compound is the molecule F_2. The same reaction is shown in shorthand notation on the right side of the diagram. Note that shared-electron or covalent bonds can also form between dissimilar atoms.

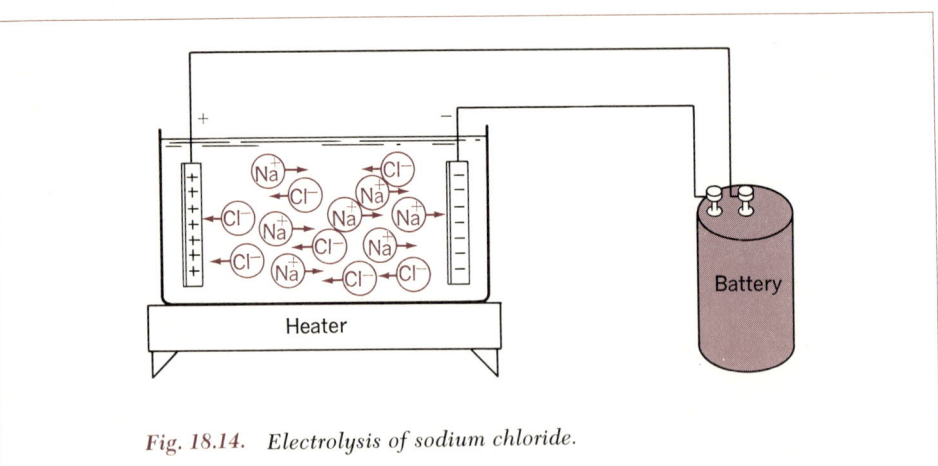

Fig. 18.14. Electrolysis of sodium chloride.

trates the manner in which the sodium ion and the chloride ion separate when in solution and become drawn to opposite electrodes under the influence of an electric field. In this example of *electrolysis*, the two electrodes may be of either metal or carbon, the choice being in part determined by whether the salt is to be melted at high temperature (above its melting point of 980°C) or simply dissolved in water. In either case the ions tend to separate and to become individually mobile. When the electrodes are connected to a battery as shown, the ions move separately toward the two electrodes, according to their electric charge. The positive sodium ions (*cations*) move toward the negative electrode (the *cathode*) and the negative chloride ions (anions) move toward the positive electrode (the *anode*). When the cations and anions reach the surface of the electrodes, they experience the following reactions in the case of the molten salt (that is, without the presence of water): At the positive electrode (anode),

$$2Cl^- \rightarrow \underset{\text{(chlorine gas)}}{Cl_2} + 2e^- \quad (18\text{-}12)$$

The sodium (positive) ions become drawn to the negative electrode (cathode),

$$2Na^+ + 2e^- \rightarrow \underset{\text{(sodium metal)}}{2Na} \quad (18\text{-}13)$$

The electrons from reaction (18-12), which constitute the electric current, pass from the anode through the wires and battery to the cathode, to make reaction (18–13) possible. The cathode collects the sodium metal, and the chlorine escapes as a gas at the anode. Here again we note that the chloride *ion* will join the sodium ion to form an ionic compound but it can also join another chloride ion to form a covalent compound, Cl_2.

Many of the properties of atoms and molecules can be explained on the basis of the Pauli exclusion principle. Table 18.1 lists the distribution of electrons in the orbital shells of atoms, based on the exclusion principle. The table lists the atomic number Z, the symbols and names of the elements, the ionization potential required to remove the outermost electron from each atom, and the distribution of electrons in the various subshells. The observations will be useful for understanding the characteristics of each of the atoms. Note that the properties tend to repeat as one moves from the first group ($Z = 3$ through $Z = 10$) to the second group ($Z = 11$ through $Z = 18$). . . .

18.10 Summary

In the discussion of the kinetic theory of gases, atoms were assumed to have no internal structure. By 1900 the conviction grew that the atom must be more complicated than had been assumed. In the first place, it had long been known that each element carried its own fingerprint —a unique series of spectral lines. Such line spectra hinted at something going on within the atom, which caused it to radiate at specific frequencies and no other. In the second place, J. J. Thomson had identified the electron, showing that it had a mass less than the mass of an atom, leaving scientists with the problem of identifying the origin of such particles. It was assumed that somehow they "come out" of the atom under certain conditions.

For the element hydrogen the pattern of spectral line distribution was finally fitted with a mathematical equation by Balmer. This was a purely empirical formula in agreement with observed regularities, but it presented a specific challenge to investigators: How was this pattern related to the internal structure of the atom? J. J. Thomson tried the simplest explanation or model, assuming that the small negative electrons were embedded in a large positive mass—like "raisins in a pudding" was the analogy used. But this hypothetical model did not work. It led to consequences not in accordance with the Balmer series.

In the meanwhile, the great English physicist Rutherford was trying to probe into the atom, using high velocity particles emitted by radioactive substances, which had recently been discovered. In this investigation, Rutherford was using a method that now has become a most fruitful aspect of modern research. Instead of merely observing phenomena by depending upon natural signals (light, sound, and so on) radiated or reflected by the objects under investigation, scientists use appropriate "external" probes. For example, X-rays, radar, laser beams, electrons, and protons shot out at high velocity by accelerators, shock waves, and satellites may be used for probing into regions that do not of themselves give off or reflect natural signals adequate to yield information about their structure.

Rutherford showed that the mass of an atom was heavily concentrated in a very small central portion. He suggested a planetary model, the electrons revolving about the central nucleus much as the planets revolve about the sun. For the reasons we have discussed, this model also would not work without some additional assumptions. Bohr provided the key, reaching into the ideas of quanta that had been developed by Planck and Einstein. He provided an elegant model of the atom, precisely fitting the most prominent lines of the hydrogen spectrum. The Bohr model was indeed an ingenious logical triumph—and a bold one because of the manner in which he extended quantum principles to atomic structure.

But it was not long before the Bohr model was subjected to two serious criticisms:

(1) One of the postulates (that 2π times the angular momentum is exactly equal to an integral number times Planck's constant h) seemed artificial (an *ad hoc* postulate is the technical term, borrowed from legal tradition). Why this particular form of equation?

(2) The Bohr model did not account for all the observed spectral lines, particularly those of atoms heavier than hydrogen.

The story of how the first objection was resolved is taken up in the next chapter. As for the second criticism, Sommerfeld and others modified the Bohr model by adding the other quantum numbers.

The Bohr and Sommerfeld models, plus the fourth quantum number associated with the spin of the electron, seemed to provide the components and combinations of atomic structure through which to explain practically all the details of line spectra. Pauli discovered another characteristic, namely, that within an atom no two electrons can have exactly the same four quantum numbers or energy state. That is, the quantum numbers for each electron must differ in at least one respect from those of any other electron in that atom. This led to the idea of "shells" in atoms. It seemed also that every atom tries to simulate a closed-shell structure by joining with other atoms and sharing electrons for that purpose. This tendency of atoms gives a clue, in fact, to very many chemical properties of atoms.

Eventually the model became more and more complex and cumbersome (compare the history of the Ptolemaic system), and today physicists have abandoned it except as a simplified model to summarize (more or less visually) some of the regularities found in atomic phenomena. Much can be learned from this. Just as the Ptolemaic system was a necessary precursor to the Copernican system, or Newtonian physics to relativity physics, so was the Bohr model necessary to further advances in atomic physics. Systematic scientific models, representing or "mapping" the regularities of nature, rarely remain unmodified for very long, but while they last they are extremely useful in pointing the way to other generalizations and in suggesting the right questions. In addition to this, the older models often can still be used for many purposes, *if not taken too literally*. Navigators continued to use the Ptolemaic system for over a century after Copernicus, and engineers still use Newtonian physics because it is simple and sufficiently accurate for the purposes at hand. (Can we not also add that the same use is made of "models" as guides to action in many fields other than the physical sciences?) A model is therefore not likely to be rated on the basis of whether it is altogether "correct" or "incorrect," but on the basis of how useful it can be.

Questions/Discussions

1. Explain the differences that you would observe in the spectrum if you were to do the following:

(a) Point the spectrometer toward the sun.
(b) Burn ordinary table salt in a bunsen burner and aim the spectrometer toward the flame.
(c) Aim the spectrometer toward a glowing discharge tube that is filled with hydrogen gas.
(d) Aim the spectrometer toward a carbon arc source of light, and interpose the discharge tube of question (c) in the path of the beam while the tube is not glowing,

2. How confident can we be that the lines observed in a spectrum represent light emitted from within atoms?

3. Using the Bohr model, explain how a single electron of the atom can give so many spectral lines.

4. What determines the frequency of light emitted by an atom?

5. List the order of transitions between the state of an atom, beginning with the light of shortest wavelength and going to the longer wavelength (that is, such transitions as from $n = 3$ to $n = 2$, $n = 50$ to $n = 40$, $n = \infty$ to $n = 1$).

6. Why do incandescent solids and liquids give a continuous spectrum?

7. Did the Bohr model of electron behavior in atoms negate the idea that a charged particle radiates energy when being accelerated? How can we reconcile the two?

8. The energy absorbed by a hydrogen atom in order to go from $n = 1$ to $n = 2$ is much greater than the energy required to go from $n = 2$ to $n = 3$ state. Why?

9. What is the distinction between ionization and excitation of an atom? How do the energies compare for hydrogen?

10. Having in mind the Pauli exclusion principle, list the energy states and the "shells" of the organization of electrons for neon.

11. Draw an orbital plan for the carbon atom.

12. How do the features of atomic structure as given in the Bohr model compare to planetary motion?

13. The maximum wavelength of light that will remove an electron from the magnesium atom is 3700 Å. Express this in electron volts.

14. Compare the concepts and the models of the atom of the Greeks, of Thomson, of Rutherford, and of Bohr.

15. Explain and contrast covalent and ionic bonds.

16. Using the covalent bonding or the ionic bonding as an example, what would you say are the characteristics of an *assembly* of parts as compared with the characteristics of individual parts? (That is, to what extent do the behavioral characteristics of individual parts also define the characteristics of the combination?)

References

The various references given in Chapter 17 all contain material pertinent to this chapter. In addition the literature listed below is highly recommended.

Boorse, H. A., and Motz, L. (eds.). *The World of the Atom*. New York: Basic Books, 1966. (*Includes prefatory comments to all papers*) (1) Thomson: "The Arrangement of Corpuscles in the Atom," p. 613. (*Discusses the "Thomson" atom*); (2) Geiger and Marsden: "On a Diffuse Reflection of the α-particle," p. 693, and Rutherford: "The Scattering of α- and β- Particles by Matter and the Structure of the Atom," p. 701 (*These two papers discuss the Rutherford scattering experiment and the nuclear atom;* (3) Bohr: "On the Constitution of Atoms and Molecules, p. 734 (*Bohr's original presentation of his atomic model*); (4) Stern and Gerlach: "Experimental Proof of Space Quantization in a Magnetic Field," p. 930; (5) Goudsmit and Uhlenbeck: "Spinning Electrons and the Structure of Spectra," p. 940; (6) Pauli: "Exclusion Principle and Quantum Mechanics," p. 953; (7) Balmer: "Note on the Spectral Lines of Hydrogen," p. 363.

Companion, A. L., *Chemical Bonding*. New York: McGraw-Hill, 1964.

Sisler, H. H., "Electronic Structure, Properties and the Periodic Law," in *Selected Topics in Modern Chemistry*. New York: Reinhold, 1963.

CHAPTER NINETEEN

Wave Mechanics

THE BOHR MODEL of atomic structure represented a great step forward in that it seemed to give clear explanation for the locations of the lines of the elements that were observed in the spectrum. Each line had a well-defined wavelength and frequency (from the simple relation that the frequency is the velocity of light divided by the wavelength). From the quantum concepts of Planck's radiation laws, light of frequency f contained energy hf. As developed in the Bohr model this amount of energy could be represented as the *difference* between energy levels (or energy states) of an atom. In fact, light of particular energy hf is emitted when the electron in an excited atom jumps back from a higher energy level, the differences between the energy levels being the value hf for the line observed in the spectrum.

19.1 Difficulties with the Bohr model

As more spectral data became available and the elements were studied more intensely, there seemed to be difficulty in using the Bohr model to explain observations. The model gave wrong answers to observations made on atoms containing more than one electron, and in some cases it offered no answers at all. Some lines that were predicted by the model were absent altogether. Also, the spectral lines varied greatly in intensity (brightness), in a manner for which the model offered only poor answers. Clearly there was need for something different. Something "different" did come very soon. But

The sort of phenomena with which quantum theory is concerned teach the same lesson as relativity theory, namely, that the world is not constructed according to the principles of common sense. However, the way in which common sense fails is somewhat different in the case of quantum phenomena. The unfamiliar world of relativity theory was the world of high velocities; the new world of quantum theory is the world of the very small.

PERCY W. BRIDGMAN†

† From an essay in the *Scientific Monthly*, June 1954. This essay appears also in *Great Essays by Nobel Prize Winners*, edited by Leo Hamalian and E. L. Volpe, The Noonday Press, New York, 1960.

564 *Wave Mechanics*

before discussing it, let us first take stock of the situation as it existed in the early 1920's. For this we must backtrack to the earlier ideas about what constitutes mass and energy.

19.2 The early concepts of matter versus energy

What are mass and energy? The work of Galileo, Newton, and others had described the energy content for mechanical systems in terms of mass. Mass seemed to be present with every evidence of potential or kinetic energy. When heat was identified with molecular motion, the mass of the atoms was again present. But this was not so in the case of radiation: Maxwell had clearly explained that light is an electromagnetic wave and without mass. Light has energy, nevertheless, and the magnitude of its energy is given as the square of the amplitude of the electric or magnetic fields that rapidly rise and fall as light travels. All experiments up to the 1900's seemed to confirm that light is energy and that it is constituted entirely of waves.

Until 1900, this was the understanding with respect to the relationship of mass, energy, and waves. Energy and mass were both conserved, independently. Neither could be destroyed. Energy could be present, with or without the presence of mass. When mass was present, the movements of the mass were (according to Newton's laws) always in a straight line unless it was forced into other motion. But when mass was absent, as in light waves, the massless variations were altogether of wave form. This duality in the character of energy and the conservation of mass, appeared to be firmly established in nature, at least until the experimental work of this century changed these ideas.

Einstein's theory of relativity proposed the equivalence of mass and energy—that is, the one can be converted into the other. We shall learn more about this in the next chapter. But this was not all. Einstein's extension of Planck's quantum hypothesis declared that light did not have only a diffused wavelike spread. It had also a concentration of energy, hf, along with a finite or minimum quantity (or minimum subdivision) that the energy could assume. In the photoelectric effect these *photons demonstrated some of the same properties that had always been attributed to particles with mass*. The Compton effect further demonstrated that if the photon was assumed to have momentum, hf/c, as well as energy, hf, the collision or interaction phenomena between photons and particles (such as electrons) seemed to follow the same laws for conservation of energy and conservation of momentum as had been applied to mechanical systems with mass.

Thus, within a span of two decades, the long-standing distinction between particles and light waves or photons was altered, with the photon taking over some of the properties that had been attributed only to mass. But this encroachment of photons on the domain of particles was destined to be modified before long, with a move in a reverse direction as well.

19.3 Wave properties of particles

It was to this newly altered picture of the relationships among energy, mass, and the photon that Prince Louis V. de Broglie (born 1892-) turned his attention in 1920. Prince de Broglie had pursued formal degree studies in history, but became interested in science to the point that he took a degree in physics in 1913. He was soon conscripted for military service in World War I. In 1920 he concentrated on the study of atomic and radiation physics and on the contradictions that seemed to surround the quantum concept and the Bohr model of the atom. Within three years he succeeded in proposing a new concept, which has remained with us to this time. In 1929, he was awarded the Nobel Prize.† Let us now try to follow his reasoning in an approximate way to see how he was driven to some ideas that were rather hard to accept at first, but which seemed to give a more correct picture of atomic processes. First, de Broglie focused his attention on the motion of the orbital electron of the atom.

What do we know about the motion of the orbital electron? We have had to recognize that, contrary to the ideas of classical physics, the electron does not pursue an infinite variety of motions. Planck introduced the idea that the absorption and emission of radiation is quantized, that it is proportional to a constant h times frequency f of the radiation. This required that the "oscillators" (the atoms) which emit radiation have quantized states. This simply meant, at least in the Bohr model, that the electron stays in certain orbits. Even when the atom becomes "excited" by the addition of energy to "kick" its electron into another permitted energy state, the electron returns in due time to the original state with emission of light of specific frequency, the frequency being one of the series that characterizes that atom.

But if Planck's idea was correct, why was the electron permitted only a restricted number of paths in an atomic assembly? Could there be some integer (whole number) requirement? One of the features of the wave motion of light is that it involves integers (or numbers) to express frequency per second of its undulations. The interference of light, which we observe, is understandable in terms of such integer and phase relationships. Could it be that the electron also has some "number" characteristic, which might explain something like a standing-wave† situation to exist, in connection with its corpuscular motion? *But such a notion would require that the electron have the attribute of frequency, or wavelength, along with a primarily corpuscular or particle character.*

But why should not every particle (corpuscle) have also a frequency or wavelike character, just as the waves that make up light have a corpuscular character? Why should there not be, as Bohr had already suggested for his model, *both a corpuscular character and a wave character that were complementary aspects of the reality applicable to all particles and all waves?* These are the questions de Broglie may have asked.

In other words, perhaps the electron, which seems to be a particle that follows Newtonian mechanics, has also a wave character in its motion, which permits it to behave like a wave and thereby to experience something resembling standing waves in its orbit. If such were the case, the electron would assume only those orbits in which it could have whole numbers of waves, since only whole number divisions would allow "standing" waves. Figure 19.1 illustrates an experiment that the reader can easily perform, and Fig. 19.2 illustrates the difference between traveling wave and standing wave patterns. Figure 19.3 shows how these would appear when applied to orbital motion of electrons.

---

† Prince de Broglie's address given "on the occasion of his receiving the Nobel Prize" is reproduced in Boorse and Motz, pp. 1048–1059. The treatment here has borrowed from his own account. The chapters of Boorse and Motz beginning with page 1041 contain an expanded treatment of the topics of this chapter.

† See Chapter 5.

If such complementarity did exist, there would be demonstrated a beautiful symmetry and parallelism with respect to the properties of matter, waves, and the energy content associated with them.

Let us see what this new idea would mean when considered in terms of the wave aspect of a photon, and then in terms of a wave motion ascribed to an electron. In the case of light, when there is energy E in the light, there is associated with it a frequency $f = E/h$ (from $E = hf$); or the wavelength of the light is†

$$\lambda = \frac{hc}{E} \quad (19\text{-}1)$$

Now suppose there is a wave motion (and therefore wavelength) associated with a particle such as an electron; what is its wavelength? For this particle, or electron, de Broglie gave a relation similar to that of Eq. (19-1), namely,

$$\lambda(\text{particle}) = \frac{h}{p} \quad (19\text{-}2)$$

where p is the momentum of the particle.‡ If the electron is in an orbit, λ is of the order of 10^{-10} m, or one angstrom unit.§

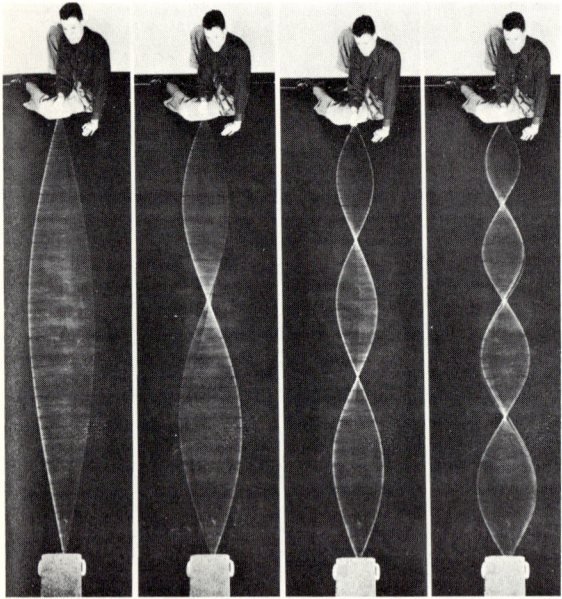

Fig. 19.1. Photograph of standing waves on a vibrating string. The number of standing waves is increased by decreasing the tension on the string. (Courtesy Physical Science Study Committee and J. Orear, Fundamental Physics. Wiley, 1967.)

† Since
$$\lambda = c/f.$$

‡ For light,
$$p = \frac{hf}{c} = \frac{E}{c};$$

thus
$$\lambda = \frac{hc}{E} = \frac{h}{p},$$

which is Eq. (19-2).

§ If the electron has a velocity of 10^7 m/sec, $h = 6 \times 10^{-34}$ J sec, $m = 9 \times 10^{-31}$ kg, then if

$$\lambda = \frac{h}{p} = \frac{h}{mv}$$

$$\lambda = \frac{6 \times 10^{34}}{9 \times 10^{-31} \times 10^7}$$

$$= \tfrac{2}{3} \times 10^{-10} \ m$$

$$\sim 0.6 \ \text{Å}$$

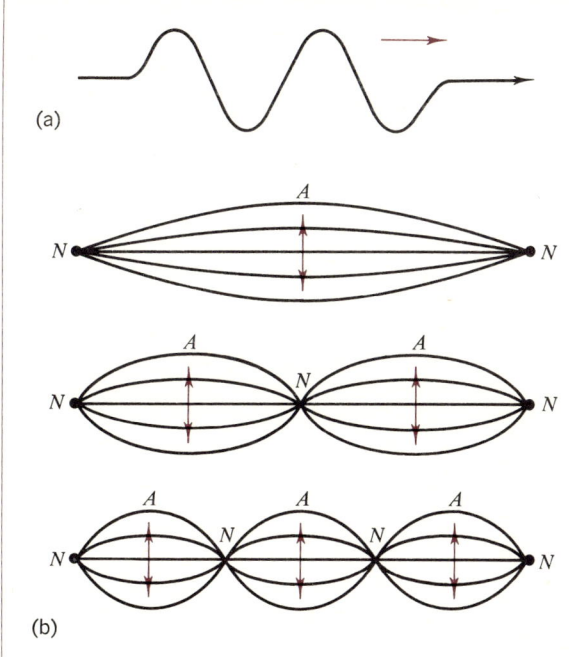

Fig. 19.2. The behavior of a traveling wave compared with the characteristic patterns of standing waves. (a) A traveling wave on a taut string. (b) Standing waves on a string of finite length firmly held at each hand.

Question: Why do we not observe a wave form in large particles of the kind we can see?

When in 1924 de Broglie presented his ideas in the form of a thesis to the faculty of sciences of Paris University, the ideas were strange indeed and difficult to accept without experimental confirmation. Experimental proof did come in 1928 in the form of the experiments with electrons by Davisson and Germer, which we shall discuss presently. In the interim, Einstein's praise of the de Broglie hypothesis encouraged a Viennese physicist who was then at the University of Zurich, Erwin Schroedinger (1887–1961), to make careful study of the dynamics of particles that do have this character. Indeed the theoretical work of de Broglie, Pauli, Heisenberg, and Schroedinger during the mid-1920's was exceedingly fruitful of ideas, ideas that are still with us today. Their work exemplified the way in which theoretical and experimental pursuits some-

---

Recalling that $E = \frac{1}{2}mv^2 = eV$, Eq. (19-2) can be written in a little different form:

$$\lambda = \frac{h}{\sqrt{2m_0 eV}}$$

or

$$\lambda = \frac{h}{\sqrt{2m_0 E}}$$

where m_0 is the rest mass of the electron, e the charge on the electron, and V the voltage through which the electron is accelerated. With $V = 150$ V, $\lambda = 1.2 \times 10^{-8}$ cm.

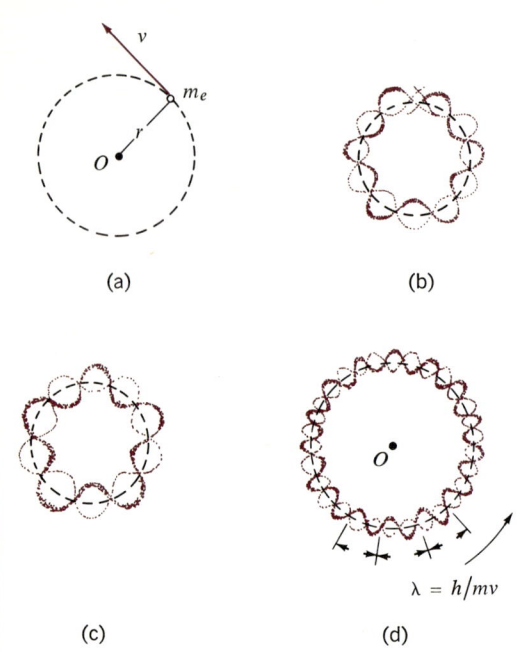

Fig. 19.3. Comparison of the de Broglie wave model with the earlier Bohr model. (a) Electron circular orbit as envisaged by original Bohr model. (b) According to de Broglie, the electron has a wave motion, with a wavelength corresponding to its velocity, or $\lambda = h/mv$. In this drawing the six waves do not quite fit into this circumference; therefore this orbit is not possible to the electron. (c) The five wavelengths meet exactly to form standing waves. This orbit is possible. (d) This orbit accommodates 16 wavelengths exactly. The waves can propagate in either sense, but they must reinforce each other on successive rotations.

times go quite far independently, but ultimately must join and support one another to achieve significant results.

19.4 "Eigenvalues" for atomic processes

Schroedinger did not fully accept the idea that the electron moves in quantum jumps, nor did he completely disregard classical mechanics. It had been possible, by classical means, to analyze the discrete wave forms that develop in a violin string and in "jelly-like" substances that do not have granular structure. The modes of standing waves that one observes in a vibrating violin string, for example, are the result of certain *eigenvalues* (characteristic values) of discrete wave forms that are permitted by the dynamics of the situation. Each combination of string mass, string tension, and string length permits a characteristic pattern of "eigenvalues" for standing waves or for transitory forms of traveling waves.

For example, Fig. 19.2 illustrates a traveling wave and three standing waves, the N's designating nodes for the standing waves. Could an electron in an atom, asked Schroedinger, have similar standing wave possibilities (or eigenvalues) in each of its possible orbits? If an electron has a velocity that corresponds to its velocity in an atom, it also has [according to de Broglie and Eq. (19-2)] a wave motion with wavelength of the order of one angstrom. When the electron is given a kick to increase its velocity, the wavelength of its wave motion becomes reduced† inversely in proportion

---

† According to Eq. (19-2), when the kinetic energy (which is proportional to p) is *increased* by four times, the wavelength is *reduced* by a factor of 2. The faster the electron, the shorter

to the change in velocity. Does this not mean, he asked, that for each velocity and associated wavelength the electron occupies only orbits in which it can have an exact number of wavelengths (Fig. 19.3)?

According to Schroedinger, the transition of an electron was nothing more than going from an orbit or path in which it could accommodate a whole number of the wavelengths associated with its velocity, v_1, to another orbit in which it could again accommodate a whole number of its new wavelengths corresponding to another velocity, v_2. His work† earned him the Nobel Prize in 1933, and it was in that same year that Schroedinger, an "acceptable" Aryan, left the high post he held as successor to Planck at the University of Berlin, in protest against the treatment of the Jews by Hitler and Nazi Germany. He was at Oxford until 1936, when he left to return to his beloved Austria. When Austria was annexed by the Germans in 1938, he fled to Italy, then to Princeton, then finally to Dublin, where President DeValera of Ireland founded the Institute of Advanced Studies in order that Schroedinger might be director of its School of Theoretical Physics. He retired in 1955.‡

---

its associated wavelength. Since kinetic energy varies as the square of velocity, show that the associated wavelength varies inversely with the velocity (that is, not with a square root relationship).

† See Boorse and Motz, pp. 1060 ff. Schroedinger had considerable interest in philosophical aspects of physics and biology, as evidenced by his fine little book, *What is Life?*, Cambridge Univ. Press, 1946 (New York: Macmillan).

‡ Schroedinger's work is highly mathematical and cannot be presented here. A brief summary of his approach is included in Sec. 19.6.

19.5 Confirmation of the wave nature of electrons

We shall return presently to modify the de Broglie-Schroedinger hypotheses on the atom. But we must now point to a brilliant confirmation of the thesis that there are wave characteristics associated with particles. The work was that of Clinton J. Davisson (1881–1958), a physicist who was working at the Bell Telephone Laboratory, assisted by L. H. Germer.

Davisson and Germer were engaged in the study of the scattering of electrons from the smooth surface of a strip of nickel. A fortunate accident, which broke the vacuum tube that contained the nickel strip, required that the strip be heated and slowly cooled. The heating and cooling caused the atoms of nickel to take on the regular pattern of a crystal, which made it similar to the grating that we discussed earlier in connection with the experiment† on spectroscopes. Now, when the experiment with electrons was resumed, the electrons reaching the nickel strip could interact with regular rows and layers of atoms of nickel. They did not at the time know of de Broglie's wave hypothesis for electrons, but they did observe patterns for maxima and minima, which varied with electron energy and with the angle of approach to the crystal of nickel. Davisson sent the results of this work to Max Born for interpretation. Born, in consultation with his associates, suggested a new experiment that might verify or negate the de Broglie-Schroedinger theory. The new experiment, begun in 1926, completely verified that theory, leading to the Nobel Prize for Davisson in 1937. According to him, the results of the experiments came

† Laboratory and Mathematics Supplement.

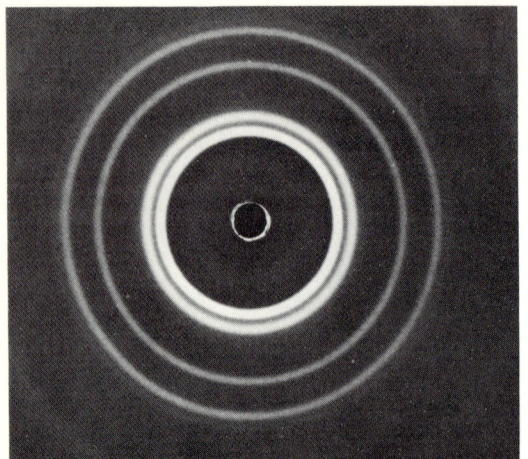

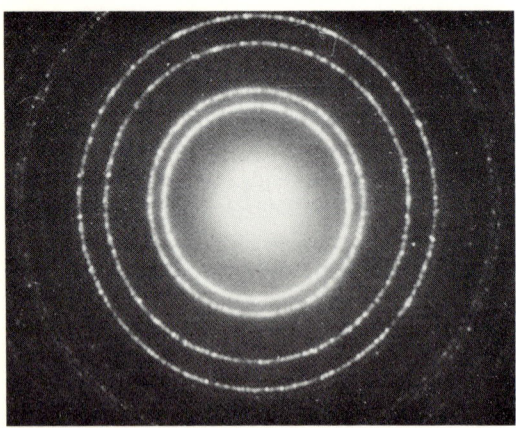

Fig. 19.4. A direct comparison of electron diffraction and X-ray diffraction. The wavelengths are the same in both cases. At the top, the diffraction pattern is obtained by X-rays passed through aluminum foil. In the bottom view, the pattern through the foil is produced by a beam of electrons. (From the P.S.S.C. film Matter Waves.)

"purely by accident," but fortune has a way of favoring with accidents those who are alert to the unexpected.

Very similar results were obtained independently by G. P. Thomson of Aberdeen, who used crystalline powder as the scatterer instead of a single crystal. At the present time the phenomenon is so well established that the technique is used in both research and industrial laboratories for examining the structure of materials. Neutrons or alpha particles are also used. When these "particles" bombard a thin sheet of metal or other forms of crystalline materials, the particles that are scattered by the atoms of the metal show diffraction patterns that are similar to those produced by X-rays or other "electromagnetic" radiation. See Fig. 19.4.

19.6 Some further considerations of matter waves

Although the following material is not necessary for the understanding of the nature of wave mechanics, it will interest readers who do not mind a little mathematics.

Consider a particle of mass m_0 traveling with a velocity v. According to the de Broglie relation, it can be thought of as being associated with a wave of wavelength $\lambda = h/m_0v$. These associated waves were called *pilot waves* by de Broglie. Let u be the speed of propagation of such a pilot wave and f its frequency. Let us consider analytically what happens when we superimpose two such waves, which differ in frequency and hence in wavelength, by an infinitesimal amount. Let us consider both waves to be traveling in the x direction with amplitudes $y_1(x,t)$ and $y_2(x,t)$ at any moment. We can write equations expressing the functional relationship of y_1 and y_2 on x and t as follows:

$$y_1(x, t) = \sin 2\pi(kx - ft) \quad (19\text{-}3)\dagger$$

$$y_2(x, t) = \sin 2\pi[(k + \Delta k)x - (f + \Delta f)t] \quad (19\text{-}4)$$

where the wave number $k = 1/\lambda$ and Δk and Δf are the small differences in wave number and frequency between the two waves. Let us add $y_1 + y_2$ to get

$$y = y_1 + y_2$$
$$= 2 \cos 2\pi \left[\frac{\Delta k}{2} x - \frac{\Delta f}{2} t\right]$$
$$\times \sin 2\pi \left[\frac{2k + \Delta k}{2} x - \frac{2f + \Delta f}{2} t\right] \quad (19\text{-}5)$$

Since $\Delta k \ll k$ and $\Delta f \ll f$, we finally get

$$y = 2 \cos 2\pi \left[\frac{\Delta k}{2} x - \frac{\Delta f}{2} t\right] \sin 2\pi(kx - ft) \quad (19\text{-}6)$$

† The reader can satisfy himself that this indeed represents a traveling wave, by plotting $y(x)$ for several different t.

Figure 19.5 is a plot of Eq. (19-6). Each of the original waves moves to the right with velocity $u = f\lambda = f/k$. The nodes (zeroes) of the pattern shown in Fig. 19.5 propagate at a velocity

$$u' = \frac{\Delta f}{\Delta k}$$

Let us consider these two velocities u and u' a little more closely. From the de Broglie and Planck relations we have

$$k = \frac{1}{\lambda} = \frac{p}{h}$$

$$f = \frac{E}{h}$$

and therefore

$$u = \frac{f}{\lambda} = \frac{E}{p} \quad (19\text{-}7)$$

It might be noted that, according to relativistic dynamics, E may be expressed as

$$E = \sqrt{p^2 c^2 + (m_0 c^2)^2}$$

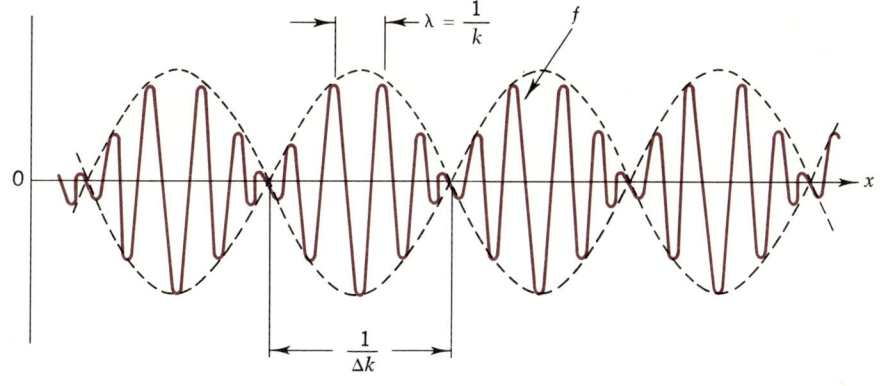

Fig. 19.5. The sum of two sinusoidal waves of slightly different frequencies and wave numbers. The wave denoted by the dotted line moves to the right with a velocity u', called the group velocity.

and thus

$$u = \frac{E}{p} = c\sqrt{1 + \left(\frac{m_0 c^2}{p}\right)^2} \quad (19\text{-}8)$$

is always greater than c! However, this does *not* violate the postulate of the theory of relativity, since the pilot wave is simply a construct and does not represent a physical energy transport.

Turning now to u', we may write

$$\Delta f = \frac{\Delta E}{h} \quad \text{corresponding to} \quad f = \frac{E}{h}$$

and

$$\Delta k = \frac{\Delta p}{h} \quad \text{corresponding to} \quad k = \frac{p}{h}$$

Thus

$$u' = \frac{\Delta f}{\Delta k} = \frac{\Delta E}{\Delta p}$$

From the equation relating E and p, given above, and using the rules of the calculus for differentiation, we find

$$\frac{\Delta E}{\Delta p} = c^2 \frac{p}{E}$$

Therefore

$$u' = c^2 \frac{p}{E} = c^2 \frac{m_0 v}{m_0 c^2} = v \quad (19\text{-}9)$$

and the "wave packets" shown in Fig. 19.5 travel with a group velocity equal to the velocity of the material object.

Although we have treated the superposition of only two waves, the results hold even when a large number of pilot waves are superposed.†

Schroedinger sought an equation that would describe atomic processes in

† The resolution of a wave packet into its component pilot waves involves a mathematical technique known as Fourier (or harmonic) analysis.

terms of de Broglie's matter waves. He found that the classical wave equation that describes the motion of a rope does not account for de Broglie waves. He was forced to *assume* a slightly different form for the wave equation that now bears his name. The Schroedinger equation involved the derivatives of a function $\psi(x)$† and Planck's constant. It also included the imaginary number i (see Sec. M.10 of the Laboratory and Mathematical Supplement). The solutions $\psi_n(x)$ of the Schroedinger equation are not real functions, but rather are complex functions. Schroedinger's equation is, however, under certain boundary conditions, an eigenvalue equation. That is, for example, if we consider a particle restricted to move in a specified range of values for x (the particle is in a box), then $\psi(x)$ will be identically zero for all values of x that are not in the specified range. This is a boundary condition. Each solution $\psi_n(x)$ of Schroedinger's equation, with its appropriate boundary conditions, corresponds to some energy E_n, of the particle. Thus, once again, energy is quantized.

The function (usually denoted y or s) that occurs in the classical wave equation is immediately identifiable with a physical characteristic of the system, namely, the amplitude of the wave motion. There is no analogous interpretation of $\psi(x)$, however. Max Born noted that the quantity $|\psi(x)|^2 \, \Delta x$ can be interpreted as being *the probability that the particle is within the interval Δx about the position x.*

19.7 The uncertainty principle

The decade following de Broglie's publication of his idea of matter waves saw the growth of the analytic formalism of

† Where ψ is the Greek letter psi.

present-day wave mechanics. Two independent developments of this formalism were made in the two years following de Broglie's work. One (which we have just mentioned), due to Erwin Schroedinger (1887–1961), sought the underlying equations of motion of de Broglie's pilot waves. The other, the so-called *uncertainty principle*, was annunciated by Werner Karl Heisenberg (1901–   ), who has been a leading figure in German physics since before and after the fearsome days of Nazi domination. While serving as Max Born's assistant at Göttingen, in 1925 he wrote a paper that gave a wholly different approach to the quantum mechanics of atoms. We cannot investigate this aspect more than to say that he introduced into his explanation of atomic phenomena a form of algebra in which $p \times q$ is not equal to $q \times p$.† The new approach, under the general title of *matrix mechanics*, was of tremendous

† Such algebras are said to be *noncommutative*. For example, consider

$$\begin{pmatrix} 1 & 2 \\ 3 & 4 \end{pmatrix}$$

This array of numbers is called a *matrix*. To multiply two matrices

$$\begin{pmatrix} 1 & 2 \\ 3 & 4 \end{pmatrix} \times \begin{pmatrix} 5 & 6 \\ 7 & 8 \end{pmatrix}$$

multiply the top row of the first by the left column of the second to get $(1 \times 5) + (2 \times 7) = 19$, and put this number in upper left. The multiplication gives

$$\begin{pmatrix} 19 & 22 \\ 43 & 50 \end{pmatrix}$$

Note that

$$\begin{pmatrix} 5 & 6 \\ 7 & 8 \end{pmatrix} \times \begin{pmatrix} 1 & 2 \\ 3 & 4 \end{pmatrix} = \begin{pmatrix} 23 & 34 \\ 31 & 46 \end{pmatrix}$$

and thus

$$\begin{pmatrix} 1 & 2 \\ 3 & 4 \end{pmatrix} \times \begin{pmatrix} 5 & 6 \\ 7 & 8 \end{pmatrix}$$

importance to the work of Born, Dirac,† and others in the further development of quantum concepts of the atom.

Our interest of the moment is not in matrix mechanics, however. Rather it is in the conclusion that Heisenberg reached while trying to analyze the significance of *an* electron in *an* orbit. How can one determine the location of a moving particle? One method is to introduce a light beam that interacts with the particle. The nature of the interaction products will then presumably tell where the electron is, and perhaps something of its nature and of its velocity or momentum. Certainly we have little difficulty in determining both position and momentum of moving objects in everyday experience.

The situation becomes quite different when the particle is as small as the electron and also has a wave character, wavelength, or "wave packet" that is very much larger than the particle itself. The wavelength is an intimate character of the momentum or velocity of the particle. That is, we cannot know the velocity or momentum without taking into account the wave packet, which may be large in comparison with the static dimensions of the particle. But if we should focus attention on measuring the velocity or *momentum* of the particle, measurement of the exact *position* of the particle would be meaningless.

If then we let Δp_x be the accuracy with which we measure the momentum p in the x direction, and let Δx be the ac-

is *not* equal to

$$\begin{pmatrix} 5 & 6 \\ 7 & 8 \end{pmatrix} \times \begin{pmatrix} 1 & 2 \\ 3 & 4 \end{pmatrix}$$

† Dirac showed that $(p \times q) - (q \times p) = ih/2\pi$, thus introducing the imaginary number i, which was necessary in the Schroedinger approach also.

curacy with which we know the position along the x-axis, the *uncertainty principle* of Heisenberg states that we cannot know the product of the two to a higher degree of accuracy than the quantity h, Planck's constant. That is

$$\Delta x \cdot \Delta p_x \geq h \quad (19\text{-}10)$$

Thus, if we wish to know the position to within an accuracy that approaches zero, the error in the measurement of the velocity becomes very large. Similarly, if we should want to know the momentum with very high accuracy, we would have to concentrate on ways to measure the velocity without any regard for the position of the particle during that velocity measurement.

The uncertainty relation can be written in terms of energy and time as:

$$\Delta E \cdot \Delta t \geq h \quad (19\text{-}11)$$

This form is useful for explaining certain phenomena such as the finite width of spectral lines.

19.8 Probabilities and atomic structure

In developing his wave representation of electron motion, Schroedinger used the symbol ψ to find solutions to his wave equations.† For example, if we consider ψ to represent the presence or motions of a particle in a box of length L, ψ must be zero outside the box, as a *boundary condition*. A solution for each function ψ_n, corresponding to a particle energy of E_n, can characterize the motions of the particle in the box. (We should note that Schroedinger adopted his wave equation approach because *it seemed to fit some situations.*)

† Section 19.6 noted some of these concepts.

While de Broglie and Schroedinger were busily developing wave concepts for particles, another group of scientists was engaged in a totally different analysis, a probability approach, for describing atomic structure. A principal figure in this development was Max Born (1882–), from whom we have quoted abundantly in earlier chapters.

The Schroedinger extensions of the de Broglie and Bohr models gave correct energy levels for the electrons in atoms. It was difficult to accept the picture of the electrons in actual wave motion, however. For one thing, waves and wave packets not only extended over large volumes of space, but also dissipated their energy in the course of time. There was another difficulty in that the famous Schroedinger equations, which accurately described the wave motion of electrons, had in them imaginary mathematical quantities to which it was difficult to assign physical meaning. Even the concept that the waves represented not physical electrons, but clouds of charge that had the character of waves, did not seem to explain the stability of atomic structure.

Born proposed that the wave representations of de Broglie and especially of Schroedinger could not be interpreted as having the meaning of a wave in the usual sense. The function ψ (psi) had been used by Schroedinger to represent these waves or wave packets. Born proposed that the square of the absolute value of the function, or $|\psi|^2 \Delta x$, gave the *probability* that the electron would be found in the interval Δx about some point of space. That is, where the amplitude of the de Broglie or Schroedinger wave is large, there is high probability for finding an electron near that spot. Where the amplitude is less, there is less probability of finding an electron. At the nodes there

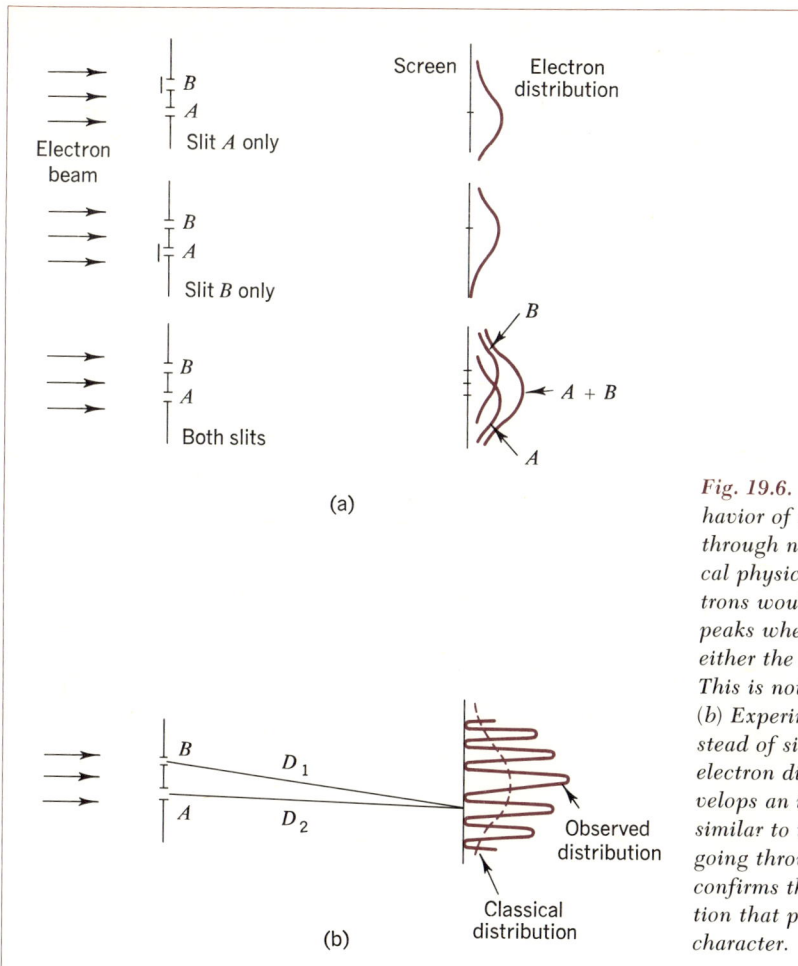

Fig. 19.6. Comparison of behavior of electrons passing through narrow slits. (a) Classical physics predicted that electrons would give the single peaks when passing through either the single or double slit. This is not observed, however. (b) Experiments reveal that instead of single peaks for the electron distribution, there develops an interference pattern similar to that produced by light going through the slits. This confirms the de Broglie prediction that particles have wave character.

was zero likelihood of finding an electron.

A most interesting experiment, to which we alluded in Chapter 16, is the following: Suppose we have the source of electrons and the two slits arranged as in Fig. 19.6(b). We would obtain the distribution pattern of electrons shown. This resembles the pattern obtained for light in a similar experiment. We can explain this by saying that the waves of the various electrons that pass through the separate slits interfere with each other to give that pattern.

But now suppose we allow only one electron to go through at a time. Speaking in terms of classical physics, we would say that each electron goes through only one of the two slits, and that at the instant of passing through, there cannot be any interference or influence from another electron. Yet we find that after many such single electrons *individually* pass through one or the other slit, the observed end effect is to produce *exactly the same pattern as was obtained with the entire beam of electrons using both slits.*

Let us now close off one of the slits [Fig. 19.6(a)] and repeat the experiment, first with the whole beam and then with the individual electrons. The patterns produced by the two would be identical; but they are totally different from the pattern we obtained with the two slits open. In other words, the mere *existence* of a second slit seems to influence the direction that an electron pursues as it passes through the first slit. What is the explanation of this apparent paradox? If we are to think of the resulting pattern as being an interference pattern, we are driven to the conclusion that the single electron interferes with itself by acting as though it were passing through *both* slits at the same time! Perhaps this means that the wave train which represents the electron passes through both slits, and the portions that pass through the different slits interfere with each other.

We may approach this problem from the point of view of the Heisenberg theory. When an electron passes through a narrow slit of width Δx, its position is known within an error no larger than Δx. Thus, according to the uncertainty principle, its known velocity in the direction of the slit width (vertical direction in Fig. 19.6) is no better than $\Delta v = h/(m \, \Delta x)$ (that is, the vertical velocity is uncertain) and hence we may speak only of a "probability" of the electron having a given velocity v_x. The smaller Δx (that is, the narrower the slit), the greater the uncertainty in v_x. A rigorous analysis of this idea yields the probability distribution of electrons passing through the slit, similar to that of a diffraction pattern of a light wave of the same wavelength as the electron (in the de Broglie theory) passing through this slit.

The preceding discussion refers to an electron passing through a single narrow slit. What happens when two wider slits are present? The uncertainty now arises because we cannot say through which slit the electron passes. Any experiment that we may design to determine this will result in changing the electron's motion after it passes the slit. The explanation was put forth by Bohr in his concept of *complementarity:* Whenever you can trace the path followed by a particle, it acts like a particle; when you can't, it acts like a wave! Thus when one slit is closed, you can follow the electron; you know it must have gone through the open slit. It therefore acts like a particle. When both slits are open, however, you can no longer follow the electron and hence it acts like a wave.

The probability model can be extended to atomic structure. Let us first, however, try a probability analogy with the results we might expect when throwing darts at a target. Suppose that, after throwing a large number of darts, we obtain the pattern shown in Fig. 19.7. We cannot say with certainty where the next throw will land, but we can estimate the probability of its landing in a given square inch of area on the board by counting the number of holes left by previous throws in this area and dividing by the total number of throws. (*Question:* If no previous holes exist in our chosen area after 1000 throws, does this mean definitely that the next (1001) throw will not land in this area?) Thus we may find a much higher probability of throwing a dart near the bull's-eye than of throwing it far away. However, the probability never becomes strictly zero except at an infinite distance from the bull's-eye.†

Likewise, for an electron in the lowest energy state of hydrogen, we cannot specify where the electron is at any moment, but we can give the probability of its being near some position, say, x. We do this by evaluating $|\psi_1(x)|^2$, where ψ_1 is the solution of the Schroedinger equation for the lowest energy state. When we do this, we get a probability distribution (Fig. 19.8) that is spherical in shape and increases in density toward the nucleus (that is, the electron will be more likely found near the nucleus than far away). The distribution extends to infinity, meaning that the electron *may*, at times, be very far from the nucleus. The average radius at which the electron will be found is approximately the Bohr radius for that energy state. Hence we may no longer speak of orbits in an atom. Instead we shall call them *orbitals*, which implies some knowledge of the electron's energy but makes no reference to its actual position.

19.9 Summary

In this chapter we have briefly introduced some of the most puzzling aspects of atomic phenomena. The rapid

† Assuming, of course, we could throw a dart an infinite distance.

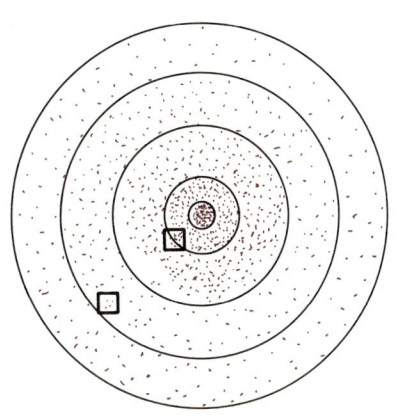

Fig. 19.7. Representation of probabilistic distribution of hits on a dart board after considerable use. Note that we are more likely to find a dart hit nearer the vicinity of the bull's-eye than much farther away from it.

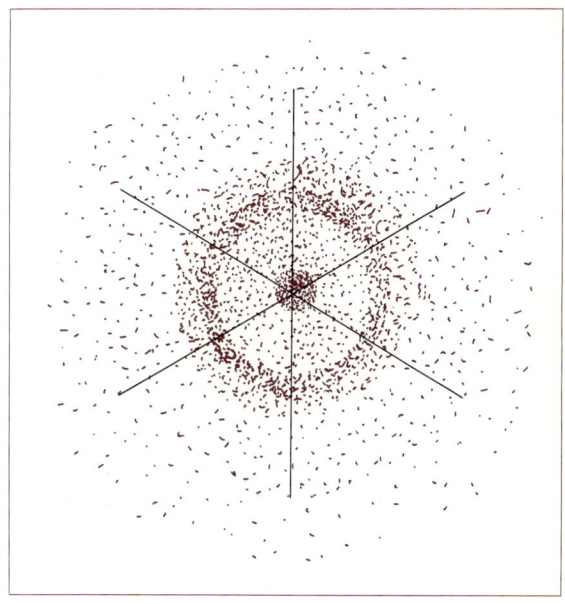

Fig. 19.8. Schematic representation of probility distribution of an electron. Although the probability is high that the electron will be found near the nucleus, the probability that the electron will be found far away from the nucleus still exists.

changes from one hypothesis to another have brought us to the paradoxes of our own day with respect to the duality, or complementarity, of the corpuscular and wave character of both photons and particles. It has been a considerable achievement to preserve this symmetry of properties between photons and particles. Atomic science has found great value in symmetry principles when investigating such things as matter and antimatter,† electrons and electron holes, and various other areas that are not appropriate to discussion here.

We have seen that de Broglie's relatively simple but brilliant theories on the wave property of particles found experimental proof in the Davisson-Germer experiments and also reached a high degree of refinement and sophistication in the work of Schroedinger. Schroedinger's mathematical analysis of the properties of standing and traveling waves seemed to say quite clearly that electrons find their proper *eigenvalues,* or proper states, by being only in orbits and energy states in which an electron can have a whole number of its associated wave motion. The quantum jumps of electrons are nothing more, therefore, than the passage of the electrons from one eigenstate to another.

Not so, said Born and Heisenberg and others. The wave feature is certainly a part of the motion of a particle, but it is wrong to think of an electron as actually experiencing such motion. Instead, the waves are related to the *probability* of finding electrons in points of space. In fact the square of the amplitude of Schroedinger's wave function, $|\psi|^2$, really gives this probability.

We cannot venture to predict future explanations of atom or electron phenomena, for the experimental means available to us thus far seem to have reached the limit of their sensitivity. When we shine a photon on an electron to see where it is located, we push the electron away with the force of its interaction with the photon. The uncertainty principle seems to have identified an upper limit to the accuracy with which we can determine both the position and momentum of a particle at the same time.

We are now approaching our own decade in atomic science, but the paradoxes are still with us. Despite these conceptual difficulties quantum mechanics has been able to account for the various atomic phenomena better than any other model suggested up to the present time.

† The concept of antimatter will be discussed in Chapter 20.

Questions/Discussions

1. How does the de Broglie model "explain" the stable energy states of atoms?

2. Should our atomic models emphasize the nature of *energy states* in atoms or the nature of *electron motion* in atoms, to be more nearly in agreement with experimental results?

3. What is the momentum of a photon that has a frequency of 4.5×10^{17} sec^{-1}? (Recall that $p = hf/c$.) If a photon has a momentum of 3×10^{-26} kg m sec^{-1}, what is its wavelength?

4. What is the de Broglie wavelength of

an electron that has kinetic energy of 1 eV? Neglect relativistic corrections.

5. A golf ball weighs 45.6 g and is given a velocity of 76 m/sec; what is its de Broglie wavelength? Compare this with the wavelength of an electron that has a velocity of $\frac{1}{100}$ that of light.

6. Does the concept of stable states in atoms, when applied to the electron, contradict the earlier experience that accelerated electrons radiate energy?

7. If an electron has a velocity of 1×10^6 m/sec, what is its de Broglie wavelength?

8. A photon and an electron each has kinetic energy of 1 eV. Show that the photon has the longer wavelength.

9. What similarities and what basic differences do you see in the way in which the concepts of the laws of thermodynamics emerged, and the way in which the de Broglie wave concepts emerged? Identify the role of experimentation, of induction, of deduction, and intuition. (Note to instructor: This question may be introduced here and pursued in connection with the discussions of Chapter 23.)

10. As in Question 9, compare the emergence of the de Broglie concepts with the emergence of Newtonian mechanics.

References

See References at the end of Chapters 16, 17, and 18, especially Boorse and Motz, pp. 1041 ff. and de Broglie, pp. 165 ff.

CHAPTER TWENTY

Radioactivity and the Atomic Nucleus

Perhaps the most central problem in theoretical physics during the last twenty years has been the search for a description of the elementary particles and of their interaction. . . . So far we have discussed particles whose basic properties are known in great detail. But there are other particles whose existence is known or suspected and whose properties are in several cases only conjectured.

ENRICO FERMI[†]

RUTHERFORD'S scattering experiments (Sec. 18.2) demonstrated that the bulk of the mass of the atom is concentrated in a small nucleus. It is time to examine the composition of that nucleus in some detail.

20.1 Introduction

The nucleus, being made up of neutrons[‡] and protons, has an electric charge that is equal to the number of protons. Around this nucleus there are negatively charged electrons, equal in number to the number of protons in the nucleus. The electrons constitute a cloud of negative charge around the nucleus, although (as we saw in Chapter 18) each electron retains a quite definite energy state. Our studies thus far have been concerned largely with the phenomena that accompany the *transition* of electrons from one energy state to another. When an electron absorbs energy, it is raised to a state of higher energy; when it can do so, the electron will jump back into a lower energy state with emission of its excess energy in the form of a photon.

We have also seen that when atoms combine to form molecules, the binding energy is derived from interaction and rearrangement of the electronic states of the atoms. All chemical reactions and the formations of all compounds derive from the interactions of these "orbital" electrons. Thus, when two hydrogen atoms combine to form the hydrogen molecule (H_2), or two hydrogen atoms

---

[†] From Fermi, *Elementary Particles*, Yale University Press, New Haven, 1951. We might note that the situation has not changed substantially, nearly two decades later, with respect to looking for new particles.

[‡] See Sec. 20.11.

combine with an oxygen atom to form the water molecule (H_2O), the interactions involve only the "orbital" electrons. By utilizing absorption and emission spectrometers of many types and frequency ranges, it has been possible to accumulate a wealth of information on many aspects of both the electron configurations and the energy states of these atoms and molecules.

What about the nucleus of the atom, where nearly all its mass is concentrated? Does the nucleus have structure or parts that in any way compare with the complex structure we observe in the atom as a whole? What are nuclei made of? What holds the parts of a nucleus together? How do nuclei differ from each other? Are there internal motions in the nucleus? How can a nucleus be transformed to another type?

The study of the nucleus began only about 70 years ago. In this brief span of time it has been possible to accumulate a great deal of information about various nuclei. More than that, we have learned how to transform atoms (by changing their nuclear structure) to produce a wholly new series of atoms that have the property of being *radioactive*. The culmination of this research, as we shall see, was the discovery of a method to convert atomic nuclei in such a way that we could derive tremendous new energies from the atoms—energies a million times those obtainable from chemical changes! The new knowledge has given us new sources of energy for useful purposes, and has yielded better understanding of how the great energy of the stars is created. It has also, unfortunately, placed the world in jeopardy because of the tremendous destructiveness of atomic bombs.

20.2 Phosphorescence; discovery of X-rays

Before proceeding to the study of the nucleus, we should mention a phenomenon called *phosphorescence*, which is useful for detecting the presence of ionizing radiation and for that reason was instrumental in the discovery of X-rays and of radioactivity.

We are all familiar with the emission of light that takes place when objects become very hot, such as burning wood in a fireplace or a hot wire filament in an electric lamp. There are some phenomena, one of which involves *phosphorescence* and another *fluorescence*, in which light is emitted without the requirement of high temperature (or even of temperature that is sensibly above room temperature). The element phosphorus, for example, when exposed to an external light, has the property of emitting light (phosphorescing) after the external light is removed. In contrast, some materials will emit light (fluoresce) when subjected to external radiation, but will not continue to emit after the external radiation is removed. Some living organisms such as certain fungi and bacteria, and inert materials such as decaying wood, exhibit phosphorescence as a result of a slow oxidation process. In fluorescence, the emitted radiation usually has longer wavelength than that of the initial radiation. (The term *luminescence* is used for any emission of light that occurs at low temperature and which is not directly ascribable to incandescence, that is, heating to glowing.)

When William Crookes was studying the behavior of gases subjected to electric discharges, he observed that where the charged particles struck the glass walls of the tube, the glass emitted a

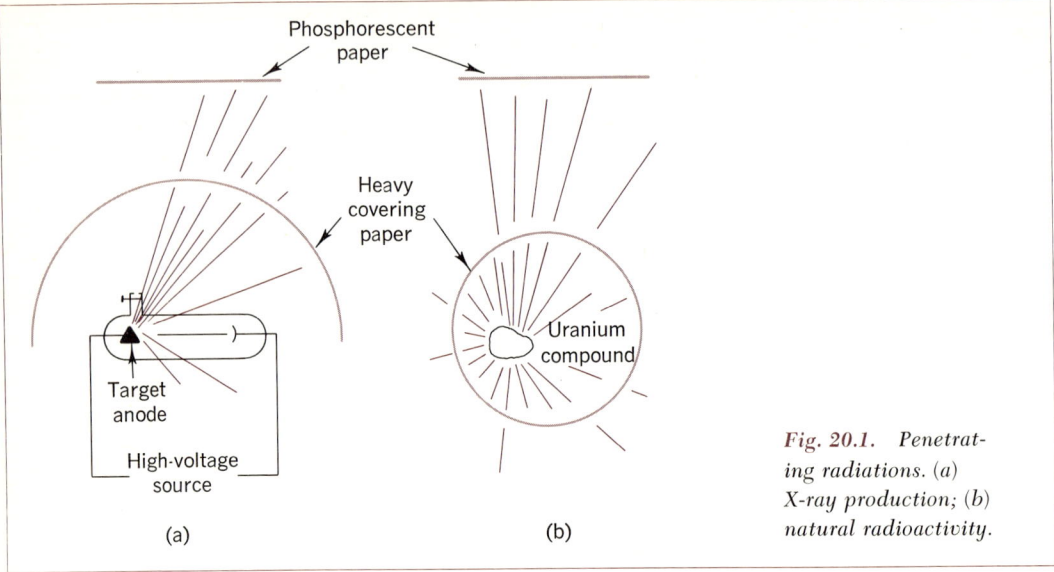

Fig. 20.1. Penetrating radiations. (a) X-ray production; (b) natural radioactivity.

green or greenish-yellow light. This emission became very useful for tracing the movements and presence of the particles. Others who followed Crookes also found the observation useful, including the German experimentalist Wilhelm Konrad Roentgen (1845–1932), who was fifty-one years of age at the time of his discovery, and held the chair in physics at the University of Würzburg. Roentgen was experimenting with cathode rays at high voltages when he observed that although the cathode ray tube was completely enclosed in a shield of black cardboard, some form of radiation seemed to emanate from the tube and penetrate the cardboard to cause fluorescence in a nearby screen. Further tests proved that he had discovered the presence of a new, highly penetrating radiation, which he called X-rays. The rays could penetrate paper, thin metal sheets, and even the hands and other parts of the human body. Magnetic fields did not cause deflection of the new rays; therefore he assumed that they were not charged particles. They seemed to be like ordinary light in many respects except for the penetrating power and except that they did not show interference patterns or refraction such as are observed with ordinary light.† [Fig. 20.1(a)]

Roentgen chose to announce his discovery (in 1895) through a local publication, the proceedings of the Physico-Medical Society of Würzburg. It is thought that this established his priority of discovery, while also giving him time to

---

† Later experiments did reveal slight refraction. Also, when X-rays are made to pass through metal or other crystals in which the atoms have regular and close spacings, the X-rays do indeed interfere and show very remarkable diffraction patterns. In fact, X-ray diffraction techniques have become of very great importance for the study of the internal atomic arrangement of many materials.

check his own findings before the scientists of Europe could pounce on it. When news did spread to the general public, it was accompanied by gross exaggerations of how the new discovery had forever destroyed all privacy of individuals; for could not the new rays penetrate the thickest wall of a building to reveal the activities of the people within? Indeed, advertisements in some major newspapers even offered X-ray proof undergarments for ladies!

In time the public came to realize that the new rays could not be made to emanate from pocket devices; that the production of X-rays required heavy installations, high voltage, and considerable shielding to protect the user of X-rays. (We see such installations in hospitals and in the offices of dentists.) In time it also was revealed that X-rays are electromagnetic waves like ordinary light except that they have wavelengths that are roughly a thousand times shorter than the wavelength of visible light. Still later it was explained that when swift charged particles such as electrons strike the heavy atoms of a target and cause the *inner orbital electrons* of those atoms to escape, the *return* of electrons to fill the inner energy levels produces X-rays. Their production is therefore not dissimilar to the production of ordinary light when the electrons of outer orbits return to states of lower energy, except that much higher energies become involved when the inner electrons of heavy atoms become disturbed. For his discovery, Roentgen was awarded the first Nobel Prize in Physics in 1901.

20.3 Discovery of radioactivity

Such were the activities and interests surrounding radiation and phosphorescence when in 1896 a new phenomenon was observed, this time by the Frenchman Antoine Henri Becquerel (1852–1908). Henri was the third generation of a family of scientists, and his son became the fourth in due time. Previous experiments had shown that commercial grade calcium sulfide, which shows phosphorescence on exposure to light, emits radiation that can pass through opaque substances (such as thick black paper) and cause photographic film to blacken. Becquerel found that the salts of *uranium* are especially active in emitting such radiation. There was confusion at first as to whether the phosphorescence (produced by incident radiation) caused the blackening of the film. It soon became clear, however, that the phosphorescence and incident light had nothing to do with the penetrating radiation emitted by the materials. In fact, Becquerel soon learned that, whether it was phosphorescent or not, the uranium emitted the radiation regardless of the presence of outside radiation. He observed [Fig. 20.1(b)] that the radiation emitted by uranium salts had properties that were quite similar to those of the X-rays discovered by Roentgen. It puzzled him that the uranium continued to emit the radiation over the months without diminishing in strength, even when locked in a tight lead box and kept in the dark. The radiation caused electrified bodies to lose their charge, and was quite penetrating.

In time it became clear that the emission was due not to any particular chemical compound of uranium, but to the uranium atom itself. Was the emitted radiation therefore due to an internal change within the uranium atom? Did the phenomenon offer a means for exploring the internal structure and workings of an atom?

584 Radioactivity and the Atomic Nucleus

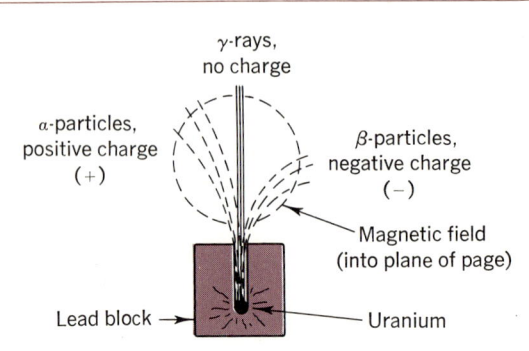

Fig. 20.2. *When a narrow beam of "rays" emitted by uranium is passed through a magnetic field (into page), the beam is split into three parts.*

20.4 Alpha, beta, and gamma radiation

A sense of urgency pervaded the scientific community and physicists were eager to learn the nature of the radiation emitted by the uranium. Was the newly discovered radiation made up of high-energy charged particles? Or was it similar to the X-rays found by Roentgen, which could not be deflected by a magnetic field? A simple test would tell the story. As shown in Fig. 20.2, a small piece of uranium placed inside a deep hole in a lead block will emit a narrow beam of radiation. If this beam is passed through a magnetic field, any bending of the beam in either direction can be easily detected, either by its effect on a photographic film or by use of charge-discharge detection techniques.

When the magnetic field was applied, it was discovered that the beam broke into three distinct components. Part of the beam continued in a straight line without any deflection. This behavior is similar to that of X-rays and visible light (that is, electromagnetic waves). But they were even more penetrating than X-rays, and were later called gamma (γ) rays.

The other two components of the beam were bent in opposite directions, showing clearly that one of them was made up of negatively charged particles, the other of positively charged particles. The negatively charged particles, called beta (β) particles, were later found to be none other than the negatively charged electrons. The particles that were positive and were bent in the opposite direction were later found to be the nuclei of helium atoms, and are called alpha (α) particles.

It was the New Zealander Ernest Rutherford (1871–1937), working with J. J. Thomson at the Cavendish Laboratory, Cambridge, England, who succeeded in identifying the electron and α-particle components of the radiation emitted by uranium. We shall have much more to say about Rutherford, whose name is associated with very many studies of the nuclear atom, including the scattering experiments noted in Sec. 18.2.

20.5 The Curies

The discovery of radioactivity brought a most unusual pair of scientists, husband and wife, to pursue the secrets of the atom. Pierre Curie (1859–1906) was a French physicist who had been quite productive in studies of the behavior of certain crystals that develop an electric charge when subjected to pressure or distortion (piezoelectricity).† A hard worker, idealist, and dreamer, determined to shun recognition or awards, he sought only to advance knowledge for its own sake, in his capacity as chief of the meager laboratory of the School of Physics and Chemistry of the city of Paris.

The story of Pierre's romance and devoted life with Marie Sklodovska Curie (1867–1934) has been celebrated in books and cinema, as it well deserves to be. The daughter of Polish schoolteachers, Marie left Poland, where higher education for women was forbidden, to study physics and mathematics at the Sorbonne in Paris. Only her intense love for learning and her exceptional ability and intelligence helped her to endure privations and inadequate nourishment, to the point of sometimes fainting from hunger. She met Pierre, they were married, and thereafter worked together

---

† Crystals of this type are now used very commonly for maintaining a fixed frequency in radio broadcasting and find applications in many industrial processes.

continously in his laboratory except for an interval of time when their daughter Irene was born.

The revelations of Becquerel's discoveries intrigued Marie, who found that thorium emits radiation similar to that of uranium. But the tests on certain uranium ores (pitchblende and carnotite) seemed to show a presence of an activity that was greater than any revealed by uranium or thorium alone. What was this more powerful emitter? Could it be isolated from the ores? Would a systematic search and careful chemical separation discover the nature of this unknown substance?

Husband and wife joined in a search that lasted nearly four years, required endless tests and chemical processing and handling of a ton of pitchblende, and brought almost endless disappointments. In 1902 they announced their discovery of *radium*, an almost negligible quantity of radium chloride which they had finally separated from the ton of ore. They measured the atomic weight of the new element, and found that it was indeed a most powerful source of penetrating radiation. Marie was finally granted a doctorate in 1903, and only six months later the two Curies shared the Nobel Prize in Physics with Henry Becquerel.

A second child, Eve, was born. Pierre died in a street accident at the age of 46, and Marie succeeded him in the Chair at the Sorbonne which had been created for Pierre in 1905. Marie's work brought her the unprecedented honor of being granted the Nobel Prize in Chemistry in 1911. Her daughter Irene joined her in the work, and later became the wife of Frédéric Joliot. In 1935, the Nobel Prize in Chemistry was granted to Irene and her husband for the work they did together in induced radioactivity, which we shall discuss presently. But Marie had died the year before of pernicious anemia which had been brought on by the years of hard work, self-denial, and exposure to the ionizing radiation of the radioactive materials with which she had worked.

20.6 The nature of radioactive decay

Up to this point it was known only that certain heavy elements, notably uranium, thorium, and the newly discovered elements polonium and radium, emit radiations that include both charged particles and the newly named gamma radiation. It seemed likely that these emanations came from within the nucleus of the atom, but there was no basis for presuming that the nucleus changed significantly in the process. The research involving uranium and thorium compounds involved processing, and as these elements were processed from one compound to another, there did appear some precipitates which, while also radioactive, had somewhat different characteristics. A uranium-x was discovered. At about the same time Ernest Rutherford, who had come to McGill University, Canada, discovered that some of these products gained from chemical processing and reprocessing of thorium produced a gas that seemed to have "excited radioactivity," or "active deposit," which could be traced to the thorium as its source. Even more important, the radioactive emanation from this "active deposit" did not continue undiminished, but *decreased exponentially with time*. His capable chemist-associate, Frederick Soddy (1877–1956), and he came to the conclusion that radioactivity represents a *change* of the nucleus of an atom from one element to another element. That is, when an atom emits an α- or β-particle (or both), the atom itself changes in the process and becomes a different element. *Radioactivity is therefore a spontaneous transformation of an element into another element.* For the first time it was revealed that atoms are not immutable entities subject only to changes in chemical state, but that some of them may even change spontaneously to other forms.

Rutherford and Soddy were able to reveal several other characteristics of radioactivity, among them the following:†

(1) The rate with which a radioactive element emits radiation is not at all influenced by environmental conditions (such as changes in temperature or pressure) or by the chemical compound in which it exists.

(2) A radioactive element may produce a "daughter" product, which is also radioactive, but which has a different rate of emanation of radiation. There can be, in fact, a long chain of such radioactive elements and radioactive daughter products of both uranium and thorium. The end product of these chains is common lead (Pb), which is a stable element.

(3) Each radioactive element emits radiations, which may include α-particles, β-particles, or γ-radiation, or some combination of two or of all three. The particles or the γ-rays may vary considerably in energy, but all three are strongly ionizing.

(4) Each radioactive element has its own characteristic rate of spontaneous transformation, or "decay." In effect, this

---

† The listing given here includes also information revealed somewhat later than the first publications of Rutherford and Soddy in 1902. See Boorse and Motz, p. 449 and succeeding pages for excerpts from Rutherford's and Soddy's early publications.

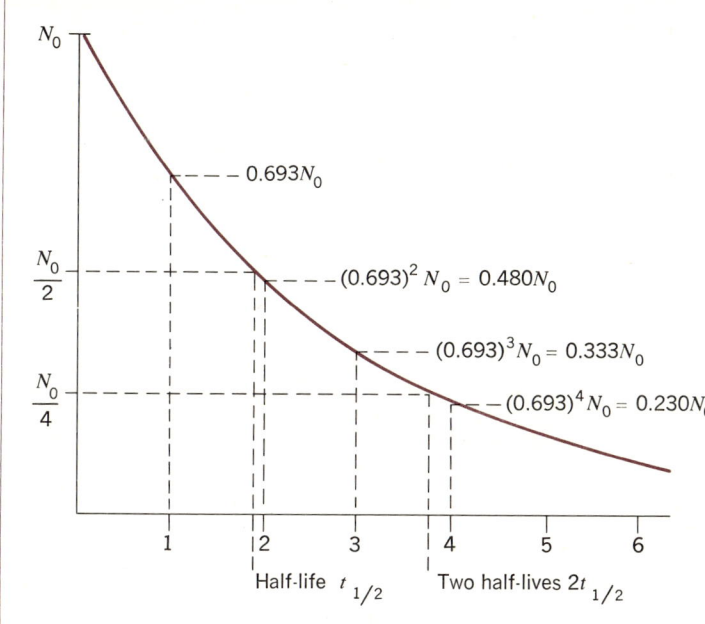

Fig. 20.3. Exponential nature of radioactive decay. The ordinate shows the number of atoms that remain untransformed, as a function of time. The half-life values are indicated, and the decay process is followed through a little over three half-lives.

gives each radioactive element a characteristic "half-life" (which may be measured in microseconds, seconds, hours, years, or thousands of years) within which time a measured quantity of that element decays to leave only half the original quantity.

(5) In a compound that includes a chain of intermediate radioactive elements, each daughter element experiences a *rate of accumulation*, which is due to the transformation of its "parent" element, and a *rate of decay*, which is due to its own spontaneous transformations. The amount of that element present in the compound is therefore determined by an *equilibrium* between the rate of production and rate of decay of that element. (For example, if a "daughter" element has a rate of decay that is very much faster than the rate of decay of the "parent" element, there will be very little concentration of that "daughter" element in the mixture at any one time.†

(6) The transformations (or decay) of naturally occurring radioactive elements seems to be entirely spontaneous, random, and unpredictable as far as any individual atom is concerned. That is, if we could "watch" a single atom, it would be impossible to predict whether that atom would change within a second, a year, or at all. When one deals with very large numbers of atoms, however (such as with numbers that are in a milligram or more of the radioactive element) the decay follows statistical laws that are very well defined with respect to the

† *Question:* Explain with a diagram or with assumed numbers why this is so. What would be the relative concentration of a succeeding "daughter" element if that element has a very long half-life?

half-life and the rate of decay during any significant interval of time. But whatever the rate may be for a specific element, *the numbers that decay within any significant interval of time is proportional to the numbers that remain at that time.* Figure 20.3 illustrates the nature of the decay by showing the number of atoms that remain at any time t.

20.7 The laws of radioactivity transformations

Suppose that there are N_0 radioactive atoms present in a batch at time t_0. How many atoms will decay in an interval of time Δt? The basic equation is simply the equation for the curve of Fig. 20.3, and was essentially presented by Rutherford and Soddy in their 1903 paper in the *Philosophical Magazine,* entitled "Radioactive Change." They reported that in all the investigated cases of radioactive material, the activity (that is, the total number of disintegrations per unit time) decreased with time in geometrical progression. Thus, if the activity of a radioactive substance decreased to one-half its value in a certain time, then after a further equal period of time the activity would be halved again. Then they deduced the law that explains this geometrical decrease in the form

$$N_t = N_0 e^{-\lambda t} \quad (20\text{-}1)$$

where N_0 is the initial number of radioactive atoms present, N_t is the number of atoms after time t, and λ is a constant which they called the "radioactive constant." In present-day literature we refer more frequently to λ as the disintegration constant or the decay constant. The constant λ is characteristic of the radioactive material and hence is different for different radioactive substances.

The half-life concept introduced by Rutherford in 1904 is related to the decay constant. The half-life of a radioactive element is the time required for it to decay from a given activity to half its activity. Since we have already mentioned that the activity decreases with time in a geometrical progression, it is obvious that the half-life of a radioactive substance is constant, irrespective of the initial activity or initial time.

Let us assume that the number of radioactive atoms present at time $t_{1/2} = 0$ is N_0. We shall designate the half-life of these radioactive atoms by the symbol t. By definition of half-life, we would expect the number of radioactive atoms at time $t = t_{1/2}$ to be $\frac{1}{2}N_0$. Using this information in the radioactive law equation, Eq. (20.1), we obtain

$$\frac{\frac{1}{2}N_0}{N_0} = e^{-\lambda t_{1/2}}$$

or

$$\frac{1}{2} = e^{-\lambda t_{1/2}}$$

Taking logarithms on both sides,

$$\ln\left(\tfrac{1}{2}\right) = -\lambda t_{1/2}$$

or

$$t_{1/2} = \frac{\ln 2}{\lambda} \quad (20\text{-}2)$$

or

$$t_{1/2} = \frac{0.693}{\lambda}$$

Equation (20-2) gives the relationship between decay constant and half-life for a radioactive sample.

The half-lives of the known radioactive elements cover an enormously wide range in time, from a nanosecond (10^{-9} sec) to a billion (10^9) years.

We have considered only the disintegration of one radioactive element. A radioactive element can decay into another radioactive element, which in turn can decay into a third radioactive element, and so on. Such radioactive chains have been observed; for example, the uranium-radium series in which, through successive decays, uranium-238 becomes lead-206.† (See Fig. 2.4 in Chapter 2.) Mathematical analysis of the decay processes in these radioactive chains are possible with the use of differential calculus.

20.8 Measurement and applications of radioactivity

The measurement of radioactivity of a sample is based on the number of disintegrations of nuclei occurring in the sample per unit time. The activity of a sample that decays at the rate of 3.7×10^{10} disintegrations per second is referred to as 1 curie. More frequently used samples in undergraduate laboratories have strengths of the order of millicuries (3.7×10^7 disintegrations per second) and microcuries (3.7×10^4 disintegrations per second).

One of the interesting applications of radioactivity (as noted in Chapter 2) is its use in dating techniques. For example, rubidium-87 decays into strontium-87 with a half-life of 4.7×10^{10} years. Therefore, by determining the ratio of rubidium-87 to strontium-87 in rocks, one can determine the age of the rock. The age determination is based on the assumption that there has been no loss of parent or daughter by leaching, between the formation time of the rock and the measurement time. Potassium-40 and argon-40, uranium-238 and lead-206 are examples of some other pairs of elements used in this type of work.

Carbon-14 is another radioactive isotope used in dating the age of carbon-bearing specimens which were at one time parts of living organisms. Atmospheric nitrogen captures neutrons produced by cosmic radiation, resulting in the formation of carbon-14 according to the following nuclear reaction:

$$_7N^{14} + {_0n^1} \rightarrow {_1H^1} + {_6C^{14}} \qquad (20\text{-}3)$$

The rate of exchange between the atmosphere, the oceans, and living matter is fairly rapid, and therefore it is assumed that all carbon in living organisms (which at any particular time exchanges freely† with the carbon of the atmosphere) must have had the same activity per gram of weight. Further, if one assumes that the intensity of cosmic-ray neutron showers has been constant for the past several tens of thousands of years and also that the specimen is free from contamination of carbon either older or newer than itself, then with the knowledge of half-life of carbon-14 and the measurement of carbon-14 activity per gram of the specimen, one can arrive at the age of the specimen.

Hydrogen-3 (or tritium) is another radioactive isotope used in dating the age of distilled spirits. Hydrogen-3 has a half-life of 12.3 years and is also produced in the atmosphere when nitrogen-14 is bombarded with cosmic ray

---

† We shall see that the nucleus of any atom consists of a number Z protons (equal to the number of outer electrons in the atom) and a number $(A - Z)$ neutrons, where A is the mass number of the atom. Thus, uranium-238 refers to a uranium nucleus possessing 92 protons and 146 neutrons ($A = 238$). This may also be written $_{92}U^{238}$. $_{92}U^{235}$ is another form of uranium, with $Z = 92$ and $A = 235$; that is, it has 92 protons and 143 neutrons. Likewise, $_7N^{14}$ possesses 7 protons (hence, also 7 outer electrons) and 7 neutrons (to give $7 + 7 = 14$ for A). The number at the lower left (7) determines the chemical properties of the atom.

† The exchange can be by breathing and by eating plants that grow in the atmosphere and thereby contain a proportional amount of the carbon-14 through the photosynthetic process.

neutrons:

$$_7N^{14} + {_0}n^1 \rightarrow {_1}H^3 + {_6}C^{12} \quad (20\text{-}4)$$

By measuring the abundance of hydrogen-3 in samples of distilled spirits, any discrepancy between the actual ages of the samples and their stated ages can be detected.

The availability of very many radioactive isotopes in recent years has promoted their use as tracers in investigating important problems in the fields of chemistry, biology, and medicine. For example, through use of this technique, it has been possible to identify many of the intermediate steps involved in photosynthesis by plants. The use of carbon-14 and radioactive species of salts has clarified the understanding of many biological processes involved in human blood flow, the diffusion of salts across body membranes, and metabolic activity.

Radioactive tracing is simple in principle. It is made possible by the fact that a radioactive atom, before disintegration, enters into the same chemical reactions as do the corresponding stable atoms. For example, radioactive carbon atoms can be incorporated in fluids or medicines and injected into the body. The injected materials are dispersed into the body fluids and tissue, and intermingle with the other carbon atoms that make up the body. But the presence of radioactive carbon atoms can be detected by the radiation they emit, by means of detection devices such as thin-window geigermüller counters, stable gas-flow counters, and films.

20.9 What is the structure of the atom?

Rutherford found himself more and more at the forefront of research that aimed to discover the secrets of the atomic nucleus. He also found himself in demand to lecture at European laboratories. His exposition given before the Royal Society of London on the chain of radioactive products that he had identified, and whose half-lives he had measured, demonstrated that the earth was much more than a few million years old (contrary to the proposal by Lord Kelvin, which was based on the assumption that sun energy was derived from gravitational contraction).†

By 1907, the young Rutherford had received many high honors, and was established as Professor of Physics at the University of Manchester, England, where he enjoyed an interesting and varied life with many friends, associates, hobbies, and promising students. While his work on the decay processes continued, his interest began to center more on questions about the structure of matter, and especially on the atomic nuclei that experienced transformations. It was known that the particles emitted by radioactive transformation have considerable energy and penetration.‡ In particular the fact that relatively heavy α-particles can pass through thin sheets suggested that possibly matter (and atoms) were made up largely of space. Could experiments with α-particles reveal details of atomic structure?

For example, suppose one were to send α-particles into a very thin sheet of matter such as gold foil. Some of the particles would pass through, while more would be deflected to the side. A few, it turned out, were hurled back toward the direction from which they came, as though repelled by very considerable repulsive

---

† See Boorse and Motz, pp. 701 ff.
‡ Gamma rays were the most penetrating, beta rays less so, while the heavy alpha particle could be stopped by a sheet of paper.

forces.† What could repel the positively charged α-particle with such force? Obviously the repulsive force may occur from charges that are alike. This clearly suggested that the atomic portions (or nuclei of atoms) which did the repelling were also of positive charge. But under what conditions could the atomic charge be large enough to hurl back the α-particles? It will be recalled that Coulomb's law states that repulsive forces between charged particles is proportional to the magnitude of the charges and inversely proportional to the square of the distances that separate their centers. It was readily acceptable that the atom could have a fairly large value for the atomic charge, but unless the physical dimensions of the charges themselves were small enough, they could not approach each other close enough to make a large repulsive force possible. Through this reasoning, Rutherford concluded that both the nucleus of the atom and the α-particle had to be very small and dense, occupying a very small fraction of the total volume of the atom, if the distance between their centers was to become small enough to produce the strong repulsion. Indeed, the atom did appear to be made up mostly of space and not a jelly-like mass. The picture began to resemble that of planetary motion, with the sun corresponding to the nucleus and the planets to the electrons, with vast separation between them.

20.10 Isotopes, and transmutation of an element

We have observed that when a heavy radioactive atom such as uranium or thorium becomes transformed, the process involves emission of γ-rays, negatively charged electrons, and/or positive α-particles. In attempting to construct a model, a first reasonable guess would be that the nucleus is made up of α-particles and electrons. This guess turned out to be wrong, however. For one thing, the nucleus of the hydrogen atom consists of a single proton. Secondly, since the α-particle is doubly charged, the formation of succeeding elements in the periodic table (which differ by one unit of nuclear charge) would necessitate the addition of one α-particle plus one electron. This would change the mass of succeeding nuclei by four units (that is, the mass of the α-particle, since the mass of the electron is small enough to be neglected in this discussion). However, for the lighter elements at least, successive stable nuclei differ by only two mass units.

While the search to find the composition of the nucleus seemed to be up against a stone wall, careful measurement of the products of radioactive transformations revealed some remarkable details. Soddy, in 1910, showed that atoms may sometimes have identical atomic charge and chemical properties, but have different atomic weights. These were called *isotopes*. The Englishman Francis William Aston (1877–1945) utilized the mass spectrometer† to reveal (in 1920) that many of the elements found in nature are in fact mixtures of isotopes. For example, even the element oxygen, which was used as reference for assigning weights to all other elements, was itself a mixture of three isotopes of masses 16, 17, 18. The fact that its natural weight was close to 16 was the result of the O^{16} isotope's being by far the most abundant. There

---

† This topic was discussed in Chapter 18, but is given again in part in this new, nuclear relationship. (See Fig. 18.4.)

† The mass spectrometer will be discussed in Sec. 20.12.

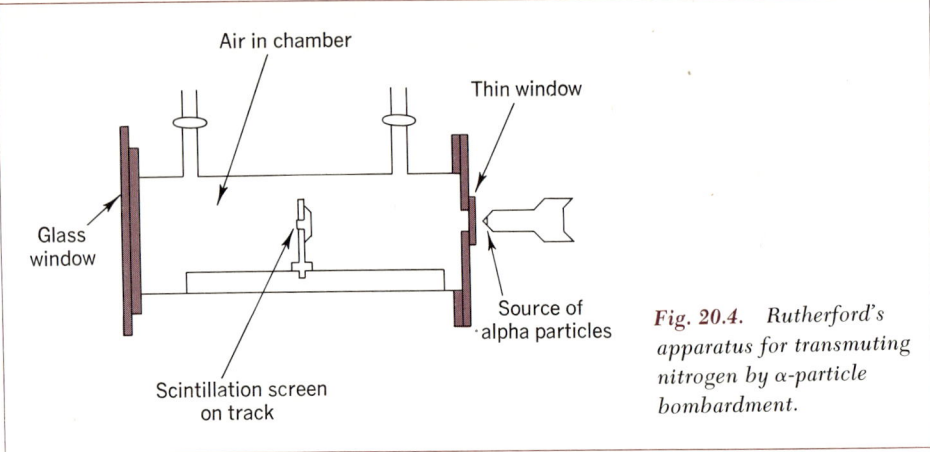

Fig. 20.4. Rutherford's apparatus for transmuting nitrogen by α-particle bombardment.

was more and more suggestion that the building block of the nucleus should include a particle that is neutral and which has a mass of approximately 1 unit, close to the mass of hydrogen. This was eventually to be called the *neutron*.

But something even more exciting had been announced the year before, in 1919. Rutherford had, despite the difficulties brought on by World War I, continued research with α-particle bombardment. His equipment was extraordinarily simple (Fig. 20.4): A metal cylinder with a thin window at one end and a glass plate at the other end contained a gas such as hydrogen or nitrogen at variable pressures. Alpha particles, from a source placed near the window, entered the thin window. A phosphorescent or scintillation screen, located on a track inside the cylinder, could be used to detect the striking of the screen by the α-particles. The screen could be moved away until the α-particles could no longer reach the screen when there was oxygen gas in the cylinder.

A most surprising thing was observed when the gas was changed to ordinary air, which is mostly nitrogen gas. Instead of ceasing when the screen was moved away, the scintillations continued to occur at long distances, distances that were clearly too long for the heavy α-particles to penetrate. The individual scintillations were also less bright, as though made by a lighter particle such as hydrogen.

Rutherford deduced what was happening: *The nitrogen atoms were being transformed (transmuted) by the energetic α-particle bombardment, and protons were being produced from the reaction.* The equation is

$$_2He^4 + {_7}N^{14} \rightarrow {_8}O^{17} + {_1}H^1 \quad (20\text{-}5)$$

This was the first demonstration that an element can be transformed, in this case by bombardment with α-particles. As we shall see, since that demonstration, every element has been transformed to one or more other forms by bombardment, and the number of isotopes has increased enormously. As a by-product of such transformations (or transmutations), radioactive forms of every element are now available with which to pursue radioactive tracer research.

20.11 Discovery of the neutron

We mentioned earlier that another particle, a *neutron*, was needed to explain nuclear structure. In 1924 James Chadwick (1891–), working in the Cavendish Laboratory, wrote to Rutherford: "I think we shall have to make a real search for the neutron. I believe I have a scheme which may just work, but I must consult Aston first."†

The scheme involved bombarding the light metal beryllium with α-particles.

† Quoted in Boorse and Motz, p. 1293.

Frederic and Irene Joliot-Curie had reported that such bombardment of beryllium seemed to produce some new kind of radiation, which carried no electric charge but which could knock protons out of paraffin. Was this new radiation an electromagnetic photon? Or was it a neutral particle? Chadwick set out to determine whether it was the latter, and possibly the sought-for neutron. In 1932 he succeeded in demonstrating that it was indeed the neutron, and was honored for this discovery by receiving the Nobel Prize in Physics in 1935. Figure 20.5 illustrates the scheme of his experiment.†

† The student is referred to Secs. 5.4 and 5.5 to show the utilization of the conservation laws in this experiment.

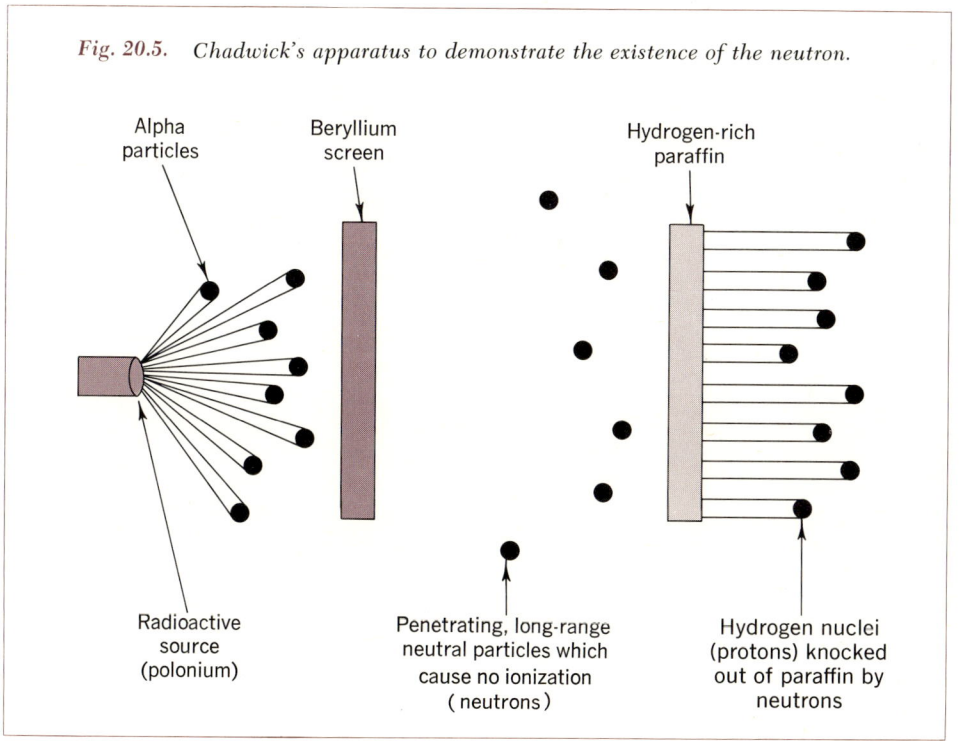

Fig. 20.5. Chadwick's apparatus to demonstrate the existence of the neutron.

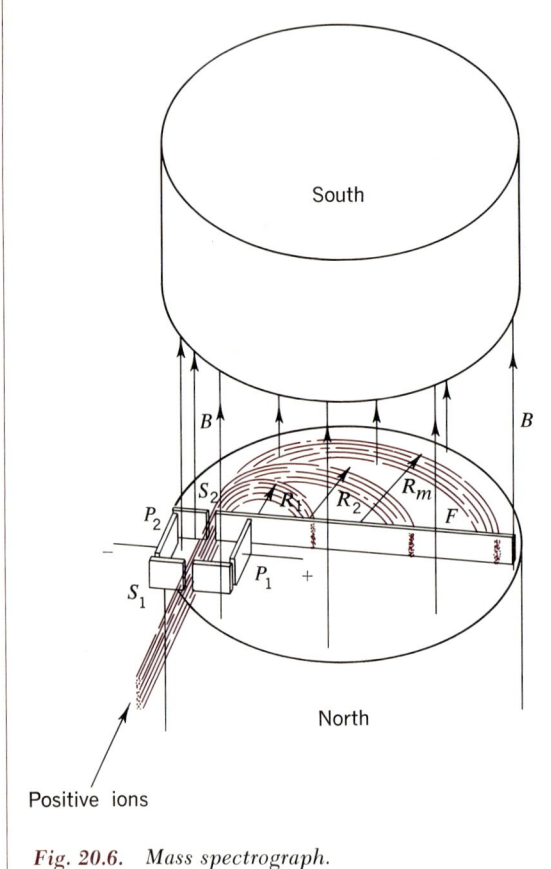

Fig. 20.6. Mass spectrograph.

20.12 The mass spectrograph and nuclear masses

We have frequently referred to the *mass* of the nucleus. It is worthwhile at this point to digress for a moment and indicate how nuclear masses may be measured.

A device for measuring the masses of atomic nuclei was developed by J. J. Thomson in 1911 and improved by F. W. Aston in the years immediately following World War I. This device, known as a *mass spectrograph*, is shown schematically in Fig. 20.6. A beam of positively charged particles (of charge Q) enters the spectrograph through a slit (S_1) and passes between a pair of charged plates P_1 and P_2. During this passage the particles are in an electric field, E, which, according to Eq. (11.7) exerts a force $F_E = QE$ on each particle. The charge thus moves on a path that curves towards P_2, and will in general miss slit S_2. If, however, a magnetic field B is impressed perpendicular to the plane of motion of the charged particle (as shown in Fig. 20.6), a second force will act on the particles. According to Eq. (11-8), this force will be perpendicular to both B and the direction of motion (v) of the particle; hence it is in the plane of the motion and at right angles to the path. It can also be shown that it is in the opposite direction to the force exerted by the electric field E. The magnitude of this magnetic force at right angles to v is $F_B = QvB$. Only those particles for which the electric force F_E is almost exactly balanced by the magnetic force F_B can pass through the second slit S_2. Equating these two forces shows that only particles having a velocity of

$$v = \frac{E}{B} \qquad (20\text{-}6)$$

can pass through S_2. The crossed electric and magnetic field therefore act as a

TABLE 20.1

ISOTOPIC MASSES OF THE LIGHTER ELEMENTS (AND MASS OF THE ELECTRONS) BASED ON $C^{12} = 12.00000000$

Symbol	Atomic mass (amu)	Symbol	Atomic mass (amu)
$_{-1}e^0$	0.000 549	O^{16}	15.994 915
		O^{17}	16.999 133
$_1p^1$	1.007 276	O^{18}	17.999 160
$_0n^1$	1.008 665	O^{19}	19.003 577
		O^{20}	20.004 071
$_1H^1$	1.007 825		
H^2	2.014 103	$_9F^{17}$	17.002 098
H^3	3.016 049	F^{18}	18.000 950
		F^{19}	18.998 405
$_2He^3$	3.016 030	F^{20}	19.999 986
He^4	4.002 604	F^{21}	20.999 972
He^5	5.012 296		
He^6	6.018 900	$_{10}Ne^{18}$	18.005 715
		Ne^{19}	19.001 892
$_3Li^5$	5.012 541	Ne^{20}	19.992 440
Li^6	6.015 126	Ne^{21}	20.993 849
Li^7	7.016 005	Ne^{22}	21.991 385
Li^8	8.022 488	Ne^{23}	22.994 475
Li^9	9.027 300	Ne^{24}	23.993 597
$_4Be^6$	6.019 780		
Be^7	7.016 931	$_{11}Na^{20}$	20.008 890
Be^8	8.005 308	Na^{21}	20.997 638
Be^9	9.012 186	Na^{22}	21.994 435
Be^{10}	10.013 535	Na^{23}	22.989 773
		Na^{24}	23.990 967
$_5B^8$	8.024 612	Na^{25}	24.989 920
B^9	9.013 335		
B^{10}	10.012 939	$_{12}Mg^{23}$	22.994 135
B^{11}	11.009 305	Mg^{24}	23.985 045
B^{12}	12.014 353	Mg^{25}	24.985 840
		Mg^{26}	25.982 591
$_6C^{10}$	10.016 830	Mg^{27}	26.998 346
C^{11}	11.011 433	Mg^{28}	27.983 880
C^{12}	12.000 000		
C^{13}	13.003 354	$_{13}Al^{24}$	24.000 090
C^{14}	14.003 242	Al^{25}	24.990 414
C^{15}	15.010 600	Al^{26}	25.986 900
		Al^{27}	26.981 535
$_7N^{12}$	12.018 709	Al^{28}	27.981 908
N^{13}	13.005 739	Al^{29}	28.980 442
N^{14}	14.003 074		
N^{15}	15.000 108	$_{14}Si^{27}$	26.986 701
N^{16}	16.006 089	Si^{28}	27.976 927
N^{17}	17.008 449	Si^{29}	28.976 491
		Si^{30}	29.973 761
$_8O^{14}$	14.008 597	Si^{31}	30.975 349
O^{15}	15.003 072	Si^{32}	31.974 020

velocity discriminator and ensure that the particles passing through slit S_2 are all traveling at the same velocity. The magnitude of this velocity can be varied by changing either the magnitude of E or B (or both).

After passing through slit S_2, the beams of all particles are traveling at a velocity v. They are still acted on by the magnetic field B as shown. Thus F_B still acts to cause their paths to bend. The radius of curvature, R, of this bending is given by Eq. (11-9) as

$$R = \frac{mv}{BQ} \qquad (20\text{-}7)$$

The charge Q on the particles can be determined experimentally. The magnitude of the magnetic field B is known, as is the velocity v [see Eq. (20-6)]. R is obtained by measuring the positions where the various charged particles strike the strip of film (F) in Fig. 20.6. Thus, only m is unknown, and this can be found from Eq. (20-7).

By use of the mass spectrograph, the masses of the various isotopes of all known elements have been measured. Although these masses may be expressed directly in grams or kilograms, the standard practice is to express them in terms of atomic mass units (amu). The amu has been defined in two ways: In the first, the common isotope of oxygen was arbitrarily chosen to have a mass of exactly 16.0000 . . . amu. In the second and more recent system, the common isotope of carbon was chosen to have a mass of 12.0000. . . amu. We shall use the system based on carbon-12, in which

$$1 \text{ amu} = 1.660 \times 10^{-27} \text{ kg}$$

Table 20.1 gives the masses of all isotopes of the lighter elements as well as of the three fundamental particles (electron, proton, and neutron) discussed so far.

The reader should note that the masses given include the masses of the outer electrons as well as the nucleus. Hence hydrogen ($_1H^1$), which consists of a nuclear proton (mass 1.007 276) and an outer electron (mass 0.000 549) is listed as mass 1.007 825, the sum of the masses of proton and electron.

20.13 Structure of nuclei; binding energy

With the discovery of the neutron there seemed to be (in combination with protons) the building blocks out of which to construct atomic nuclei. Every nucleus has an *atomic number* (Z), which represents the number of protons in the nucleus. Each proton having a positive charge of $+1$ gives the nucleus a charge of $+Z$. Thus, oxygen of mass 16 ($_8O^{16}$) has a nucleus that contains 8 protons. The remainder of the nucleus is made up of 8 neutrons to give a total atomic mass A of 16 atomic units. The 8 positive protons attract an equal number of orbital electrons to complete the oxygen atom.

Can there be more than 8 neutrons in this nucleus? Yes. There can be 9 neutrons, to make the isotope $_8O^{17}$ ($A = 17$) and 10 neutrons to make the isotope $_8O^{18}$ ($A = 18$). The orbital electrons remain unchanged. All three forms of nuclei are *stable* and do not need to change to other forms. An additional neutron would not make for a stable assembly, however.

But let us ask a more basic question: How can positive protons, which repel each other, exist in the same close assembly? The answer is that protons repel each other when they are farther apart than 10^{-14} m. But as they are forced to come closer than this distance, *the repulsive force suddenly changes to an attractive force* and holds the protons and neutrons very tightly together.† In fact it becomes necessary to supply considerable energy, of the order of millions of electron volts, to tear them apart again.

Where do these binding energies come from? They come from a conversion of mass into energy, in agreement with Einstein's mass-energy equivalence equation. Let us see how this works out in the case of the buildup of the α-particle, which constitutes the nucleus of the helium atom. We shall add up the masses of the two individual protons and the two individual neutrons and see how the addition or sum of masses corresponds with the mass that is actually measured for the alpha particle.

For two protons:

Rest mass = 2(1.007276) amu
= 2.014552 amu

For two neutrons:

Rest mass = 2(1.008 665) amu
= 2.017330 amu
Sum = 4.031882 amu

That is, taking the sum of the individual *separated* masses of the four particles adds up to 4.031882 amu. When the helium nucleus (α-particle) is measured by means of the mass spectrometer or by collision experiments, the mass is found to be only 4.001 506 amu.‡ This is 0.030376 amu *less* than the sum computed above. That is, there is a mass deficiency, or *mass defect*, of this amount. What happened to this mass? Where did it go? The answer is that this amount of mass was converted to *binding energy* to hold the four particles (we now call them nucleons) together.

20.14 Mass-energy equivalence

Let us see what this small amount of mass defect amounts to in binding energy. We shall use Einstein's equation:

$$E = mc^2 \qquad (20\text{-}8)$$

where m is the mass, c is the velocity of light (3×10^8 m/sec). We have seen that 12 g of carbon (1 molar weight) contains 6.02×10^{23} carbon atoms. (That is, a molar weight of any element or compound contains this number, called *Avogadro's number*, of atoms or molecules.) Each carbon atom weighs

$$\frac{12 \text{ g}}{6.02 \times 10^{23}} = 2 \times 10^{-23} \text{ g}$$

An atomic mass unit would be $\frac{1}{12}$ of this, or 1.66×10^{-24} g, or 1.66×10^{-27} kg. Then

$$E = 1.66 \times 10^{-27} \text{ kg} \times (3 \times 10^8)^2$$
$$= 14.94 \times 10^{-11} \text{ J†}$$
$$= 931 \text{ MeV}$$

That is, each atomic mass unit would give, if converted, 931 MeV of energy, or

$$1 \text{ amu} = 931 \text{ MeV} \qquad (20\text{-}9)$$

With this conversion factor we can easily determine what the mass defect of 0.030376 amu amounts to:

(931 MeV/amu) (0.030376 amu)
= 28.280 MeV

In other words, we would have to give the α-particle this much energy if we

---

† Even though this short-range nuclear force is measurable, it remains an area in which much theoretical and experimental research is needed.

‡ This is the mass given in Table 20.1 minus 2 electron masses.

† The joule (J) is the energy expended when 1 amp flows for 1 sec through a resistance of 1 ohm (Ω) and is equal to 10^7 erg. The electron volt (eV) is a more useful energy unit for many calculations: 1 eV = 1.59×10^{-22} Btu = 1.602×10^{-12} erg = 1.602×10^{-19} J; 1 MeV = 10^6 eV.

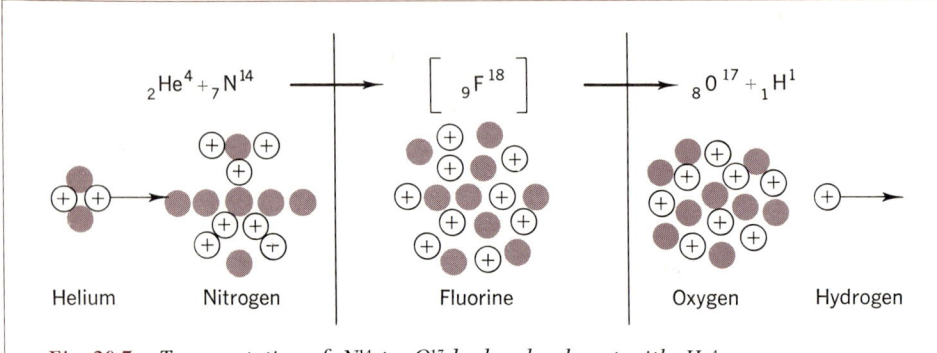

Fig. 20.7. Transmutation of $_7N^{14}$ to $_8O^{17}$ by bombardment with $_2He^4$.

wanted to separate the four nucleons. *The addition of this much energy would also provide the additional mass to restore them to their individual, separated mass values.* The 28.280 MeV is then the total *binding energy* of the α-particle, and is also the additional energy required to break up the assembly. To remove only one nucleon from an atomic nucleus requires from 5 to 8.5 MeV.

20.15 Some nuclear reactions

Before leaving this chapter we might look at some representative processes of nuclear reactions. Figure 20.7 illustrates graphically the transmutation that was expressed by Rutherford in the form of Eq. (20-5). The figure now illustrates the presence of protons and neutrons, and also shows that the α-particle and the nitrogen nucleus first form an intermediate compound nucleus (which is assumed to be a fluorine nucleus) for a very brief instant before the reaction proceeds to its final products.

The reaction observed by Chadwick when he discovered the neutron is

$$_4Be^9 + {_2He^4} \rightarrow [{_6C^{13}}] \rightarrow {_6C^{12}} + {_0n^1}$$
(20-10)

In this case the beryllium and helium nuclei combine to form an intermediate carbon compound of mass 13, which is unstable and rapidly ejects a neutron to become stable.

One of the early results of the neutron discovery was the realization that since the neutron carries no electric charge, it is not repelled as are the proton and α-particle. Therefore the neutron should be unusually capable of inducing transmutations. This is indeed so, as is illustrated by the reactions

$$_{13}Al^{27} + {_0n^1} \rightarrow [{_{13}Al^{28}}] \rightarrow {_{11}Na^{24}} + {_2H^4}$$
(20-11)

and

$$_7N^{14} + {_0n^1} \rightarrow [{_4N^{15}}] \rightarrow {_6C^{14}} + {_1H^1}$$
(20-12)

The product $_6C^{14}$ of reaction (20-12) is itself radioactive and has been especially useful for biological tracer research. A reaction that we shall take up in detail in the next chapter involves the neutron and uranium:

$$_{92}U^{238} + {_0n^1} \rightarrow [{_{92}U^{239}}] \rightarrow {_{92}U^{239}} + \gamma$$
(20-13)

It should be noted that in all these reactions, as indeed in all nuclear reactions, the sum of the Z values (subscripts) on the left-hand side equals that of the products on the right-hand side. This is simply the conservation of charge noted in Chapter 11. Also, the sums of the atomic mass numbers (superscripts) are equal on both sides, showing a conservation of nucleons. Let us consider the reaction given in reaction (20-12) from the point of view of energy conservation. The total energy on the left-hand side of reaction (20-12) is the sum of the rest energies of $_7N^{14}$ plus $_0n^1$ plus the kinetic energy of the neutron (as seen by the $_7N^{14}$ nucleus), K_n.

The total rest mass is

Rest mass $_7N^{14}$ = 14.003 074 amu
Rest mass $_0n^1$ = 1.008 665 amu
Total   15.011 739 amu

Total rest energy
= 15.011 739 amu × 931 MeV/amu
= 13,975.929 009 MeV

The total energy represented by the left-hand side of reaction (20-12) is therefore (about $14.000 + K_n$) MeV.

Similarly, the total energy given by the right-hand side of reaction (20-12) is the sum of the rest energies of the carbon and hydrogen atoms plus the sum of their kinetic energies (taken again with respect to the original $_7N^{14}$ at rest).

The rest masses are found to be

Rest mass $_6C^{14}$ = 14.003 242 amu
Rest mass $_1H^1$ = 1.007 276 amu
Total rest mass = 15.010 518 amu

Total rest energy

= 15.010 518 amu × 931 MeV/amu
= 13,974.792 258 MeV
= ~14,000 MeV

The total energy represented by the right-hand side of reaction (20-12) is therefore $(14,000 + K_C + K_H)$ MeV, where K_C and K_H are the kinetic energies of the carbon and hydrogen atoms, respectively. Equating the left- and right-hand sides gives

$$K_C + K_H - K_n = (15.011739 - 15.010518) \times 931 \text{ MeV}$$
$$= 1.14 \text{ MeV}$$

For thermal neutrons (such as are needed for this reaction), K_n is only a fraction of an electron volt; the kinetic energy of the neutron may be neglected. Thus, after the capture of a thermal neutron by a nitrogen-14 atom, the reaction products (carbon-14 and hydrogen-1) carry away over 1 MeV of kinetic energy.

Questions: (1) Use the law of conservation of linear momentum to show that

$$\frac{K_H}{K_C} = \frac{M_C}{M_H} \sim 14$$

(2) How do you interpret the above result in terms of the motion of the reaction products?

20.16 Beta decay; the positron and the neutrino

We have previously noted that one form of natural radioactive decay involves the emission of a negatively charged electron. This process is called beta (β) decay. In some decay processes, however, a single *positively* charged particle is emitted. For example,

$$_6C^{11} \rightarrow {}_5Be^{11} + {}_1e^0$$

The emitted particle has a mass, determined by experiment, equal to that of the electron. This "positive electron" is called the *positron*. It was first discovered by C. D. Anderson in 1932 in his study of cosmic rays.

The positron has the interesting property that upon collision with a normal electron, both are "annihilated" and a γ-ray having an energy equal to the sum of the rest energies of the positron plus

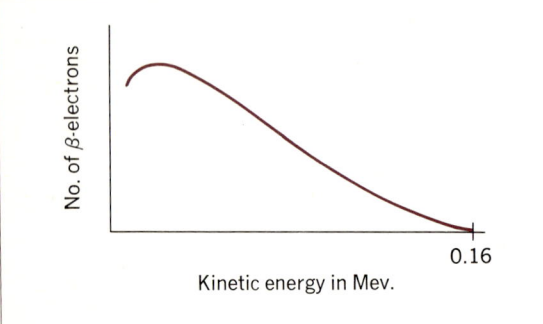

Fig. 20.8. Energy spectrum of β-decay electrons from the reaction $_6C^{14} \to {}_7N^{14} + {}_{-1}e^0$.

the electron (that is, over 1 MeV) is created. The inverse process—the decay of a γ-ray into a positron and electron—also takes place when a γ-ray of sufficient energy passes close to a massive nucleus. This latter process is termed *pair-formation*. Both pair-annihilation and pair-formation are excellent examples of the mass-energy conversion implied by Einstein's equation, $E = mc^2$.

Another interesting phenomenon of β-decay concerns the kinetic energy of the emitted electron. Consider a typical β-decay reaction:

$$_6C^{14} \to {}_7N^{14} + {}_{-1}e^0$$

The difference in rest energy between $_6C^{14}$ and $_7N^{14}$ is (according to Table 20.1) $(14.003\ 242 - 14.003\ 074) \times 931$, or about 0.16 MeV. This should appear as the kinetic energy of the emitted electron. However, when the kinetic energies of a large number of β-electrons emitted by $_6C^{14}$ are measured, they are not all equal to 0.16 MeV; rather they show a distribution similar to that in Fig. 20.8.

Although no electron is emitted with more than 0.16 MeV of kinetic energy, most are emitted with *less* than this amount. Does this mean that the conservation of energy does not hold?

In 1931 Wolfgang Pauli suggested that perhaps a second particle is emitted from the decaying nucleus along with the electron. This particle would have no charge (since our original equation already accounted for all the charge) and little or no mass (since the total energy = rest energy plus kinetic energy, which can vary down to almost zero). Pauli's work was refined by Enrico Fermi who named this particle the *neutrino*. Abundant experimental evidence now confirms Pauli's prediction of its existence.

The neutrino can also explain another apparent failure of a classical conservation law—that of angular momentum. In

Chapter 18 we saw that the electron possesses an internal angular momentum called *spin*, which is quantized in half-integer units of Planck's constant. Similarly the nucleus has a spin, which remains the same during β-decay. The neutrino must therefore have half-integer spin, as does the electron, and in the *opposite sense* so that the total angular momentum of the neutrino plus that of the electron amounts to zero.

The positron, like the electron, is accompanied in its emission by a neutrino. This neutrino, it has been found, must spin in the opposite sense to that of the emitted positron. In more recent terminology, the particle emitted with the positron is called the *neutrino;* that emitted with the electron is called the *antineutrino.* Symbolically, these are represented as ν and $\bar{\nu}$, respectively. (See Fig. 20.9.)

20.17 High-energy physics and elementary particles

We saw in Sec. 20.16 that when physicists were faced with the alternatives of discarding a conservation principle or accepting the existence of a particle that defied detection, the preference was to hold on to the conservation principle. The reason why physicists are so reluctant to abandon the conservation principle is that these relationships are founded on some rather basic assumptions about the nature of space and time. Among them there is ascribed a symmetry to space and time, which in turn implies an invariance of physical laws with respect to that symmetry. For example, the symmetry of space (which we call its *homogeneity*) implies the invariance of experimental results when the apparatus is moved to another location. A collision experiment should give the same results whether it is lo-

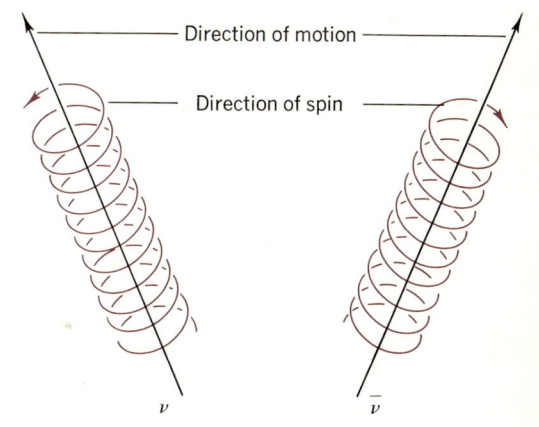

Fig. 20.9. Neutrino and antineutrino. Note that the neutrino is "right-handed," the antineutrino "left-handed." A neutrino cannot be transformed into an antineutrino, or vice versa.

cated at one corner of a room or at another corner of the room. A container of gas located on a corner of a table does not move, despite the fact that its constituent atoms are in a state of considerable motion. The conservation laws are all satisfied in the course of the multitudinous collisions of its gas molecules, without in any way affecting the center of mass of the container and the contained molecules. If the center of mass is to remain immobile, the particles of the atom will have equal and opposite momenta, that is, the total momentum will be constant or conserved.

Other new particles became evident as physicists continued to probe the structure of the nucleus within the concepts of conservation laws. The bombardment of nuclei occasionally reveals a collision phenomenon that requires new explanation for its cause. A new particle may also be postulated on the basis of theoretical analysis, such as the *meson theory* put forth in 1935 by the Japanese physicist Yukawa. He proposed that when two nucleons interact, a field is associated with the system and the mechanism of the interaction involves the exchange of a "quantum" of the field between the two nucleons. In such an interaction the momentum carried by the "quantum" is transferred from one nucleon to the other, and is the source of the force acting between nucleons. Note that this theory is analogous to the case of Coulomb interaction, in which the field is the electromagnetic field and the quantum of the field is the photon.

In 1947 Powell and collaborators observed *mesons* as a component of cosmic radiation, and shortly thereafter mesons were produced artificially in collisions of high energy (~300 MeV) nucleons. Mass measurements are in agreement with theory and the observed particle is assumed to be the same as the meson of the theory. Experiments show strong interactions characteristic of nuclear force for some mesons; these are called π-mesons and are the ones involved in Yukawa's theory. Some mesons undergo weak interactions characteristic of β-decay; these are called μ-mesons and are associated with electrons because they have identical spin, symmetry and magnetic moment. The π- and μ-mesons are classified under the same heading, since the term *meson* simply indicates a mass of "middle" value, that is, between the rest mass of an electron and a proton.

It is interesting to note that if it were not for the strength of nuclear binding forces, the world would contain only hydrogen. The existence of every other substance in the world depends upon the fact that the nuclear forces are powerful enough to stabilize the normally unstable neutron and to make it also available for use as a universal building block.

Theory again preceded experiment when Dirac predicted the existence of the *positron* (see Sec. 20.16), which had the same mass and energy as the electron but had opposite charge. However, the distinguishing characteristic of the positron is its ability to combine with an electron, in the process in which it and the electron are annihilated. For this reason the positron is termed the *antiparticle* corresponding to the electron. Dirac's theory predicts the existence of an antiparticle for every particle and therefore, in principle, *antimatter* is possible (that is, atoms with a negatively charged nucleus surrounded by positrons).

Having discussed the roles of the neutrino, π- and μ-mesons, and the positron in the drama of the subatomic world, we now turn to the general classification

20.17 High-Energy Physics and Elementary Particles

of the elementary particles. The purpose of this listing is to indicate what criteria are used to classify the particles and to show the extent and complexity of the interrelations among them.

(1) Massless bosons: photon, graviton.
(2) Lepton: neutrino associated with electron, neutrino associated with μ-mesons, the electron, the μ-meson (and their antiparticles).
(3) Mesons: π-meson, k-meson, η (eta).
(4) Baryons: nucleons, that is, proton, neutron, and their antiparticles; the hyperons.

Bosons have zero spin and are not subject to the Pauli exclusion principle (see Chapter 18). Leptons, or the electron family, are grouped because they obey a conservation law of lepton number in the reactions in which they are involved. The mesons are bosons that have mass, but which obey no conservation law of meson number. The baryons have spin ($\frac{1}{2}$), as do the leptons, and obey a conservation of baryon number.

As an application of the law of conservation to these criteria, consider the conservation of baryon number in re-

Fig. 20.10(a). Production of an eta "resonance." The antiproton entering from the right annihilates a proton in the bubble chamber to form two pions and an eta-one particle. After a time of about 10^{-22} sec, the eta-one decays into three more pions. The transitory existence of the eta-one is inferred only from studies of the pion tracks. (Notice that one of the positive pions is also observed to undergo decay into a muon through the reaction, $\pi^+ \rightarrow \mu^+ + \nu_\mu$, and the muon, in turn, decays into a position according to $\mu^+ \rightarrow e^+ + \nu_\beta + \bar{\nu}_\mu$. The neutrinos are of course unseen.) (Photograph by Lawrence Radiation Laboratory of the University of California.)

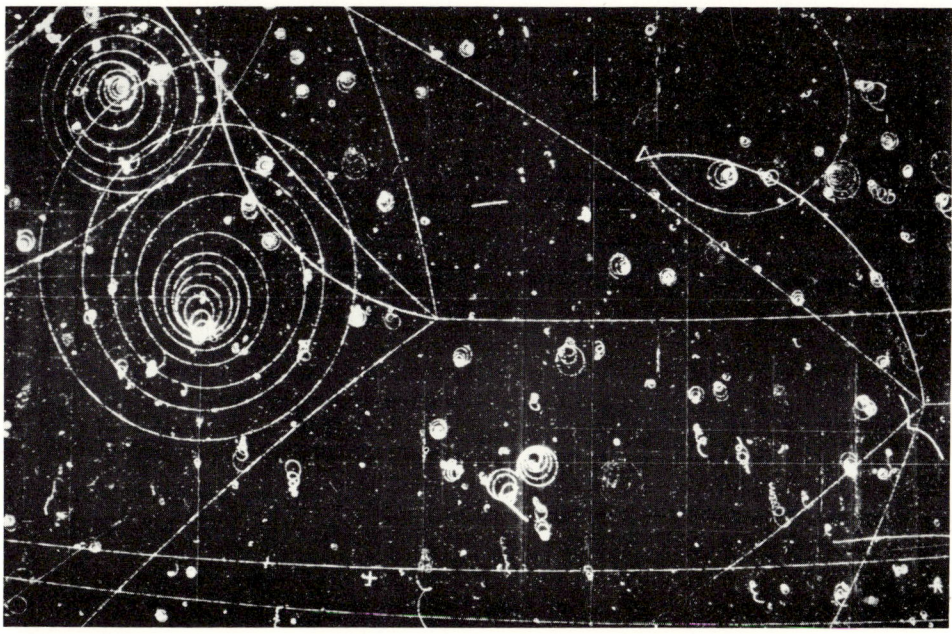

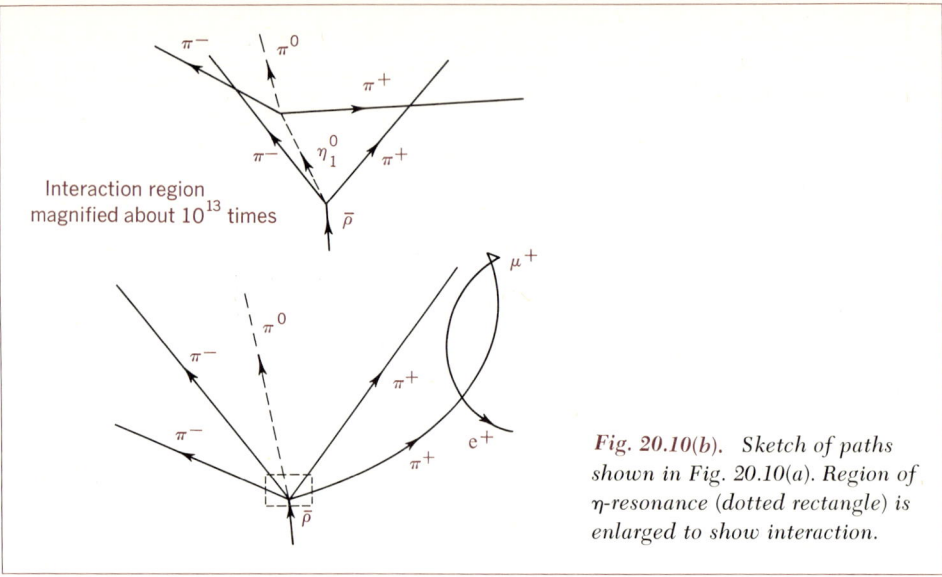

Fig. 20.10(b). Sketch of paths shown in Fig. 20.10(a). Region of η-resonance (dotted rectangle) is enlarged to show interaction.

lation to proton (a baryon) decay. Since the proton is the lightest of the baryon family, it cannot produce any other baryon upon decaying. However, the law of baryon conservation requires that a baryon be produced for every baryon that disappears. Therefore the decay of a proton is seen to be at least highly improbable.

It is important to note that there is no theoretical basis for holding so rigidly to these conservation laws other than that they have been found to hold thus far.

Until recently the principal source of high-energy particles has been the cosmic radiation from space, which constantly bombards our atmosphere. During the past two decades the development of various types of particle accelerators (that is, linear accelerators, cyclotrons, and similar ones) has made it possible to study particles with energies as high as 32 billion electron volts (BeV), and even larger accelerators are now under construction. Several techniques exist for the detection and analysis of these particles and their reaction products. Among these are the various forms of cloud and bubble chambers, which work on the principle that as a high-energy charged particle passes through a column of air saturated with water vapor, it produces ions as a result of collisions with air molecules. These ions serve as condensation nuclei about which water droplets form; hence the passage of a high-energy particle through a cloud or bubble chamber leaves a track of water droplets. These "trails" can be photographed (see Fig. 20.10(a)) and the pictures analyzed. In this way, 36 "elementary" particles have been discovered. Most of these are very short-lived, decaying in a fraction of a second to another more stable particle.

In addition a large number of so-called

resonances have been discovered. These may be thought of as "particles," with lifetimes of the order of 10^{-22} sec! Figure 20.10(b) shows an η(eta)-resonance.

Extension of our ideas about symmetry into the realm of high-energy physics by Gell-Mann and others has led to an entirely new view of the subnuclear physical world. As a consequence of this new approach, it has been suggested that a set of three truly fundamental particles—the *quarks*—may be the building blocks of which all the others are made. The quarks, still undiscovered, are hypothesized to have a third-integer charge! (All known particles have integer charge.)

20.18 Summary

It was once thought that the atoms of elements had no internal structure. The discovery and identification of electron emission, together with Rutherford's scattering experiments, led to the concept of the atom composed of a central, positively charged nucleus surrounded by electrons. Similarly, the study of radioactivity, with its evidence that α-particles come from the nucleus of atoms, led to the conception of a complex nuclear structure.

There are a great many puzzles still connected with the way the nucleus is organized. It is hoped that the newest high-energy accelerators will yield some of the sought-for answers. But a good working model of the nucleus, its component parts, and some understanding of the forces that bind the parts together has been achieved. The essential features can be briefly summarized:

(1) The nucleus, containing nearly all the mass of the atom, is composed of neutrons and protons, the number of protons determining the positive charge of the nucleus. That number also determines the number of outer electrons, the chemical properties, and the atomic number of the element.

(2) Protons can have a stable existence outside the atom, whereas neutrons decay in a brief period, splitting usually into an electron and a proton.

(3) Holding the particles packed into the tiny nucleus is a rather mysterious force called the *binding force*, stronger by far than the Coulomb forces of repulsion between the protons.

(4) In certain of the heaviest elements, where the number of protons and neutrons packed together is large and where as a result the Coulomb forces are very great, the binding force is insufficient to hold the nucleus together immutably, and there is a chance that at any instant of time the nucleus will change spontaneously. Among billions of atoms of such an element, constant changes occur, and this process is known as *radioactive decay*.

(5) When radioactive decay occurs, a new element or elements are produced as *daughter products*. The resulting products are less massive than the parent nucleus, but the law of conservation of mass remains valid if Einstein's equation for the equivalence of mass and energy is applied.

(6) Artificial radioactivity of the lighter elements can be produced if the nuclei of such elements are bombarded by particles having adequate velocity.

A few points in this picture need emphasis. The binding force referred to is a force of very different nature than any we have met before. It does not follow the inverse square law, but is a very short-range and still quite mysterious force. It is stronger by far than Coulomb forces which, in turn, are immensely more powerful than gravitational forces. The binding forces exist only between particles in a nucleus and hence are very small, in absolute terms, between individual particles. The resulting binding energies are also small within any single atom and are most conveniently measured in million electron volts (MeV) instead of joules (J). But obviously, because in any mass of recognizable size there are billions upon billions of atoms present, the total binding energy of a gram may be enormously large. Hence small amounts of material can be used as fuel for nuclear reactors (discussed in the next chapter).

The decay that spontaneously takes place in radioactive substances is governed by statistical laws. We cannot predict *when* a given atom will decay, but we can predict the proportionate *number* of atoms that will decay in the next second out of say a billion atoms present. The exponential decay equation and the half-life of any element can therefore be defined with precision. The variation of these decay rates is of tremendous range—some radioactive substances have a half-life of less than a billionth of a second; others have a half-life of hundreds of thousands of years.

Discovery of radioactivity has led to the uncovering of many of nature's secrets hidden within matter itself, and in addition has resulted in practical applications of vast consequence. A tiny speck of radioactive substance will reveal its presence to instruments sensitive to the rays it emits. Some of the uses of radioactive tracers were discussed in this chapter. Additional uses will be described in the next chapter.

Undoubtedly the employment of this radioactivity technique has only begun. The application of our knowledge and control of nuclear processes to provide energy for cities and for industries is of even greater significance for the future of civilization. We shall turn to this in the next chapter.

Questions/Discussions

1. Explain the reasons for assuming that the internal structure of an atom consists mostly of space.

2. How many protons are there in 12 g of $_6C^{12}$?

3. How many nucleons are there in 12 g of $_6C^{12}$?

4. How many electron volts are equivalent to 1 atomic mass unit (amu)?

5. Show that the density of a nucleus is roughly 10^{17} kg/m³, or roughly 10^9 tons/in.³. (If a proton has a diameter d of 3×10^{-15} m, its volume is $(4\pi/3)(d/2)^3$, or $\sim(d^3/2)$, or $(27/2) \times 10^{-45}$ m³. It weighs approximately 1.6×10^{-24} g $= 1.6 \times 10^{-27}$ kg. Its density value is

$$\frac{\text{Mass}}{\text{Volume}} = \frac{1.6 \times 10^{-27}}{27/2 \times 10^{-45}} = (1+) \times 10^{17} \text{ kg/m}^3$$

6. Describe how one would analyze the radiations emitted by uranium.

7. Give reasons why it was not proper to assume that a nucleus is made up of α-particles and electrons.

8. Explain what happens when an α-particle makes glancing collision with a nucleus.

9. Explain what is likely to happen when an α-particle of moderate energy heads directly for the nucleus of an atom.

10. Why is it desirable to assume intermediate compound nuclei in certain nuclear reactions?

11. What is meant by half-life of a radioactive element?

12. Describe the tracer technique that utilizes radioactive atoms.

13. Give some reasons for postulating before its discovery that there had to be a neutron.

14. What is the mass defect of the deuteron ($_1H^2$)?

15. Balance the following nuclear reactions

(a) $_1H^1 + {}_9F^{19} = $ _____ $ + {}_2He^4$.
(b) $_1H^1 + {}_5B^{11} = {}_6C^{12} + $ _____.
(c) $_1H^2 + {}_4Be^9 = {}_5B^{10} + $ _____.
(d) _____ $ + {}_8O^{16} = {}_7N^{14} + {}_2He^4$.
(e) $_2He^4 + {}_7N^{14} = {}_8O^{16} + $ _____.

References

Atkins, K. R., *Physics*. New York: Wiley, 1965.

Boorse, Henry A., and Motz, Lloyd, *The World of the Atom*, Vols. 1 and 2. New York: Basic Books, 1966. *These two volumes contain a wealth of technical and biographical information and excerpts from the original papers of scientists. The student is urged to refer to the volumes for any and all topics mentioned in this chapter.*

Cook, C. S., *The Structure of The Atomic Nucleus*. Princeton, N.J.: Van Nostrand (Momentum Book No. 8), 1966.

Curie, Eve, *Madame Curie*. New York: Doubleday, 1937.

Ford, K. W., *The World of Elementary Particles*. Waltham, Mass.: Blaisdell, 1963.

Frisch, O. H., and Thorndike, A. M., *Elementary Particles*. Princeton, N.J.: Van Nostrand (Momentum Book No. 1), 1964.

Heisenberg, W., *Nuclear Physics*. New York: Philosophical Library, 1953.

Mann, W. B., and Garfinkel, S. B., *Radioactivity and Its Measurement*. Princeton, N.J.: Van Nostrand (Momentum Book No. 10), 1966.

Swartz, C. E., *The Fundamental Particles*. Reading, Mass.: Addison-Wesley, 1965.

White, H. E., *Introduction to Atomic and Nuclear Physics*. Princeton, N.J.: Van Nostrand, 1964.

CHAPTER TWENTY-ONE

The New World of Nuclear Power

The unleashed power of the atom has changed everything except our ways of thinking.

ALBERT EINSTEIN†

PRIOR TO 1945 the nuclear reactions that were discussed in Chapter 20 would have appeared to be somewhat removed from the daily life of the citizen; and indeed nuclear interests were primarily the concern of specialists. Suddenly these phenomena were precipitated at the very center of social and political decisions that meant the life or death of nations, and which today greatly influence the lives of average citizens. The catalyst in this development was the event of World War II, which began in 1939 and engulfed most of the world before its conclusion in 1945. This event ushered in a new era: an era that encompasses the fearful atomic bomb as well as very great potential benefits from atomic power. The effects of this new source of energy on international relations, on science and technology, and on the pattern of governmental functions—and indeed on the future of mankind—have yet to be fully evaluated.

We shall introduce the subject of nuclear power (or atomic power, as it is usually called) within the context of the political events of the period, for it was political events that gave direction to science and technology as they had never done before.

21.1 The world of 1939

What was the political and social background of the half-century within which atomic science developed? World War I (1914–1918) witnessed the rise and fall of the political ambitions of Germany. The Versailles Treaty, which confirmed the fall of the central powers of Europe, appeared to assure that Germany would not again menace the peace of Europe. The inefficient Czarist Regime of Russia

† *The New York Times*, May 25, 1946.

had collapsed in 1917, and a very new kind of government ultimately came into being in the form of the Communist Party and the Union of Soviet Socialist Republics (U.S.S.R.). The war had introduced the airplane, the tank, poison gas, refined weaponry mechanisms, and more deadly explosives.

The years that followed this "war to end wars" were far from peaceful. The League of Nations, established in 1919 to substitute arbitration for war as a means of resolving international disputes, struggled helplessly to meet the new problems arising from the peace. The treaties that ended the war seemed to be totally inadequate in a new environment created by inequities and the threat of Communist expansion. The absence of the United States among League memberships was a blow to that organization. The rise of Italian Fascism seemed to give inspiration to its German counterpart, which grew rapidly by feeding hungrily on the unemployment and inflation that wrought havoc in the German economy. Adolph Hitler assumed the leadership of the National Socialist (Nazi) Party, and in 1933 became Chancellor of the Third Reich.

At the time of the discovery of the neutron in 1932, the United States itself was deeply immersed in an economic depression that brought Franklin Delano Roosevelt to the presidency and the New Deal to the nation's fallen economy. By 1939 the U.S. economy had recovered, but the remote war clouds hovering over the Asian continent were drifting toward the near-horizon. In 1937 Japan attacked China. In March, 1938, Hitler dared to annex Austria, and in March of 1939 Czechoslovakia was sacrificed to him by the Western Powers as a price for continued peace. Slowly there emerged the need to prepare for a new international emergency. During the next few months of 1939, President Roosevelt initiated substantial increase in defense appropriations, although the nation was firm in its resolve not to be dragged into another world conflict. On September 1, 1939, Hitler attacked Poland. Great Britain and France, in accordance with treaty promises, declared war on Germany and World War II began.

This was the political situation when, in January 1939, the Danish physicist Niels Bohr, whose explanations of the atom we have studied, brought word of a new atomic discovery, which seemed to have enormous consequences in the environment of these political conflicts. He reported that late in 1938 two German scientists working at Kaiser Wilhelm Institute for Chemistry in Berlin had discovered an unusual phenomenon involving the heavy element uranium, atomic number 92, which was the last element that was then listed on the atomic chart.

Question: What does this atomic number signify with respect to the number of protons in the nucleus of uranium?

The two scientists, Otto Hahn and Fritz Strassman, irradiated uranium with neutrons. This in itself was simply like other tests with neutrons, which the Italian physicist Enrico Fermi and his associates had begun years earlier and for which Fermi had received the Nobel Prize. It seemed that most elements could be transformed by neutron bombardment, and now it was uranium that was being bombarded. But something unusual was taking place. The two scientists seemed always to find the lighter element barium (which has atomic number $Z = 56$ and weight A of 135 to 140) with the uranium after bombardment. Where was the barium coming from?

Hahn suspected that the heavy uranium

atom was experiencing not the usual transformation but a complete split (later called *fission*) into lighter atoms; the barium that always seemed to come into the picture was apparently one of the products of the split. On communicating the findings to Lise Meitner in Austria (she had been forced to flee the Nazis because she was a Jewess) and to her nephew Otto R. Frisch, the suspicions seemed confirmed. They, in turn, communicated their conclusions to Bohr, and together they outlined an experiment that might confirm or alter their conclusions. With the experiments still to be tried, Bohr started out for a stay at Princeton University and on the way reported the developments to Fermi and others in the United States.

The news initiated great excitement and several other experiments to determine whether the fission of uranium was accompanied by significant energy release. When the bombardment was conducted in an ionization chamber with the chamber connected to an oscilloscope, the oscilloscope revealed very large bursts of energy release from fission. Reports of experiments followed each other in rapid succession in the technical journals; a January, 1940 article (Turner) listed over 100 published articles on the subject.

The picture that evolved over these months seemed to be: When a heavy atom such as uranium (thorium and protactinium as well) is bombarded by neutrons, it sometimes splits into two smaller atoms and gives up considerable energy. *Moreover, some fast neutrons are also emitted by the process.* With this last information a distressing question took shape in the minds of many scientists here and abroad, namely: *If one neutron can cause fission with emission of other neutrons, could a system be so designed that the emitted neutrons, on reaching other uranium atoms, would repeat the fission process in an explosive chain reaction?* If an atomic bomb could be developed, it seemed likely that the German government would be the first to capitalize on the development to bring other nations of the world to their knees. In view of the horrors that even then had been attributed to the Nazis, the world at large and especially those who had fled Nazi Germany had everything to fear.

The evidence to that point gave only the assurance that fission of each uranium atom would produce a lot more heat energy than could the chemical reaction between a pair of molecules. The possibilities for the fission process to be useful for making a superbomb depended on meeting one additional condition, namely, the achievement of a chain reaction. That is, if, when a neutron causes fission of a uranium atom, other neutrons are also emitted, and each of these new neutrons initiates fission of another uranium atom. If this chain of reactions took place within a fraction of a second, it would cause a whole avalanche of fissions that could accumulate tremendous energy of explosive proportions (see Fig. 21.2 on page 615).

The immediate question that required early answer was: Can a chain reaction be achieved in an uranium assembly?

21.2 The fission process

Several subsidiary questions required closer examination of the fission process. Natural uranium has three isotopes: U-238, U-235, and U-234. The ratio is about 138 atoms of U-238 for every atom of U-235, with the third negligible. *Which isotope was the one that experienced fission?* Small quantities of separated isotopes were obtained through the mass

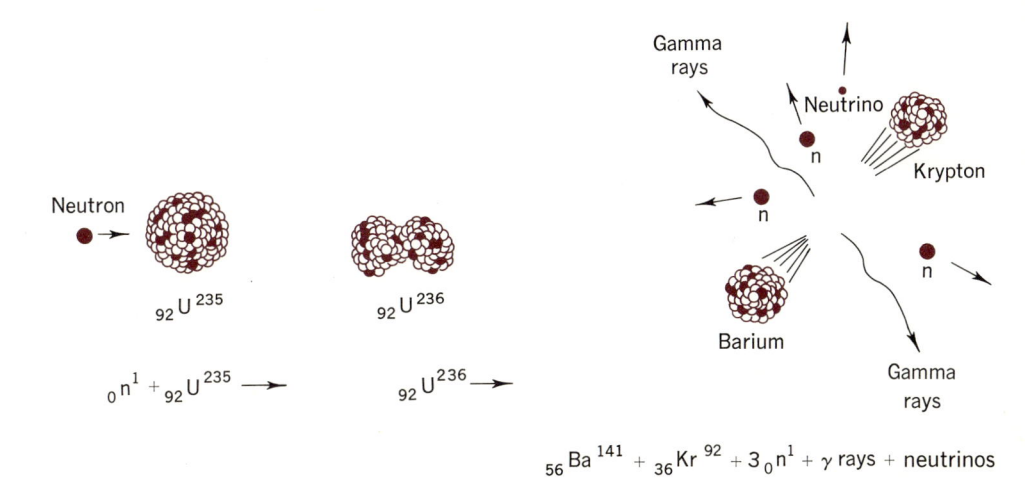

Fig. 21.1. One example of how uranium experiences fission.

spectrometer technique, and tests revealed that while U-238 captures neutrons easily, it fissions only when struck by a fast neutron. The U-235 isotope, on the other hand, splits quite easily *especially with very slow (thermal) neutrons.* One mode of fission (it turned out later that there are over 30 modes) is indicated by the following equation:

$$_{92}U^{235} + {}_0n^1 \underset{\text{(slow)}}{\rightarrow} {}_{56}Ba^{141} + {}_{36}Kr^{92} + (3)_0n^1 + Q \cdots \quad (21\text{-}1)$$

where the right side of the reaction indicates that 3 neutrons are emitted plus energy Q (Fig. 21.1).†

How much energy is represented by Q? One way for determining this would

† One or more neutrons are usually produced in each of the other fission modes noted above. When all these modes are occurring in a sample of uranium, each fission produces, *on the average,* about 2.5 neutrons.

be to collect the (heat) energy that is given off by the reaction. Tests indicate the energy given up to be about 189 MeV per fission. Where does this large energy come from? It must come from the transformation of mass into energy, according to the relationship given by Einstein, namely,

$$E = mc^2 \quad (21\text{-}2)$$

This must mean that as a result of fission into smaller fragments, *the total mass of the fragments is less than the mass of the uranium plus-neutron with which the reaction begins* [Eq. (21.1)].

The mass of the original U-235 (=235.044 atomic mass units) plus the mass of the neutron (=1.009 amu) totals 236.053 amu. When the masses of the final product atoms and emitted neutrons are summed up, the total is only 235.938 amu, which is 0.215 amu less than we started with. This much mass, when converted into energy according to the relationship of Eq. (21-2), represents an energy of

$$(0.215 \text{ amu}) \times (931 \text{ MeV/amu}) = 200 \text{ MeV}$$

(Notice that this is larger than the 189 MeV mentioned earlier. The reason will be given soon.)

21.3 The power of nuclear devices

What is the significance of this 200 MeV of energy in more conventional units? The 200 MeV amounts to $200 \times 4.45 \times 10^{-20} = 8.9 \times 10^{-18}$ kilowatt-hour (kwh) per fission. But in a gram molecular weight of U-235 (a gram molecular weight in this case would be 235 g, a little over half a pound), there are 6.02×10^{23} uranium atoms. If all these atoms experienced fission, the energy output would be the product of the two numbers, or 5,300,000 kwh of energy. A pound of U-235 has the energy equivalent of burning about 1200 tons of coal! That is, nuclear mass transformations give several million times greater energy than that given by ordinary chemical reactions.

In natural uranium, of course, where the U-235 content is only one part to 138 of U-238, the amount of energy available from the U-235 alone is less, by a factor of about 138.

In what form does this energy appear? The largest part is in the form of kinetic energy of the fragments of fission; that is, when the large atom splits, the pieces fly apart with great velocity. An explanation of the fission process offered by Neils Bohr and John A. Wheeler pictured the uranium atom as resembling a liquid drop that experiences internal motions and changes in shape but which normallly manages to hang together. When a neutron enters this "liquid drop," however,

TABLE 21.1

APPROXIMATE ENERGY RELEASE IN THE FISSION OF U-235

Nature of Energy	Energy in MeV
Kinetic energy of fission fragments	167
Kinetic energy of fission neutrons	5
Prompt γ-rays	7
β-ray decay energy	5
γ-ray decay energy	5
Neutrino energy	11
Total	200

a more violent state of agitation and elongation develops and finally overcomes the binding forces. The parts of the "liquid drop" then fly apart with great velocity. The energy Q is therefore in large part revealed as kinetic energy of the two lighter atoms that result from the fission. Table 21.1† lists the several categories in which the energy appears.

The first two items of Table 21.1 are understandable on the basis of the explanation just given. The third, prompt gamma-rays, indicates that γ-rays are also emitted at the instant of fission. What about the fourth and fifth items? The gamma-ray and beta-ray decay energies represent latent energy, which is embodied in the radioactive fission fragments and given up later as radioactivity. The atoms barium-141 and krypton-92 of Eq. (21-1) are in fact radioactive atoms that in time change to other forms by giving off γ- or β-radiation.

Barium-141 has a half-life of 18 min, decaying by β-emission to lanthanum-141, thence (by emission of a second β-particle) to cesium-141, and finally (with emission of a third β-particle) to the stable isotope praesodymium-141.

This type of *successive emission and change from one radioactive atom to another is typical of fission products*. About 250 different radionuclides have been identified as products of fission. While most of the 250 decay to stable forms in a matter of months, some of them are very slow to change and may be dangerous sources of radiation for hundreds of years.

Question: The last item of Table 21.1, neutrino energy, amounts to 11 MeV. In the earlier paragraph we said that if we were to *measure* all the energy emitted by fission we would obtain about 189 MeV. Explain the discrepancy between the 189 MeV and the 200 MeV given in the table.

What was the significance of this added information? The fact that the energy Q is so large was very promising with respect to the *power* possibilities of a nuclear device. The emission of an average of 2.5 neutrons with each fission was also very encouraging (feasibility of chain reaction). But easy fission was limited to the U-235 isotope instead of the much more abundant U-238 and this was very discouraging. It became especially discouraging when further measurements revealed that the U-238 atoms were as likely to absorb neutrons as were the U-235 atoms, except when the neutrons were very slow (called *thermal energy level*). Neutrons of thermal energy are much more likely to react with U-235 than with U-238 atoms, but before the neutrons slow down to thermal energies, they can be gobbled up by the much more abundant U-238 atoms.

The answer seemed to require that the uranium be enriched in its U-235 content so that there would be more U-235 than U-238 atoms in a quantity of uranium. But this required very unusual processes for concentrating the U-235, which was lighter than the more abundant U-238 by only three parts in 238. We shall presently discuss the separation processes that became feasible. But first let us turn to a new direction that opened up, which in fact ultimately gave an alternative path to an atomic bomb by means of the brand new man-made element plutonium.

21.4 Discovery of plutonium

To this point we have referred to the U-238 isotope, which is the most abun-

---

† In this table we include data that actually came later than 1941.

dant of the uranium isotopes, as "unproductive" of fission and very much in the way of achieving a chain reaction. In the course of the early studies it was discovered that the capture of a neutron by U-238 develops an interesting series of changes. The capture is accompanied by the emission of a γ-ray:

$$_{92}U^{238} + {_0}n^1 \rightarrow {_{92}}U^{239} + \gamma \quad (21\text{-}3)$$

But presently this new, heavier uranium loses an electron from its nucleus. The loss of negative charge leaves the nucleus with one additional positive charge ($92 + 1 = 93$), making it a new element, which was called *neptunium*:

$$_{92}U^{239} \xrightarrow[\text{23 min half-life}]{} {_{93}}Np^{239} + {_{-1}}\beta^0 \quad (21\text{-}4)$$

Question: Why doesn't the weight of Np change significantly?

But the new atom is highly radioactive, with a half-life of 2.3 days, and emits still another β-particle, to become a new atom, *plutonium*:

$$_{93}Np^{239} \xrightarrow[\text{2.3 days half-life}]{} {_{94}}Pu^{239} + {_{-1}}\beta^0 \quad (21\text{-}5)$$

Plutonium, while also radioactive, has a very long half-life, finally emitting an α-particle. But the interesting characteristic of Pu-239 is that it experiences the *same easy fission* and *emission of neutrons and energy* that U-235 does. That is, Pu-239 is just as good for use in bombs as is U-235.

21.5 Getting a chain reaction

The properties of Pu-239 made it desirable to explore how this new atom could be produced in quantity. Could it be produced in a nuclear reactor made up of natural uranium? That is, could natural uranium make up a controlled, critical (chain reacting) assembly through the fissioning of the U-235 alone? If so, enough neutrons would be captured by U-238 atoms to produce significant amounts of the new Pu-239.†

A reactor is said to be *critical* when it sustains a *chain reaction;* that is, when the number of uranium atoms that fission per unit time continues without decreasing to zero. When the chain reaction tends to die down, the reactor is said to be *subcritical*. But when the number of fissions per unit time increases rapidly, the reactor is called *supercritical*.

Question: How would you describe the three types of criticality in terms of a system and feedback influences?

Power reactors must be of the controlled critical category, whereas a bomb must be highly supercritical to be effective (Fig. 21.2).

For the proposed production of plutonium all that was needed was to have the reactor of the *critical* type, holding the power level at a constant value. But the power level had to be high if it was to produce sufficient plutonium with which to make a bomb. This meant that the reactor had to operate as a huge "furnace" in which there would be a dense cloud of neutrons, involving very many fissions each second. The fission energy heats the uranium metal, and a fluid heat transfer system must be incorporated to remove this heat to exchangers, which convert the energy to steam energy that may be used to drive turbines and generators.

---

† The reader will notice that in developing the idea of a chain reaction we have begun with a "furnace" or power reactor instead of with a bomb. Historically, the emphasis on getting the bomb was only because of the urgency of military needs.

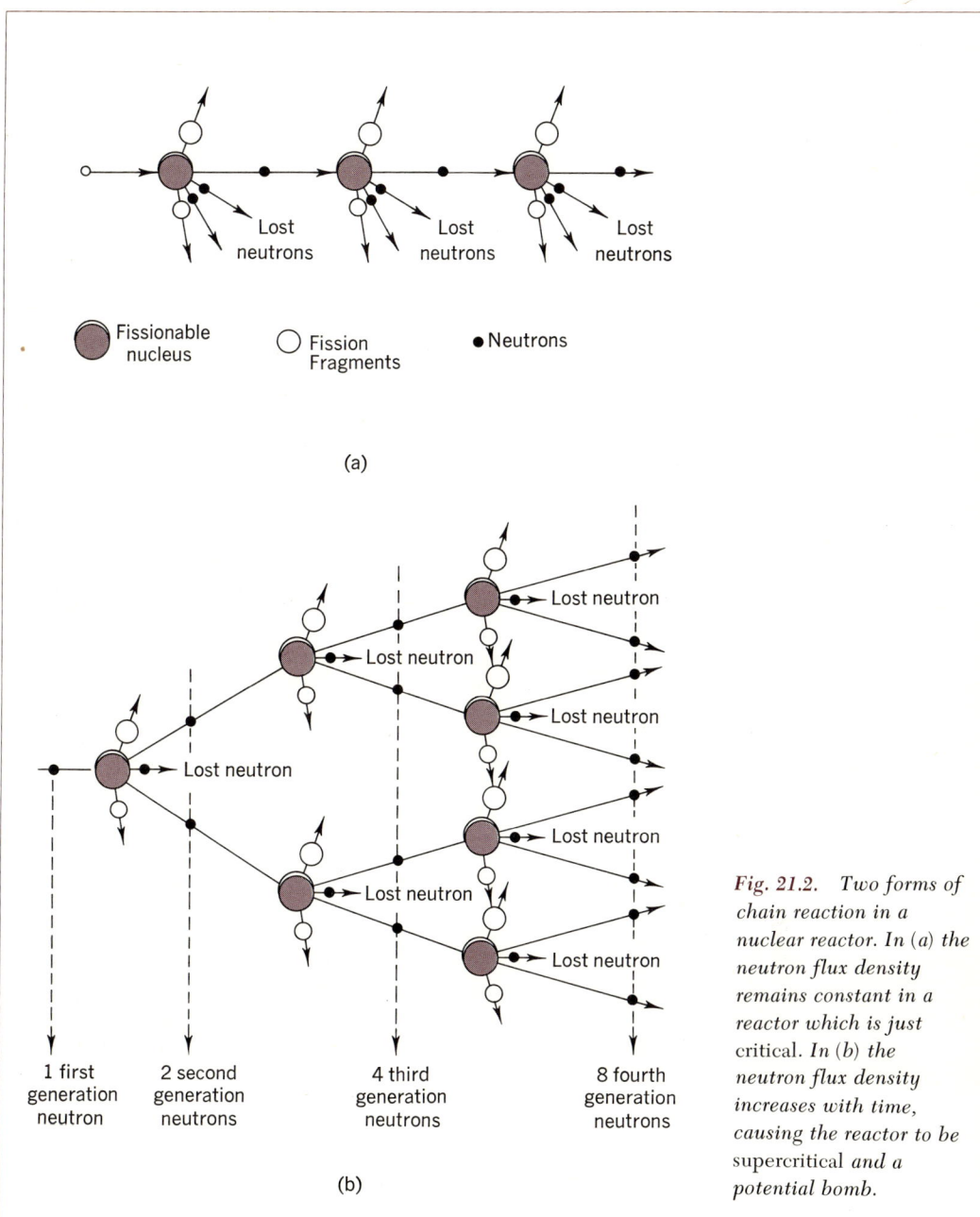

Fig. 21.2. Two forms of chain reaction in a nuclear reactor. In (a) the neutron flux density remains constant in a reactor which is just critical. In (b) the neutron flux density increases with time, causing the reactor to be supercritical *and a potential bomb.*

But how was it going to be possible to build a critical reactor using normal uranium, in which the useful U-235 isotope was almost like a needle in a haystack and all other metals (for example, U-238, the impurities in the metal, and the materials of construction) gobbled up the precious neutrons? Of course the uranium could be processed to be quite free of contaminants, and all other materials could be selected to be of the kinds that would have as little as possible reaction with neutrons. (This is not easy to do, for as we saw in Chapter 20, neutrons will interact with most elements.)

One encouraging step was in finding that when the neutron is very slow (that is, has thermal energy), the likelihood of a fission interaction with U-235 increases by a large factor. Perhaps the uranium could be made up in the form of rods of such diameter (say, 1 in.) that when fission occurred, the fast neutrons would escape from that rod. If then the neutrons could encounter collisions with nonabsorbing (moderator) material outside the uranium rods and thereby lose energy, they could thereafter wander among the uranium rods with increased likelihood of meeting U-235 atoms.

This was indeed the case. The first critical assembly was successfully tested on December 2, 1942, under the general direction of Fermi. The "pile," assembled in the stadium of the University of Chicago, achieved a controlled power level of 200 watts (W). This simple statement of the test belies the drama, the fears, and the hopes that attended this first criticality test at a time when it was far from certain that the assembly would not go up in fiery failure.

Although the Chicago "pile" had a power level of only 200 W, and a practical production reactor had to have a power level of nearly half a million times larger, it seemed reasonable to pursue both the plutonium path and the enriched U-235 path toward an atomic bomb.

21.6 "Manhattan district" and the bomb

We cannot trace all the steps and alternatives that were pursued by these scientists who saw the possibility of German dominance over the world through atomic power. They struggled feverishly to obtain the necessary measurements through which to propose the courses of action that appeared to be favorable. They struggled just as hard to persuade the government to assign funds to the work. They adopted their own secrecy measures when it appeared that the possibilities for an atomic bomb were very real. Many months elapsed until adequate funding was available and teams could be organized to attack the multitudinous tasks. A vast organization to complete the job ultimately came into being by September 1942 under the title of "Manhattan District" of the U.S. Corps of Engineers.

The job of concentrating the U-235 isotope became one of the main thrusts of the project as one possibility for obtaining an uncontrolled (explosive) chain reaction for an atomic bomb. A diffusion process turned out to be most successful for this concentration. New processes and equipment were developed for converting the uranium into a gas (uranium hexafluoride, UF_6) and diffusing the gas through barriers that were developed for this purpose (Fig. 21.3). The $U^{235}F_6$ molecule finds it a little easier to pass through the tiny holes (whose diameter is of the order of 0.0000004 in.) than does the molecule

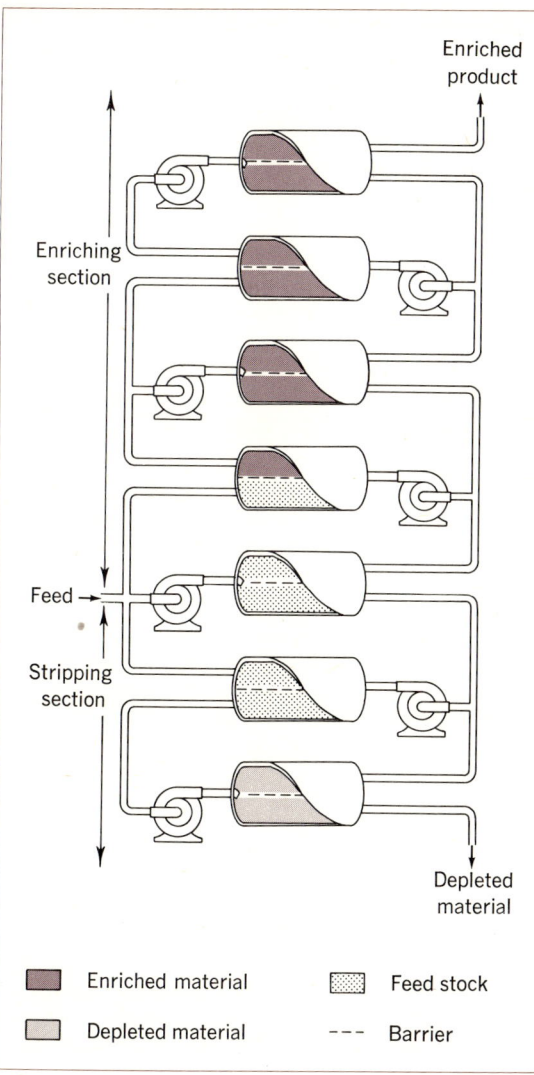

The uranium hexafluoride (UF_6), in gaseous state, enters at the "feed" point partly up the cascade, and is pumped through the barrier material. The $U^{235}F_6$ molecules diffuse slightly more rapidly than the $U^{238}F_6$ through the barrier material, so that there is a slow enrichment of the $U^{235}F_6$ concentration at the upper end of the cascade. The cascade must include many stages until the desired enrichment is achieved. The product may be drawn off at any point. The process is carried out at lower than atmospheric pressure in order to lengthen the mean free path of the colliding molecules.

Fig. 21.3. *Schematic diagram of the gaseous diffusion process. (Courtesy of U.S. Atomic Energy Commission.)*

with the heavier isotope ($U^{238}F_6$). After many cycles and recycles through long barriers, the product gas obtained is very much enriched with the U^{235} isotope. This concentrated gas can then be converted to uranium metal for use in a bomb. The bomb itself must be designed to hold the uranium parts separated until the instant of explosion. At that instant the uranium parts are hurled together to make a supercritical chain-reacting assembly. With proper containment it becomes possible to assure that most of the uranium atoms will fission before the assembly is blown apart. Each kilogram of fissioning uranium can produce an explosive release equivalent to that of about 20,000 tons of TNT.

In view of the uncertainties with respect to separation of the U-235 isotope, it was decided that the plutonium path should also be pursued to make available an alternative material for use in bombs. This required building large uranium reactors for which the main purpose would be the production of plutonium.

The design of such large nuclear reactors introduced a host of new problems. There were uncertainties as to the necessary and optimum conditions for establishing criticality in a natural-uranium assembly, to keep it chain-reacting through fission of the U^{235} isotope while also producing plutonium by the capture of some neutrons in U^{238}. Also, American industry had to devise processes by which to produce many tons of uranium and graphite and other materials in a state of purity that was of laboratory grade and far beyond earlier production experience.

Three reactors were built at Hanford, Washington, each made up of over 100 tons of natural uranium in the form of rods encased in aluminum and arranged in a lattice within a graphite moderator, and a heat transfer system that utilized the cooling waters of Columbia River to keep the uranium temperatures down to a safe level. The first reactor began operation in September 1944, and all three achieved power levels of millions of watts. They produced plutonium in quantities that were measured in grams per ton of uranium fuel, and huge reprocessing plants were constructed to recover this precious plutonium. Because fuel elements become intensely radioactive from the presence of fission products, the entire reprocessing operation was conducted by remote manipulation behind concrete shielding walls that were 10 ft or more thick. The Hanford project represented an amazingly successful combination of new science, new engineering techniques, and new boldness in production techniques.

The diffusion barrier approach for obtaining U^{235} concentrate was pursued by building huge diffusion plants at Oak Ridge, Tennessee, and was equally successful. Processes that utilized thermal diffusion, centrifuge principles, and electromagnetic separation were also under development as alternatives.

A vast research and development program at Hanford and Oak Ridge supported these two main thrusts. At the end of 1941 there were available only a few grams of pure natural uranium. This had to be increased to many tons within a couple of years, using wholly new production techniques. The "Metallurgical Laboratory" at the University of Chicago (later Argonne National Laboratory) and a laboratory in Ames, Iowa, attacked many of these problems with the support of industrial groups. The University of California group developed the chemistry of plutonium, while Columbia University and a few other schools developed much of the experimental nuclear data basic to uranium and plutonium technology. A great many pieces of special equipment such as pumps, valves, and remote manipulating devices had to be devised along the way.

By late 1942 there had developed enough confidence in the program to justify establishing a group at the University of California to develop bomb designs. The Los Alamos Laboratory was established in early 1943 to undertake the design, construction, and test of the bomb, working with gram quantities of materials until more became available.

The culmination of all this work was the successful test of July 16, 1945, at Alamogordo, New Mexico. The official War Department informational release on that test describes the great drama that surrounded every aspect of this vast operation as the components of the bomb were formally transferred from the scientists to the Army:

The entire cost of the project, representing the erection of whole cities and radically new plants spread over many miles of countryside, plus unprecedented experimentation, was represented in the pilot bomb and its parts. Here was the focal point of the venture. No other country in the world had been capable of such an outlay in brains and technical effort.

The full significance of these closing moments before the final factual test was not lost on these men of science. They fully knew their position as pioneers into another age. They also knew that one false move would blast them and their entire effort into eternity. . . . †

In time this test was followed by the uranium-235 bomb dropped on Hiroshima (August 6, 1945), and by a plutonium bomb dropped on Nagasaki two days later. Each had the effectiveness of about 20,000 tons of high explosive. The uranium bomb ("little boy") measured about 28 in. in diameter and 120 in. long and weighed about 9000 lb. The plutonium bomb ("fat man") was 60 in. in diameter, 128 in. long, and weighed about 10,000 lb. Of course the U-235 and the plutonium content of these amounted to only a few pounds of weight.

The decision to unleash the bombs on civilian populations was not arrived at easily. The scientists who had created the bomb were the most distressed over the consequences of unleashing it. There were many agonizing discussions among them and many diverse opinions expressed. There was little doubt that such attack would end the war quickly and save tens of thousands of American lives. The alternative, to demonstrate the power of the bomb in a remote area or on the field of battle, seemed less effective for ending the war. The decision to drop the bomb on Hiroshima was finally made by President Harry S. Truman, who had assumed office on the untimely death of Franklin D. Roosevelt. But the decision was fraught with fears on the part of many others that the precedent would bring such stigma as to greatly weaken the ability of the United States to promote international control of the atom.

Japan capitulated on August 14, 1945, when her emperor, anxious to prevent further destruction of cities, intervened to force the military to accept defeat.

21.7 Following Hiroshima

The brief account we have given cannot begin to convey the intensity of the drama that began in 1939 and brought death to Hiroshima. Never before had the fever of war effort so impressed itself on the scientists and technologists of any nation. Never before had the scientific community been so willing to step down from the ivory towers of academic life to the brutal business of winning a war. Never before had the products of science appeared to be the final arbiter of salvation or death for so many nations. Science seemed to have been suddenly transformed from a benign seeker of knowledge to an instrument of power in support of national policy.

---

† Appendix 6 of Smyth; see References.

Never before had the work of a brief period of five years so completely transformed the balance of power and the techniques of science and of war to the extent that in truth there emerged a new world. But this new world was a troubled one. The flush of victory was soon abated by the fears and threats that loomed just over the horizon. The Manhattan District project was carried out under very tight secrecy controls, with only some nationals of Great Britain and Canada participating in addition to United States personnel and the principals who were political refugees from Germany and Italy. Every individual had been carefully investigated and considered acceptable with respect to character and loyalty. In fact there became established the pattern of clearances, which from that time on was going to divide the people of the nation into millions who were "cleared" and considered trustworthy on one side, and many more millions who were not "cleared" and therefore not to be given any information of security importance. This great "Secret" loomed as much a handicap as an asset in assessing the future. Germany and Japan were badly beaten, and relationships with the Soviet Union were fairly cordial at this time; but there was little reason to assume that there would not be new threats to international peace. The secret of the atomic bomb seemed to be in reasonably safe hands, but the scientists knew that other nations would soon produce atomic bombs through their own work. There seemed to be a most urgent need to establish international control over this new fantastic weapon, to prevent the peace of the world from becoming the victim of blackmail at the hands of other restless nations. There was especial concern as to how the Soviet Union would react toward the new alignment with the preponderance of power that atomic energy brought to the United States and Great Britain.

There were domestic problems as well. The atomic work was still in the hands of military agencies. Both the scientists and the agencies were anxious to transfer responsibilities for the future of the work to a civilian agency, although there were many differing views on how much responsibility this new civilian agency should command over weapons aspects of atomic development. After a year of struggle in the halls of Congress and in meetings of many groups of scientists and laymen, the Atomic Energy Act of 1946 established the U.S. Atomic Energy Commission to take over this work. Neither the secrecy requirements nor government monopoly control over the operations was relaxed, however. Some people referred to the event as the transfer from the uniform into a straitjacket. Others referred to the atomic program as an island of socialism in a capitalistic society. But at least the long trek had begun for integrating the new developments into the science and industry of the nation.

The triumph of this atomic bomb project was mixed with strong revulsion and despair among scientists who feared that the unleashed forces would now become political tools of uncertain and fearful consequence. Many exerted effort to seek a balance in the peacetime uses of atomic energy. There was, of course, the assurance that nuclear energy of vast magnitudes would become available to power the industries of the nations and thus compensate the depletion of coal and oil reserves that were bound to come. Hope was expressed that this new source of energy would somehow provide the means with which the underdeveloped countries might succeed in raising their productivity and living standards. Perhaps the dream of transforming the deserts of the world into habitable areas, and of eliminating hunger and poverty, would finally be realized through this abundant source of energy.

There were other aspects. The decade preceding the war had witnessed progress in nuclear transformations, by which new radioactive atoms could be produced by bombardment with nuclear particles. As noted earlier, bombardment with neutrons was especially effective for transforming elements, but these uses remained in the area of scientific interest because neutron sources were scarce and expensive. But now nuclear reactors promised to make available very intense and low-priced sources of neutrons, since each fission in a chain-reacting, "critical" reactor requires the presence of at least one neutron in this nuclear furnace. (In a modern "test" reactor, as many as 10^{18} neutrons will pass through each square centimeter of area each second of time.) The nuclear reactor promised to become the most important scientific tool of the century.

21.8 The decades since Hiroshima

Before proceeding with specific topics that require more detailed attention, we might attempt a bird's-eye view of developments in the intervening period since Hiroshima. The end of the war brought not peace but a "cold" war (which continues to this day) in which the United States and western Europe faced the communist nations to the east. For this reason, the development and production of nuclear bombs has had dominant interest and funding, both in the United States and in a number of other nations. The Korean war, which began in 1950, brought a decision to press the development of an even more powerful bomb—the H-bomb—in which the fusion of light atoms added to the energy of the fission bomb. (Fusion will be discussed in Sec. 21.15.) The Soviet Union tested its first nuclear bomb in August 1949, and since then has tested bombs having destructive power of over 50 megatons (50×10^6 tons) of TNT equivalent. This latter is 2500 times more powerful than the Hiroshima bomb, and we are assured that the figure can be exceeded many times at will. The total quantity of fissionable material available to be hurled at potential enemies is said to be enough to kill every enemy person several times over.† Atomic weaponry has become very sophisticated as to types, sizes, methods of delivery. Bombs or missiles can be dropped from airplanes, shot from cannon, rocketed from land bases or submarines. At this writing, five nations have tested nuclear bombs of their own manufacture.

This period of development of the fission bomb and the fusion bomb called for a deep searching of the scientific conscience. The pure scientific interest of scientists in atomic phenomena seemed suddenly to have degenerated into projects designed to bring death to people on a monstrous scale. There seemed to be no alternative to the course of action that the governments followed, and yet it was difficult for individual scientists or groups of scientists to consider the alternatives without severe pangs of

† Using unclassified information, Ralph Lapp estimated in March 1964 that there were available in the United States the equivalent of 140,000 atom bombs, each twice as powerful as the Hiroshima bomb.

conscience. There came also a period of "McCarthyism,"† when anyone who dared to appear friendly or conciliatory toward communism was suspect. It was a period when the former director of the Los Alamos Laboratory of the Manhattan District, J. Robert Oppenheimer, was suddenly tried and discredited by the government he had served so well.‡ However, the Atomic Energy Commission did honor him with high award before his death in 1966.

Unfortunately, this spread of nuclear armament has not had parallel development of international control to keep the bomb threat within bounds. Atomic power has become an instrument for international threat and competition. It may be that the weapons problem will worsen as nuclear plants become more common for supplying the electric power needs of more nations, since each power reactor is a potential producer of fissionable materials that can be converted to bomb usage. Or possibly this very saturation with weapon materials will somehow bring its own cure, like a disease that defeats itself by excessive growth or finds some new equilibrium with its environment.

At this writing, the world is enjoying a ban on atmospheric and underwater testing of nuclear bombs, achieved after long effort.† It is fortunate that the ban is now in operation, for there was evidence that both the Soviet Union and the United States were becoming a little "test happy" and were contaminating the atmosphere and disturbing the natural blankets of charged particles that surround the earth.

The other international aspect of atomic energy, namely, its exploitation on behalf of the underdeveloped nations, has scarcely begun. These hopes have receded in the face of the technical and financial realities of atomic energy. We know now that each nation must first strengthen its educational system and achieve substantial progress in nonnuclear industry before it can exploit the fruits of atomic energy. Nations that have not acquired enough knowledge to exploit the steel plow and the combustion engine cannot expect to jump over this phase into atomic energy. Moreover, we appreciate the fact that the first costs of developing economical nuclear systems can be paid for only by the few nations that are well advanced. Later, these systems can be copied or purchased by nations not now able to use them.

In pushing the frontiers of other technologies, we have learned that education in atomic phenomena is part of the new approach toward the study of such things as fluid flow, metal behavior, properties of materials, chemical reactions, and soil mechanics. We have gained, as byproducts of atomic power, very many new types of radioactive atoms or radioisotopes. The chart of nuclides has blos-

---

† After Senator Joseph McCarthy, who conducted his own campaign of identifying and weeding out so-called Communist sympathizers from all positions of any consequence in government or universities.

‡ The reader who is interested in pursuing the subject is urged to obtain an Atomic Energy Commission report entitled: "In the Matter of J. Robert Oppenheimer; Transcript of Hearing before Personnel Security Board, Washington, D.C., April 12, 1954, through May 6, 1954." Available from the U.S. Government Printing Office, Washington, D.C.

† The Chinese government of Peking, which is not party to this agreement, exploded its first atomic bomb on October 16, 1964. Likewise, France is not a signatory to this agreement and has carried on several series of atmospheric tests since 1964.

somed from a list of a few hundred to over 1300. The new atoms, mostly produced in nuclear reactors, give off characteristic ionizing radiation. This is useful in either simply detecting or tracing the presence of the radioisotope itself or (when there are many of them) in utilizing it to bring about some chemical change.

Tracer techniques, using radioisotopes, have become very important to the study of chemical processes, biological and medical processes, mechanical wear, atomic displacement in metals, and hundreds of other applications. The technique is incomparably powerful and sensitive for many types of experiments. For example, through use of this technique it has been possible to identify many of the intermediate mechanisms involved in the photosynthesis of matter by plants. The use of radioactive carbon (C^{14}) and radioactive species of salts has clarified understanding of many of the biological processes involved in human blood flow, the diffusion of salts across body membranes, and metabolic activity. Industry has found activation analysis to be particularly sensitive for identifying contaminants in metals or other materials.

We mentioned that the fission products are very radioactive and that many materials can be made radioactive. The ionizing radiation given off by radioisotopes is highly concentrated and intense. Since this radiation can penetrate and ionize and can induce changes in chemical and biological systems, it promises to become significant for chemical processing, for destroying living organisms (sterilizing or pasteurizing), and even for utilizing as energy in battery form. Thickness gauges have become common in paper, plastic sheet, and rubber sheet manufacture. These gauges are based on the principle that one can detect small changes in material thickness by selecting radioisotopic radiation that barely passes through an absorbing sheet.

On the legislative side, the severely restrictive provisions of the Act of 1946, which retained to the federal government complete monopoly and control over all aspects of atomic energy, finally gave way to the revision of 1954. The new Act permitted industry to participate in the development of atomic energy, although still under the complete domination of government license and direction. The tight secrecy that had characterized atomic work from the beginning was relaxed somewhat to permit this participation. The slow pace of transfer of atomic energy information from secret files to open literature has been a troublesome issue. The precedent set for development of atomic energy has had considerable influence on the pattern by which the federal government has moved into other areas of new technology, such as activities in outer space travel. Students who are interested in government, political science, economics, and management of projects that involve industrial and governmental cooperation are urged to study the atomic program in some detail.†

21.9 Future of atomic power

The economic, social, and technological progress of a nation is intimately dependent on the amount of electric and other forms of energy that the nation can bring to the support of its economy. The United States enjoys an excess of food supply through the efforts of only 20 percent of its population, as compared with 80 percent and higher of the population of

---

† See Parsegian in References.

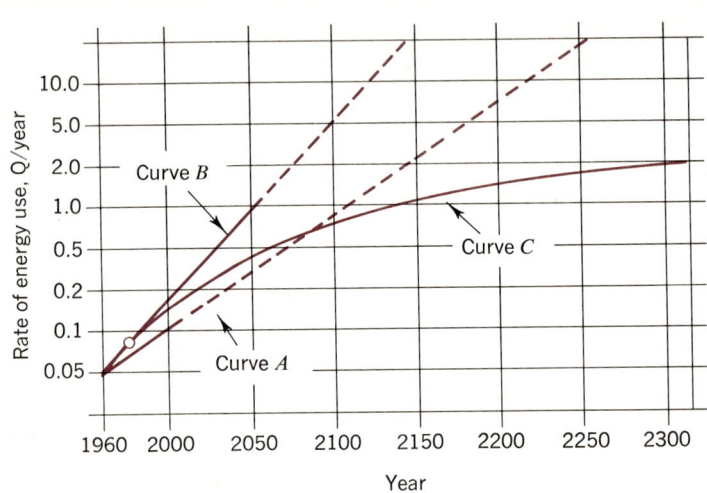

Fig. 21.4. Estimates of future rates of energy consumption in the United States. Curve A is based on Fig. 15 in the 1962 report "Energy Resources," prepared for the National Academy of Sciences Committee on Natural Resources. Curve B from the September 21, 1962, "Report of the National Fuels and Energy Study Group, on an Assessment of Available Information on Energy in the United States, to the Senate Committee on Interior and Insular Affairs." Curve C has the same initial rate as curve B but has a downward trend. (Courtesy of U.S. Atomic Energy Commission.)

undeveloped countries, simply because so much of the farming is performed with machines and gasoline energy. The automatic machines of industry that turn out vast quantities of consumer and other products are themselves the products of technological skill that commands large energy resources to fashion new metals, new designs, new science. The energy utilization per person continues to increase, as do population numbers. Since the world's resources in coal, oil, natural gas, and hydraulic power are not unlimited, there is need to examine at what point of time the energy resources of the world will fail to meet the growing needs of man.

In 1962 the U.S. Atomic Energy Commission undertook a study, at the request of the President, to determine the future energy needs of the nation in relation to available energy resources. The purpose of the study was to obtain some guide lines for the rate at which atomic energy must pick up the energy load.

Figure 21.4, taken from the report of that study, illustrates the trend in power consumption for the nation, extrapolated to the year 2300. Three curves are offered for this projection, each subject to many unknowns because the rapid rise in power consumption over the past few decades may not continue at the same rate. If the same pace is assumed to continue, one might expect an increase according to curve A (extrapolation of experience of past 50 years). If one fixes on a power value for 1980 estimated by the Department of the Interior, and extrapolates from there, a very sharp increase continues (curve B, an exponential curve). But if one estimates a downward trend in population growth rate and in the power required per individual, curve C may be the result. The ordinates express energy in unit Q per year (Q for quintillion), where one Q equals 10^{18} British thermal units (Btu). This is the energy contained in about 40 billion tons of average high-grade coal. As indicated on the curves, current power consumption is at the rate of roughly $\frac{1}{20} Q$ per year. It is increasing at a rate of 2.04 percent per year and doubling every 30 years. It is assumed that the population in A.D. 2000 will not be more than 320 million.

With an increase in yearly power consumption of at least four times during the next half-century, what may our grandchildren expect to use for energy sources?

Figure 21.5, also taken from the report to the President, illustrates the total cumulative energy consumption in terms

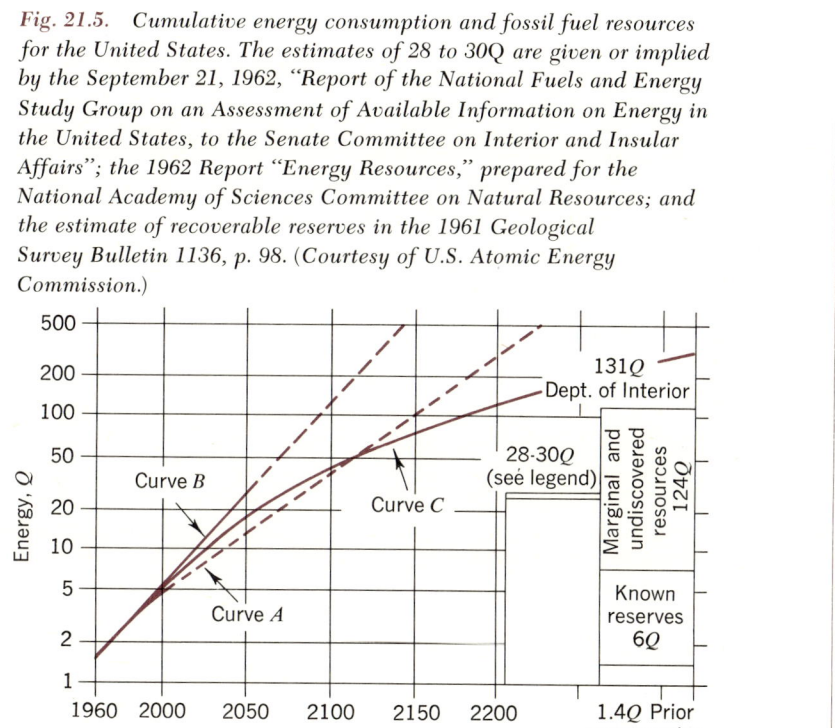

Fig. 21.5. Cumulative energy consumption and fossil fuel resources for the United States. The estimates of 28 to 30Q are given or implied by the September 21, 1962, "Report of the National Fuels and Energy Study Group on an Assessment of Available Information on Energy in the United States, to the Senate Committee on Interior and Insular Affairs"; the 1962 Report "Energy Resources," prepared for the National Academy of Sciences Committee on Natural Resources; and the estimate of recoverable reserves in the 1961 Geological Survey Bulletin 1136, p. 98. (Courtesy of U.S. Atomic Energy Commission.)

TABLE 21.2

FISSION ENERGY CONTENT OF DOMESTIC NUCLEAR RESOURCES[a]

Cost range, $/lb of oxide[b]	Energy in U^{235}, Q		Total energy content, Q	
	Reasonably assured resources	Estimated total resources[c]	Reasonably assured resources	Estimated total resources[c]
I Uranium				
0–10	0.16[d]	0.4[d]	22[d]	50[d]
10–30	0.17	0.3	24	40
30–100[e]	5	10	700	1,400
100–500[e]	220	900	30,000	120,000
II Thorium				
0–10			6[f]	25[f]
10–30			6[f]	13[f]
30–100[e]			700	2,200
100–500[e]			63,000	190,000

SOURCE: Courtesy, U.S. Atomic Energy Commission. *Civilian Nuclear Power, A Report to the President*, 1962.

[a] The magnitude of the resources has been estimated by the USAEC. The energy unit, Q, equals one billion billion Btu, or 0.252 billion billion kilocalories. The fission energy content is presented on the basis that all the resource material will ultimately fission after being recycled through reactor cores in refabricated fuel. The figures do not take account of losses during fuel recycling and other relatively minor losses.

[b] Present Commission contracts call for a price of $8.00 per pound of uranium oxide. Its present open market price would be somewhat less. Market prices have not been established for thorium oxide on a significance scale.

[c] Includes geologic estimates of future discoveries.

[d] Includes uranium already mined, most of which still exists as uranium.

[e] Cost based on recovery of both uranium and thorium from granite, and only uranium from shale and phosphate rock.

[f] Incomplete estimate because of lack of data.

of Q, which we may anticipate on the basis of curves A, B, and C of Fig. 21.4. Curve C may again be looked on as a compromise, conservative estimate.

On the right of Fig. 21.5 the rectangular blocks indicate various estimates for fossil fuel reserves. The "known reserves" of 6Q represent an amount that can probably be mined at current costs, according to the U.S. Department of the Interior. Further the Department has estimated informally that 124Q undiscovered and marginal resources exist, and that a possible 24Q can be recovered with improved technology at costs of 10 to 15 percent above present levels. Thereafter, costs would increase by amounts that would depend on the advances in technology. It seems clear that the fossil fuel sources cannot comfortably supply our needs beyond 75 to 100 years from now.

The world situation for fossil fuels is less encouraging because the rate of consumption in undeveloped countries is increasing very rapidly. The United States has roughly 30 percent of the world's fossil fuel reserves, but only 6 percent of the population. The report also notes that a detailed analysis of future energy sources must take into account the special advantages possessed by fossil fuels and hydrocarbons for small power plants, for mobile units, and for the chemical and metallurgical industries, advantages not now offered by nuclear devices.

What about sources of energy other than coal or nuclear sources? Coal and oil now provide about 81 percent of the power being used in the world, with natural gas adding another 12 percent, and hydraulic energy just below 2 percent. Coal and lignite reserves in North America amount to about 120 times the total reserves of oil, natural gas, and hydraulic energy. Therefore the future needs are understandably matched against coal sources.

All this suggests that *supplementary energy sources and substitutes for fossil fuel systems* will become necessary to our economy *within a few decades* and should therefore be sought *even now*.

What about energy resources from nuclear fission?

Table 21.2 lists the estimates prepared by the U.S. Atomic Energy Commission for uranium and thorium. The estimates are in two categories, depending on whether one looks to the heat energy derivable from easy fission of the U^{235} isotope of uranium or to the total energy that can be derived from uranium and thorium by fission and breeding processes. As shown in the column under "Energy in U^{235}," the total energy contained in resources that can be mined and purified for under $30 per pound of oxide amounts to less than $1Q$, which is enough for little more than one generation to come. But if, through conversion and breeding processes, the full energy of uranium and thorium are utilized, the same cost range for pure metal oxide could provide over $110Q$. If higher costs for mining and purifying can be condoned (as they easily might be with converters), the reserves amount to many times the known fossil reserves.

The Commission looked into the future role of atomic energy as supplementary source of power, and its report produced the estimates of Fig. 21.6. The graph, based on the compromise curve C of Fig. 21.4, produces the total energy estimate represented by curve A. It is assumed that by the year 2000, nuclear stations will generate about half the total electric power of the country and that new construction will be largely of nuclear plants. Curve C of Fig. 21.6 represents the total fuel consumption other than nuclear energy, most of which goes for uses other than for producing electric power. (Curve B shows fuel uses other than for electric power.) The difference between curves A and C represents nuclear power, which in A.D. 2000 will equal that produced by fossil fuels. Therefore the nuclear portion increases rapidly, with fossil fuels being conserved for applications other than producing electric power. It should be noted that the need for coal continues to increase even while nuclear energy takes on a major part of the power load.

Efficiencies in the use of coal have increased markedly in recent decades by the use of higher plant temperatures and pressures. Whereas 7 lb of coal were required in 1920 to produce a kilowatt

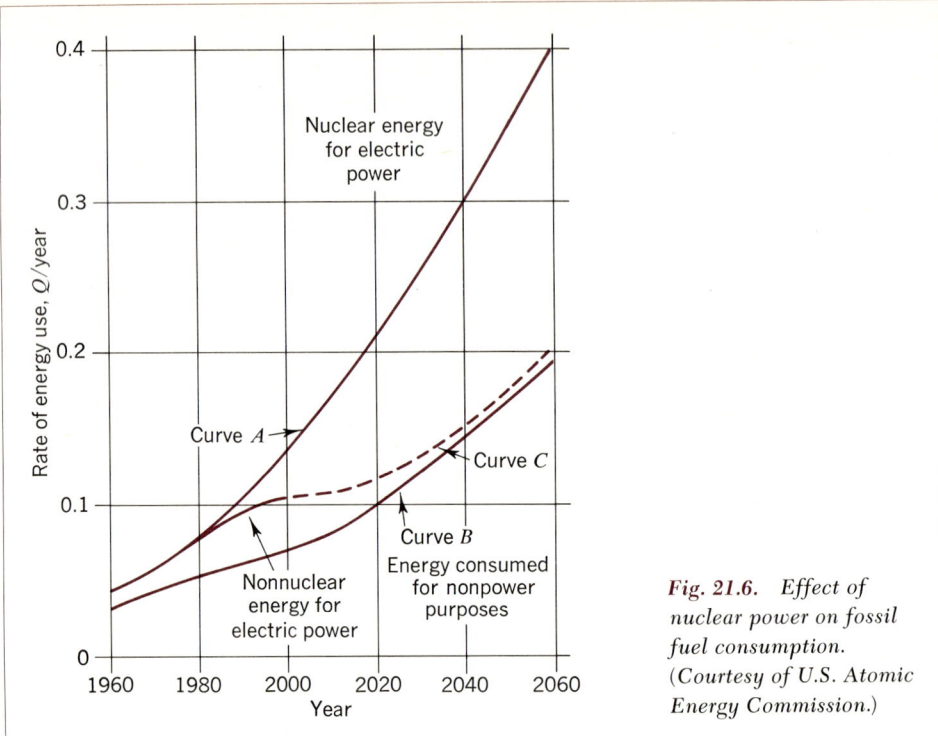

Fig. 21.6. Effect of nuclear power on fossil fuel consumption. (Courtesy of U.S. Atomic Energy Commission.)

hour of electricity, the figure was reduced to 1.1 lb in 1953. It is doubtful that such improvement in efficiencies of power plants can continue to a degree that will make up for the increases in costs that result from having to work poorer deposits. Some interest has developed recently concerning the merits of reducing coal to slurries or other forms that may be transported by pipe lines, and which could be consumed more efficiently. Therefore it is safe to assume that nuclear sources of electric power may have somewhat stiffer competition from coal and other sources in the immediate years to come. During 1964 the price of coal dropped back to the 1954–1955 levels as a direct result of this competition.

21.10 Nature of nuclear power reactors

Figure 21.7 shows the relationship of the nuclear reactor to the various mining, milling, processing, and fabrication operations that make up the nuclear power complex. A nuclear reactor is essentially a furnace that consumes fissionable material in place of oil or coal or chemicals to produce heat. Figure 21.8 compares coal burning with nuclear fission plants for producing electric power. Conventional furnaces have a number of designs and methods for converting heat into mechanical motion or into electricity. Nuclear furnaces permit much wider variety of designs, as is shown schematically in Fig. 21.9. But the

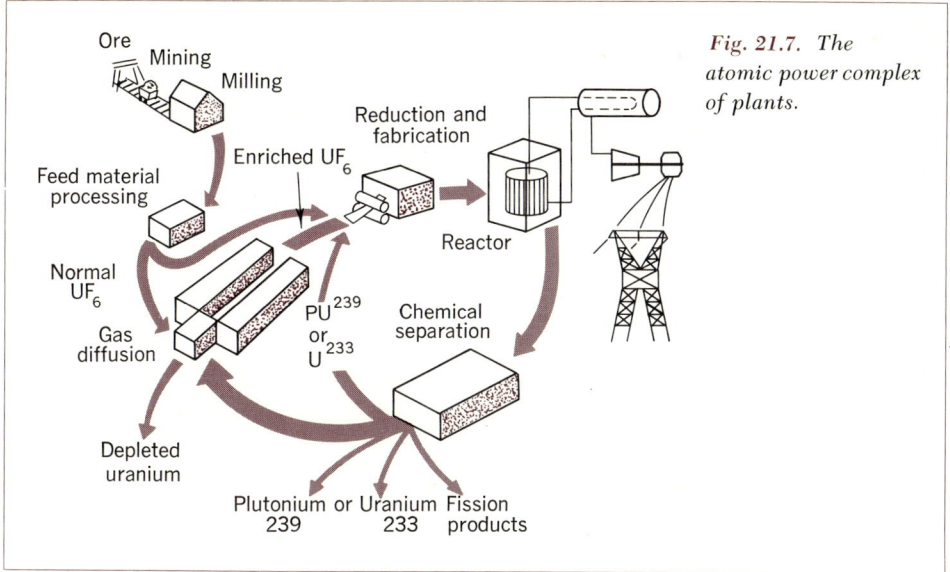

Fig. 21.7. The atomic power complex of plants.

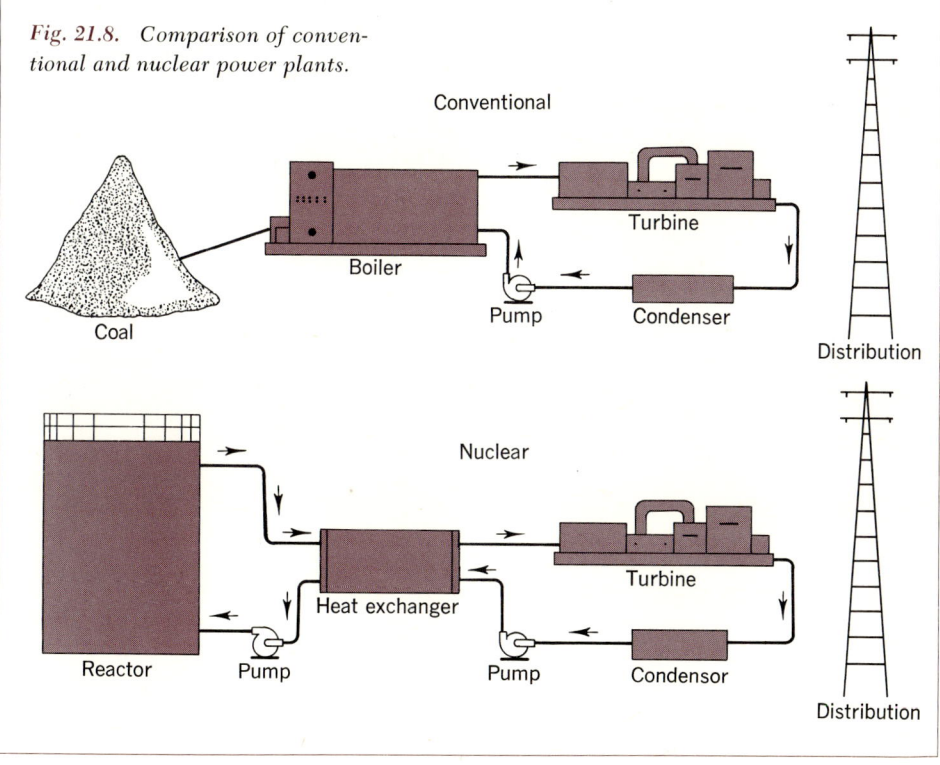

Fig. 21.8. Comparison of conventional and nuclear power plants.

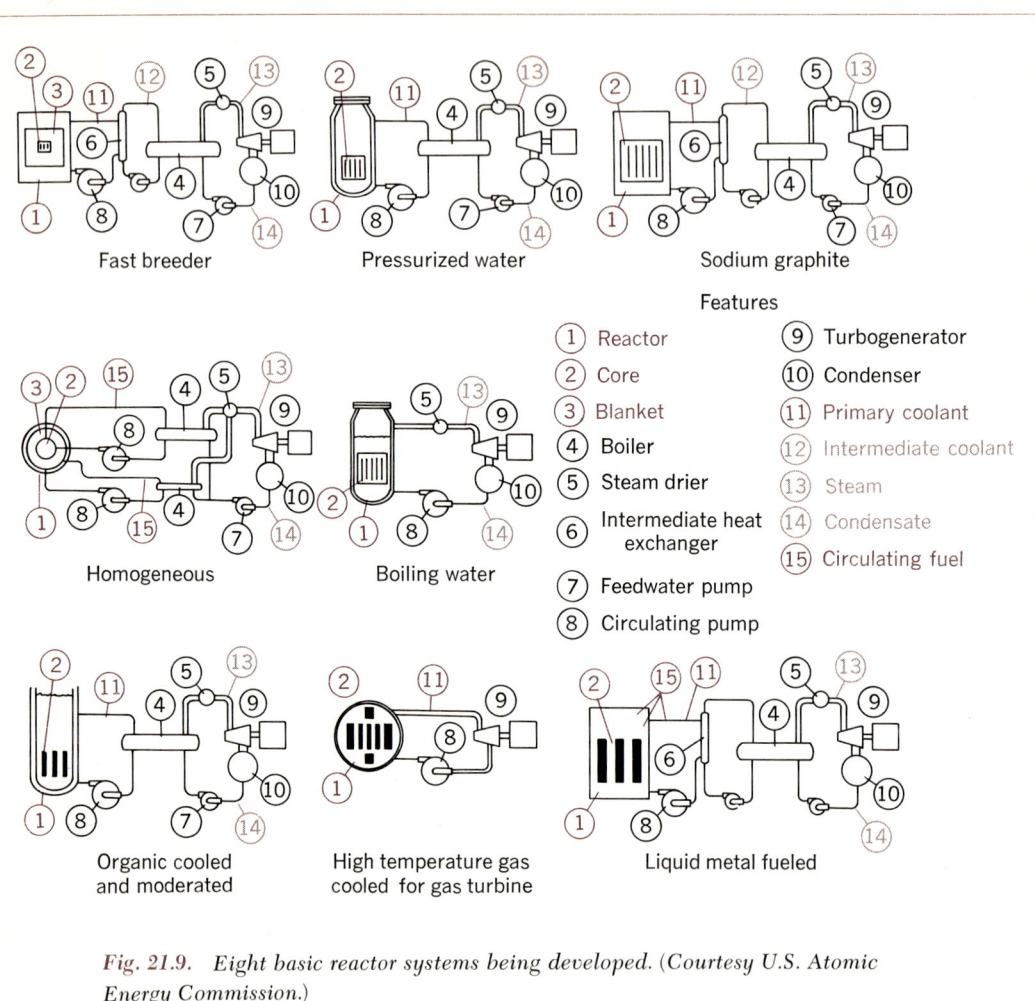

Fig. 21.9. Eight basic reactor systems being developed. (Courtesy U.S. Atomic Energy Commission.)

design of an optimum system requires a very sophisticated analysis of the required performance, of the specific experience and technology on which the design must be based, of the environmental and contractual considerations, and often of unusual circumstances within which the system must function effectively. (See Figs. 21.10 through 21.12.)

The easily consumed nuclear fuels are the three fissionable isotopes U^{235} (found in natural uranium in a concentration of only one part U^{235} to 138 parts U^{238}), Pu^{239} (produced in converter or breeder reactors from the plentiful U^{238}), and U^{233}

(produced similarly from the plentiful thorium). Of these three, U^{235} is the current basis for most reactor designs, although in time the other two will assume more importance as breeder and converter reactors become more common. In fact, the Commission report discussed earlier stressed the importance of advancing breeder and converter technology to take advantage of nuclear energy resources. Reactor design, therefore, usually involves optimum location and use of U^{235} and U^{238} atoms to support a chain reaction that is controllable with respect to neutron flux, temperature excursions, and heat removal under conditions that permit high thermal efficiency and long life.

From the mining and milling operations, the concentrated uranium oxide (U_3O_8) is sent to processing plants for purification and conversion to "orange oxide" (UO_3)†, then to "green salt" (UF_4).‡ The "green salt" then goes directly to reduction and fabrication plants, to be converted to metal ingots and fuel elements for reactors using natural uranium or to be converted to the hexafluoride gas (UF_6) and sent through diffusion plants to concentrate the U-235 isotope. The enriched UF_6 then proceeds to reduction and fabrication plants, to be converted to fuel elements for use in reactors using enriched fuel. The spent fuel is periodically given chemical separation to remove fission products and prepare fissionable materials for reuse in fuel elements.

In an assembly using natural uranium, the total weight of uranium has to be many tons—of the order of 10 to 100 tons or more—to become critical, depending on the moderator used. The heat gen-

† Uranium trioxide, called "orange oxide" because of its characteristic color.
‡ Uranium tetrafluoride.

Fig. 21.10. *Examining a fuel element assembly for the Yankee reactor before shipment. (Courtesy of Yankee Atomic Electric Company.)*

Fig. 21.11. *Fuel element assemblies loaded in Yankee reactor. (Courtesy of Yankee Atomic Electric Company.)*

Fig. 21.12. The Yankee reactor vessel head in place. (Courtesy of Yankee Atomic Electric Company.)

erated, or power, is therefore distributed over a large volume, and the power density (kilowatts per unit volume) will be much lower than it would be if the same total power were generated in a smaller total volume.

Since U^{235} can be concentrated, it becomes possible to enrich uranium fuels all the way from the normal enrichment (0.7 percent) up to the very nearly 100 percent U^{235}. Even slight enrichment, up to 2 percent, can very markedly decrease the weight of uranium required for criticality, thus increasing the power density. With high enrichment, one may locate the same total number of U^{235} atoms in a reactor assembly of small size. But this enrichment would generate the same total heat in such a small volume as to make controlled and useful heat transfer impossible. The unit would melt down. Usually, precautions must be taken to make sure that hot spots do not develop around any single fuel element, either because of poor heat transfer through its containing sheath or because of local disturbance to the heat transfer system.

When natural uranium is used as fuel, it is necessary to exercise great care in conserving neutrons to establish criticality. Construction materials must not have too high an absorption coefficient for neutrons. Fuel element geometry and dimensions must be selected to improve the chances that one of these product neutrons will reach a U^{235} atom and cause another fission. U^{235} is more likely to catch a neutron (that is, it has a higher-capture cross section) than will U^{238} if the neutron has very slow speed (referred to as thermal energy). Just above the thermal energies, in the epithermal or resonance energy region, U^{238} atoms have a very high probability of capturing neutrons without producing fission. That is favorable for producing plutonium, but makes it more difficult to maintain chain reaction if the uranium has only the normal enrichment of U^{235}. The emitted neutrons have high speed at fission, corresponding to kinetic energies of the order of millions of electron volts, and must be slowed down rapidly to avoid capture in U^{238} when there is heavy dependence on U^{235} content. To help this escape from U^{238}, the uranium rods are made of selected diameter and are carefully spaced to improve the chances for the emitted neutrons to pass through the rods without capture by U^{238}. When the neutrons leave the rods, they enter moderator material, which helps to slow them down to thermal energies with a minimum of capture losses. Light materials such as deuterium (in heavy water), beryllium, and graphite are useful as moderators.

21.11 The world of new atoms and of ionizing radiation

We have gained, as by-products of atomic power, very many new types of radioactive atoms or radioisotopes. There are now about 1100 nuclides that are new and man-made. Each is unstable, but changes in its own time to a more stable form. The change is accompanied by the emission of radiation, either in the form of a γ-ray photon, β-ray, sometimes positron, an α-particle, or some other form or combination. Each nuclide has the chemical properties of a stable, conventional atom, but in addition each also emits radiation of a type and energy that is characteristic of that nuclide. Also, each unstable nuclide (radioisotope) has a particular time rate or half-life for its change of form.

The early forms of Mendeleev's Periodic Table of the atoms listed up to 92 elements. Within the limited science and technology revolving around the chemistry of these elements, there were built up vast chemical industries. The chart of over 1300 nuclides now offers a much larger variety of atoms and building blocks out of which to develop an understanding of atomic behavior.

For example, consider the isotopes of carbon. Two stable forms of carbon are found in nature, one of mass 12 (C^{12}) and one of mass 13 (C^{13}). When nitrogen-14 (N^{14}) is bombarded by neutrons, it captures a neutron and emits a proton, leaving a new atom which has six protons and which therefore behaves chemically like carbon. This is the isotope C^{14}, which is unstable and eventually emits a weak β-particle as it reverts back to the original stable N^{14}. The half-life for this transition is very long, about 5700 years, and the β-ray energy is 0.156 MeV.

These C^{14} atoms become important for several purposes.† They may be incorporated into drugs that contain carbon. When the drug is injected into man or animal (or incorporated into carbon dioxide gas, which may be absorbed by a plant), it becomes possible to follow the course of the carbon in these systems simply by "tracing" the behavior of the C^{14} components; this is done by measuring the radiation they emit. Both time rate and distribution of the drug (or CO_2) in these complex systems can then be determined even though the systems themselves are already full of carbon atoms. This process has made it possible to identify a long series of intermediate steps in the photosynthesis of carbon dioxide for plant growth. The use of radioactive carbon (C^{14}) and radioactive

---

† We have already discussed the use of C^{14} in radioactivity dating techniques in Chapters 2 and 20.

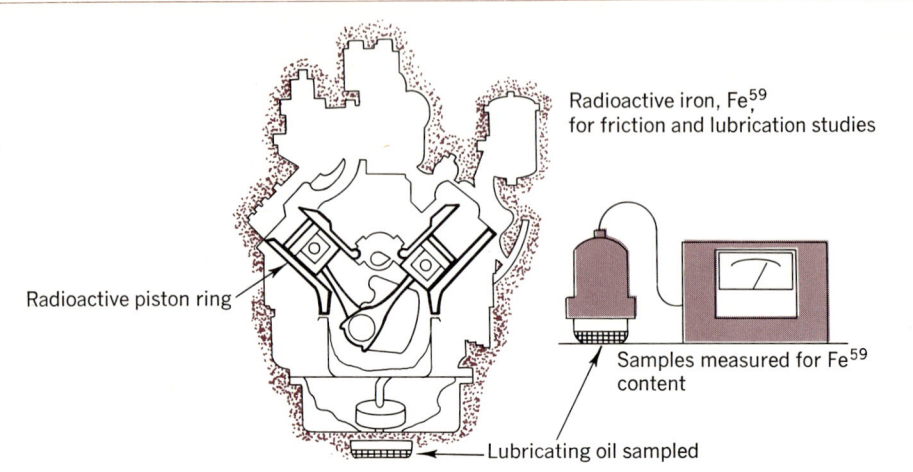

Fig. 21.13. A common application for use of radioisotope iron-59 to measure wear of metal parts. The piston rings are first made radioactive by exposing them to neutrons in a nuclear reactor, then installed in a motor which is under test for wear characteristics. As the piston ring loses metal to the oil, the presence of radioactivity in the oil gives a measure of the wear while the motor is running. When the motor is disassembled, the transfer of metal to the cylinder wall can also be measured accurately. Advantages: (1) transfer of metal measured to $\frac{1}{1,000,000}$ oz.; (2) oil sampled during operation of motor; (3) rapid, simple, economical. (Courtesy of U.S. Atomic Energy Commission.)

species of salts has clarified the understanding of many of the biological processes involved in human blood flow, the diffusion of salts across body membranes, and metabolic activity. Industry has found activation analysis to be particularly sensitive to contaminants in metals or other materials and has used it for identifying these contaminants. Considerable literature has been written about the characteristics and uses of radioisotopes. Many useful publications and references are available through the AEC.

Figures 21.13, 21.14, 21.15, 21.16, and 21.17 illustrate some applications involving radioisotopes.

Radiotracer and dating techniques require relatively weak concentrations of C^{14}, of the order of microcuries. In such applications all that is required of the emitted radiation is that it be measurable, either with Geiger (or similar) counters or with photographic film.

The various types and energies of radiation have penetrating power of differing orders. For example, α-particles can be stopped by a sheet of paper; β-particles may require from several sheets of paper to inches of solid material to stop them, depending on their energy. Gamma rays can penetrate inches of lead. By selecting suitable radiation, one may easily construct gauges for industrial applications that may be used for a wide range of thicknesses.

As noted earlier, the analytic technique called *activation analysis* has become important for industry as well as for research.† If a specimen has a very small amount or trace of impurities and is placed in the neutron flux of a nuclear reactor, the trace impurities (as well as the main body of the specimen in some cases) become radioactive. In many cases the type and amount of the impurity can be determined by comparing the results of irradiation of the unknown sample with the results one obtains by irradiating specimens with known impurities.

The sensitivity of activation analysis is illustrated by the following case: Ordinary arsenic, arsenic-75, on capturing a neutron, becomes radioactive arsenic,

† The term *activation analysis* refers to the process of making a material (which may be a contaminant) radioactive by bombardment with suitable nuclear radiation.

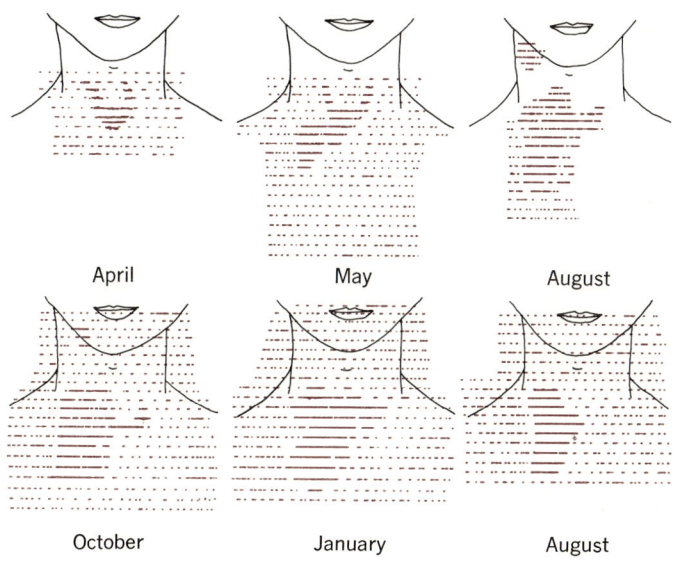

Fig. 21.14. Thyroid cancer. This is a series of six radioiodine scans of the neck and chest of a patient with cancer of the thyroid, made over a period of 16 months at the Oak Ridge Institute of Nuclear Studies. The initial scan (top, left) shows the pattern of the normal thyroid tissue (dark lines) and the presence of the tumor is questionable. With subsequent therapeutic doses of radioiodine, the normal thyroid is progressively fainter and the tumor becomes more apparent as it takes up the radioiodine. Finally, shrinkage in the size of the tumor begins (lower, right scan) as a result of the radioiodine therapy. (AEC Report for 1965.)

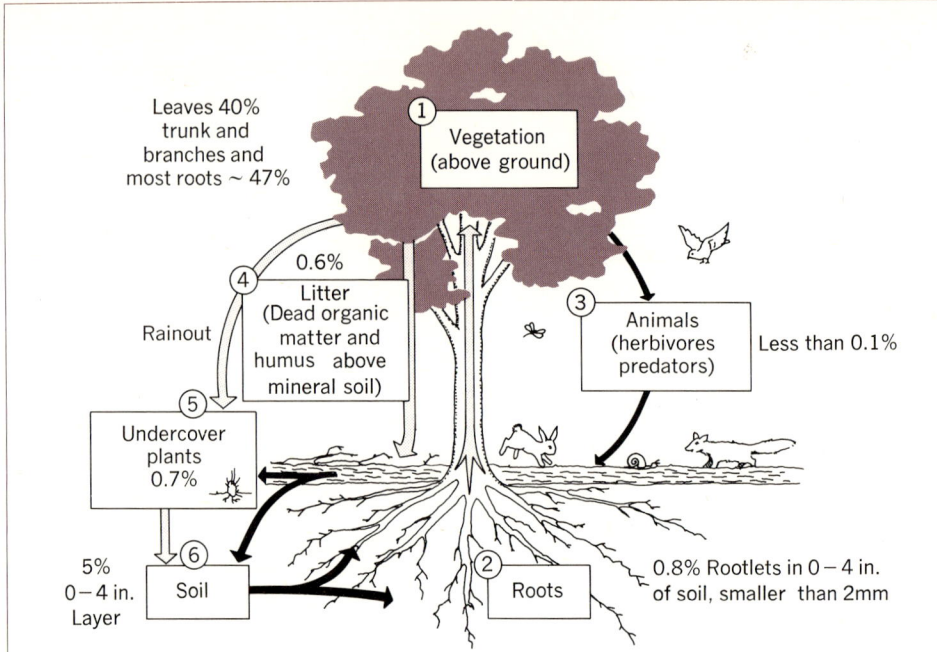

Fig. 21.15. Ecological cycle. The rapid ecological movement of a radioisotope such as a cesium-134 is illustrated in the above drawing of a white oak tree whose trunk was tagged with two millicuries of the radioisotope. Within 165 days, the tracer had become redistributed in different parts of the ecological system and was again entering the tree, this time through the root system. Use of radioisotopes in such studies in the forests at Oak Ridge National Laboratory helps ecologists understand the basic processes that maintain our forest resources. (AEC Report for 1965.)

Fig. 21.16. (Facing page) Treatment of leukemia by irradiation of blood. A patient at the Medical Research Center at Brookhaven National Laboratory is shown in the photo top undergoing treatment for leukemia by extracorporeal irradiation of his blood. The nurse is about to connect the arteriovenous shunt in the patient's forearm to the tubing leading into a shielded container where the gamma-ray source is located. The technique, as diagrammed below, was applied to the study and treatment of human leukemia following extensive studies of the origin, function, and turnover rates of cells and other blood constituents of normal and leukemic cows. The purpose of this form of treatment is to destroy leukemic white cells in the blood without injuring other cells or organs in the body; the red blood cells are much more resistant to radiation damage than the leukemic cells. A semipermanent external arteriovenous shunt, which may last for many months, is inserted in the patient's forearm. Arterial blood is propelled by the action of the heart through plastic tubing into the shielded container, past an intense source of gamma rays, and back into the patient's arm. As the blood passes through the gamma source (4000 curies of cesium-137) it receives a radiation dose of from 250 to 900 rads, depending upon its flow rate (900 rads would be a lethal dose of radiation if applied to the whole body). The treatment can be repeated as necessary to reduce the numbers of leukemic cells in the blood. (Courtesy Brookhaven National Laboratory.)

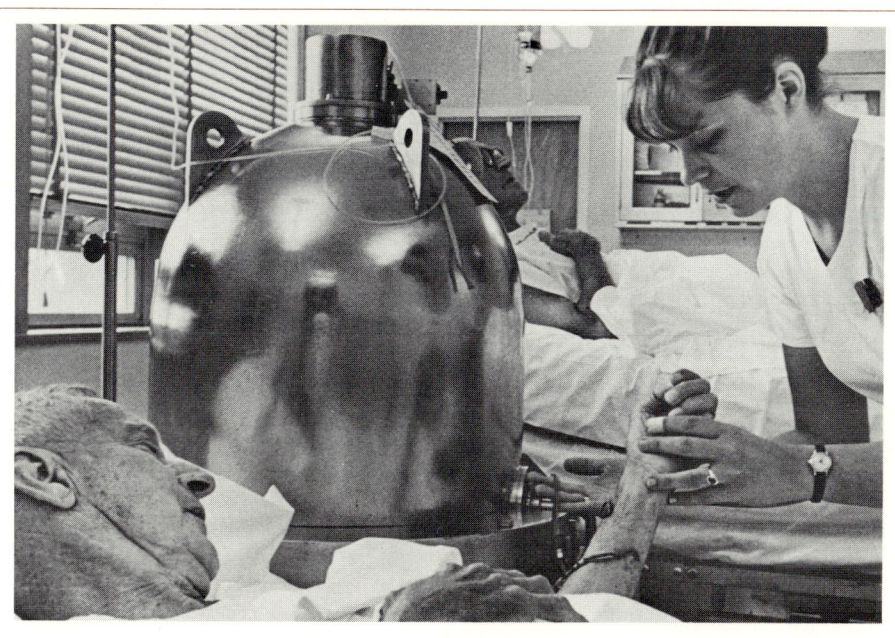

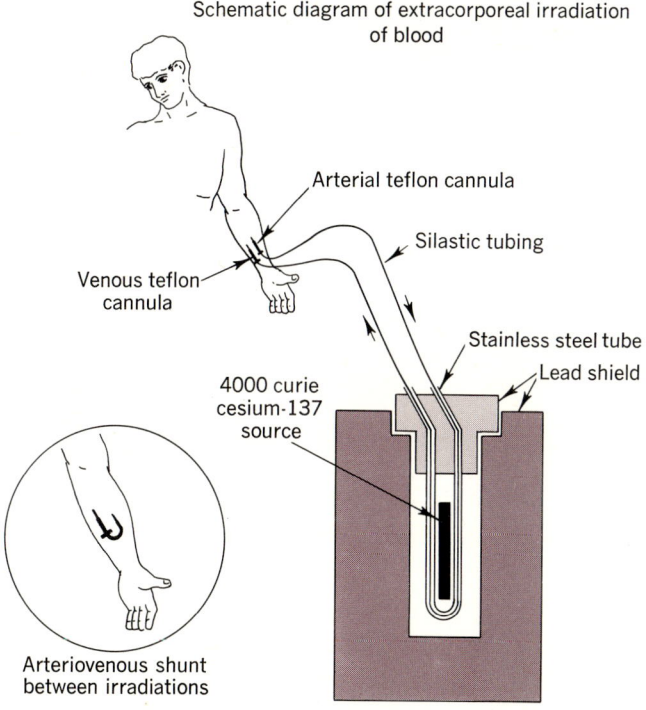

Schematic diagram of extracorporeal irradiation of blood

Arterial teflon cannula
Silastic tubing
Venous teflon cannula
Stainless steel tube
Lead shield
4000 curie cesium-137 source

Arteriovenous shunt between irradiations

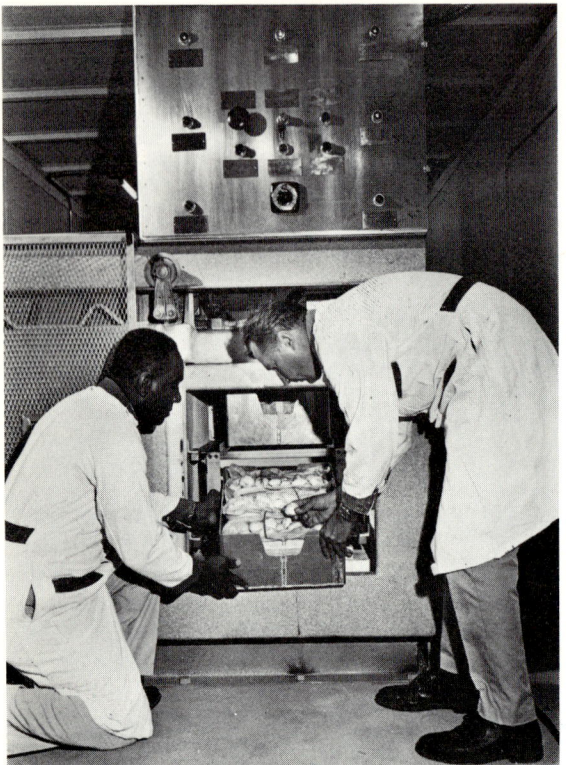

Fig. 21.17. *Irradiation of food with ionizing radiation to increase shelf life against spoilage. (Courtesy Brookhaven National Laboratory.)*

arsenic-76, which emits beta and gamma radiation on decay. Therefore, by radioactive assay, one can determine the concentration of arsenic in a sample. In 1961 a group of Scottish and Swedish scientists subjected a few strands of hair, cut from the head of Napoleon at his death in 1821, to neutron irradiation and found arsenic to be present in thirteen times normal concentration, thus suggesting that Napoleon might have been poisoned. Closer investigation indicated a definite pattern of the variation of arsenic concentration in the hair. This pattern, when compared with the record of Napoleon's sickness, revealed a correlation with his periods of severest pain. It seems arsenic was in the medicine given to relieve his pain and it may have had untoward effects as well.

21.12 Effects and products of ionizing radiation

The ionizing radiation given off by radioactive isotopes can be concentrated and intense. Since this radiation is highly penetrating and ionizing, and induces changes in biological and chemical systems, it promises to become significant in chemical processing and in destroying unwanted bacteria (such as in milk) and tissues (such as in tumors, cancers). But this promise is a mixed blessing and curse, for overexposure to radiation is a health hazard. It has been found to cause leucopenia (decrease in number of white cells in blood), epilation (loss of hair), sterility, cancer, mutations (altered heredity of offspring), bone necrosis (destruction and death of bones), and eye cataracts.

In conventional processes, chemical reactions proceed as a result of atomic collision, favorable valence combinations, excitement of atom systems by heating, Coulomb attraction, free radical intermediates, and other similar activators. The energy exchanges are likely to be of the order of a few electron volts or less per atom (or molecule).

When swift, charged particles (such as α-particles, protons, or β-particles) pass through matter, they leave tracks of ionized and excited atoms and molecules, which undergo vigorous reorganization. The concentration of energy can be hundreds or more times the intensity of conventional processes, especially with heavy charged particles and toward the end of particle tracks in the material. As a result, radiation effects are often deleterious to the properties of the material.

There are, however, applications wherein the destructiveness of radiation is desirable, such as for killing insects that infest grain or microbe systems in medical supplies. There are also cases where the reorganization of atoms and molecules following irradiation results in improved physical properties or produces desired chemical changes. Radiation induces such widely different reactions that it becomes a very versatile research tool. Processing by irradiation also appears to have very real possibilities of competing with some conventional industrial processes and of inducing reactions that cannot be produced by other means.

The activities involving radiation and radiation chemistry may be grouped under six categories: food preservation, sterilization, chemical processing, radiography and medical therapy, radioisotope power sources, miscellaneous.

Since ionizing radiation can be lethal to living organisms and microorganisms, one of the early programs sought to sterilize foodstuffs and thus give them longer shelf life. Early efforts concentrated on sterilizing meats and other foods by radiation dosage ranging from 2 to 5 megarads (million rads†). The results of these early years were not successful because the heavy dose caused changes in the taste and appearance of foods. More recent work has been much more encouraging. In 1963 the Food and Drug Administration (FDA) approved sterilization of bacon by gamma radiation (up to 2.2-MeV energy) and by electron beams (up to 5-MeV energy) from accelerators. The sterilization of ham, chicken, and beef appears promising.

When the radiation dose is kept well below the doses required for sterilization, down to values of 500,000 rads or less, the effect is to "pasteurize" foods in a way that often permits longer shelf life. For example, a dose of 250,000 rads will extend the shelf life of haddock fillets to 21 days at 32° to 33°F. Crabmeat treated with 200,000 rads had its shelf life increased from 7 days to 35 days when held at 33°F. Fruits (strawberries, cherries, citrus, pears, tomatoes) show similar gain. Insects in grains and wormy (helminthic) parasites such as those associated with trichinosis from pork are killed by 30,000 rads. Sprouting of potato tubers can be inhibited with doses from 10,000 to 15,000 rads. But dosages in excess of 10 million rads appear to be needed to inactivate some enzymes.

Radiation does not raise the temperature of the processed materials at these dosages. Furthermore, with γ-radiation, the whole process can be mechanized and the foods can be irradiated in the packaged state. The main difficulty is the cost of the radiation, whether one uses radioisotopic sources or an accelerator. The irradiation of fish adds from 1 to 3 cents per pound, which is probably acceptable. Because strawberries may cost about 50 cents per pound, they can stand

† A rad represents the absorption of 100 ergs of radiation energy per gram of absorbing material.

an irradiation expense of an additional 1¾ cents per pound. But for other fruits and for grains, the cost probably must remain at ¼ cent per pound, to be economically acceptable. To help this industrial development, the Commission has reduced the selling price for certain radioisotopes such as cobalt-60 (Co^{60}, which emits strong γ-rays of 1.1 and 1.3 MeV and has a half-life of 5.3 yr) and cesium-137 (Cs^{137}, which emits gammas of 0.66 MeV with a half-life of about 33 yr).

Radiation costs come down sharply as the radiation intensity of the facility is increased, in terms of kilowatt capacity, for either radioisotopic sources or accelerator sources. But it is difficult to find many geographic sites where one can provide high enough production quantities to bring the cost of radiation pasteurization down to 1 cent per pound.

How about irradiation to sterilize materials that are not foodstuffs, such as medical supplies, sutures, bandages, and drugs? While there are limitations in this area also, there are some real advantages to radiation processes as compared with the use of heat, chemicals, or ultraviolet light. When penetrating radiation is used, sutures or other supplies can be packaged in conventional work areas and then irradiated while in sealed state.

21.13 Radiography and medical therapy

These two subjects may be treated together because they depend on similar sources and techniques. Gamma rays are very penetrating—more so than X-rays from conventional machines. A cobalt-60 source can therefore be used effectively for penetrating metal parts, castings, tank walls, and the human body. As in X-rays, the radiation that passes through the target or body can be recorded on photographic film or on a fluorescent screen, to give a faithful picture of the variations of matter through which it passes. Flaws, cracks, cavities will show up as clearly as with X-rays.

The advantages of radioisotopic gamma sources over X-ray machines are three:

(1) These sources can be made portable and do not require electric power for their operation.

(2) Radioisotopes emit radiation in all directions, which makes it possible to obtain radiographs all around a vessel into which the source is placed.

(3) Radioisotopes can provide higher penetrating power without requiring excessively large installations.

Very many industrial firms make use of such radioisotopes as Co^{60}, which is equivalent to 2.5-MeV X-rays and can be used for steel of 2- to 5-in. thickness. For lesser penetrability, iridium-192 (Ir^{192}), cesium-134 (Cs^{134}), and Cs^{137} are the equivalent of up to 1400-keV X-rays and are useful for radiographing steel plates from ½- to 2½-in. thickness (or an equivalent density of other materials). Thulium-170 (Tm^{170}), europium-155 (Eu^{155}), and certain isotopes of americium (Am) provide still lower penetration.

For many fixed installations, X-ray machines may be preferred. Some industrial firms engaged in the production and testing of tanks, ships, and transmission pipe in the field have found the radioisotopic sources to be much more practical than X-ray machines.

We have noted that radiation kills living organisms. Malignant disease in body tissue can often be arrested by exposure to penetrating, ionizing radiation. But since healthy tissue also suffers, radiation must be applied carefully and restrictively to the tissues to be treated. This has given rise to very many designs

21.14 The hazards from nuclear bombs

that use radioisotopic sources in the form of tiny needles that are inserted into tissue; or the sources may be contained in a housing that directs a well-collimated beam onto the tissue.

Radioisotopic sources offer portability and considerable choice in the type and energy of radiation that they emit. Also they can be fabricated into very many shapes and sizes.

21.14 The hazards from nuclear bombs

It would have been fortunate if, nearly three decades after the bombing of Hiroshima, we could have completed this chapter with only the useful peacetime aspects of atomic energy. Unfortunately this is not the case. At this writing more money is being devoted to the weapons aspects than to the constructive utilization aspects of nuclear energy in the United States, in the Soviet Union, and in other nations. Because the problem of the nuclear bomb is a responsibility of civilian populations and of civilian governments, and because the control of the problem relates to the very topics that we have just discussed within a new context, we shall briefly review the subject of nuclear bombs and of the damage that they cause.

The power or energy release of a nuclear weapon is expressed in terms of its equivalent in explosion of ordinary explosive TNT (trinitrotoluene). Thus the explosion of the bomb over Hiroshima released energy equivalent to that which 20,000 tons (20 kilotons) of TNT would release when it exploded. This amount of energy corresponds to the energy release of only about 1 kg (2.2 lb) of U-235 undergoing fission. The power increases with the amount of U-235 or plutonium incorporated in the bomb.

The early bombs depended only on the release of energy from the *fission* of U-235 or Pu-239. It was realized even during the war days that there could also be energy release from the *fusion* of very light elements. For example, when two atoms of deuterium (which is the "heavy" hydrogen isotope that has a proton plus a neutron for its nucleus) combine, the product is a helium atom; but since the helium atom has a smaller mass than do the two deuterium atoms, some of the mass becomes converted to energy. It is very difficult to bring about this deuterium-deuterium interaction, to give the two positive nuclei enough kinetic energy (this is often referred to as the atoms being at high enough *temperature* to overcome their repulsion forces). During the explosion of the fission bomb, the temperature rises, to an order of a million degrees.† This is sufficiently high so that when deuterium is included in the fission bomb the atoms develop enough kinetic energy to interact. There can be similar reactions on the part of other light atoms (helium, tritium, lithium) with substantial energy gain. In fact, pound for pound, the fusion process offers more energy than does the fission process—and since there is a vast supply of light atoms in nature, the possibilities of power production are almost infinite.

Using this principle, it has been possible to develop very powerful fusion bombs (H-bombs) that utilize the fission process to initiate the high temperatures required for fusing the light atoms. The largest bomb tested had energy equivalent to that of 50 million tons (50 megatons) of TNT, but there seems to be no limitation on the size that can be produced. The relatively small Hiroshima bomb destroyed the city. The death toll is estimated by the Japanese to have

† When expressed in this approximation does it matter very much whether the units are °F, °C, or °K?

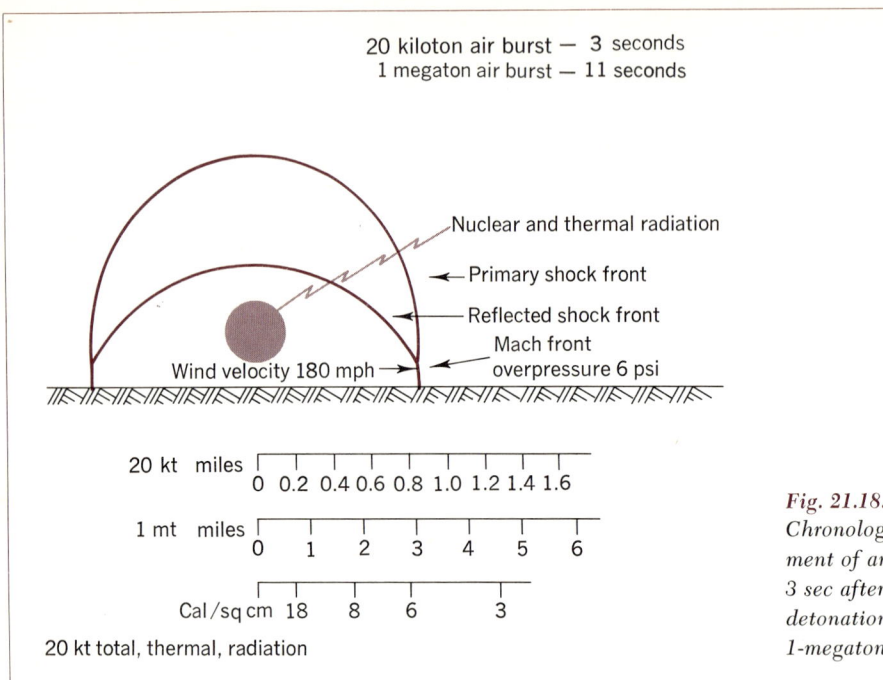

Fig. 21.18. Chronological development of an air burst: 3 sec after 20-kiloton detonation; 11 sec after 1-megaton detonation.

been from 153,500 minimum to 211,500, of a total 310,000 to 340,000 people who were in the city at the time. A United States AEC report† lists 70,000 killed and missing and 70,000 injured out of a total population of 255,000 and 417 square miles destroyed. This same bomb is now capable of producing a casualty rate of 260,000 on the basis of a population density of 1 per 1000 sq ft.

Causes of injuries to the people were classified into three main categories: blasts due to shock waves, falling buildings and flying objects; burns from burning buildings; and burns and other damage from nuclear radiation. In the case of Hiroshima victims, only 30 percent of the people exposed to the nuclear radiation survived, whereas the survival

† See Glasstone in References.

rate for blast and common burn injuries was between 65 and 85 percent.

Figures 21.18 and 21.19 illustrate the nature of a nuclear bomb burst in air for a 20-kiloton and a 1-megaton nuclear bomb. Immediately following the detonation of a nuclear bomb in the air, an intensely hot and luminous (gaseous) ball of fire is formed. Owing to its extremely high temperature, it emits thermal (or heat) radiation capable of causing skin burns and starting fires in flammable material at a considerable distance. The nuclear processes that cause the explosion and the radioactive decay of the fission products are accompanied by harmful nuclear radiations (γ-rays and neutrons) that also have a long range in air. Very soon after the explosion, a destructive shock (or blast) wave develops in the air and moves rapidly away from the fireball.

When the primary shock (or blast) wave from the explosion strikes the ground, another shock (or blast) wave is produced by reflection. At a certain distance from ground zero, which depends upon the height of burst and the energy of the bomb, the primary and reflected shock fronts fuse near the ground to form a single, reinforced Mach front (or stem).

As time progresses, the Mach front (or stem) moves outward and increases in height. The overpressure (pressure above normal atmospheric pressure) at the Mach front is 6 psi and the blast-wind velocity immediately behind the front is about 180 mph.

Nuclear radiations continue to reach the ground in significant amounts. About 3 sec after the detonation of a 20-kiloton bomb, the fireball, although still very hot, cools to such an extent that the thermal radiation is no longer important. The total accumulated amounts of thermal radiation, expressed in calories per square centimeter, received at various distances from ground zero after a 20-kiloton air burst, are shown on the scale at the bottom of Fig. 21.9. Appreciable amounts of thermal radiation continue to be emitted from the fireball at 11 sec after a 1-megaton explosion; the thermal radiation emission is spread over a longer time interval than for an explosion of lower energy yield.

At 10 sec after a 20-kiloton explosion

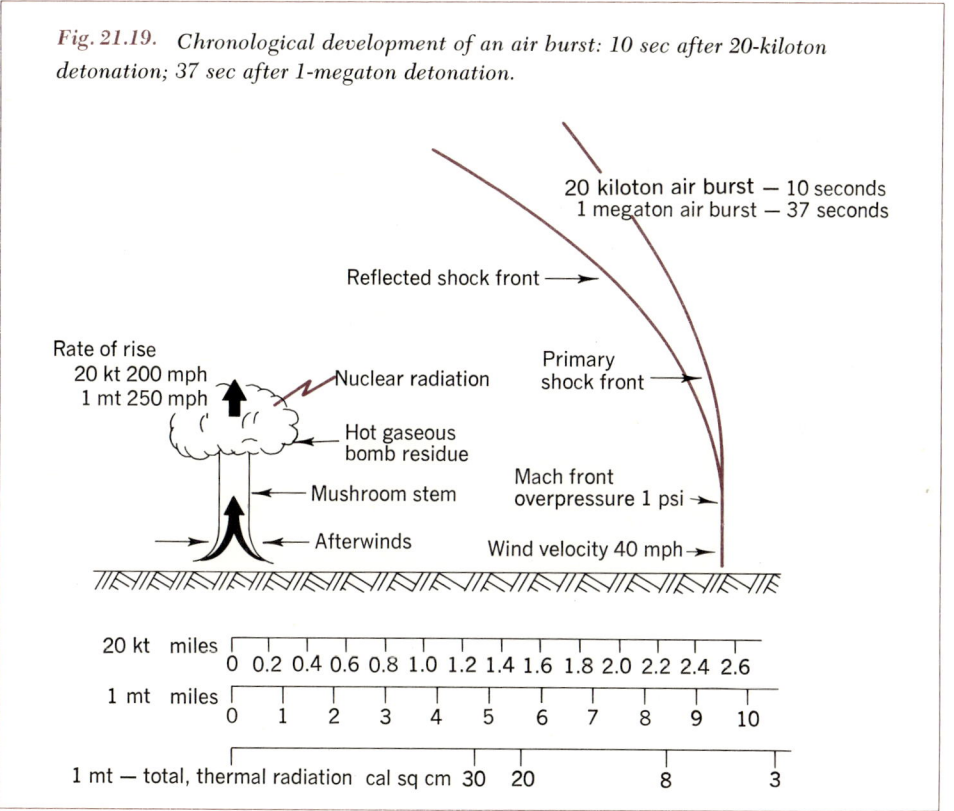

Fig. 21.19. Chronological development of an air burst: 10 sec after 20-kiloton detonation; 37 sec after 1-megaton detonation.

the Mach front is over 2½ miles from ground zero, and 37 sec after a 1-megaton detonation it is nearly 9½ miles from ground zero. The overpressure at the front is roughly 1 psi in both cases, and the wind velocity behind the front is 40 mph. Apart from plaster damage and window breakage, the destructive effect of the blast wave is essentially spent at this time. Thermal radiation is no longer important, even for the 1-megaton burst; the total accumulated amounts of this radiation at various distances are indicated on the scale at the bottom of Fig. 21.19. Nuclear radiation of an appreciable extent can still reach the ground, however, consisting mainly of γ-rays from the fission products. At that point the ball of fire is no longer luminous but it is still very hot and behaves like a hot-air balloon rising at a rapid rate. As it ascends, it causes air to be drawn inward and upward, somewhat similar to the updraft of a chimney. This produces strong air currents, called *afterwinds*, which raise dirt and debris from the earth's surface to form the stem of what will eventually be the characteristic mushroom cloud.

The hot residue of the bomb continues to rise, expand, and cool. As a result, the vaporized fission products and other bomb residues condense to form a cloud of highly radioactive particles. The afterwinds, having velocities of 200 or more miles per hour, continue to raise a column of dirt and debris, which will later join with the radioactive cloud. Within about 10 min, the bottom of the mushroom head will have attained an altitude of 5 to 15 miles, depending on the energy yield of the explosion. Ultimately, the particles in the cloud will be dispersed by the wind, and will eventually fall to earth in local or distant areas according to the pattern (see Fig. 21.20) that develops for wind and for precipitation (radioactive fallout).

21.15 The sun and fusion reactions

We have seen that nuclear fission holds promise of providing a vast energy source for the future use of mankind—a source that we are beginning to use quite effectively even now. We still have not considered our principal source of energy— the sun. In this section we consider how the sun and stars generate their energy.

Astronomically speaking, the sun is a typical star. Modern measurements show us that the sun is a sphere 1.5 million kilometers in diameter, located approximately 150 million kilometers from the earth. Using the mechanical laws of planetary motion introduced earlier, we may deduce that the sun has a mass some 330,000 times that of the earth, or about 2×10^{30} kg. Accurate measurements of the amount of solar radiation received per unit area per unit time at the top of the earth's atmosphere† yield a figure of about 1.36×10^3 J/m²/sec.

Consider a sphere of radius 150 million kilometers (1.50×10^{11} m) surrounding the sun. This sphere has a surface area ($4\pi R^2$) of approximately 3×10^{23} m². If we assume that through each square meter an amount of energy equal to that received on earth (that is, 1.36×10^3 J) passes in each second, then through the entire sphere we find 4×10^{26} J passing in each second. This same amount of energy will pass through any spherical surface centered at the sun's center. In fact that is exactly the amount of energy leaving the surface of the sun.

---

† We refer the solar energy to the top of the earth's atmosphere because the atmosphere reflects and absorbs much of the radiation incident on it, especially in the ultraviolet and infrared regions of the spectrum. When measurements of the incident solar energy are made from the earth's surface, suitable corrections must be made for the reflection and absorption rates. Recently, satellite measurements made outside our atmosphere have eliminated this problem.

21.15 The Sun and Fusion Reactions   645

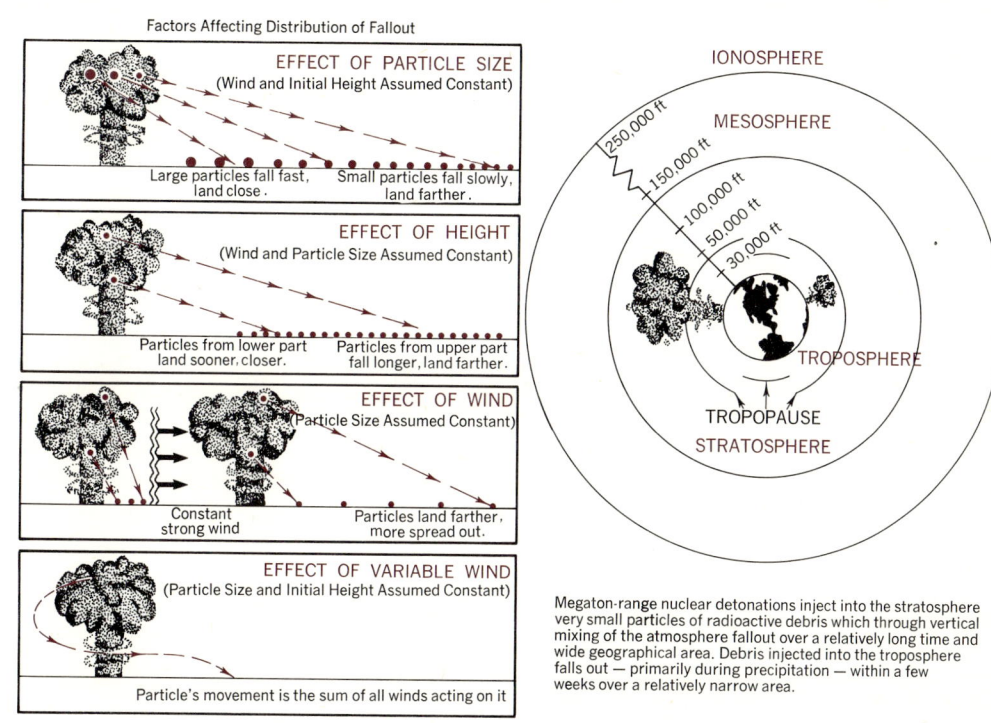

Megaton-range nuclear detonations inject into the stratosphere very small particles of radioactive debris which through vertical mixing of the atmosphere fallout over a relatively long time and wide geographical area. Debris injected into the troposphere falls out — primarily during precipitation — within a few weeks over a relatively narrow area.

On left in the drawing are illustrated the primary factors which affect the distribution patterns of fallout. On the right are shown the various atmospheric zones of particular interest. The troposphere rises to between about 35,000 and 50,000 feet and the stratosphere extends out to about 140,000 feet. The tropopause boundary between the two zones suffers a break in the temperate latitudes which is associated with the strong "jetstream" currents of air. Nuclear debris injected into the low-lying trophosphere generally returns to earth within a few weeks—in the same latitudes or zones of the earth's surface in which the detonation occurred. Rain or snow helps "wash" the fallout from the atmosphere. If the nuclear detonation is very large, it will send its debris into the stratosphere where the light dust particles may remain for months or years before working their way down to the tropospheric "weather zone" and eventually descending to any part of the earth.

Fig. 21.20. This illustration has been taken from the AEC Supplemental Report for 1965. It illustrates the various factors that affect damage and fallout of radioactive particles developed by a nuclear explosion. (AEC Report for 1965.)

The total *flux* of radiation emitted (per second) by the sun is called its *luminosity* and is denoted by L_0. Its value, again, is

$$L_0 = 4 \times 10^{26} \text{ J/sec}$$

Where does this energy come from? A simple calculation, using the Stefan-Boltzmann law (see Chapter 17), allows us to show that if the sun were a perfect blackbody, of radius 7×10^8 m (R_0) and luminosity L_0, then its temperature would have to be almost 6000°K. At such temperatures all the fossil fuels (coal, oil) used on earth would burn up quickly. Let us assume the sun to be one vast lump of coal, for example. Then, using the known energy released per kilogram of coal, we would find that our coal-sun would continue to emit energy at the rate L_0 for only a few thousand years. But geologic evidence points to the fact that L_0 has not changed appreciably over several *billion* years, nor has the sun's mass changed appreciably over this period. Chemical burning, therefore, can be completely eliminated as the source of the sun's energy.

The chemical binding energy released during burning amounts to a few electron volts per molecule. But we noted that the fission of a single U^{235} atom produces some 200 million electron volts (or nearly 10^8 times as much energy as released in chemical processes). If burning would sustain the sun 1000 years, nuclear fission would sustain it for 100 billion years! But this assumes that the sun is all U^{235}, a very improbable assumption. Yet it seems that the energy released in nuclear reactions could sustain the sun for the requisite length of time.†

---

† It can also be seen from Einstein's energy-mass relation ($E = mc^2$) that if we convert the sun's entire mass to energy, we obtain

$$E = (2 \times 10^{30})(3 \times 10^8)^2 = 18 \times 10^{46} \text{ J}$$

An emission rate of 4×10^{26} J/sec would require 4.5×10^{20} sec. (or 1.5×10^{13} years) to use

In 1938 Hans Bethe suggested a new type of nuclear reaction to account for the sun's energy production. Since spectroscopic studies indicate that the chemical composition of the sun is approximately 70 percent hydrogen and 27 percent helium, Bethe suggested that instead of using the breakup of heavy nuclei (such as uranium) as an energy source, we should look to the buildup of light nuclei into heavier ones. In particular, he suggested the fusion of four hydrogen nuclei into normal helium.

To see that this process yields energy, note that the mass of one hydrogen atom is 1.007276 amu (thus four hydrogens have mass 4.029504 amu). Normal helium ($_2\text{He}^4$) has a mass 4.002604 amu. Thus, if we can fuse four $_1\text{H}^1$ into one $_2\text{He}^4$, we have 0.0269 amu of mass converted into energy (0.0269 amu is equivalent to about 25.7 MeV of energy per fusion).

Having seen that fusion also yields energy, what is the mechanism for fusion reaction? Bethe proposed the following series of nuclear reactions:

Step 1:

$$_6\text{C}^{12} + {_1\text{H}^1} \rightarrow {_7\text{N}^{13}}$$

by which a normal carbon nucleus captures a proton and becomes an isotope of nitrogen, which is radioactive.

Step 2:

$$_7\text{N}^{13} \rightarrow {_6\text{C}^{13}} + \beta^+ + \nu$$

wherein the nitrogen isotope of step 1 decays, emits a positron and a neutrino, and forms isotope of carbon.

Step 3:

$$_6\text{C}^{13} + {_1\text{H}^1} \rightarrow {_7\text{N}^{14}}$$

---

up this energy. This is the life of the sun for total mass annihilation. Nuclear processes, which generally use up only a fraction of 1 percent of the original mass, could still sustain the sun for billions of years.

by which the newly formed carbon isotope captures a second proton and becomes normal nitrogen.

Step 4:

$$_7N^{14} + {}_1H^1 \rightarrow {}_8O^{15}$$

where the nitrogen captures a third hydrogen nucleus and becomes a radioactive oxygen isotope.

Step 5:

$$_8O^{15} \rightarrow {}_7N^{15} + \beta^+ + \nu$$

because the oxygen is highly unstable and decays to another nitrogen isotope with the emission of a positron and neutrino.

Step 6:

$$_8O^{15} + {}_1H^1 \rightarrow {}_6C^{12} + {}_2He^4$$

In this last step the nitrogen captures a fourth proton, but instead of becoming $_8O^{16}$ (stable oxygen) it usually splits into two parts, giving back the original carbon nucleus and a helium atom. Thus, since the carbon reappears, the entire series of reactions has the effect of converting four hydrogen atoms into one helium atom. In this process, however, *it liberates over 25 MeV of energy.* This energy is emitted as γ-rays during the various reactions.

It was suggested at about the same time (by von Weizsäcker) that a direct fusion of hydrogen was possible. He suggested the following series of nuclear reactions:

Step 1:

$$_1H^1 + {}_1H^1 \rightarrow {}_1H^2 + e^+ + \nu$$

in which two hydrogen atoms fuse to form deuterium, or heavy hydrogen, and a positron and neutrino.

Step 2:

$$_1H^2 + {}_1H^1 \rightarrow {}_2He^3$$

wherein the newly formed deuterium captures still another hydrogen to form the light isotope of helium.

Step 3:

$$_2He^3 + {}_2He^3 \rightarrow {}_2He^4 + {}_1H^1 + {}_1H^1$$

Here the two $_2He^3$ atoms fuse to give a $_2He^4$ (α-particle) plus two hydrogen nuclei. Since each $_2He^3$ was produced by the fusion of three hydrogens, this step effectively converts six hydrogens into a $_2He^4$ and two hydrogens, or in other words, again four hydrogen atoms fuse to produce a helium nucleus (α-particle). This is a reaction similar to step 2, which forms the basis of the fusion (or hydrogen) bomb.

It appears, according to modern theories of stellar structure, that both Bethe's and Weizsäcker's processes probably occur in stars.

It should be noted that when four hydrogen nuclei with a mass of over 4 amu react to form helium, about 0.027 amu is transformed into energy, or about 0.7 percent of the reacting mass. If 0.7 percent of the total mass of the sun were converted to radiant energy, the sun could continue to shine at its present rate for 100 billion years! Actually, the current estimate of the lifetime of the sun is about 20 billion years, since the entire sun does not take part in the reactions described, but rather only the inner 10 percent of its radius (or about 20 percent of its mass).

The reader may immediately ask: How does the $_6C^{12}$ nucleus capture its first proton? Since both are positively charged, they must repel one another. In order to enter the carbon nucleus, the incident proton would seem to need sufficient kinetic energy to overcome this repulsion. Such a high kinetic energy corresponds to temperatures of billions of degrees Kelvin. The internal temperatures are believed to be high by our everyday standards (tens of millions of degrees), but the average kinetic energy of the protons is still insufficient to overcome the repulsive forces.

We are reminded, however, that, according to the quantum mechanical picture, even a much lower-energy proton has a finite probability of *tunneling*† into the carbon nucleus.

Thus, even at lower temperatures, some captures can be anticipated. But lowering the temperature (and hence kinetic energy of the hydrogen nuclei) does greatly reduce the probability of a reaction's occurring. It can be shown that at 6000°K (the surface temperature of the sun), almost no fusion reactions occur. However, with temperatures in the range of 10 to 20 million degrees Kelvin, sufficient numbers of captures should occur each second to account for the energy output of the sun.

A brief calculation shows that temperatures of the order of 10 million degrees Kelvin must exist at the sun's center. Without actually carrying out this calculation, we may consider the line of reasoning. The sun does not seem to be changing in size, and geologic data seem to indicate that the size, mass, and distance to the sun have not changed measurably during the past several billion years. Since each particle that makes up the solar material is gravitationally attracted to the sun's center, why then, hasn't the sun collapsed? The answer is that the gas pressure inside the sun is sufficiently high to stop any tendency toward collapse. This implies a very high central pressure for the sun. (Numerical results give the central pressure of the sun at over 10^{12} times that of the earth's atmosphere at sea level.) If the sun is gaseous throughout and acts as a perfect gas (a seemingly arbitrary assumption, which appears to be substantiated by more elaborate theories), then the pressure is proportional to the temperature. This proportion yields a temperature of about 15 million degrees Kelvin at the central portion of the sun. Very sophisticated models, in which we account in detail for the physical effects that occur, do not significantly change this value.

Thus the sun (and by similar reasoning, almost all stars) derive their energy from one of the two possible processes for fusing hydrogen into helium. Central temperatures of the order of 10 million degrees Kelvin are needed for these processes, and theoretical calculations based only on gas dynamics show that it is reasonable to expect such central temperatures.

It should be noted before ending our discussion of nuclear fusion that several attempts are presently being made to harness the fusion reaction as an energy source. The biggest problem seems to be the high temperatures involved. Although temperatures of the order of millions of degrees Kelvin have been reached in laboratory experiments in the United States, the United Kingdom, and Russia, they have been sustained only for times measured in microseconds. What material can we use for a vessel to hold a fluid with a temperature in excess of 1 million degrees Kelvin? The answer may lie in using strong magnetic fields as a "container," for certainly any common material container would melt. When fusion energy can be produced commercially it will offer a nearly unlimited source of energy.

21.16 Solar energy utilization

It is appropriate that we conclude this chapter with a very brief discussion of the utilization of solar energy. Until the

---

† According to the wave concept (Chapter 19), the proton may be thought of as a probability wave. This wave does not become actually zero inside the nucleus; hence the proton has a small probability of being found there. This process is called *tunneling*.

beginning of the "atomic age," almost all energy came indirectly from the sun. The burning of coal or oil, for example, depends on the fact that at some earlier epoch, living plants stored solar energy as chemical energy. The burning process simply releases this chemical energy in the form of heat and light.

With the growth of population and technology, the demands on our energy resources are continually increasing. Our supply of fossil fuels is not unlimited and cannot continue to supply this demand for very much longer. The use of fission energy (nuclear reactors) will undoubtedly play a large role in solving this problem, as will fusion energy when it becomes available. However, since the sun still provides the earth with so much energy, it is worth inquiring as to whether this energy can be utilized more efficiently.

Consider an average house whose dimensions are, say 40 ft (13 m) by 30 ft (10 m). If we consider the 130 m² of flat roof and the total solar energy hitting this roof (using the numbers given in Sec. 21.15), we see that about 200,000 J strike the roof each second! Or, in more common units, 200 kw of power. If the sun were to shine every day for 10 hours (h), the roof would receive 2000 kwh of power per day. The average American family needs about 1500 kwh of power per day to maintain our present level of comfort. Therefore, if we could convert and store the solar energy available to us, our power problems would be solved.

How can this energy be utilized? One way is to build reflectors that will concentrate this heat; and indeed this has been done successfully. But the cost remains high for such units because the reflectors are large and expensive, and the capital costs make the cost per unit of heat far more than the cost of heat obtained through other means (for example, coal, oil, and nuclear energy).

We shall not discuss here the various other attempts to harness solar energy directly, such as solar heaters and solar batteries. It is inevitable that much future research will be directed toward ways of including solar radiation among our energy supplies.

21.17 Summary

The fission process is distinguished from natural radioactivity in three ways:

(1) Radioactive decay occurs spontaneously, whereas the fission process is induced by bombarding the fissionable atom with a neutron of appropriate velocity.

(2) Whereas in radioactivity only the light particles are given off, the heaviest of these being an α-particle, in the fission process (as the name implies) the parent atom is split into heavy daughter atoms.

(3) In the process of fissioning, more than one neutron may be shot out of the nucleus, a fact that makes chain reactions possible.

Although most of the heavy elements are radioactive, few elements are suitable for sustaining a chain reaction; these are two of the isotopes of uranium, U^{235} and U^{233}, and plutonium. (Strictly speaking, thorium and uranium-238 are also fissionable under a more complicated process. Higher velocity neutrons must be used.) Plutonium can be produced artificially, but is not found in nature; this is also true for uranium-233. In nature, uranium-235 is found mixed in a very small proportion (less than 1 percent) with uranium-238. Since isotopes cannot be separated by any chemical means, the separation of uranium-235, required for reactor fuels or atomic bombs, is both difficult and expensive. But there are no secrets as to how it can be done and any country has the capacity, if it desires to do so, to manufacture uranium-235.

The principles governing chain reactions, whether in a bomb or in a nuclear reactor, are simple to understand, even if the arrangement and control of the processes involve many engineering difficulties. Any chain reaction to be sustained requires that more than one neutron on the average be emitted by the fission of one atom and that these neutrons strike and produce fission in a neighboring atom. If the average number of *effective* neutrons (that is, neutrons effective in causing another fission) emitted per fission is maintained at exactly 1, a controlled reaction results. If the number is greater than 1, an explosion results. To achieve either of these results, two minimal things are required: a fission process in which, on the average, more than one neutron is emitted by the fuel element, and a *critical mass* of the fuel. (Below the critical mass, too many of the emitted neutrons escape without striking any neighboring atoms.) How these two requirements are met has been presented in this chapter.

Obviously the importance of nuclear processes for modern civilization, for good or for ill, lies in the uniqueness of the reaction. Because of the mass-energy relationship, very little fuel is required to produce a vast quantity of energy. Furthermore, although the sources of supply of uranium-235 may be exhausted, just as the fossil fuels are now being exhausted, it is possible to use the much more common uranium-238 or thorium in "breeder reactors," in which more fuel is produced than is consumed, even while the "breeding" reactor is generating power.

An additional advantage of nuclear fuel over fossil fuel, and one growing in significance, is that carbon dioxide is not constantly being exhausted into the atmosphere. Somewhat offsetting this advantage is, of course, the problem of disposing of radioactive wastes.

In addition to power production, nuclear reactions have provided a new source of useful radioactive materials that can be used for a growing variety of purposes, illustrations of which have been given. These radioactive product elements of such a wide variety result from the fact that when a large number of uranium atoms are undergoing fission, they do so in many different ways. At times they split into two elements almost equal in mass; at other times into a heavy and a light element; and sometimes even into three or more elements. Hence a tremendous number of artificial radioactive isotopes are made available for the use of man. Accompanying the possible usefulness is, of course, the danger of misuse. When an atomic bomb explodes, the same variety of dangerous radioactive products is released into the atmosphere and the whole world may be contaminated to some degree. The story of Prometheus takes on a new guise.

One other future possibility should be kept in mind when sources of energy are being considered—the possibility of using fusion reactions in a controlled manner. Fusion (or thermonuclear) bombs have been produced; how to control such a reaction has so far escaped the scientist's ingenuity. In a fusion reaction, light elements are brought together under the condition of extraordinarily high temperatures. This sets off the fusion process, which develops an even greater production of energy than can be obtained by the fissioning processes. Furthermore, the source of supply of the fuel—an isotope of hydrogen—is almost limitless, so vast are the oceans of the world. If scientists could find another way of starting the fusion process, other than through high temperatures, and means of controlling it after it starts, the energy problems of man would be forever solved.

Questions/Discussions

1. Define the atomic mass unit.

2. What is the energy produced when a gram (not gram-mole) of U-235 experiences fission? Express it first in million electron volts and then in kilowatt-hours.

3. What would be the energy obtained from a gram of U-235 if all the mass could be converted to heat?

4. What energy would be released if a gram of helium became converted entirely to heat energy?

5. What energy is released when hydrogen atoms are fused to produce 1 g of helium?

6. What is the function of "moderating" material in a critical nuclear reactor?

7. Give the reasons for assembling uranium fuel elements in rod form, spaced away from each other and with moderator material between them.

8. Why is a nuclear reactor sometimes referred to as being a most valuable scientific tool?

9. The development of the nuclear bomb has had impact on very many areas of human activity. There can be observed many "feedback" influences, of both positive and negative character, of the kind we employ for systems analysis. Analyze the impact and feedback influences related to this new science and technology with respect to:

   (a) The promotion of war and peace.
   (b) Industrial progress.
   (c) Aid or hindrance to the effectiveness of the scientific endeavor.
   (d) Social progress of underdeveloped societies.
   (e) The problem of overpopulation of the world.
   (f) The problem of inadequate sources of energy.
   (g) The problem of smog and pollution of the atmosphere.
   (h) Issues involving morality and conscience of scientists with respect to the products of their efforts.
   (i) Issues involving morality and conscience of society.
   (j) Your own attitude toward world government.
   (k) Your attitude toward the significance of the scientific endeavor in the United States.
   (l) The precedence established by the United States in using the first nuclear bomb on populations.

   (*Note:* Read Question 10 also before giving answers to this Question 9.)

10. Refer again to the impact of the development of nuclear bomb technology and to each of the items under Question 9. Having in mind the long range (evolutionary) trends (as contrasted with the more immediate effects that we can observe in our own lifetime);

    (a) What do you see to be the long-range *gains*?
    (b) What do you see to be the long-range *losses*?

11. One of the accusations against J. Rober Oppenheimer was that he did not favor proceeding with the development of the more powerful H-bomb. It was implied that he feared the escalation of war potential that would result from this new development. At the time, he was chairman of the important General Advisory Committee of the U.S. Atomic Energy Commission. The international situation was tense, with reason to expect the conflict in Korea to become extended.

How would you have judged Oppenheimer in this situation?

12. Describe how radioactive iodine may be used to fight cancer of the thyroid glands.

13. Suppose you are accused by your farmer neighbor of having impaired the water supply from his wells because you have dug wells of your own. It is your contention that your wells do not draw from the same underground waters. How would you go about proving your case?

14. Government authorities have recently authorized the use of ionizing radiation for pasteurizing certain meats and vegetables. Having in mind the fact that some types of radiation, and especially radiation at high energies, can cause foods to become radioactive, what might be the government specifications for the irradiation process to make sure that the product is completely safe for human consumption? (Identify the types of radiation and radiation energies that are permitted.)

References

Bishop, A. S., *Project Sherwood—The U.S. Program in Controlled Fusion.* New York: Doubleday (Anchor books), 1960.

Fermi, Laura, *Atoms in the Family.* Chicago: Univ. of Chicago Press (Phoenix Book P58), 1954.

Glasstone, Samuel (ed.), *The Effects of Nuclear Weapons.* Washington, D.C.: U.S. Govt. Printing Office. (U.S. Dept. of Defense), 1957.

Hewlett, Richard A., and Anderson, Jr., Oscar E., *The New World, 1939–1946,* **2.** University Park: Pennsylvania State Univ. Press, 1962. *This history of the U.S. Atomic Energy Commission, written in 1962, provides a fascinating review of the story of the transition of atomic energy authority from military to civilian control.*

Jungk, Robert, *Brighter Than a Thousand Suns* (A Personal History of the Atomic Scientists). New York: Harcourt, Brace, 1958. (Trans. from the German by James Cleugh.)

Parsegian, V. L., *Industrial Management in the Atomic Age.* Reading, Mass.: Addison-Wesley, 1965. *This volume offers a brief history of the atomic development program from 1939 to the time of its publication; also an analysis of the patterns, restraints, opportunities, and trends that characterize new technological developments involving cooperation of industry and government.*

Smyth, Henry De Wolf, *Atomic Energy for Military Purposes.* Princeton, N.J.: Princeton Univ. Press, 1945. *This was the first account of the "Manhattan District" project that developed the atomic bomb. It is a fascinating account, and was the subject of much debate as to whether it should have revealed so much of the details of the project.*

Turner, L. A., *Nuclear Fission, Rev. of Modern Physics* (Jan. 1940).

U.S. Atomic Energy Commission. *The Commission maintains a substantial information and library service in principal cities and through its office at Washington, D.C.*

CHAPTER TWENTY-TWO

What is Scientific Method?

Science is an attempt to make the chaotic diversity of our sense-experience correspond to a logically uniform system of thought.

ALBERT EINSTEIN
Out of My Later Years

A GREAT DEAL of human history has been glossed over in the chapters to this point. We have reviewed briefly the earliest stages of organized society, when tools and techniques brought assurance of food and comfort and even brought safety (except from other societies of human beings). Over long centuries, the curtains of knowledge moved rather slowly in reluctant response to the gropings of man, revealing the secrets of nature. Within recent decades the search for the secrets of nature has become more intense, systematic, sophisticated, and varied than ever before. Each discovery, each new skill, and every application of these discoveries and skills contributes more or less toward the birth of new ideas, new discoveries, and new applications of knowledge.

22.1 Limitations of models

Our early mathematical forms, which said that $2 \times 3 = 6$ and gave its symbolic form as $a \times b = c$, has blossomed into computers that pursue this logic with unimaginable speed and complexity. More than that, there are now varieties of mathematical logic, in some of which $a \times b$ is not equal to $b \times a$ as prescribed by everyday experience. Surprisingly enough, while this particular variety is different from its counterpart in common experience, it seems to be remarkably useful for application to atomic phenomena, which we neither see nor hear. For describing other physical phenomena we have developed abstract concepts of fields, probability, and entropy, which seem to hold the future of the universe to a vaguely foreseeable course. Often we are compelled to resort to the use of models, to picture things that apparently must forever remain unseen to human eyes and beyond the reach of the senses.

The models, though clearly limited in what they can reveal, are convenient to work with because they give shapes and visible action to near-unreal situations. But sometimes this very convenience causes models to obstruct further progress into areas that do not so easily lend themselves to such representation.

An example illustrating the failure of visual models is found in atomic physics, which in its initial stages (1913-1925) was greatly aided by an attractive picture introduced by Niels Bohr. It involved a planetary model of the hydrogen atom, with the proton at its center forming the (nearly) stationary sun and the lightweight electron revolving about it in some specified stationary orbit. All properties of the system were calculable from the basic laws of mechanics and electromagnetism, and the consequences to which they led, in particular with regard to the frequencies of the lines in the H-spectrum, were beautifully verified in observations. Rarely did a model satisfy the scientists' craving for understanding and predicting as thoroughly as did Bohr's picture.

22.2 Failure of the Bohr model

There was consternation when the Bohr model failed, not so much in the domain of observations as in the theoretical domain by implying things that were irrational and contradictory in a basic theoretical sense. They were irrational because some features of the model were incapable of being observed, and many scientists do not operate with concepts that have no operational counterparts. They were contradictory because some of the implications could not be simultaneously true, to wit:

Bohr's model, in assuming the electron to pursue a path around the proton, assigned to it a definite place in space at any given time. The electron's speed of revolution, when calculated in accordance with Bohr's assumptions, is such that it revolves about 10^{16} times in 1 sec. Despite this terrific speed, the electron's "position at an instant" was thought to be meaningful because it could be measured by allowing a signal (for example, a gamma ray) to be reflected from it while it was revolving.† The direction of the reflected ray would then allow an inference as to the electron's position. To be sure, an experiment of this kind could hardly be performed upon a single electron in an atom, but as a "thought experiment"—and atomic physics bristles with thought experiments—it seemed to be a perfectly meaningful method for determining the electron's position.

Yet a simple consideration invalidates it altogether. For by the laws of electromagnetism on which Bohr's model is in part founded, a certain time interval is required if an electromagnetic signal like a gamma ray is to be reflected by a charged particle, and this interval, the time during which the photon and the electron are in collision, turns out to be of the order of 10^{-9} sec. Where, then, was the electron at the time of reflection? The answer is "anywhere in its orbit, for it revolved a million times during this position measurement." Thus, when the Bohr model suggests that the electron should have a measurable position while at the same time admitting that the position cannot be measured, it contradicts itself.

† The reader will easily realize that we have no way of "measuring" an object except as we touch that object with a ruler, micrometer, or with a light photon. In this case we mention a gamma ray as being the photon of the shortest possible wavelength, to offer the greatest accuracy for the measurement.

The brief sketch given here is not wholly convincing, for it leaves open the possibilities that (1) there may be other ways in which the electron's position can in fact be determined at every instant, or (2) the electron *has* a position, even though it is not accessible to measurement. The facts, however, are that possibility 1 has been virtually ruled out because an exhaustive search has failed to reveal such ways. Possibility 2 has been abandoned by most scientists for the reason that science is not content to operate with concepts that are wholly barred to observation and measurement.

Our intention here is merely to show that a model may lead to erroneous conclusions, against which the scientist must be on guard. It also confronts us with a certain basic presupposition of science, namely, its unwillingness to accept matters that are not open to inspection. Such peculiarities, together with accepted procedures, make up what is called the *methodology* or *the logic of science*, to which the remainder of this chapter is devoted. For the sake of completing the account of our electron, whose position was left in mid-air, let us first briefly note how the difficulty was in fact resolved.

A feasible experiment allowing a consistent inference with respect to the electron's place is one in which *many* photons (for example, X-rays) are reflected from many atoms in one collective act. It is then found that the reflected rays form a diffraction pattern, which suggests that the electrons are in all sorts of places, mostly on or near their respective Bohr orbits. If a pattern were drawn in which the density of shading indicates the likelihood that an electron might be found at the place in question, a diagram like Fig. 22.1 would result. The Bohr orbit marks the most likely

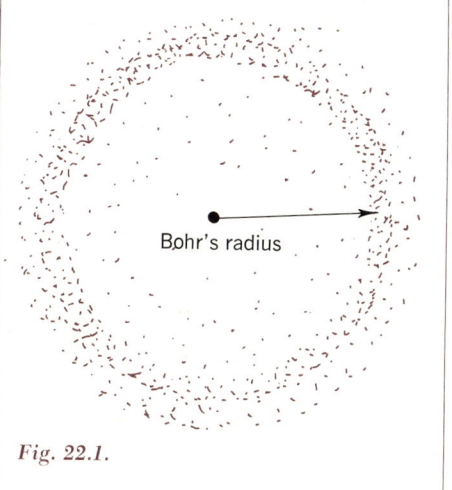

Fig. 22.1.

locus of all electrons, each with respect to its own proton. Yet many can be found some distance away from this "normal" path, quite in contrast with Bohr's visual scheme.

22.3 Deterministic versus probability approach

What interests the physicists is no longer the electron's path or its location in space. In fact many of them deny that it has a path and deny that its motion can be visualized at all. They take seriously the density smear of Fig. 22.1 which in mathematical parlance is a *probability*. The quantum theory, which was designed for understanding and predicting the behavior of electrons, no longer speaks of their instantaneous position but only of their location probabilities. And this theory has been so successful that the Bohr model is now taught only as a useful, qualitative stepping stone to the more adequate theory. Of course this improved theory is itself a model, but a model that is less dependent on difficult mechanistic explanations. It is a probabilistic, mathematical model, more modest in its claim to portray the atom; and it has been fruitful in the extreme, for thus far everything it says about atomic electrons has been verifiably correct.

Considerations of this kind go contrary to the claims that have been made for science over the past 300 years. Beginning with the philosopher Descartes and on through the nineteenth century, the claim was made that every scientific concept had to satisfy a requirement, often designated as Cartesian clarity or "clarté Cartesienne." It demanded that concepts must be capable of being pictured referable to points, lines, and figures of three-dimensional space; that models must be mechanisms, perhaps intricate and complex, in which points or figures move in regular fashion, in which a location is assigned to every part at every time. Bohr's model was the last attempt in the atomic domain to do justice to Descartes' seemingly reasonable demand, and it failed.

Rather strangely, physics, which is often called the most refined and successful science and which has made extensive use of probability concepts over the years, finally decided to abandon the use of mechanical models in favor of the statistical approach for atomic research. The early success of life insurance companies is to be ascribed largely to the fact that the expert in actuarial statistics does not worry over Cartesian clarity with respect to individual deaths; he constructs anonymous survival charts and operates with probabilities. Some of the greatest accomplishments throughout recent science, behavioral as well as natural, may be traced to the use of the notions of random events instead of detailed trajectories, of ensembles rather than specific occurrences. This is one reason why the concepts of probability as applied to systems, statistical behavior, and feedback were stressed in earlier chapters of this text. But, of course, statistical tabulation alone does not explain what causes the electron to stay with the nucleus, nor does it identify the factors that make for life or death in human society.

22.4 Science is a human enterprise

What, we ask, are the characteristic features of the so-called scientific method?

What is the role of science within the larger context of human experience? How does science differ, for instance, from the humanities and the arts? How does science build meaningful superstructures on human experience when the superstructures are abstract concepts for which our senses provide no direct guidance? In seeking answers, we shall first make clear the meaning of science as the word is used in the present text. We shall let it refer to all the physical, biological, and behavioral sciences. The word is taken in the typical, limited Anglo-Saxon sense, because elsewhere (notably in France and in Germany) scholars speak of the science of literature, the science of history, and so on (the German word *Geisteswissenchaft* is almost synonymous with the humanities).

Whatever else its character, we must recognize that science is an *ever evolving human enterprise* which adheres to no static norms. Therefore, whatever is said of science at this time may not be true at some future time. But this potentially tentative understanding of science is nevertheless the understanding we need for making the decisions that apply to our day. The circumstance that the method of science cannot be pinned down and stated in cookbook fashion might well induce an attitude of healthy skepticism with respect to finality. However, it is clear to every observer that the progress of science has followed certain patterns, which perhaps may be too vague and flexible to be called rules but which are worth our closest attention nevertheless.

To emphasize by means of a simple allegory the time- and space-bound character of scientific knowledge, let us recall our human limitations as investigators. We are like passengers in a transcontinental railroad train. Being inside the train affords good opportunity to analyze the details of the interior of the train and the individual experiences and attitudes of the passengers. We can achieve some sense of time and of motion from the turning of the wheel, the lurching of the cars, and the repetitive functions of eating and sleeping. But our knowledge is essentially confined to the interior, plus such events and phenomena as we can view from looking out the windows. As our train moves from the east coast to the west, we could easily develop the impression at first that the universe is made up of rolling hills. We might even find some equations or laws that seem to express the rolling character of this world. In a matter of days the character of the world would suddenly change to become huge flatlands as we traverse the Midwestern plains, and there would be important changes again when the great plains give way to the Rocky Mountains and finally to the ocean.

But individuals who never pass from their region to another would scarcely agree in their descriptions of the terrain. In our gropings we must then make the most of the freedom to move around inside the train and of our ability to look through the windows; but we must never forget that *we are limited to the inside of the train*. Moreover, the experiences and impressions of the world we gather from the train with each day's passing do not assure that the new day will not reveal a wholly different universe. That is, the experiences of the past do not necessarily define the pattern for the future.

22.5 *Experience, knowledge, and the domain of science*

Science is a part of human knowledge, and knowledge is a part or product of experience. Thus experience is more extensive than knowledge, and knowledge is more extensive than science in the sense that the latter represents the refined or selected portions of the former. Let us begin then by explaining the meaning of "experience," a word that needs careful examination. In the original Latin it meant all forms of ordinary perception, such as feeling, seeing, and hearing. The notion of experience likewise included sentiments, attitudes, decisive acts, esthetic appreciation—anything that expressed man's awareness. During the history of philosophy, however, especially during the seventeenth and eighteenth centuries, in the writings of the British empiricists the word "experience" took on a somewhat narrower significance. It referred not to *all* experience but rather to the kinds of experiences that enter the mind through the senses. Thus arose the philosophy of empiricism, which claimed as fundamentally real only that of which one becomes aware through the sense organs. Henceforth we shall not refer to experience in this narrow sense; instead, we shall restore the word to its general earlier meaning.

But having restored this broad scope of what constitutes experience, we might then ask: "Do science and scientific study also encompass this full scope?" The question is significant because clearly such experiences as enjoyment, shame, and frustration are more difficult to describe and to study than are the physical experiences with which we have dealt in this volume. We can illustrate the difference through the experience of a technologist; say, an architect who spent years in learning the constructional and architectural details of many buildings. It occurred to him once to examine the dwelling in which he lived, from the point of view of his technology. When he did this he discovered that his technological competence for analyzing the physical details of the house was, by itself, totally inadequate for describing the structure as a *home*. It seemed quite clear that the terminology and approach for defining the characteristics of a *home* are different and, to a large extent, less easy to quantify and measure than were those required for defining the *house*. It was equally clear that there was no hard and fast boundary between the concepts and techniques that entered the analysis of the house as compared with those that revealed the characteristics of the home. A feature of a home may be intangible and undefinable today, but may tomorrow become more tangible and may be included as part of the professional education of the architect. There can be no absolute boundaries between the physical and the life sciences, where in the term *life* we include every interest of man. In this connection we are reminded that at one time there was great difficulty in moving the concepts of space and time out of the exclusive realm of philosophical speculation.

22.6 Perceptual and conceptual planes

At any rate we recognize the existence of a wide range of cognitive experiences. According to the British empiricists, cognitive experience begins with the things we see, hear, feel, smell, and taste, these finally giving impressions that are called *sensations*. But cognitive experience also includes explanatory ideas, the sort of things which logicians call *concepts*. The differences between the two are not always easily identified because they tend to fuse into each other. In a general way, however, we may distin-

22.6 Perceptual and Conceptual Planes

guish them by saying that *percepts*† enter the mind through sensory awareness; *concepts* are constructed by the mind. These two types of human experiences, percepts and concepts, include what is traditionally conceived to be cognitive experience.

Thus a concept begins with a certain kind of knowledge to which one often applies the word "immediate." Immediate experience consists mainly of sense data; it includes observations and the results of experiments in science. It furnishes a set of data or facts which, as long as they stand alone and remain unmodified by rational interpretation, make little sense of the world. If all experiences we had were simply bits of sensory data here and there, ever emerging and perishing particulars that assail us from without, if we were incapable of thoughts about our sensory experiences which place them in relationships to one another, then our world would be chaotic. Immediate sensations, as long as they remain uninterpreted and uncorrelated, make a senseless world. The philosopher Kant called them "rhapsodies of perception," alluding to the fact that they are beautiful in their immediate appeal, charming and suggestive, but never self-explanatory.

The differences between the perceptual and the conceptual planes are sometimes illustrated by the experiences one has when sitting at a desk. We can, through sensory reception, identify the characteristics of the desk in terms of its size, shape, color, composition. This has been referred to as *cognitive* experience at the perceptual or sensory level. But the physicist recognizes that the material of the desk is made up of atoms and molecules, which he cannot "see" through the sensory receptors. He must therefore accept this latter desk as existing in the conceptual plane only, not supported by direct sensory information.‡

The reader must by now have recognized (or should have recognized) the great difficulty of cataloging human experiences that involve sensory receptors and interpretation of sensory information. There are few clearly drawn boundaries. We know that sensory information begins with a stimulus that causes electric discharges from certain nerve cells in the sense organs. Of itself, an electric pulse carries little information except as early experience begins to identify the stimulus related to each type of electric pulse. Where and when does this identification or interpretation begin? We noted, in connection with the discussion of vision in Chapter 9, that the retina of the eye itself appears to do some of this identifying of form even before sending the message along to the brain. On receiving the electrical message in this partially organized form, the brain proceeds to further organize the message to give it fuller significance than the eye alone could produce. For this organization and interpretation the brain utilizes its memory of earlier experiences and its earlier interpretations of *any and all kinds*. Some of the earlier experiences may be of simple sensory variety, but much of it can be the product of book-learning, philosophical deliberation, imagination, and of earlier interpretations

---

† Of course both percepts and concepts involve integration of past experiences. We must warn the reader that there are likely to be as many definitions for percept and concept as there are psychologists. Our choice of definition is for heuristic purposes only. See Von Fieandt in the References.

‡ Eddington, in his book *The Nature of the Physical World*, made this comparison between the desk of the positivist-philosopher and the desk of the physicist.

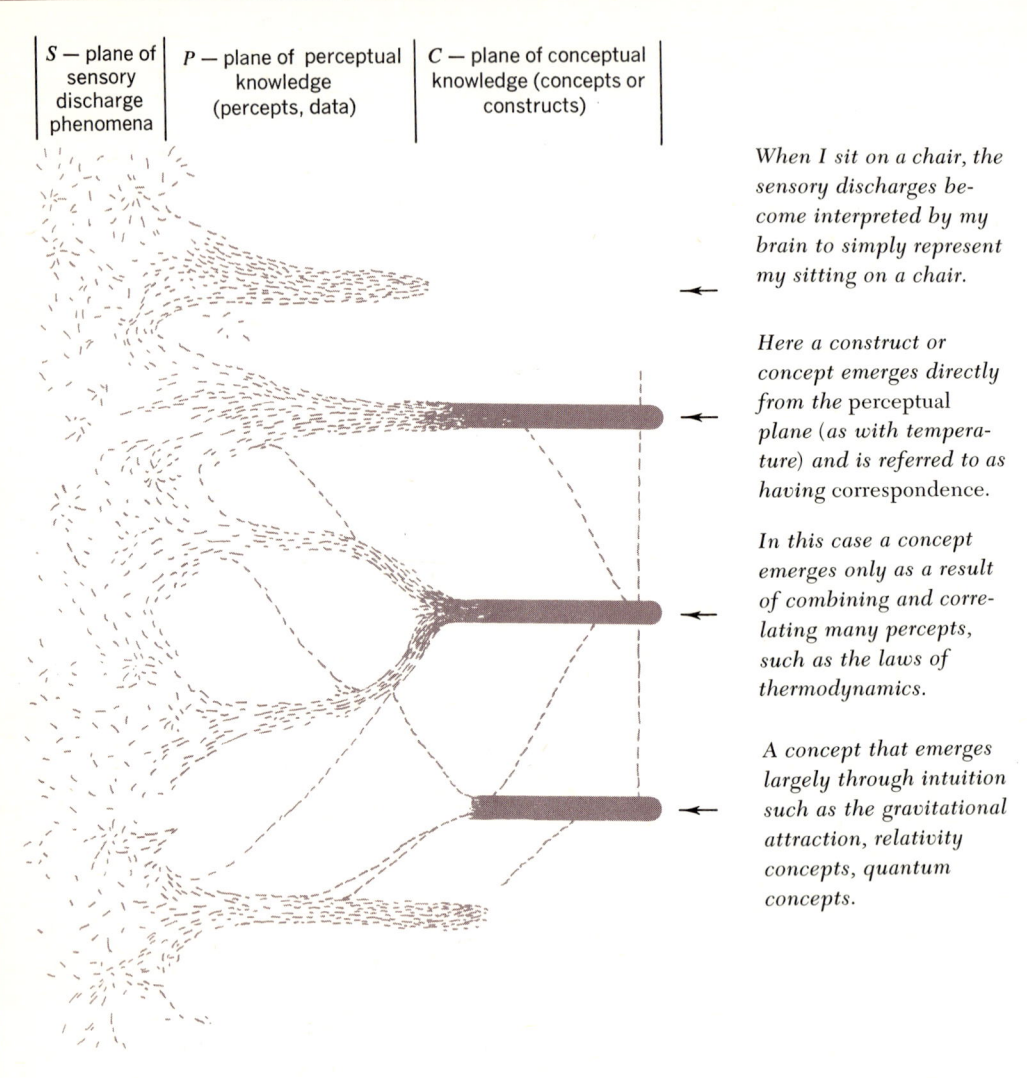

Fig. 22.2. Relationship between the sensory receptor, the perceptual and the conceptual planes. The nerve discharges that originate in the S-plane may, through some organization and correlation, become data, percepts, or perceptual knowledge in the P-plane. Through further organization these latter may become concepts in the conceptual or C-plane. As illustrated, some limited organization of random sensory discharges can be initiated even at the S-plane. There can be some percepts that readily become concepts, while more often concepts evolve from many interconnections among percepts or among concepts and percepts.

that were themselves several steps removed from direct visual observation.

The explanation given above says that we cannot offer clear-cut distinctions between perceptual and conceptual knowledge, that their differences are more in degree than in kind. If we are careful to avoid tight boundaries, we can nevertheless represent broad categories of differences. Figure 22.2 illustrates these categories in terms of the *stimulus plane*, a perceptual (or protocol perceptual) plane, and a conceptual plane. The stimulus plane contains bits of sensory discharges, which in most cases are quite random, but in a few cases tend to show a (flow) pattern. When these reach the perceptual or *P*-plane the random bits become organized to become meaningful observations (sometimes referred to as the immediate observations) or data. This may also be called *protocol experience*, from the Greek work *protokollon* meaning "first glue" which referred to the first leaf of a Greek book. When an ancient Greek author wanted to write a book, he would enumerate upon a sheet of papyrus the individual items with which the book was to deal. The writing of the book would then be nothing more than an elaboration and a correlation of all the items that the protokollon contained. By protocol experience we mean the raw experience with which the scientist, or for that matter the man of everyday life, starts—that is, the items that he wishes to weave into a texture of multiple understandings.

The *P*-(for protocol) plane of experience comprises such things as immediate sensations, the scientist's observations, and all the measurements made by physicists, chemists, psychologists, and economists. Their character, however, tends to be unsatisfying to the human mind, for these experiences lack the cohesive, logical bonds that the scientist as well as the man in the street invariably seeks. As mentioned before, if our world were composed of nothing but these spontaneous sense impressions, it would make no sense. Protocol experience has this important positive feature: Its incidences and even the quality is rarely doubted. *P*-experience reaches the scientist from without in such a way that he cannot deny it. Therefore he uses it as his last instance of appeal, as the final testing ground for his ideas and theories.

Because the scientist is not satisfied with the lack of coherence, the unrelatedness of the items in the *P*-plane, he uses ideas, concepts, or constructs as counterparts of them in the *C*-field. Let us use the word *construct*, since it defines accurately the manner in which an idea is produced—the mind has made it, man is responsible for it because he has opposed it to that for which nature is responsible, that which is coercive and indubitable, vivid but irrational. Scientists introduce constructs because these enable them to reason. To be sure, the inchoate items of the protocol plane can be collected and arranged in many ways and inductive inferences can be based upon them, but one cannot reason in their terms alone. For example, individual apples and pears, fingers and toes are not objects of reason unless you have the idea of number and know the laws of arithmetic. Number, however, is not something like toes or apples or pears; it is not encountered in the *P*-plane, but is a construct. It becomes necessary for the scientist and also for the layman to associate *P*-experiences with each other in such a way as to produce constructs in the *C*-field. The question, therefore, arises as to how this association is effected. The answer is quite simple: One does this by means of some sort of measurement, as will now be illustrated.

22.7 Experience and concept of temperature

When we stick a finger into a bath of hot water we experience a quality that we have learned to call hotness. This is a P-fact. It is a subjective feeling of temperature, with perhaps only qualitative information to qualify it. The physicist and the chemist, however, who wish to include a concept with the idea of temperature, are unable to readily insert that kind of experience of temperature into an equation or a law that would allow them to reason about it. In order to obtain an observation that is more amenable to logical analysis, they measure the temperature instead of depending on feeling with a finger.

In this particular case, one employs a device called a *thermometer*, which is inserted into the bath. The top of the meniscus of its liquid defines a number on a scale that is no longer entirely a subjective feeling and may be taken to be an objective measure of temperature. In this instance we have started at a certain point on the P-plane and have gone from there by an act called *measurement* to a more objective concept of the quantity in question, a concept that we have constructed from earlier work and which we therefore call *a construct*. This passage from P to C, from a protocol fact to a construct, is called a *rule of correspondence* in analytic philosophy. It simply represents any act within experience which takes us from the P-plane to the C-field, and by this act it establishes a construct (in this case temperature) to which one can now assign a number, to be the same for all observers.

Numbers are useful things because they can be placed in a context of logical and mathematical relations. One can reason with numbers, for numbers are the ingredients of arithmetic. By means of numbers via operations called measurement or rules of correspondence, immediate P-facts are made objective; that is to say that they are subsumable under the laws of thought, and these laws give coherence and logical connectivity to the constructs that correspond to the P-facts.

There is a great deal that is arbitrary about the way in which the notion of temperature was objectively stabilized. There are many ways of measuring temperature, many rules of correspondence, each corresponding to an instrument of a different sort. For instance, if we use a mercury thermometer we obtain one number; an alcohol thermometer will give another number; an ideal gas thermometer would give still another number; and a constant pressure thermometer a different result again. Which number is right? Which number can the scientist accept? Which rule of correspondence is to be considered as valid? Although we seem to have resolved certain problems by going from the P- to the C-level, we now face new questions. Perhaps this illustrates first that there is no royal road to absolute knowledge, and perhaps there is no such thing as absolute knowledge.

22.8 Experience and concept involving light

Let us take another example, that of light. The yellow patch we see is P-fact; the physical entity called *light* is the construct. The passage from one to the other, indeed the conception of what that physical entity is, has changed in time. The ancient Greeks thought that a light beam was composed of material particles; they thought that the sensation of light arose when a beam of particles emanating from the light source impinged upon the eye. A later Greek theory involved two beams, one emitted by the eye and another by coming from the light source. When the two make contact the sensation of vision was thought to occur. Much later, during the Middle Ages, people continued to believe that light was made of particles, but particles more clearly specified in their nature. Today we sometimes assume that light consists of electromagnetic vibrations, or sometimes, in the quantum theory, as a particle or photon. Thus we see that there is a great range of choice available in making that passage from the P-level to the C-field, and also that in the history of science many of these choices have alternately been made. The question as to which of them is ultimately acceptable to science has no immediate answer because science does not stand still. It looks for further and further refinements in its ideas about the world, man's knowledge becomes more and more finely structured, and there is no telling where the scientific quest will end. Nevertheless, although we do not know the final goal, it is proper for us to ask how science proceeds in its choice of rules of correspondence, how science establishes reality even if that reality is not final. Knowing this, we may understand more fully how man relates himself to reality in nature.

22.9 What makes a concept acceptable?

In the course of scientific history certain procedures that guide the passage from the P-plane to the C-field have evolved. The multiplicity of passages is limited by certain regulative principles, which for want of a better name we may call *metaphysical requirements*. The adjective "metaphysical" is chosen because these principles have little to do with the P-plane; they stand behind the empirical facts of science much in the way conveyed by the sense of the word "metaphysical." Let us emphasize once more, however, that these principles are not the innate ways of categorizing experience; for example, the categories of the mind postulated by the philosopher Kant in his *Critique of Pure Reason*. They are assumed to be simple requirements, which the history of ideas and of science has recommended to the knower as being most fruitful in the process of acquiring knowledge.

These metaphysical requirements are not specific or discrete items that can be easily put on a list. They form a sort of continuous spectrum, with a fine gradation between them all. One of them may partially overlap another, and in listing them as separate constituents we emphasize a central idea in each.

LOGICAL FERTILITY

First among these metaphysical requirements is one called *logical fertility*. It requires that no construct shall be introduced into a system of scientific ideas unless it sets itself into fertile logical relations with others, unless it allows consequences to be deduced from its acceptance. The scientist insists that all constructs used in the explanation of protocol experience shall be of the fertile kind, of the kind that leads to specific suggestions which can be verified in the factual P-domain. To the physicist, the requirement of logical fertility might be judged on whether the explanation lends itself to confirmation by a critical experiment.

EXTENSIBILITY

Secondly, the scientist would like the constructs he uses to have a *maximum measure of extensibility*. This means that a set of constructs shall be usable not only for the explanation of a small portion of the P-plane, but for a large area of it. This principle is used mainly to discriminate between rival theories, bestowing preference upon that which is more extensible.

It is said that Newton, while resting in the orchard at Woolsthorp, saw an apple fall and that this incident suggested to him the law of universal gravitation. If this law had succeeded in explaining only the fall of the apple, it would have commanded very little attention. The law of universal gravitation, on the other hand, explains not only the fall of this apple, or of all apples, but the fall of all objects near the surface of the earth as well as the motions of the moon about the earth and the planets about the sun. This is what is meant by *extensibility*.

RELATEDNESS TO OTHER CONCEPTS

Thirdly, *acceptable constructs must be multiply connected to a set of other constructs*. This requirement of multiple connections between constructs is very closely related to extensibility, although it places the emphasis in a slightly different place. About 30 years ago the physicist Fermi conceived the neutrino, a massless particle postulated to circumvent the apparent contradiction of the laws of conservation of energy and of angular momentum in certain processes of atomic decay. The neutrino was to have the properties of a small particle, yet it escaped observation. For a while it served only the purpose for which it had been invented, namely, that of conserving energy and angular momentum in beta decay. Had it done nothing more than that, the concept of the neutrino would have been singly connected to that set of P-plane facts (namely, beta decay) that it was designed to regularize. Dissatisfaction on the part of physicists with this single purpose led them to search for further manifestations of the neutrino, and finally these were found. Thus the construct, which was at first insular, entered into relations with other constructs, led to coincidence with other parts of the P-plane, and the neutrino was accepted as being real. A larger degree of reality was thus gained.

SIMPLICITY AND ELEGANCE

The fourth metaphysical requirement is one of *elegance* and *simplicity*. One might wonder whether these items may be appropriately extended from the domain of art into that of science, yet it is becoming increasingly clear that a certain reliance upon elegance and simplicity is utterly necessary in the whole of science. To expand here upon the precise meaning of elegance and simplicity in science is beyond the scope of this text. Suffice it to say, however, that a theoretical scientist is likely to sense when a theory is simple enough to be accepted and when it is complex, ugly, and burdensome enough to be rejected. In our day, simplicity and elegance are more and more identified with the mathematical property of invariance with respect to transformations that theories must display. This was discussed in Chapter 12.

The great theoretical physicist Dirac pointed out that a scientific theory, a set of constructs, that does not partake initially of the highest measure of elegance and simplicity, is not worthy of being considered by scientists. In fact, as modern science develops, we find ourselves more and more forced to apply such esthetic criteria as elegance and simplicity.

EMPIRICAL VERIFICATION

At a certain stage, science converts constructs into real entities. Physical reality, therefore, depends in large measure upon whether the constructs corresponding to reality satisfy this set of metaphysical requirements. This, however, is not enough. There is another set of criteria, equally important, related to *empirical verification.*

When a construct is introduced and is deemed to satisfy the metaphysical requirements, it must be further related, directly or indirectly, to the P-plane. It must be possible by actual observation to find something in the world of observations, that is, in P-experience, which corresponds to the construct. This process of empirical verification usually takes the form of a circuit in the following sense:

The scientist begins with an observation in the P-plane and proceeds from there via rules of correspondence to the C-field, where he can reason logically and solve mathematical equations. Actually he rarely takes a single step from one plane to the other; progress is made by repeatedly going back and forth between the two planes. In the course of this he hopefully comes to a set of constructs that enable him to make a prediction. This prediction is a passage from the C-field to the P-plane. If that prediction and very many other predictions seem to be borne out, the theory can be considered to be feasible, at least until further notice. These are the conditions that tend to establish reality in science, to the extent that the scientist dares to speak of reality.

Question: With the five criteria given above in mind, to what extent would you say their applicability is limited to science?

22.10 A "systems" approach to concepts

We have, quite properly, emphasized how difficult it is to assign boundaries to the sensory, perceptual, and conceptual planes, and how complex are the processes by which we learn. It must not be assumed that concepts take shape only in the mind of the adult scientist. There is no reason to assume that it is not a significant factor in the experience of a child, or possibly even of the infant. Any model that attempts to describe the processes by which stimuli to sensory organs finally produce concepts is certainly likely to be oversimplified and faulty. Nevertheless, there are a few relationships present in these epistemic processes which seem to be nicely describable in terms of the "systems" diagrams and concepts. Figure 22.3 illustrates such an attempt. The figure may apply to a child, to an adult, and indeed to society as a whole. It begins at the left with a box that represents the body at its simple operational level. This box is in intimate contact with its environment. In some respects this is the lowest level of animal existence, wherein the reflexes move the body in response to push and pull, cold and hot, hunger, and intense stimuli to the sensory receptors. There is from the very first, however, a progression through the upper box which represents the intermediate and higher nerve centers. This brings into play memory of experiences, and reason and logic which give additional meaning to the crude body experiences represented by the first box. The discomforts suffered by the body from the heat and cold find some adaptation to environment in the form of shelters. The stimuli affecting the eyes and ears and the other sensory organs gradually develop adaptation to environment—more efficient, less frightened, less at the mercy of the elements, more able to interrelate phenomena toward meaningful purposes. There is one other major achievement, namely, a sense of time, of progression, of looking ahead to

666 *What Is Scientific Method?*

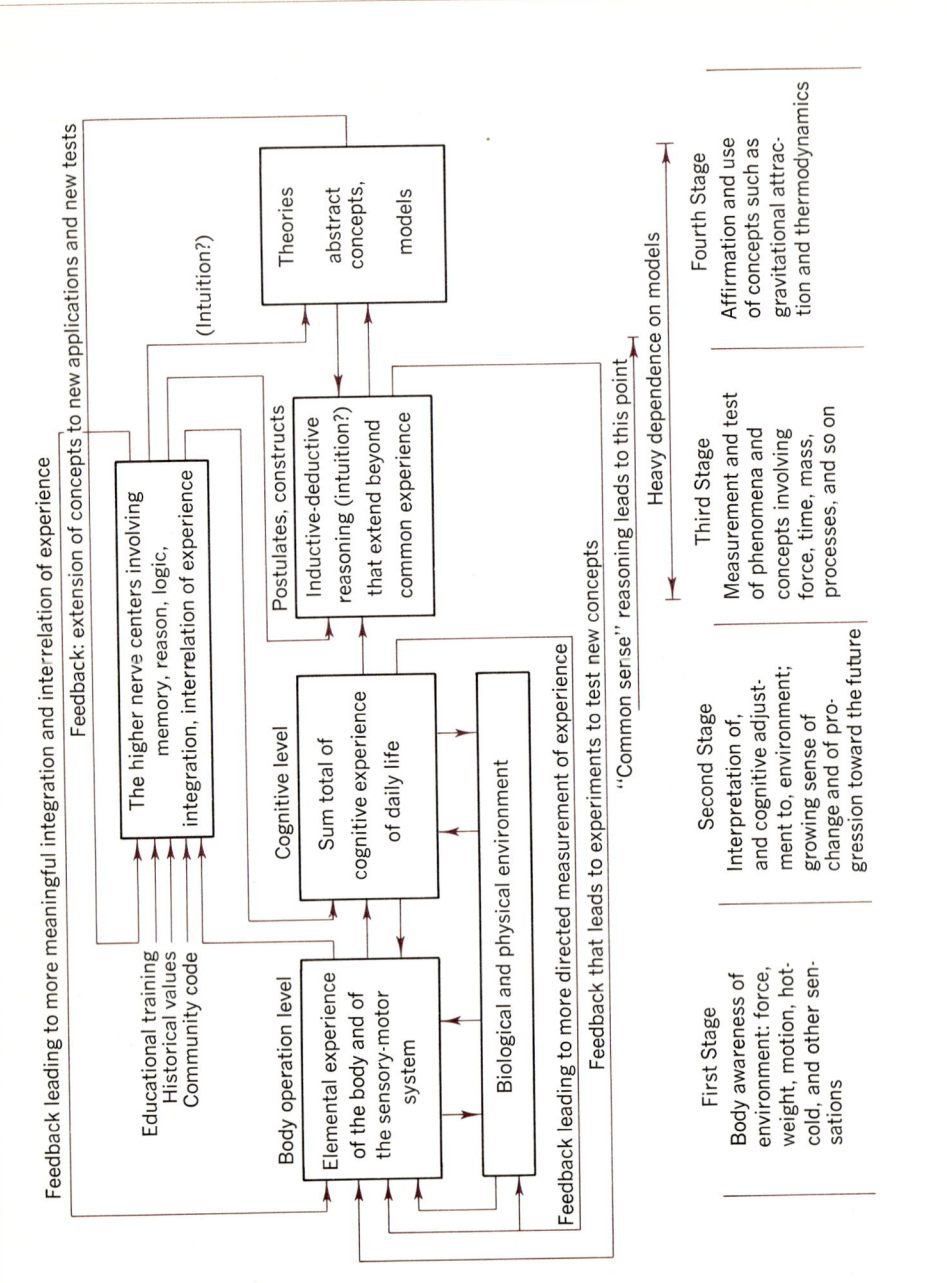

Fig. 22.3. *Progression of human experiences to useful concepts.*

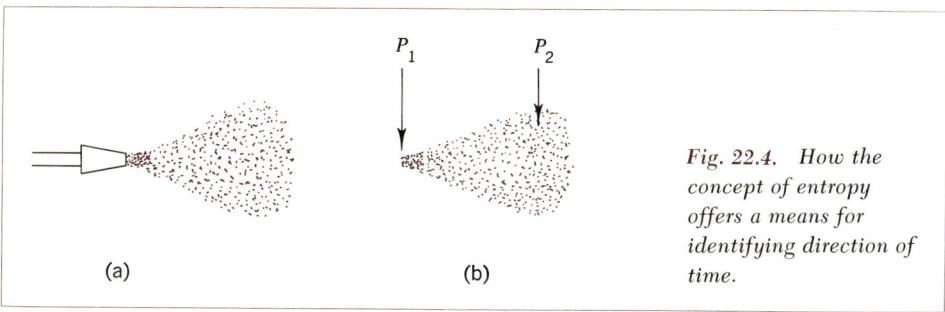

Fig. 22.4. How the concept of entropy offers a means for identifying direction of time.

a future. We may call this the level of simple constructs.†

Figure 22.3 shows feedback influences from the higher nerve centers and from the box of simple constructs, back to the initial experiences. An idea that is the product of brain action can thus give rise to succeeding ideas, and indeed to a whole chain of developments that give the appearance of positive feedback and amplification. On the other hand, the idea might, as a result of a test or further contemplation, be found to be of little value and a negative feedback can suppress its further pursuit.

We can assume that up to this point our subject (child, or man, or society) has found reasonable adjustment to environment to afford a comfortable existence. It is not likely that this is the end, however. Even an infant may exercise effort to discover some general rules that seem to spell pleasure or pain. There can easily develop something of an hypothesis that associates hunger, crying, and the mother's attention in a combination that clearly goes beyond simple reflex action. A little later there will be a period of testing to see how far the child can go against instructions before he is punished. He may test his capacity

† See Margenau in References.

against the elements. The adult person may pursue a more sophisticated program of measurement and of reasoning. He will develop some ideas as to *how* things happen, and perhaps even *why* they happen. His analysis is likely to suggest some tentative ideas on causality, which may encourage him to erect a statue to his god or to offer burnt offerings to placate the gods.

There can be very many other *assumptions* and new *constructs* as a result of this reasoning and measurement process. The quantitative experiments involving force and matter can lead to identification of a more abstract characteristic of matter, which we call *mass*. The constant presence of a weight property associated with mass might, with the flash of intuition and genius of a Newton, reveal a universal law of gravitation. The testing and retesting of this law (as indicated in Fig. 22.3 by the close interaction between the hypothesis and the theories boxes, and by the feedback to the original boxes) may continue either to strengthen this concept if there are no contradictions or to suppress it if contradictions develop.

There can be concepts such as entropy which are of even more abstract nature. We said that according to this concept, nature tends toward greater disorder. Suppose we watch a stream of vapor or of dust particles emanating from a nozzle (Fig. 22.4) *without actually seeing the nozzle itself.*† We observe the vapor stream to be small and well directed at one end. We see also that at another point, the stream volume expands and the vapor trails lose precise forward direction in favor of a more disordered state. Experience tells us that the nozzle is always at the end where the vapor stream is at its smallest size, and that the *larger and more disordered state of the vapor occurs later in time.* Experience thus seems to give us the basis for a hypothesis (suggested by repeated observations) that *order tends to give way to disorder.*

This hypothesis, which forms the beginnings of a postulate, can become the basis for a more directed effort, through feedback, for experimental tests and for more critical analysis. Does the hypothesis that order tends toward disorder with time conflict with other observations in other areas? Have we ever observed uninfluenced random particles to converge into an ordered stream? If so, the same feedback and test processes can very quickly suppress or kill the hypothesis or reduce it to qualified terms, useful only within a smaller domain of application. But if every experience tends to confirm the hypothesis, we may find ourselves stating a very basic theory, namely, that of *entropy increase.* The theory is very abstract in its simplicity, yet effective for predicting what is likely to be the next observation of vapor issuing from nozzles.

22.11 The concept of time

It was noted in Chapter 9 that Newton thought of time as a stream that ". . . of itself and from its own nature flows equably without regard to anything external," providing us with a reference coordinate to which all events could be referred as function of time. It carried the notion of *absolute time* that exists and persists independently of any other change, or absence of change in nature.†

But within the past few decades we have come to understand that time is meaningless unless it is accompanied by changes. That is, time "moves" only because things change from one position to another, from one state to another state; the flow of time is meaningful only because it is identified with the progression of states. To be sure, everybody has an immediate awareness, a subjective knowledge of the progression of time in the sense of the *P*-experience of Sec. 22.6. This is often spoken of as our *stream of consciousness.* It is involved in the succession of pictures, thoughts, or moods that pass through our minds. But it is not measurable as such. As we have seen, to make it so, we must employ operational definitions using clocks, and these invariably refer to changes in our external milieu.

Thus it is physical changes such as the rotation of the earth, the swing of the pendulum, or the frequency emissions from atoms that establish our time base.

But what of the *direction* of time? What is the significance of time in nature? The laws of Newton defined processes that could go in either direction. In a collision process the initial and final states could just as readily be reversed. Does nature have a *forward* direction, and if so, what constitutes going in the forward direction versus backward direction? This so-called time-reversibility, which is indifference with respect to the direction of time, is also apparent from the form of Newton's laws. The first and third laws make no

† See Bronowski in References.

† See Schlegel in References.

reference to time and are therefore indifferent to its course. The second law, force equals mass times acceleration, involves the time variable t because acceleration is dv/dt, while the velocity $v = ds/dt$, s being distance traveled. Now reversal of the time direction means replacement of t by $-t$. When this is done, v changes sign. Hence, if the laws of Newton involved velocities instead of accelerations, they would be sensitive to the sign of t, and one could by reference to them tell the future from the past. But they are cast in terms of accelerations, which change their sign twice (and hence not at all) upon replacement of t by $-t$. Therefore Newton's laws are time-reversible; they do not distinguish the future from the past. For every phenomenon in nature that satisfies them, we can find another perfectly possible one that results when t is traced backward. Take a movie of any sequence of mechanical processes satisfying Newton's laws (they must be frictionless because friction is proportional to v!) and run it backward! What you see is another possible sequence of mechanical processes.

We said that the passing of time is represented by a succession of states. Thus all the states that have occurred are in the past, while the states that have yet to come into being are still in the future. But is there some characteristic that will distinguish the states that will come from the states that have passed?

It appears that the concept of entropy offers us the means for making this distinction of future events from past events. To see this, let us take a movie of a tennis game, supposing that all motions satisfy Newton's laws and that friction is absent. Now we run the reel backward. You see a curious series of happenings, not in accord with Wimbledon rules. The ball will bounce in normal fashion (if friction were operative its height of bounce would increase in time!), but the racket, instead of propelling will be running backward (more often than forward)—all these occurrences violate the rules of the game but not the laws of nature. We know they could happen in our world.

But suppose our picture includes a spectator on the side lines who is smoking a cigarette. In the reverse picture he would seem to be drawing smoke out of the thin air as he inhales, exhibiting an occurrence that is precisely the contrary of what is illustrated in Fig. 22.4. This is unbelievable; it does not happen naturally in our world. As the smoke stream in Fig. 22.4 progresses away from the nozzle, or in this case the smoker's mouth, it expands and mixes with the air. The passage of time brings greater disorder, and we can readily distinguish which part of the stream must have come into existence first as a stream. *The change in entropy gives the direction of time, in that successive later states of a system are states with successively greater total entropy values.*†

We can carry this a little further. The first law of thermodynamics says that the energy content of the universe is constant, and we can transform energy only from one form to another. The second law [on which the German physicist Rudolf Clausius (1822–1888) based the entropy concept] says that with *most* energy transformations, an increase in entropy reduces our future ability to convert and utilize the energy of the universe. We therefore face the bleak prospect of ultimately finding the same total energy in the universe, but without any prospect for utilizing that energy. In other words, the flow of time is a measure of our approach to that time of "heat death" envisioned by Clausius, when the whole

† See Schlegel in References.

universe will have reached a dead level of uniform temperature. But, of course, we remember again that this is the way it *appears* to us, and we are only passengers in a passing train!

22.12 Induction, deduction, intuition in birth of concepts

Thus far we have stressed the importance of progressing from sense experience to concepts; but we have not said much about the specific steps by which the transition is accomplished. Widely accepted at one time as the representative of science was the so-called *method of induction* of Francis Bacon. Bacon suggested† some very specific guides by which one might progress from particular experiences to generalizations through a sequence of steps involving induction and some use of deduction and of tests. It seems, however, that neither the rules nor the implications of his approach have applied completely in the birth of the more important concepts of science. We shall discuss a few of these concepts to identify the roles of induction, deduction, and intuition in their development.

THE EXAMPLE OF THE THEORY OF GRAVITATIONAL ATTRACTION

The contemporaries of Newton were cognizant of many common experiences involving force, momentum, and the idea of the conservation of momentum. It was a period of ferment for new theories and for experimental testing of theories. Mathematical techniques developed by Descartes, Leibnitz, Fermat, and Newton himself gave additional impetus to the search for the laws of nature. Two thousand years of theorizing had brought the ideas of planetary motion to roughly the third or fourth stages represented in Fig. 22.3, but without the benefit of the feedback in the form of tests. Now the Greek theories regarding basic properties of matter did not seem to stand up under careful inquiry. Ptolemy's ideas seemed inadequate in the face of the alternative suggestions of Copernicus. Experience with moving bodies and with interaction between bodies had given birth to the concepts of mass and energy. It was possible to make fairly accurate measurement of some of these phenomena. From these observations, when viewed in the light of simply related constructs, it became possible to develop the three laws of motion introduced by Newton; that is, to go from specific experiments to general ideas. This is technically called the process of *induction*.‡

The general laws that developed from this inductive process could be tested. In fact we saw in an earlier chapter that the three laws of Newton apply quite generally. We can, having these laws, deduce what should be the consequences of the laws under new situations. If tests confirm our *deductive*‡ *reasoning* that arrives at these consequences, we say that the general laws are true.

Let us assume that this was the situation when Newton looked more deeply into the subject of gravitational attraction, the reason why the planets execute elliptic orbits with the sun at the focus of the ellipse. By what steps of logic

---

† See *Novum Organum* of Bacon in References for his "tables of discovery."

† Inductive reasoning is said to be from specific phenomena to the general, from the individual case to the universal law of theory, or from the part to the whole.

‡ Deductive reasoning, in contrast with inductive, is reasoning that applies general laws or axioms to specific situations. Mathematical deduction solves specific problems according to the axioms and rules that are given for the mathematics in question, an example being the elements of geometry as presented by Euclid.

could he make transition to *the inverse square law of gravitation?* What experiences or experiments did he have at his command that might offer a *logical* progression to the fundamental law of gravity, which required among other things that distances and forces be related to the *center of the earth?*

The answer is that there were no experiences or formal steps of logic by which to go from a finite set of observations to a universal inverse square law. It was, in fact, a stroke of genius, a kind of *intuition*, which leaped over the steps of logic and of experience to give to the world the gravitational law. Once achieved, the theory permitted calculation of planetary motion, and the calculations gave excellent agreement with the orbital positions that had been observed by Tycho Brahe. The observed facts and the new concept of gravitational attraction seemed to fit well to describe the world of nature and of man.

The validity of the gravitational law also lay partly in the beautiful way in which that law complemented other laws of motion. Here is where the metaphysical principles of Sec. 22.6 enter the scene. The law completed the package that was needed to explain the mechanics of nature. The new law was simple. It was seemly that the heavenly bodies should attract one another. It was somehow felt to be proper that all masses, large or small, should feel the force of attraction in proportion to their masses. The simplicity and beauty of the whole scheme promised that God was in His heaven, and all would be well with the world system. And yet the picture could not be considered to be complete. For, although the law of gravity held true, and holds true to our day, we are forced to question its validity as absolute truth. Indeed it is only a half-truth, for it does not tell the whole story. Newton himself was left dissatisfied because he could not fathom the "cause" of gravitational attraction. He was aware that there were other kinds of forces that became involved in friction, cohesion, and chemical behavior, which he had not defined. He was aware of the immensity of space and of time, and had a deep sense of concern for the importance of identifying the role of God in that immensity as well as in the humbler operations of his laws. But he sensed, as we must all sense, that absolute truth is not within reach. Despite the magnificence of its accomplishments, we remain keenly aware that the achievements of science are enmeshed within very human emotions and convictions. The achievement of absolute truth, therefore, is encumbered not simply because we are inside a train; the handicaps are even more confining within ourselves as human beings. There is, however, some comfort in the thought that while we suffer limitations in reasoning power, there may be available to us on occasion the talent of *intuition*. This is not to be considered as a mystical sense, but understood as a sudden glimpse of an idea—an act of creative imagination that is difficult to analyze and which sometimes, as in Newton's case, affords us a view of partial truth in a manner that is almost unscientific, if we define science in terms of formal pursuit.

22.13 *Deduction, induction, intuition in the progress of science*

Let us continue with comparisons of the roles of deduction, induction, and intuition in the development of science. We are all familiar with the process of *deduction* as it is used in geometry and in algebra. There are axioms, or premises, and rules of the game by which we reason our way to the solution of specific problems. Among other things, Euclidean geometry says that the sum of the angles of a triangle totals 180 deg. With this proposition we are able to solve very many problems of geometry by simply following the prescribed rules of the game. The axioms of geometry and those of algebra turn out to fit the physical world in a way that makes them of tremendous importance. From them we deduce a host of conclusions and solutions that apply well to the details of our daily life. Their practical value would break down, however, if the deductions did not fit our world of experience. An axiom that permits the sum of the angles to be greater than 180 deg, which is the case in non-Euclidean geometries, could yield quite different deductions and consequences.†

At any rate, general principles are of enormous value for application to specific cases to get specific answers. The goal of science is to find these general principles. It is a *reaching* for an absolute truth, even though there is awareness that absolute truth seems to be well beyond reach; indeed every "law" is, strictly speaking, something less than a law. For example, we discuss the laws of the perfect gas, and the behavior of a body in motion which is not subject to effects of friction, while knowing full well that there are no ideal gas situations or moving bodies that are not subject to some frictional force. Nevertheless we find the laws of the perfect gas and of frictionless motion to be very important because, with suitable corrections, they approach real-life situations that are meaningful and useful to us. This is demonstrated by the fact that science and technology depend on these laws for explaining past experiments and for giving direction to new experiments. Of course every extension of a concept, law, or theory to a new experimental situation carries with it a risk that something unexpected may be revealed, raising a question as to the validity of the law. Strictly speaking, the so-called *laws* of nature are man-made and hence fallible. They are introduced to *describe* how nature appears to act. Despite this they are responsible for most of the progress in science through the ages.

But how do general laws develop to such proportions that they become useful bases from which to deduce other information? It may be worthwhile to delve into this subject a little further.

We noted earlier that all knowledge and science find source in human experience. We observe that the sun rises each day (allowing for the fact that we do not always see the sun), and that our bodies require daily food to exist. These are the facts, the empirical data, or *P*-experiences of life. Such experience gives rise to impressions and to ideas or hypotheses through a process of induction, which involves at least in rudimentary fashion a passage into the *C*-field. We noted earlier that inductive reasoning looks for general laws or concepts through which to explain the rising of the sun and the hunger pangs, perhaps even to seek a relationship between the two phenomena. The general law we think up must be of a kind that can be valid to other phenomena as well, subject to confirmation by others through test. *Indeed this inclination to subject every idea to test was the feature that characterized the new science that came with the Renaissance.* Prior to that period, science was highly speculative, bordering on the

---

† It must be noted that the sum of the angles of a large triangle on the surface of the earth does *not* add up to 180 deg. It was this fact that led Gauss to introduce spherical geometry into surveying (early nineteenth century).

inductive and requiring a minimum of experimental proof along the way.

But development and test of a hypothesis usually involves use and test of the hypothesis, and this itself may involve a form of deductive reasoning. That is, before a formal hypothesis is born, it is subjected (in modern science at least) to test and confirmation (or rejection) through experiments that deal with very specific situations; but when the tests cease to be tests and fall into the category of deductive may not be clear.

The progress of science is therefore determined not by inductive reasoning alone or deductive reasoning alone, but by a combination of both working back and forth and often involving many steps. But sometimes the stepping back and forth is not enough. Sometimes there is needed the influence of *intuition*, or creative imagination.

What do we mean by intuition? Margenau represents it as an uncanny ability that some people have for contemplating a set of facts or observations and from there to take an "inductive leap" over intervening steps of logic or experiments to a valid set of "constructs" or ideas that are suggested by the observations. Sometimes intuition is of a kind that may be classified as a successful conjecture that can be demonstrated by test to have value. On occasion some thinkers have claimed for intuition a more deep-seated, metaphysical role approaching revelation and leading to capacity for understanding the laws of the universe; but there is coupled with it the qualification that this understanding does not permit testing for validity. A concept that can find neither test nor confirmation has limited value in science.†

† There are some who see intuition as simply a manifestation of chance phenomena that accompany logical processes. Others see the existence of patterns in discovery. See Hanson in References.

22.14 A creed for the scientist and the layman

Perhaps this chapter can best end with the following quotation from Margenau (p. 76):

"Were I to set down a creed which expresses the new faith of science, such a catechism would include the following articles:

"I believe that the search for truth is a never-ending quest; yet I pledge myself to seek it.

"I will not recognize or accept any kind of truth that pretends to be ultimate or absolute. I will consider and weigh all claims as provisional conclusions. If examination shows them to be stop signs on the road of inquiry, I will ignore them; if they are signposts, I will note them and move on.

"I recognize no subjects and no facts which are alleged to be forever closed to inquiry or understanding; a mystery is but a challenge.

"I believe that new principles of understanding are constantly created through the efforts of man, that a philosophy which sees the answers to all questions already implied in what is now called science is presumptuous and contrary to the spirit of science.

"I believe in the convergence of the scientific laws upon principles that are all embracive, though they may never be completely within our reach.

"And finally, it should be added that scientific illumination of human experience is not confined to the area of inquiry about nature. As later chapters of the book may suggest, scientific understanding can be made to penetrate, not only the dealings of man with nature, but likewise the dealings of man with man."

Questions/Discussions

1. In Chapters 2 and 14 we presented the general premises of the theory of uniformitarianism. Discuss the strength and weakness of that theory in the light of the discussions of this chapter.

2. Repeat the analysis of Question 1 for the theory of evolution of organisms.

3. Suppose that you have discovered in a laboratory experiment that a certain chemical substance kills bacteria of a serious human disease, and you market this chemical without delay for use on human beings. Suppose also that use of the drug restores some patients to health but kills other patients.
  Should you be tried for homicide?
  Explain the strong points and the weaknesses of your position which you might present in court.
  Can you claim the drug was fairly tested?
  What should have been done to determine the effects of age, medical history, hereditary background, and other influences?
  What about experimentation on quantities, method of administering the drug, and similar factors?
  What about analysis of the chemical properties, impurities, and other characters of the drug?
  What about the need for adequate knowledge of the progress of the disease in the patients, with or without administering the drug?
  How would you establish a control group for purposes of comparison?

4. Having in mind the questions and difficulties noted in connection with Question 3, trace the steps you would pursue for the development of the drug.

5. Two approaches have been proposed for determining the governing forces of the world. In the one, the exponents of *vitalism* expect that it is necessary to include the influence of something special or "vital forces" (which some may call god or spirit or the supernatural) to explain the workings of nature. Exponents of the alternative approach, called *mechanism*, claim that "natural laws," such as those of physics and chemistry, will ultimately explain all phenomena of the universe. In view of the difficulty in confirming all laws for the past and the future, and the difficulty of attaining absolute truth, what might be a proper "scientific attitude" toward the two concepts?

6. Discuss what might constitute controlled experiments for testing the *vitalistic* point of view and for testing the *mechanistic* point of view.

7. With the *mechanistic* point of view, design a controlled experiment that might test: (a) the usefulness of a drug, (b) the theory of gravitational attraction, and (c) the concept of entropy.

8. When a nucleus emits an electron (β-decay), the electron does not gain as much energy as the nucleus loses. To save the principle of conservation of energy, Pauli advanced, in 1931, the hypothesis that along with the electron a second invisible particle is emitted, which carries away the excess energy. This hypothetical particle was called the *neutrino*. Physicists were reluctant to accept the reality of the neutrino until it manifested itself in other nuclear phenomena, leaving its trace, for example, in the recoil momentum of decaying nuclei. Explain this initial reluctance by reference to the principles in Sec. 22.4.

9. All objects in our known universe have velocity smaller than that of light (c). It has been conjectured that a universe in which all velocities are greater than c might exist, but if it did, there could be no physical interaction between it and our actual universe. Can the hypothesis that a superluminary universe exists be ruled out by means of the principles that constitute the method of science?

10. A person claims to read someone else's mind. The situation involves two individuals; the mind reader and the person whose mind is being read. The experiment appears to be successful. Suppose the two individuals placed themselves at your disposal and allowed you to take steps in order to verify in rigorous scientific fashion the reality of telepathic communication; what steps would you take?

11. Give an operational definition of the following terms: mass, force, temperature, entropy, probability, feedback. Are there also operational definitions for the concepts one encounters in the behavioral sciences; for example, supply of a commodity, demand for a commodity, preference between two political candidates, intelligence?

12. How would you distinguish between a postulate, a scientific principle, a scientific law, a hypothesis, and a mere working formula? (*Note:* There is no clear unanimity among scientists in the use of these terms.) State what you think the meaning of these terms ought to be.

13. Is probability a reliable scientific quantity? How would you establish the probability that a good die will fall with a two on its top surface?

14. At the end of the nineteenth century, many scientists believed that the universe was moving toward a state of stagnation called "heat death." In this state all physical objects in the universe are at the same temperature, so that no useful work could possibly be done. The entropy content of the universe would have reached a maximum. It was believed that the heat death was inescapable, in view of the second law of thermodynamics. Comment on the cogency of this conclusion. (Remember that the second law of thermodynamics is a probability statement and that it holds only for closed systems.)

References

Bacon, Francis, *Novum Organum*, edited by T. Fowler. New York: Oxford Univ. Press, 1878 edition.

Beveridge, W. I., *The Art of Scientific Investigation*. New York: Random House (Vintage Book V-129).

Bronowski, J., *The Common Sense of Science*. New York: Random House (Vintage Book).

Hanson, N. R., *Patterns of Discovery*. Cambridge: Cambridge Univ. Press, 1965.

Margenau, Henry, *Open Vistas*. New Haven: Yale Univ. Press, 1961. (Available in paperback.)

Schlegel, Richard, *Time and the Physical World*. East Lansing: Michigan State Univ. Press, 1961.

Von Fieandt, K., *The World of Perception*. Homewood, Ill.: Dorsey Press, 1966.

Weisz, Paul B., *The Science of Biology*, 2d ed. New York: McGraw-Hill, 1963.

CHAPTER TWENTY-THREE

Science and the Progress of Man

Concern for man himself and his fate must always form the chief interest of all technical endeavors. . . . Never forget this in the midst of your diagrams and equations.

ALBERT EINSTEIN†

IT IS TIMELY to assess the significance of science for the progress of man. We have dwelt at some length on the history of science, from its uncertain beginnings in the Neolithic Age and the early civilizations, through its maturing into the empiricism of the centuries after Roger Bacon and Galileo, to the present-day scientific methods that utilize both deduction and induction. The intervening centuries have witnessed vast changes in the *content* and *methodology* of what is called science, and have brought equally large changes in social patterns that are directly attributable to the progress of science and to its by-product in technology.

23.1 The panorama before us

We are in possession of much more information about natural phenomena and about the relationship of man to nature than was ever dreamed of in the days of the Sumerians, the Egyptians, the Babylonians, and the Greeks. We also seem to know more about those early civilizations, in historical perspective, than was available to their own scholars. There is an interesting contrast in the *attitudes* of the two periods with respect to the progress of man, however: Some early Sumerian clay tablets reported that "in the beginning" mankind had enjoyed a Golden Age of universal peace, a world of plenty without fear of want or fear of animals. But the world, man, and society had *degenerated.* This view was dominant among the Greeks and persisted in some theological form even into the seventeenth century. But the "Philosophés" of The Age of Reason saw things differently: Man has not degenerated but

† Quoted in Robert S. Lynd, *Knowledge for What?* Princeton, N.J.: Princeton Univ. Press, 1939.

is improving and advancing toward a Golden Age.

When one regards the very great and rapid changes that have come to our way of life in the form of goods, services, and community activities that derive directly or indirectly from the fruits and by-products of science, it is easy to think of science as the great force that may lead toward that Golden Age. The *rate* of change has increased to the point that the spirit of the times calls for change and is impatient with whatever is old. This emphasis on *Becoming* contrasts strongly with the Greek search for *Being*. Many people view contemporary man's activities to be a sophisticated approach that is capable of systematically revealing the basic laws and the order that underlie the surging flux of events. There is the impression that we now have the *Science of Science* by which to do this. There are others, however, who fear that the pursuits of modern man are not so orderly; that they reflect the frenzy of change for the sake of change more than they reflect plan and purpose. They fear, moreover, the rapid growth of science in the form of vast amounts of data, very complex instrumental techniques, and well nigh esoterically abstract theories. They point to our overpowering technological environment that now encompasses man and science, and see an uncertain future because of the absence of comparable progress in the study of man's purpose in nature and of nature's requirements for man. It is interesting that it has been in the period of the greatest achievements of science that there has developed the most intense reservations with respect to the purposes and even the foundations of science. At any rate, science and its by-products are no longer being generally regarded as the panacea for mankind.

But while we may wonder at the marvels of scientific progress, we cannot but be aware that *the past continues to live side by side with the present.* Remnants of each stage of the development of man and of his society, almost every form of religious practice, social tradition, agriculture, and community living that have existed since Neolithic days, seem to persist in some corner of the globe along with the most modern gadgets. While some societies are devising electronic and photographic methods to relieve the heavy load on libraries that have become overcrowded with books, other societies have yet to discover writing. Along with the concepts that give rise to space projects and the search for the outer limits of space, there still persist all the fears of natural phenomena that ever troubled primitive man. The evidences of the history of man and of nature are within this panorama for us to view, to study, and through which to develop the perspective of history. Along with study of the impact of science on the society of the more technologically advanced countries we are permitted to make comparison with societies that have been barely touched by technology. There are, of course, all the intermediate categories as well.† Truly, one may wonder whether any age has been more blessed with such abundant research source material for the study of society and of man, and may also ponder to what extent we are taking advantage of these opportunities.

While we can easily see *differences* between societies, it is far less easy to achieve a *perspective with respect to what constitutes progress.* This would not be difficult to do if there were universally accepted criteria by which to evaluate the progress of society; for ex-

---

† The reader may recall similarities in this situation with the premises of the theory of uniformitarianism, which was presented in Chapter 2.

ample, in terms of kilowatt hours of energy utilization per individual, or of number of books or newspapers published, or in terms of gadgetry. These "measures" are important and should not be underestimated. Nevertheless, when discussing the progress of science and technology, and the progress of man, care must be taken not to confuse the two or to fuse one into the other.

For example, if we call it "progress" to use improved tools and agriculture so that a society no longer lacks food, what can one say when further "progress" in agriculture produces gluttony and waste of food, especially while many in the same nation and in other nations suffer from insufficient food? If we say that both science and society progress when motor power relieves man and beast of heavy tasks and makes extensive transportation possible, what can we say when it is pointed out that the automobile has become an instrument of destruction, a major source of air pollution as well as a prodigal waster of precious and limited oil resources? It is not easy to ridicule those who say that science has built machine behemoths that can destroy the men and the society that created them.

All this points to the need for better definitions and guide lines than we now have for evaluating the progress of society. There is little agreement among people about either definitions or guide lines. There are those who would glorify gadgets, whose conception of progress is quite different from those of people who talk with equal vigor (and often with equal narrowness of understanding) about their ideals, using such phrases as Liberty, Fraternity, Equality, Freedom, Democracy.†

The emphasis of this volume has been on the history and progress of science rather than on the history of man because this is a text on natural science. Progress in science has been viewed in terms of the theories and problems, but not in terms of human objectives or values. Attempt has been made, however, to develop the story within the context of the men and societies that produced the science. We now turn to the question of what has been the influence (feedback) of science on man and on society. We shall explore the question within each of several categories of human interests and activities, with the realization that the exploration will reveal more questions than answers.

23.2 Science and technology in national affairs

In Sec. 23.1 we noted that there is great diversity of societies and culture among the peoples that populate the earth. There are vast differences even among the nations that meet regularly as members of the United Nations.† These differences go beyond those of geography, climate, and natural resources, but include among other things the differences in attitude toward social values, government, economics, religion, and culture. The differences exist even though each nation has access to the useful knowledge possessed by any other nation. Each nation is likely to defend its own way of life with respect to its values, religion, and national customs, and we certainly do not imply that they should do otherwise. There has tended to be general agreement, however, that there is far too

---

† An entertaining example of the difficulty of evaluating human culture is given in *Are We Civilized?* by R. H. Lowie.

† It should be noted that of the roughly 110 member nations of the United Nations, about a third were colonies that gained independence in the period since 1945.

much disparity in the economic levels of the nations in terms of such factors as individual income, food supply, and standard of living (where the "standard" has to do with health needs and necessities of life). For example, in the mid-1960's the average income of an individual in nations that were termed undeveloped (or "have-not") was *between $70 and $110* a year, against an average of about $1700 for Europe and $2300 for the United States, the latter groups being referred to as the more developed (or "have") nations.

This disparity, which is deplored by the "have" as well as the "have-not" nations, is generally attributed to the low status (or absence) of science and technology in the "have-not" nations. Large sums of money and much effort have, therefore, been given (especially by the United States, the United Nations, and the Soviet Union) to improve this situation by providing education, especially in science and technology, to the regions that lack it. The hope has been that this will almost automatically improve the economic level of the region. At this writing, however, the results have been much less than satisfactory. It also appears that the economic disparity has tended to increase rather than decrease because the advanced nations have a higher rate of technological growth than do the less advanced nations. The reasons are many. Much of the "aid" given to the undeveloped countries has been in the form of military equipment. Much of it has been used by privileged individuals and vested interests rather than being used to establish a firmer base for ordinary industrial and agricultural production. There has also been waste of funds and effort because of a lack of clear understanding of what was best to do. Confusion and misdirection have often resulted because the nature of the "help"

given by the United States and the U.S.S.R. tended to be guided more by the objectives of their political competition than by the needs of the countries that were being "helped." In some areas the progress in agriculture has been nullified by a greater increase in population.

It appears that the progress of science and technology, and therefore of economic productivity, is intertwined with the values and cultural features of a people and with the philosophy of their government. The industrial development of Europe required three centuries to reach its present state. That of the United States began in the early 1700's, although the development of American science has largely taken place during the present century. Under the pressures of the Communist revolution, science and technology in the U.S.S.R. developed rapidly and became competitive with those of western Europe and of the United States within a half-century. The technological development of China, which began later, may catch up within even fewer years. Most other underdeveloped nations have had less success in changing their status, however, the reasons being, among others, the values and cultural preferences of their societies. It cannot be said that the undeveloped nations are completely enamored with the way of life that characterizes the "advanced" nations. To some rulers of these nations, the ideas of a democratic form of government, which call for some effort to satisfy the needs of the people in order to be elected to office, is far from attractive. Some people of these countries fear the spread of irreligion and of "materialism," and the substitution of technological efficiency for their traditional way of life. At any rate, there are many questions that these nations ask; for example:

How can a society incorporate the new scientific technology that produces more goods and food for the nation, without destroying the dominant values of its cultural pattern?

How can the technology that brings such great material benefits be kept subservient to, and influenced by, the cultural pattern of a people rather than reducing the culture to a mere by-product of technology?

The transition from old customs to the new technology is taking place in these nations at a time of history wherein the political and social climates are quite different from those of the eighteenth and nineteenth centuries. The industrial revolution of England and particularly of the United States of America occurred at a time when free enterprise and private initiative were permitted full reign, whereas the political climate of the past several decades has tended to favor socialism and considerable governmental direction of the economy and the activities of citizens.† It is interesting that although in the United States of America there has been a strong trend toward government direction of the nation's economy, individual Americans nevertheless are ready to change anything that does not measure up to specifications or that can be exchanged for something more efficient, more convenient, and newer. In contrast, some other peoples find their traditional way of living to be desirable to them, to the extent that they are less prepared to sacrifice their ways to the comforts of technological change. Only time will reveal which of these attitudes is better for mankind, in view of limited energy and other resources of nature. Meanwhile the United States continues to be the foremost industrial nation on earth because tradition is not valued above technological progress.

† See References for the analysis of P.M.S. Blackett and a reply by V. L. Parsegian.

It is unfortunate that there has not yet been revealed a pattern of economic and technological development that can be taken on by a "developing" society without complete loss of its own character in favor of the "European" type of society. Each "successful" adoption of new technology seems to sacrifice some of the individuality of a society's culture, and to that extent to reduce the richness of variety in the world's cultures.

At any rate, it is generally recognized that science and technology provide the basis for the economic progress of nations and for the maintenance of a high standard of living. They also provide the modern war machine with nuclear bombs, germ warfare, poison gas, napalm, airplanes, rockets, and endless electronic and mechanical gear for pursuing defensive and offensive warfare. The importance of this relationship is reflected in the national budgets of the advanced nations. For example, the United States expenditure for research and development in the mid-1960's was close to 30 billion dollars, of which about 16 billion dollars was spent through the federal government budget,† the latter being roughly 100 billion dollars at the time. Of the 16 billion dollars, from 8 to 10 percent was devoted to basic research and the remainder to applied research. (*Basic research* may be said to include studies that have "knowledge for its own sake" as its only objective, without its application in mind, while *applied research* is concerned with solving specific problems or with finding applications for new phenomena.) Of the federal government's expenditures for research and

† The 1939 government budget for research and development had been only about 100 million dollars. The 16 billion dollars represents about 3 percent of the gross national product of the mid-1960's.

development, about 31 percent was for the life sciences (biological, agricultural, medical), 64 percent for the physical sciences and engineering, and only 5 percent for the social sciences. Of the total, roughly 70 percent was for defense and space activities. It has been estimated that activity in the area of research and development doubles every 15 years in the technologically advanced nations. (*Question:* What is the present doubling time for population increase?) The most tangible evidence of this rate of increase (which seems to have some exponential features) is related to the number of scientific societies and journals of the world. At the turn of the nineteenth century these numbered about 100. By 1850 the number was 1000, and over 100,000 in 1950. It has often been said that because of this rapid rate of increase, between 80 to 90 percent of all scientists who have lived since the beginning of history are alive today!

How are research and development funds spent? Because national and international expenditures for science and technology represent such large fractions of a nation's resources, the question has become a continuing issue in the halls of government. What portion of the expenditures should be for basic research? For applied research? What types of laboratories are best suited for each? In the United States the work is conducted in four categories of organizations. These are university laboratories, industrial laboratories, government laboratories, and those that are generally called "not-for-profit" laboratories. Research activities of the "basic" variety have been conducted at universities for several centuries. During the early decades of this century, especially as a consequence of World War I (1914–1918), there developed a need for a few government laboratories to work on military problems and to expand the work of the National Bureau of Standards. Industry also found it necessary to organize research and development divisions to support production activities and to devise new products.

With the coming of World War II and of the events that followed that war, all four categories of organizations increased in number and in size. The atomic bomb program gave rise to many large government laboratories, nearly all of which continue to this day under one name or another. With the birth of the "space age" came many additional government laboratories and additions to the list of not-for-profit laboratories. Colleges and universities were required to greatly expand their enrollment as well as research activities that could provide experience to potential scientists and engineers. Industrial firms had for some time realized that only through adequate research and development activities could they hope to compete with one another and with government and other laboratories.

What kinds of activities engage these laboratories? A large portion of the income and effort for all these organizations is administered (and often directed) through government agencies such as the Department of Defense (DOD), the Atomic Energy Commission (AEC), the National Aeronautical and Space Administration (NASA), the National Institute of Health (NIH), and the National Science Foundation (NSF). The Department of Defense budget alone accounts for roughly half of the total federal budget. The larger portion of the budgets of the Atomic Energy Commission is devoted to military aspects. As noted earlier, about 70 percent of all government expenditures for research and development have military or space interests in mind. The expenditures of civilian and military applications cannot be easily identified or segregated.† It is fairly easy to see that the lion's share of all federal expenditures for research and development are geared to military needs and interests. The situation is comparable in the U.S.S.R. and in most other "advanced" nations.

Nor are laboratory activities confined to federal organizations. The problems attending safe use in laboratories of radiation-producing equipment, industrial health control, waste disposal, water pollution, and similar hazards all require that states and communities establish many laboratories and sufficient inspection and licensing divisions to maintain proper control that will assure the safety of communities.

Science and research, therefore, have become very important aspects of the life of the most "advanced" or "progressive" nations. For that reason they become distressing problems to every Congress of the United States of America; and because of taxes, to every citizen.

23.3 The contributions of science and of technology

What aspects of science and technology are the most influential within the life of a modern society? For an answer to this question we might note first the distinctions between basic science and technology. The two are often difficult to separate clearly because they merge and overlap to a considerable extent. *Basic science* is considered to be knowledge

---

† For example, how much of the approximately 5 billion dollars of the yearly budget of NASA can be said to relate to civilian versus military interests?

than need not be associated with any particular field of application. Basic research is said to be a scientific search for knowledge for its own sake without having in mind any use or application for what may come out of the research. It is not uncommon for scientists to resent the intrusion of the question of "practicality" or "usefulness" of what they are doing in the realm of "pure research." *Technology*, on the other hand, is the application of knowledge or science for solving some problem, or to find some new application for the knowledge. The terms *engineering* and *applied science* are often used interchangeably with technology. The engineering professions are almost entirely devoted to the application of science and of technology to various uses. Many of the members of disciplines such as physics, chemistry, biology, geology, astronomy, and mathematics are also more likely to be engaged in applied work rather than basic research, perhaps because of the monetary rewards and social acclaim that await them there.

The distinctions between basic research and technology have changed substantially over the past two centuries. After Faraday and others revealed some of the basic facets of electricity, some decades elapsed before the ideas were put to practical use in the form of the electric motor and generator, the telegraph and telephone, the electric light, and radio broadcasting. In the nineteenth century, scientific research tended to be separated from technology. Nevertheless as time went on, the time interval between a discovery and its application in technology also tended to become reduced. This was true in the case of the discovery or invention of the vacuum tube and its wide use in radios and amplifiers. The more recent discovery of the transistor and its extensive use in electronic circuitry and in computers all took place within one decade. As noted in Chapter 21, the discovery of fission and its application in the critical reactor and in the explosion of the first bomb required little more than half a decade. These examples illustrate how the application of new knowledge now follows hard on the discovery of that knowledge, and for that reason how important the pursuit of science has become to our way of life. Even when a discovery passes too quickly into use, as sometimes happens in the discovery of a new drug when it causes harm as a result of inadequate testing, the solution lies in doing *more* research rather than less.

If we were to examine recent interchanges between science and technology, there would be revealed the converse, namely, that basic science and basic research techniques depend heavily on the progress of technology. The discoverers of the transistor, while doing basic research, were expert in their use of electronic devices and in the techniques for producing crystals. The discovery and utilization of nuclear fission required equipment that could be produced only by an advanced technology. The huge accelerators that today probe the innermost secrets of the atom are made possible only because they are the product of fine engineering design, involving the most complex calculations, special materials and techniques, and construction methods.

In short, technology not only "feeds on" basic science; it also "feeds back" new equipment and techniques that make new science possible. It is this close relationship that has given the "have" nations the ability to increase their *rate* of progress, leaving the "have-nots" farther behind.

One noteworthy example of the close interrelationship of research and technological progress is found in the computer. A modern electronic computer is able to calculate exceedingly fast while including a very large number of variables in the process. The cost for such computers may reach a million dollars; their speed and capability are such, however, that for the equivalent of one dollar of their operating cost, problems can be solved which if performed by human beings and desk calculators could cost tens of thousands of dollars.† Such computers now perform innumerable types of service. They maintain inventory control for many industries. They have been used successfully to plan the economy of nations with respect to agricultural versus other needs. They render incomparable service for computing the complex relationships of the large organic molecules; they systematically organize and analyze the data produced from the accelerators, and contribute enormously to the development of new materials and techniques, which in turn will produce the more powerful computer techniques of the future. The feedback and systems relationship go in many directions. Along with this computing advantage, we must note that the "mechanical" operations of a computer offer little creativeness except as the one who feeds data into the computer can exercise creativeness. Technology alone offers little that can substitute for human judgment and creativity.

23.4 Problems created by advancing technology

There are not only gains for science and society from advancing technology, but

† *Question:* What are the necessary conditions that justify such a large financial investment in a large computer?

problems as well, and these become more severe with each new decade. Some of these problems can be easily identified; for example, those that attend limitations in natural oil and other fossil fuel resources, the spoilage of natural waters by industrial and community wastes, the growing toxicity of city atmospheres, city crowding, and transportation. There are also the hazards and threats to the survival of civilization itself if a nuclear war should be unleashed.

There are subtler aspects as well to the problems, which are less easy to identify because they depend on human reactions that do not lend themselves to easy analysis. It will be useful to note some of these, without attempting to give a complete analysis of them.

Foremost are the problems of government. It seems quite clear that the magnitude and complexity of such activities as space engineering, atomic power development, major expansion of laboratory capabilities of nations, defense and military preparedness, transportation, all require strong support from, if not complete domination by, federal governments. For these reasons the managerial functions within most nations have moved toward more and more detailed and centralized direction from the federal government. This, coupled with the trend to direct the lion's share of research expenditures toward military interests, threatens to condition the mentality and pursuits of scientists and engineers and make them more and more in tune with military interests. Since each government agency must be prepared to justify before Congress its reasons for selecting the scientific research programs it chooses to support, such justification often requires identification of the research with the primary mission of the agency (which is usually of a military nature or which requires awareness of military interests). Under these conditions, there is the risk of concentrating the scientific resources of advanced nations primarily on military goals and of neglecting the society's welfare. This trend is enhanced when programs are undertaken on a *competitive* basis, with each nation in military competition with all other nations. The resources of the science and scientists of advanced nations (and even of the advancing nations) are therefore too often being developed into an arsenal that can be used in the event they become pitted against each other in a life and death struggle. It seems likely that when military interests dominate research to that extent, there is less emphasis on scientific endeavor as a search for truth.

A contributing problem is that, within the population of every advanced nation, there exist extremist elements who hold widely divergent ideas with respect to such matters as religion, political and social structure, ethnic and racial relationships, and economic theories. They live within technological communities that are centers of all the capabilities and hazards involved in modern chemical processes, nuclear bombs, fast transportation, and "fire" power. As a result, each community and nation has within itself the capabilities for self-destruction as as well as for progress, and under some circumstances the danger of self-destruction can be accentuated by the activity of the extremist elements.

Undesirable influences from the technological changes may affect the mental state of individuals. It is no longer possible to lead a comfortable life without involvement in the fast pace of learning, growing, and adjusting that is required by modern living conditions. How these affect the nervous system and the attitude of a race of people only time can reveal. There is speculation that the very rapid changes in social habits that are brought about by changing technology tend to produce anxiety and alienation, a feeling of "normlessness" and of anomie among people. It has become less easy to take roots or to feel the stability of firm footing.† If this is true, and if it does indeed conflict with the tendency of man to be a creature of custom and tradition, there may be developing a new force toward modifying some of the characteristics of society and possibly of man himself.

There is still another source of concern, which some scientists feel, having to do with questions of their personal responsibility for the proper use of their discoveries. What is the responsibility of a scientist toward the proper or constructive use of his work? The question has troubled some of the leading scientists, especially those who have been involved in the development of nuclear bombs and other militarily important material.‡

23.5 Science and human values

The view of the significance of science as presented in the earlier sections of this chapter are likely to be disquieting and even upsetting. The disquietude is unavoidable if one has a picture of science as being an impersonal, precise, and objective probing of nature by an inhuman machine instead of by thinking

† Based on certain correlations they observe, some ecological sociologists speculate that schizophrenia is often present within social structures that are in rapid transition and which for that reason do not allow individuals to find root within an environment.
‡ The journal called *Bulletin of the Atomic Scientists* has been a very responsible medium for the discussion of issues such as these. The reader is urged to make use of this publication.

and feeling human beings. Some view science as concerned with physical nature and with man only insofar as physics and chemistry govern his biological processes. This view would relegate all aspects relating to the mind and consciousness of man as being outside the realm of science and therefore outside the scope of scientific interests except as they can be reduced to explanatory terms of chemical or physical processes.

The centuries have brought many changes and many efforts to prevent science from being the esoteric domain of a privileged few. More recently, mass education, reduced illiteracy, and daily involvement with the technological products of science have tended to invite greater interest in the relationship of science to other interests. There were major steps in this direction from the time of the priests of Sumeria to the philosophers of Greece, and then to the noblemen and their protégés of the Middle Ages. The dramatic advances that came with the development of the modern empirical approach and with Newtonian mechanics provided basis for even closer relationship. Those who were concerned with ethics, philosophy, and even theology sought to explain human behavior and beliefs in terms of the ideas of science, although not always successfully. For example, the physiocrats and some political economists used Newtonian mechanics to justify the Laissez-Faire economic and political doctrines which they proposed. Sometimes a concept would be used both to justify and to attack a religious idea; for example, Newton developed a *natural theology*, which did not conflict in his mind with the science of mechanics, while those with a materialistic attitude found the science of mechanics to be contradictory to religion and to traditional beliefs.

The dependence of science on the empirical approach did not lead to a sense of humility with respect to its limitations. Rather, there developed the conviction in many quarters that whatever could not be studied and confirmed experimentally was of little consequence. The attitude was common in the seventeenth to the nineteenth centuries, when it was assumed that science had the answers to all social and personal problems, and that man had to be explainable entirely in terms of machine concepts. On this basis, concepts relating to *mind, conscience,* and *values* represented unscientific notions that were better left with the poet and novelist. Others who did study them tried to do it "scientifically," to formulate and to explain these concepts in terms of the laws of physics (Hobbes, Condillac, Bentham, for example). The experimental approach had brought such success and prestige to physical science as to make its methods exemplary. But objections were raised by those who saw man as a being who is much more than a mechanism. Some of them (Dilthey) went so far as to insist that the scientific piece-by-piece analytic approach of the natural sciences could not be used to study human and cultural phenomena. Thus there arose a philosophical movement which claimed that man and society require quite different methods than those used in the natural sciences. The qualities we associate with man and his society are qualities of *wholeness* that represent the integration of many characteristics, while the empirical, analytic approach to social and human problems can only "destroy" this quality of wholeness. (Smuts, Goldstein, and von Berthalanfy were among those who pointed out these disadvantages of the analytic approach.)

In this connection it is of interest to refer back to the illustration of the rider and the bicycle of Chapter 1. It was observed that when an assembly (a system, if you will) of parts is designed to perform a function, the resulting function is often wholly new and not likely to be in any sense an extension of the characteristics of the parts themselves. The uniqueness of riding the bicycle is not at all a natural consequence of anything contained in the individual characteristics of the metal or rubber or in man himself. As someone has noted, the most extensive study of the approximately 20 amino acids that make up the human body will not reveal the functions of which the body is capable.

If, then, organization brings about new features that cannot be often guessed at or described by a study of the parts of the organization, does it mean that man cannot comprehend the totality of the nature of which he is a part, so long as his methods of study are suited only for the study of the parts of the system? And if this is so, does this raise an insurmountable barrier to the study and the understanding of mental processes and consciousness? The answer probably is that while some barriers do exist beyond which we cannot at present go (as we saw to exist even in physics), there is not complete exclusion from understanding. According to this view, the progress of science is not helped by denial of the existence or importance of aspects of life that are part of human experience but which at present may not be measured or analyzed in detail. There was a time, not too many centuries ago, when even the inanimate was quite out of reach of science. Indeed, even today, the laws of physics and chemistry are far from able to explain the basic phenomena of the physical world. In contrast with the boldness of the beliefs of the early mechanicists, the present-day physicist is likely to feel considerable humility with respect to what he knows or does not know of the fundamental phenomena of the physical world. Nor is there ground for less humility when one acknowledges that there are many levels of greater complexity as one proceeds to build up from the inanimate to the animate and up to the level of human reason that makes use of abstract concepts and symbols.

It is of interest to note some of the so-called *subjective* and immeasurable phenomena of human behavior which are at present being studied with some success:

(1) Until about 75 years ago the field of *learning* was considered to be too personal and subjective to be ever reduced to experimentation and mathematical analysis, while today some aspects of it are being studied experimentally.

(2) The field of *emotions* was also considered to fall in the realm of speculative philosophy, but today physiological psychology has helped to unravel some of the neurological mechanisms that produce and change the emotional responsivity of people and animals.

Such research gives assurance that a "breakthrough" of method will come to help us to understand, predict, and perhaps control many more of man's behavioral characteristics.

It appears, therefore, that while the methods and approach of present-day science do not permit it to encompass the thinking processes that go on within us, it is shortsighted (if not absurd) to deny the significance to natural phenomena of the intensity of the emotions of love, hate, trust, fear, and hope that bear so heavily on us. This is turn implies that the scientific process must allow room for human judgment, human likes and dislikes, and for a sense of values with respect to science itself and its role in society. This is not contrary to the statements given in this volume on the human aspect of science and the scientific effort. There is a judgment factor, an ethical aspect, and evaluation of values to which the scientific effort has always been subject and which modern science must not deny.† Perhaps the disquietude to which we referred at the beginning of this section would be greatly reduced if there could be acceptance of the idea in science, and among scientists generally, that science invokes judgment, ethics, and consideration for the rights of a society of human beings. This is important for their own purposes as well as when they promote understanding of the less anthropomorphic features of scientific progress.

Indeed science shares a number of common facets with other social and human endeavors. The desire to know, to learn, to understand is common to them all. Each activity begins with a curiosity, followed by a questioning that may lead to the development of new skills, arts, and crafts. Each activity gives nurture to aesthetic, intellectual, and emotional appreciation, which is as intense in the poet, the novelist, the philosopher, or the theologian as ever it is in the scientist. Each activity is concerned as much with what is invisible and out of reach as it is with tangibles that can be touched and measured. Each is on occasion concerned and fearful of its premises, assumptions, methods, and conclusions. In short, each is a *human* endeavor with all that the name implies. It is only when one becomes enamored by the successes of the moment, or discouraged by the failures of the moment, that it becomes difficult

† See Michael Polanyi in References.

to maintain perspective and to recognize how each human effort complements the others within the total human endeavor.

A discussion of values comes dangerously close to requiring some thoughts on purpose in the teleological sense. It would be wrong to close this volume without at least taking cognizance of the turmoil that has often evolved at this boundary of the scientific effort.

23.6 The impact of science on religion and philosophy

The progress of science has from time to time brought considerable pressure to bear on the way of life and thought of a people, pressures to change their technology, their economic system, and their social life as well. These particular influences do not often stir up as much long-term difficulty or conflict as do pressures on religious thought, however. For one thing, any inconvenience that science may cause labor, industry, or government is often counteracted by the new products or services that usually come with change. In contrast, because theology and religious practices often spring from a combination of high authority and deep individual convictions, and because religion tends to hold to tradition, an attack or even the appearance of contradiction of religion by scientific discovery usually invites strong repercussions that do not quickly disappear.

Religious beliefs and practices have been among the most strongly felt of all human experiences from time immemorial. An inclination toward religious thought does not spring merely from fear of the unknown. It springs as easily from a sense of happiness, awe, beauty, or gratefulness. Both science and theology seek to understand the nature of physical and social reality and the relationship of man to that reality. Both probe as deeply into the invisible as into the visible cosmos to find that reality and that relationship. Some individuals have pointed out a number of similarities between science and religion. Both strive to develop a logical base and system of ideas to elucidate their tenets and concepts. Despite these common elements, the methods they use for defining their premises and concepts tend to channel each along its own way. Each may become guilty of an "imperialistic" attitude that on occasion tries to deny the existence of the other, to the point that the complementarity of their interests becomes lost in the shuffle. For this reason the concepts and ideas that might normally constitute boundary regions between religion and science more often tend to become walls that isolate them.

There were conflicts of this nature in ancient Greece, when Anaxamenes preached the concept of an Eternal Unknowable as he attacked the anthropomorphic gods of the Greeks. Epicurus and Lucretius, in particular, made an open attack on religion as superstition. The Age of Reason and of Enlightenment brought even severer attacks on religion, for which it is claimed the humanistic paganism of the Renaissance had prepared the way.

One aspect of the attacks on religion is often overlooked: Any attack on religion by a scientist, or attack on science by a theologian, is too often treated as though *all* scientists or *all* theologians participate in the attack. The fact is that there exist a great variety of "heretical" opinions among scientists and among theologians. Opinions regarding scientific concepts are not confined to scientists; nor are theologians the only ones who are committed to theological concepts. It appears that nearly all the great ideas, whether derived from Aristotle, Newton's mechanics, or Darwin's theory of evolution, have been used in one form or another either to inspire attacks on religion or to support certain religious beliefs. Often the attacks were not so much on the central beliefs of a religion as they were on specific concepts that were derived from outworn scientific ideas or social practices, which religion had incorporated within its system of institutionalized beliefs and rituals. An example of this was Galileo's attack, which appeared to be an attack on church dogma only because Aristotelian physics had been incorporated in that dogma.†

The Age of Reason aimed to do away with old beliefs and traditions and to establish a world based on the fruits of science, since science seemed to have an answer to all problems of man. It was believed that man's reason, if allowed to operate freely, could discern the "laws of nature"; and since man was a part of nature, the laws would apply to him as well. It has been necessary, however, for scientists to give up the reliance on the simple mechanistic laws of the Age of Reason because the laws of nature turned out to be neither simple nor mechanistic. The negative attitude of the Age of Enlightenment toward religion and tradition nevertheless persists, and science is still envisioned as made up of deterministic laws that govern both nature and man. There has come into being a new age of faith that derives from constant use of gadgets and ideas drawn from science. Each of us, whether housewife, laborer, or office executive, is surrounded by electronic or mechanical gadgets that give tribute to man's "scientific" ingenuity for conceiving and producing machines that do the work of man and almost think for man. Moreover, it seems that man penetrates to the "essence" of life when the biologist and the medical research scientists apply electronic instruments to the human body and the brain, for the nerves and the muscles produce electric signals that are similar to those that can be generated within the instruments themselves. Although to the scientists these methods and signals represent only a beginning for the study of man, in the minds of others they assume larger importance. It becomes "common sense" to see man and nature as having the features of material mechanisms that can be understood. It becomes easy for the large questions that face science to become fused and "lost" within technology, and *technology* becomes the ever present framework within which religious beliefs and institutions must function. It has not been easy for traditional religious symbols, customs, and beliefs to function or to find a *modus vivendi* or comfortable integration within this materialistic framework with a sense of perspective.

Religions that are based on theology† are therefore handicapped by being a "nonscience" activity of man, in a world in which science is glorified. The rapid progress of the technological sciences and their economic and military counterparts has preempted the interest of people and of competing governments to the disadvantage of other human interests, which have become peripheral features of our contemporary world.

† *Question:* It has been said that the physics of yesteryear is the metaphysics and theology of today. Do you agree?

† There are religions that are not based on theology and in fact do not have either a formal doctrine or concern with developing a doctrine of the nature of God.

Religious institutions are seeking some conceptual and functional niche that is compatible both with the scientific and technological interests of modern life as well as with the socioeconomic transformations that are taking place. The search for a *modus vivendi* with technology is reminiscent of the attempts by leaders of the past to reconstruct religious concepts in terms of new philosophical orientation. Among them were Philo in ancient times, Maimonides and St. Thomas in the Middle Ages, and Newton in a more recent century.

It has been proposed by some that science be entrusted with those aspects of nature and of man that can be "seen and heard and measured," while theology may concern itself with the "unknown" or "unseeable." The division becomes rather arbitrary and difficult, however. To begin with, science finds itself more and more involved with the features of nature and of man that cannot be seen or defined. The most basic aspects of the physical world (for example, gravitational attraction, electric fields, entropy, and probabilistic features of nature) are not seen or understood except for their effects. The medical doctor knows how important the thought processes and attitudes of a patient are for determining the state of his health and recovery from illness. Nor is it possible for the theologian to delve into the spiritual aspects of man while disregarding the physical and socioeconomic environment that bears on the man's existence. Einstein and Barbour have pointed out the necessity to disregard such division of spheres of influence; to recognize that to a certain extent both science and religion seek the *Eternal Unknowable* and to that extent they play complementary roles in the life of man.

There have been somewhat similar rivalries affecting the progress of science and philosophy. In fact many of the earlier topics that have been presented in this volume were once topics of philosophical thought. Philosophy has often sought to develop new systems of thought by utilizing the facts and theories of science. The philosophies of the eighteenth and nineteenth centuries were very much influenced by the prevailing ideas of science. One sees the impact of Galileo's observations on motion in the thinking of Hobbes, of the discoveries in physics and astronomy on the thinking of Descartes, Leibnitz, Spinoza, and Kant, and the impact of biological theorizing on the work of Nietzsche and Bergson. Modern philosophers continue to reflect and to quote the concepts of modern science, just as theologians and philosophers of past ages used to quote church doctrine or biblical texts.

That there should be interaction between philosophy and science is not surprising, since it was not too long ago (until about 1850) that science was Natural Philosophy and the philosopher and scientist were members of the same university department and often were one and the same person. The association gave philosophy some of the axioms and principles on which to base speculations, while philosophy took on the task of analyzing the concepts and methods of science for the purpose of providing a firmer logical foundation for science. The interaction was mutually beneficial in that it made some scientists more aware of the significance of methodology, and altered some of them to the metaphysical assumptions of their concepts. The growth of analytic philosophy tended to lead philosophy (at least in the United States) away from the metaphysical and speculative philosophy that was so common in the nineteenth century, and which contributed also to the elucidation and to the clarification of so many of the concepts and methodology of science.†

To the extent that science is able to explain or describe the workings of man and of nature, to that extent the mystique and the sacredness of man appears to be minimized. This effect of science is considered to be unfortunate by the humanistic psychologists and by some philosophers. They may be justified in some cases. There is a great tendency to draw conclusions, which often go far beyond that which can be justified on the basis of the evidence at hand. For example, we shall see in the chapters of the succeeding volume how the sensory and nerve systems send their signals to the brain, but no one has yet identified exactly how the electric signals become knowledge and how they become involved in the thinking processes and the passions that often accompany reception of such signals. Since physical science has not yet succeeded in studying the whole man, or even in identifying the total complexity of the interrelationships that exist among even a limited number of the organs of the body, it seems risky to assign too many machine-like attributes to man. This does not mean that we should escape into obscurantism. It does suggest, however, that it is advisable to withhold judgment on what man *is*.

It has not been our intent to attempt any conclusions with respect to the relationship of science, religion, and philosophy, but simply to note some of the questions and problems that persist in their relationship. In view of this posture on our part, perhaps the best conclusion for this section would be a series of questions over which the reader may ponder†:

(1) It is said that in both scientific and religious thinking there are certain inherently undefined terms or axioms. Try to list some of assumptions of science and religion. Compare and contrast them. Do they have the same function in science as they do in religion?

(2) It has been said that theoretical science and religion are opposed to each other. Can you identify the religious ideas that are clearly opposed to scientific ideas? In what religion, and in what time and place were certain scientific ideas opposed? In what science, and in what time and place was an idea rejected by science as belonging to "religion" or being mystical?

(3) "Science believes in determinism and religion in free will," is often asserted by behavioral scientists. Do you know of religious sects that believe in determinism? Do you know of sects that believe in free will? Does it make them any more or less scientific if they believe in one or the other? Is the issue of free will and determinism really a crucial issue between religion and science?

(4) "Scientists have destroyed the beauty of nature in their attempt to analyze it." Analyze the criticism in terms of:

(a) Study of the human body.
(b) Development of the laws of government.
(c) Modification of landscapes.

To what extent is the criticism justified by the unique action of science as contrasted with social or industrial influences? Is damming the Colorado River an act of science or of technology?

† See White in References.

† The reader is also urged to turn back to Chapter 13 and read again the quotation from Max Born.

(5) Comment on the following sentences:

(a) Science proceeds from qualitative observations to quantification. It attempts to make precise the intuitions of the poet.

(b) "It is more important to seek the truth than to assert that one has found it" is an assertion that can be made by a religious person as well as a scientist.

(c) Science is a kind of experimental philosophy.

23.7 Impact of science on education

There have developed rather dramatic changes in educational patterns since the days when schooling consisted largely of following the words of the philosophers of ancient Greece. Developments in the sciences, in technology, agriculture, the arts, industry, commerce, the humanities, and in government have contributed toward making the world a little more complex. This complexity has demanded modification of educational curricula to reflect what has become the culture and technological society of our age. We noted in the earlier sections of this chapter how much like an explosion the sciences and technology have expanded in recent decades. This expansion has created major problems of devising educational concepts that are effective for preparing competent scientists and technologists, as well as for preparing the rest of the citizenry to live in a technological world. It has been difficult to resolve the problems of education, partly because the total amount of information that now exists is so vast and difficult to sift out and categorize, and partly because the information becomes modified so rapidly through new research.

A century ago all the sciences were under one roof, in the philosophy department under the title of Natural Philosophy. It became necessary to differentiate and subdivide this into various departments (physics, chemistry, biology, and other life sciences). The process of further differentiation and specialization with ever new departments continues to this day. Specialization has been unavoidable, since it permits a scientist to focus on specific problems of his chosen field. Unfortunately, specialization also tends toward separation of disciplines to a degree that makes mutual understanding and cooperative research difficult. When educational curricula do not recognize and take steps to reduce this tendency, students emerge from universities with such narrow outlook that specialization takes on the function of "blinders" rather than developing the ability to probe in depth into a specific area while retaining perspective with respect to the whole.

The weaknesses that are inherent in specialization have been the cause for concern on the part of many people. The splintering of disciplines goes in all directions. Many individuals see the separations to be most serious (disastrous in its effects) between the physical sciences and the social sciences and the humanities. (More recently the biological sciences have had closer liaison with chemistry and physics in such combinatory efforts as biochemistry and biophysics, which tends to merge biological sciences with the physical sciences.) There has been extensive discussion of this as the cause for a trend toward a two-culture society within a nation.†

There has emerged also, partly as a result of fears of the harmful effects of specialization and partly because of the

---

† See Snow in References.

need to evaluate existing scientific effort, still other specialties that are variously called the *Science of Science*, or *Philosophy of Science*, and *History of Science*. Their purpose is to trace historical trends in ideas, or to identify underlying logical themes that are common and basic to the various sciences, sometimes with emphasis on the relationship of science to the humanities. These approaches have helped to show some unity among the sciences and even to promote an interdisciplinary outlook. The feeling persists, however, that educational curricula can do much more than they are now doing to reduce the harmful effects of specialization by cultivating awareness of the *interrelatedness* of the research work in which each is engaged to other research. (The existence of this volume, and the course of study of which it is a part, represent an interdisciplinary, integrative effort in this direction.) The handicaps to making major improvements are sometimes posed by the specialists themselves, by their university departments, and by their professional societies, which are ever on guard to defend their "domains." Also, the heavy emphasis on military research tends to leave some of the social sciences (particularly the humanities) at the "short end of the stick." The net effect is to create more rivalry than cooperation among disciplines.

Explicit in our previous sections was the complaint that there is a conflict of interests between technology and the human values of our society. It reflects contemporary discontent with the trends that appear to be inherent to technology and science. Ironically, there was more faith in science and technology *before* the appearance of the abundant fruits of technology than many people are inclined to credit to them at this time.

Although these fruits have given us much more than men dared dream possible a hundred years ago, there is concern about whether we really got what we expected from science. The two World Wars gave science an unexpected boost, but this was largely in connection with military technology. Science reflects the pressures and the human limitations of its creators, and is highly susceptible to the added pressures and the bias that develop as a result of the close ties between technology and national needs of the moment.

We noted that the controversy is often represented in terms of a conflict between two cultures: science and the humanities, with the latter being considered more concerned with the welfare of man. History, on the other hand, suggests that the propagation of justice and mercy toward man and nature is not the monopoly of either the humanistic or the scientific "culture." The ivory-towered scholars of the humanistic or scholastic tradition have not been more humane in the social area than the scholars of the scientific traditions have been. Nor did their work make man happier or more religious, if we use Micha's ideal of what God wants of man, namely, to deal righteously with mercy and justice, and to walk humbly with one's God. Perhaps the difference lies not in basic differences between the sciences and the humanities as much as in the differences of attitude between the literal-minded magicians and craftsmen and the seekers of truth who are present in both "camps." The "field" within which people with these attitudes function may be art, science, religion, or technology. These attitudes reflect, moreover, a difference of ideas on how to develop an educated person. The differences pertain not merely to the content of education, but also to the attitudes and values developed by education. Of course this brings up the question: What is education?

(1) Education is seen by some as the inculcation of habits and skills. Learning becomes a process of memorizing by rote in order to repeat what was presented by the text, teacher, teaching machine.

(2) Education is seen by others as the opening of the "mind's" eyes, to inspire a search for understanding. Not memorizing but *understanding* is stressed. Not repetition of the text alone, but investigation beyond what the book or teacher has set as the goal. The purpose of such education is to create a person who will be able to add to knowledge and who will be oriented toward seeking the True, the Beautiful, and the GOOD.

The first type of education is a form of indoctrination, whether its content is the "liberal arts" or the sciences, or logic or psychology. It is as often seen in routine study of literature as in a laboratory course in science.

The second concept of education is premised on a need to go beyond the limitations of one's understanding. "He is wise who knows that he knows not." The Delphic oracle said to an inquirer that *the wisest of all men is Socrates because he knows that he does not know*. His approach to education has come down to us in the idea of the *liberal education*, an education that liberates the mind from the belief that mere sense experiences constitute real truths. Perhaps we have to be liberated from the belief that gadgets are gods and that the scientist's and technologist's task is to outwit nature or at least to shift the balances more nearly in favor of the technological man.

This concludes the first volume. Its emphasis has been on the historical beginnings of science and the scientific method, and on the current status of modern physical science. In the course of the development we have discussed some of the philosophical and social implications of the physical sciences. It is time to ask what we wish the reader will have gained from this volume. The volume contains, of course, a great deal of scientific information and information on the scientific method, past and present. More than this information, however, we hope the reader has developed greater appreciation for science as a very *human* enterprise, a *dynamic* and *changing* activity of man, and as a study that has personal implications for all of us, whether we plan careers in science or in other fields.

More than anything else, we hope that the reader has been encouraged to maintain an open but analytic and discerning mind toward the basic strengths and limitations of science, and toward the values and hazards to society that accompany the utilization of science in the form of technology. For the sciences, the humanities, and religious thought retain within themselves the seeds that can grow, albeit slowly, to produce self-correction and progress—if only we can maintain an open mind.

Questions/Discussions

1. Comment on the following statements:
   (a) Science is the objective, dispassionate study of nature.
   (b) Science is power; it gives man control over nature.
   (c) The progress of man rests on his

ability to widen the application of science to life.
(d) The technologically advanced nations are the superior societies.
(e) Religious beliefs belong to the prescience stages of human cultural development.
(f) Since the technological way of life is the most "efficient," it is therefore the "best" way of life.
(g) There is very little that a technologically advanced country may learn from an underdeveloped society.
(h) The scientific spirit is one of questioning. This is unique to science.

2. Niels Bohr once said (in 1933): "The existence of life must be taken as a basic fact for which no specific reason can be given and which must be accepted as the starting point of biology in the same way in which the quantum together with the existence of the elementary particles form the basis of physics." Do you agree or disagree? Why?

3. It has been said that the attitude of scientists is to put question marks where there are periods. Please identify about ten sentences of this chapter which in your opinion should become questions instead of statements.

4. What do you regard to be the present responsibility of science and technology with respect to the needs of the generations to come? For how many generations should we have concern? With respect to what aspects (for examples resources)?

5. What should be the attitude of science with respect to uniformity of culture around the world?

6. What are the criteria for judging the *efficiency* of any society?

7. What do you consider to be necessary to include (in decreasing order of importance) in the college level education of a scientist from the following fields? (List these separately for (a) a physicist; (b) a biologist; (c) a psychologist.):

philosophy, theology, literature, mathematics, art, current events, economics, political science, history.

8. Identify and discuss any *changes* that you may have experienced as a result of study of this volume, with respect to:

(a) Your understanding of the concepts and methods of science.
(b) Your understanding of the strengths and limitations of the scientific endeavor.
(c) Distinctions between science and technology.
(d) The responsibility of the "non-scientist" citizen with respect to the *direction* and *magnitude of support* that the scientific endeavor should be given by public agencies.
(e) The field of science that you consider to be the most important to support and promote.

References

While we cannot endorse all the contents of the following references, we can recommend them for reading and discussion.

American Academy of Arts and Sciences, "Evolution and Man's Progress," *Daedalus* (Summer 1961). (*The entire issue is devoted to articles that are relevant to the theme of this chapter.*)

Barbour, Ian G., *Issues in Science and Religion*. Englewood Cliffs, N.J.: Prentice-Hall, 1966.

Blackett, P. M. S., "The Ever Widening Gap," *Science,* **155,** No. 3765 (Feb. 24, 1967).

Boyko, Hugo (ed.), *Science and the Future of Mankind.* Bloomington, Ind.: Indiana Univ. Press, 1961. (*Contains a number of articles by scientists under the general titles of* "The Need," "The Means," *and* "The Goal.")

Colodny, Robert G. (ed.), *Beyond the Edge of Certainty.* Englewood Cliffs, N.J.: Prentice-Hall, 1965. (*A series of interesting chapters, including one by Paul K. Feyerabend on* "Problems of Empiricism.")

Crombie, A. G. (ed.), *Scientific Change.* New York: Basic Books, 1963. (*Symposium on the history of science held at University of Oxford, July 9–15, 1961.*)

Gillispie, Charles C., *The Edge of Objectivity; An Essay in the History of Scientific Ideas.* Princeton, N.J.: Princeton Univ. Press, 1960.

Kapp, K. William, *Toward a Science of Man in Society.* The Hague, Netherlands: Martinus Nighoff, 1961.

Lindsay, Robert Bruce, *The Role of Science in Civilization.* New York: Harper & Row, 1963.

Marsak, Leonard, M., *The Rise of Science in Relation to Society.* New York: Macmillan, 1964.

Parsegian, V. L., "Letters," *Science,* **155,** No. 3767 (April 21, 1967).

Polanyi, Michael, "Science and Man in the Universe," in Woolf, Harry, edit, *Science as a Cultural Force.* Baltimore, Md.: The Johns Hopkins Press, 1964.

Snow, C. P., *The Two Cultures: A Second Look.* New York: Cambridge Univ. Press, 1964.

White, W., *The Age of Analysis.* New York: New American Library (a Mentor book), 1955.

Appendix

1. Atomic and other constants (including miscellaneous constants and ratios pertaining to the planets)
2. Metric prefixes
3. Greek alphabet
4. Equivalents and conversion factors
5. Isotopes
6. Partial list of radio-isotopes
7. Relative atomic weights

APPENDIX 1
ATOMIC AND OTHER CONSTANTS

(The values given are within ±0.1 percent of the most precise values known at present.)[°]

		cgs	mks
G	Gravitational constant	6.67×10^{-8} dyne cm²/gm²	6.67×10^{-11} nt m²/kg²
F	Faraday	9.65×10^3 emu·(gm-equiv)⁻¹ (phys)	9.65×10^7 coul·(kgm-equiv)⁻¹ (phys)
N_0	Avogadro's number	6.023×10^{23} (gm-mole)⁻¹ (phys)	6.023×10^{26} (kgm-mole)⁻¹ (phys)
h	Planck's constant	6.625×10^{-27} erg·sec	6.625×10^{-34} j·sec
c	Velocity of light in vacuum	3.00×10^{10} cm·sec⁻¹	3.00×10^8 m·sec⁻¹
e	Electronic charge	$\begin{cases} 1.602 \times 10^{-20} \text{ emu} \\ 4.80 \times 10^{-10} \text{ esu} \end{cases}$	1.602×10^{-19} coul
m	Electron rest mass	$\begin{cases} 9.11 \times 10^{-28} \text{ gm} \\ 5.49 \times 10^{-4} \text{ amu} \end{cases}$	9.11×10^{-31} kgm ; 5.49×10^{-4} amu
e/m	Specific electronic charge	1.760×10^7 emu·gm⁻¹	1.760×10^{11} coul·kgm⁻¹
h/mc	Compton wavelength, electron	2.426×10^{-10} cm	2.426×10^{-12} m
M_0	1 amu = $\tfrac{1}{12}$ mass $C^{12} = 1/N_0$	1.6603×10^{-24} gm	1.6603×10^{-27} kgm
M	Proton rest mass	1.6724×10^{-24} gm	1.6724×10^{-27} kgm
H	H-atom rest mass	1.6733×10^{-24} gm	1.6733×10^{-27} kgm
n	Neutron rest mass	1.6747×10^{-24} gm	1.6747×10^{-27} kgm
M/m	Ratio, proton to electron mass	1.836×10^3	1.836×10^3
mc^2	Energy equiv. of electron rest mass	0.511 Mev	0.511 Mev
$M_0 c^2$	Energy equiv. of 1 amu	931 Mev	931 Mev
σ	Stefan-Boltzmann constant	5.67×10^{-5} erg·cm⁻²·(°K)⁻⁴·sec⁻¹	5.67×10^{-8} j·m⁻²·(°K)⁻¹·sec⁻¹
k	Boltzmann constant	1.380×10^{-16} erg·(°K)⁻¹	1.380×10^{-23} j·(°K)⁻¹
J	Joule equivalent	4.185×10^7 ergs/cal	4.185 j/cal
0°K	Absolute zero	−273.15°C	
$\lambda_{max} T$	Wien displacement law constant	0.290 cm·°K	0.290×10^{-2} m·°K
R_∞	Rydberg const for infinite mass	1.097×10^5 cm⁻¹	1.097×10^7 m⁻¹
R	Gas constant	8.32×10^7 erg(gm-mole·°K)⁻¹	8.32×10^3 j·(kgm-mole·K°)⁻¹
V_0	Standard volume of ideal gas	2.242×10^4 cm³(gm-mole)⁻¹	22.42 m³(kgm-mole)⁻¹

[°] When solving textbook problems the reader is encouraged to limit the constants to two significant figures. (For example, when working with Planck's constant, use 6.6×10^{-27} erg sec instead of 6.625×10^{-27} erg sec.)

MISCELLANEOUS CONSTANTS AND RATIOS PERTAINING TO THE PLANETS

	Mercury	Venus	Earth	Mars	Jupiter	Saturn	Uranus	Neptune	Pluto
Sidereal period	87.97^d	224.7^d	365.26^d	687.0^d	11.86^y	29.46^y	84.02^y	164.8^y	247.7^y
Synodic period	115.88^d	583.9^d	—	779.9^d	1.092^y	1.035^y	1.012^y	1.006^y	1.004^y
Mean distance,									
10^6 km	57.94	108.27	149.68	228.06	778.73	1,427.7	2,872.4	4,500.8	5,914.8
Astron-units	0.387	0.723	1.000	1.524	5.203	9.539	19.19	30.07	39.46
Orbital speed,									
km/sec	47.9	35.0	29.8	24.1	13.6	9.6	6.8	5.4	4.8
Orbital eccentricity	0.2056	0.0068	0.017	0.093	0.048	0.056	0.047	0.0086	0.249
Orbital inclination	7°0'	3°24'	—	1°51'	1°18'	2°29'	0°46'	1°47'	17°19'
Mean diameter, km	5,000	12,400	12,742	6,870	139,760	115,100	51,000	50,000	12,700?
Earth diameters	0.39	0.973	1.000	0.532	10.97	9.03	4.00	3.90	0.46
Volume (earth									
volumes)	0.06	0.92	1.00	0.15	1,318	736.	64.	39.	0.10
Mass (earth									
masses)	0.04	0.82	1.00	0.11	318.3	95.3	14.7	17.3	1.0?
Density (earth									
densities)	0.69	0.89	1.00	0.70	0.24	0.13	0.23	0.29	?
in g/cc	3.8	4.86	5.52	3.96	1.33	0.71	1.26	1.6	?
Surface gravity									
(earth's)	0.27	0.86	1.00	0.37	2.64	1.17	0.92	1.44	?
Velocity of escape,									
km/sec	3.6	10.2	11.2	5.0	60.	36.	21.	23.	11?
Maximum surface									
temperature °F	770.	140.	140.	86.	−216.	−243.	−300?	−330?	−348?
Length of day	88^d	30^d?	1^d	37^{m}23^s	9^{h}55^m	10^{h}38^m	10^h.7	15^h.8	?
Inclination of									
equator to orbit		0°?	23°27'	25°12'	3°7'	26°45'	8°0'	29°	?
Oblateness	0.00	0.00	$\frac{1}{298}$	$\frac{1}{192}$	$\frac{1}{15.4}$	$\frac{1}{9.5}$	$\frac{1}{14}$	$\frac{1}{45}$	?
Albedo	0.07	0.59	0.29	0.15	0.44	0.42	0.45	0.52	0.04?
Atmosphere	none	CO_2	(see p. 21)	H_2O?	CH_4, NH_3	CH_4, NH_3	CH_4, NH_3	CH_4, NH_3	none
Moon (known)	none	none	1	2	12	9	5	2	none

Mass of Earth	M_e	5.977×10^{24} kgm
Radius (equatorial) of Earth	R_e	6.378×10^6 m
Mass of Sun	M	1.991×10^{30} kgm
Radius of Sun	R	6.960×10^8 m
Luminosity of Sun	L	3.86×10^{26} joules
Astronomical unit	au	1.49598×10^{11} m
Parsec	pc	206,265 au
		= 3.262 light years
		= 3.086×10^{16} m

APPENDIX 2
METRIC PREFIXES

Prefix	Abbreviation	Meaning	Typical examples
kilomega	KM	$\times 10^9$	1 kilomegacycle/sec (radar frequency) = 10^9 cycles/sec
mega	M	$\times 10^6$	1 megaton (equivalent TNT strength of nuclear weapon) = 10^6 tons
kilo	K, k	$\times 10^3$	1 kilogram = 1000 g
deci	d	$\times 10^{-1}$	1 decimeter = 0.1 m
centi	c	$\times 10^{-2}$	1 centimeter = 0.01 m
milli	m	$\times 10^{-3}$	1 milliampere = 0.001 amp
micro	μ	$\times 10^{-6}$	1 microinch = 10^{-6} in.
millimicro	mμ	$\times 10^{-9}$	1 millimicrovolt = 10^{-9} v
micromicro	$\mu\mu$	$\times 10^{-12}$	1 micromicrofarad = 10^{-12} farad
tera	T	$\times 10^{12}$	
giga	G	$\times 10^9$	
nano	n	$\times 10^{-9}$	1 nanosecond = 10^{-9} sec
pico	p	$\times 10^{-12}$	

APPENDIX 3
GREEK ALPHABET

A	α	alpha	H	η	eta	N	ν	nu	T	τ	tau
B	β	beta	Θ	θ	theta	Ξ	ξ	xi	Y	υ	upsilon
Γ	γ	gamma	I	ι	iota	O	o	omicron	Φ	ϕ	phi
Δ	δ	delta	K	κ	kappa	Π	π	pi	X	χ	chi
E	ϵ	epsilon	Λ	λ	lambda	P	ρ	rho	Ψ	ψ	psi
Z	ζ	zeta	M	μ	mu	Σ	σ	sigma	Ω	ω	omega

APPENDIX 4
EQUIVALENTS AND CONVERSION FACTORS

Length
1 ft = 30.48 cm
1 mi = 5280 ft = 1.609 km
1 yd = 0.9144 m
1 in. = 2.540 cm
1 angstrom (A) = 10^{-8} cm = 10^{-10} m
1 micron (μ) = 10^{-4} cm = 10^{-6} m = 10^4 A
1 light-year = 5.88×10^{12} mi = 9.464×10^{12} km

Area
1 ft² = 929.0 cm² = 0.09290 m²
1 in.² = 6.452 cm² = 645.2 mm²

Volume
1 liter (l) = 1000 cm³ = 1.0576 qt = 61.03 in.³
1 ft³ = 7.481 gal = 28.32 l
1 m³ = 1000 l = 10^6 cm³ = 1.308 yd³

Velocity 60 mi/hr = 88 ft/sec = 26.82 meters/sec

Mass 1 slug = 14.59 kg

Mass density 1 slug/ft³ = 0.5154 g/cm³ = 515.4 kg/m³

Weight density 1 lb/ft³ = 16.02 kgf/m³ = 0.01602 gf/cm³

Force
1 newton (nt) = 10^5 dynes = 0.2248 lb = 102.0 gf
1 lb = 453.6 gf = 4.448 nt
1 kgf = 2.205 lb = 9.81 nt
1 ton = 2000 lb = 907.2 kgf
1 metric ton = 1000 kgf = 2205 lb

Pressure
1 atm = 14.70 lb/in.² = 76.00 cm Hg = 1.013×10^6 dyne/cm²
= 1.013×10^5 nt/m²
1 lb/in.² = 6.89×10^4 dynes/cm² = 6.89×10^3 nt/m²

Work and energy
1 joule (j) = 10^7 ergs = 0.239 cal = 0.7376 ft·lb
1 cal = 4.18 j = 3.087 ft·lb
1 Btu = 252 cal = 778 ft·lb = 1055 j
1 kilowatt·hour (kwh) = 3.60×10^6 j
1 electron volt (ev) = 1.60×10^{-19} j

Power
1 horsepower (hp) = 0.746 kw = 550 ft·lb/sec
1 watt (w) = 1 j/sec = 0.738 ft·lb/sec

Heat 1 Btu/lb = 0.556 cal/g

APPENDIX 5
PARTIAL LIST OF ISOTOPES

Atomic no. Z	Element	Symbol	Mass no., A	Isotopic mass, u	Relative abundance, %	No. of isotopes Stable	No. of isotopes Radioactive
0	Neutron	n				0	1
			1 (R)	1.008665			
1	Hydrogen	H				2	1
			1	1.007825	99.985		
		D					
			2	2.01410	0.015		
		T					
			3 (NR)				
2	Helium	He				2	3
			3	3.01603	0.00013		
			4	4.00260	100		
3	Lithium	Li				2	3
			6	6.01513	7.42		
			7	7.01601	92.58		
4	Beryllium	Be				1	6
			9	9.01219	100		
5	Boron	B				2	4
			10	10.01294	19.78		
			11	11.00931	80.22		
6	Carbon	C				2	6
			12	12.00000	98.89		
			13	13.00335	1.11		
7	Nitrogen	N				2	5
			14	14.00307	99.63		
			15	15.00011	0.37		
8	Oxygen	O				3	5
			16	15.99491	99.759		
			17	16.99914	0.037		
			18	17.99916	0.204		

The isotopic masses are given on the assumed base of 12.00000 for the carbon-12 isotope. Naturally occurring radioactive isotopes are indicated by (NR). The mass numbers given for the radioactive elements are those of the longest-lived isotopes.

The data for this table were obtained from the Chart of the Nuclides, 8th edition, revised to March, 1965 by David T. Goldman. (Courtesy of Knolls Atomic Power Laboratory, Schenectady, N.Y., operated by the General Electric Company for the United States Atomic Energy Commission.)

Appendix 5

Atomic no. Z	Element	Symbol	Mass no., A	Isotopic mass, u	Relative abundance, %	No. of isotopes Stable	No. of isotopes Radioactive
9	Fluorine	F				1	5
			19	18.99840	100		
10	Neon	Ne				3	5
			20	19.99244	90.92		
			21	20.99395	0.257		
			22	21.99138	8.82		
11	Sodium	Na				1	6
			23	22.98977	100		
12	Magnesium	Mg				3	5
			24	23.98504	78.70		
			25	24.98584	10.13		
			26	25.98259	11.17		
13	Aluminum	Al				1	7
			27	26.98153	100		
14	Silicon	Si				3	5
			28	27.97693	92.21		
15	Phosphorus	P				1	6
			31	30.97376	100		
16	Sulfur	S				4	6
			32	31.97207	95.0		
17	Chlorine	Cl				2	7
			35	34.96885	75.53		
			37	36.96590	24.47		
18	Argon	Ar				3	6
			40	39.96238	99.60		
19	Potassium	K				2	9
			39	38.96371	93.10		
			40 (NR)		0.0118		
			41	40.96184	6.88		
20	Calcium	Ca				6	8
			40	39.96259	96.97		
			44	43.95594	2.06		
21	Scandium	Sc				1	11
			45	44.95592	100		
22	Titanium	Ti				5	5
			48	47.94795	73.94		
23	Vanadium	V				1	9
			50 (NR)	49.9472	0.24		
			51	50.9440	99.76		

(cont'd)

Atomic no. Z	Element	Symbol	Mass no., A	Isotopic mass, u	Relative abundance, %	No. of isotopes	
						Stable	Radioactive
24	Chromium	Cr				4	7
			52	51.9405	83.76		
			53	52.9407	9.55		
25	Manganese	Mn				1	8
			55	54.9381	100		
26	Iron	Fe				4	6
			56	55.9349	91.66		
			57	56.9354	2.19		
27	Cobalt	Co				1	10
			59	58.9332	100		
28	Nickel	Ni				5	7
			58	57.9353	67.88		
			60	59.9303	26.23		
29	Copper	Cu				2	9
			63	62.9298	69.09		
			65	64.9278	30.91		
30	Zinc	Zn				5	8
			64	63.9291	48.89		
			66	65.9260	27.81		
			68	67.9249	18.57		
31	Gallium	Ga				2	12
			69	68.9256	60.4		
			71	70.9247	39.6		
32	Germanium	Ge				4	10
			70	69.9242	20.52		
			72	71.9221	27.43		
			74	73.9212	36.54		
33	Arsenic	As				1	14
			75	74.9216	100		
34	Selenium	Se				6	11
			78	77.9173	23.52		
			80	79.9165	49.82		
35	Bromine	Br				2	16
			79	78.9183	50.54		
			81	80.9163	49.46		
36	Krypton	Kr				6	16
			82	81.9135	11.56		
			83	82.9141	11.55		
			84	83.9115	56.90		
			86	85.9106	17.37		

(cont'd)

					Relative	No. of isotopes	
Atomic no. Z	Element	Symbol	Mass no., A	Isotopic mass, u	abun- dance, %	Stable	Radio- active
37	Rubidium	Rb				1	16
			85	84.9117	72.15		
			87		27.85		
38	Strontium	Sr				4	12
			88	87.9056	82.56		
39	Yttrium	Y				1	14
			89	88.9056	100		
40	Zirconium	Zr				5	9
			90	89.9047	51.46		
			92	91.9050	17.11		
			94	93.9063	17.40		
41	Niobium (or Columbium, Cb)	Nb	93	92.9064	100	1	13
42	Molybdenum	Mo				7	10
			92	91.9068	15.84		
			95	94.9058	15.72		
			96	95.9047	16.53		
			98	97.9054	23.78		
43	Technetium	Tc				0	14
			99 (R)				
44	Ruthenium	Ru				7	9
			102	101.9043	31.61		
			104	103.9054	18.58		
45	Rhodium	Rh				1	14
			103	102.9055	100		
46	Palladium	Pd				6	12
			105	104.9051	22.23		
			106	105.9035	27.33		
			108	107.9039	26.71		
47	Silver	Ag				2	14
			107	107.9051	51.82		
			109	108.9047	48.18		
48	Cadmium	Cd				8	11
			110	109.9030	12.39		
			111	110.9042	12.75		
			112	111.9028	24.07		
			113	112.9046	12.26		
			114	113.9034	28.86		
49	Indium	In				1	18
			113	112.9043	4.28		
			115 (NR)	114.9039	95.72		

(cont'd)

Atomic no. Z	Element	Symbol	Mass no., A	Isotopic mass, u	Relative abundance, %	No. of isotopes Stable	No. of isotopes Radioactive
50	Tin	Sn				10	15
			116	115.9017	14.30		
			118	117.9016	24.03		
			119	118.9033	8.58		
			120	119.9022	32.85		
51	Antimony	Sb				2	22
			121	120.9038	57.25		
			123	122.9042	42.75		
52	Tellurium	Te				8	16
			123 (NR)	122.9043	0.87		
			126	125.9033	18.71		
			128	127.9045	31.79		
			130	129.9062	34.48		
53	Iodine	I				1	22
			127	126.9045	100		
54	Xenon	Xe				9	16
			129	128.9048	26.44		
			131	130.9051	21.18		
			132	131.9042	26.89		
55	Cesium	Cs				1	20
			133	132.9051	100		
56	Barium	Ba				7	14
			137	136.9056	11.32		
			138	137.9050	71.66		
57	Lanthanum	La				1	20
			138 (NR)	137.9068	0.089		
			139	138.9061	99.911		
58	Cerium	Ce				3	16
			140	139.9053	88.48		
			142 (NR)	141.9090	11.07		
59	Praseodymium	Pr				1	14
			141	140.9074	100		
60	Neodymium	Nd				6	8
			142	141.9075	27.11		
			144 (NR)	143.9099	23.85		
			146	145.9127	17.22		
61	Promethium	Pm				0	14
			145 (R)				
62	Samarium	Sm				4	14
			147 (NR)	146.9146	14.97		
			148 (NR)	147.9146	11.24		

(cont'd)

Atomic no. Z	Element	Symbol	Mass no., A	Isotopic mass, u	Relative abundance, %	No. of isotopes Stable	No. of isotopes Radioactive
			149 (NR)	148.9169	13.83		
			152	151.9195	26.72		
			154	153.9209	22.71		
63	Europium	Eu				2	16
			151	150.9196	47.82		
			153	152.9209	52.18		
64	Gadolinium	Gd				6	12
			152 (NR)	151.9195	0.20		
			156	155.9221	20.47		
			158	157.9241	24.87		
			160	159.9271	21.90		
65	Terbium	Tb				1	17
			159	158.9250	100		
66	Dysprosium	Dy				6	13
			156 (NR)	155.9238	0.052		
			162	161.9265	25.53		
			163	162.9284	24.97		
			164	163.9288	28.18		
67	Holmium	Ho				1	18
			165	164.9303	100		
68	Erbium	Er				6	12
			166	165.9304	33.41		
			167	166.9320	22.94		
			168	167.9324	27.07		
69	Thulium	Tm				1	17
			169	168.9344	100		
70	Ytterbium	Yb				7	10
			172	171.9366	21.82		
			174	173.9390	31.84		
71	Lutetium	Lu				1	15
			175	174.9409	97.41		
			176 (NR)		2.59		
72	Hafnium	Hf				5	13
			174 (NR)	173.9403	0.18		
			178	177.9439	27.14		
			180	179.9468	35.24		
73	Tantalum	Ta				2	13
			181	180.9480	99.988		
74	Tungsten (Wolfram)	W				5	10
			182	181.9483	26.41		

(cont'd)

Atomic no. Z	Element	Symbol	Mass no., A	Isotopic mass, u	Relative abundance, %	No. of Isotopes	
						Stable	Radio-active
			184	183.9510	30.64		
			186	185.9543	28.14		
75	Rhenium	Re				1	14
			185	184.9530	37.07		
			187 (NR)	186.9560	62.93		
76	Osmium	Os				7	8
			190	189.9586	26.4		
			192	191.9612	41.0		
77	Iridium	Ir				2	15
			191	190.9609	37.3		
			193	192.9633	62.7		
78	Platinum	Pt				5	16
			190 (NR)	189.9600	0.0127		
			194	193.9628	32.9		
			195	194.9648	33.8		
			196	195.9650	25.3		
79	Gold	Au				1	18
			197	196.9666	100		
80	Mercury	Hg				7	14
			199	198.9683	16.84		
			200	199.9683	23.13		
			202	201.9706	29.80		
81	Thallium	Tl				2	18
			203	202.9723	29.50		
			205	204.9745	70.50		
			207 (NR)				
82	Lead	Pb				3	18
			204 (NR)	203.9731	1.48		
			206	205.9745	23.6		
			207	206.9759	22.6		
			208	207.9766	52.3		
83	Bismuth	Bi				1	18
			209	208.9804	100		
			210 (NR)				
84	Polonium	Po				0	27
			210 (NR)	209.9829			
85	Astatine	At				0	20
			210 (NR)				
			211 (NR)	210.9875			
86	Radon	Rn				0	18
			222 (NR)	222.0175			

(cont'd)

Atomic no. Z	Element	Symbol	Mass no., A	Isotopic mass, u	Relative abundance, %	No. of isotopes Stable	No. of isotopes Radioactive
87	Francium	Fr				0	18
			223 (NR)	223.0198			
88	Radium	Ra				0	13
			226 (NR)	226.0254			
89	Actinium	Ac				0	11
			227 (NR)	227.0278			
90	Thorium	Th				0	13
			232 (NR)	232.0382			
91	Protactinium	Pa				0	12
			231 (NR)	231.0359			
92	Uranium	U				0	14
			234 (NR)	234.0409	0.0057		
			235 (NR)	235.0439	0.72		
			238 (NR)	238.0508	99.27		
93	Neptunium	Np				0	11
			237 (R)	237.0480			
94	Plutonium	Pu				0	15
			239 (R)	239.0522			
			242 (R)	242.0587			
			244 (R)				
95	Americium	Am				0	10
			241 (R)	241.0567			
			243 (R)	243.0614			
96	Curium	Cm				0	13
			243 (R)	243.0614			
			247 (R)				
97	Berkelium	Bk				0	8
			247 (R)	247.0702			
98	Californium	Cf				0	11
			251 (R)				
99	Einsteinium	Es				0	11
			254 (R)	254.0881			
100	Fermium	Fm				0	11
			257 (R)				
101	Mendelevium	Md				0	2
			256 (R)				
102	Nobelium	No				0	3
			255 (R)				
103	Lawrencium	Lw				0	1
			257 (R)				
104						0	1
			260 (R)				

APPENDIX 6
PARTIAL LIST OF RADIOISOTOPES

Element	Nuclide	Half-life, T	Decay constant λ, s^{-1}	Principal particle energy, MeV
Antimony	$_{51}Sb^{122}$	2.80 d	2.87×10^{-6}	β^-, 1.40; γ, 0.56, 0.70
Argon	$_{18}Ar^{37}$	35.1 d	2.29×10^{-7}	K-capture
Arsenic	$_{33}As^{76}$	26.5 h	7.26×10^{-6}	β^-, 2.97; γ, 0.56, 1.21
Barium	$_{56}Ba^{140}$	12.8 d	6.26×10^{-7}	β^-, 1.02; γ, 0.03, 0.54
Bismuth	$_{83}Bi^{210}$ (NR)	5.0 d	1.60×10^{-6}	β^-, 1.16
Cadmium	$_{48}Cd^{115}$	2.3 d	3.49×10^{-6}	β^-, 1.11; γ, 0.52
Calcium	$_{20}Ca^{45}$	165 d	4.87×10^{-8}	β^-, 0.25
Carbon	$_{6}C^{14}$	5730 y	3.83×10^{-12}	β^-, 0.156
Cerium	$_{58}Ce^{141}$	32.5 d	2.47×10^{-7}	β^-, 0.44; γ, 0.15
Cesium	$_{55}Cs^{134}$	2.1 y	1.05×10^{-8}	β^-, 0.65; γ, 0.60
	$_{55}Cs^{137}$	30 y	7.32×10^{-10}	β^-, 0.51; γ, 0.66
Chlorine	$_{17}Cl^{36}$	3.0×10^5 y	7.32×10^{-14}	β^-, 0.71
Chromium	$_{24}Cr^{51}$	27.8 d	2.89×10^{-7}	γ, 0.32; K-capture
Cobalt	$_{27}Co^{58}$	71 d	1.13×10^{-7}	β^+, 0.48; γ, 0.81, 1.65
	$_{27}Co^{60}$	5.26 y	4.17×10^{-9}	β^-, 0.31; γ, 1.17, 1.33
Gold	$_{79}Au^{198}$	64.8 h	2.97×10^{-6}	β^-, 0.96; γ, 0.41, 0.67
Hafnium	$_{72}Hf^{181}$	45 d	1.78×10^{-7}	β^-, 0.41; γ, 0.48
Hydrogen	$_{1}H^{3}$	12.26 y	1.79×10^{-9}	β^-, 0.018
Iodine	$_{53}I^{131}$	8.05 d	9.96×10^{-7}	β^-, 0.61; γ, 0.36, 0.72
Iron	$_{26}Fe^{59}$	45 d	1.78×10^{-7}	β^-, 0.46; γ, 1.10, 1.29
Krypton	$_{36}Kr^{85}$	10.76 y	2.04×10^{-7}	β^-, 0.67; γ, 0.52
Lanthanum	$_{57}La^{140}$	40.2 h	4.79×10^{-6}	β^-, 1.34; γ, 1.60, 0.49
Mercury	$_{80}Hg^{203}$	47 d	1.71×10^{-7}	β^-, 0.21; γ, 0.28
Molybdenum	$_{42}Mo^{99}$	66 h	2.92×10^{-6}	β^-, 1.23; γ, 0.04, 0.14
Neptunium	$_{93}Np^{237}$	2.14×10^6 y	1.03×10^{-14}	α, 4.50; γ, 0.03, 0.09
	$_{93}Np^{239}$	2.35 d	3.42×10^{-6}	β^-, 0.72; γ, 0.05, 0.33
Nickel	$_{28}Ni^{63}$	92 y	2.39×10^{-10}	β^-, 0.07
Phosphorus	$_{15}P^{32}$	14.3 d	5.61×10^{-7}	β^-, 1.71
Plutonium	$_{94}Pu^{239}$	2.44×10^{-4} y	9.01×10^{-13}	α, 5.15; γ, 0.013, 0.038
Potassium	$_{19}K^{40}$ (NR)	1.3×10^9 y	1.69×10^{-17}	β^-, 1.32; γ, 1.46
	$_{19}K^{42}$	12.4 h	1.55×10^{-5}	β^-, 3.53; γ, 1.52
Selenium	$_{34}Se^{75}$	120 d	6.70×10^{-8}	γ, 0.27; K-capture
Silver	$_{47}Ag^{111}$	7.5 d	1.07×10^{-6}	β^-, 1.05; γ, 0.34
Sodium	$_{11}Na^{24}$	15.0 h	1.28×10^{-5}	β^-, 1.39; γ, 1.37, 2.75
Strontium	$_{38}Sr^{89}$	50.4 d	1.59×10^{-7}	β^-, 1.46
	$_{38}Sr^{90}$	28 y	7.85×10^{-10}	β^-, 0.54
Sulfur	$_{16}S^{35}$	86.7 d	9.25×10^{-8}	β^-, 0.17
Tantalum	$_{73}Ta^{182}$	115 d	6.98×10^{-8}	β^-, 0.51; γ, 0.10, 1.12
Xenon	$_{54}Xe^{135}$	9.2 h	2.09×10^{-5}	β^-, 0.91; γ, 0.25, 0.61
Zinc	$_{30}Zn^{65}$	245 d	3.28×10^{-8}	β^-, 0.33; γ, 1.12

APPENDIX 7
RELATIVE ATOMIC WEIGHTS
(Based on the atomic mass of $^{12}C = 12$)

Name	Symbol	Atomic number	Atomic weight	Name	Symbol	Atomic number	Atomic weight
Actinium	Ac	89	—	Osmium	Os	76	190.2
Aluminum	Al	13	26.9815	Oxygen	O	8	15.9994[a]
Americium	Am	95	—	Palladium	Pd	46	106.4
Antimony	Sb	51	121.75	Phosphorus	P	15	30.9738
Argon	Ar	18	39.948	Platinum	Pt	78	195.09
Arsenic	As	33	74.9216	Plutonium	Pu	94	—
Astatine	At	85	—	Polonium	Po	84	—
Barium	Ba	56	137.34	Potassium	K	19	39.102
Berkelium	Bk	97	—	Praseodymium	Pr	59	140.907
Beryllium	Be	4	9.0122	Promethium	Pm	61	—
Bismuth	Bi	83	208.980	Protactinium	Pa	91	—
Boron	B	5	10.811[a]	Radium	Ra	88	—
Bromine	Br	35	79.909[b]	Radon	Rn	86	—
Cadmium	Cd	48	112.40	Rhenium	Re	75	186.2
Calcium	Ca	20	40.08	Rhodium	Rh	45	102.905
Californium	Cf	98	—	Rubidium	Rb	37	85.47
Carbon	C	6	12.01115[a]	Ruthenium	Ru	44	101.07
Cerium	Ce	58	140.12	Samarium	Sm	62	150.35
Cesium	Cs	55	132.905	Scandium	Sc	21	44.956
Chlorine	Cl	17	35.453[b]	Selenium	Se	34	78.96
Chromium	Cr	24	51.996[b]	Silicon	Si	14	28.086[a]
Cobalt	Co	27	58.9332	Silver	Ag	47	107.870[b]
Copper	Cu	29	63.54	Sodium	Na	11	22.9898
Curium	Cm	96	—	Strontium	Sr	38	87.62
Dysprosium	Dy	66	162.50	Sulfur	S	16	32.064[a]
Einsteinium	Es	99	—	Tantalum	Ta	73	180.948
Erbium	Er	68	167.26	Technetium	Tc	43	—
Europium	Eu	63	151.96	Tellurium	Te	52	127.60
Fermium	Fm	100	—	Terbium	Tb	65	158.924
Fluorine	F	9	18.9984	Thallium	Tl	81	204.37
Francium	Fr	87	—	Thorium	Th	90	232.038
Gadolinium	Gd	64	157.25	Thulium	Tm	69	168.934
Gallium	Ga	31	69.72	Tin	Sn	50	118.69
Germanium	Ge	32	72.59	Titanium	Ti	22	47.90
Gold	Au	79	196.967	Tungsten	W	74	183.85
Hafnium	Hf	72	178.49	Uranium	U	92	238.03
Helium	He	2	4.0026	Vanadium	V	23	50.942
Holmium	Ho	67	164.930	Xenon	Xe	54	131.30
Hydrogen	H	1	1.00797[a]	Ytterbium	Yb	70	173.04
Indium	In	49	114.82	Yttrium	Y	39	88.905
Iodine	I	53	126.9044	Zinc	Zn	30	65.37
Iridium	Ir	77	192.2	Zirconium	Zr	40	91.22
Iron	Fe	26	55.847[b]				
Krypton	Kr	36	83.80				
Lanthanum	La	57	138.91				
Lead	Pb	82	207.19				
Lithium	Li	3	6.939				
Lutetium	Lu	71	174.97				
Magnesium	Mg	12	24.312				
Manganese	Mn	25	54.9380				
Mendelevium	Md	101	—				
Mercury	Hg	80	200.59				
Molybdenum	Mo	42	95.94				
Neodymium	Nd	60	144.24				
Neon	Ne	10	20.183				
Neptunium	Np	93	—				
Nickel	Ni	28	58.71				
Niobium	Nb	41	92.906				
Nitrogen	N	7	14.0067				
Nobelium	No	102	—				

The values for atomic weights given in the table apply to elements as they exist in nature, without artificial alteration of their isotopic composition, and, further, to natural mixtures that do not include isotopes of radiogenic origin.

[a] Atomic weights so designated are known to be variable because of natural variations in isotopic composition. The observed ranges are:

Hydrogen	±0.00001	Oxygen	±0.0001
Boron	±0.003	Silicon	±0.001
Carbon	±0.00005	Sulfur	±0.003

[b] Atomic weights so designated are believed to have the following experimental uncertainties:

Chlorine	±0.001	Bromine	±0.002
Chromium	±0.001	Silver	±0.003
Iron	±0.003		

Index

(Bold face page numbers refer to definitions in the text.)

A

Abbott, L., 222
Abell, G., 98, 264
Ablation, **242**
Abrahams, A., 222
Abscissa, 395
Absolute motion, 404
Absolute scale of temperature, 340
Absolute space, 395
Absolute time, 395, 668
Absolute zero, 698
Absorption of energy by atom, 542
Absorption spectrum, 538
Academia dei Lincei, 127
Academie des Sciences, 127
Academies (19th century), *see* by name
Accelerating charges, 385
Acceleration, 109, 111, **128**
  gravity and, 415
Accelerators, 604
Accuracy in measurement, 309
Acoustic waves, 172
Action, **133**
  at distance, 180, **377**
  of electron, 547
Activation analysis, **635**
Actualism, 477
Adrian, E. D., 439
Aerial, 387
After-winds from nuclear bomb detonation, 644
Age of Reason, 177, 429
Air, as primordial substance in Greece, 45
Akkad, 35
Alamogordo, New Mexico, 619
Alchemy, 80
Alexander the Great, 78
Alexandria, Museum at, 78
Algae, 490
Allen, G. E., 222
Alpha radiation, 584
Alphabet, Greek, 700
Alloys, bronze age, 35
Almagest, 74
Alpha particle bombardment, 592
Amber, 363
Ammonia, 488
Amonton's law, 343
Ampere, **374**
Ampère, André Marie, 374
Amplification gain, 217
AMU (atomic mass unit), 596, 698
Analytic geometry, 124
Anatolia, 35
Anderson, C. D., 599
Anderson, D. L., 516, 579
Anderson, D. W., 393
Anderson, O. E., Jr., 652
Andrade, E. N. de C., 183
Andromeda Nebula, 258
Angrist, S. W., 223
Angstrom, **318**
Angular momentum, **145**
Angular velocity, **142**
Anode, 373, 559
  of battery, 509
Anthropogeography, **483**
Antimatter, 602
Antineutrino, 601
Antiparticle theory, 602
Apogee, **233**
Applied research, **680**
Applied science, **682**
Aptitude, testing of, 317
Aquinas, St. Thomas, 81, 82, 429
Arab science, 81
Archimedes of Syracuse, 78
Area, 701
Argon-40, 589
Argonne National Laboratory, 618
Aristarchus, 78
Aristotle, 10, 45, 46
Arithmetic mean, **288**
Armenia, 80
Armitage, A., 98
Army alpha tests, 317
Army beta tests, 317
Ashby, W. R., 212, 222
Astronomical unit, **247**, **249**, 699
Asteroids, **249**
Aston, F. W., 591, 594
Astrology, 80
Astronauts, 240
Astronomy
  beginnings of, 40
  development of, 54
Atkins, K. R., 122, 183, 362, 393, 607
Atmosphere, 449
  formation of, 476
  theory of origin, 485
Atom(s)
  Bohr's model of, 535, 542, 544
  bombardment of, 598
  collision frequency of, 340
  Dalton's conception of, 333, 504
  energy states of, 543
  excitation of, 548
  ground state of, 543
  ionization of, **548**
  kinetic energy of, 342
  normal state of, 543
  Rutherford's model, *see* Rutherford
  Schrödinger's model, 567
  size of, 338
  velocity of, 340
  weight of, 338
Atomic bomb project, 616
Atomic bombs, 619
Atomic chart, 506
Atomic constants, 698
Atomic energy, feedback influences of, 651
Atomic Energy Act of 1946, 620
Atomic mass, 595
  units (amu), **596**
Atomic model, energy states, 544
Atomic number, 559
  dimensions related to heat and, 338
  listing of, 702
  Z, **596**
Atomic picture of matter, 333
Atomic power, 608
  complex of plants, 629
  future of, 623

Atomic spectra, 318, 535, 536
Atomic theory
 in early 19th century, 503
 Greek version, 47
Atomic weight(s), 504
 classification of elements according to, 505
AU, astronomical unit, 247
Augustine, St., 81
Autotrophs, **486**
Average, **102**
 deviation, 290
 of distribution, **288**
Avicenna, 81
Avogadro, Amadeo, 334, 505,
Avogadro's law, 334
Avogadro's number, **338**, **373**, 507, 597, 698
Azimuthal quantum number, 553
Azote, **507**

B

Bacon, Sir Francis, 123, 177, 427, 670, 675
Bacon, Roger, 82
Balmer, Johann Jacob, 538
Balmer lines, 538
Balmer series, 544
Balsley, H. L., 300
Bancroft, A., 12
Barium, 609
Barnett, L., 425
Baryons, **603**
Basalt, volcanic origin of, 466
Basic research, **680**
Battery, electric, 509
Baker, J. J., 222
Becquerel, Antoine Henri, 583
Beiser, A., 393
Bellman, R., 223
Benade, A. H., 183
Bentham, J., 686
Beta decay, 599
Beta rays, 584
Bethe, H., 646
Beveridge, W. I., 675
Bimetal, **206**, **311**
Binding energy of nuclei, **596**
Binet, A., 317

Biological clock in relativity, 419
Biosphere, **481**
 as environment for populations, 481
Biscovich, R. J., 504
Bishop, A. S., 652
Bitter, F., 393
"Black box" approach, 213
"Blackbody" cavity, principle of, 519
Blackett, P. M. S. 696
Boase, R. L., 300
Bohr, Niels, 540, 609
Bohr atom, 535, 542
Bohr model of hydrogen atom, 544
 difficulties with, 563
 failure of, 654
Boltzmann, L., 339
Boltzmann's constant, **340**, **343**, 698
Bomb detonation, effects of, 642
Bonner, F. T., 393
Bonnor, W., 425
Boorse, H. A., 393, 516, 534, 562, 579, 607
Borgstrom, G., 501
Born, Max, 425, 436, 441, 516, 531, 534, 579
Bosons, **603**
Boundary condition in atomic structure, 574
Boyko, H., 696
Brackett series, 544
Brahe, Tycho, 86
Bresler, J. B., 501
Bridgman, P. W., 563
Brightness of stars, 252
Broad, C., 222
Bronowski, J., 15, 441, 675
Bronze-age civilization, 36
Brooks, N., 223
Brown, H., 499, 501
Brown, R., 336
Brown, S. C., 362
Brownian motion, 336
Bruno, G., 86
Bubble chamber, **604**
Bulletin of the Atomic Scientists, 685

By-products of atomic power, 622
Byzantium, 80

C

C-plane (conceptual plane), 660
Calcium sulfide, 583
Calculus, 108, 127
 application of, 140
Calendar, beginnings of, 40
Calibration of instruments, 313
Caloric theory (fluid theory) of heat, 334
Calorie (unit), 335, **352**
Candlepower, **319**
Cannizzaro, S., 507
Cantzlaar, G. L., 480
Carbon-14, 25, 589
Carbon cycle, 485
Carbon dioxide in atmosphere, 476
Cardano, G., 275
Carnotite, **585**
Cartesian clarity, 656
Cartesian coordinates, 124, 395
Cartesian determinism, 275, **433**
Carthy, J. D., 501
Casey, E. J., 222
Catastrophic changes, geological, 19, 477
Catastrophists, 465
Cathode, **373**
 of battery, **509**
Cathode ray tube, 373, 582
Cathode rays, **510**, 559
Catholic Church, 81
 relations between science and, see de Chardin
Cations, 559
Causality, **438**
 concept of, 511
Cause and chance phenomena, philosophy of, 436
Cause and effect, 433
 concepts, weaknesses of, 435
Cavendish, Henry: on mass of earth, 149
Cavendish laboratory of Cambridge University, 540
Ceiling population density, 493

Celestial equator, **61**
Celestial meridian, **58**
Celestial pole, **58**
Celestial sphere, **56**
Celsius, 313
Cementation, **473**
Centi, 700
Centigrade temperature scale, 313
Centimeter-gram-second system (cgs), 373
Central limit theorem, 293
Centrifugal force, **145**
 in atom, 546
Centripetal acceleration, **144**
Centripetal force, **144**
Cepheid variable, **252**
Cesium isotope, 640
Cgs system, **132**
Chadwick, J., 593
Chain reaction, **610**
 nuclear, 614
Chance errors in measurement, 308
Chance phenomena, 6
Changeux, J. P., 223
Charge conservation, 375
Charged particles, 382
 effects of moving, 518
Charges in motion, 369
Charles' law, 343
Chauncey, H., 331
Chemical properties of atoms, explanation of, 552
Childe, V. G., 53
Chinese science, 36
Christianity, 80
Chronological age, 317
Chronological table, 500 B.C. to 1500 A.D., Middle Ages, events and people, 94
Circumpolar, **59**
Civilization, beginnings of, 34
Clark, A. C., 264
Classical determinism, 515
Clausius, R., 669
Climate, influence of on people, 482
Cline, B. L., 534, 579
Clocks, slowing down of, 419

Closed-loop amplifier system, 215
Closed system, 356
Cloud chamber, **604**
Coal reserves, 627
Coal seams, **473**
Cobalt-60 isotope, 640
Coefficient of elasticity, 172
Cognitive experience, 658
Coins, tossing of, 279
 of three coins, 281
Colburn, R., 222
"Cold" war, 621
Collimator, 521
Collision
 elastic, **139**
 inelastic, **139**
Collision of bodies, 137
Collision experiment, atomic, 540
Colloidal particle, 336
Colodny, R. G., 441, 696
Colors of light, 169
Combination of events (in probability theory), 281
Combinatorial analysis, 297
Combustion
 Lavoisier on process, 335
 phlogiston theory, 334
Commission on Ground Water, "Ground Water Basin Management," ASCE, 480
Common sense, 10
Communism, coming of, 609
Compaction in geology, **459**
Companion, A. L., 562, 579
Complementarity in atomic theory, 576
Compound nucleus, **598**
Compression waves in earth, 447
Compton, A. H., 529
Compton wavelength, 529
 election, 698
Computer technology, 683
Concept of time, 667
Concepts, **658**
 metaphysical requirements for acceptability of, 663
 from observations, 9
Conceptual plane(s), 658, 660
Condenser, electric, 368

Condillac, E., 686
Conjunction, **67**
Conservation of angular momentum, 145
Conservation of electric charge, 374
Conservation of energy
 principle of, 145, 357
 in simple harmonic motion, 157
Conservation of matter, principle of, 130
Constant(s)
 atomic and other, 698
 of proportionality, 104
 ratios pertaining to planets and, 699
Constantinople, 80
 capture of by Turks, 81
Constellations, **62**, 255
Construct, **661**
Constructive interference, **166**
Control
 anticipatory, 197
 proportional, 197
Control system, **186**
 driving automobile, 198
 elements of, 205
 on-off, 192, 200
 proportional type, 202
Conversion factors, 701
Cook, C. S., 607
Copernican system, 85
Copernicus, Nicolas, 84
Corpuscular theory of light, 512, 526
Correspondence, rule of, in logic, 662
Cosmic radiation, 602, **604**
Cosmic ray, 589
Cosmogany, **17**
Cosmology, atomists, **17**
Coulomb, 366, 373
Coulomb, C., 364
Coulomb's law, 366
Covalent bonding of atoms, **554**
Covalent compounds, 557
Cowling, T. G., 362
Creed for scientist and layman, 673
Crescent, **65**

Critical mass of nuclear reactor, 614
Critical reactor, **614**
Critique of Pure Reason, 429, 663
Crombie, A. G., 696
Crookes, Sir William, 510, 581
Cross product, example of, 384
Crystalization, 474
Crystallography, 346
Cuneiform script, 38
Curie, **589**
Curie, Eve, 607
Curie, Irene, 585
Curie, Marie Skodovska, 585
Curie, Pierre, 585
Current density, **375**
Curtis, H. D., 258
Curvature of space, 420
Cushing, S. W., 501
Cybernetics, **5, 207,** 210
Cyclotherm, in geology, **473**

D

Dalton's atomic theory, 333, 504
Dampier, Sir W. C., 53, 98, 183
Damrin, D. E., 331
Dante's "The Divine Comedy," 83
Darwin, Charles: genetics and, 19, 477, 492
Dating techniques, 589
Daughter products of radioactivity, 605
David, E. E., 184
da Vinci, Leonardo, 83
Davis, K., 501
Davisson, C. J., 567, 569
Day, **60**
Dead zone
  in control systems, **197**
  in measurement, 316
DeBenedetti, S., 425
de Broglie, L., 534, 564, 579
Decay constant, 588
  of radioisotopes, 710
Deceleration, 109
deChardin, P. T., 2, 11, 15
Deci, 700

Decimal system, 132, 373
Deduction
  in birth of concepts, 670
  induction, intuition in progress of science, 671
Deductive process, 10
Deductive reasoning, **670**
Deevey, E. S., 223
Deferent, **76**
Degree of freedom, **340**
de LaPlace, P. S., 275
Delta (Δ), **104**
De Morgan, W. F., 275
Dependent variable, **102**
Derivative, 119
Descartes, René, 6, 124, 177, 395
Destructive interference in waves, **166**
Determinacy, 7
Determinism
  to indeterminacy, 426
  in macroscopic and microscopic phenomena, 515
  in nature, 433
Deterministic versus probability approach, 656
Devonian period, 29
Diatomic, 505
Dice, rolling of, in probability theory, 283
Diffraction, of waves, 161
Diffusion process, for separation of isotopes, 616
Dilthey, 686
Dimensions of objects, 4
Dirac, 602
Di Santillana, G., 98
Discovery and application of, 682
Disintegration constant in radioactivity, 588
Displacement, current, 103, **386**
Diurnal motion, **58**
Dobbin, J. E., 331
Doppler effect, **166**
Dorn, H. F., 223
Drake, S., 98
Dreyer, J. L. E., 98
Duality, of matter and waves, 513
Du Nouy, L., 15

Duveen, A., 264
Dyne, **132, 335**

E

Ear, frequency range of, 173
Early Bird communications satellite, 390
Earth
  age of, 21
  atmosphere, 443
  crust of, 449
  formation of, 26
  interior of, 445
  magnetic field, 243
  mass, 149, 699
  outer layers, 446
  as primordial substance, 45
Earthquake, Alaskan (1964), 467
Eastern Roman Empire, 80
Ebel, R. R., 331
Ebert, J., 222
Ebling, F. J., 501
Eccentric hypothesis, **76**
Eclipse of moon, 72
Ecliptic, **62**
Ecological system, **481**
Ecology, **481**
Economic disparity among nations, 679
Ecosystem, **481**
Eddington, A., 659
Education, 694
  science and, 692
"Eigenvalues" for atomic processes, 568
Einstein, Albert, 394, 403, 425, 437, 653, 676
Einstein's equation, 597
Einstein's photoelectric effect, 526
Einstein's special theory of relativity, 402
Elastic vibrations, 172
Electric charge
  characteristics of, 364
  discovery of, 363
Electric currents, 369
  magnetic effects of, 370

Index 717

Electric field, 377
  magnetic field and, 382
Electric motor, origin of, 371
Electrical conductivity, **376**
Electrical resistivity, 312
Electricity
  discovery of, 363
  induced current, 371
Electrochemical cell, 508
Eectrochemistry, Faraday's work, 509
Electrolysis, 509, 556, 559
Electromagnetic field, 381
Electromagnetic induction, **386**, 510
Electromagnetic radiations, 389
Electromagnetic spectrum, 389, 390
Electromagnetic theory, 512
Electromagnetic waves, 387
Electromagnetism, 370
Electron(s), **373**, 511
  annihilation of, 599
  atomic theory, 370
  in Bohr model of atom, 542
  charge of, 374
  diffraction of, 514, 570
  forces on in atom, 546
  interference patterns of, 575
  mass of, 595
  passing through slits, 575
  rest mass of, 698
  spin in atom, 551
  volt, **597**
  wave motion in orbits, 568
  wave properties of, 565
Electronic charge, 698
Electroscope, 364
  principle of, 366
Electrostatic force in atom, 546
Electrostatics, 364
Elektron (Greek name), 363
Element(s)
  abundance of, 702
  atomic chart, 506
  chemical, periodic table, 506
  earth, water, air, and fire, 45
  transformation of, 586
Elementary particles, 603

Elliptical orbit, 87
  of atoms, 548
Emission of radiation by charged particle, 387
Emotions in scientific endeavor, 687
Empirical verification of concepts, 664
Empiricism, 664
Energy, 187
  conversions in earth, 450
  as matter or motion, 513
  modern civilization and, 430
  nuclear resources, 626
  potential, gravitational, 229
  requirements for life, 490
  simple harmonic motion, 157
Energy consumption in United States, 624
  cumulative, 625
Energy equivalent
  of 1 amu, 698
  of electron rest mass, 698
Energy states in atoms, 543
Engine, steam, 431
Engineering, **682**
Entropy, 7, **355**
  in chance phenomena, 435
  change
    direction of time and, 668
    examples in photosynthesis, 357
  in living organisms, 490
  concept, 354, 490, 669
  related to direction of time flow, 358
  of time, 667
Environmentalism, **483**
Eosphoros, 67
Epicenter, referred to earthquakes, 468
Epicyclic hypothesis, **76**
Equant, **76**, 84
Equation of state of ideal gas, 343
Equilibrium concepts in gases, 351
Equilibrium in radioactive decay, 587
Equinox, **61**

Eratosthenes, 79
Erech, 35
Erg, **352**
Erosion, 454
  rates of, 466
Escape velocity, 227
Eternal unknowable, 688, 690
Ether concept, 386, 401
Euclid, 43, 44, 78
Eudoxus of Cnidus, 74
Evaporation
  from earth, 462, 463
  of molecular crystal, 349
Evolution, *see also* Hutton, Darwin
  geological, 19
  theories of, 476
  idea of progress, 677
Expanding universe, **260**
Experience, knowledge, and domain of science, 657
Experimental approach to science, 428
Experimental method, *see* Bacon
Explorer I, 225
Extensibility of concepts, 663
Extrapolation, **103**
Eye
  physiology of, 319
  structure of, 319

F

Fahrenheit temperature scale, 313
Failure rate, **272**
  determination, 273
  in statistical analysis, 267
Faraday, M., 371, 508
Faraday constant, 698
Farrington, B., 53
Fascism, coming of, 609
Faul, H., 53
Feedback, 2–5, 185, 186, **204**
  in form of information, 196
  negative versus positive, 194
Fender, D., 223
Fermi, Enrico, 580, 600
Fermi, L., 652
Field, electric, 377

Fine structure, 549, 551
   of spectral lines, 549
Finnigan, R. E., 222
Fire, as primordial substance, 45
Fission
   energy content of domestic nuclear resources, 626
   energy release in, 612
   nuclear, **610**
   products, 613
Fission process, 610
   liquid drop model of, 612
Fitzgerald, G. F., 402
Fluorescence, **581**
Food, problem of, 491
Food chain in nature, 490
Food and Drug Administration, 639
Force, **128**, 701
Ford, K. W., 607
Fossil fuel resources for United States, 625
Foster, R. J., 480
Fourth state of matter, 508
Frames of reference, 395
Frank, P., 441
Franklin, Benjamin, 364
Fraunhofer lines, 538
Free enterprise, role in national development, 680
Free fall of bodies (gravitation), 126, 148
Free will, existence of, 433
Frequency, 173
   distribution, tabulation of, 270
   in simple harmonic motion, 155
   transformations, 523
Fresnel, A. J., 518
Frisch, O. H., 607
Frisch, O. R., 610
Fuel element assemblies, 631
Function, representation of, 103
Functional relationship, **102**
Functions of whole versus those of parts, 686
Fusion
   energy processes in sun, 646
   of light atoms, 621
      for energy release, 641

Fusion bomb, 641
Fusion reactions, 644
Fusion temperature, 641

G

Gagarin, Y., 225
Gain in amplifier systems, 217
Galaxies, 252
Galilean transformations, **398**
Galilei, Galileo, 10, 89, 92, 98, 100, 122, 124, 180
Galvani, L., 369
Galvanometer, **369**
Gamblers' manual, 275
Gamma radiation, 584
Gamow, G., 53, 183, 300, 362, 425, 516, 534, 579
Garfinkel, S. B., 607
Gas constant, 698
Gas engine, pressure of, 341
Gaseous diffusion process, 617
Gaseous state, 504
Gauss, K. F., 269
Gaussian curve, **269**, 287, 292
Gay-Lussac, J. L., 334, 505
Gay-Lussac's law, 343
Geiger, 540
Geikie, A., 480
General relativity, 414
Geodesic lines, 421
Geodesy, 244
Geography (early exploration)
   influence on man, 482
   on political and economic development, 484
Geologic cyclic processes, 473
   of rock change, 475
Geological processes, theories of, 464, 473
Geology
   effect on society, 456
   estimates on age of earth, 21
   technology and, 457
Geometry, beginnings of, 39
Geosynclinal theory, 469
Geosyncline, **469**
Germanium, 376
Germer, L. H., 567, 569

Giga, 700
Gilbert, C. N., 480
Gilbert, W., 367
Gillispie, C. C., 53, 696
Gilluly, 480
Glacial periods, theory of cyclic phenomena, 188
Glaciers, 21
Glasstone, S., 264, 652
Glenn, Jr., J. H., 225
Globular cluster, 255
Gneiss, in rock formation, **459**
Goddard, R. H., 225
Goldstein, 686
Goodfield, J., 98
Goudsmit, S. A., 551
Grabbe, E. M., 222
Grabens, 471
Graham, C. H., 331
Gram-atom, 373
Gram (measure) molecular weight, 338
Granite(s), **460**, 473
   formation of, 474
Granitization, **475**
Gravel, 460
Gravitation, 127, 228
   Einstein's theory, *see* Relativity theory
Gravitational attraction, development of, concept of, 670
Gravitational constant, 698
Gravitational deflection of light, 417
Gravitational field(s), **379**, 416
Gravitational forces, 148
   constant G, 148
   in space work, 228
Gravitational red shift, 419
Graviton, **603**
Gravity, 126
Great spiral in Andromeda, 257
Greek alphabet, 700
Greeks and beginnings of science, 41
"Green salt" (UF_4), **631**
Greenaway, F., 516, 579
Greenberg, D. A., 122
Greenhouse effect, **488**
Group velocity, **571**

H

Guidance system for spacecraft, 238
Gutsell, J. S., 501

H-bomb (hydrogen bomb), 621
Hahn, O., 609
Half-life, in radioactive decay, 23, 587, 588
Hall, Sir James, 466, 469
Halliday, D., 122, 183, 393, 425
Hanford reactors, 618
Hanson, N. R., 15, 675
Hardin, G., 222, 501
Harmonics, **173**
Harriott, P., 222
Hawkes, J., 33, 53
Hayek, J., 222
Heat, 701
   caloric theory of, 334
   early concepts of, 332
   mechanical equivalent of, 335
   thermodynamics and, 332
   of vaporization, 313, 348, 462
   versus temperature, 341
"Heat death," 675
Heat energy, source of in earth, 452
Heat transfer process, 357
Hebb, D. O., 8, 15
Heisenberg, W. K., 573, 607
Heliocentric theory of universe, 84
Hellenistic period, 77, 78
Henry, J., 371
Hepatia, 80
Hepler, L. G., 362
Herbivores, **486**
Herschel, W., 150, 252
Hertz, H., 388, 524
Hesperos, 67
Heterotrophs, **486**
Hewlett, R. A., 652
High-energy physics, 601
High-fidelity, 207
Hill, W., 122
Hindu science, 36
Hipparchus, 78
Hiroshima, 619
Histogram, example, 267
History of science, 693
Hitler, Adolf, 609
Hoagland, H., 223, 501
Hobbes, T., 686
Hodgson, J. H., 480
Hoffmann, B., 534, 579
Holton, G., 183, 362, 393
Holmes, A., 480
Homeostatic equilibrium, **495**
Homo neanderthalensis, 30
Homo sapiens, 28, 30
Homogeneity of space, 601
Homogenizing processes, **475**
Horizon, 56
Horsepower, **335**
Horst, **471**
Hubble, E., 258
Hunger, problem of, 491
Hunting, control systems and, 196, 200
Huntington, E., 501
Hutchings, E., 222
Hutton, J., 18, 52, 464, 480
Huttonian theory of geological formations, 464
Huygens, Christian: wave theory of light, 124, **512**
Hydrogen-3, 589
Hydrogen atom, rest mass, 698
Hydrogen line spectrum, 538
Hydrogen molecule, 345, 554
Hydrologic cycle, 461
Hyperons, **603**
Hypotheses, development of, 667

I

Ibn Habib, 81
Ice
   melting point of, 340
   structure of, 346
Igneous processes, **475**
Igneous rocks, 474
Ihde, A. J., 516, 579
Impulse, **133**, **135**
Independent variable, **102**
Indeterminacy, 7
   in nature, 515
Index Expurgatorius, 91
India, science of, 36
Induction, **670**
   deduction, intuition in birth of concepts, 670
   in electricity, 364
   example of, 294
Inductive approach, 9
Inductive method, 670
Inductive processes in probability considerations, 295
Industrial Revolution, 431
Industry and geologic materials, 458
Indus Valley civilization, 36
Inert gases, 554
Inertia, 130
Inertial frames, **398**
   reference, **399**
Infeld, L., 425
Inference, from statistical distributions, 294
Information as feedback, 193, 196
Instantaneous rates of change, 116
Instruments, calibration of, 313
Insulators, **376**
Integration, 117
Intelligence quotient, **317**
Interference phenomena in light waves, 168
Interference of waves, 161
Interferometer, 401
Internal energy, **353**
Interpolation, **103**
Interstellar dust, 27
Intuition, 10, **673**
   in birth of concepts, 670
Invariance
   of physical laws, 601
   related to time, 358
Inverse square law
   of charged particles, 366
   in gravitational attraction, 127, 228
Ion, **557**
Ionian school of cosmologists, 43
Ionic bond, 556
Ionization energy, **548**

Ionization potential, 559
Ionizing radiation, 633
  effects and products of, 638
Iron (tools), 42
Irradiation of food, 638
Islam, *see* Arab science
Isolated system, **352**, **356**
Isostasy, **467**
Isostatic, **470**
Isotopes, **591**
  table of, 702
Isotopic masses, 595
  of nuclides, 702

J

Jeans, Sir James, 523
Jericho, 34
Joliot, F., 585
Joule, **335**, **352**, **374**, 597
Joule, J. P., 335
Joule equivalent, 698
Jungk, R., 652
Junishapur, 80
Jupiter (planet), 90

K

Kac, M., 300
Kaempffer, F. A., 393
Kant, Immanuel, 429, 663
Kaplan, A., 222
Kapp, K. W., 222, 696
Katz, R., 425
Keller, Helen, 12
Kelvin, Lord (William Thomas), 313, 467
Kelvin temperature scale, 313, 340
Kepler, Johannes, 86, 180, 224
Keynes, J. M., 275, 300
Kilo, 700
Kilogram-mass, **130**
Kilomega, 700
Kilotons, **641**
Kinematics, **100**
Kinetic energy, **136**
Kinetic theory
  of gases, 339, 351, 360
  of heat, 339

King, C., 480
King, P. B., 480
King-Hele, D., 264
Klemm, F., 442
Kleppner, D., 122
Kock, W. E., 183
Koenig, S., 222
Korean War, 621
Kuhn, T. S., 98
Kummel, B., 52
Kybernes (cybernetics), **5**

L

Laboratories, types of, 681
Lagemann, R., 122
Laissez-Faire economic and political doctrines, 685
Langbein, W. B., 480
Language
  beginning of, 35
  problem of, 13
Laplace, 6, 300, 335
Lapp, R., 621
Latent heat of fusion, 313, 314
Latitude (early exploration), **70**
Lattice structure of ice, 346
Lavoisier, Antoine, 335
Law of diminishing marginal returns, 113
League of Nations, 609
Learning in science, 687
Lee, A. E., 222
Length
  measurement of, 306
  units of, 701
Lens of eye, 322
Lenz, H. F. E., 372
Leopold, L. B., 480
Lepton, **603**
Levinson, H. C., 300
Leyden jar, **367**
Libby, W., 25, 52
Life expectancy, 491
Light
  absorption of by atoms, 542
  corpuscular theory of, 169, 512, 526
  curvature of, 416

  emission of by atoms, 542
  nature of, 168
  rays in gravitational field, 416
  red shifts in, 259
  spectrum, 169, 535
  speed of, 390
  velocity of, 399, 698
Limestone, **459**, **469**
Lindsay, R. B., 222, 696
Line of nodes, **73**
Line spectrum, **536**
Lines of force, **379**
Lines of induction, 384
Lithification, in geology, 475
Lithium, 702
Locke, J., 180
  on heat, 334
Lodestone, 367
Loefsack, T., 501
Logic of science, 655
Logical fertility of concepts, 663
Lorentz, H. A., 402
Lorentz contraction, 410
Lorentz transformations, 405
Los Alamos Laboratory, 618
Luchins, A. S., 15, 53, 331, 442
Luchins, E. H., 15, 53, 442
Luminescence, **581**
Luminosity of sun, 645
Lunar eclipses, **66**
Lundquist, C. A., 264
Luther, Martin, 86
Lyell, Sir Charles, 53, 477
Lyman series, 544

M

"McCarthyism," 622
Mach, Ernst, 403
Mach front from nuclear bomb, 643
Mach number, **175**
Magellanic clouds, 253
Magie, W. F., 393
Magma, in rock formation, **473**
Magnetic needle, 370
Magnetism, induced, 368
Magnetostatics, 367
Magnets and magnetic fields, 367
Malkus, J. S., 223

Malthus, T. R., 491
Man and environment, 443
"Manhattan district" project, 616
Mann, W. B., 607
Marble, **459, 473**
Margenau, H., 442
Mars, photo of, 246
Marsden, 540
Mason, B., 480
Mason, S., 52, 480
Mason, S. F., 53, 98, 183
Mass, **128**
  change with velocity, 412
  Einstein theory, *see* Relativity theory
  standard of, 130
  units of, 132
Mass defect, **597, 701**
Mass density, 701
Mass-energy equivalence, 413, 597, 611
Mass spectrograph, 594
Mass spectrometer, 591
Mathematics, beginnings of, 38
  Plato and, 46
  Pythagoreans, 44
Mather, K., 52, 480
Matrix mechanics, 573
Matter
  electronic properties of, 375
  versus energy, concepts of, 564
  wave-particle duality of radiation and, 564
Maturity of chances, 277
Maxwell, J. C., 339, 379, 393, 508, 518
Maxwell-Boltzmann velocity distribution of molecules, 520
Mean free path of electrons in metal, 376
Measurement
  chance errors in, 308
  errors of judgment, 304
  of observation, 304
  limitations of, 303
  role of, 301, 662
  systems of, 39
  techniques, elements of, 315
Measures and weights, 701

Mechanical theory of heat, *see* Heat
Mechanics, 128
  science of, 180
Mechanistic view of nature, 674
Median of distribution, **289**
Medical therapy with ionizing radiation, 640
Medicine, beginnings of, 41
Mega, 700
Megahertz (MHz), 388
Megarads, **639**
Megatons, 642
Meinzer, O. E., 480
Meitner, L., 610
Mendeléev, D. I., 507
Mendeléev periodic table of elements, 505, 633
Mental age, 317
Mesolithic Age, 34
Meson theory, 602
Mesons, **602**
Mesopotamia, 35
Metal reduction plants, 631
"Metallurgical Laboratory" at University of Chicago, 618
Metamorphic rocks, 474
Metaphysical conclusions on cause and chance, 437
Metaphysical requirements for acceptability of concepts, 663
Meteoric matter, 243
Meteorites, 449
Meter, 101
Meter-kilogram-second system (mks), 373
"Method of fluxions," 127
Methodology of science, 655
  inductive, deductive, intuitive, 670
Metric prefixes, 700
Michelangelo, 83
Michelson, A. A., 399
Michelson-Morley experiment, 399
Micro, 700
Microcurie, **589**
Micrometer, 307
Micromicro, 700
Micron, **318**

Microscope, 325
Microscopic versus macroscopic analysis, 337
Middle Ages, 81
Milky Way, 252
  age of, 261
Miller, F., 393
Milli, 700
Millicurie, **589**
Millikan, R. A., 374, 393, 516, 529, 579
Millimeters, 101
Millimicro, 700
Mineral production in New York State, 458
Mining and milling of uranium, 629
Miracle Worker, 12
Mks system, **366**
Mode of statistical distribution, **289**
Models, 213
  limitations of, 653
Mössbauer, R. L., 419
Mohammed, 80
Molecules
  internal modes of motion in, 344
  rotational energy of, 345
  vibrational energy of, 345
Moment of inertia, **145**
  of masses, 146
Momentum, **132, 135**
  conservation of, 135
Monatomic gas, ideal, 341
Moon
  movements of, 64
  satellite photographs of, 244
Moors, 81
Moroney, M. J., 300
Moshor, R., 223
Motion
  of particle, 101
  significance of, 100
  in straight line, 102
Motor transform devices, in control systems, 206
Motz, L., 264
Mountain formation, 467

Moving charges and magnetic field, 381
Mueller, C. G., 331
Muon, 603
Museum of Alexandria, 78, 81
Music, frequency characteristics of, 173

N

Nagasaki, 619
Nagel, E., 300
Nano, 700
Napoleon's hair, studies of, 638
Nash, L., 516, 579
Natural philosophy, **1**, 690, 692
  of cause and chance, 436
Natural science, **1**
Natural theology, 685
Nazi Germany, 610
Nebulae, **257**
Neolithic Age, 34
Negative charges, 375
Neoplatonism, 79
Neptune, discovery of, 150
Neptunist school, 465
Neptunium, discovery of, 614
Neugebauer, O., 53
Neutrino, 600, **603**
Neutron(s), 592
  discovery of, 593
  rest mass, 698
  of thermal energy, 613, **616**
  wave properties of, 570
Newman, J. R., 15, 300
Newton, Isaac, 6, 108, 124, 180, 503
  laws of, 129, **131**
  Principia (1686), 183
Newtonian mechanics, limitations of, 433
Ney, E. F., 425
Nitrate ions, 488
Nitrifying bacteria, 488
Nitrite ions, 488
Nitrogen
  transformations of, by cosmic rays, 589
  transmutation of, 592

Nitrogen-containing living matter, 488
Nitrogen cycle in nature, 488
Nobel Prize, 583
Noble gases, 554
Noise, **199**
  in measurement, 316, 324
Noncommutative algebra, 573
Non-Euclidean space, 421
Normal distribution, 269, **270**, 287
North Star, 58
Notz, L., 393, 516, 534, 562, 579, 607
Novum Organum, 670
Nuclear bombs, 621
  energy and damage from, 641
  explosion, nature of, 642
  feedback influences of, 651
  hazards from, 641
Nuclear devices, power of, 612
Nuclear energy, rate of development, 627
Nuclear fission, **610**
Nuclear particles, 603
Nuclear "pile," 616
Nuclear power, 608
Nuclear power plants, 629
Nuclear power reactors, 628
Nuclear reactions, 598
Nuclear reactors, production of plutonium, 618
Nucleons, **597**
Nucleus
  of atom, size of, 541
  binding energy, **596**
  structure of, 591
Nuclides
  man-made, 633
  radioactive
    decay constants of, 710
    half lines of, 710
    table of, 710
  stable
    atomic numbers of, 711
    atomic weights of, 711
    table of, 711
  symbols of, 702
  table of, 702
  unstable, 633

Numbers
  Arabic, 39
  beginnings of, 38
  role of in analyzing experience, 662

O

Oak Ridge, 618
Objectivity, principle of, 438
Observation, limitations of, 304
Occam's Razor, 82–83
Odum, E. P., 501
Oersted, Hans Christian, 370
Ohm, **374**
Omega, 142
  as angular velocity, 142
On-off control, **192**
Oparin, A. I., 222
Opik, E. J., 223
Oppenheimer, J. R., 622
Opposition, of planets, **67**
Optical pyrometer, 313
Optner, S. L., 222
"Orange oxide" (UO_3), **631**
Orbital (atomic) angular momentum, **550**
  quantum number, 548
Orbital (atomic) magnetic quantum number, **550**
Orbital motion of planet Mercury, 418
Orbits of electrons in atoms, 549
Orbits of satellites, 231
Order and disorder, concepts of, 353, 355
Ordinate, 395
Ore, O., 122, 300
Origin of Species, 19, 492
Oxygen in atmosphere, 486
Oxygen cycle, 485

P

P-(percept) plane, 660
Pair-annihilation (nuclear), **600**
Pair-formation (nuclear), **600**
Paleolithic period, 34
Paleontology, 476
Pannekoek, A., 98

Parallax, 87, **249**
    in depth perception, 323
Parameters, 119
Parsec, **251**, 699
Parsegian, V. L., 331, 652, 696
Particle concept of light, 517
Particles, wave properties of, 564
Paschen series, 544
Pasteurizing with radiation, 638
Paul, J., 222
Pauli, Wolfgang, 552, 600
Pauli exclusion principle, **552**
Payne-Gaposchkin, C., 264
Peirce, C. S., 300
Pendulum, 157
Percepts, **659**
Perceptual planes, 658
Perigee, **233**
Period in simple harmonic motion, 155
Period-luminosity relation, **254**
Periodic table of elements, explanation of, 552
Perpillon, A. V., 501
Perrin, F., 338
Persian Empire, 42
Perturbation of planetary orbits, 150
Phase in simple harmonic motion, 155
Phases of moon, 64
Phillips, M., 393
"Philosophés," 676
Philosophy, **1**
    of science, 693
        relationship, 690
Phosphorescence, **581**
Phosphorus, 581
Photocell, 527
Photoelectric effect, 514, 526
Photon(s), 319, 514, 526
    duality of, 531
    energy of, 529
    momentum of, 529
Photosynthesis, 357, **486**
Phyla, 476
Pico, 700
Pictographs, 37
Pierce, J. H., 184
Piezoelectricity, **585**

Pilot waves, 570
Pitchblende, **585**
Planck, Max, 524
Planck's constant, **529**, 698
Planck's postulate, 543
Planck's radiation law, 547
Planetary motion, models of, 73
Planets, 54
    motion of, 66
    periods of, 247
    various constants on, 699
Plasma, **510**
Plato, 427
    Aristotle and, 45
    Bacon and, compared, 427
Playfair, J., 52
Plücker, J., 510
Plutonists, 465
Plutonium
    discovery of, 613
    production of, 618
Poincaré, H., 300, 402
Polanyi, M., 687, 696
Polaris, 58
Pollution, 482
Population controls in nonhuman societies, 493
Population frequency curve, **268**
Population growth, 491
Population increase in human societies, 495
Positive charges (protons), 375
Positron, 599, 602
Potassium-40, 589
Potential energy, 157, 187
Power, 377, 701
Pragmatic approach to science, 428
Precision in measurement, 309
Pressure, 701
Prime mover (of universe), **77**
Principal quantum number, 548, 553
Principia, 127, 180
Principle of equivalence in relativity, 415, 416
Prism, 536
Prism spectrometer, 521

Probability, 7, **277**
    applied to electron position, 656
    axioms of, 279
    combinations, 279, 284
    concepts of, 275
    considerations in atomic theory, 574
    density function, **271**
    distribution of electron, 577
    early beginnings, 274
    of failure, **271**
    statistical concepts and, 265
Probable error in measurements, 308
Progress, measure of, 677
Proportional control, 197, **202**
Protocol experience, **661**
Proto-Earth, 27
Protokollon, 661
Proto-Sun, **26**
Proton
    mass of, 595
    rest mass, 698
Prout, W., 507
Prout's hypothesis, 507
Psi probability function (ψ), 572
Psychological measurements, 316
Ptolemaic system, description of, 74
Ptolemy, 10, 74
Puritan spirit, 430
Pythagoras, 44
Pythagorean concept of numbers, 44

Q

Quadrature, **67**
Quality control, **267**
Quanta, 319
Quantum concept of atom, 547
Quantum conditions in atoms, 548
Quantum of energy, 543
Quantum hypothesis, birth of, 524
Quantum numbers of atoms, 552
Quark, **605**

Quasar, 260
Quasistellar radio sources, 260

R

Radian, **142**
Radiation
  from charged particles, 517
  versus frequency at different temperatures, 522
  from hot bodies, 519
Radiation pyrometers, 312
Radiation therapy, 636
Radio astronomy, 256, 260
Radioactive carbon, C^{14}, 25
Radioactive chains, 589
Radioactive constant, 588
Radioactive dating techniques, 21, 589
Radioactive decay, 605
  half life of, 587
  nature of, 586
Radioactive disintegrations, 589
Radioactive tracer techniques, applications of, 634
Radioactive tracing, 590
Radioactivity, 22
  atomic nucleus and, 580
  daughter products of, 587
  discovery of, 583
  equilibrium in, 587
  as source of heat in the earth, 467
  transformations, laws of, 588
Radiography, 640
Radioiodine, 635
Radioisotopes, table of, 710
Radium, discovery of, 585
Rads (radiation unit), **639**
Rain, formation of, 462
Rainbow, 170
Ramsey, H., 122
Randall, J. E., 222
Random errors in measurement, 308
Range of distribution, **288**
Rankama, K., 480
Rapport, S., 53

Rate of change, 102
Ratio, proton to electron mass, 698
Rationalism, growth, 429
Ratzel, F., 483
Rayleigh, Lord, 523
Reaction, **133**
Reactor, nuclear
  critical, 614
  subcritical, 614
  supercritical, 614
Reaumur, 313
Red shift in light spectrum, 259
Reflection of waves, 161
Refraction
  of waves, 161
  of x-rays, 582
Regulatory systems, 220, *see also* Systems, control
Relatedness of concepts, 664
Relationship of time and space, 411
Relative frequency in testing operations, 268
Relativity theory, 394
Relativity of simultaneity, 406
Reliability, **272**
Religion and philosophy, impact of science on, 688
Renaissance, 82, 83
Research
  basic versus applied, 680
  military and civilian, 681
Resistance, electrical, **375**
Resistivity, **375**
Resnick, R., 122, 183, 393, 425
Resonances as nuclear particles, 605
Restoring force, 187
Retrograde motion, 88
  of planets, 67
Reversibility of time and of processes, 358
Reversible process, 357
Rhodopsin, 321
Ripley, J. A., Jr., 331, 362, 393, 425
Ripple tank, 161
Ritow, I., 122

Ritter, K., 483
Ritz, 538
Rock cycle, 474
Rocket propulsion, early history, 225
  escape velocity of, 227
Rockets
  principles of, 233
Rocks, 460
Roemer, O., 390
Roentgen, W. K., 510, 582
Rogers, E. M., 98, 122, 183
Roller, D., 183, 362, 393
Roman Catholic Church, 86
Roman Empire, 79
Roosevelt, Franklin Delano, 609
Root mean square deviation, **290**
  applied to atomic motion, 343
Rosenbleuth, A., 207
Rotational motion, 141
Rourke, 300
Royal Observatory, 151
Royal Society, 127
Royal Society of London, 123
Rubidium-87, 589
Rule of correspondence, 662
Rumford, Count (Benjamin Thompson), 334
Rush, J. H., 393
Rutherford, E., 540, 584
Rutherford scattering experiments, 539, 541, 590
Rydberg, 538
Rydberg constant, **538**
  for infinite mass, 698

S

S-plane, 660
Sahama, T. G., 480
Sampling, **266**
  errors, 273
Samuelson, P. A., 222
Sandstone, **459**, 469, 473
Satellite orbits, 231
Sawyer, W. W., 122
Sax, K., 501
Scalar, **105**

Scattering experiments, use of, 560
Schlegel, R., 331, 362, 425, 675
Schroedinger, 567
Schwab, C. E., 480
Science, **1, 16**
  beginnings of, 41
  as human enterprise, 656
  human values and, 685
  humanities and, 693
  impact of on education, 692
    on religion, philosophy, 688
  progress of man and, 676
  of science, 677, 693
  scope, 658
  as search, 13
  technology and
    compared, 682
    contributions of, 682
    feedback between, 683
    in national affairs, 678
Scientific Method, 653
Scintillations, **540**
Secrecy clearances, 620
Secrecy controls in atomic work, 620
Sedimentary layers, 459
  formations of, 20
Sedimentary rocks, 474
Sedimentation, rate of, 465
Seebeck effect, 311
Seeding for rain, 463
Segregation processes, 475
Seismic data, 447
Semiconductors, **376**
Sensations, 658
Sensor-transducer, **219**
Sensors, in control systems, 206
Sensory data, perceptual knowledge, conceptual knowledge, 660
Sensory experience, 658
Sensory interpretations and limitations, 439
Sensory system, extension of through instruments, 324
Setlow, 222
Sexagesimal system of numbers, 39

Shale, **459, 469**
Shapley, H., 252, 258
Shear waves in earth, 447
Shell concept of atomic structure, 552, 553
Shock front from nuclear detonation, 642
Shock waves, 175
Sidereal day, **63**
Sidereal period, **67**
Silicates, **449**
Silliman, R. P., 501
Silver iodide seeding for rain, 463
Simple harmonic motion, 151
  equations for, 155
Simplicity and elegance of concepts, 664
Simultaneity, concept of, 403, 406
Singer, C., 183
Sisler, H. H., 562, 579
Slug, **132**
Smith, J. H., 425
Smith, K. V., 222
Smith, W. S., 477
Smuts, 686
Smyth, H. De W., 652
Snow, 692
Snow, C. P., 15, 696
Society, panorama before us, 676
Societies, diversity of, 676
Socrates, 45
Soddy, F., 586
Solar corona, **66**
Solar day, **63**
Solar eclipse, **65**
Solar energy, 644
  utilization, 648
Solar parallax, **248**
Solstice, **61**
Sommerfeld, A., 548
Sonic barrier, **176**
Sonic boom, **176**
Sophists, 45
Sound barrier, **176**
Sound waves, 172
Space, structure of, 420
Space exploration, history of, 224
Space projects, results of, 243

Space quantization of an electron's orbit, 550
Space sciences, 224
Spacecraft
  guidance and control, 234
  launching pattern, 237
  life support in, 239
  listing of, 226
  reentry problems, 241
Specialization, effects of, 692
Species, man, 30
Specific electronic charge, 698
Specific heat, **341**
Spectra of elements, 536
Spectrometer, 536
Spectrum
  of hydrogen, 536
  of light, 535
  of visible light, 318
Speed, **102**
  of light, 390, 403
Spherical geometry, 672
Spheroidal weathering, 455
Spin quantum number, **551**
Spinoza, 6
Sputnik I, 225
Standard deviation, **288, 290**
  example of, 308
Standard volume of ideal gas, 698
Standing wave, **173,** 565, 567
Starr, V., 223
State variables, 353
Static, **199**
Statistical approach to nature, 437
Statistical behavior, 7
Statistical equilibrium, 351, 352
Statistical inference, 294
Stefan-Boltzmann constant, 698
Step function, 199
Stellar parallax, 251
Sterilization
  of bacon, 639
  with radiation, 638
Stern, W., 317
Stimulus plane, 660
Stochastic variable, **278**
Stoicism, 79

Stone, R., 223
Stoney, G. J., 373
Strassman, F., 609
Stroboscope, 161
Strontium-87, 589
Struve, O., 264
Sublimation, 349
Sub-system, 5, **186**
Sulfides, **449**
Sumer, 35
Sun
  compositions of, 646
  energy source of, 644
  luminosity, 699
  mass, 699
  movements of, 59
  radiation energy on earth, 463
Superconductivity, **377**
Supernovae, 252
Superposition of waves, 164
Supersonic waves, 175
Swartz, C. E., 607
Symmetry, 601
Synodic period, **67**
Syria, 80
System(s), 3–5, **185**
  closed or conservative, **199**
  examples of, 209
    of biological systems, 218
  feedback, cybernetics, 185
  stability of, 190
Systems aspects of nature, 3
"Systems" approach to concepts, 665
Systematic errors in measurement, 309

T

Tacoma Narrows suspension bridge, 189
Tangent line for determining rates of change, 109
Taylor, E. F., 425
Taylor, L. W., 184
Technology, **682**
  beginnings of, 82
  impact of on religious thinking, 689
  problems created by, 683
  rationalism, and Industrial Revolution, the, 429
Telescope, 90, 325
Temperature, **340**
  absolute scale of, 340
  control in room, 201
  early concepts of, 332
  experience and concept of, 662
  measurement, 310
  scale, 313
Tepper, M., 223
Tera, 700
Terman, L. M., 317
Tetrahedrons, 347
Thales, 42, 363
Theophilus, Bishop, 80
Theory of relativity, 394
Thermal energy neutrons, **616**
Thermal equilibrium, 351
Thermocouple, 311
Thermodynamics, 351
  first law of, 352
  second law of, 354
  zeroth law of, 352
Thiel, R., 98
Thirring, H., 442
Thompson, B. (Count Rumford), 334
Thomson, G. P., 570
Thomson, J. J., 373, 511, 513, 540, 584, 594
Thomson model of atom, 539
Thomson, Sir William (Lord Kelvin), 53
Thorium, 585, 586
Thorndike, A. M., 607
"Thought experiment," 654
Thrust of rockets, 232
Time
  concepts of, 327, 668
  dilation, 408
  direction of, 329, 668
  geological processes and, 464
  measurement of, 309, 327
Tiryakian, E., 222
TNT equivalent of nuclear bombs, 641
Tools, significance of in history of man, 30, 32
Toulmin, S., 98
Transducer, **210**
Transfer (transform) functions, 194, 205
Transistor, **376**
Transmutation
  of element, 591
  by neutrons, 598
  of nitrogen, 598
Traveling wave, 567
Tritium, 589
Truman, Harry S., 619
Tsunami, **454**
"Tunneling" in nuclear reactions, **647**
Turks, coming of, 81
Turner, F. J., 480
Turner, L. A., 652

U

Ubbelohde, A., 223
Uhlenbeck, G. E., 551
Ultrasonic, 173
Uncertainty principle, 515, 572
Uniform motion, **103**
Uniformitarianism, **19**, 477
Union of Soviet Socialist Republics, 609
United Nations, 678
Units
  CGS system, 132
  English system, 101, 132
  standard of length, 101
  system of, 39, 373
Ur, 35
Uranium
  enrichment process, 617
  enrichment in U^{235}, 632
  isotopes of, 610
Uranium hexafluoride, 616
Uranium oxide (U_3O_8), **631**
Uranium salts, 583
Uranium tetrafluoride, 631
Uranium trioxide (UO_3), 631
Uranium-238 radioactive series, 24
Uranus, discovery of, 150
Urban Revolution, 35

V

Van Allen radiation belts, 226, 243
Van Bergeÿk, W. A., 184
Van der Waals force, 348, 349
Variable
　dependent, 205
　independent, 205
Varves, **465**
Vectors, **105**
Velocity, v, **105**, 701
　average, **107**
　constant, **107**
　instantaneous, **107**, 108
　of light in vacuum, 698
　measurement of, 309
Venus, 68
Versailles Treaty, 608
Vision, characteristics of, 318
Visual acuity, accommodation, intensity discrimination, depth perception, 322
Vitalistic view of nature, 674
Volt, **374**
Volta, Alessandro, 369
Volume, units of, 701
von Berthalanfy, 686
Von Fieandt, K., 675
von Humboldt, A., 482
von Leibniz, G. W., 6, 108
von Weizsäcker, 647
Vulcanists, 465

W

Waldron, R. A., 184
Water
　boiling point of, 340
　density of, 348
　earth, and man, 460
　freezing point, 313
　heat of vaporization, 348
　molecule, 345
　phases, 345
　properties of, 498
　structure of, 346
Watt, J., 335, 430
Watt, unit, **374**
Watt, K. E. F., 223
Wave concept of light, 517
Wave number, **538**
"Wave packets," 572
Wave properties
　of electrons, 575
　of particles, 564
Wave transmission in earth, 447
Wavelength, **173**
Waves
　characteristics of, 566
　group velocity of, 570
　longitudinal, 172
　mechanical systems, 158
　seismic, 172
　sound, 172
　standing, 565
　transverse, 172
Weaver, W., 300
Weber, E. H., 316
Weber fraction in vision, 322
Weber/meter, **384**
Weber's law, 316
Wechsler, D., 317
Weight density, 701
Weighted arithmetic mean, 289
Weisskopf, V. F., 15, 442
Weisz, P. B., 223, 675
Wertheimer, M., 331
Wesley, J., 430
Wexlor, H., 223
Wheeler, J. A., 425, 612
White, H. E., 607
White, W., 696
White light, **535**
Whitehead, A. N., 15
Whitrow, G. J., 425
Whittlesey, D., 484, 501
Wien's displacement law, 522
　constant, 698
Wiener, N., 207, 223
Wilks, S. S., 300
Williams, H., 480
Wilson, E. B., Jr., 331
Winchester, A. M., 223
Windleband, W., 53
Wolf, A., 331
Woodcock, A., 223
Woodford, 480
Woolley, Sir L., 53
Work, **136**
　energy and, 701
　function in photoelectric effect, 528
　by method of calculus, 140
World War I, 608
World War II, 609
Wright, H., 53
Writing, beginning of, 36
Wynne-Edwards, V. C., 223, 501

X

X-rays, **510**
　discovery of, 581
　nature of, 583
X-ray diffraction, 570, 582
X-ray production, 582

Y

Yankee reactor, 631
Year, **61**
Young, H. D., 331
Young, T., 518
Yukawa, 602

Z

Zeeman effect, 550
Zelburgs, V., 264
Zenith, **56**
Zeroth law of thermodynamics, 352

Q
160
P48

AUG 29 1974

RAYMOND H. FOGLER LIBRARY
DATE DUE

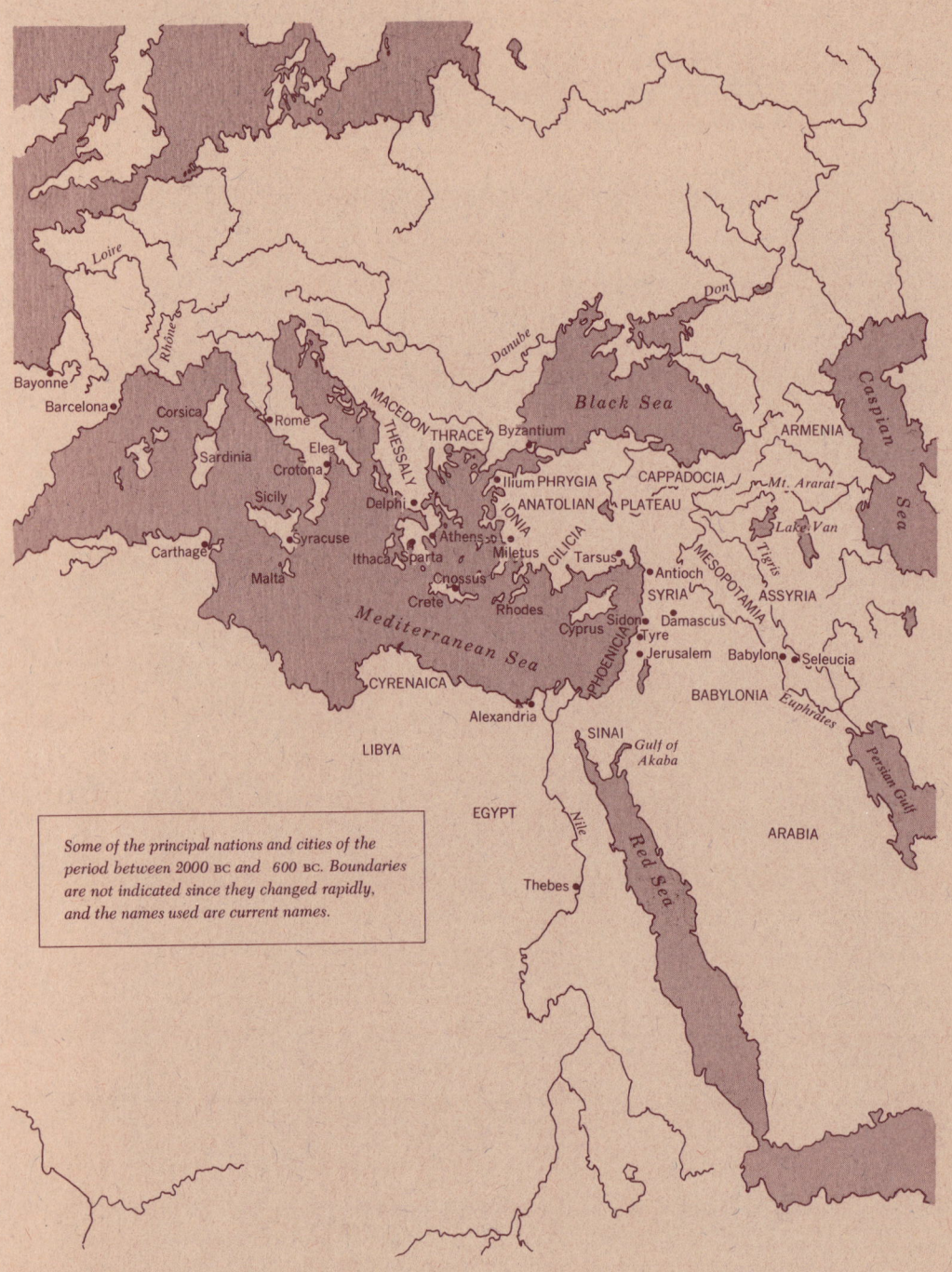

Some of the principal nations and cities of the period between 2000 BC and 600 BC. Boundaries are not indicated since they changed rapidly, and the names used are current names.